George Bebis   Richard Boyle   Bahram Parvin
Darko Koracin   Ronald Chung
Muhammad Hussain   Tan Kar-
Daniel Thalmann   David Kao

Advances in
Visual Computing

6th International Symposium, ISVC 2010
Las Vegas, NV, USA
November 29 – December 1, 2010
Proceedings, Part I

 Springer

Volume Editors

George Bebis, E-mail: bebis@cse.unr.edu

Richard Boyle, E-mail: richard.boyle@nasa.gov

Bahram Parvin, E-mail: parvin@hpcrd.lbl.gov

Darko Koracin, E-mail: darko@dri.edu

Ronald Chung, E-mail: rchung@mae.cuhk.edu.hk

Riad Hammoud, E-mail: riad.hammoud@dynavoxtech.com

Muhammad Hussain, E-mail: mhussain@ccis.edu.sa

Tan Kar-Han, E-mail: karhan.tan@hp.com

Roger Crawfis, E-mail: crawfis@cse.ohio-state.edu

Daniel Thalmann, E-mail: daniel.thalmann@epfl.ch

David Kao, E-mail: davidkao@nas.nasa.gov

Lisa Avila, E-mail: lisa.avila@kitware.com

Library of Congress Control Number: 2010939054

CR Subject Classification (1998): I.3, H.5.2, I.4, I.5, I.2.10, J.3, F.2.2, I.3.5

LNCS Sublibrary: SL 6 – Image Processing, Computer Vision, Pattern Recognition, and Graphics

ISSN	0302-9743
ISBN-10	3-642-17288-1 Springer Berlin Heidelberg New York
ISBN-13	978-3-642-17288-5 Springer Berlin Heidelberg New York

springer.com

© Springer-Verlag Berlin Heidelberg 2010
Printed in Germany

Typesetting: Camera-ready by author, data conversion by Scientific Publishing Services, Chennai, India
Printed on acid-free paper   06/3180

Lecture Notes in Computer Science 6453

Commenced Publication in 1973
Founding and Former Series Editors:
Gerhard Goos, Juris Hartmanis, and Jan van Leeuwen

Editorial Board

David Hutchison
  Lancaster University, UK
Takeo Kanade
  Carnegie Mellon University, Pittsburgh, PA, USA
Josef Kittler
  University of Surrey, Guildford, UK
Jon M. Kleinberg
  Cornell University, Ithaca, NY, USA
Alfred Kobsa
  University of California, Irvine, CA, USA
Friedemann Mattern
  ETH Zurich, Switzerland
John C. Mitchell
  Stanford University, CA, USA
Moni Naor
  Weizmann Institute of Science, Rehovot, Israel
Oscar Nierstrasz
  University of Bern, Switzerland
C. Pandu Rangan
  Indian Institute of Technology, Madras, India
Bernhard Steffen
  TU Dortmund University, Germany
Madhu Sudan
  Microsoft Research, Cambridge, MA, USA
Demetri Terzopoulos
  University of California, Los Angeles, CA, USA
Doug Tygar
  University of California, Berkeley, CA, USA
Gerhard Weikum
  Max Planck Institute for Informatics, Saarbruecken, Germany

Preface

It is with great pleasure that we present the proceedings of the 6th International, Symposium on Visual Computing (ISVC 2010), which was held in Las Vegas, Nevada. ISVC provides a common umbrella for the four main areas of visual computing including vision, graphics, visualization, and virtual reality. The goal is to provide a forum for researchers, scientists, engineers, and practitioners throughout the world to present their latest research findings, ideas, developments, and applications in the broader area of visual computing.

This year, the program consisted of 14 oral sessions, one poster session, 7 special tracks, and 6 keynote presentations. The response to the call for papers was very good; we received over 300 submissions for the main symposium from which we accepted 93 papers for oral presentation and 73 papers for poster presentation. Special track papers were solicited separately through the Organizing and Program Committees of each track. A total of 44 papers were accepted for oral presentation and 6 papers for poster presentation in the special tracks.

All papers were reviewed with an emphasis on potential to contribute to the state of the art in the field. Selection criteria included accuracy and originality of ideas, clarity and significance of results, and presentation quality. The review process was quite rigorous, involving two – three independent blind reviews followed by several days of discussion. During the discussion period we tried to correct anomalies and errors that might have existed in the initial reviews. Despite our efforts, we recognize that some papers worthy of inclusion may have not been included in the program. We offer our sincere apologies to authors who contributions might have been overlooked.

We wish to thank everybody who submitted their work to ISVC 2010 for review. It was because of their contributions that we succeeded in having a technical program of high scientific quality. In particular, we would like to thank the ISVC 2010 Area Chairs, the organizing institutions (UNR, DRI, LBNL, and NASA Ames), the government and industrial sponsors (Air Force Research Lab, Intel, DigitalPersona, Equinox, Ford, Hewlett Packard, Mitsubishi Electric Research Labs, iCore, Toyota, Delphi, General Electric, Microsoft MSDN, and Volt), the international Program Committee, the special track organizers and their Program Committees, the keynote speakers, the reviewers, and especially the authors that contributed their work to the symposium. In particular, we would like to thank *Air Force Research Lab, Mitsubishi Electric Research Labs*, and *Volt* for kindly sponsoring four "best paper awards" this year.

We sincerely hope that ISVC 2010 offered opportunities for professional growth.

September 2010   ISVC 2010 Steering Committee and Area Chairs

Organization

ISVC 2010 Steering Committee

Bebis George   University of Nevada, Reno, USA
Boyle Richard   NASA Ames Research Center, USA
Parvin Bahram   Lawrence Berkeley National Laboratory,
                           USA
Koracin Darko   Desert Research Institute, USA

ISVC 2010 Area Chairs

Computer Vision

Chang Ronald   The Chinese University of Hong Kong,
                           Hong Kong
Hammoud Riad   DynaVox Systems, USA

Computer Graphics

Hussain Muhammad   King Saud University, Saudi Arabia
Tan Kar-Han   Hewlett Packard Labs, USA

Virtual Reality

Crawfis Roger   Ohio State University, USA
Thalman Daniel   EPFL, Switzerland

Visualization

Kao David   NASA Ames Research Lab, USA
Avila Lisa   Kitware, USA

Publicity

Erol Ali   Ocali Information Technology, Turkey

Local Arrangements

Regentova Emma   University of Nevada, Las Vegas, USA

Special Tracks

Porikli Fatih   Mitsubishi Electric Research Labs, USA

ISVC 2010 Keynote Speakers

Kakadiaris Ioannis	University of Houston, USA
Hollerer Tobias	University of California at Santa Barbara, USA
Stasko John	Georgia Institute of Technology, USA
Seitz Steve	University of Washington, USA
Pollefeys Marc	ETH Zurich, Switzerland
Majumder Aditi	University of California, Irvine, USA

ISVC 2010 International Program Committee

(Area 1) Computer Vision

Abidi Besma	University of Tennessee, USA
Abou-Nasr Mahmoud	Ford Motor Company, USA
Agaian Sos	University of Texas at San Antonio, USA
Aggarwal J. K.	University of Texas, Austin, USA
Amayeh Gholamreza	Eyecom, USA
Agouris Peggy	George Mason University, USA
Argyros Antonis	University of Crete, Greece
Asari Vijayan	University of Dayton, USA
Basu Anup	University of Alberta, Canada
Bekris Kostas	University of Nevada at Reno, USA
Belyaev Alexander	Max-Planck-Institut fuer Informatik, Germany
Bensrhair Abdelaziz	INSA-Rouen, France
Bhatia Sanjiv	University of Missouri-St. Louis, USA
Bimber Oliver	Johannes Kepler University Linz, Austria
Bioucas Jose	Instituto Superior Tecnico, Lisbon, Portugal
Birchfield Stan	Clemson University, USA
Bourbakis Nikolaos	Wright State University, USA
Brimkov Valentin	State University of New York, USA
Campadelli Paola	Università degli Studi di Milano, Italy
Cavallaro Andrea	Queen Mary, University of London, UK
Charalampidis Dimitrios	University of New Orleans, USA
Chellappa Rama	University of Maryland, USA
Chen Yang	HRL Laboratories, USA
Cheng Hui	Sarnoff Corporation, USA
Cochran Steven Douglas	University of Pittsburgh, USA
Cremers Daniel	University of Bonn, Germany
Cui Jinshi	Peking University, China
Darbon Jerome	CNRS-Ecole Normale Superieure de Cachan, France
Davis James W.	Ohio State University, USA

Debrunner Christian	Colorado School of Mines, USA
Demirdjian David	MIT, USA
Duan Ye	University of Missouri-Columbia, USA
Doulamis Anastasios	National Technical University of Athens, Greece
Dowdall Jonathan	510 Systems, USA
El-Ansari Mohamed	Ibn Zohr University, Morocco
El-Gammal Ahmed	University of New Jersey, USA
Eng How Lung	Institute for Infocomm Research, Singapore
Erol Ali	Ocali Information Technology, Turkey
Fan Guoliang	Oklahoma State University, USA
Ferri Francesc	Universitat de Valencia, Spain
Ferryman James	University of Reading, UK
Foresti GianLuca	University of Udine, Italy
Fowlkes Charless	University of California, Irvine, USA
Fukui Kazuhiro	The University of Tsukuba, Japan
Galata Aphrodite	The University of Manchester, UK
Georgescu Bogdan	Siemens, USA
Gleason, Shaun	Oak Ridge National Laboratory, USA
Goh Wooi-Boon	Nanyang Technological University, Singapore
Guerra-Filho Gutemberg	University of Texas Arlington, USA
Guevara, Angel Miguel	University of Porto, Portugal
Gustafson David	Kansas State University, USA
Harville Michael	Hewlett Packard Labs, USA
He Xiangjian	University of Technology, Sydney, Australia
Heikkilä Janne	University of Oulu, Filand
Heyden Anders	Lund University, Sweden
Hongbin Zha	Peking University, China
Hou Zujun	Institute for Infocomm Research, Singapore
Hua Gang	Nokia Research Center, USA
Imiya Atsushi	Chiba University, Japan
Jia Kevin	IGT, USA
Kamberov George	Stevens Institute of Technology, USA
Kampel Martin	Vienna University of Technology, Austria
Kamberova Gerda	Hofstra University, USA
Kakadiaris Ioannis	University of Houston, USA
Kettebekov Sanzhar	Keane inc., USA
Khan Hameed Ullah	King Saud University, Saudi Arabia
Kim Tae-Kyun	University of Cambridge, UK
Kimia Benjamin	Brown University, USA
Kisacanin Branislav	Texas Instruments, USA
Klette Reinhard	Auckland University, New Zealand
Kokkinos Iasonas	Ecole Centrale Paris, France
Kollias Stefanos	National Technical University of Athens, Greece

Papanikolopoulos Nikolaos	University of Minnesota, USA
Pati Peeta Basa	First Indian Corp., India
Patras Ioannis	Queen Mary University, London, UK
Petrakis Euripides	Technical University of Crete, Greece
Peyronnet Sylvain	LRDE/EPITA, France
Pinhanez Claudio	IBM Research, Brazil
Piccardi Massimo	University of Technology, Australia
Pietikäinen Matti	LRDE/University of Oulu, Finland
Porikli Fatih	Mitsubishi Electric Research Labs, USA
Prabhakar Salil	DigitalPersona Inc., USA
Prati Andrea	University of Modena and Reggio Emilia, Italy
Prokhorov Danil	Toyota Research Institute, USA
Prokhorov Pylvanainen Timo	Nokia, Finland
Qi Hairong	University of Tennessee at Knoxville, USA
Qian Gang	Arizona State University, USA
Raftopoulos Kostas	National Technical University of Athens, Greece
Reed Michael	Blue Sky Studios, USA
Regazzoni Carlo	University of Genoa, Italy
Regentova Emma	University of Nevada, Las Vegas, USA
Remagnino Paolo	Kingston University, UK
Ribeiro Eraldo	Florida Institute of Technology, USA
Robles-Kelly Antonio	National ICT Australia (NICTA), Australia
Ross Arun	West Virginia University, USA
Salgian Andrea	The College of New Jersey, USA
Samal Ashok	University of Nebraska, USA
Sato Yoichi	The University of Tokyo, Japan
Samir Tamer	Ingersoll Rand Security Technologies, USA
Sandberg Kristian	Computational Solutions, USA
Sarti Augusto	DEI Politecnico di Milano, Italy
Savakis Andreas	Rochester Institute of Technology, USA
Schaefer Gerald	Loughborough University, UK
Scalzo Fabien	University of California at Los Angeles, USA
Scharcanski Jacob	UFRGS, Brazil
Shah Mubarak	University of Central Florida, USA
Shi Pengcheng	The Hong Kong University of Science and Technology, Hong Kong
Shimada Nobutaka	Ritsumeikan University, Japan
Singh Meghna	University of Alberta, Canada
Singh Rahul	San Francisco State University, USA
Skurikhin Alexei	Los Alamos National Laboratory, USA
Souvenir, Richard	University of North Carolina - Charlotte, USA

Su Chung-Yen	National Taiwan Normal University, Taiwan
Sugihara Kokichi	University of Tokyo, Japan
Sun Zehang	Apple, USA
Syeda-Mahmood Tanveer	IBM Almaden, USA
Tan Tieniu	Chinese Academy of Sciences, China
Tavakkoli Alireza	University of Houston - Victoria, USA
Tavares, Joao	Universidade do Porto, Portugal
Teoh Eam Khwang	Nanyang Technological University, Singapore
Thiran Jean-Philippe	Swiss Federal Institute of Technology Lausanne (EPFL), Switzerland
Tistarelli Massimo	University of Sassari, Italy
Tsechpenakis Gabriel	University of Miami, USA
Tsui T.J.	Chinese University of Hong Kong, Hong Kong
Trucco Emanuele	University of Dundee, UK
Tubaro Stefano	DEI, Politecnico di Milano, Italy
Uhl Andreas	Salzburg University, Austria
Velastin Sergio	Kingston University London, UK
Verri Alessandro	Universitá di Genova, Italy
Wang Charlie	The Chinese University of Hong Kong, Hong Kong
Wang Junxian	Microsoft, USA
Wang Song	University of South Carolina, USA
Wang Yunhong	Beihang University, China
Webster Michael	University of Nevada, Reno, USA
Wolff Larry	Equinox Corporation, USA
Wong Kenneth	The University of Hong Kong, Hong Kong
Xiang Tao	Queen Mary, University of London, UK
Xue Xinwei	Fair Isaac Corporation, USA
Xu Meihe	University of California at Los Angeles, USA
Yang Ruigang	University of Kentucky, USA
Yi Lijun	SUNY at Binghampton, USA
Yu Kai	NEC Labs, USA
Yu Ting	GE Global Research, USA
Yu Zeyun	University of Wisconsin-Milwaukee, USA
Yuan Chunrong	University of Tuebingen, Germany
Zhang Yan	Delphi Corporation, USA
Zhou Huiyu	Queen's University Belfast, UK

(Area 2) Computer Graphics

Abd Rahni Mt Piah	Universiti Sains Malaysia, Malaysia
Abram Greg	IBM T.J.Watson Reseach Center, USA
Adamo-Villani Nicoletta	Purdue University, USA

Agu Emmanuel   Worcester Polytechnic Institute, USA
Andres Eric   Laboratory XLIM-SIC, University of Poitiers, France

Artusi Alessandro   CaSToRC Cyprus Institute, Cyprus
Baciu George   Hong Kong PolyU, Hong Kong
Balcisoy Selim Saffet   Sabanci University, Turkey
Barneva Reneta   State University of New York, USA
Bartoli Vilanova Anna   Eindhoven University of Technology, The Netherlands

Belyaev Alexander   Max Planck-Institut fuer Informatik, Germany

Benes Bedrich   Purdue University, USA
Berberich Eric   Max-Planck Institute, Germany
Bilalis Nicholas   Technical University of Crete, Greece
Bimber Oliver   Johannes Kepler University Linz, Austria
Bohez Erik   Asian Institute of Technology, Thailand
Bouatouch Kadi   University of Rennes I, IRISA, France
Brimkov Valentin   State University of New York, USA
Brown Ross   Queensland University of Technology, Australia

Callahan Steven   University of Utah, USA
Chen Min   University of Wales Swansea, UK
Cheng Irene   University of Alberta, Canada
Chiang Yi-Jen   Polytechnic Institute of New York University, USA

Choi Min   University of Colorado at Denver, USA
Comba Joao   Univ. Fed. do Rio Grande do Sul, Brazil
Cremer Jim   University of Iowa, USA
Culbertson Bruce   HP Labs, USA
Debattista Kurt   University of Warwick, UK
Deng Zhigang   University of Houston, USA
Dick Christian   Technical University of Munich, Germany
DiVerdi Stephen   Adobe, USA
Dingliana John   Trinity College, Ireland
El-Sana Jihad   Ben Gurion University of The Negev, Israel
Entezari Alireza   University of Florida, USA
Fiorio Christophe   Université Montpellier 2, LIRMM, France
Floriani Leila De   University of Genoa, Italy
Gaither Kelly   University of Texas at Austin, USA
Gao Chunyu   Epson Research and Development, USA
Geist Robert   Clemson University, USA
Gelb Dan   Hewlett Packard Labs, USA
Gotz David   IBM, USA
Gooch Amy   University of Victoria, Canada

Gu David	State University of New York at Stony Brook, USA
Guerra-Filho Gutemberg	University of Texas Arlington, USA
Habib Zulfiqar	National University of Computer and Emerging Sciences, Pakistan
Hadwiger Markus	KAUST, Saudi Arabia
Haller Michael	Upper Austria University of Applied Sciences, Austria
Hamza-Lup Felix	Armstrong Atlantic State University, USA
Han JungHyun	Korea University, Korea
Hao Xuejun	Columbia University and NYSPI, USA
Hernandez Jose Tiberio	Universidad de los Andes, Colombia
Huang Mao Lin	University of Technology, Australia
Huang Zhiyong	Institute for Infocomm Research, Singapore
Joaquim Jorge	Instituto Superior Tecnico, Portugal
Ju Tao	Washington University, USA
Julier Simon J.	University College London, UK
Kakadiaris Ioannis	University of Houston, USA
Kamberov George	Stevens Institute of Technology, USA
Kim Young	Ewha Womans University, Korea
Klosowski James	AT&T Labs, USA
Kobbelt Leif	RWTH Aachen, Germany
Kuan Lee Hwee	Bioinformatics Institute, ASTAR, Singapore
Lai Shuhua	Virginia State University, USA
Lakshmanan Geetika	IBM T.J. Watson Reseach Center, USA
Lee Chang Ha	Chung-Ang University, Korea
Lee Tong-Yee	National Cheng-Kung University, Taiwan
Levine Martin	McGill University, Canada
Lewis Bob	Washington State University, USA
Li Frederick	University of Durham, UK
Lindstrom Peter	Lawrence Livermore National Laboratory, USA
Linsen Lars	Jacobs University, Germany
Loviscach Joern	Fachhochschule Bielefeld (University of Applied Sciences), Germany
Magnor Marcus	TU Braunschweig, Germany
Majumder Aditi	University of California, Irvine, USA
Mantler Stephan	VRVis Research Center, Austria
Martin Ralph	Cardiff University, UK
McGraw Tim	West Virginia University, USA
Meenakshisundaram Gopi	University of California-Irvine, USA
Mendoza Cesar	NaturalMotion Ltd., USA
Metaxas Dimitris	Rutgers University, USA
Myles Ashish	University of Florida, USA
Nait-Charif Hammadi	University of Dundee, UK

Nasri Ahmad	American University of Beirut, Lebanon
Noma Tsukasa	Kyushu Institute of Technology, Japan
Okada Yoshihiro	Kyushu University, Japan
Olague Gustavo	CICESE Research Center, Mexico
Oliveira Manuel M.	Univ. Fed. do Rio Grande do Sul, Brazil
Ostromoukhov Victor M.	University of Montreal, Canada
Pascucci Valerio	University of Utah, USA
Peters Jorg	University of Florida, USA
Qin Hong	State University of New York at Stony Brook, USA
Razdan Anshuman	Arizona State University, USA
Reed Michael	Columbia University, USA
Renner Gabor	Computer and Automation Research Institute, Hungary
Rosenbaum Rene	University of California at Davis, USA
Rushmeier	Holly, Yale University, USA
Sander Pedro	The Hong Kong University of Science and Technology, Hong Kong
Sapidis Nickolas	University of Western Macedonia, Greece
Sarfraz Muhammad	Kuwait University, Kuwait
Scateni Riccardo	University of Cagliari, Italy
Schaefer Scott	Texas A&M University, USA
Sequin Carlo	University of California-Berkeley, USA
Shead Tinothy	Sandia National Laboratories, USA
Sorkine Olga	New York University, USA
Sourin Alexei	Nanyang Technological University, Singapore
Stamminger Marc	REVES/INRIA, France
Su Wen-Poh	Griffith University, Australia
Staadt Oliver	University of Rostock, Germany
Tarini Marco	Università dell'Insubria (Varese), Italy
Teschner Matthias	University of Freiburg, Germany
Tsong Ng Tian	Institute for Infocomm Research, Singapore
Umlauf Georg	HTWG Constance, Germany
Wald Ingo	University of Utah, USA
Wang Sen	Kodak, USA
Wimmer Michael	Technical University of Vienna, Austria
Wylie Brian	Sandia National Laboratory, USA
Wyman Chris	University of Iowa, USA
Yang Qing-Xiong	University of Illinois at Urbana, Champaign, USA
Yang Ruigang	University of Kentucky, USA
Ye Duan	University of Missouri-Columbia, USA
Yi Beifang	Salem State College, USA
Yin Lijun	Binghamton University, USA

Yoo Terry   National Institutes of Health, USA
Yuan Xiaoru   Peking University, China
Zabulis Xenophon   Foundation for Research and
                                   Technology - Hellas (FORTH), Greece
Zhang Eugene   Oregon State University, USA
Zhang Jian Jun   Bournemouth University, UK
Zordan Victor   University of California at Riverside, USA

(Area 3) Virtual Reality

Alcañiz Mariano   Technical University of Valencia, Spain
Arns Laura   Purdue University, USA
Balcisoy Selim   Sabanci University, Turkey
Behringer Reinhold   Leeds Metropolitan University UK
Benes Bedrich   Purdue University, USA
Bilalis Nicholas   Technical University of Crete, Greece
Blach Roland   Fraunhofer Institute for Industrial
                                   Engineering, Germany
Blom Kristopher   University of Hamburg, Germany
Borst Christoph   University of Louisiana at Lafayette, USA
Brady Rachael   Duke University, USA
Brega Jose Remo Ferreira   Universidade Estadual Paulista, Brazil
Brown Ross   Queensland University of Technology,
                                   Australia
Bruce Thomas   The University of South Australia,
                                   Australia
Bues Matthias   Fraunhofer IAO in Stuttgart, Germany
Chen Jian   Brown University, USA
Cheng Irene   University of Alberta, Canada
Coquillart Sabine   INRIA, France
Craig Alan   NCSA University of Illinois at
                                   Urbana-Champaign, USA
Cremer Jim   University of Iowa, USA
Egges Arjan   Universiteit Utrecht, The Netherlands
Encarnacao L. Miguel   Humana Inc., USA
Figueroa Pablo   Universidad de los Andes, Colombia
Fox Jesse   Stanford University, USA
Friedman Doron   IDC, Israel
Froehlich Bernd   Weimar University, Germany
Gregory Michelle   Pacific Northwest National Lab, USA
Gupta Satyandra K.   University of Maryland, USA
Hachet Martin   INRIA, France
Haller Michael   FH Hagenberg, Austria
Hamza-Lup Felix   Armstrong Atlantic State University, USA
Hinkenjann Andre   Bonn-Rhein-Sieg University of Applied
                                   Sciences, Germany

Hollerer Tobias	University of California at Santa Barbara, USA
Huang Jian	University of Tennessee at Knoxville, USA
Julier Simon J.	University College London, UK
Klinker Gudrun	Technische Universität München, Germany
Klosowski James	AT&T Labs, USA
Kozintsev	Igor, Intel, USA
Kuhlen Torsten	RWTH Aachen University, Germany
Liere Robert van	CWI, The Netherlands
Majumder Aditi	University of California, Irvine, USA
Malzbender Tom	Hewlett Packard Labs, USA
Mantler Stephan	VRVis Research Center, Austria
Meyer Joerg	University of California, Irvine, USA
Molineros Jose	Teledyne Scientific and Imaging, USA
Muller Stefan	University of Koblenz, Germany
Paelke Volker	Leibniz Universität Hannover, Germany
Pan Zhigeng	Zhejiang University, China
Papka Michael	Argonne National Laboratory, USA
Peli Eli	Harvard University, USA
Pettifer Steve	The University of Manchester, UK
Pugmire Dave	Los Alamos National Lab, USA
Qian Gang	Arizona State University, USA
Raffin Bruno	INRIA, France
Reiners Dirk	University of Louisiana, USA
Richir Simon	Arts et Metiers ParisTech, France
Rodello Ildeberto	University of Sao Paulo, Brazil
Santhanam Anand	MD Anderson Cancer Center Orlando, USA
Sapidis Nickolas	University of Western Macedonia, Greece
Schulze	Jurgen, University of California - San Diego, USA
Sherman Bill	Jurgen, Indiana University, USA
Slavik Pavel	Czech Technical University in Prague, Czech Republic
Sourin Alexei	Nanyang Technological University, Singapore
Stamminger Marc	REVES/INRIA, France
Srikanth Manohar	Indian Institute of Science, India
Staadt Oliver	University of Rostock, Germany
Swan Ed	Mississippi State University, USA
Stefani Oliver	COAT-Basel, Switzerland
Sun Hanqiu	The Chinese University of Hong Kong, Hong Kong
Varsamidis Thomas	Bangor University, UK
Vercher Jean-Louis	Université de la Méditerrane, France
Wald Ingo	University of Utah, USA

Yu Ka Chun	Denver Museum of Nature and Science, USA
Yuan Chunrong	University of Tuebingen, Germany
Zachmann Gabriel	Clausthal University, Germany
Zara Jiri	Czech Technical University in Prague, Czech Republic
Zhang Hui	Indiana University, USA
Zhao Ye	Kent State University, USA
Zyda Michael	University of Southern California, USA

(Area 4) Visualization

Andrienko Gennady	Fraunhofer Institute IAIS, Germany
Apperley Mark	University of Waikato, New Zealand
Balázs Csébfalvi	Budapest University of Technology and Economics, Hungary
Bartoli Anna Vilanova	Eindhoven University of Technology, The Netherlands
Brady Rachael	Duke University, USA
Benes Bedrich	Purdue University, USA
Bilalis Nicholas	Technical University of Crete, Greece
Bonneau Georges-Pierre	Grenoble Université , France
Brown Ross	Queensland University of Technology, Australia
Bühler Katja	VRVIS, Austria
Callahan Steven	University of Utah, USA
Chen Jian	Brown University, USA
Chen Min	University of Wales Swansea, UK
Cheng Irene	University of Alberta, Canada
Chiang Yi-Jen	Polytechnic Institute of New York University, USA
Chourasia Amit	University of California - San Diego, USA
Coming Daniel	Desert Research Institute, USA
Dana Kristin	Rutgers University, USA
Dick Christian	Technical University of Munich, Germany
DiVerdi Stephen	Adobe, USA
Doleisch Helmut	VRVis Research Center, Austria
Duan Ye	University of Missouri-Columbia, USA
Dwyer Tim	Monash University, Australia
Ebert David	Purdue University, USA
Entezari Alireza	University of Florida, USA
Ertl Thomas	University of Stuttgart, Germany
Floriani Leila De	University of Maryland, USA
Fujishiro Issei	Keio University, Japan
Geist Robert	Clemson University, USA
Goebel Randy	University of Alberta, Canada

Gotz David	IBM, USA
Grinstein Georges	University of Massachusetts Lowell, USA
Goebel Randy	University of Alberta, Canada
Gregory Michelle	Pacific Northwest National Lab, USA
Hadwiger Helmut Markus	VRVis Research Center, Austria
Hagen Hans	Technical University of Kaiserslautern, Germany
Hamza-Lup Felix	Armstrong Atlantic State University, USA
Heer Jeffrey	Armstrong University of California at Berkeley, USA
Hege Hans-Christian	Zuse Institute Berlin, Germany
Hochheiser Harry	University of Pittsburgh, USA
Hollerer Tobias	University of California at Santa Barbara, USA
Hong Lichan	Palo Alto Research Center, USA
Hotz Ingrid	Zuse Institute Berlin, Germany
Jiang Ming	Lawrence Livermore National Laboratory, USA
Joshi Alark	Yale University, USA
Julier Simon J.	University College London, UK
Kohlhammer Jörn	Fraunhofer Institut, Germany
Kosara Robert	University of North Carolina at Charlotte, USA
Laramee Robert	Swansea University, UK
Lee Chang Ha	Chung-Ang University, Korea
Lewis Bob	Washington State University, USA
Liere Robert van	CWI, The Netherlands
Lim Ik Soo	Bangor University, UK
Linsen Lars	Jacobs University, Germany
Liu Zhanping	Kitware, Inc., USA
Ma Kwan-Liu	University of California-Davis, USA
Maeder Anthony	University of Western Sydney, Australia
Majumder Aditi	University of California, Irvine, USA
Malpica Jose	Alcala University, Spain
Masutani Yoshitaka	The University of Tokyo Hospital, Japan
Matkovic Kresimir	VRVis Forschungs-GmbH, Austria
McCaffrey James	Microsoft Research / Volt VTE, USA
McGraw Tim	West Virginia University, USA
Melançon Guy	CNRS UMR 5800 LaBRI and INRIA Bordeaux Sud-Ouest, France
Meyer Joerg	University of California, Irvine, USA
Miksch Silvia	Vienna University of Technology, Austria
Monroe Laura	Los Alamos National Labs, USA
Morie Jacki	University of Southern California, USA

ISVC 2010 Special Tracks

1. 3D Mapping, Modeling and Surface Reconstruction

Organizers

Nefian Ara	Carnegie Mellon University/NASA Ames Research Center, USA
Broxton Michael	Carnegie Mellon University/NASA Ames Research Center, USA
Huertas Andres	NASA Jet Propulsion Lab, USA

Program Committee

Hancher Matthew	NASA Ames Research Center, USA
Edwards Laurence	NASA Ames Research Center, USA
Bradski Garry	Willow Garage, USA
Zakhor Avideh	University of California at Berkeley, USA
Cavallaro Andrea	University Queen Mary, London, UK
Bouguet Jean-Yves	Google, USA

2. Best Practices in Teaching Visual Computing

Organizers

Albu Alexandra Branzan	University of Victoria, Canada
Bebis George	University of Nevada, Reno, USA

Program Committee

Bergevin Robert	University of Laval, Canada
Crawfis Roger	Ohio State University, USA
Hammoud Riad	DynaVox Systems, USA
Kakadiaris Ioannis	University of Houston, USA, USA
Laurendeau Denis	Laval University, Quebec, Canada
Maxwell Bruce	Colby College, USA
Stockman George	Michigan State University, USA

3. Low-Level Color Image Processing

Organizers

Celebi M. Emre	Louisiana State University, USA
Smolka Bogdan	Silesian University of Technology, Poland
Schaefer Gerald	Loughborough University, UK
Plataniotis Konstantinos	University of Toronto, Canada
Horiuchi Takahiko	Chiba University, Japan

Program Committee

Aygun Ramazan	University of Alabama in Huntsville, USA
Battiato Sebastiano	University of Catania, Italy
Hardeberg Jon	Gjøvik University College, Norway
Hwang Sae	University of Illinois at Springfield, USA
Kawulok Michael	Silesian University of Technology, Poland
Kockara Sinan	University of Central Arkansas, USA
Kotera Hiroaki	Kotera Imaging Laboratory, Japan
Lee JeongKyu	University of Bridgeport, USA
Lezoray Olivier	University of Caen, France
Mete Mutlu	Texas A&M University - Commerce, USA
Susstrunk Sabine	Swiss Federal Institute of Technology in Lausanne, Switzerland
Tavares Joao	University of Porto, Portugal
Tian Gui Yun	Newcastle University, UK
Wen Quan	University of Electronic Science and Technology of China, China
Zhou Huiyu	QueenŠs University Belfast, UK

4. Low Cost Virtual Reality: Expanding Horizons

Organizers

Sherman Bill	Indiana University, USA
Wernert Eric	Indiana University, USA

Program Committee

Coming Daniel	Desert Research Institute, USA
Craig Alan	University of Illinois/NCSA, USA
Keefe Daniel	University of Minnesota, USA
Kreylos Oliver	University of California at Davis, USA
O'Leary Patrick	Idaho National Laboratory, USA
Smith Randy	Oakland University, USA
Su Simon	Princeton University, USA
Will Jeffrey	Valparaiso University, USA

5. Computational Bioimaging

Organizers

Tavares João Manuel R. S.	University of Porto, Portugal
Jorge Renato Natal	University of Porto, Portugal
Cunha Alexandre	Caltech, USA

Program Committee

Santis De Alberto	Università degli Studi di Roma "La Sapienza", Italy
Reis Ana Mafalda	Instituto de Ciencias Biomedicas Abel Salazar, Portugal
Barrutia Arrate Muñoz	University of Navarra, Spain
Calvo Begoña	University of Zaragoza, Spain
Constantinou Christons	Stanford University, USA
Iacoviello Daniela	Università degli Studi di Roma "La Sapienza", Italy
Ushizima Daniela	Lawrence Berkeley National Lab, USA
Ziou Djemel	University of Sherbrooke, Canada
Pires Eduardo Borges	Instituto Superior Tecnico, Portugal
Sgallari Fiorella	University of Bologna, Italy
Perales Francisco	Balearic Islands University, Spain
Qiu Guoping	University of Nottingham, UK
Hanchuan Peng	Howard Hughes Medical Institute, USA
Pistori Hemerson	Dom Bosco Catholic University, Brazil
Yanovsky Igor	Jet Propulsion Laboratory, USA
Corso Jason	SUNY at Buffalo, USA
Maldonado Javier Melenchón	Open University of Catalonia, Spain
Marques Jorge S.	Instituto Superior Tecnico, Portugal
Aznar Jose M. García	University of Zaragoza, Spain
Vese Luminita	University of California at Los Angeles, USA
Reis Luís Paulo	University of Porto, Portugal
Thiriet Marc	Universite Pierre et Marie Curie (Paris VI), France
Mahmoud El-Sakka	The University of Western Ontario London, Canada
Hidalgo Manuel González	Balearic Islands University, Spain
Gurcan Metin N.	Ohio State University, USA
Dubois Patrick	Institut de Technologie Médicale, France
Barneva Reneta P.	State University of New York, USA
Bellotti Roberto	University of Bari, Italy
Tangaro Sabina	University of Bari, Italy
Silva Susana Branco	University of Lisbon, Portugal
Brimkov Valentin	State University of New York, USA
Zhan Yongjie	Carnegie Mellon University, USA

6. Unconstrained Biometrics: Advances and Trends

Organizers

Proença Hugo	University of Beira Interior, Portugal
Du Yingzi	Indiana University-Purdue University Indianapolis, USA

Scharcanski Jacob	Federal University of Rio Grande do Sul Porto Alegre, Brazil
Ross Arun	West Virginia University, USA
Amayeh Gholamreza	EyeCom Corporation, USA

Program Committee

Júnior Adalberto Schuck	Federal University of Rio Grande do Sul, Brazil
Kwolek Bogdan	Rzeszow University of Technology, Poland
Jung Cláudio R.	Federal University of Rio Grande do Sul, Brazil
Alirezaie Javad	Ryerson University, Canada
Konrad Janusz	Boston University, USA
Kevin Jia	International Game Technologies, USA
Meyer Joceli	Federal University of Santa Catarina, Brazil
Alexandre Luís A.	University of Beira Interior, Portugal
Soares Luis	ISCTE, Portugal
Coimbra Miguel	University of Porto, Portugal
Fieguth Paul	University of Waterloo, Canada
Xiao Qinghan	Defense Research and Development Canada, Canada
Ives Robert	United States Naval Academy, USA
Tamir Samir	Ingersoll Rand Security, USA

7. Behavior Detection and Modeling

Organizers

Miller Ron	Wright-Patterson Air Force Base, USA
Bebis George	University of Nevada, USA
Rosen Julie	Science Applications International Corporation, USA
Davis Jim	Ohio State University, USA
Lee Simon	Army Research Laboratory, USA
Zandipour Majid	BAE Systems, USA

Organizing Institutions and Sponsors

Table of Contents – Part I

ST: Behavior Detection and Modeling

ST: Low-Level Color Image Processing

Feature Extraction and Matching

Visualization I

Motion and Tracking

ST: Unconstrained Biometrics: Advances and Trends

ST: Computational Bioimaging II

Computer Graphics II

ST: 3D Mapping, Modeling and Surface Reconstruction

Virtual Reality I

Table of Contents – Part II

Calibration, Pose Estimation, and Reconstruction

Segmentation

Registration

Medical Imaging

ST: Low Cost Virtual Reality: Expanding Horizons

ST: Best Practices in Teaching Visual Computing

Applications

Visualization II

Video Analysis and Event Recognition

Poster Session

Table of Contents – Part III

Poster Session

Ontology-Driven Image Analysis for Histopathological Images

Ahlem Othmani, Carole Meziat, and Nicolas Loménie*

CNRS - French National Center for Scientific Research
IPAL joint lab - UMI CNRS
Institute for Infocomm Research, A*STAR
nicolas.lomenie@mi.parisdescartes.fr
http://ipal.i2r.a-star.edu.sg/

Abstract. Ontology-based software and image processing engine must cooperate in new fields of computer vision like microscopy acquisition wherein the amount of data, concepts and processing to be handled must be properly controlled. Within our own platform, we need to extract biological objects of interest in huge size and high-content microscopy images. In addition to specific low-level image analysis procedures, we used knowledge formalization tools and high-level reasoning ability of ontology-based software. This methodology made it possible to improve the expressiveness of the clinical models, the usability of the platform for the pathologist and the sensitivity or sensibility of the low-level image analysis algorithms.

1   Introduction

Usually in medical imaging, after the acquisition step, computer vision researchers propose new algorithms dedicated to a specific task like the segmentation of the liver out of MRI images or the counting of cells over stained images. For macroscopic natural images also, dedicated softwares for face recognition for instance have already been successfully delivered on the market so far. But, when considering new devices from satellite to microscopy imaging systems, the resolution and size at which images are acquired provide huge amount of biological and natural data to process in a parallel way, within a limited frame time and more or less on a pervasive mode in the near future [1].

For instance, the digitalization of biopsy images is raising new issues due to the exploration of what is called Whole Slide Images (WSI). For one patient, the amount of visual data to process over this WSI is about eight gigabyte. Various biological objects must be detected and segmented in order to infer any aid to the pathologist for the diagnosis. The spatial relationships between these different objects must be used as well to improve the efficiency of the automatic analysis of the data. As a matter of fact, if a WSI in histopathology is about a 50 000 by 40 000 pixels size image, it is now common to produce satellite images

---

* Corresponding author.

G. Bebis et al. (Eds.): ISVC 2010, Part I, LNCS 6453, pp. 1–12, 2010.

at very high resolution of about 30 000 by 30 000 pixels size (see the Pleiades satellite resolution).

It is not reasonable to systematically plan the design of image analysis algorithm on the fly as new needs are required. In any case, it will take time and money to improve the capacity of automatic annotation of these images. Ontology-driven interface and processing can be an alternative to this engineering constraints. First, systematically involving ontological descriptions on a platform improves the interaction standards with the novice end-user and also within the software designer team by modeling the knowledge and the objects in a formal way. Second, high-level reasoning based on the formalized concepts can provide an alternative way to detect biological objects. Last, it provides the end-user (like the pathologist) a semantic way to specify a query based on the results of the image analysis modules currently available in the system. From a pattern recognition point of view, it can help to lower the false alarm rate by adding high-level constraint rules or to improve the correct detection rate within a fixed time frame constraint by triggering the image analysis algorithms only on specific areas in the image defined by high-level spatial relationships rules for instance.

The ontology contribution is particularly relevant in the field of pathology and clinical imaging where a mental database is constantly used by the physician either coming from books or from his/her acquired experience over years of visual inspection of clinical data. This is the reason why our research work -even though aimed at being quite generic however- is however dedicated to a specific application and a platform we designed to automatically grade breast cancer out of histopathological images [2].

This work proposes to leverage the high-level reasoning and knowledge formalization ability of ontology-based softwares to make annotation of high-content images more efficient and interactive. Few works have operationally explored this kind of idea among which we can mention [3,4].

Section 2 focuses on the low-level image analysis modules currently available in our system. Section 3 elaborates on the ontology part of the system and illustrates the use of the reasoning capability to infer new results or control the low-level image engine. Section 4 gives and discusses elements of quantitative assessment of the ontology-driven strategy before drawing a conclusion in the last section.

2   Low-Level Image Annotation

The low-level image processing aims at outlining and describing general biological objects in the histopathological image. The current platform uses images from breast cancer biopsies. Three characteristics are used in breast cancer grading according to the Nottingham Grading System [5]:

- tubular formation of cells around lumina: the better formed the tubular formations are, the lower the cancer grade is;

- nuclear pleomorphism, that comes from nuclei features (area, mean and standard deviation intensity and circularity or roundness features): the bigger the nuclei are, the less regular their shape is and the less uniform their texture is, the higher the cancer grade is.
- mitosis number: the more mitoses are present in the image, the higher the cancer grade is.

Currently in our application, three different kinds of biological objects are detected to be able to provide an image with a cancer grade: the nuclei, the lumina and the invasive areas.

Nuclei segmentation. The nuclei detection module proceeds in two steps. First nuclei seeds are identified and then each detected nucleus is automatically segmented to extract geometric and radiometric features about it. The nuclei seeds extraction follows two processing steps: the regions of interest detection and then the nuclei identification (see Figures 1(b) and (c)).

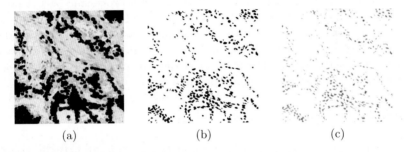

(a)   (b)   (c)

Fig. 1. Nuclei identification - (a) Regions of interest detection, (b) nuclei identification (coarse nuclei separation), (c) nuclei identification based on a distance map

The region of interest detection step locates the part of the images that contains the nuclei. Usually, cells are grouped together all around lumina and form what is called tubules. This step creates a mask to locate nuclei clusters that contain useful information. The following processing chain is performed: (1) Automatic image thresholding in order to distinguish the nuclei from the image background; (2) Morphological closure in order to group close nuclei together; (3) Removal of small objects not useful for ulterior processing or studies (see Figure 1(a)). The nuclei identification step proceeds by similar morphological filtering operators before drawing a distance map over which points within the nuclei area being the furthest from the boundaries are identified as the nuclei seeds (see Figure 2(a)). The nuclear boundaries are extracted using a snake-based method described in [6]. Patches of images that contain nuclei are extracted and are subjected to a polar transform of the coordinate system. After a first processing that constructs the first lace close to the real nuclei boundaries, the iterative snake algorithm outlines the nuclei boundary (see Figure 2(b)). Then geometric and radiometric features can be extracted over each detected nucleus.

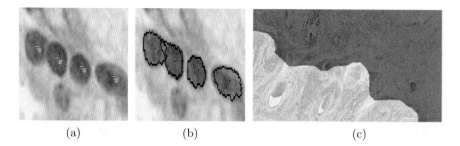

(a)   (b)   (c)

Fig. 2. An example of (a) seeds detection and (b) nuclei segmentation at high magnification x40 (c) invasive area detection at low magnification x1.2

Lumina and Invasive Area segmentation. The low-level detection of the lumina uses mathematical morphology tools . The invasive ROI detection is currently casted as a classification problem whereby we exploited the relationship between human vision and neurosciences [7]. As the low-level processing part is not the core of this paper, we just give an illustration of the obtained results in our platform for the low-level detection of the invasive areas (see Figure 2(c)). The idea is now to exploit these biological landmarks to perform reasoning and knowledge management over the microscopy slide, as the extraction of all the biological concepts in an exhaustive way is not possible.

3   Ontology-Driven Image Analysis

The algorithms briefly described in the previous section are actually standard low-level ones based on signal analysis: they allow a "black-box" detection of biological structures useful to draw a diagnosis based on a medical protocol. Yet, medical protocols or knowledge are constantly evolving or refining and can be very specific to an expert mental database as well, so that developing a complete image analysis platform in a complex field as histopathology is very costly and versatile in terms of engineering.

One way to overcome this issue and to facilitate the design of such complex, evolving platform as well is to work at a higher semantic level. For that purpose, the ontology framework constitutes a powerful tool for formalizing knowledge and for reasoning. In fact, the ontology is used to make the engineering of our heterogeneous knowledge database: medical knowledge but also image processing results coming from the image engine.

With this in mind, a few applications have been developed so far on our grading platform to experiment the benefits of articulating ontology capabilities with image processing outcomes. They correspond to two generic objectives of our research work:

- Consistency-checking annotation: to improve the specificity rate;
- Image Analysis Engine Triggering Control: to improve the sensibility rate within a limited response time.

Before illustrating these two concepts, the next subsection draws a brief technical description of the core anatomical ontology for breast cancer grading we already built up.

3.1 Anatomical Ontology: OWL

An ontology is a system of knowledge representation of a domain in the form of a structured set of concepts and relationships between these concepts. An ontology is expressed in the form of a XML graph and produces reasoning through a rule language. Our Breast Cancer Ontology (BCO) is based on two languages: OWL-DL (Web Ontology Language Description Logics) to describe the ontology and SWRL (Semantic Web Rule Language) to write and manage rules for the reasoning part. Technically, OWL and SWRL are specifications of the W3C[1], OWL is an extension of RDF (Resource Description Framework) used in the description of classes and types of properties, SWRL combines OWL and RuleML (Rule Markup Language) to produce the rules for the reasoning. The annotated images are described with the Wide Field Markup Language (WFML)[2] specific to the histopathology field (see Figure 4). Finally, the query language SPARQL (Simple Protocol And RDF Query Language) is used for querying in Java. SPARQL has been chosen for its ease of use and the very good integration of the API in Java. A thorough description of this ontology-based platform can be found in [8,9].

3.2 Rules and Reasoning

Once the anatomical and medical core concepts are formalized, we can feed our WFML database with new annotations based on a reasoning process.

Consistency Checking Annotation. Usually in the bio-medical field, the objects of interest are described by the biologists with high-level descriptions. However, the image analyzers use signal-based definition of these concepts. Subsequently, it is not uncommon to have the opportunity to cross both ways of defining a biological structure, like for the mitosis for instance. On an ideal platform, we will get two ways for defining mitoses:

- a low-level - in a sense implicit - signal-based extraction providing a set of results $\mathcal{R}^{signal}$, usually by statistical learning;
- an explicit high-level description corresponding to a SWRL rule like the one expressed in the Protégé[3] platform in Figure 3 and potentially providing a set of results $\mathcal{R}^{knowledge}$, and where Circularity and Roundness are the standard shape features.

---

[1] World Wide Web Consortium.
[2] A XML language produced by the company TRIBVN for its platform ICS Framework.
[3] http://protege.stanford.edu/

⟶ Nucleus(?x) ∧ hasIntensity(?x, ?value) ∧ swrlb:lessThan(?value, 110.0) ∧ hasCircularity(?y, ?cir)
∧ swrlb:lessThan(?cir, 0.75) ∧ hasRound(?z, ?round) ∧ swrlb:lessThan(?round, 0.65) ⟶ Mitosis(?x)

Fig. 3. A SWRL rule for mitosis description in our BCO (Breast Cancer Ontology) within the Protégé platform

In the case that $\mathcal{R}^{signal}$ is currently available on the platform, we can check the consistency of this result set by the semantic rule expressed in Figure 3 in the way Mechouche et al. proceeded for brain annotation issues [3]. This semantic checking will provide a set of results $\mathcal{R}^{signal \times knowledge}$ lowering the false alarm rate and subsequently improving the specificity rate of the image engine.

In the case that an image analysis module detecting the nuclei is currently available but not a mitosis detector, the platform can use the semantic rules defined by the pathologist on the fly to enrich the annotation WFML file by reasoning and providing a knowledge-based result set $\mathcal{R}^{knowledge}$ to this kind of semantic query. The basic principle of this case study is the following: from an original WFML file containing annotations about nuclei, we seek those corresponding to mitoses based on the semantic rule in Figure 3 and enrich the annotation file whenever it detects a mitosis. Technically, the WFML is parsed to retrieve information for each nucleus. The OWL file in the Protégé platform is powered by the list of nuclei in order to use the logic reasoning engine. This is the reason why we need a matching procedure between the WFML Files (specific to our application) and generic OWL Files (to benefit from the reasoning capability) as described in Figure 4(a). A reasoning procedure is then performed with the SPARQL query language which defines the syntax and semantics necessary to express queries on RDF type Database (Figure 4(b)).

Then the nuclei that are recognized as mitoses are modified in the WFML file by changing the annotation from a nucleus NP (standing for Nuclear Pleomorphism) into Mitosis (see Figure 5 for the global overview of the annotation updating process and Figure 6 for the WFML-based annotated resulting images[4]).

Image Analysis Engine Triggering Control. Another issue of high-content image annotation is the limited response time we must fit in. The ability to control the image analysis triggering over the While Slide Image can help to improve the sensibility rate of the platform under the time constraint.

From rule R1 for instance defined in first order logic in Equation 1, we can trigger the image analysis algorithms to detect the neoplasm as an invasive area, then locate formally the border of the neoplasm using a formalization of the spatial relation *Border* (see Figures 7 and 8).

$$R1 : Mitosis(X) \rightarrow \exists NeoplasmN/X \in Border(N) \qquad (1)$$

---

[4] In the ICS^{TM} Technology interface from TRIBVN S.A., the image format is a SVS format involving both a pyramidal TIFF multiscale description and WFML description file for the annotations currently available in the database (http://www.tribvn.com/

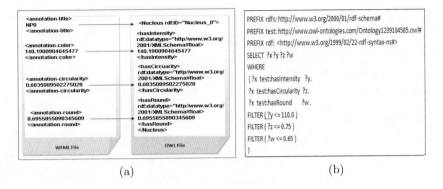

(a)   (b)

Fig. 4. (a) Matching between WFML and OWL files (b) SPARQL query sample

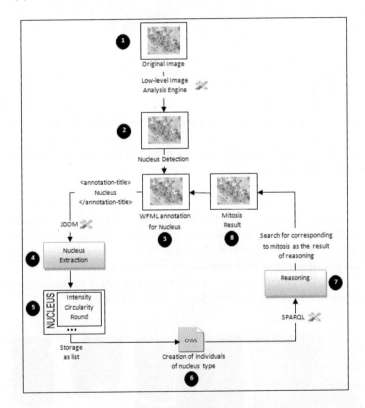

Fig. 5. Mitosis detection process

For the location of the border, we model the spatial relationship " Around" as a landscape resulting from mathematical morphology operators such as dilations [4,10] (see Figure 8). If a request is sent to the system to detect mitoses, we can scan the WSI image by image or does trigger that detector where this is relevant regarding the spatial relations constraints we are currently formalizing in the

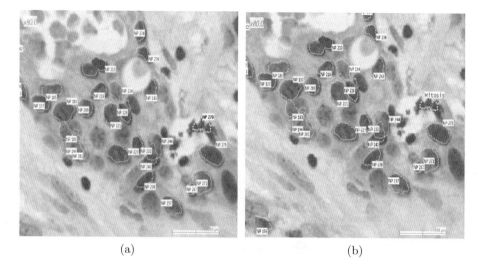

|(a)|(b)|

Fig. 6. (a) WFML before Mitosis Detection corresponding to step 3 in Figure 5 (b) WFML after Rule-based Mitosis Detection corresponding to step 8 in Figure 5

$$\rightarrow Mitosis(?x) \wedge Neoplasm(?y) \wedge hasNeoplasmPeriphery(?y, ?x) \rightarrow sqwrl{:}select(?x, ?y)$$

Fig. 7. A SWRL rule for the expression of the spatial relationship constraint of Eq. 1 within the Protégé platform

knowledge base according to the pathologist experience. The processing of the request is the result of a reasoning step that can evolve as new rules are added. For example, it triggers the rule R1 which is linked to mitoses. By doing this, we save between five and ten-fold increase in processing time which is of dramatic importance for WSI exploration. In addition, this kind of spatial relationship rules can help to check the consistency of $\mathcal{R}^{signal}$ results.

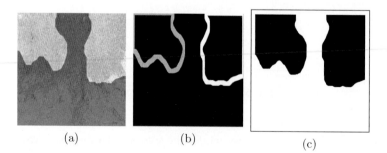

|(a)|(b)|(c)|

Fig. 8. An example of (a) invasive area detection at low magnification and (b) of its instantiated "Around" region after various morphological based filtering like (c) dilation and erosion of the invasive area

4   Results and Discussion

The preliminary assessment of the ontology-driven annotation was achieved by quantifying the improvement of the grading platform specificity and sensibility rate related to the mitosis detection based on the previous ideas. The database is made of several histopathological samples as listed in Figure 9.

Query with intensity. As quantified in Figure 9(a), all mitoses (TP=9) are detected but the detection is not very specific, there are a lot of false alarms. For 253 cells on average per frame, the algorithm returns 5 mitoses on average, sensitivity is equal to 1, all true mitoses were detected among the 5 mitoses returned. This could be a good diagnosis aid for histopathologists who can focus on the study of detected mitoses instead of having to analyze all the cells. This will reduce the workload and save time, which is of utmost importance in this field regarding the size of the images. The pathologist's task is complex and requires a lot of experience and we noticed an important point: real mitoses were detected by the semantic procedure but were not identified by the pathologist.

Query with intensity and geometrical constraints. A normal cell has a regular shape almost round or oval while a mitosis has an irregular shape and tends to divide. The test consists in reducing the false alarm rate and increasing the specificity by adding geometric constraints (see Figure 9(b)). In half the cases the algorithm becomes more specific, false alarms are reduced but the sensitivity may decrease in some cases. Of course, this result is very dependent of the low-level image processing algorithms. If the outlines of cells are correctly detected specificity gets better and sensitivity is maintained, the algorithm is efficient. In addition the algorithm works better on images with a grade NGS (Nottingham Grading System) equals to 1 than image with a NGS equals to 3. The reason is that cells are less deformed for grade 1, the algorithm is less dependent on the quality of segmentation and just need to locate each cell.

A detailed test. The image in Figure 10 contains 258 cells and 3 mitoses. A pathologist detects three mitoses. By testing with the intensity without geometric constraints only six mitoses (N194, N170, N14, N82, N210, N224) are detected among which three true mitoses (N210, N224, N14). By testing with geometrical constraints like Circularity ≤ 0.75 and Roundness ≤ 0.65, only 2 mitosis are detected, one of the three true mitosis is not detected. The algorithm detects the mitosis N224 and N210 but not the mitosis N14. These results show that the algorithm with geometric constraints is more specific but that it decreases sensibility.

Furthermore, a qualitative assessment about usability by the novice end-user (that is the pathologist in our case) remains to be drawn. However, the formalization of the knowledge is a definite asset that the clinical world requires (see the European-based virtual physiological human project) for sharing and reusing tools that are developed worldwide [11]. In addition, for internal development requirement, the need for knowledge engineering in clinical and medical imaging

Query with intensity

Image	Frame	Nb of cells	Correct Nb of mitosis	Nb of mitosis detected	TP	TN	FP	FN	Specificity	Sensitivity
IMG001	f001	257	3	6	3	251	3	0	0,988	1,000
IMG001	f003	314	2	4	2	310	2	0	0,994	1,000
IMG001	f008	296	2	3	2	293	1	0	0,997	1,000
IMG002	f001	244	0	8	0	236	8	0	0,967	1,000
IMG002	f004	242	0	18	0	224	18	0	0,926	1,000
IMG003	f002	297	2	4	2	293	2	0	0,993	1,000
IMG004	f002	261	0	6	0	255	6	0	0,977	1,000
IMG004	f003	214	0	1	0	213	1	0	0,995	1,000
IMG004	f011	229	0	5	0	224	5	0	0,978	1,000
IMG004	f016	234	0	5	0	229	5	0	0,979	1,000
		2588	9	60	9	2528	51	0	0,980	1,000

Query with intensity and geometrical constraints

Image	Frame	Nb of cells	Correct Nb of mitosis	Nb of mitosis detected	TP	TN	FP	FN	Specificity	Sensitivity
IMG001	f001	257	3	2	2	255	0	1	1,000	0,667
IMG001	f003	314	2	3	2	311	1	0	0,997	1,000
IMG001	f008	296	2	3	2	293	1	0	0,997	1,000
IMG002	f001	244	0	0	0	244	0	0	1,000	1,000
IMG002	f004	242	0	0	0	242	0	0	1,000	1,000
IMG003	f002	297	2	3	2	294	1	0	0,997	1,000
IMG004	f002	261	0	5	0	256	5	0	0,981	1,000
IMG004	f003	214	0	1	0	213	1	0	0,995	1,000
IMG004	f011	229	0	3	0	226	3	0	0,987	1,000
IMG004	f016	234	0	0	0	234	0	0	1,000	1,000
		2588	9	20	8	2568	12	1	0,995	0,889

(a) Ontological intensity constraints   (b) Ontological intensity and geometry constraints

Fig. 9. Query results

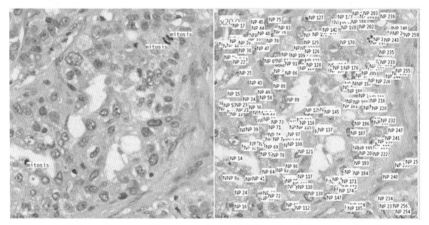

(a) Manual annotation   (b) Low-level image annotation

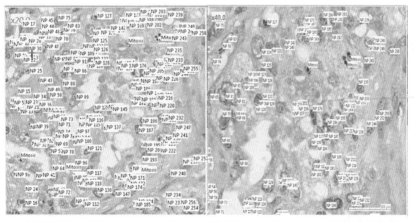

(c) High-level image annotation   (d) A zoomed-in area of the (c) image

Fig. 10. Detailed test image

fields is gaining momentum in order to be able to share issues and experience between the various key players of the platform design, from the imaging researcher to the clinician expert.

5   Conclusion

We experimented and made preliminary assessment of the articulation between ontology-based platforms and image analysis engine in a field where images contain a lot of complex, documented information, partly in the form of a mental database acquired by experience over years of practice. We showed that formalizing the knowledge can lead to improvement of image analysis results both in terms of specificity and sensibility of the pattern recognition system by involving semantic reasoning procedures. The new amount of visual data available in fields like satellite or bio-clinical imaging definitely calls for new paradigms whereupon knowledge engineering and computer vision issues must cooperate for the end-user benefits. In particular, digitized pathology is a growing market like digitized radiology has been over the previous couple of decades but with more than ever increased requirement for interoperability and expressiveness in the modeling [12]. Ontology-driven and reasoning based software engineering should play a key role for these new issues within the visual computing paradigm. In the next phase of our research program, we will consolidate the spatial relation reasoning part for the image analysis engine control.

References

1. Wu, C., Aghajan, H.: Using context with statistical relational models – object recognition from observing user activity in home environment. In: Workshop on Use of Context in Vision Processing (UCVP), ICMI-MLMI (2009)
2. Dalle, J.R., Leow, W.K., Racoceanu, D., Tutac, A.E., Putti, T.C.: Automatic breast cancer grading of histopathological images. In: Proceedings of the 30th International Conference of the IEEE Engineering in Medicine and Biology Society, Vancouver, BC, Canada, pp. 3052–3055 (2008)
3. Mechouche, A., Morandi, X., Golbreich, C., Gibaud, B.: A hybrid system using symbolic and numeric knowledge for the semantic annotation of sulco-gyral anatomy in brain MRI images. IEEE Transactions on Medical Imaging 28, 1165–1178 (2009)
4. Hudelot, C., Atif, J., Bloch, I.: A spatial relation ontology using mathematical morphology and description logics for spatial reasoning. In: ECAI Workshop on Spatial and Temporal Reasoning, pp. 21–25 (2008)
5. Frkovic-Grazio, S., Bracko, M.: Long term prognostic value of nottingham histological grade and its components in early (pT1N0M0) breast carcinoma. Journal of Clinical Pathology 55, 88–92 (2002)
6. Dalle, J.R., Li, H., Huang, C.H., Leow, W.K., Racoceanu, D., Putti, T.C.: Nuclear pleomorphism scoring by selective cell nuclei detection. In: IEEE Workshop on Applications of Computer Vision, Snowbird, Utah, USA (2009)
7. Miikkulainen, R., Bednar, J.A., Choe, Y., Sirosh, J.: Computational Maps in the Visual Cortex. Springer, Berlin (2005)

8. Tutac, A.E., Racoceanu, D., Leow, W.K., Dalle, J.R., Putti, T., Xiong, W., Cretu, V.: Translational approach for semi-automatic breast cancer grading using a knowledge-guided semantic indexing of histopathology images. In: Microscopic Image Analysis with Application in Biology, MIAAB, third MICCAI Workshop, 11th International Conference on Medical Image Computing and Computer Assisted Intervention, New-York, USA (2008)

9. Roux, L., Tutac, A., Lomenie, N., Balensi, D., Racoceanu, D., Leow, W.K., Veillard, A., Klossa, J., Putti, T.: A cognitive virtual microscopic framework for knowlege-based exploration of large microscopic images in breast cancer histopathology. In: IEEE Engineering in Medicine and Biology Society - Engineering the Future of Biomedecine, Minneapolis, Minnesota, USA (2009)

10. Lomenie, N.: Reasoning with spatial relations over high-content images. In: IEEE World Congress on Computational Intelligence - International Joint Conference on Neural Networks, Barcelona, Spain (2010)

11. Gianni, D., McKeever, S., Yu, T., Britten, R., Delingette, H.: Sharing and reusing cardiovascular anatomical models over the web: a step towards the implementation of the virtual physiological human project. Philosophical Transactions of The Royal Society A 368, 3039–3056 (2010)

12. Klipp, J., Kaufman, J.: Adoption trends in digital pathology. Laboratory Economics 5, 3–10 (2010)

Attribute-Filtering and Knowledge Extraction for Vessel Segmentation

Benoît Caldairou[1,2], Nicolas Passat[1], and Benoît Naegel[3]

[1] Université de Strasbourg, LSIIT, UMR CNRS 7005, France
[2] Université de Strasbourg, LINC, UMR CNRS 7191, France
[3] Université Nancy 1, LORIA, UMR CNRS 7503, France
{caldairou,passat}@unistra.fr, benoit.naegel@loria.fr

Abstract. Attribute-filtering, relying on the notion of component-tree, enables to process grey-level images by taking into account high-level *a priori* knowledge. Based on these notions, a method is proposed for automatic segmentation of vascular structures from phase-contrast magnetic resonance angiography. Experiments performed on 16 images and validations by comparison to results obtained by two human experts emphasise the relevance of the method.

Keywords: vessel segmentation, mathematical morphology, component-trees, magnetic resonance angiography.

1   Introduction

For a long time, mathematical morphology has been involved in the design of vessel segmentation methods[1] by only considering low-level operators (see, *e.g.*, [7,5]). The ability of high-level operators to be efficiently considered for medical image processing has been pointed out in recent works [6], especially in the context of vessel segmentation [12,16,1]. The usefulness of such mathematical morphology operators –including those based on component-trees– is justified by their intrinsic capacity to model morphological information, and then to enable anatomical knowledge-guided approaches.

The notion of *component-tree* [15,9,4] associates to a grey-level image a descriptive data structure induced by the inclusion relation between the binary components obtained at the successive level sets of the image. In particular, it has led to the development of morphological operators [2,15].

Thanks to efforts devoted to its efficient computation [2,15,13], component-trees have been considered for the design of various kinds of grey-level image processing methods, including image filtering and segmentation [9,19,18], some of them being devoted to the analysis of (bio)medical images: CT/MR angiography [21], confocal microscopy [14], dermatological data [11]. Some of these methods are automatic, can filter complex objects in 3-D images [21,14], or take into account complex anatomical knowledge [11]. However none of them fuses all these virtues. The challenges to be faced towards the development of efficient medical image segmentation methods based

---

[1] A whole state of the art on vessel segmentation is beyond the scope of this article. The reader may refer to [10] for a recent survey.

G. Bebis et al. (Eds.): ISVC 2010, Part I, LNCS 6453, pp. 13–22, 2010.

on component-trees then consists in simultaneously dealing with *automation* and *complexity* requirements.

Based on advances related to the use of component-trees in this difficult context [17,3], a method has been developed for the segmentation of phase-contrast magnetic resonance angiography (PC-MRA). Indeed, such data –non invasive, non-irradiant, and then harmless for the patients– are often considered in clinical routine, but generally not in the literature devoted to vessel segmentation. The low SNR and resolution of PC-MRAs however justify the use of segmentation methods in order to simplify their analysis.

The remainder of this article is organised as follows. Section 2 provides background notions on component-trees. Section 3 describes the proposed vessel segmentation method. Section 4 provides experimental results and validations. Section 5 summarises the contributions, and describes further works.

2   Component-Trees

Let $I : E \rightarrow V$ (we also note $I \in V^E$) be a discrete grey-level image (with $E \subset \mathbb{Z}^n$ and $V = [\![a, b]\!] \subset \mathbb{Z}$), as illustrated in Fig. 1(a).

Let $X \subseteq E$ be a binary image. The connected components of X are the equivalence classes of X w.r.t. a chosen adjacency relation. The set of the connected components of X is noted $C[X]$.

Let $v \in V$. We set $\mathcal{P}(E) = \{X \subseteq E\}$. Let $X_v : V^E \rightarrow \mathcal{P}(E)$ be the thresholding function defined by $X_v(I) = \{x \in E \mid v \leq I(x)\}$ for all $I : E \rightarrow V$ (see Fig. 1(b–f)).

Let $v \in V$ and $X \subseteq E$. We define the cylinder function $C_{X,v} : E \rightarrow V$ by $C_{X,v}(x) = v$ if $x \in X$ and a otherwise. A discrete image $I \in V^E$ can then be expressed as $I = \bigvee_{v \in V} \bigvee_{X \in C[X_v(I)]} C_{X,v}$, where $\bigvee$ is the pointwise supremum for the sets of functions.

Let $\mathcal{K} = \bigcup_{v \in V} C[X_v(I)]$. The inclusion relation $\subseteq$ is a partial order on $\mathcal{K}$. Let $v_1 \leq v_2 \in V$. Let $B_1, B_2 \subseteq E$ be the binary images defined by $B_k = X_{v_k}(I)$ for $k \in \{1, 2\}$. Let $C_2 \in C[B_2]$ be a connected component of B_2. Then, there exists a (unique) connected component $C_1 \in C[B_1]$ of B_1 such that $C_2 \subseteq C_1$ (see Fig. 1(b–f)). In particular, we necessarily have $B_2 \subseteq B_1$.

Based on these properties, is can be easily deduced that the Hasse diagram of the partially ordered set $(\mathcal{K}, \subseteq)$ is a tree (*i.e.*, a connected acyclic graph), the root of which is its supremum $X_a(I) = E$. This tree is called the *component-tree of I* (see Fig. 1(g)).

Definition 1 (Component-tree). *Let $I \in V^E$ be a grey-level image. The component-tree of I is the rooted tree $(\mathcal{K}, L, R)$ such that: $\mathcal{K} = \bigcup_{v \in V} C[X_v(I)]$ (namely the nodes); $R = \sup(\mathcal{K}, \subseteq) = X_a(I)$ (namely the root); L (namely the set of edges) is composed of all pairs $(X, Y) \in \mathcal{K} \times \mathcal{K}$ verifying (i) $Y \subset X$ and (ii) $\forall Z \in \mathcal{K}, Y \subseteq Z \subset X \Rightarrow Y = Z$.*

In Fig. 1(g), $\mathcal{K}$ is the set of white rectangles, R is the one located at the highest level, and L is visualised by the set of black lines (linking the elements of each pair).

Component-trees enable the storage, at each node, of *attributes*, *i.e.*, elements of information related to the binary connected components associated to the nodes. For instance, in Fig. 1(g), the size of the connected component has been added at each corresponding node. In this (simple) example, the considered attribute is a single numerical

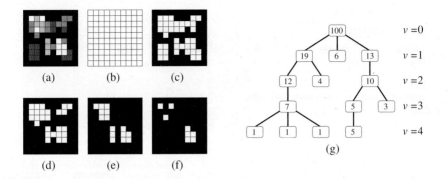

Fig. 1. (a) A grey-level image $I : [\![0,9]\!]^2 \rightarrow [\![0,4]\!]$. (b–f) Threshold images $X_v(I)$ (white points) for v varying from 0 (b) to 4 (f). (g) The component-tree of I. Its levels correspond to increasing thresholding values v. The root ($v = 0$) corresponds to the support ($E = [\![0,9]\!]^2$) of the image.

value. It is however possible to consider any kinds of –quantitative or qualitative– attributes, provided they can be conveniently modelled. It is also possible to store several elements of information, no longer leading to scalar attributes but to vectorial ones [17].

3   Segmentation Method

The proposed method (summarised in the flowchart of Fig. 2) consists in (*i*) determining the characteristic properties of the structures of interest (vessels) thanks to a supervised learning process (Subsection 3.1), and (*ii*) using this knowledge to automatically process images *via* their component-tree (Subsection 3.2).

3.1   Learning Process

This first step (see Fig. 2(a)) enables to extract from one (or possibly several) ground-truth data (*i.e.*, correctly segmented images) a set of characteristic parameters chosen from a given set of criteria.

The learning step takes as input (*i*) a ground-truth image $I_g \in V^E$, (*ii*) its segmentation $B_g \subset E$, and (*iii*) a function $A : \mathcal{P}(E) \rightarrow \Omega$, associating to each possible node of the component-tree of I_g, a feature vector in the parameter space Ω induced by a chosen set of criteria. It provides as output a subset $\omega \subset \Omega$ of the parameter space, characterising the nodes of the component-tree of I_g which enable to fit at best the segmentation B_g.

Let $(\mathcal{K}, L, R)$ be the component-tree of I_g. Let $\mathcal{S} = \{\bigcup_{X \in C} X\}_{C \subseteq \mathcal{K}}$ be the set of all the binary images which can be generated from the set of nodes $\mathcal{K}$. We need to determine the "best" binary image which may be computed from $\mathcal{K}$ w.r.t. B_g. This requires to define a distance d on $\mathcal{P}(E)$ enabling to compare B_g and the binary images of $\mathcal{S}$. The best binary image $\widehat{B}$ can be set as $\widehat{B} = \arg\min_{B \in \mathcal{S}}\{d(B, B_g)\}$. We define such a distance d by $d(B, B_g) = \alpha.|B \setminus B_g| + (1 - \alpha).|B_g \setminus B|$, with $\alpha \in [0, 1]$. It aims at finding a best compromise (parametrised by α) between the amount of false positives/negatives of B w.r.t. B_g.

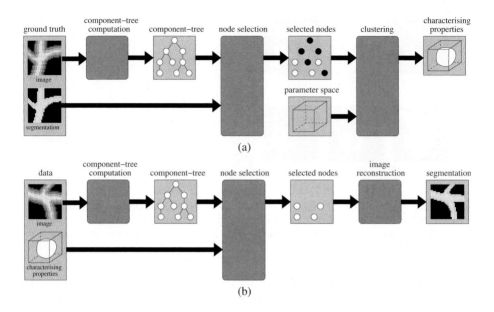

Fig. 2. Summary of the method (dark grey: processes; light grey: data). (a) Learning process. (b) Segmentation process.

Based on these definitions, a minimal set $\widehat{B}$ can be extracted from S (in linear time $O(|\mathcal{K}|)$). Then, an adequate set of nodes $\widehat{K} \subseteq \mathcal{K}$ associated to $\widehat{B}$ (i.e., such that $\bigcup_{X \in \widehat{K}} X = \widehat{B}$) has to be determined. Let $\widehat{C} = \{X \in \mathcal{K} \mid X \subseteq \widehat{B}\} \subseteq \mathcal{K}$ (note that the nodes of $\widehat{C}$ generate a set of subtrees of the component-tree $(\mathcal{K}, L, R)$ of I_g). The set $\widehat{B}$ can be generated by any set of nodes $\widehat{K} \subseteq \widehat{C}$ verifying $\bigcup_{X \in \widehat{K}} X = \bigcup_{X \in \widehat{C}} X = \widehat{B}$. Two main strategies can be considered: by setting $\widehat{K}^+ = \widehat{C}$, any node included in $\widehat{B}$ is considered as a useful binary connected component, while by setting $\widehat{K}^- = \{X \in \widehat{C} \mid \forall Y \in \widehat{C}, X \not\subset Y\}$, only the roots of the subtrees induced by $\widehat{C}$ are considered as useful. The first (resp. second) strategy is the one considering the largest (resp. smallest) possible set of nodes/connected components among $\widehat{C}$.

Once a set of nodes $\widehat{K}$ has been defined, the determination of the subset of characterising knowledge $\omega \subset \Omega$ has to be performed. The determination of ω can be expressed as a clustering problem consisting in partitioning Ω into two classes thanks to the samples $A(\widehat{K}) = \{A(N)\}_{N \in \widehat{K}}$ (corresponding to the attributes of the structures of interest) and $A(\mathcal{K} \setminus \widehat{C}) = \{A(N)\}_{N \in \mathcal{K} \setminus \widehat{C}}$. This process can, for instance, be carried out by usual classification tools, such as the Support Vector Machine (SVM) [20].

3.2   Segmentation Process

This second step (see Fig. 2(b)) enables to segment the structures of interest from an image, based on the characterising properties modelled by the set of knowledge $\omega \subset \Omega$.

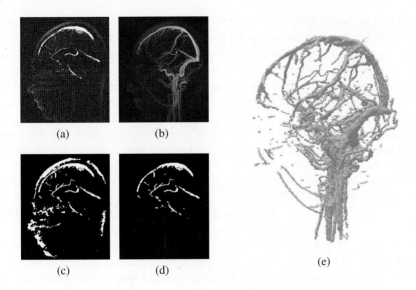

Fig. 3. (a,b) Cerebral PC-MRA (this image has been used as ground-truth): (a) sagittal slice, (b) maximum intensity projection. (c–e) Ground-truth segmentation of (a): vessels and artifacts $(B_v \cup B_a)$ (c); vessels only (B_v) (d), 3-D visualisation of vessels (B_v) (e).

The segmentation step takes as input an image $I \in V^E$ and a subset $\omega \subset \Omega$ characterising the structures to be segmented in I. It provides as output a segmentation $B \subseteq E$ of these structures of interest.

Let $(\mathcal{K}, L, R)$ be the component-tree of I. We define $\mathcal{K}_f = \{N \in \mathcal{K} \mid A(N) \in \omega\}$. The set $\mathcal{K}_f$ is composed of the nodes which satisfy the characterising properties modelled by ω, and are then considered as the parts of the image to be preserved by the segmentation process. We can finally reconstruct the segmentation result as $B = \bigcup_{N \in \mathcal{K}_f} N$.

4   Experiments and Results

4.1   PC-MRA Segmentation

The proposed methodology has been considered for the segmentation of cerebral PC-MRAs. Such images (see Fig. 3(a,b)) are composed of three kinds of semantic elements: low-intensity background, high-intensity artifacts and high-intensity vascular signal.

Input/Output. The method processes PC-MRA images $I \in V^E$ (with $E = [\![0, 255]\!]^3$ and $V \subset \mathbb{N}$). The learning step requires a ground-truth image $I_g \in V^E$ (Fig. 3(a,b)), and its segmentation $B_v \subset E$ (vessels) and $B_a \subset E$ (artifacts) (Fig. 3(c–e)). The segmentation step provides as output a fuzzy segmentation $S \in [0, 1]^E$ of the vessels visualised in I.

Multiscale Approach. In order to deal with the complexity of the structures of interest, the segmentation method can be applied in a multiscale fashion. The image is then processed as a collection of subimages obtained from an octree decomposition, thus enabling to "break" complex structures into smaller –and easier to characterise– ones. In order to improve the behaviour of the method at the border of these subimages, it can be convenient to consider two shifted octree decompositions of the image. In such a context, if the octrees have d levels, any point $x \in E$ of the image belongs to $2d$ subimages, the sizes of which vary from 256^3 to $(256/2^{d-1})^3$. (Note that this octree decomposition also has to be considered in the learning step, resulting in a distinct set ω for each one of the d subimage sizes.) This value $2d$ can be seen as a redundancy factor of the multiscale segmentation method. In our case, the value d has been set to 4.

By opposition to the initially proposed strategy, the variant proposed here does not provide a binary result $B \subseteq E$. Indeed, overlaps induced by multiscale and redundancy may lead to ambiguous results for any point $x \in E$. A grey-scale segmentation $S \in \mathbb{R}^E$ can however be obtained by setting $S(x) = s(x)/2d$, where $s(x) \in \mathbb{N}$ is the number of nodes which contain x (among the component-trees induced by the $2d$ subimages where x appears) and which have been classified as being vascular. This segmentation S (which can be assimilated to a fuzzy segmentation, although not normalised) provides, for each point $x \in E$, a value which can be seen as a "vesselness" score.

Learning Step: Presegmentation. In order to perform the learning step, it is first required to choose the best segmentation results w.r.t. the ground-truth segmentations B_v and B_a and the distance d. Several results have then been computed, for various values of α (which determines the authorised ratio between false positives/negatives) sampled in $[0, 1]$. The most satisfying sets $\widehat{B_v}$ and $\widehat{B_a}$ have then been chosen by a human expert, based on a visual analysis. (Note that in the current experiments, the corresponding value of α was 0.9 for B_v and 0.4 for B_a.) The associated sets of nodes $\widehat{K_v}$ and $\widehat{K_a}$ have been defined as $\widehat{K_v^-}$ and $\widehat{K_a^-}$, respectively. Indeed, it has been chosen to give a higher importance to the shape of the structures than to their grey-level profile (since only the roots of the component-trees induced by the best segmentation results are then considered, see Subsection 3.1).

Learning Step: Parameter Estimation. In order to determine a set ω characterising the high-intensity vascular signal from the background noise and the high-intensity artifacts, a three class SVM classification has been applied on the component-trees of I_g, using the parameter space Ω induced by the following attributes: moment of inertia, flatness (computed from the eigenvalues of the inertia matrix of I), intensity, size, volume (related to the grey-level profile of the image at the node), contrast (distance between the node and the closest leaf in the component-tree), distance to the head (computed thanks to the morphological image associated to I). Among these attributes, some "simple" ones (*e.g.* intensity, size) are considered in order to enable the classification between the background noise and the high intensity structures (vessels and artifacts), while some more sophisticated ones (*e.g.* moment of inertia, flatness) are assumed to enable the discrimination between vessels and artifacts, since they are dedicated to shape characterisation [8].

Segmentation Step. The obtained set $\omega \subset \Omega$ has then been used to determine the desired set of nodes $\mathcal{K}_s(k, l) \subseteq \mathcal{K}$ from each set of component-trees of I, for each octree decomposition $k = 1, 2$, and at each level $l = 1$ to d. From these sets of nodes, it has been possible to build a segmented grey-level image $I_s : E \to \mathbb{R}$ defined by

$$I_s(x) = \frac{1}{2d} \sum_{k=1}^{2} \sum_{l=1}^{d} |\mathcal{K}_s(k, l) \cap \{X \mid x \in X\}| \qquad (1)$$

In particular, $I_s(x) = 0$ (resp. $\neq 0$, resp. $\gg 1$) means that x never matches a vessel (resp. matches at least once a vessel, resp. often matches a vessel) in the segmentation result.

4.2 Results

Technical Details. The method has been applied on a database composed of 17 PC-MRAs (Philips Gyroscan NT/INTERA 1.0 T, TR = 10 ms, TE = 6.4 ms), of millimetric resolution (see Fig. 3(a,b)). One image has been considered for the learning step (I_g) while the other 16 ones have been considered for quantitative validations.

With a standard PC (Intel Quad Core i7 860, 4GB RAM), the average computation time for the method is 3 min/image for the learning step (which only requires to be carried out once), and 4 min/image for the segmentation step.

Validations. The 16 MRAs have also been interactively segmented by two human experts. These results $S_e^1, S_e^2 \subset E$ are generally slightly different. Indeed, the mean of the maximal sensitivity (resp. minimal sensitivity) between S_e^1 and S_e^2 is 87.2% (resp. 68.3%). This tends to mean that the accurate segmentation of such images (presenting a very low SNR) is a complex and error-prone task, even for experts. Based on this uncertainty, we set $V = S_e^1 \cap S_e^2$, $\overline{V} = E \backslash (S_e^1 \cup S_e^2)$, and $V_? = (S_e^1 \backslash S_e^2) \cup (S_e^2 \backslash S_e^1)$, assumed to be the vascular, non-vascular, and ambiguous areas, according to these ground-truths.

For each MRA, let $S \subset E$ be a visually satisfactory segmentation obtained from I_s (by interactive thresholding of I_s, which requires only a few seconds; we provide in Tab. 1 the threshold value λ such that $S = I_s(\lambda)$). Let TP = $|V \cap S|$, AP = $|V_? \cap S|$, FP = $|\overline{V} \cap S|$, the true, ambiguous and false positives of S w.r.t. the ground-truth, respectively. Let TN = $|\overline{V} \backslash S|$, AN = $|V_? \backslash S|$, FN = $|V \backslash S|$, the true, ambiguous and false negatives of S w.r.t. the ground-truth, respectively. In the standard case (*i.e.*, when $V_? = \emptyset$), the formulae for sensitivity (Sen) and specificity (Spe) are given by Sen = TP/(TP + FP) and Spe = TP/(TP + FN). Here, since $V_? \neq \emptyset$, we can only compute intervals [Sen$^-$, Sen$^+$], [Spe$^-$, Spe$^+$] $\subseteq [0, 1]$ providing the potential sensitivity and specificity values induced by the uncertainty on the ground-truth. The bounds of these intervals are straightforwardly given by

$$\text{Spe}^+ = (\text{TP} + \text{AP})/(\text{TP} + \text{FN} + \text{AP}) \qquad (2)$$
$$\text{Spe}^- = \text{TP}/(\text{TP} + \text{FN} + \text{AN}) \qquad (3)$$
$$\text{Sen}^+ = (\text{TP} + \text{AP})/(\text{TP} + \text{FP} + \text{AP}) \qquad (4)$$
$$\text{Sen}^- = \text{TP}/(\text{TP} + \text{FP} + \text{AP}) \qquad (5)$$

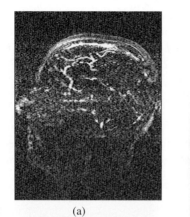

(a)

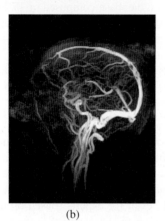

(b)

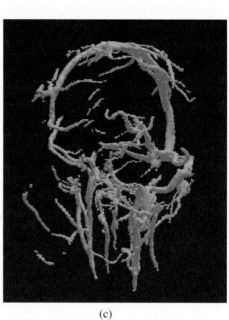

(c)

Fig. 4. (a) MRA (I) and (b) its grey-level segmentation (I_S) viewed as a MIP (line 15 in Table 1). (c) Binary segmentation (S) viewed as a 3-D object (line 11 in Table 1, $\lambda = 40.3$).

Table 1. Quantitative analysis of the segmentation results (see text)

	λ	Spe (%)	Sen (%)	$d(S, V)$		λ	Spe (%)	Sen (%)	$d(S, V)$
1	100	[49.3, 79.8]	[77.6, 85.0]	2.8 ± 7.7	9	80	[39.9, 57.7]	[75.8, 77.3]	2.0 ± 4.3
2	50	[51.7, 77.5]	[80.4, 85.5]	1.5 ± 5.2	10	150	[38.8, 62.9]	[63.7, 68.7]	3.8 ± 7.6
3	120	[47.7, 66.6]	[79.3, 77.0]	4.0 ± 10.8	11	50	[48.7, 74.8]	[80.6, 87.9]	1.7 ± 5.1
4	80	[46.8, 66.0]	[88.5, 87.3]	1.8 ± 5.0	12	50	[43.2, 68.9]	[83.1, 89.5]	1.6 ± 4.9
5	80	[33.3, 64.1]	[83.3, 87.4]	2.3 ± 5.9	13	80	[33.9, 65.2]	[62.9, 73.1]	4.8 ± 10.3
6	100	[34.6, 58.9]	[86.4, 89.0]	3.6 ± 11.2	14	100	[35.7, 63.4]	[71.2, 74.3]	2.5 ± 6.5
7	80	[28.5, 46.6]	[71.2, 75.7]	2.1 ± 4.2	15	150	[36.6, 65.2]	[67.1, 77.8]	2.7 ± 6.2
8	80	[43.3, 65.2]	[68.4, 74.5]	3.5 ± 8.1	16	100	[36.0, 65.4]	[62.1, 66.7]	5.5 ± 11.5

From a quantitative point of view (see Tab. 1 for intervals [Sen⁻, Sen⁺], [Spe⁻, Spe⁺] and mean point-to-set distance between S and V, in mm), the measures can appear as low. If this can be partially explained by possible segmentation errors (the main errors generally result from the loss of the smallest vessels, composed of small connected components less accurately modelled by the considered attributes), it also results from the quality of the ground-truths (which contain several errors, as aforementioned). Moreover, the proposed segmentations are grey-level ones (see Fig. 4(a,b)), which can be thresholded to favour either sensitivity or specificity. The binary segmentation obtained by such a thresholding (see Fig. 4(c)) also emphasises the ability of the method to provide qualitatively satisfactory results. Finally, one has to note that the experts provided results in approximately one hour by using interactive tools, *vs.* a few minutes for the method.

5   Conclusion

Based on the notion of component-tree, a method has been proposed for the segmentation of angiographic data. It takes advantage of *a priori* knowledge, thanks to a learning process enabling to define characteristic properties of vessels. This knowledge is, in particular, embeddable in the component-tree structure of the images to be processed.

The validations performed on phase-contrast images tend to emphasise the relevance of the proposed methodology, in particular by comparison to segmentations performed by human experts (from both qualitative and time-consumption points of view). It has to be noticed that vessel segmentation method, although being fast and automated, however still presents a few weaknesses, especially in the detection of small vessels.

In order to improve its robustness, further works will focus on the following points: (*i*) automatic choice of the pertinent parameters enabling to determine the Ω space (enabling to initially consider a larger set of potential parameters), (*ii*) incremental improvement of the learning process (by taking into account the segmentations performed by the method) and (*iii*) use of parameters of high level (for instance shape descriptors). Points (*i*) and (*ii*) will require to develop solutions to reinject the evaluation of the segmentations in the learning/segmentation process, while point (*iii*) will require to develop solutions to compute (or at least approximate) complex attributes with a satisfactory algorithmic cost.

References

1. Bouraoui, B., Ronse, C., Baruthio, J., Passat, N., Germain, P.: 3D segmentation of coronary arteries based on advanced Mathematical Morphology techniques. Computerized Medical Imaging and Graphics (in press) doi: 10.1016/j.compmedimag.2010.01.001
2. Breen, E.J., Jones, R.: Attribute openings, thinnings, and granulometries. Computer Vision and Image Understanding 64(3), 377–389 (1996)
3. Caldairou, B., Naegel, B., Passat, N.: Segmentation of complex images based on component-trees: Methodological tools. In: Wilkinson, M.H.F., Roerdink, J.B.T.M. (eds.) International Symposium on Mathematical Morphology- ISMM 2009, 9th International Symposium, Proceedings. LNCS, vol. 5720, pp. 171–180. Springer, Heidelberg (2009)

4. Chen, L., Berry, M.W., Hargrove, W.W.: Using dendronal signatures for feature extraction and retrieval. International Journal of Imaging Systems and Technology 11(4), 243–253 (2000)
5. Cline, H.E., Thedens, D.R., Meyer, C.H., Nishimura, D.G., Foo, T.K., Ludke, S.: Combined connectivity and a gray-level morphological filter in magnetic resonance coronary angiography. Magnetic Resonance in Medicine 43(6), 892–895 (2000)
6. Cousty, J., Najman, L., Couprie, M., Clément-Guinaudeau, S., Goissen, T., Garot, J.: Segmentation of 4D cardiac MRI: Automated method based on spatio-temporal watershed cuts. Image and Vision Computing (in press), doi: 10.1016/j.imavis.2010.01.001
7. Gerig, G., Koller, T., Székely, G., Brechbühler, C., Kübler, O.: Symbolic description of 3-D structures applied to cerebral vessel tree obtained from MR angiography volume data. In: Barrett, H.H., Gmitro, A.F. (eds.) IPMI 1993. LNCS, vol. 687, pp. 94–111. Springer, Heidelberg (1993)
8. Hu, M.K.: Visual pattern recognition by moment invariants. IRE Transactions on Information Theory 8(2), 179–187 (1962)
9. Jones, R.: Connected filtering and segmentation using component trees. Computer Vision and Image Understanding 75(3), 215–228 (1999)
10. Lesage, D., Angelini, E.D., Bloch, I., Funka-Lea, G.: A review of 3D vessel lumen segmentation techniques: Models, features and extraction schemes. Medical Image Analysis 13(6), 819–845 (2009)
11. Naegel, B., Passat, N., Boch, N., Kocher, M.: Segmentation using vector-attribute filters: methodology and application to dermatological imaging. In: International Symposium on Mathematical Morphology - ISMM 2007, Proceedings, vol. 1, pp. 239–250. INPE (2007)
12. Naegel, B., Passat, N., Ronse, C.: Grey-level hit-or-miss transforms - Part II: Application to angiographic image processing. Pattern Recognition 40(2), 648–658 (2007)
13. Najman, L., Couprie, M.: Building the component tree in quasi-linear time. IEEE Transactions on Image Processing 15(11), 3531–3539 (2006)
14. Ouzounis, G.K., Wilkinson, M.H.F.: Mask-based second-generation connectivity and attribute filters. IEEE Transactions on Pattern Analysis and Machine Intelligence 29(6), 990–1004 (2007)
15. Salembier, P., Oliveras, A., Garrido, L.: Anti-extensive connected operators for image and sequence processing. IEEE Transactions on Image Processing 7(4), 555–570 (1998)
16. Tankyevych, O., Talbot, H., Dokládal, P., Passat, N.: Direction-adaptive grey-level morphology. Application to 3D vascular brain imaging. In: International Conference on Image Processing - ICIP 2009, Proceedings, pp. 2261–2264. IEEE Signal Processing Society (2009)
17. Urbach, E.R., Boersma, N.J., Wilkinson, M.H.F.: Vector attribute filters. In: Proceedings of International Symposium on Mathematical Morphology - ISMM 2005, Proceedings. Computational Imaging and Vision, vol. 30, pp. 95–104. Springer SBM, Heidelberg (2005)
18. Urbach, E.R., Roerdink, J.B.T.M., Wilkinson, M.H.F.: Connected shape-size pattern spectra for rotation and scale-invariant classification of gray-scale images. IEEE Transactions on Pattern Analysis and Machine Intelligence 29(2), 272–285 (2007)
19. Urbach, E.R., Wilkinson, M.H.F.: Shape-only granulometries and gray-scale shape filters. In: International Symposium on Mathematical Morphology - ISMM 2002, Proceedings, pp. 305–314. CSIRO Publishing (2002)
20. Vapnik, V.: Statistical Learning Theory. Wiley-Interscience, New York (1998)
21. Wilkinson, M.H.F., Westenberg, M.A.: Shape preserving filament enhancement filtering. In: Niessen, W.J., Viergever, M.A. (eds.) MICCAI 2001. LNCS, vol. 2208, pp. 770–777. Springer, Heidelberg (2001)

A Human Inspired Local Ratio-Based Algorithm for Edge Detection in Fluorescent Cell Images*

Joe Chalfoun[1], Alden A. Dima[1], Adele P. Peskin[2],
John T. Elliott[1], and James J. Filliben[1]

[1] NIST, Gaithersburg, MD 20899
[2] NIST, Boulder, CO 80305

Abstract. We have developed a new semi-automated method for segmenting images of biological cells seeded at low density on tissue culture substrates, which we use to improve the generation of reference data for the evaluation of automated segmentation algorithms. The method was designed to mimic manual cell segmentation and is based on a model of human visual perception. We demonstrate a need for automated methods to assist with the generation of reference data by comparing several sets of masks from manually segmented cell images created by multiple independent hand-selections of pixels that belong to cell edges. We quantify the differences in these manually segmented masks and then compare them with masks generated from our new segmentation method which we use on cell images acquired to ensure very sharp, clear edges. The resulting masks from 16 images contain 71 cells and show that our semi-automated method for reference data generation locates cell edges more consistently than manual segmentation alone and produces better edge detection than other techniques like 5-means clustering and active contour segmentation for our images.

1   Introduction

Optical microscopy is an important technique used in biological research to study cellular structure and behavior. The use of fluorescence stains and other labeling reagents enables fluorescence microscopy by allowing cell components to be clearly visible on a darker background [1]. The Cell Systems Science Group at NIST has developed a high contrast cell body fluorescence staining technique that highlights cell edges [2]. The cell morphology (i.e. spread area) can then be measured and used to determine regions-of-interest (ROI) for quantifying other fluorescent probes that report on cellular behavior. This technique is often combined with automated microscopy which provides the ability to collect cell images from a large number of fields in an unbiased fashion [3]. Data generated from these images represent a sample of the distribution of responses from a cell population and provide a signature for that cell population. These distributions of

---

* This contribution of NIST, an agency of the U.S. government, is not subject to copyright.

G. Bebis et al. (Eds.): ISVC 2010, Part I, LNCS 6453, pp. 23–34, 2010.
© Springer-Verlag Berlin Heidelberg 2010

responses are generated from individual cell measurements and provide information about the noise processes that are inherent in biological mechanisms. This information is critical for developing accurate biological mathematical models.

Automated segmentation routines are required to extract cell response data from the large number of cell images typically acquired during a study. To ensure that segmentation routines are robust and accurate, their results can be compared to reference data that invariably must come from expert manual segmentation. However, manual cell edge detection (i.e. segmentation) cannot be used to generate the large amount of reference data needed to thoroughly evaluate automated algorithms due to the tedious nature of the task. Although it is well known that such hand-selection leads to some level of error, this error has not been critically analyzed and the extent of the error has not been quantified. Furthermore, because a certain amount of error is expected due to the tedious nature of the task, a method more rigorous than one dependent upon the continuous application of human attention and hand-eye coordination would eliminate much of this error if it can be proven to be as just as effective. In this paper we examine the differences between the two sets of manually segmented masks of 71 cells. We present an analysis of the differences between these data sets. We then present our semi-automated method for computing reference data and compare our results to manual segmentations. The data reveal the cell morphologies most at risk of error during manual segmentation. Our results suggest that this approach can be used to generate cell masks that are similar to the manually segmented masks but appear to track the cell edges in a more consistent fashion. As a result, human effort is shifted from the tedious manual identification of the entire cell boundary to the correction of the automated identification of much smaller regions of the cell boundary. This new technique is useful for generating reference quality data from a large number of images of cells seeded at low density for high-quality comparison and analysis of segmentation algorithms.

2   Data Description

Images of two different cell lines whose cells differ in size and overall geometric shape were prepared. These images consist of A10 rat smooth vascular muscle cells and NIH3T3 mouse fibroblasts stained with a Texas Red cell body stain [1]. We examined 16 fixed cell images, 8 images from each cell line that represent the variability observed in a culture. Each image is comprised of multiple individual cells and cell clusters (typically less than 3 cells) that were treated as single cells. Exposure times of 0.15 s for the A10 cells, and 0.3 s for the NIH3T3 cells were used with Chroma Technology's Texas Red filter set (Excitation 555/28, #32295; dichroic beam splitter #84000; Emission 630/60, #41834).[1] Three exposure times were selected to produce a large intensity to background ratio and

---

[1] Certain trade names are identified in this report only in order to specify the experimental conditions used in obtaining the reported data. Mention of these products in no way constitutes endorsement of them. Other manufacturers may have products of equal or superior specifications.

to produce a similar ratio for the two cell lines. Optimal filters and high exposure conditions result in maximizing signal to background ratios and saturating the central region of each cell. The sample preparation for this data is designed to minimize ambiguity in segmentation. Example images of A10 cells and NIH3T3 cells are shown in Figure 1.

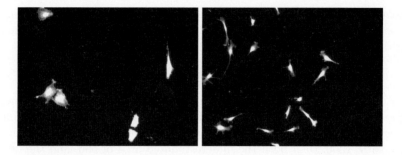

Fig. 1. Example images of A10 cells (left) and NIH3T3 cells (right) used to find reference data

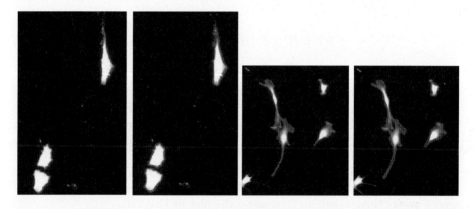

Fig. 2. Comparisons of manual segmentations: A10 (left), NIH3T3 (right), zoomed in on a few cells of each image of Figure 1, to show differences in hand selections of cell edges

3 Manual Segmentation Procedure

For each of our 16 images, two individuals independently segmented the images by hand using ImageJ [4]. Manual segmentation was performed as follows: Using the zoom function for maximum resolution, the edges of cells are marked pixel by pixel with the pencil tool set at width = 1. After the edge is manually selected a hole filling routine creates the masks. Cells touching the border of an image are disregarded. The estimation of the cell edges is different for each individual and depends on how a person sees and interprets the edge pixels. These differences are visible when comparing the two sets of masks in Figure 2 and will be quantified in the following section.

4   Comparison of Hand-Selected Data Sets

Various metrics have been used to evaluate segmentation algorithm performance. The commonly used Jaccard similarity index [5], compares a reference mask T (truth) with another mask E (estimate), and is defined as:

$$S = |T \cap E|/|T \cup E|, \tag{1}$$

where $0.0 \leq S \leq 1.0$. If an estimate matches the truth, $T \cap E = T \cup E$ and $S = 1$. If an algorithm fails, then $E = 0$ and $S = 0$. However, S cannot discriminate between certain underestimation and overestimation cases. For example, if the true area = 1000, then both the underestimated area of 500, and the overestimated area of 2000 yield the same value for the similarity index $S = 500/1000 = 1000/2000 = 0.5$. For our algorithm evaluation work we use a pair of bivariate metrics that can distinguish between underestimation and overestimation.

We define these metrics as follows. To compare the reference mask T, with a estimate mask E:

$$TET = |T \cap E|/|T|, 0.0 \leq TET \leq 1.0 \tag{2}$$

$$TEE = |T \cap E|/|E|, 0.0 \leq TEE \leq 1.0 \tag{3}$$

Each similarity metric varies between 0.0 and 1.0. If the estimate matches the reference mask, both TET and TEE = 1.0. TET and TEE were designed to be independent and orthogonal. This bivariate metric divides performance into four regions: Dislocation: TET and TEE are small; Overestimation: TET is large, TEE is small; Underestimation: TET is small, TEE is large; and Good: both TET and TEE are large. Figure 3 shows a plot of TET vs. TEE for the manually segmented 71 cells. On this plot, a perfect agreement between the two sets of masks corresponds to a point on the plot at (1.0,1.0).

Figure 3 shows that the two hand-selected data sets agree better for the A10 cells than for the NIH3T3 cells. To identify edge features that may be responsible for the reduced agreement between manual segmentation masks of the NIH3T3 cells, we use a metric which we call the extended edge neighborhood, outlined more fully in an accompanying paper [6]. This metric describes the fraction of cell pixels at the edge of the cell that are at risk for being selected differently by a segmentation method. It is a product of the pixel intensity gradient at the edge of a cell and the overall geometry of the cell. We use the pixel intensity gradient at the cell edge to determine a physical edge thickness on the image, and determine a band of pixels surrounding the cell within this band. The ratio of the number of pixels in this band around the cell to the total number of cell pixels describes the ratio of pixels at risk during segmentation. Briefly, the edge thickness is determined by estimating the number of physical pixels on the image

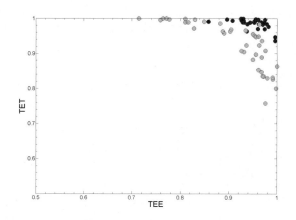

Fig. 3. TET vs. TEE for two manually segmented data sets, cell line 1 in blue, cell line 2 in cyan

that represent a cell edge, calculated using a quality index (QI) calculation [7]. The quality index ranges from 0.0 to 2.0, with a perfectly sharp edge at a value of 2.0. We define the thickness of a perfectly sharp edge to be equal to 1.0 pixel unit, or in this case 2.0/QI, which is how we define the edge thickness in general:

$$Th = 2.0/QI \qquad (4)$$

We approximate the number of pixels at the edge by multiplying the edge thickness, Th, by the cell perimeter, and then define our new metric, the ratio of pixels at the edge to the total number of pixels, the extended edge neighborhood (EEN), as:

$$EEN = (P \times Th)/area \qquad (5)$$

The extended edge neighborhoods for all 71 cells are shown in Figure 4, where again the A10 cells and NIH3T3 cells are plotted separately. Populations of cells with a lower extended edge neighborhood value (A10 cells) correspond to the cells where comparisons of manual segmentation results have lower variability. This includes larger rounder cells, with a smaller perimeter-to-area ratio. The thin, more spindly NIH3T3 cells, with a larger perimeter-to-area ratio, have higher extended edge neighborhood values for cells with the same edge characteristics as the A10 cells.

5   Selection of Reference Data for Segmentation Comparison

Dependable reference data is needed to be able to compare segmentation methods. In order to develop a semi-automatic method to trace the cell edges that mimics manual segmentation, we analyzed the pixel characteristics under the

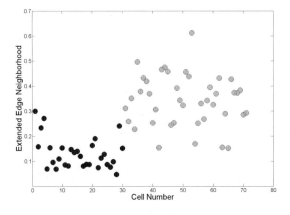

Fig. 4. Extended edge neighborhoods for the 71 cells in this study, A10 cells in blue, NIH3T3 cells in cyan

human-generated outlines. We assume that a background pixel will be similar in intensity to its surrounding background pixel neighbors, i.e. ratios of neighboring background pixel intensities should be close to 1.0. A larger difference will exist between a background pixel value and an adjacent cell edge pixel; i.e., we expect to find pixel intensity ratios well below 1.0. In the case of the fluorescent microscopy images, the background has relatively low pixel intensities and the pixels at the edge of the cells have higher intensities. A low ratio of neighboring pixel intensities will thus imply that two pixels are very different and that the higher intensity pixel might be an edge pixel.

Figure 5 shows an example of how we isolate background pixels to compare neighboring pixel values. We use a 2-step process to eliminate all pixels from the image that are not in the background. First, we threshold with a low value to overestimate the cell edge, and thus include the entire cell, shown in Figure 5 on the left. Then we dilate the edges so that the background pixels are completely separated from the cells, shown on the right of Figure 5. Also shown are the small clusters of brighter pixels that are also eliminated from the background. Background pixels include all pixels outside of the blown up boundaries. For each pixel intensity, $u[j]$, we find the minimum value of the ratio of any neighboring intensity $u[j]$ to $u[i]$, $\min(u[j]/u[i])$. We use 8-pixel neighborhoods for the ratio calculations. These ratio values, shown on the top row of Figure 6, are consistently above 0.7 and are close to 1.0, as expected.

We now compare the plot of ratios from background pixels with a plot a ratios collected from pixels selected as edge pixels in the manual segmentations. Figure 7 shows the set of pixels marked in red that were used for the lower plot of Figure 6. For the hand selected edge pixels, most of the minimum neighbor ratio values are less than 0.7. Occasionally, pixels are selected by hand that do not appear to align with the cell edge. These pixels will appear as a point on the plot above the 0.7 level. This is due to the manually generated outlines not being consistent all over the edges. The inset of Figure 7 shows some pixels that were outlined as part of the edge but the surrounding eight neighbors have similar pixel intensities (ratio higher than 0.7).

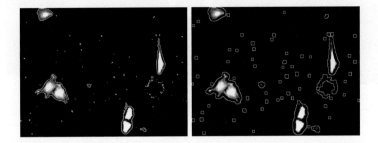

Fig. 5. Edge detection by threshold (left), Dilated edges including cells that touch the borders (right)

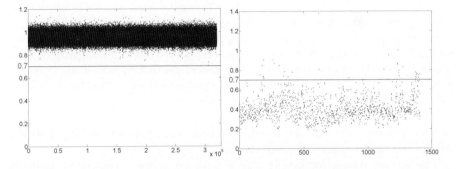

Fig. 6. Ratio between every background pixel from Figure 5 and its background pixel neighbors on the left; Minimum of eight-neighbor-ratio for all edge pixels in the manual outline on the right

Our localized ratio segmentation technique takes advantage of this intensity difference at the cell edge. It is a very simple algorithm with only one parameter (the threshold value for the minimum pixel ratio) and has a very fast execution time. Our algorithm tests all pixels in the image with an intensity below a certain threshold, including all background pixels and pixels at the cell edge and inside the cell edge. The cutoff intensity we use is based on a 5-means clustering segmentation. Five-means clustering consistently under-estimates the cell area, as described in a paper currently in process [8]. To ensure that we include a broad band of pixels around the cell edge, we use an upper cutoff value that is 1 standard deviation above the 5-means clustering lowest centroid value, assuming that the lowest centroid represents the image background. The masks resulting from our new method were surprisingly good at representing the actual cell images, as we will show below.

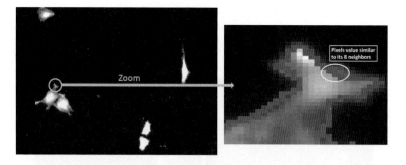

Fig. 7. Human manual outlines under study

There is much literature on the ability of the human visual system to discern changes in brightness in the visual field. The method described here, where edge pixels are identified by comparing ratios of neighboring pixel intensities, may be related to the Weber-Fechner Law which describes human detectible intensity changes over a background level. Our results suggest that manual segmentation produces edges at pixels where an immediate neighboring pixel exists with an intensity that is lower by a factor of approximately 0.7 or more and that our method simulates some parts of the human visual system for identifying edge pixels [9] [10] [11]. The algorithm for this technique that identifies edge pixels is given in Appendix A. We describe our method as a semi-automated procedure for two reasons. The first is that it is expected that for a given image set, a human user would adjust the image properties, such as brightness and contrast, of a few representative images to ensure concurrence between the segmentation results and the perceived edges of the cells.

The second reason for calling this a semi-automatic technique is that the resultant segmentation masks must be human verified to assure accuracy. This verification, however, can be done much more rapidly than manual segmentation and consists of a rapid visual inspection. Segmentations that do not pass visual inspection can be discarded or manually edited. The intent is that our method will be embedded within an overall human-supervised laboratory protocol that generates reference data for evaluating fully automatic segmentation techniques which would not need such human intervention.

6   Comparison of Individual Cell Masks

We compared the semi-automated reference masks with masks from manual segmentation, using two sets of bivariate metrics. This allowed us to compare our semi-automated results with the range of results from the hand-selected masks. The first pair of bivariate metrics compares our method, set E, to the intersection of the manual segmentation masks $\cap M_i$, for the manual sets, M_i:

$$TEE_\cap = \cap M_i \cap E / E \tag{6}$$

$$TET_\cap = \cap M_i \cap E / \cap M_i. \tag{7}$$

The second pair of bivariate metrics compares the semi-automated data, E, to the union of the manual segmentation masks, $\cup M_i$:

$$TEE_\cup = \cup M_i \cap E / E \tag{8}$$

$$TET_\cup = \cup M_i \cap E / \cup M_i. \tag{9}$$

These two sets of metrics serve to bracket the algorithm within the range of manual segmentation variability. A segmentation mask should fall in the range of the manual segmentations if it overestimates the intersection and underestimates the union of these sets. This is the case, seen in the plots of $TEE_\cap$ vs $TET_\cap$ and $TEE_\cup$ vs. $TET_\cup$, shown in Figure 8. The results clearly show that the semi-automated reference data lie within the bracket that represents the hand-generated data. When masks from this new technique are compared with the intersection of the hand-selected masks, the fraction $TET_\cap$ is close to 1.0, or most of the pixels of the new mask are present in the intersection. When the masks are compared with the union of the hand-selected sets, the fraction $TEE_\cup$ is close to 1.0, or most of the pixels in the new mask are present in the union of the hand-selected masks.

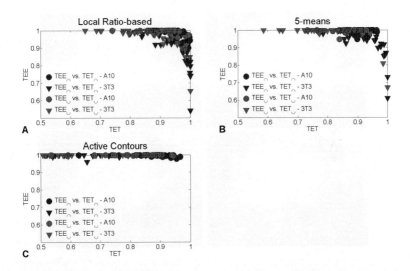

Fig. 8. Results of comparing segmentation masks with both the intersection and the union of the two manually segmented cells, for our local ratio-based method, for 5-means clustering, and for active contours

We compare these results to a comparable analysis looking at the differences between the hand-selected masks and a 5-means clustering segmentation of the 16 images. The results for this analysis are shown in Figure 8B. When the 5-means clustering masks are compared with the intersection or with the union of the hand-selected masks, the fractions and are both close to 1.0, or most of the pixels of the 5-means clustering masks are contained within the intersection and within the union of these sets. The 5-means clustering masks are seen to underestimate both the intersection and union of the two hand-selected sets, whereas our semi-automatic method was larger than the intersection and smaller than the union of the hand-selected sets. Our new method produced masks that lie inside the bracket of hand-selected data, unlike the 5-means clustering sets. We show the same type of analysis for masks created using a sparse field active contour segmentation [12], which underestimate with respect to both the intersection and union of the manually segmented masks.

Figure 9 also visually compares the segmentation of our new method with a manual selection, the result of a 5-means clustering segmentation, and a segmentation performed using a Canny edge method. By examining the small details of each mask, the consistency of our new method is clear. The mask from the Canny edge method fails to capture the hole seen in the upper left of the other three masks. The bottom right section of cell edge is missing a segment in both the 5-means and Canny edge masks. In the manually segmented cell edge, bright pixels are clearly seen outside of the right border, while our new method consistently selects pixels that appear to be on the edge.

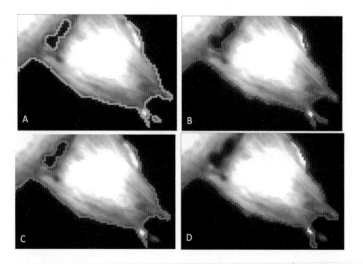

Fig. 9. Comparison of four segmentations: A: our new method; B: manual segmentation; C: 5-means clustering; D: Canny edge detection

7   Conclusions

Evaluating algorithms for biological cell segmentation depends upon human created reference data which is assumed to be close to the truth. Manually selecting pixels for a reference image is a long and tedious process, and it is error prone for that reason. We have found that the expected precision in hand-selecting reference data varies both with the pixel intensity gradient at cell edges and with the size and shape of the imaged cell. For large round cells, hand-selection is a fairly repeatable process. For smaller, more spindly, cells with complex geometries, or for any cell where the thickness of the cell edge is large and the clarity not sharp, there can be large differences in hand-selected segmentation masks over repeated assessments. For very small cells, for which the number of pixels near the edge of the cell is a large fraction of the entire set of cell pixels, these differences can be large enough to confuse and complicate the issues of segmentation analysis.

Manually generated reference data is too time-consuming to be used in large scale segmentation studies. Our new semi-automated method for determining reference masks from cells with very sharp edges consistently reproduces the geometry expected by human observers and can be used to collect reference data sets for large-scale segmentation algorithm performance studies. It relieves much of the tedium of selecting individual edge pixels and makes more effective use of human supervision in selecting reference data. The masks generated are a better approximation to the hand-selected reference data than those created using 5-means clustering.

Acknowledgements

Anne Plant and Michael Halter, of NIST's Biochemical Science Division, provided helpful discussion and input. We thank our students Asheel Kakkad and Sarah Jaffe for creating the manual segmentation masks.

References

1. Suzuki, T., Matsuzaki, T., Hagiwara, H., Aoki, T., Takata, K.: Recent Advances in Fluorescent Labeling Techniques for Fluorescence Microscopy. Acta Histochem. Cytochem. 40(5), 131–137 (2007)
2. Elliot, J.T., Tona, A., Plant, A.L.: Comparison of reagents for shape analysis of fixed cells by automated fluorescence microscopy. Cytometry 52A, 90–100 (2003)
3. Langenbach, K.J., Elliott, J.T., Tona, A., Plant, A.L.: Evaluating the correlation between fibroblast morphology and promoter activity on thin films of extracellular matrix proteins. BMC-Biotechnology 6(1), 14 (2006)
4. ImageJ, public domain software, http://rsbweb.nih.gov/ij/
5. Rand, W.M.: Objective criteria for the evaluation of clustering methods. Journal of the American Statistical Association 66(336), 846–850 (1971)
6. Peskin, A.P., Dima, A., Chalfoun, J.: Predicting Segmentation Accuracy for Biological Cell Images (in process)
7. Peskin, A.P., Kafadar, K., Dima, A.: A Quality Pre-Processor for Biological Cells. In: 2009 International Conference on Visual Computing (2009)
8. Dima, A., Elliott, J.T., Filliben, J., Halter, M., Peskin, A., Bernal, J., Stotrup, B.L., Marcin, Brady, A., Plant, A., Tang, H.: Comparison of segmentation algorithms for individual cells. Cytometry Part A (in process)
9. Hecht, S.: The Visual Discrimination of Intensity and the Weber-Fechner Law. J. Gen. Physiol. 7(2), 235–267 (1924)

10. Jianhong, S.: On the foundations of visual modeling I. Weber's law and Weberized
    TV restoration. Physica D 175, 241–251 (2003)
11. Schulman, E., Cox, C.: Misconceptions about astronomical magnitudes. Am. J.
    Phys. 65(10) (October 1997)
12. Lankton, S.: Active Contour Matlab Code Demo (2008),
    http://www.shawnlankton.com/2008/04/active-contour-matlab-code-demo

Appendix A: Matlab Code for Segmentation Algorithm

```
for j = 2:nb_columns-1
  for i = 2:nb_rows-1
    % if the pixel(i,j) is definitely a background pixel, skip it
    if I2(i,j) < 6000, continue, end

    % if pixel(i,j) is definitely a cell, fill it
    if I2(i,j) > 10500, segmented_image(i,j) = 1; continue, end

    % Check the 8 neighbors of pixel(i,j)
    % if they meet the selected criterion.
    % If one neighbor does ==> that pixel(i,j) is a cell pixel

    if I2(i-1, j-1)* Ratio < I2(i,j), segmented_image(i,j) = 1;
       continue, end
    if I2(i, j-1)* Ratio < I2(i,j), segmented_image(i,j) = 1;
       continue, end
    if I2(i+1, j-1)* Ratio < I2(i,j), segmented_image(i,j) = 1;
       continue, end
    if I2(i-1, j)* Ratio < I2(i,j), segmented_image(i,j) = 1;
       continue, end
    if I2(i+1, j)* Ratio < I2(i,j), segmented_image(i,j) = 1;
       continue, end
    if I2(i-1, j+1)* Ratio < I2(i,j), segmented_image(i,j) = 1;
       continue, end
    if I2(i, j+1)* Ratio < I2(i,j), segmented_image(i,j) = 1;
       continue, end
    if I2(i+1, j+1)* Ratio < I2(i,j), segmented_image(i,j) = 1;
       continue, end
  end
end
```

A Non-rigid Multimodal Image Registration Method Based on Particle Filter and Optical Flow

Edgar Arce-Santana, Daniel U. Campos-Delgado, and Alfonso Alba

Facultad de Ciencias, Universidad Autónoma de San Luis Potosí,
Av. Salvador Nava Mtz. S/N, Zona Universitaria, 78290,
San Luis Potosí, SLP, México
Tel.: +52 444 8262486 x 2907
arce@fciencias.uaslp.mx, ducd@fciencias.uaslp.mx, fac@fc.uaslp.mx

Abstract. Image Registration is a central task to many medical image analysis applications. In this paper, we present a novel iterative algorithm composed of two main steps: a global affine image registration based on particle filter, and a local refinement obtained from a linear optical flow approximation. The key idea is to iteratively apply these simple and robust steps to efficiently solve complex non-rigid multimodal or unimodal image registrations. Finally, we present a set of evaluation experiments demonstrating the accuracy and applicability of the method to medical images.

1   Introduction

One of the most important stages in medical image analysis is the registration (alignment) of images obtained from different sources [1], [2]. For instance, image registration is a common pre-processing stage during the development of atlases, the analysis of multiple subjects, and the study of the evolution of certain diseases or injuries. Generally speaking, image registration is the problem of finding a geometrical transformation which aligns two images. Solutions to this problem can be divided in two classes: rigid and non-rigid. Rigid image registration methods assume that a single global transformation is applied equally to each pixel in the image, reducing the problem to finding only the few parameters which describe the global transformation. While rigid or affine registration is widely used, it presents serious drawbacks when one of the images shows complex deformations with respect to the other, for example, when both images belong to different subjects. On the other hand, non-rigid (also called elastic) registration methods estimate a transformation for each pixel, incorporating only weaker smoothness assumptions in order to make the problem well-posed. These methods are more general than rigid methods; however, they are also more computationally demanding, and difficult to implement and calibrate. An extensive and comprehensive survey can be found in [3], [4].

Registration methods can also be classified as unimodal, where it is assumed that pixel intensities between both images are similar, and multimodal, which

G. Bebis et al. (Eds.): ISVC 2010, Part I, LNCS 6453, pp. 35–44, 2010.

are capable of aligning images coming from different types of sensors (e.g., MRI and PET), or with inhomogeneous illumination and contrast. The most popular multimodal registration methods are based on the maximization of the mutual information (MI) between both images, a method which was originally proposed by Viola and Wells [5], and also by Collignon and Maes [6]. These mehods, however, were originally described for rigid transformations; hence, recent research has mostly focused on multimodal elastic registration [7], [8], [9], [10]. Most of these methods deal with two difficulties: (1) find a model for the joint probability between both images (e.g., Parzen windows), which is required for the computation of the MI, and (2), propose an optimization algorithm (usually a complex one) capable of dealing with the non-linearities of the MI when expressed in terms of the joint probability model. Other approaches use landmarks and radial basis functions (RBF) to model local deformations as in [11], which require to detect some characteristic as edges and evaluate RBF.

A recent work [12] proposes a very different approach for MI-based rigid registration methods: instead of modeling the MI with a differentiable function (so that it can be optimized with continuous methods), a stochastic search is performed, where the MI is simply seen as a measure of similarity for each solution in a relatively large population. In this case, the authors employ a technique called Particle Filter (PF), which is commonly used for parameter estimation in dynamical systems. This new approach results in a simple yet robust method to perform multimodal affine registration with great accuracy.

This work builds from [12] to present a new approach to solve the non-rigid image registration problem. The key idea behind the proposed method lies in a two-step iterative algorithm, where we first compute the global affine registration between both images (using the PF approach), and then refine this transformation locally for each pixel using a very simple optical flow approximation. The aligned candidate image is then used as the input candidate image in the following iteration until convergence is reached. This iterative process produces a multi-stage algorithm where each stage is efficient, and easy to implement and tune, yet powerful enough to achieve complex non-rigid unimodal and multimodal registrations. The paper is organized as follows: Section 2 describes the methodology and some mathematical properties; Section 3 demonstrates the methods through some experiments and results; and finally, in Section 4, our conclusions are presented.

2   Methodology

This section describes the proposed algorithm in detail, and provides some mathematical insight about the convergence of the method. For the rest of this paper, we will use the following notation: $I_1(\mathbf{x})$ and $I_2(\mathbf{x})$ are, respectively, the reference and candidate images, both of which are observed in a finite rectangular lattice $L = \{\mathbf{x} = (x, y)\}$. For simplicity's sake, we will only deal with the 2D case, but the method can be easily extended to 3D volumes. The problem consists in finding a smooth transformation field $T_{\mathbf{x}}$ that aligns I_2 with I_1, i.e., so that

$\tilde{I}_2(\mathbf{x}) = I_1(\mathbf{x})$ where $\tilde{I}_2(\mathbf{x}) = f(I_2(T_\mathbf{x}(\mathbf{x})))$, and f is an unknown tone transfer function. Here we propose a solution based on a two-step iterative approach where, at each iteration t, a high-precision affine registration technique is used to find a global transformation matrix $M^{(t)}$, and then an efficient optical flow algorithm is applied to refine the transformation locally for each pixel; specifically, the desired transformation $T_\mathbf{x}^{(t)}$ is obtained as

$$T_\mathbf{x}^{(t)}(\mathbf{x}) = M^{(t)}\mathbf{x} + \mathbf{v}^{(t)}(\mathbf{x}), \tag{1}$$

where $\mathbf{v}^{(t)}$ is the optical flow field.

This method is generalized to the multimodal case by using Mutual Information (MI) as similarity function during the affine registration stage. Once the affine transformation is found, an adequate tone transfer function $f^{(t)}$ is applied to match the grayscale levels between both images, so that the optical flow constraint holds.

Our method can be summarized in the following steps:

1. Let $t = 0$.
2. Initialize $I_2^{(0)}$ with the candidate image I_2.
3. Find an affine transformation matrix $M^{(t)}$ that aligns $I_2^{(t)}$ with I_1 using a (possibly multimodal) affine registration method.
4. Estimate $\hat{I}_2(\mathbf{x}) = I_2^{(t)}(M^{(t)}\mathbf{x})$ using, for instance, bicubic interpolation.
5. Find a pixel-wise tone transfer function $f^{(t)}$ that adequately maps gray levels in I_1 to their corresponding gray values in $\hat{I}_2$. Let $\tilde{I}_2 = f(\hat{I}_2)$.
6. Estimate the optical flow field $\mathbf{v}^{(t)}$ between I_1 and $\tilde{I}_2(\mathbf{x})$.
7. Obtain $I_2^{(t+1)}(\mathbf{x})$ as $I_2^{(t)}(M^{(t)}\mathbf{x} + \mathbf{v}^{(t)}(\mathbf{x}))$.
8. Increase t and go back to step 3 until a convergence criteria is met.

Steps 3, 5, and 6 are described below in more detail.

2.1   Algorithm Details

The key idea behind the proposed method is to efficiently perform a complex elastic registration from simple and robust building blocks, which can be tested, tuned, and optimized independently. Each particular block has been chosen to achieve a good balance between simplicity, efficiency, and accuracy. Here we discuss some of our particular choices.

Affine registration by particle filter: Registration methods based on MI have proven to be very robust for both uni-modal and multi-modal applications. One difficulty, however, is that the inherently non-linear definition of the MI results in algorithms which often require complex optimization techniques which are either computationally expensive (e.g., Markov Chain Monte Carlo (MCMC) methods), or sensitive to the initial parameters (e.g., gradient descent methods). A recent approach, introduced in [12], employs instead a stochastic search known as Particle Filter (PF) [13] to estimate the optimal parameters of the affine transformation. An overview of this algorithm is as follows:

1. Generate a random initial population $S^1 = \{s_j^1\}$ of N_s parameter vectors (particles). We specifically use seven affine parameters: rotation angle, scaling along X and Y, translation along X and Y, and shearing along X and Y.

2. Let $k = 1$. Follow the next steps until convergence is met:

   (a) For each particle $s_j^k \in S^k$, compute the *likelihood* function $\gamma(s_j^k \mid I_1, \hat{I}_2)$, which in this case is given by

   $$\gamma(s \mid I_1, \hat{I}_2) = \frac{1}{\sqrt{2\pi}\sigma} \exp\left\{-\frac{\left(H(I_1) - \mathrm{MI}(I_1, \hat{I}_2)\right)^2}{2\sigma^2}\right\}, \qquad (2)$$

   where $H(I_1)$ is the entropy of I_1, $\mathrm{MI}(I_1, \hat{I}_2)$ is the MI between I_1 and $\hat{I}_2$, and $\hat{I}_2(x) = I_2(M_s x)$, with M_s the transformation matrix defined by parameter vector s.

   (b) Normalize the likelihoods to obtain the weights $w_j^k = \gamma(s_j^k \mid I_1, \hat{I}_2)/Z$, where $Z = \sum_j \gamma(s_j^k \mid I_1, \hat{I}_2)$, so that $\sum_j w_j^k = 1$. Note that w^k can now be seen as a distribution over S^k.

   (c) Obtain an intermediate population $\hat{S}^k$ by resampling from population S^k with distribution w^k [13].

   (d) Generate the next population S^{k+1} by stochastically altering each particle in the intermediate population. In this case, each particle follows a random walk profile given by:

   $$s_j^{k+1} = \hat{s}_j^k + \nu_j \qquad j = 1, \ldots, N_s, \qquad (3)$$

   where $\hat{s}_j^k \in \hat{S}^k$, and $\nu_j \sim \mathcal{N}(0, \Sigma^k)$; here, Σ^k is a diagonal matrix of standard deviations of the noise applied to each parameter. To enforce convergence, we apply simulated annealing to Σ^k by letting $\Sigma^{k+1} = \beta\Sigma^k$ at each iteration with $0 < \beta < 1$.

   (e) Increase k.

3. Based on the normalized weights $\{w^k\}$ for the last population, obtain the final parameter estimators s^* as the population average: $s^* = \sum_j w_j^k s_j^k$.

Some of the advantages of the PF approach are:(1) it allows complex and nonlinear likelihood functions to be used, without seriously affecting the complexity of the method, (2) it is robust to its initial parameters, and (3) it can be easily generalized to the 3D case.

Estimation of tonal transfer function: Once the global transformation matrix M is known, optical flow can be used to refine the transformation at each pixel in order to achieve an elastic registration. In the multi-modal case, however, the brightness consistency constraint (also known as optical flow constraint) [14] may be violated, preventing the optical flow algorithm to estimate the correct displacements. To deal with this obstacle, one may choose an adequate tone transfer function $f(i)$ to match the gray levels between image I_1 and the registered candidate $\hat{I}_2(\mathbf{x}) = I_2(M\mathbf{x})$. In particular, we compute $f(i)$ as a maximum likelihood estimator from the joint gray-level histogram of I_1 and $\hat{I}_2$, given by

$$h_{i,j} = \sum_{\mathbf{x} \in L} \delta\left(I_1(\mathbf{x}) - i\right) \delta\left(\hat{I}_2(\mathbf{x}) - j\right), \tag{4}$$

where δ is the Kronecker delta function. The tone transfer function is thus computed as

$$f(j) = \arg\max_i p(I_1(\mathbf{x}) = i \mid \hat{I}_2(\mathbf{x}) = j) \approx \arg\max_j \{h_{i,j}\}. \tag{5}$$

Note, however, than in the unimodal case, it is possible that the estimated transfer function differs significatively from the identity, introducing spurious correlations which may affect the performance of the optical flow estimation method. When the input data is known to be unimodal, we suggest to simply use the identity function.

Optical flow estimation: Our approach for optical flow estimation is based on a discrete reformulation of the classical Horn-Schunck method [14]. Consider a first order Taylor expansion of $\tilde{I}_2(\mathbf{x} + \mathbf{v}(\mathbf{x}))$ around $\mathbf{x}$:

$$\tilde{I}_2(\mathbf{x} + \mathbf{v}(\mathbf{x})) \approx \tilde{I}_2(\mathbf{x}) + \nabla_{\mathbf{x}}\tilde{I}_2(\mathbf{x})\mathbf{v}'(\mathbf{x}), \tag{6}$$

where $\mathbf{v}'$ denotes the transpose of $\mathbf{v}$, so that one can write the optical flow constrain as

$$I_1(\mathbf{x}) = \tilde{I}_2(\mathbf{x}) + \nabla_{\mathbf{x}}\tilde{I}_2(\mathbf{x})\mathbf{v}'(\mathbf{x}). \tag{7}$$

Equation 7 is ill-posed; therefore, additional regularization constraints must be introduced to obtain a unique solution. Both constraints are typically expressed by means of an energy function $U(\mathbf{v})$ that must be minimized. For efficiency reasons, we chose a quadratic penalty function for both constraints, so that $U(\mathbf{v})$ is given by

$$U(\mathbf{v}) = \sum_{\mathbf{x} \in L} \left\| \nabla_{\mathbf{x}}\tilde{I}_2(\mathbf{x})\mathbf{v}'(\mathbf{x}) + \tilde{I}_2(\mathbf{x}) - I_1(\mathbf{x}) \right\|^2 + \lambda \sum_{<\mathbf{x},\mathbf{y}>} \|\mathbf{v}(\mathbf{y}) - \mathbf{v}(\mathbf{x})\|^2, \tag{8}$$

where the sum in the second term ranges over all nearest-neighbor sites $< \mathbf{x}, \mathbf{y} >$ in L, and λ is a regularization parameter which controls the smoothness of the resulting flow field. The gradient in the first term is computed by applying an isotropic Gaussian filter to $\hat{I}_2$ and computing the symmetric derivatives with the kernel $(0.5, 0, -0.5)$.

Since $U(\mathbf{v})$ is quadratic in the unknowns, it can be solved very efficiently by the Gauss-Seidel method [15].

2.2   Mathematical Properties of the Problem Formulation

In general, from its proposed formulation, the nonlinear transformation field $T_{\mathbf{x}}$ that aligns I_2 with I_1 is sought under the following optimality condition

$$\max_{T_{\mathbf{x}}} \mathrm{MI}\left(I_1(\mathbf{x}), I_2(T_{\mathbf{x}}(\mathbf{x}))\right). \tag{9}$$

This previous formulation has important robustness properties and capability to handle multimodal registration. However, the optimization problem in (9) is highly nonlinear and its solution involves a complex multivariable search. Hence, as it was mentioned at the beginning of this section, the transformation field $T_{\mathbf{x}}$ is assumed with the special structure in (1). Furthermore, it is considered that the optical flow field $\mathbf{v}^{(t)}$ can be represented as a linear combination of unknown vector fields $\hat{\mathbf{v}}^{(j)}$ $j = 1, \ldots, N$, i.e.

$$\mathbf{v}^{(t)}(\mathbf{x}) = \sum_{j=1}^{N} \alpha^j \hat{\mathbf{v}}^{(j)}(\mathbf{x}) \qquad \alpha^j \in \mathbb{R}, \ \forall j. \tag{10}$$

Therefore, in order to solve (9), an iterative algorithm is proposed to construct $T_{\mathbf{x}}^{(t)}$:

$$T_{\mathbf{x}}^{(t)}(\mathbf{x}) = \mathbf{M}^{(t)} T_{\mathbf{x}}^{(t-1)}(\mathbf{x}) + \mathbf{v}^{(t)}(\mathbf{x}) \qquad t = 1, 2, \ldots \tag{11}$$

Let us assume that $T_{\mathbf{x}}^{(-1)}(\mathbf{x}) = \mathbf{x}$, then it is satisfied in the iterative process

$$T_{\mathbf{x}}^{(1)}(\mathbf{x}) = \mathbf{M}^{(1)}\mathbf{x} + \mathbf{v}^{(1)}(\mathbf{x}) \tag{12}$$
$$T_{\mathbf{x}}^{(2)}(\mathbf{x}) = \mathbf{M}^{(2)} T_{\mathbf{x}}^{(1)}(\mathbf{x}) + \mathbf{v}^{(2)}(\mathbf{x}) \tag{13}$$

$$\vdots$$

$$T_{\mathbf{x}}^{(t)}(\mathbf{x}) = \underbrace{\left[\prod_{j=1}^{t} \mathbf{M}^{(t-j+1)}\right]}_{\mathbf{M}(t)} \mathbf{x} + \sum_{j=1}^{t} \underbrace{\left[\prod_{i=1}^{t-j} \mathbf{M}^{(t-i+1)}\right] \mathbf{v}^{(j)}(\mathbf{x})}_{\alpha^j(t)\hat{\mathbf{v}}^{(j)}(\mathbf{x})} \tag{14}$$

where $\prod_{i=a}^{b} \mathbf{M}^{(i)} = \mathbf{M}^{(b)} \mathbf{M}^{(b-1)} \cdots \mathbf{M}^{(a)}$ for $a \leq b$, and $\prod_{i=a}^{b} \mathbf{M}^{(i)} = \mathbf{I}$ for $a > b$. Therefore, at t-th iteration, the resulting transformation field $T_{\mathbf{x}}^{(t)}(\mathbf{x})$ has the structure highlighted in (1) and (10). In addition, note that the iterative construction of $T_{\mathbf{x}}(\mathbf{x})$ in (11) can be interpreted as the response of a linear time-varying system under the input vector field $\mathbf{v}^{(t)}(\mathbf{x})$. Consequently, in order to guarantee the convergence (*stable response*) of (11) [16], it is sufficient to satisfy

$$\mathbf{M}^{(t)} \to \mathbf{I} \quad \& \quad \mathbf{v}^{(t)} \to \mathbf{0}, \quad \text{as} \quad t \to \infty \tag{15}$$

where $\mathbf{I}$ denotes the identity matrix and $\mathbf{0}$ the zero matrix, in order to guarantee

$$\|\mathbf{M}(t)\| \leq \gamma \quad \forall t \tag{16}$$

and $\gamma > 0$. Therefore, we may define as convergence conditions:

$$\|\mathbf{M}^{(t)} - \mathbf{I}\| < \epsilon_1 \quad \& \quad \|\mathbf{v}^{(t)}(\mathbf{x})\| < \epsilon_2, \tag{17}$$

given $\epsilon_1, \epsilon_2 > 0$.

Hence the proposed algorithm suggests to solve iteratively the two-stage optimization problems:

I.- The rigid transformation $\mathbf{M}^{(t)}$ through the optimization process

$$J_1^{(t)} = \max_{\mathbf{M}^{(t)}} \text{MI} \left(I_1(\mathbf{x}), I_2 \left(\mathbf{M}^{(t)} T_\mathbf{x}^{(t-1)}(\mathbf{x}) \right) \right). \tag{18}$$

II.- And the linear optical flow $\mathbf{v}^{(t)}(\mathbf{x})$ by the following minimization problem

$$J_2^{(t)} = \min_{\mathbf{v}^{(t)}(\mathbf{x})} \sum_{\mathbf{x} \in L} \left\| \nabla_\mathbf{x} \tilde{I}_2 \left(\mathbf{M}^{(t)} T_\mathbf{x}^{(t-1)}(\mathbf{x}) \right) \mathbf{v}'^{(t)}(\mathbf{x}) + \tilde{I}_2(\mathbf{M}^{(t)} T_\mathbf{x}^{(t-1)}(\mathbf{x})) - \right.$$

$$\left. I_1(\mathbf{x}) \right\|^2 + \lambda \sum_{<\mathbf{x}, \mathbf{y}>} \left\| \mathbf{v}^{(t)}(\mathbf{y}) - \mathbf{v}^{(t)}(\mathbf{x}) \right\|^2. \tag{19}$$

Note that if $\{J_1^{(t)}\}$ and $\{J_2^{(t)}\}$ represent increasing and decreasing convergent sequences with respect to t, respectively, then conditions in (15) will be satisfied, since there is a continuity property of the parameters of $T_\mathbf{x}^{(t)}(\mathbf{x})$ with respect to the cost functions in the optimization processes.

Assume that the iterative algorithm stops at $t = N$, then the optimal (*locally*) transformation is given by

$$T_\mathbf{x}^*(\mathbf{x}) = \mathbf{M}^* \mathbf{x} + \mathbf{v}^*(\mathbf{x}) \tag{20}$$

where

$$\mathbf{M}^* = \prod_{j=1}^{N} \mathbf{M}^{(N-j+1)} \tag{21}$$

$$\mathbf{v}^*(\mathbf{x}) = \sum_{j=1}^{N} \left[\prod_{i=1}^{N-j} \mathbf{M}^{(N-i+1)} \right] \mathbf{v}^{(j)}(\mathbf{x}), \tag{22}$$

which are carried out by the method proposed here.

3   Results

In this section, we present two different kinds of results. The first one corresponds to the registration of images having similar gray values, so that the transfer function $f(.)$ (see Section 2) is the identity. These kinds of experiments are important in the medical field, since for example tumors or vital organs may be monitored. Rows in Fig. 1 show a set of TC-images, these rows correspond to the

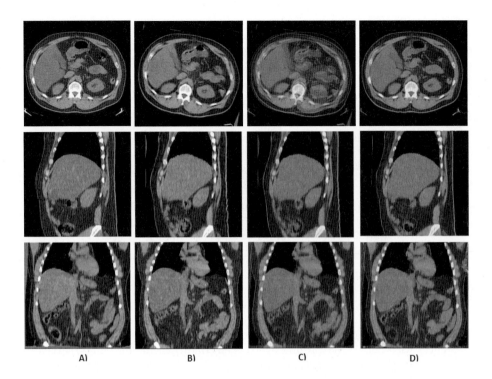

A)   B)   C)   D)

Fig. 1. Chest TC-images: *A*) Reference Image; *B*) Moving Image; *C*) Superimposed Non-registered Images; *D*) Superimposed Registered Images

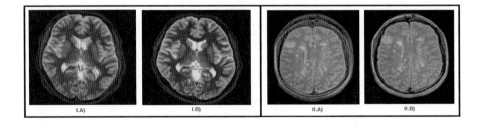

I.A)   I.B)   II.A)   II.B)

Fig. 2. Superimposed image registration: I.A) MRI-images before registration; I.B) MRI-images after registration; II.A) T1-GdDTPA and T2-images before registration; II.A) T1-GdDTPA and T2-images after registration

axial, sagittal, and coronal views from the chest of the same subject. The images in column *A*) are the reference; meanwhile, column *B*) shows the moving images. The difference between these two columns corresponds to posture and breathing in two different intervals of time. We can see in the superimposed images in column *C*) the difference between the images to register; the reference image is displayed using the red channel, while the moving image uses the green channel;

therefore, those pixels where the images coincide should look yellow. Column D) shows the non-rigid image registration obtained by the proposed method. Notice how some structures have been deformed in order to achieve a good matching, reducing the pixels in red or green, and incrementing the matching pixels in yellow explained by a good elastic registration; the image size (in pixels) for the axial, sagittal, and coronal views are respectively 247×242, 268×242, and 268×244; and the registration times were 214, 231, and 231 seconds. All the these experiments were performed on a PC running at 3.06 GHz.

The following set of experiments concern to the case of multimodal image registration. Panel I in Fig. 2 shows the registration of anatomical (spin-echo) MRI images of two different subjects; observe how the different structures in the center of the Fig. 2-I.A) are aligned after the registration (2-I.B)). Similar results are shown in Panel II in Fig 2, which corresponds to the alignment of a T1-weighted MRI-image, with a contrast agent (gadolinium diethylene triamine pentaacetic acid), with a T2-weighted image from the same subject.

In all the experiments described above , we used the same set of parameters; the values of the main parameters are defined in Table 1.

Table 1. Main parameter values of the non-rigid registration algorithm

Particle Filter						
Particles	Likelihood-σ	PF-Iterations	Displacement-σ	Angle-σ	Scale-σ	Shearing-σ
200	$H(I_1)/6$	150	2	0.2	0.02	0.02
Optical Flow						
Iterations		λ		Smoothing-σ		
200		2000		0.5		

4   Conclusions

We presented a new iterative algorithm which is mainly based on two building blocks: 1) a global affine registration, and 2) a linear optical flow. The key idea behind this approach is to efficiently perform a complex elastic registration from these simple and robust building blocks. The affine registration was carried out by an accurate algorithm based on the particle filter [12], meanwhile the second block can be solved very efficiently by a Gauss-Seidel method. We provided the mathematical properties which support the problem formulation, and establish some convergence conditions. Finally, we presented experiments where the algorithm proved to be efficient, and accurate to estimate complex 2D elastic unimodal and multimodal registrations.

Acknowledgements. This work was supported by grant PROMEP/UASLP/-10/CA06.

References

1. Hajnal, J., Hawkes, D., Hill, D.: Medical image registration. CRC, Boca Raton (2001)
2. Modersitzki, J.: Numerical Methods for Image Registration. Oxford University Press, Oxford (2004)
3. Zitova, B., Flusser, J.: Image registration methods: a survey. Image and Vision Computing 21, 977–1000 (2003)
4. Brown, L.G.: A survey of image registration techniques. ACM Computing Survey 24, 326–376 (1992)
5. Wells III, W.M., Viola, P.A., Atsumi, H., Nakajima, S., Kikinis, R.: Multi-modal volume registration by maximization of mutual information. Medical Image Analysis 1, 35–51 (1996)
6. Maes, F., Collignon, A., Vadermeulen, D., Marchal, G., Suetens, P.: Multimodality image registration by maximization of mutual information. IEEE Trans. Med. Image. 16, 187–198 (1997)
7. Zhang, J., Rangarajan, A.: Bayesian Multimodality Non-rigid image registration via conditional density estimation. In: Taylor, C.J., Noble, J.A. (eds.) IPMI 2003. LNCS, vol. 2732, pp. 499–512. Springer, Heidelberg (2003)
8. Periaswamya, S., Farid, H.: Medical image registration with partial data. Medical Image Analisys 10, 452–464 (2006)
9. Arce-Santana, E., Alba, A.: Image registration using markov random coefficient and geometric transformation fields. Pattern Recognition 42, 1660–1671 (2009)
10. Bagci, U., Bai, L.: Multiresolution elastic medical image registration in standard intensity scale. Computer Graphics and Image Processing 29, 305–312 (2007)
11. Yang, X., Zhang, Z., Zhou, P.: Local elastic registration of multimodal image using robust point matching and compact support RBF. In: International Conference on BioMedical Engineering and Informatic, BMEI-(2008)
12. Arce-Santana, E., Campos-Delgado, D.U., Alba, A.: ISVC 2009. LNCS, vol. 5875, pp. 554–563. Springer, Heidelberg (2009)
13. Arulampalam, M.S., Maskell, S., Gordon, N., Clapp, T.: A Tutorial on Particle Filters for Online Nonlinear/Non-Gaussian Bayesian Tracking. IEEE Transactions On Signal Processing 50, 174–188 (2002)
14. Horn, B.K.P., Schunck, B.G.: Determining Optical Flow. Artificial Intelligence 17, 185–203 (1981)
15. Barrett, R., Berry, M., Chan, T.F., Demmel, J., Donato, J., Dongarra, J., Eijkhout, V., Pozo, R., Romine, C., van der Vorst, H.: Templates for the Solution of Linear Systems: Building Blocks for Iterative Methods, 2nd edn. SIAM, Philadelphia (1994)
16. Rugh, W.J.: Linear System Theory. Prentice-Hall, Englewood Cliffs (1996)

Stitching of Microscopic Images for Quantifying Neuronal Growth and Spine Plasticity

SooMin Song[1], Jeany Son[1], and Myoung-Hee Kim[1,2,*]

[1] Department of Computer Science and Engineering
Ewha Womans University, Seoul, Korea
`{smsong,filia00}@ewhain.net`
[2] Center for Computer Graphics and Virtual Reality
Ewha Womans University, Seoul, Korea
`mhkim@ewha.ac.kr`

Abstract. In neurobiology, morphological change of neuronal structures such as dendrites and spines is important for understanding of brain functions or neuro-degenerative diseases. Especially, morphological changes of branching patterns of dendrites and volumetric spine structure is related to cognitive functions such as experienced-based learning, attention, and memory. To quantify their morphologies, we use confocal microscopy images which enables us to observe cellular structure with high resolution and three-dimensionally. However, the image resolution and field of view of microscopy is inversely proportional to the field of view (FOV) we cannot capture the whole structure of dendrite at on image. Therefore we combine partially obtained several images into a large image using image stitching techniques. To fine the overlapping region of adjacent images we use Fourier transform based phase correlation method. Then, we applied intensity blending algorithm to remove uneven intensity distribution and seam artifact at image boundaries which is coming from optical characteristics of microscopy. Finally, based on the integrated image we measure the morphology of dendrites from the center of cell to end of each branch. And geometrical characteristics of spine such as area, location, perimeter, and roundness, etc. are also quantified. Proposed method is fully automatic and provides accurate analysis of both local and global structural variations of neuron.

1 Introduction

Morphological analysis of neurons is not only important for studying the normal development of brain functions, but also for observing neuropathological changes. Therefore considerable attempts have been made to identify and quantify neural changes at a cellular level. However, to date, most biologists in the neuroscience field still use semi-automatic methods to identify and quantify neural changes. Even in the recent studies conducted in the field of cellular biology [1, 2], quantification of dendrite was carried out manually. In [1], a clear acetate sheet with concentric rings,

---

* Corresponding author.

G. Bebis et al. (Eds.): ISVC 2010, Part I, LNCS 6453, pp. 45–53, 2010.
© Springer-Verlag Berlin Heidelberg 2010

was placed over the printed datasets; it was oriented such that the cell soma and the innermost rings, were aligned over each other. The points of intersection between the dendritic branches and the concentric circles (60 µm apart) were used to identify the number of dendritic segments.

In neuroscience, dendrites and spines are the most frequently observed structures in a neuronal cell. Dendrites are the tree-like structure and they grow up to several micrometers in length. Spines are small protrusions formed on the surface of a dendrite. They are smaller in size than dendrites, with their head volume ranging from 0.01 to 0.8 µm. Therefore high sampling rate and magnification are required to capture them. However, it is found that the field of view (FOV) decreases with an increase in the optical magnification. Therefore to get a complete view of a specimen, a number of image tiles are acquired and they are then reconstructed into one large image. This image combining technique is called tiling, mosaicing, or stitching.

Previously developed techniques or existing software for image mosaicing were mostly focused on stitching of 2D photographs for panorama image generation nevertheless of increasing need for 3D image stitching. Furthermore these skills were used to correct perspective and lens distortion concerning high-order transformation models [5]. In case of microscopic images, however, optics is mechanically fixed on equipment and there is no need to correct lens distortion or scale transformation. Rather, due to the optical characteristics, intensity attenuation problem should be concerned which intensity signal weakens along depth. It is not feasible to simply apply 2D photograph stitching techniques because of the several reasons mentioned above. Non-commercial software VIAS (Volume Integration and Alignment System) creates a single volume from multiple image volumes,; however, in uses olnly the projected images along the X-, Y- and Z-axes to deal with 3D volume images. This makes the stitching procedure fast but provides erroneous results. Moreover a user manually positions adjacent images to approximate alignment, and then, an auto-alignment process is applied for fine-tuning. This procedure can lead an increase in the number of intra-, and inter-observer variability errors. With HCA-vision local information can be missing because of the application of the binarization process before mosaicing.

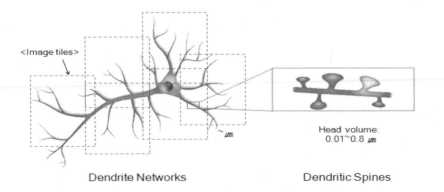

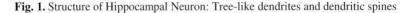

Fig. 1. Structure of Hippocampal Neuron: Tree-like dendrites and dendritic spines

In this paper, we explain the manner in which several images can be integrated into one large mosaic image; we also describe an automatic quantification method for analyzing the morphology of dendrites and spines.

2   Deformable-Model-Based Segmentation of Dendritic Spines

2.1   Phase-Correlation-Based Image Stitching

We assume that all the image tiles to be stitched overlap each other, and we only consider the translation transform between them because they are acquired only moving the microscope stage horizontally. To compute translational offsets between images we use the Fourier transform based phase correlation method.

Applying the phase correlation method to a pair of images produces image which contains a single peak. The location of this peak corresponds to the relative shift between the images.

Real images, however, contain several peaks indicating different translational offsets with high correlation. Moreover, each peak describes many possible translations due to the periodicity of the Fourier space. To determine the correct shift, we select the n highest local maxima from the inverse Fourier transform and evaluate their possible translations by means of cross correlation on the overlapping area of the images. Then, the peak with the highest correlation is selected to indicate the translation between two images.

When two images $I_a(x, y)$, and $I_b(x, y)$ are given and they are related by translational transform about a horizontally and b vertically, the result of the Fourier transform is denoted as $A(u,v)$, and $B(u,v)$, respectively, and their relationship is expressed as $B(u,v) = A(u,v)\exp\{-i(au+bv)\}$ as follows:

Then phase correlation matrix is computed (eq.1)

$$Q(u,v) = \frac{A(u,v)B(u,v)^*}{\left| A(u,v)B(u,v)^* \right|} \tag{1}$$

$$Q(x, y) = \delta(x-a, y-b) \tag{2}$$

It is theoretically represented by a single peak whose location defines the correct translation vector between the images. Phase correlation uses the fast Fourier transform to compute the cross-correlation between the two images, generally resulting in large performance gains. Applying the phase correlation method to a pair of images produces an image that contains a single peak. The location of this peak corresponds to the relative translation between the two images. Translational shift in the image domain corresponds to linear phase change in the Fourier domain.

2.2   Intensity Blending on Image Boundaries

Even after tiling, images have uneven intensity distributions, shading, and seam artifacts in overlapping regions, because confocal microscopy uses an optical source.

Therefore, to compensate for the brightness difference between tiles, we apply an intensity blending technique to all overlapping areas of adjacent tiles. The simplest way to achieve this compensation is to average the intensity values in the transition zone. However, this produces a blurred result. Therefore, we use an inverse-weighted blending method. This method involves the use of an exponentially weighted factor that is determined on the basis of the distance of the current pixel coordinate from its own tile boundary. On the basis of the distance from the pixel position, we create a weighting map for each tile that has high influence on the central pixels and no influence on the pixels at the edge of the image.

3   Quantitative Analysis of Dendritic Arbor, Spine Density and Spine Morphology

Using an integrated stitching image, we construct a dendrogram which is a graphical structure of dendrite networks by analyzing the branching features of dendrites. Then, in magnified mosaic image, we measure various shape features of spines, such as their area, diameter, perimeter, and length, etc. and we classify them into three categories.

The structure of both the neuron backbone and the dendritic spines show that they exhibit synaptic and neurological functions [10, 11]. For example, the dendrogram increases in size and exhibits a complicated structure when neural signal transmission is active and vice-versa. Furthermore, morphological changes in the spine are key indicators of brain activities; for example, the spine transforms into a mushroom shape when it transmit a synaptic signal.

3.1   Graph Reconstruction of Dendrite Networks

For measuring the neuronal growth rate, we construct a graph model to concisely capture the neuron geometry and topology for a dendrite and quantify the morphology of a single spine. Dendrites are tree-like structures that are represented in terms of their total length and the total number of branches they contain. First, we apply unsharp mask filtering [12] to enhance the boundary of the images, because the optical source of fluorescence microscope causes blurred artifacts and uneven intensity distributions. Then we convert the filtered image to a binary one using a local adaptive thresholding method. On this binary image, we extract the skeleton of a dendrite area by applying a thinning method. After extracting the skeleton of the dendrite area we construct a tree-like graph structure. We simply identify each skeletal point as a branch point, end point, and other skeletal points by analyzing 26 neighboring voxels. A branch point is defined a point that has more than three neighbouring skeletal points; and end point is defined as a voxel with only one neighbouring point. Each end point and branching point of a dendrite are assigned to a graph node. This graph model is called a dendrogram; it is used to analyze the structure of the neuron backbone. The neuronal growth is expressed in terms of the dendritic length and the total number of branches using the dendrogram.

3.2 Quantitative Representation of Single Spine Profile and Its Classification

Once the dendrites are detected in the previous step, we distinguish spines from them. To extract the boundaries of the spines, we use a deformable-model-based segmentation approach. The end points of each branch are used as seed points to generate the initial curve, which then evolves using a geodesic active contour method.

However, it is difficult to separate the individual spine are when spines are either too close or they overlap each other. We apply a watershed algorithm to overcome this difficulty; this algorithm is efficient in separating or merging adjacent objects. Then, each spine is represented quantitatively with various morphological features such as area, location, perimeter, roundness (shape factor), and spine length. Using these parameters, we can classify spines into three general types, namely, thin-, stubby-, and mushroom-shaped spines.

For each spine, we compute the following features:

- Area: The number of pixels in the object. To calculate the area of object i, we use the connected component labeling method. Connected components labeling scans an image and groups its pixels into components based on pixel connectivity
- Location: The X, Y coordinates of the geometric center of mass of the spine. The center of mass is represented by a relation of area size and first moments of object.

$$L_i = \frac{M_i}{A_i},$$ where M_i is first moments of area A_i of object i.

- Perimeter: The length of the boundary of the object. Chain coding is used to follow the outer edge of the object. It follows the contour in a clockwise manner and keeps track of the directions as it goes from one contour pixel to the next.
- Roundness (shape factor): The ratio of the area of the shape to the area of a circle having the same perimeter. For a circle, the ratio is one; for an infinitely long and narrow shape, it is zero.

$$R_i = \frac{4\pi \times A_i}{P_i^2},$$ where A_i is the area and P_i is the perimeter of object i.

- Spine length: For continuous skeletons, the spine length can be measured by counting the number of pixels between each endpoint and branch point. For discontinuous skeletons, the endpoint of the branch is extracted and then tracked along the branch. If the branch is connected to the backbone, then the number of pixels from endpoint to the branch point will be calculated. When the branch point cannot be identified because the skeletons are discontinuous, spine length is calculated as the distance between the endpoint and backbone (Fig 2).

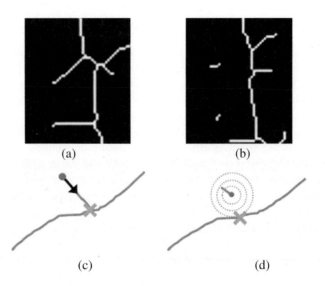

Fig. 2. Spine length calculation; (a) and (b) are images of the skeleton of dendrites and spines: (a) when all skeletons are connected, (b) spine's skeleton is unconnected, (c) and (d) illustrate the calculation methods of spine length for each case

Finally, we classified spines into three general types: thin, stubby, and mushroom using following method.

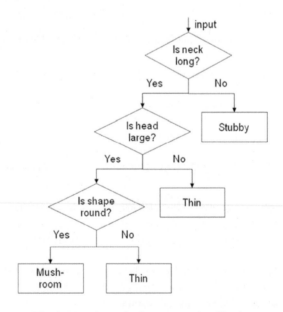

Fig. 3. Flowchart of spine pattern classification

4 Experimental Results

The cells used in this study are dendritic spines in a hippocampal neuron of CA1 in an E18 rat. Images of the neurons were labeled with green fluorescent protein (GFP) and obtained with an Olympus IX-71 inverted microscope with a 60x 1.4 N.A. oil lens using a CoolSNAP-Hq CCD camera. We used 8 data sets for the experiments, each consisting of 7 or more partially overlapping image tiles approximately 512 ×512×24 voxels at a resolution of 0.21×0.21×1.0 μm.

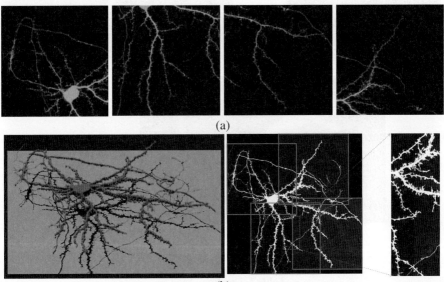

(a)

(b)

Fig. 4. Acquired images and stitching result

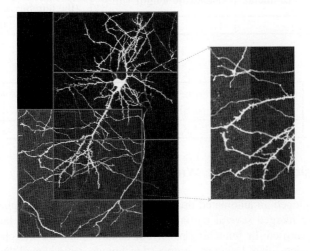

Fig. 5. Stitching Results and magnified view on image boundary

To validate stitching accuracy we observe image boundaries of integrated image. As shown in Fig. 5. dendrite branches and spines are all well connected.

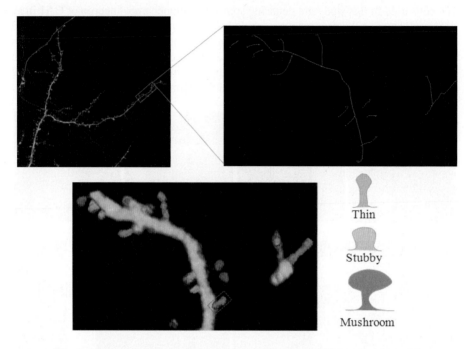

Thin

Stubby

Mushroom

Fig. 6. Dendrogram and result of spine classification in ROI (purple rectangle)

Table 1. Dendrom Analysis

Total branch Numbers	Dendritic length (μm)	Density (the number of spines/μm)
121	18	9

Table 2. Single Spine's Morphological Analysis

Area (μm^2)	Diameter (μm)	Perimeter (μm)	Roundness	Length (μm)	Class
0.3	0.08	1.3	0.4	1.5	thin

5   Conclusion and Future Works

We have developed an automated image stitching and neuronal morphology analysis technique. We merged several partial microscopic images into one large integrated image and quantified the branching pattern of dendrites and the morphological characteristics of each spine. Using the proposed technique, we widened the field of view to capture the entire structure of a neuron, from soma to its branch tips. This makes it possible to carry out a follow-up study of live neurons from their genesis. We also provided the quantitative morphological characteristics of spines located in every dendritic branch. Simultaneous observation of the branching pattern of a neuron and the spines' shape is beneficial to neurobiologists who study mechanisms of cognitive function or development of neurodegenerative diseases. Thus far, our method was applied to control data so that the shapes of the spines did not change dynamically. In future, we intend to use cells with various simulations such as electrical stimulation or drugs that speed up or regress the growth of neuron.

Acknowledgement

This work is financially supported by the Korea Science & Engineering Foundation through the Acceleration Research Program. And we would like to thank Prof. Sunghoe Chang of Seoul National University for providing confocal microscopic images of neurons.

References

1. Hao, J., et al.: Interactive effects of age and estrogen on cognition and pyramidal neurons in monkey prefrontal cortex. PNAS, 104(27) (2007)
2. Kim, H., et al.: δ-catenin-induced dendritic morphogenesis. The Journal of Biological Chemistry 283(977-987) (2008)
3. Vollotton, P., et al.: Automated analysis of neurite branching in cultured cortical neurons using HCA-Vision. Cytometry 71A, 889–895 (2007)
4. Wearne, S.L., et al.: New techniques for imaging, digitization and analysis of three-dimensional neural morphology on multiple scales. Neuroscience 136, 661–680 (2005)
5. Emmenlauer, M., et al.: XuvTools: free, fast and reliable stitching of large 3D data-sets. Journal of Microscopy 233(1), 42–60 (2009)
6. Ideker, T., et al.: Globally optimal stitching of tiled 3D microscopic image acquisi-tions. Bioinformatics Advance Access 25(11), 1463–1465 (2009)
7. Brown, M., Lowe, D.G.: Automatic panoramic image stitching using invariant features. International Journal of Computer Vision 74(1), 59–73 (2002)
8. Bai, W., et al.: Automatic dendritic spine analysis in two-photon laser scanning microscopy images. Cytometry, 818–826 (2007)
9. Rittscher, J., Machiraju, R., Wong, S.T.C.: Microscopic image analysis for life science applications. Artech House, Norwood
10. Weaver, C.M., Hof, P.R., Wearne, S.L., Lindquist, W.B.: Automated algo-rithms for multiscale morphometry of neuronal dendrites. Neural Computation 16(7), 1353–1383 (2004)
11. Zito, K., Knott, G., Shepherd, G., Shenolikar, S., Svoboda, K.: Induction of spine growth and synapse formation by regulation of the spine actin cytoskeleton. Neuron 44(2), 321–334 (2004)
12. Castleman, K.R.: Digital Image Processing. Prentice Hall, Englewood Cliffs (1995)

Semi-uniform, 2-Different Tessellation of Triangular Parametric Surfaces

Ashish Amresh and Christoph Fünfzig

School of Computing, Informatics and DS Engineering, Arizona State University, USA
Laboratoire Electronique, Informatique et Image (LE2I), Université de Dijon, France
amresh@asu.edu, c.fuenfzig@gmx.de

Abstract. With a greater number of real-time graphics applications moving over to parametric surfaces from the polygonal domain, there is an inherent need to address various rendering bottlenecks that could hamper the move. Scaling the polygon count over various hardware platforms becomes an important factor. Much control is needed over the tessellation levels, either imposed by the hardware limitations or by the application. Developers like to create applications that run on various platforms without having to switch between polygonal and parametric versions to satisfy the limitations. In this paper, we present SD-2 (Semi-uniform, 2-Different), an adaptive tessellation algorithm for triangular parametric surfaces. The algorithm produces well distributed and semi-uniformly shaped triangles as a result of the tessellation. The SD-2 pattern requires new approaches for determining the edge tessellation factors, which can be fractional and change continuously depending on view parameters. The factors are then used to steer the tessellation of the parametric surface into a collection of triangle strips in a single pass. We compare the tessellation results in terms of GPU performance and surface quality by implementing SD-2 on PN patches.

1 Introduction

Rendering of parametric surfaces has a long history [1]. Early approaches consist of direct scanline rasterization of the parametric surface [2]. Modern GPUs have the ability to transform, light, rasterize surfaces composed of primitives like triangles and more recently also to tessellate them. Tessellation in particular can save bus bandwidth for rendering complex smooth surfaces on the GPU.

The proposed triangle tessellation approaches can be distinguished either by the hardware stage they employ or by the geometric tessellation pattern. Approaches originating from triangular subdivision surfaces use a 1-to-4 split of the triangle, as seen in [3] and [4]. For greater flexibility of the refinement, factors are assigned to the edges of each triangle, which results in three different tessellation factors for an arbitrary triangle. Tessellation patterns for this general case go back to several authors after [5], for example, in [6] optimized for a hardware implementation, and in [7] for a fast CUDA implementation. Fractional tessellation factors target the discontinuity problems with integer factors during

G. Bebis et al. (Eds.): ISVC 2010, Part I, LNCS 6453, pp. 54–63, 2010.

animations, whose abrupt changes are visible in the shading and other interpolated attributes. With fractional tessellation factors and a tessellation pattern continuously depending on them, the changes are less abrupt with the view.

The interior patch is tessellated uniformly with one of the tessellation factors and the two other edges have to be connected accordingly. This can be achieved by a gap-filling strip of certain width, see Figure 1. We call this process as *stitching* the gaps.

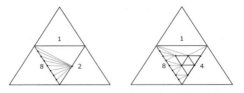

Fig. 1. Gap-filling strips connect the two border curves resulting from different tessellation factors [7]

Shading artifacts can be visible as the tessellation is strictly regular in the interior part and highly irregular in the strips. This can be mitigated by choosing not too different tessellation factors. Also the recently presented DirectX 11 graphics system [8] contains a tessellation stage with a tessellator for triangular and quadrangular domains, steered by edge and interior fractional tessellation factors. The tessellator's architecture is proprietary though.

Non-uniform, fractional tessellation [9] adds reverse projection to the evaluation of the tessellation pattern to make it non-uniform in parameter-space but more uniform in screen-space.

In a recent work, [10] keeps a uniform tessellation pattern and snaps missing vertices on the border curve to one of the existing vertices at the smaller tessellation factor. This is a systematic way to implement gap-filling but it is restricted to power-of-two tessellation factors and 1-to-4 splits of the parameter domain. Figure 2 gives an illustration of the uniform tessellations with snapped points in black. For largely different tessellation factors, the resulting pattern can be quite irregular.

In this work, we restrict the edge tessellation factors in such a way that only two different tessellation factors can occur in each triangle. For such cases, a tessellation pattern can be used, which is much simpler and more regular than in the three different case (Section 3). The tessellation code can output the resulting triangles as triangle strips, which makes this approach also suitable for an inside hardware implementation. We will demonstrate this advantage by implementing the pattern in a geometry shader (Section 3.3). In Section 3.2, we analyze which assignments are possible with two different factors and show that many important cases are contained, i.e., factors based on the distance to the camera eye plane and to the silhouette plane. We show visual results and GPU metrics obtained with our implementation in Section 4. Finally, we give conclusions and future work in Section 5.

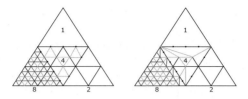

Fig. 2. Semi-uniform tessellation by snapping edges of larger tessellation factor (black points) to the nearest point of the smaller tessellation factor [10]

2 Locally Defined Triangular Parametric Surfaces

Vlachos et al. [11] propose *curved PN triangles* for interpolating a triangle mesh by a parametric, piecewise cubic surface. This established technique generates a C^0-continuous surface, which stays close to the triangle mesh and thus avoids self-interference.

A parametric triangular patch in Bézier form is defined by

$$p(u, v, w) = \sum_{i+j+k=n} \frac{n!}{i!j!k!} u^i v^j w^k b_{ijk} \tag{1}$$

where b_{ijk} are the control points of the Bézier triangle and $(u, v, w = 1 - u - v)$ are the barycentric coordinates with respect to the triangle [12]. Setting $n = 3$ gives a cubic Bézier triangle as used in the following construction.

At first, the PN scheme places the intermediate control points $\bar{b}_{ijk}$ at the positions $(ib_{300} + jb_{030} + kb_{003})/3$, $i + j + k = 3$, leaving the three corner points unchanged. Then, each b_{ijk} on the border is constructed by projecting the intermediate control point $\bar{b}_{ijk}$ into the plane defined by the nearest corner point and its normal.

Finally, the central control point b_{111} is constructed moving the point $\bar{b}_{111}$ halfway in the direction $m - \bar{b}_{111}$ where m is the average of the six control points computed on the borders as described above. The construction uses only data local to each triangle: the three triangle vertices and its normals. This makes it especially suitable for a triangle rendering pipeline.

3 Semi-uniform, 2-Different Tessellation

Semi-uniform 2-Different (SD-2) is our proposed pattern for adaptive tessellation where the tessellation factors on the three edges of a triangle are either all same or only two different values occur. As parametric triangular surfaces are composed of patches, it is necessary to do a tessellation of patches. In order to change the tessellation based on various criteria, the computation of suitable tessellation factors and an adaptive tessellation pattern for them is necessary. In the following sections, we describe both these components.

3.1   Edge Tessellation Factors Based on Vertex/Edge Criteria

For adaptive tessellation of a triangular parametric, normally tessellation factors are computed per vertex or per edge. If computed per vertex then they are propagated to tessellation factors per edge. Given the three edge tessellation factors (f_u, f_v, f_w), the subdivisions of the three border curves into line segments are given by $p(0, i/f_u, (f_u - i)/f_u)$, $i = 0 \ldots \lfloor f_u \rfloor$, $p(j/f_v, 0, (f_v - j)/f_v)$, $j = 0 \ldots \lfloor f_v \rfloor$, and $p(k/f_w, (f_w - k)/f_w, 0)$, $k = 0 \ldots \lfloor f_w \rfloor$. A gap-free connection to the mesh neighbors is guaranteed by a tessellation that incorporates these conforming border curves.

An assignment of edge tessellation factors can be based on criteria like vertex distance to the camera eye plane, edge silhouette property, and/or curvature approximations using normal cones [4].

3.2   Edge Tessellation Factors for the 2-Different Case

For our following tessellation pattern, we need that in the edge tessellation factor triple (f_u^*, f_v^*, f_w^*) only two different values $f_u^* = f_v^*$ and f_w^* occur.

Approximation of an arbitrary tessellation factor assignment (f_u, f_v, f_w), $f_u \neq f_v$, $f_v \neq f_w$, $f_u \neq f_w$ by a 2-different one (f_u^*, f_v^*, f_w^*) is a non-local problem. Therefore, we avoid the general case and guarantee that the tessellation factor calculation never produces 3-different factors for the edges of a triangle. Then SD-2 can be used for a faster and simpler tessellation.

Let $d : V(M) \to \mathbb{R}$ be a *level function* on the vertices of the mesh, which means d is strictly monotone increasing on shortest paths $\{v_0 = x \in I, \ldots, v_l = y\}$, i.e. $d(v_{i-1}) < d(v_i)$, from a vertex $x \in I$ in the set $I = \{x \in V(M) : \forall y \ d(x) \leq d(y)\}$ of minimum elements. Given a level function d, it is easy to derive a tessellation factor assignment f^*, which is only 2-different, as follows $f^*(\{s, e\}) := g(\min\{d(s), d(e)\})$ with a normally monotone, scalar function g. Note that the level function is only used to impose an order on the mesh vertices, which is easy to compute based on the vertex coordinates.

We give examples of tessellation factor assignments below, which are constructed with the help of a level function as described.

Distance from the camera eye plane. The smallest distance d_p to a plane, for example, the camera eye plane, naturally is a level function, as defined above. A semi-uniform edge tessellation factor assignment (f_u^*, f_v^*, f_w^*) for a triangle then is $f_{\text{edge}}^* := g(\min\{d(s_{\text{edge}}), d(e_{\text{edge}})\})$ with a linear function $g_1(d) := f_{\max} \frac{d_{\max} - d}{d_{\max} - d_{\min}} + f_{\min} \frac{d - d_{\min}}{d_{\max} - d_{\min}}$ or a quadratic function $g_2(d) := \frac{f_{\max} - f_{\min}}{(d_{\min} - d_{\max})^2}(d - d_{\max})^2 + f_{\min}$ mapping the scene's depth range $[d_{\min}, d_{\max}]$ to decreasing tessellation factors in the range $[f_{\min}, f_{\max}]$.

Silhouette refinement. Silhouette classification is usually done based on a classification of the vertices into front-facing $(n^t(e - v) \geq 0)$ and back-facing $(n^t(e - v) < 0)$ using the vertex coordinates v, vertex normal n and the camera eye point e. The distance function d_{silh} to the silhouette plane is a level function

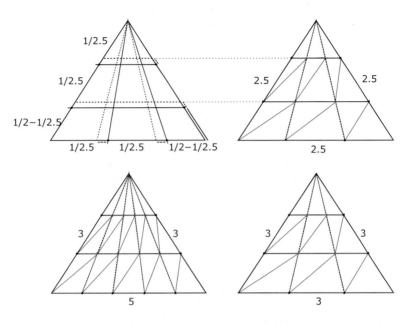

Fig. 3. Tessellation pattern for 2-different factors, which is composed of a bundle of parallel lines, intersected by a second bundle of radial lines towards the tip vertex. In the top row, for a fractional example: $f_u = f_v = 2.5$, $f_w = 2.5$; in the bottom row, for integer examples: $f_u = f_v = 3$, $f_w = 3$ and $f_u = f_v = 3$, $f_w = 5$.

as defined above. But the function $n^t(e - v)$ is easier to compute by just using the vertices and vertex normals of the mesh. It is not a level function though, but an edge is crossed by the silhouette plane in case the two incident vertices are differently classified. Each triangle can have exactly 0 or 2 such edges. An edge tessellation factor assignment with just two values, a for an edge not crossed by the silhouette, and b for an edge crossed by the silhouette, can be used to refine the silhouette line and present full geometric detail at the silhouette.

The edge tessellation factor assignments based on a level function can be directly computed inside a GPU shader. We show examples of this in Section 4. On the contrary, the edge tessellation factors obtained by a curvature approximation in the patch vertices can not be made 2-different for arbitrary meshes easily. Computing an approximation is possible though on the CPU.

3.3 Adaptive Tessellation Pattern with 2-Different Factors

In case of triangles with only two different edge tessellation factors $f_u = f_v$ and f_w, it is possible to tessellate in an especially simple way. Our tessellation pattern is composed of a bundle of parallel lines $w = i/f_u := w_i$, $i = 0, \ldots, \lfloor f_u \rfloor$, which are intersected by a second bundle of radial lines from the tip vertex $(0, 0, 1)$ to $((f_w - j)/f_w, j/f_w, 0)$, $j = 0, \ldots, \lfloor f_w \rfloor$. The intersections with the parallel line i are in the points $((1 - w_i)(f_w - j)/f_w, (1 - w_i)j/f_w, w_i)$, $j = 0, \ldots, \lfloor f_w \rfloor$. This

```
// reorder control points/normals (b000/n000, b030/n030, b003/n003)
// so that fu=fv on u,v isolines
void calcUVValues(float diff, float same) //fu=fv=same, fw=diff
{
    float line;   // counter that traverses along the diff edge
    float line1;  // counter that traverses along the same edge

    float incS = 0.5*(1.0/same - (same - floor(same))/same); // remainder for same fractional
    float incD = 0.5*(1.0/diff - (diff - floor(diff))/diff); // remainder for diff fractional

    vec2 diff1, diff2; // pair of u,v values on the diff edge
    float u, v;
    float par; // keeps track of the current location on the same edge

    for (line=0; line < diff; ++line)
    {
        diff1.u = (line -incD)/diff;
        if (diff1.u < 0.0)
            diff1.u = 0.0;
        diff1.v = 1.0 - diff1.u;
        diff2.u = (line+1.0 -incD)/diff;
        if (diff2.u > 1.0)
            diff2.u = 1.0;
        diff2.v = 1.0 - diff2.u;

        u = 0.0;
        v = 0.0;
        //Use PN evaluation code to calculate the point and the normal
        //In GLSL use EmitVertex()
        for (line1=0; line1<same; ++line1)
        {
            /* evaluate the point on the second radial line */
            par = (line1+1.0 -incS)/same;
            if (par > 1.0)
                par = 1.0;
            v = diff2.y*par;
            u = diff2.x*par;
            //Use PN evaluation code to calculate the point and the normal
            //In GLSL use EmitVertex()

            /* evaluate the point on the first radial linee */
            v = diff1.y*par;
            u = diff1.x*par;
            //Use PN evaluation code to calculate the point and the normal
            //In GLSL use EmitVertex()
        }
        //Finish the triangle strip
        //In GLSL call EndPrimitive()
    }
}
```

Fig. 4. Pseudo code used for generating the barycentric coordinates in SD-2 tessellation

pattern is very flexible as it works also with fractional factors $f_u = f_v$ and f_w. In the fractional case, the remainders $(f_u - \lfloor f_u \rfloor)/f_u$ and $(f_w - \lfloor f_w \rfloor)/f_w$ can be added as additional segments. In case the subdivision is symmetric to the mid-edge, it can be generated in arbitrary direction. We achieve this by shrinking the first segment and augmenting the last segment by the half fractional remainder $0.5(1/f_u - (f_u - \lfloor f_u \rfloor)/f_u)$ and $0.5(1/f_w - (f_w - \lfloor f_w \rfloor)/f_w)$ respectively. Otherwise, it has to be generated in a unique direction, for example from the nearest to the farthest vertex, which complicates things a lot. Concerning the distribution of lines, it is also possible to place them non-uniformly by a reverse projection according to [9]: $u' = \frac{u/z_1}{(1-u-v)/z_0+u/z_1+v/z_2}$, $w' = \frac{w/z_0}{w/z_0+u/z_1+(1-u-w)/z_2}$ where z_0, z_1, z_2 are the vertex depths of the triangle.

Figure 3 shows an example of the construction for fractional values $f_u = f_v = 2.5$, $f_w = 2.5$, and for integer factors $f_u = f_v = 3$, $f_w = 3$ and $f_u = f_v = 3$, $f_w = 5$.

It is possible to output all triangles between two adjacent radial lines or two adjacent parallel lines as a triangle strip, which reduces vertex repetition considerably and is beneficial on some hardware architectures. This property becomes a great advantage at the silhouettes where the triangles have edge tessellation factors ($f_u = f_v \gg f_w$) or ($f_u = f_v \ll f_w$) and the tessellation can be emitted with a minimum number of strips. For edge tessellation factors ($f_u = f_v, f_w$), we give the pseudo code for barycentric coordinates ($u, v, w = 1 - u - v$) on triangle strips along radial lines in Figure 4.

4   Results

The method can use any triangular parametric surface, however we have chosen the PN triangles scheme. We compare SD-2 with uniform tessellation of the PN

Fig. 5. Results of SD-2 refinement for improving the silhouette

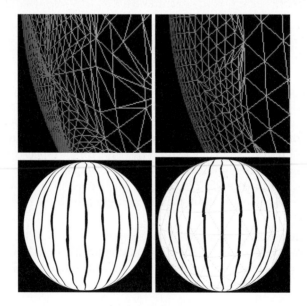

Fig. 6. Comparison of SD-2 and stitching methods using reflection lines

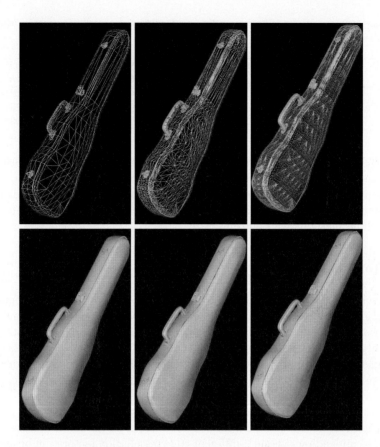

Fig. 7. Results of SD-2 refinement for continuous LOD based on the distance to the camera eye plane. From left to right, the original mesh, and adaptive tessellations generated by a linear level function with $f_{\max} = 7$, $f_{\min} = 2$, $d_{\max} = 1$ and $d_{\min} = 0$.

Table 1. Performance for silhouette refinement with max tessellation factor 7 and triangle strips

Model Name	Base Mesh Triangles	Uniform PN $f = 7$ FPS/Primitives	PN SD-2 FPS/Primitives	PN Stitch FPS/Primitives
Sphere	320	121.0/15680	153.0/12952	122.0/12184
Violin Case	2120	19.5/103880	24.5/83932	19.9/75446
Cow	5804	7.0/284396	9.5/222586	7.4/200383

patches as well as the stitching pattern methods described in [5], [6] and [7]. We compare frame rate and number of primitives generated on the GPU. For demonstration purposes, we have implemented all methods as geometry shaders. The SD-2 method clearly outperforms the other two and it can be clearly seen that there is a significant boost in frame rate for SD-2 by switching over to triangle strips. See Tables 1 and 2 for the concrete values on a PC with Windows

Table 2. Performance for silhouette refinement with max tessellation factor 7 and not using triangle strips

Model Name	Base Mesh Triangles	Uniform PN $f = 7$ FPS/Primitives	PN SD-2 FPS/Primitives	PN Stitch FPS/Primitives
Sphere	320	60.0/15680	65.0/12952	65.0/12184
Violin Case	2120	7.5/103880	16.0/83932	16.0/75446
Cow	5804	3.2/284396	4.0/222586	4.0/200383

Vista, 32bit, and NVIDIA 9800 GTX graphics. For surface interrogation, we render a series of reflection lines on the final surface and look at the smoothness of these lines. In general, smoother reflection lines indicate better surface quality. Reflection lines are much smoother for SD-2 in the adaptive region compared to the stitching method, see Figure 6.

5   Conclusions

In this paper, we have described, SD-2, a tessellation pattern for fractional edge tessellation factors with only two different values per triangle. Under continuous changes of the tessellation factors, the pattern fulfills all the requirements on the continuity of tessellation changes. This is especially important for the sampling of geometry and applied texture/displacement maps during animations. The scheme is especially simple to implement, and it is suitable for triangle output in the form of triangle strips.

In terms of adaptivity, it can cover the most important cases, where the edge tessellation factors are derived from a level function on the mesh vertices. Then, the edges can be directed and 2-different edge tessellation factors can be assigned based on the minimum (or maximum) vertex on each edge. We have shown results in terms of speed and quality with an implementation of the pattern and the edge tessellation factor assignment in the geometry shader of the GPU. This shows that the approach is also suitable for a future hardware implementation.

References

1. Sfarti, A., Barsky, B.A., Kosloff, T., Pasztor, E., Kozlowski, A., Roman, E., Perelman, A.: Direct real time tessellation of parametric spline surfaces. In: 3IA Conference (2006), Invited Lecture,
   http://3ia.teiath.gr/3ia_previous_conferences_cds/2006
2. Schweitzer, D., Cobb, E.S.: Scanline rendering of parametric surfaces. SIGGRAPH Comput. Graph. 16, 265–271 (1982)
3. Bóo, M., Amor, M., Doggett, M., Hirche, J., Strasser, W.: Hardware support for adaptive subdivision surface rendering. In: HWWS 2001: Proceedings of the ACM SIGGRAPH/EUROGRAPHICS Workshop on Graphics Hardware, pp. 33–40. ACM, New York (2001)
4. Settgast, V., Müller, K., Fünfzig, C., Fellner, D.: Adaptive Tesselation of Subdivision Surfaces. Computers & Graphics 28, 73–78 (2004)

5. Moreton, H.: Watertight tessellation using forward differencing. In: HWWS 2001: Proceedings of the ACM SIGGRAPH/EUROGRAPHICS Workshop on Graphics Hardware, pp. 25–32. ACM, New York (2001)
6. Chung, K., Kim, L.: Adaptive Tessellation of PN Triangle with Modified Bresenham Algorithm. In: SOC Design Conference, pp. 102–113 (2003)
7. Schwarz, M., Stamminger, M.: Fast GPU-based Adaptive Tessellation with CUDA. Comput. Graph. Forum 28, 365–374 (2009)
8. Gee, K.: Introduction to the Direct3D 11 graphics pipeline. In: nvision 2008: The World of Visual Computing, Microsoft Corporation, pp. 1–55 (2008)
9. Munkberg, J., Hasselgren, J., Akenine-Möller, T.: Non-uniform fractional tessellation. In: Proceedings of the 23rd ACM SIGGRAPH/EUROGRAPHICS Symposium on Graphics Hardware, GH 2008, Aire-la-Ville, Switzerland, Switzerland, Eurographics Association, pp. 41–45 (2008)
10. Dyken, C., Reimers, M., Seland, J.: Semi-uniform adaptive patch tessellation. Computer Graphics Forum 28, 2255–2263 (2009)
11. Vlachos, A., Peters, J., Boyd, C., Mitchell, J.L.: Curved PN triangles. In: I3D 2001: Proceedings of the 2001 Symposium on Interactive 3D Graphics, pp. 159–166. ACM Press, New York (2001)
12. Farin, G.: Curves and Surfaces for Computer-Aided Geometric Design — A Practical Guide, 5th edn., 499 pages. Morgan Kaufmann Publishers, Academic Press (2002)

Fast and Reliable Decimation of Polygonal Models Based on Volume and Normal Field

Muhammad Hussain

Department of Computer Science,
College of Computer and Information Science,
King Saud University, KSA
mhussain@ccis.edu.sa

Abstract. Fast, reliable, and feature preserving automatic decimation of polygonal models is a challenging task. Exploiting both local volume and normal field variations in a novel way, a two phase decimation algorithm is proposed. In the first phase, a vertex is selected randomly using the measure of geometric fidelity that is based on normal field variation across its one-ring neighborhood. The selected vertex is eliminated in the second phase by collapsing an outgoing half-edge that is chosen by using volume based measure of geometric deviation. The proposed algorithm not only has better speed-quality trade-off but also keeps visually important features even after drastic simplification in a better way than similar state-of-the-art best algorithms; subjective and objective comparisons validate the assertion. This method can simplify huge models efficiently and is useful for applications where computing coordinates and/or attributes other than those attached to the original vertices is not allowed by the application and the focus is on both speed and quality of LODs.

1 Introduction

Recent advances in technology and the quest of realism have given rise to highly complex polygonal models for encoding 3D information. Despite the enhancement in graphics acceleration techniques, fast reliable and feature preserving decimation of polygonal models is a challenging task. Based on the application requirements, existing simplification techniques can be classified into three categories: quality simplification [1], fast simplification [2], and simplification with best speed-quality trade-off [3], [4], [5], [6].

The algorithm discussed in this paper is concerned with the class of simplification algorithms where the focus is on better quality of approximations and less simplification time, and recomputing the vertex attributes such as texture coordinates, colors, etc. is not allowed or there is no straightforward method of interpolation. Though, so far QSlim [3] and Memoryless Simplification (MS) [4] are identified as the best algorithms in respect of speed-quality trade-off, both the algorithms can't preserve important shape features, especially after drastic reduction of polygon count, e.g. see the close-up of right eye of David model in Figure 1. The normal field and volume based algorithm (GeMS) [7] produces

G. Bebis et al. (Eds.): ISVC 2010, Part I, LNCS 6453, pp. 64–73, 2010.

Original   VNSIMP   MS   QSlim   GeMS

Fig. 1. Close-up of the right eye of David model (#Faces: 7,227,031) and its approximations (consisting of 50,000 faces) generated by the four methods

LODs of acceptable quality and is relatively better in preserving visually important features but it is not as efficient as QSlim and MS.

Motivated by the fact that volume based error measure [4] results in better quality approximations whereas the one based on normal field variation and volume loss [7] better preserves visually important features, we propose an algorithm - VNSIMP - that exploits both volume loss and normal field variation in a novel way. In contrast to QSlim, MS and GeMS where simplification of a model proceeds by selecting and collapsing half-edges according to their significance, VNSIMP selects vertices according to their importance, and removes them by collapsing the least significant respective incident half-edges. The importance of a vertex is defined using the normal field variation across its one-ring neighborhood. After a vertex is liable to be removed, it is eliminated by collapsing one of the outgoing half-edges, which is decided by a measure of geometric deviation that is based on volume loss caused by the collapse. A comparison with the help of standard evaluation tool - Metro - reveals that VNSIMP generates better quality LODs in terms of symmetric Hausdorff distance in less time than QSlim, MS and GeMS. Its memory over-head is small like MS and it is comparable with GeMS in preserving significant shape features automatically but better than QSlim and MS.

The remainder of this paper is organized as follows. In Section 2, we discuss in detail the new measures of geometric distortion. Section 3 describes the two phase simplification algorithm. Quality and efficiency of the algorithm is discussed and it is compared with the sate-of-the-art best algorithms in Section 4. Section 5 concludes the paper.

2   Measures of Geometric Deviation

VNSIMP selects vertices randomly for removal according to their importance; after selection, a vertex is removed by collapsing one of its outgoing half-edges. In this section, first the measure of importance of a vertex is elaborated, then detail of the error measure, which finds the outgoing half-edge resulting in minimum volume loss, is presented.

2.1   Metric for Vertex Selection

Normal field of a surface model plays fundamental role in its appearance and it has been used for constraining the geometric distortion in different geometry processing tasks [2], [8], [9], [10]. The Poincar-Wertinger- Sobolev inequality implies that minimizing the normal field distortion ensures the minimization of the geometric deviation [9]. Normal field variation truly represent the importance of a vertex. In view of this evidence in support of the strength of normal field, we use the normal field variation across one-ring neighborhood of a vertex for defining its importance. The normal field variation over NF_v, the set of triangular faces incident on v, can be defined as follows:

$$\sum_{t \in NF_v} \int_{\Delta_t} n_t(s)ds = \sum_{t \in NF_v} \Delta_t n_t$$

where n_t and Δ_t are, respectively, the unit normal and the area of the triangular face t. According to triangular inequality

$$\left\| \sum_{t \in NF_v} \Delta_t n_t \right\| \leq \sum_{t \in NF_v} \| \Delta_t n_t \|.$$

Since $\| \Delta_t n_t \| = \Delta_t$ so

$$\left\| \sum_{t \in NF_v} \Delta_t n_t \right\| \leq \sum_{t \in NF_v} \Delta_t$$

or

$$\sum_{t \in NF_v} \Delta_t - \left\| \sum_{t \in NF_v} \Delta_t n_t \right\| \geq 0. \tag{1}$$

In case v is flat i.e. all faces in NF_v are coplanar, then left-hand-side of inequality (1) is zero, otherwise it is greater than zero depending on how much the vertex v departs from being flat. This expression defines a measure of significance of a vertex as follows:

$$VC(v) = \sum_{t_i} \Delta_{t_i} - \left\| \sum_{t_i} \Delta_{t_i} n_{t_i} \right\| \tag{2}$$

where the summation is over all faces in NF_v. Note that $VC(v) = 0$ when a vertex is flat and it gets larger values depending on the degree of departure of a vertex from being flat i.e. according to its level of significance. The value of $VC(v)$ is used for selecting a vertex randomly.

2.2   Metric for Half-Edge Collapse

Once a vertex is selected for removal according to its cost (2), it is eliminated by half-edge collapse operation (see Figure 2(b)). This decimation operation is simple to implement and is easy to invert, so it is most suitable for applications like visualization of 3D information across networks. Half-edge collapse can also simplify dealing with vertex attributes that have no straightforward method of

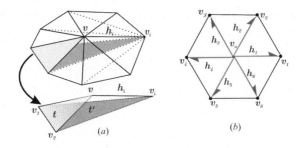

Fig. 2. (a) Half-edge collapse $h_i : (v, v_i) \rightarrow v_i$ and the volume loss caused by a typical triangle $t(v, v_1, v_2)$. (b) Half-edges with v_0 as origin.

interpolation. For instance, many meshes come with vertex normals, texture coordinates, colors, etc., and half-edge collapse frees one from having to recompute such attributes at the position of the new vertex.

A vertex can be removed by collapsing any one of the outgoing half-edges (see Figure 2(b)); we eliminate it by collapsing the half-edge $h_i \in NHE_v$ (the set of outgoing half-edges) that causes minimum geometric distortion in its local neighborhood - this half-edge is termed as *optimal half-edge* and is denoted by h_o.

For finding h_o we employ a measure, similar to the one in [4], that ensures that the loss of volume is minimum. A half-edge collapse $h_i(v, v_i) \mapsto v_i$ causes each triangle $t \in NF_v$ to sweep out a tetrahedron, see Figure 2(a). The volume of this tetrahedron represents the volume loss due to the movement of t as a result of half-edge collapse and is indicative of the geometric deviation. As such the geometric deviation introduced due to face $t(v, v_1, v_2)$ is defined as follows

$$TC(t) = \frac{1}{6}[(v_1 - v) \times (v_2 - v) \cdot (v_i - v)]^2 \qquad (3)$$

In view of this, the geometric error introduced due to half-edge collapse $h_i(v, v_i) \mapsto v_i$ is defined as follows

$$HC(h_i) = \sum_{t \in NF_v} TC(t) \qquad (4)$$

For each half-edge $h_i \in NHE_v$, we compute $HC(h_i)$ and choose the one as optimal half-edge h_o for which $HC(h_o)$ is minimum. Note the difference between our idea of using volume measure and that of Lindstrom et al. [4]; Lindstrom uses it for global ordering of edges whereas we employ it for local decision of determining the optimal half-edge once a vertex is chosen for removal.

3   Two Phase Simplification Algorithm

The proposed algorithm employs a simplification framework that is similar to multiple-choice approach [11]. The algorithm involves two degrees of freedom to

Table 1. Simplification times (in seconds to simplify to 10000 faces)

Model	Model Size		VNSIMP	MS	QSlim	GeMS
	#Vertices	#Faces				
Oilpump	570,018	1,140,048	7.711	13.156	21.312	25.454
Grago	863,210	1,726,420	11.245	18.664	27.187	35.398
Blade	882,954	1,765,388	12.875	20.347	28.750	37.128
Satva	1,815,770	3,631,628	25.957	40.547	67.438	80.484
David	3,614,098	7,227,031	61.387	93.457	151.650	164.757
Statuette	4,999,996	10,000,000	82.887	125.664	228.453	245.757

be fixed: to select a vertex for decimation and to determine the corresponding optimal half-edge. The first phase fixes the vertex to be eliminated. The second phase finds the optimal half-edge h_o corresponding to the selected vertex. The pseudo code of the algorithm is given below.

Algorithm VNSIMP
  Input: $M = (V, F)$ - Original triangular mesh consisting
        of the set of vertices V, and the set of faces F
        $numf$ - the target number of faces,
  Output: An LOD with the given budget of faces
  Processing
    Step-1 **Select a vertex randomly**
        Randomly choose m vertices $v_1, v_2, \cdots, v_m$, compute cost of each
        vertex using equation (2) and select $v \in \{v_1, v_2, \cdots, v_m\}$ such that
        $VC(v) = \min \{VC(v_1), VC(v_2), \cdots, VC(v_m)\}$.
    Step-2 **Find optimal half-edge and collapse**
        (a) For each half-edge $h_i \in NHE_v$, compute cost using equation
            (4) and select the optimal half-edge $h_o \in NHE_v$ such that
            $HC(h_o) = \min \{HC(h_i) \mid h_i \in NHE_v\}$.
        (b) Collapse $h_o : (v, v_o) \rightarrow v_o$, if it does not create foldovers, by elim-
            inating faces incident on the edge $e = \{v, v_o\}$ and substituting
            v_o for every occurrence of v in left-over faces in NF_v.
    Step-3 Repeat Steps 1 and 2 until the number of faces in the decimated
        mesh is $numf$.

For identifying foldovers, we use the usual test that checks whether the normal vector of any face in FN_v turns through an angle greater than 90° after collapse. For results presented in this paper, m is 24.

4   Results and Discussion

In this section, we discuss the results and evaluate the performance of VNSIMP by comparing it with similar state-of-the-art best decimation algorithms MS,

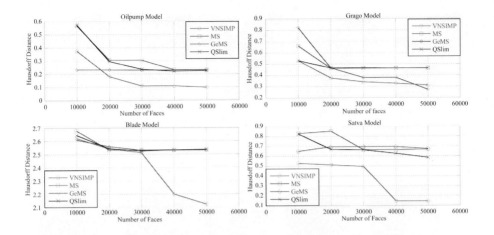

Fig. 3. Plots of Symmetric Hausdorff distance for (top-left) oilpump model, (top-right) grago model, (bottom-left) blade model and (bottom-right) Satva model

Fig. 4. Original oilpump model (#faces:1,140,048) and its LODs (each consisting of 20000 faces) generated by the four methods

QSlim, and GeMS. The four simplification algorithms (VNSIMP, MS, QSlim and GeMS) are scaled using four parameters: running time, memory consumption, quality of the generated LODs, and the preservation of salient features at low levels of detail. Huge models of varying complexities - oilpump, grago, Satva, David and statuette models - have been used as benchmark models; the statistics of these models are given in Table 1. The experimental environment used for this study consists of a system equipped with Intel Centrino Duo 2.1GHz CPU and 2GB of main memory, and C++ has been used as a programming development tool.

Table 1 lists the execution times of the four algorithms to reduce each benchmark model to 10000 faces. It indicates that VNSIMP outperforms MS, QSlim and GeMS in terms of running time. VNSIMP is about 1.5 times faster than MS and about 3 times faster than both QSlim and GeMS. The decimation framework used in MC algorithm selects r vertices randomly, computes cost of each vertex, chooses the vertex with minimum cost and removes it by collapsing the optimal half-edge. The cost of a vertex is assumed to be the cost of the

Fig. 5. Original grago model (#faces:1,726,420) and its LODs (each consisting of 20000 faces) generated by the four methods

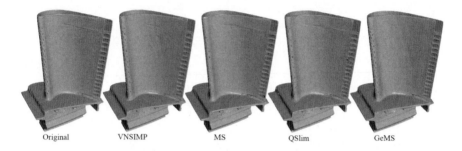

Fig. 6. Original blade model (#faces:1,765,388) and its LODs (each consisting of 20000 faces) generated by the four methods

corresponding optimal half-edge and its calculation involves the computation of the cost of 6 half-edge operations, see Figure 2(b), assuming that the valence of each vertex is 6. It means that the selection of a vertex and its removal requires the computation of the cost of $6r$ half-edge operations. In contrast, the two-phase decimation framework used in VNSIMP does not need to compute the cost of 6 half-edge operations to calculate the cost of a vertex, it is directly calculated using equation (2) and can be roughly assumed to be equivalent to the computation of the cost of one half-edge operation. So, the selection of a vertex out of r randomly selected vertices involves the computation of the cost of r half-edge operations. For removing the vertex , the selection of the optimal half-edge involves the computation of the cost of 6 half-edge operations. In this way, selection and removal of a vertex by VNSIMP involves the computation that is equivalent to the calculation cost of $r + 6$ half-edge operations. In case n vertices are removed, the computational costs of MC algorithm and VNSIMP will be $6nr$ and $n(r + 6)$ respectively.

For evaluating the quality of the LODs generated by VNSIMP, the LODs created by the four methods are compared employing Symmetric Hausdorff distance (SHD) that is widely used for thorough comparison of polygonal models in graphics community because it provides tight error bounds and does not discount local deviations. To avoid any kind of bias, symmetric Hausdorff distance has been calculated using well-known Metro tool [12]. Plots of the symmetric Hausdorff distances for 5 LODs of four benchmark models created by the four

Fig. 7. Original Satva model (#faces:3,631,628) and its approximations (each consisting of 40000 faces) generated by the four methods. Models have been shown partly.

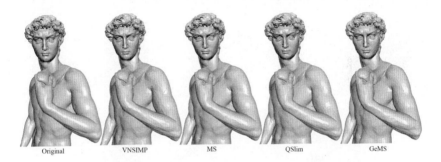

Fig. 8. Original David model (#faces:7,227,031) and its approximations (each consisting of 40000 faces) generated by the four methods. Models have been shown partly.

simplification algorithms are shown in Figure 2. On average VNSIMP improves over MS, QSlim and GeMS by 46.84% 43.64% and 24.27%, respectively, in terms of SHD for oilpump model, 24.88% 20.97% and 18.62% for grago model, and 50.32% 45.63% and 46.32% for Satva model. In case of turbine blade model, the performance of VNSIMP is similar to that of MS, QSlim and GeMS at low levels of detail but it performs better at higher levels of detail.

For visual comparison, low resolution LODs generated with the four algorithms of the benchmark models have been presented in Figures 4, 5, 6, 7, 8, and 9. A close look at these figures shows that the LODs generated by VNSIMP are comparable with those produced by the other three algorithms. VNSIMP preserves visually important high frequency information in a better way and keeps the semantic meaning of the surface model even after drastic reduction. For example, the high resolution detail like eyes of a huge David model (consisting of 7.2 million faces) is better preserved by VNSIMP (see Figure 1) even after drastic simplification (50,000 faces, 0.69% of original) whereas other algorithms fade out this information. Also have a look at Figure 10 that shows the close-up of a portion of a huge Statuettes model consisting of 10 million triangular faces and its LODs (consisting of only 100,000 triangular faces) generated by MS,

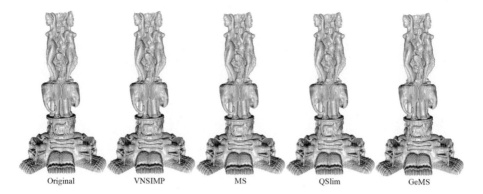

Fig. 9. Original Statuettes model (#faces:10,000,000) and its approximations (each consisting of 100000 faces) generated by the four methods

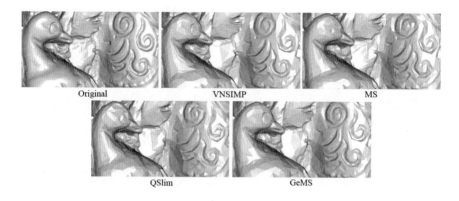

Fig. 10. Close-up of a small region of statuettes model, shown in Figure 9 and its approximations generated by the four methods

QSlim and GeMS; in this case the detail features like eye and beak of the bird, and other small detail is relatively better preserved by VNSIMP. Note that performance of GeMS in terms of preserving high resolution detail is comparable with that of VNSIMP, but its execution time is about three times of that of VNSIMP.

VNSIMP consumes almost as much memory as MS and GeMS. As such, it is enough to compare the memory overhead of VNSIMP and QSlim. VNSIMP consumes about 44% less memory than QSlim, because VNSIMP needs not to keep error quadrics like QSlim or any other kind of geometric history.

5   Conclusion

Though simplification is well-searched area, still there is space for improvement. The idea of using normal field deviation and volume measure in measuring simplification error has been around for a while, however in this paper these measures have been used in a novel way for developing a simple and reliable automatic simplification algorithm for applications where good time-accuracy trade-off is important and the presence of vertex attributes does not allow the creation of new vertices. Thorough comparison with similar state-of-the-art algorithms such as MS, QSlim and GeMS, which are known for their good time-accuracy trade-off, demonstrates that it is more efficient than MS, GeMS, and QSlim in terms of execution time. It has less memory overhead than QSlim. It creates LODs which are better than those produced by other algorithms in terms of symmetric Hausdorff distance and preserves salient shape features in a better way as compared to MS and QSlim. The ability of preserving detail features of a surface model of VNSIMP is comparable with GeMS. To sum-up, VNSIMP has better quality-speed trade-off than similar best stat-of-the-art algorithms. The proposed algorithm involves two degrees of freedom i.e. to select a vertex and to fix the corresponding optimal half-edge. Selecting a vertex dominates the simplification process. We employ normal field variation as a measure for selecting vertices. Even better measures can be further investigated for better simplification results.

References

1. Yan, J., Shi, P., Zhang, D.: Mesh simplification with hierarchical shape analysis and iterative edge contraction. Transactions on Visualization and Computer Graphics 10, 142–151 (2004)
2. Boubekeur, T., Alexa, M.: Mesh simplification by stochastic sampling and topological clustering. Computers and Graphics 33, 241–249 (2009)
3. Garland, M., Heckbert, P.S.: Surface simplification using quadric error metric. In: Proc. SIGGRAPH 1997, pp. 209–216 (1997)
4. Lindstrom, P., Turk, G.: Evaluation of memoryless simplification. IEEE Transactions on Visualization and Computer Graphics 5, 98–115 (1999)
5. Hussain, M., Okada, Y.: Lod modelling of polygonal models. Machine Graphics and Vision 14, 325–343 (2005)
6. Hussain, M.: Efficient simplification methods for generating high quality lods of 3d meshes. Journal of Computer Science and Technology 24, 604–inside back cover (2009)
7. Southern, R., Marais, P., Blake, E.: Gems: Generic memoryless polygonal simplification. In: Proc. Computer Graphics, Virtual Reality, Visualization and Interaction, pp. 7–15 (2001)
8. Cohen, D.S., Alliez, P., Desbrun, M.: Variational shape approximation. ACM Transactions on Graphics 23, 905–914 (2004)
9. Guskov, I., Sweldens, W., Schroeder, P.: Multiresolution signal processing for meshes. In: Proc. SIGGRAPH 1999, pp. 325–334 (1999)
10. Sorkine, O., Cohen-OR, D., Toledo, S.: High-pass quantization for mesh encoding. In: Proc. Symp. on Geometry Processing 2003, pp. 42–51 (2003)
11. Wu, J., Kobbelt, L.: Fast mesh decimation by multiple-choice techniques. In: Proc. Vision, Modeling, and Visualization 2002, pp. 241–248 (2002)
12. Cignoni, P., Rocchini, C., Scopigno, R.: Metro: Measuring error on simplified surfaces. Computer Graphics Forum 17, 167–174 (1998)

Lattice-Boltzmann Water Waves

Robert Geist, Christopher Corsi, Jerry Tessendorf, and James Westall

Clemson University

Abstract. A model for real-time generation of deep-water waves is suggested. It is based on a lattice-Boltzmann (LB) technique. Computation of wave dynamics and (ray-traced) rendering for a lattice of size 1024^2 can be carried out simultaneously on a single graphics card at 25 frames per second. In addition to the computational speed, the LB technique is seen to offer a simple and physically accurate method for handling both dispersion and wave reflection from obstructing objects.

1   Introduction

The goal of this effort is to provide the mathematical basis for a particularly simple, real-time computational model of deep-water waves. Computation of wave dynamics and a ray-traced rendering of the wave height field can be carried out simultaneously, in real-time, on a single NVIDIA GTX 480 graphics card.

The model is based on a lattice-Boltzmann method. Lattice-Boltzmann (LB) methods are a class of *cellular automata* (CA), a collection of computational structures that can trace their origins to John Conway's famous *Game of Life* [1], which models population changes in a hypothetical society that is geographically located on a rectangular lattice. In Conway's game, each lattice site is labeled as populated or not, and each lattice site follows only local rules, based on nearest-neighbor populations, in synchronously updating itself as populated or not. Although the rules are only local, global behavior emerges in the form of both steady-state population colonies and migrating colonies who can generate new steady-state colonies or destroy existing ones.

In a general CA, arbitrary graphs and local rules for vertex updates may be postulated, but those that are most interesting exhibit a global behavior that has some provable characteristic. Lattice-Boltzmann methods employ synchronous, neighbor-only update rules on a discrete lattice, but the discrete populations at each lattice point have been replaced by continuous distributions of some quantity of interest. The result is that the provable characteristic is often quite powerful: the system is seen to converge, as lattice spacing and time step approach zero, to a solution of a targeted class of partial differential equations (PDEs).

Lattice-Boltzmann methods are thus often regarded as computational alternatives to finite-element methods (FEMs), and as such they have have provided significant successes in modeling fluid flows and associated transport phenomena [2–6]. They provide stability, accuracy, and computational efficiency comparable

G. Bebis et al. (Eds.): ISVC 2010, Part I, LNCS 6453, pp. 74–85, 2010.

to FEMs, but they offer significant advantages in ease of implementation, parallelization, and an ability to handle interfacial dynamics and complex boundaries. The principal drawback to the methods, compared to FEMs, is the counterintuitive direction of the derivation they require. Differential equations describing the macroscopic system behavior are derived (emerge) from a postulated computational update (the local rules), rather than the reverse.

The paper is organized as follows. After discussing related work in the next section, we describe our computational model (the local rules) in Section 3. In Section 4, we derive the wave equation directly from the postulated local rules. Section 5 contains a brief discussion of dispersion and wave number spectra, and Section 6 describes initial conditions. A principal benefit of our approach is the ease with which we can handle wave reflections, and we describe this in Section 7. Finally, we provide implementation details in Section 8 and conclusions in Section 9.

2   Related Work

The graphics literature on physically-based modeling and rendering of water flow is extensive. Foundational work by Kass and Miller [7], Foster and Metaxas [8], Foster and Fedkiw [9], and Stam [10], among others, has led to numerous, visually stunning examples of water flow on small to medium scale, such as water pouring into a glass or water sloshing in a swimming pool. Large-scale, deep-water simulations appropriate for oceans or lakes, which is our focus here, usually avoid full-scale, 3D Navier-Stokes solutions and instead employ 2D spectral approaches to simulate displacement of the free surface. Mastin et al. [11] was probably the first. In this case weights in frequency space are obtained by sampling from models fitted to observed spectra, e.g., Hasselmann et al. [12], and then applying a fast Fourier transform to construct the height field. As seen in the work of Jensen et al. [13], this approach can offer real-time performance suitable for interactive gaming [14]. The principal drawback to FFT-based approaches is their inability to handle obstructions, i.e., wave/object interactions. Hinsinger et al. [15] also achieve visually impressive results in real-time using an adaptive sampling, procedural approach that includes dispersion, but again, they do not consider wave obstructions.

In a somewhat complementary approach, Yuksel, House, and Keyser [16] focus on obstructions. They use *wave particles*, which are dynamically blended cosine segments, to provide extremely effective, real-time wave-object interaction. They do not include dispersion, and so their technique is not appropriate for deep-water waves. Nevertheless, they provide impressive demonstrations of large-scale, open water simulations by augmenting their technique with those of Tessendorf [17, 18].

Applications of lattice-Boltzmann methods to water flow are also numerous. Salmon [19] provided an early application to ocean circulation modeling, in particular, a "reduced-gravity" model in which a homogeneous, wind-driven layer of fluid overlays a denser layer that remains at rest. More recently, Thürey et al. [5]

use a full, multi-phase 3D LB model to create impressive animations that include object/free-surface interaction for open water. As yet this approach is computationally-intensive, even for relatively small grids. They report 80 seconds per frame for a 120^3 grid.

Our approach is most closely related to the *iWave* system of Tessendorf [18], in that we apply 2D site updates based on local information. Our updates are based on a collision matrix, whereas *iWave* uses a convolution kernel applied over a 13×13 or larger neighborhood. The total computational effort is remarkably similar. The advantage of our approach is the flexibility it offers in handling wave-object interaction. *iWave* uses a grid bit mask to indicate object position, and its update operation simply forces wave height to zero at masked sites. A fortuitous consequence of its kernel, which controls damping of the second derivative of height, is that zeroing the height removes this damping and yields a visually effective simulation of reflection. The additional flexibility we offer includes selected directional reflection, damping energy in response to varying restitutional characteristics of the obstruction, changing wave numbers for harmonics, and simulating semi-porous surfaces.

3   The Computational Model

Although 3D grids are common in LB models, we seek to achieve real-time performance, and so we restrict our development to a 2D, rectangular grid with four, unit-length directions, c_i, $i = 1, ..., 4$, and a single zero-length direction, c_0, as shown in Figure 1. Although 2D, rectangular grids can generate anisotropic

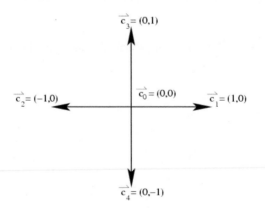

Fig. 1. Model grid

flows for certain LB models, we will see that anisotropy is avoided here through a careful choice of the site collision matrix.

We assume a lattice spacing, λ, a time step, τ, unit velocity $v = (\lambda/\tau)$, and velocity vectors $v_i = vc_i$, $i = 0, ..., 4$. We further assume that $h(r, t)$, the wave height at site r and time t, comprises 5 directional flows,

$$h(r,t) = \sum_{i=0}^{4} f_i(r,t) \tag{1}$$

where $f_i(r,t)$, represents the mass flow at location r at time t moving in direction c_i. The velocity field is then

$$u(r,t) = (\sum_{i=0}^{4} v_i f_i(r,t))/(\sum_{i=0}^{4} f_i(r,t))$$

and the momentum tensor is

$$\Pi_{\alpha\beta} = \sum_{i=0}^{4} v_{i\alpha} v_{i\beta} f_i(r,t),$$

where $\alpha, \beta \in \{x,y\}$. Note that for the limited set of directions we use, $v_{i\alpha} v_{i\beta} = v^2 \delta_{\alpha\beta}$, and so Π is diagonal.

The fundamental system update equation (basis for simulation) is given by:

$$f_i(r + \lambda c_i, t + \tau) = f_i(r,t) + \Omega_i \cdot f(r,t), \qquad i = 0,1, ..., 4 \tag{2}$$

where Ω_i is the ith row of a matrix $\Omega : \Re^5 \rightarrow \Re^5$, which is a *collision matrix* in the sense that $\Omega_{i,j}$ represents the deflection of flow f_j into the ith direction. Once Ω is specified, equation (2) is, essentially, the entire computational model. Starting with initial conditions, we apply (2) synchronously to all lattice sites and then generate the new wave height field at time $t = t + \tau$ by (1).

The choice of Ω determines the properties of the system. Some important constraints on this choice can be specified immediately. From (2) we have:

– conservation of mass: $\sum_{i=0}^{4} \Omega_i \cdot f(r,t) = 0$
– conservation of momentum: $\sum_{i=0}^{4} v_i \Omega_i \cdot f(r,t) = (0,0)$

The principal constraint is that the limiting behavior of (2) as $\lambda, \tau \rightarrow 0$ should be a recognizable wave equation.

We choose to specify

$$\Omega = \begin{pmatrix} -4K & 2-4K & 2-4K & 2-4K & 2-4K \\ K & K-1 & K-1 & K & K \\ K & K-1 & K-1 & K & K \\ K & K & K & K-1 & K-1 \\ K & K & K & K-1 & K-1 \end{pmatrix} \tag{3}$$

where $K \in (0, 1/2]$ is a parameter. We will see that this choice ultimately yields a limiting wave equation with speed (phase velocity) $v\sqrt{K}$. For now we note that 0 is a triple eigenvalue of Ω and that the eigenvectors $e_0 = (2-4K, K, K, K, K)$, $e_1 = (0, 1, -1, 0, 0)$, and $e_2 = (0, 0, 0, 1, -1)$ span the null space.

4   Derivation of the Wave Equation

In this section, we show that the limiting behavior of (2) as $\lambda, \tau \to 0$ is indeed the well-known wave equation. Intermediate results include the continuity equation, which is a statement of conservation of mass, and the Euler equation of hydrodynamics, which is a statement of conservation of momentum. It should be noted that, although this derivation is essential in verifying that our model is physically correct, model implementation does not depend upon the derivation in any way.

4.1   The Continuity Equation

We begin with a standard Chapman-Enskog expansion [2]. If we apply a Taylor expansion to the basic update equation (2) we obtain:

$$[(\lambda c_i, \tau) \cdot \nabla] f_i(r, t) + \frac{[(\lambda c_i, \tau) \cdot \nabla]^2}{2!} f_i(r, t) + \dots = \Omega_i \cdot f(r, t) \qquad (4)$$

As noted, we want to consider the limiting behavior here as $\lambda, \tau \to 0$; they can, of course, approach at different rates, but we assume they do not. Specifically, we write

$$t = \frac{s}{\epsilon} \qquad \text{where} \quad s = o(\epsilon)$$

$$r = \frac{q}{\epsilon} \qquad \text{where} \quad q = o(\epsilon)$$

and where the limit of interest is $\epsilon \to 0$. Then

$$\frac{\partial}{\partial t} = \epsilon \frac{\partial}{\partial s}$$

$$\frac{\partial}{\partial r_\alpha} = \epsilon \frac{\partial}{\partial q_\alpha} \qquad \text{for} \quad \alpha \in \{x, y\}$$

So

$$\nabla = (\partial/\partial r_x, \partial/\partial r_y, \partial/\partial t) = \epsilon(\partial/\partial q_x, \partial/\partial q_y, \partial/\partial s) \qquad (5)$$

We also assume that the solution, $f(r, t)$, is a small perturbation on this same scale about some local equilibrium, i.e.,

$$f(r, t) = f^0(r, t) + \epsilon f^1(r, t) + \epsilon^2 f^2(r, t) + \dots \qquad (6)$$

To qualify as a local equilibrium, f^0 must carry the macroscopic quantities of interest, that is,

$$h(r, t) = \sum_{i=0}^{4} f_i^0(r, t) \qquad (7)$$

and

$$u(r, t) = \left(\sum_{i=0}^{4} v_i f_i^0(r, t)\right) / \left(\sum_{i=0}^{4} f_i^0(r, t)\right) \qquad (8)$$

For the chosen Ω, these two conditions uniquely determine f^0. Since f^0 is an equilibrium, it is in the null space of Ω, and so we can write $f^0 = Ae_0 + Be_1 + Ce_2$. Then (7) and (8) together provide 3 independent equations in A, B, and C. The result is:

$$f_i^0(\boldsymbol{r},t) = \begin{cases} h(\boldsymbol{r},t)(1 - 2K) & i = 0 \\ (h(\boldsymbol{r},t)/2)\left[K + (\boldsymbol{v}_i \cdot \boldsymbol{u}(\boldsymbol{r},t))/v^2\right] & i = 1,2,3,4 \end{cases} \tag{9}$$

The continuity equation is now at hand. We insert (5) and (6) into (4), then sum (4) over $i = 0, 1, ..., 4$, divide by τ, and equate coefficients of ϵ^1. We obtain

$$\left(\frac{\partial}{\partial q_x}, \frac{\partial}{\partial q_y}\right) \cdot \sum_{i=0}^{4} \boldsymbol{v}_i f_i^0(\boldsymbol{r},t) + \frac{\partial}{\partial s} \sum_{i=0}^{4} f_i^0(\boldsymbol{r},t) = 0$$

and so, after multiplying by ϵ,

$$\partial h(\boldsymbol{r},t)/\partial t + \nabla_{\boldsymbol{r}} \cdot [h(\boldsymbol{r},t)\boldsymbol{u}(\boldsymbol{r},t)] = 0 \tag{10}$$

4.2   The Euler Equation

If we multiply (4) by $\boldsymbol{v}_i = (v_{ix}, v_{iy})$, sum over $i = 0, 1, ..., 4$, divide by τ, and again equate coefficients of ϵ^1, we obtain a pair of equations:

$$\frac{\partial}{\partial s} \sum_{i=0}^{4} v_{i\alpha} f_i^0(\boldsymbol{r},t) + \frac{\partial}{\partial q_x} \sum_{i=0}^{4} v_{i\alpha} v_{ix} f_i^0(\boldsymbol{r},t) + \frac{\partial}{\partial q_y} \sum_{i=0}^{4} v_{i\alpha} v_{iy} f_i^0(\boldsymbol{r},t) = 0 \quad \alpha \in \{x,y\} \tag{11}$$

where the right hand side vanishes due to conservation of momentum. This pair can be expressed as

$$\frac{\partial}{\partial t}[h(\boldsymbol{r},t)\boldsymbol{u}(\boldsymbol{r},t))] + \nabla_{\boldsymbol{r}} \cdot \Pi^0(\boldsymbol{r},t)) = \boldsymbol{0} \tag{12}$$

where Π^0 denotes the momentum tensor based on the local equilibrium, f^0. This is the Euler equation. We have already observed that the momentum tensor is diagonal, and now the explicit expression for f^0 in (9) allows an important simplification. We have $\Pi_{xx}^0 = \Pi_{yy}^0 = Kv^2 h(\boldsymbol{r},t)$, and so

$$\frac{\partial}{\partial t}[h(\boldsymbol{r},t)\boldsymbol{u}(\boldsymbol{r},t))] + Kv^2 \nabla_{\boldsymbol{r}} h(\boldsymbol{r},t) = \boldsymbol{0} \tag{13}$$

4.3   The Wave Equation

If we differentiate (10) with respect to t, differentiate (13) with respect to $\boldsymbol{r}$, and subtract, we obtain

$$\partial^2 h(\boldsymbol{r},t)/\partial t^2 - Kv^2 \nabla_{\boldsymbol{r}}^2 h(\boldsymbol{r},t) = 0 \tag{14}$$

the classical wave equation with wave speed $v\sqrt{K}$.

To this point, our derivation is similar in spirit to that of Chopard and Droz [2], but we have avoided the complexity of their approach by using an explicit collision matrix and a single time scale, rather than the more conventional, relaxation equation with multiple time scales. Note that although (14) gives us only a constant-speed wave, that speed is controllable by selection of the collision matrix parameter, K.

5   Dispersion and Wave Number Spectra

In the standard model of deep-water waves, wave speed (phase velocity) is given by $\sqrt{g/k}$, where g is the gravitational constant and k is the *wave number*, the spatial analogue of frequency with units m^{-1} [20]. Note that phase velocity, $\sqrt{g/k}$, yields wave frequency $\sqrt{gk}$. If $\Omega(K)$ denotes the collision matrix of (3), then given a target wave number, k, we can use $\Omega(g/(v^2 k))$ in the update equation (2) to achieve the desired wave speed. An accurate model of a large body of water is likely to have multiple wave numbers per site. Such composite waves disperse with time according to their component wave numbers. Since (2) describes only local, per site collisions, our strategy is to adjust Ω per wave number to control speeds.

If the wave numbers present in a given wave height field, $h(\boldsymbol{r})$, are not evident from the height field construction, it is straightforward to estimate them. If the underlying process is wide-sense stationary, then the lag $\boldsymbol{r}$ auto-covariance function of the wave height field is given by

$$R(\boldsymbol{r}) = E[h(\boldsymbol{x})h(\boldsymbol{x} + \boldsymbol{r})]$$

where E is the expected value operator. The *wave number spectrum* is then the Fourier transform

$$\phi(\boldsymbol{k}) = \frac{1}{(2\pi)^2} \int R(r)e^{-i\boldsymbol{k}\cdot\boldsymbol{r}} d\boldsymbol{r}$$

which carries the amount of energy in, and hence importance of, the waves at each wave vector. The wave number is the modulus of the wave vector. If the wave height field is specified on a lattice, we can use the sample auto-covariance sequence and estimate the wave number spectrum as its discrete Fourier transform (DFT).

In the absence of obstructions, water waves can maintain their speeds (and hence wave numbers) for great distances, sometimes hundreds of miles [20]. To update a composite wave at any given site, we need to apply multiple update matrices, $\Omega(K)$, one to each wave component. We thus maintain total site density, $h(\boldsymbol{r})$, in terms of its wave-number-indexed components,

$$h(\boldsymbol{r}, t) = \sum_{k}\sum_{i=0}^{4} f_{i,k}(\boldsymbol{r}, t) \tag{15}$$

and we apply $\Omega(g/(v^2 k))$ to update the $f_{i,k}(\boldsymbol{r}, t)$ as in (2). In the absence of obstructions that change wave numbers (described in Section 7), we can treat the wave-number-indexed components independently.

This opens the issue of how many wave numbers will be needed for visually accurate representation of interesting surfaces. If the height field is centered on a lattice of edge dimension N, then by symmetry alone we need at most $N(N + 2)/8$ wave numbers, one for each lattice point in a 45-degree octant. Of course, some circles about the origin will contain more than one such lattice point. The number of distinct radii among all circles through all lattice points is asymptotically $0.764 \times (N/2 - 1)^2/\sqrt{2log(N/2 - 1)}$ [21] (cited in [22]). To represent all of them would require both excessive storage and computation time.

Instead, we observe that if we restrict our reflection model (Section 7) to first and second order effects, wave numbers will either remain constant or double on each update. Thus $logN$ wave numbers (powers of 2 in lattice units) should suffice, which yields a total update effort of $O(N^2 logN)$, identical to that of the fast Fourier transform.

6   Initial Conditions

Initial conditions can be arbitrary, but if the goal is to model naturally occurring water waves, we are obliged to begin with a height field that is a reasonable representation of such. Thus, we initially ignore any wave obstructions and begin with a known solution to the general wave equation, in particular, a finite, weighted sum of cosine functions,

$$h(\boldsymbol{r}, 0) = \frac{1}{N^2} \sum_{\boldsymbol{k}} w(\boldsymbol{k})e^{2\pi i k \cdot \boldsymbol{r}} \tag{16}$$

where the weights, $w(\boldsymbol{k})$, are specified in frequency space, and the height field is given by the (inverse) DFT. For the field to be real, we must have the conjugate $w^*(\boldsymbol{k}) = w(-\boldsymbol{k})$, where positions are interpreted mod N. We follow Tessendorf [17] and enforce this constraint by taking

$$w(\boldsymbol{k}) = w_0(\boldsymbol{k}) + w_0^*(-\boldsymbol{k}) \tag{17}$$

where $k = |\boldsymbol{k}|$ and $w_0(\boldsymbol{k})$ is calculated from the targeted wave number spectrum, $\phi(\boldsymbol{k})$. The Phillips spectrum [23] is a standard choice. We use a slightly modified version and instead take

$$w_0(\boldsymbol{k}) = (C\sqrt{e^{-1/k^2}}/k^2)(N(0, 1) + iN(0, 1))((\boldsymbol{k}/k) \cdot \boldsymbol{D}) \tag{18}$$

where $N(0, 1)$ denotes a random sample from a standard normal distribution, D is the wind direction, and C is a scaling constant.

We can write (16) in terms of individual wave numbers as

$$h(\boldsymbol{r}, 0) = \sum_{k} \sum_{|\boldsymbol{k}|=k} \frac{1}{N^2} w(\boldsymbol{k})e^{2\pi i k \cdot \boldsymbol{r}} \tag{19}$$

and again treat each wave number independently. Comparing (15), we see that for each site, r, and each wave number, k, we need to specify values $f_{i,k}(r,0)$ so that

$$\sum_{|\mathbf{k}|=k} \frac{1}{N^2} w(\mathbf{k}) e^{2\pi i \mathbf{k} \cdot \mathbf{r}} = \sum_{i=0}^{4} f_{i,k}(\mathbf{r},0) \tag{20}$$

The specification of these values is otherwise open, but we find the most compelling wave action to arise if we first decompose the wind direction, $\mathbf{D}$, into its associated positive lattice directions, f_{i_1} and f_{i_2}. We then select that vector, $\mathbf{k}$, having maximum dot product, $\mathbf{D} \cdot \mathbf{k}$, and distribute the entire left hand side of (20) to f_{i_1} and f_{i_2} in proportion to the components of $\mathbf{k}$.

7   Obstructions

In addition to their computational simplicity, a widely recognized advantage of LB methods over conventional (finite element, finite difference) methods is their ability to handle complicated boundary conditions. We can represent the collision of a wave with an obstruction by simply reflecting the directional density, dissipating its amplitude, and if harmonics are desired, doubling its wave number. For example, if r is a site adjacent to an obstruction at $r + (\lambda, 0)$, then the update at r at time t, which would have routed density to $f_{1,k}(r + (1,0), t + \tau)$ will instead route a possibly reduced amount to $f_{2,k+s}(r - (1,0), t + \tau)$, where s is either 0 or k. The flow reduction, if any, represents energy dissipation.

8   Implementation

We implemented both the lattice-Boltzmann wave model and a ray-tracing renderer in OpenCL. They can be executed simultaneously on a single NVIDIA GTX 480. Individual frames from a sample animation are shown in Figure 2. For this animation, we used a 1024×1024 grid with 16 wave numbers, and we were able to render 1024×768-pixel frames at 25 frames per second.

The OpenCL kernel for the wave model is nearly trivial, as should be expected from the update equation (2). The only item of note is that the directional density storage, which requires WIDTH×DEPTH×DIRECTIONS floats, is implemented as

```
#define store(i,j,k) ((i)*(WIDTH*DIRECTIONS)+(k)*WIDTH+(j))
```

so that the WIDTH index, rather than the DIRECTIONS index, varies most rapidly in linear memory. There are 5 directions, but the width is usually a large power of 2, and so this storage alignment allows the NVIDIA architecture to fully coalesce accesses to device (card) memory, which is important to performance.

The ray-tracing renderer is based largely on the approach developed by Musgrave [24], as this algorithm lends itself well to GPU computation. Unlike kd-tree traversals of large sets of triangles, there is no control-flow based on the direction of the cast ray, which allows all rays to follow the same execution path

Fig. 2. Frames from sample animation

until a potential hit is encountered. The increased coherence allows the GPU to compute a larger number of rays in parallel, thereby enabling real-time frame rates.

The entire lattice-Boltzmann grid represents a height map. This height map is stored in the red component of a texture object, since NVIDIA's architecture caches accesses to textures. A modified Bresenham Digital Differential Analyzer (DDA) algorithm [25] is then used for the traversal. Once a potential intersection point is found, triangles representing the cell that is intersected are generated from the height map, taking advantage of the spatial locality of the texture cache. A standard ray-triangle intersection test is used. Once all rays have been tested for intersection against the water surface, intersection tests for the channel markers and the beach ball are carried out using traditional ray-tracing methods for reflection, refraction, transmission, and occlusion with respect to the water.

9   Conclusions

We have suggested a new technique for modeling deep-water waves that is based on a two-dimensional, lattice-Boltzmann method. It includes wave dispersion and offers a flexible facility for handling wave-object interaction. Modeling and rendering can be carried out simultaneously, in real-time, on a single graphics card.

Extensions currently under investigation include wave interaction with boats or other partially submerged, moving objects and wave interaction with porous materials.

Acknowledgments

This work was supported in part by the U.S. National Science Foundation under Award 0722313 and by an equipment donation from NVIDIA Corporation. Thanks to the Peabody Symphony Orchestra, Hajime Teri Murai, conductor, for use of an excerpt from the second movement of Debussy's *La Mer* in the accompanying video.

References

1. Gardner, M.: Mathematical games: John conway's game of life. Scientific American (1970)
2. Chopard, B., Droz, M.: Cellular Automata Modeling of Physical Systems. Cambridge Univ. Press, Cambridge (1998)
3. Geist, R., Steele, J., Westall, J.: Convective clouds. In: Natural Phenomena 2007 (Proc. of the Eurographics Workshop on Natural Phenomena), Prague, Czech Republic, pp. 23 – 30, 83, and back cover (2007)
4. Geist, R., Westall, J.: Lattice-Boltzmann Lighting Models. In: GPU GEMS 4, vol. 1. Morgan Kaufmann, San Francisco (2010)

5. Thürey, N., Rüde, U., Stamminger, M.: Animation of open water phenomena with coupled shallow water and free surface simulations. In: SCA 2006: Proceedings of the 2006 ACM SIGGRAPH/Eurographics Symposium on Computer Animation, Vienna, Austria, pp. 157–164 (2006)

6. Wei, X., Li, W., Mueller, K., Kaufman, A.: The lattice-boltzmann method for gaseous phenomena. IEEE Transactions on Visualization and Computer Graphics 10 (2004)

7. Kass, M., Miller, G.: Rapid, stable fluid dynamics for computer graphics. In: SIGGRAPH 1990: Proceedings of the 17th Annual Conference on Computer Graphics and Interactive Techniques, pp. 49–57. ACM, New York (1990)

8. Foster, N., Metaxas, D.: Realistic animation of liquids. Graph. Models Image Process. 58, 471–483 (1996)

9. Foster, N., Fedkiw, R.: Practical animation of liquids. In: SIGGRAPH 2001: Proceedings of the 28th Annual Conference on Computer Graphics and Interactive Techniques, pp. 23–30. ACM, New York (2001)

10. Stam, J.: Stable fluids. In: SIGGRAPH 1999: Proceedings of the 26th annual Conference on Computer Graphics and Interactive Techniques, New York, NY, USA, pp. 121–128. ACM Press/Addison-Wesley Publishing Co. (1999)

11. Mastin, G., Watterberg, P., Mareda, J.: Fourier synthesis of ocean scenes. IEEE Computer Graphics and Applications 7, 16–23 (1987)

12. Hasselmann, D.E., Dunckel, M., Ewing, J.A.: Directional wave spectra observed during jonswap 1973. Journal of Physical Oceanography 10, 1264–1280 (1980)

13. Jensen, L.: Deep-water animation and rendering (2001),
    http://www.gamasutra.com/gdce/2001/jensen/jensen_pfv.htm

14. GmbH, C.: Cryengine3 specifications (2010),
    http://www.crytek.com/technology/cryengine-3/specifications/

15. Hinsinger, D., Neyret, F., Cani, M.P.: Interactive animation of ocean waves. In: SCA 2002: Proceedings of the 2002 ACM SIGGRAPH/Eurographics Symposium on Computer Animation, pp. 161–166. ACM, New York (2002)

16. Yuksel, C., House, D.H., Keyser, J.: Wave particles. ACM Transactions on Graphics (Proceedings of SIGGRAPH 2007) 26, 99 (2007)

17. Tessendorf, J.: Simulating ocean water. In: Simulating Nature: Realistic and Interactive Techniques, SIGGRAPH 2001 Course #47 Notes, Los Angeles, CA (2001)

18. Tessendorf, J.: Interactive Water Surfaces. In: Game Programming Gems 4. Charles River Media, Rockland (2004)

19. Salmon, R.: The lattice boltzmann method as a basis for ocean circulation modeling. Journal of Marine Research 57, 503–535 (1999)

20. Kinsman, B.: Wind Waves: their generation and propagation on the ocean surface. Prentice-Hall, Englewood Cliffs (1965)

21. Landau, E.: Über die einteilung der positiven ganzen zahlen in vier klassen nach der mindestzahl der zur ihrer additiven zusammensetzung erforderlichen quadrate. Archiv. der Math. und Physik. 13, 305–312 (1908)

22. Moree, P., te Riele, H.J.J.: The hexagonal versus the square lattice. Math. Comput. 73, 451–473 (2004)

23. Phillips, O.: On the generation of waves by turbulent wind. Journal of Fluid Mechanics 2, 417–445 (1957)

24. Musgrave, F.: Grid tracing: Fast ray tracing for height fields. Technical Report RR-639, Yale University, Dept. of Comp. Sci. (1988)

25. Bresenham, J.E.: Algorithm for computer control of a digital plotter. IBM Systems Journal 4, 25–30 (1965)

A Texture-Based Approach for Hatching Color Photographs[*]

Heekyung Yang[1], Yunmi Kwon[1], and Kyungha Min[2],[**]

[1] Dept. of Computer Science, Graduate School, Sangmyung Univ., Seoul, Korea
[2] Div. of Digital Media, School of Software, Sangmyung Univ., Seoul, Korea
minkh@smu.ac.kr

Abstract. We present a texture-based approach that hatches color photographs. We use a Delaunay triangulation to create a mesh of triangles with sizes that reflect the structure of an input image. At each vertex of this triangulation, the flow of the image is analyzed and a hatching texture is then created with the same alignment, based on real pencil strokes. This texture is given a modified version of a color sampled from the image, and then it is used to fill all the triangles adjoining the vertex. The three hatching textures that accumulate in each triangle are averaged, and the result of this process across all the triangles form the output image. This method can produce visually pleasing hatching similar to that seen in colored-pencil strokes and oil paintings.

1 Introduction

Hatching, an artistic technique used to create tonal or shading effects by drawing closely spaced parallel lines, is one of the most interesting topics in NPR. Many researchers have presented various hatching techniques that can render 3D triangular meshes or re-render photographs using strokes or line segments [1–4]. The hatching patterns can also be found in researches on line illustrations [5–7], pencil drawings [8–15] and oil paintings [16–21]. Even color pencil drawings [11, 13, 14] and oil-paintings [17, 21] exhibit some hatching patterns.

In this paper, we present a texture-based hatching approach to the re-rendering of color photographs. Our scheme is inspired by Lee et al. [22] who create monochrome pencil drawings on triangular meshes. We have extended this work to the re-rendering of color photographs with hatching. Our scheme produces a color image with hatching patterns similar to those found in a color pencil drawing or an oil painting. Our basic strategy is to build a color hatching texture and to apply it at a vertex of a triangular mesh in which the input image is embedded. This mesh is constructed by the Delaunay triangulation of adaptively sampled points in the image. The color hatching textures are constructed

---

[*] This research is supported by the Ministry of Culture, Sports and Tourism (MCST) and Korea Culture Content Agency (KOCCA) in the Culture Technology (CT) Research and Development Program 2010. This research is also supported by Basic Science Research Program through the Korea Research Foundation (KRF) funded by the Ministry of Education, Science and Technology at 2010 (2010-0006807).
[**] Corresponding author.

G. Bebis et al. (Eds.): ISVC 2010, Part I, LNCS 6453, pp. 86–95, 2010.

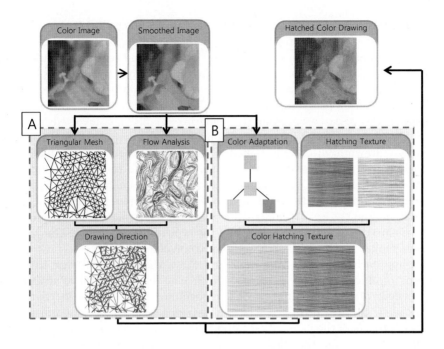

Fig. 1. An overview of our hatching algorithm. The contents of the dotted boxes **A** and **B** are explained in Sections 3 and 2 respectively.

from a base hatching texture made from real strokes. During the hatching step, we build a color hatching texture with a color selected from the input image at each vertex of the triangular mesh. We modify the sampled color to emulate the color found in actual drawings. The pixels inside the triangles which meet at each vertex are drawn using the textures. The direction in which each texture is drawn is selected with the aim of improving the understanding of the shapes in the image. An overview of our scheme is shown in Fig. 1.

We will now survey some of the more significant related work on color pencil drawing and oil painting that include hatching patterns in their results. In some work on color pencil rendering [11, 13, 14], hatching effects are generated using line integral convolution (LIC) [13], using strokes along contours [14], and paper textures may be also simulated [11]. Yamamoto et al. [13] simulated the overlapping effect of two different color pencils and hatching patterns of uniform direction inside a region. They segmented an image into several regions and generated particles of two different colors in each region. They then applied LIC to create hatching patterns, which are overlapped using the Kubelka-Munk model. However, since all the regions in the image are re-rendered using only two different pencil colors, the results are not visually pleasing. Matsui et al. [14] developed a scheme that re-renders a color photograph as a color pencil drawing by drawing strokes that follow the boundaries of regions in the image. They extracted the boundaries of regions in the image, and generated strokes along the boundary curves. The colors of the strokes are sampled from the image and again mixed using the Kubelka-Munk model. Murakami et al. [11] presented an algorithm that generates strokes that mimic pastels, charcoals or crayons using multiple illuminated paper textures. They captured the texture of paper and simulated the stroke effects on the paper in various ways. By applying colored strokes to their paper model, they produce realistic color drawing effects. But none of these techniques can produce the effects of color pencils with sharp tips.

Some of the research on painterly rendering [17, 21] has involved hatching effects. Hays and Essa [17] presented a painterly rendering scheme in which strokes are drawn in directions estimated using an RBF (radial basis function). Hatching patterns are created by capturing the stroke textures produced by artists' brushes. However, the results of this work have a limited ability to convey shape information, since the stroke-generation process does not have any shape information. Zeng et al. [21] presented a painterly rendering scheme in which an image is parsed to determine the size and the type of brushes to use. Our scheme achieves similar results, although we use a triangular mesh.

Our approach offers several advantages: First, the use of different hatching patterns produces a range of effects, which can suggest media as diverse as color pencils and oil painting. Second, embedding input images in triangular meshes allow us to control the scales of hatching patterns to conform to the shapes of objects in the image. Less important regions, such as the backgrounds, are embedded in large triangles and more important regions in small triangles; then we apply coarser hatching patterns to the less important regions and finer patterns to more important regions. Third, we can improve the effects of hatching by modifying the colors that we extract from the image. For example, reducing the saturation of the sampled color creates a texture suggestive of pencil drawing, whereas increasing the saturation suggests oil painting.

The rest of this paper is organized as follows: In Section 2 we explain how the sampled color is modified and how color hatching textures are created with the sampled color. In Section 3 we describe how a color hatching texture is applied to an input image embedded in a triangular mesh. In Section 4 we explain how our scheme is implemented and present several results. Finally, we conclude our paper and mention some plans for future research direction in Section 5.

2   Generating Color Hatching Textures

A color hatching texture is generated by rendering a base texture with the target color. We build a base texture by capturing and overlapping real strokes [22]. We define the *tone* of a hatching texture to be the average intensity of all the intensities in the texture, and its *range* to be a pair consisting of the maximum and minimum intensity across the texture. The tone and the range of a base texture are denoted by t^0, t_m^0 and t_M^0. Fig. 2 (a) illustrates a base texture with its tone and range.

2.1   Color Modification

Different artistic media tend to be used in ways which give the colors in the resulting picture a particular relationship with those in the scene. Therefore,

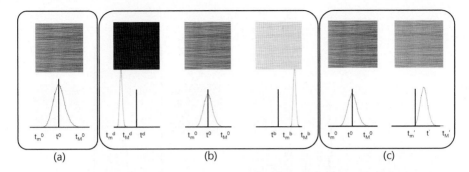

Fig. 2. Histogram control to achieve a hatching texture of a required tone: (a) base texture; (b) darkest and brightest textures; (c) required texture of tone t' and range (t'_m, t'_M)

we modify the color extracted from the input image before we build a color hatching texture. Let C_p be the color sampled at a pixel p of the input image in RGB format with components in the range of $(0, 1)$. We convert C_p to HSV format (H_p, S_p, V_p). We then modify C_p by changing S_p and V_p. We use nine different variations on C_p, selected from the combinations expressed by $\{S_p + \delta, S_p, S_p - \delta\} \times \{V_p + \delta, V_p, V_p - \delta\}$, as shown in Fig. 3 (a). Out of these, we set the modified color (H'_p, S'_p, V'_p) to $(H_p, S_p - \delta, V_p)$ to represent pencil, and to $(H'_p, S'_p, V'_p) = (H_p, S_p + \delta, V_p + \delta)$ to represent oils. The extent of the changes, δ, is in the range $(0.1, 0.3)$. Empirically, we use 0.3. Note that the modified value is clamped to keep it in the range $(0, 1)$. The RGB format of the modified color C'_p is reconstructed from (H'_p, S'_p, V'_p).

2.2 Color Hatching Textures

We build three individual monochrome hatching textures for R'_p, G'_p and B'_p, and then merge them into a color hatching texture as shown in Fig. 3 (b). If we assume $R'_p = t'$, then our target texture has a tone of t' and a range of (t'_m, t'_M), where the subscripts m and M denote the minimum and the maximum of the histogram, respectively. Whereas previous techniques build a series of textures by overlapping strokes or textures [2, 22–24], we use a histogram-based approach to create textures of different tones. The darkest and brightest hatching textures respectively have tones of t^d and t^b, and ranges (t^d_m, t^d_M) and (t^b_m, t^b_M), as shown in Fig. 2 (b). A texture of the required tone and range can be constructed by manipulating the histogram of the texture, although the range of a texture is reduced as it becomes darker or brighter. The new range (t'_m, t'_M) is estimated from the ratio between the tone and the original range. Then an intensity $t^0_i \in (t^0_m, t^0_M)$ sampled from the base texture is converted to $t'_i \in (t'_m, t'_M)$ to match the intensity of the target texture. If $t' > t_0$, then t'_m and t'_M are determined as follows:

$$t'_M = t^0_M + (t^b_M - t^0_M)\frac{t' - t^0}{t^b - t^0}, \qquad t'_m = t^0_m + (t^b_m - t^0_m)\frac{t' - t^m}{t^0 - t^m}.$$

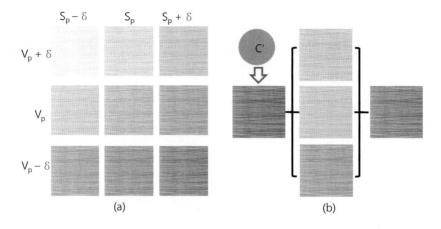

Fig. 3. (a) Eight modified colors: the center color is the original, (b) A color C' applied to a monochrome hatching texture. The corresponding color hatching texture is composed from three color components with different textures.

But if $t' < t_0$, then t'_m and t'_M are determined as follows:

$$t'_M = t^d_M + (t^0_M - t^d_M)\frac{t' - t^m}{t^0 - t^d}, \qquad t'_m = t^d_m + (t^0_m - t^d_m)\frac{t' - t^d}{t^0 - t^d}.$$

Having obtained t'_m and t'_M, other intensities $t^0_i > t^0$ or $t^0_j < t^0$ of a base texture can be converted to t'_i and t'_j in the target hatching texture using the following formulas:

$$t'_i = t' + (t_i - t_0)\frac{t'_M - t'}{t^0_M - t^0}, \qquad t'_j = t^0 - (t^0 - t_j)\frac{t' - t'_m}{t^0 - t^0_m}.$$

We illustrate the relationships between the tones and histograms in Fig. 2 (c).

3 Drawing Color Hatching Textures

Before re-rendering the input image using color hatching textures, we smooth the input image using the mean-shift scheme [25].

3.1 Adaptive Delaunay Triangulation

The first step in drawing the color textures is to embed the input image into a triangular mesh, which requires the following steps:

1. Sample n_0 points on the image using a Poisson disk distribution, and then use Delaunay triangulation to build an initial triangular mesh. We set n_0 to be 100.

2. In each triangle having at least one vertex on an important pixel, we generate a new point at its center. We decide whether a pixel is important using the DoG (difference of Gaussian) filter in [26]. If the evaluated DoG value of a pixel is greater than a threshold, then that pixel is labeled as important.

3. Apply Delaunay triangulation to the modified set of points.

4. Repeat Steps **2** and **3** until the image is triangulated appropriately. An appropriate triangulation means that the triangles are not too tiny to ignore the hatching effects and not too large to ignore shape information. In practice, the triangulation only has to be performed for three or four times.

The final result of this procedure is a triangular mesh which reflects the structure of the input image. Fig. 4 (a) shows three stages in the triangulation of a test image, and Fig. 4 (b)shows how the result is affected by the triangulation.

(a)   (b)

Fig. 4. Triangulation of images: Hatching results on a coarse uniform mesh without modifications, a mesh modified to reflect the structure of the image, and a fine mesh without modification

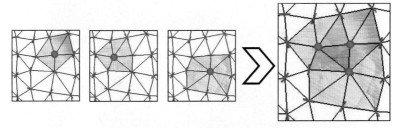

Fig. 5. Merging the textures drawn inside the triangles which share three vertices

3.2 Estimating Drawing Directions

The directions in which the hatching textures are drawn are determined at the vertices of the embedding triangular domain by performing an edge tangent flow (ETF) algorithm [26] at the pixels in which each vertex is located. This produces similar drawing directions to these seen in real pencil drawings. These are usually drawn in on of three styles: (i) along contours or feature lines, (ii) in locally uniform directions, or (iii) in random directions. The directions on important vertices follow style (i), since the ETF's computed at the matching pixels follow contours or feature lines and the vertices are located densely. Since

(a) :800 X 530 (13.8 sec.)

(b) : 800 X 530 (15.6sec.)

(c) 800 X 530 (16.8 sec.)

Fig. 6. Re-rendered images (1)

(c) 1082 X 716 (23.9 sec.)

(e) 1000 X 456 (15.1 sec.)

Fig. 7. Re-rendered images (2)

the unimportant vertices are coarsely distributed, their ETF's correspond to a locally uniform flow, and therefore follows style (ii).

3.3   Drawing Textures on a Triangular Mesh

We mimic the procedures in [22]. At each vertex v, we locate a pixel p that contains v. The color at p, which is C_p, is modified to C'_p using the scheme described in Section 2. The C'_p is then used in generating the color hatching texture. The pixels inside triangles which have v as one of their vertices are filled with the hatching texture. Therefore, every pixel inside a triangle is drawn three times using the different textures generated from each of its vertices. The colors from the three textures are averaged at each pixel. Fig. 5 shows this process.

4   Implementation and Results

We implemented our algorithm on a PC with an Intel Pentium QuadCore^{TM} Q6600 CPU and 4G Byte of main memory. The programming environment was Visual Studio 2008 with the OpenGL libraries. We selected five photographs containing the images of a child, flowers, an animal and a landscape. Each of these photographs was re-rendered using our scheme, with the level of triangulation set to 3. The original images and resulting hatched images are shown in Figs. 6 and 7. The resolutions of the images and their computation times are given in the captions of 6 and 7.

5   Conclusions and Future Plans

We have presented a texture-based hatching technique for re-rendering a color image in order to give the impression of a color pencil drawing. The input image is embedded into an adaptive triangular mesh, and color hatching textures, created by applying an modified colors from the image to a monochrome hatching texture, are drawn into the triangles of the mesh.

Our results have the favor of an artistic rendering, but it is not possible to produce clearly identifiable pencil drawing or oil painting effects by controlling the parameters currently available. We plan to solve this problem by extending the range of the strokes from the pencil strokes of various thickness and shapes used in this paper to other types of strokes, such as those made by artists' brushes and watercolor pencils.

References

1. Paiva, A., Brazil, E., Petronetto, F., Sousa, M.: Fluid-based hatching for tone mapping in line illustrations. The Visual Computer 25, 519–527 (2009)
2. Praun, E., Hoppe, H., Webb, M., Finkelstein, A.: Real-time hatching. In: SIGGRAPH 2001, pp. 579–584 (2001)

3. Salisbury, M., Wong, M., Hughes, J., Salesin, D.: Orientable textures for image-based pen-and-ink illustration. In: SIGGRAPH 1997, pp. 401–406 (1997)
4. Webb, M., Praun, E., Finkelstein, A., Hoppe, H.: Fine tone control in hardware hatching. In: NPAR 2002, pp. 53–58 (2002)
5. Hertzmann, A., Zorin, D.: Illustrating smooth surfaces. In: SIGGRAPH 2000, pp. 517–526 (2000)
6. Salisbury, M., Anderson, S., Barzel, R., Salesin, D.: Interactive pen-and-ink illustration. In: SIGGRAPH 1994, pp. 101–108 (1994)
7. Salisbury, M., Anderson, C., Lischinski, D., Salesin, D.: Scale-dependent reproduction of pen-and-ink illustrations. In: SIGGRAPH 1996, pp. 461–468 (1996)
8. Cabral, B., Leedom, C.: Imaging vector field using line integral convolution. In: SIGGRAPH 1993, pp. 263–270 (1993)
9. Li, N., Huang, Z.: A feature-based pencil drawing method. In: The 1st International Conference on Computer Graphics and Interactive Techniques in Australasia and South East Asia 2003, pp. 135–140 (2003)
10. Mao, X., Nagasaka, Y., Imamiya, A.: Automatic generation of pencil drawing using lic. In: ACM SIGGRAPH 2002 Abstractions and Applications, p. 149 (2002)
11. Murakami, K., Tsuruno, R., Genda, E.: Multiple illuminated paper textures for drawing strokes. In: CGI 2005, pp. 156–161 (2005)
12. Yamamoto, S., Mao, X., Imamiya, A.: Enhanced lic pencil filter. In: The International Conference on Computer Graphics, Imaging and Visualization 2004, pp. 251–256 (2004)
13. Yamamoto, S., Mao, X., Imamiya, A.: Colored pencil filter with custom colors. In: PG 2004, pp. 329–338 (2004)
14. Matsui, H., Johan, J., Nishita, T.: Creating colored pencil images by drawing strokes based on boundaries of regions. In: CGI 2005, pp. 148–155 (2005)
15. Xie, D., Zhao, Y., Xu, D., Yang, X.: Convolution filter based pencil drawing and its implementation on gpu. In: Xu, M., Zhan, Y.-W., Cao, J., Liu, Y. (eds.) APPT 2007. LNCS, vol. 4847, pp. 723–732. Springer, Heidelberg (2007)
16. Haeberli, P.: Paint by numbers: Abstract image representations. In: SIGGRAPH 1990, pp. 207–214 (1990)
17. Hays, J., Essa, I.: Image and video based painterly animation. In: NPAR 2004, pp. 113–120 (2004)
18. Hertzmann, A.: Painterly rendering with curved brush strokes of multiple sizes. In: SIGGRAPH 1998, pp. 453–460 (1998)
19. Litwinowicz, P.: Processing images and video for an impressionist effect. In: SIGGRAPH 1997, pp. 406–414 (1997)
20. Meier, B.: Painterly rendering for animation. In: SIGGRAPH 1996, pp. 477–484 (1996)
21. Zeng, K., Zhao, M., Xiong, C., Zhu, S.C.: From image parsing to painterly rendering. ACM Trans. on Graphics 29, 2 (2009)
22. Lee, H., Kwon, S., Lee, S.: Real-time pencil rendering. In: NPAR 2006, pp. 37–45 (2006)
23. Lake, A., Marshall, C., Harris, M., Blackstein, M.: Stylized rendering techniques for scalable real-time 3d animation. In: NPAR 2000, pp. 13–20 (2000)
24. Winkenbach, G., Salesin, D.: Computer cenerated pen-and-ink illustration. In: SIGGRAPH 1994, pp. 91–100 (1994)
25. Comaniciu, D., Meer, P.: Mean shift: A robust approach toward feature space analysis. IEEE Trans. on Pattern Analysis and Machine Intelligence 24, 603–619 (2003)
26. Kang, H., Lee, S., Chui, C.: Flow-based image abstraction. IEEE Trans. on Visualization and Computer Graphics 15, 62–76 (2009)

Camera Pose Estimation Based on Angle Constraints

Fei Wang, Caigui Jiang, Nanning Zheng, and Yu Guo

Institute of Artificial Intelligence and Robotics, Xi'an Jiaotong University, Xi'an, China
fwang@aiar.xjtu.edu.cn

Abstract. A novel linear algorithm to estimate the camera pose from known correspondences of 3D points and their 2D image points is proposed based on the angle constraints from arbitrary three points in 3D point set. Compared with Ansar's N Point Linear method which is based on the distance constraints between 3D points, due to more strict geometric constraints, this approach is more accurate. Simultaneously some strategies of choosing constraint equations are introduced so that this algorithm's computational complexity is reduced. In order to obtain more accurate estimated pose, we propose the singular value decomposition method to derive the parameters from their quadratic terms more exactly. Finally, the experiments show our approach's effectiveness and accuracy compared with the other two algorithms using synthetic data and real images.

1 Introduction

The estimation of the camera pose from 3D-to-2D point correspondences is the well-known PnP problem. The solution of the PnP problem is widely used in the field of computer vision, robot navigation, photogrammetry, augmented reality (AR) and so on. The key technique in AR is the registration of the virtual object in the real scene. It requires effective and stable camera pose estimation in the real scene, so the research on the algorithm which can determine the camera pose fast and effectively is essential.

In this paper, our approach estimates the camera pose from the angle constraints connected by arbitrary three points in 3D point set. Ansar's N Point Linear (NPL) method[1] presents a linear algorithm based on the constraints from the distances between 3D points, his basic idea is using the well known linearization technique named "relinearization" which was first presented in CRYPTO' 99 by Kipnis and Shamir[2] to solve some special overdetermined quadratic equations. The main idea of "relinearization" is to deal with the overdetermined quadratic equations as linear equation system. The solution of this equations system is weighted sum of basic solutions and the weight of each basic solution can be finally determined by the correlation between those quadratic terms. Compare with the distance constraints, the angle constraints applied are more strict geometric constraints. The overdetermined constraint equations deduced by angle constraints are also quadratic equations which can

G. Bebis et al. (Eds.): ISVC 2010, Part I, LNCS 6453, pp. 96–107, 2010.
© Springer-Verlag Berlin Heidelberg 2010

be solved by the "relinearization" method. Since the detected corresponding 2D image points are noisy, the coefficients of quadratic equation systems are also noisy and the noise can be propagated to the solution of quadratic terms. So we develop a method which uses the singular value decomposition (SVD) to reduce the parameter noise when we deduce the parameters from their quadratic terms. The application of this method effectively improves the pose estimation accuracy.

2 Relative Work

In the research field of computer vision and augmented reality, camera pose estimation is a basic and significant research area. According to the different direction of coordinate transformation, we classify the pose estimation algorithms into two categories: 3D-3D-2D and 2D-3D-3D.

The 3D-3D-2D algorithms transfer the 3D points in the world coordinate into 3D camera coordinate using rotation matrix R and translation vector T, and then produce the constraint equations between 3D points in the camera coordinate and their projected 2D image points. These constraint equations can directly solve the pose parameters (rotation matrix R and translation vector T). Most of these algorithms were nonlinear which must be solved by iterative approaches. Lowe [6] and Haralick [7] applied Gauss-Newton methods to solve the nonlinear optimization problems. Lu et al. [8] proposed an effective iterative algorithm which were proved to be globally convergent and can attain the orthonormal rotation matrix. In this category, the direct linear transform (DLT) [9] is a well known linear algorithm which required at least six points to calculate the intrinsic and extrinsic parameters simultaneously, but it ignored the orthonormality constraints on the rotation matrix.

The 2D-3D-3D algorithms directly recover the point parameters in the 3D camera coordinate and apply the known corresponding 3D points in the world coordinate to determine the pose parameters from 3D-to-3D point correspondences. In this category, Fiore [5] directly recovered the point depths in the camera coordinate, and then transferred the problem to a pure rotation problem using the SVD with at least six points. Quan and Lan [3] applied the distance constraints between 3D points and deduced a four-degree polynomials equation system. Then they considered the nonlinear overdetermined equation as linear equation system and obtained the basic solution set of the linear system by the SVD. The weight of each basic solution was finally estimated according to the correlation between quadratic form variables. The basic principle in Ansar's NPL method was similar with Quan and Lan's algorithm. Ansar derived a set of quadratic equations according to distance constraints. When the number of points $n \geq 4$, the number of equations is $n(n-1)/2$ and the overdetermined quadratic equations can be solved by the "relinearization" method. It's valuable to mention that Morenno-Noguer et al. [4] proposed an algorithm named Epnp which represented each reference point as a weighted sum of four control points both in world coordinate and camera coordinate. This algorithm's computational complexity is $O(n)$ with at least six points.

3 Our Algorithm

3.1 Notation

In our approach we define three coordinates: (a) 3D world coordinate, the space point in this coordinate can be represented as $P_{Wi} = \begin{bmatrix} x_{Wi} & y_{Wi} & z_{Wi} \end{bmatrix}^T$; (b) 3D camera coordinate, the space point in this coordinate can be represented as $P_{Ci} = \begin{bmatrix} x_{Ci} & y_{Ci} & z_{Ci} \end{bmatrix}^T$; (c) 2D image plane coordinate, the corresponding projected point on the image plane can be represented as $P_{Ii} = \begin{bmatrix} u_i & v_i \end{bmatrix}^T$. The camera intrinsic matrix is

$$A = \begin{bmatrix} f_u & 0 & u_0 \\ 0 & f_v & v_0 \\ 0 & 0 & 1 \end{bmatrix}$$

In the camera coordinate, the vector form the camera optical center to the image point can be denoted as

$$\bar{p}_i = \begin{bmatrix} \dfrac{u_i - u_0}{f_u} & \dfrac{v_i - v_0}{f_v} & 1 \end{bmatrix}^T$$

Under the ideal pinhole imaging model, the point P_{Ci}, its corresponding projected image point and the camera optical center are collinear. Then we can obtain

$$P_{Ci} = \rho_i \bar{p}_i,$$

Where ρ_i is a scale factor.

In this paper, a set of n ($n \geq 4$) correspondences of $P_{Wi} (i = 1...n)$ and $P_{Ii} (i = 1...n)$ is known and we assume the camera intrinsic matrix A was calibrated. Then $\bar{p}_i$ can be calculated. If the factor $\rho_i (i = 1...n)$ can be estimated, then $P_{Ci} (i = 1...n)$ can be recovered.

Deriving quadratic equation system in n variables $\rho_i (i = 1...n)$ from the distance constraint between points or angle constraints of arbitrary three points in 3D point set, we can apply the "relinearization" method to recover $P_{Ci} (i = 1...n)$. Finally, the camera pose can be determined from the correspondence of $P_{Wi} (i = 1...n)$ and $P_{Ci} (i = 1...n)$.

3.2 Basic Constraint

3.2.1 The Distance Constraint

Ansar's NPL and Quan's methods used the distance constraint shown as Figure 1. The distance d_{ij} between P_{Wi} and P_{Wj} is calculable and equal to the distance between P_{Ci} and P_{Cj} in the camera coordinate. Let P_{Ci} and P_{Cj} to be denoted as $\rho_i \bar{p}_i$, $\rho_j \bar{p}_j$ respectively, the constraint equation of scale factors will be as follows:

$$d_{ij} = \left| \rho_i \bar{p}_i - \rho_j \bar{p}_j \right|$$

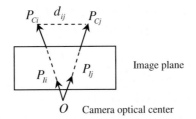

Fig. 1. The distance constraint between two point P_{Ci} and P_{Cj}. The distance d_{ij} from P_{Ci} to P_{Cj} is invariable in world coordinate and camera coordinate. P_{Ci} and P_{Cj} can be denoted as $\rho_i \bar{p}_i$ and $\rho_j \bar{p}_j$ respectively.

Let $b_{ij} = d_{ij}^2$, then,

$$b_{ij} = \left(\rho_i \bar{p}_i - \rho_j \bar{p}_j\right)^T \left(\rho_i \bar{p}_i - \rho_j \bar{p}_j\right) = \rho_i^2 \left(\bar{p}_i^T \bar{p}_i\right) + \rho_j^2 \left(\bar{p}_j^T \bar{p}_j\right) - 2\rho_i \rho_j \left(\bar{p}_i^T \bar{p}_j\right)$$

In the equation above, b_{ij}, $\bar{p}_i$ and $\bar{p}_j$ can be figured out from known information. Let $P_{ij} = \bar{p}_i^T \bar{p}_j$, we can simplify the equation as follows:

$$P_{ii}\rho_i^2 + P_{jj}\rho_j^2 - 2P_{ij}\rho_i\rho_j - b_{ij} = 0 \tag{1}$$

For n points, the number of the constraint equation will be $C_n^2 = n(n-1)/2$.

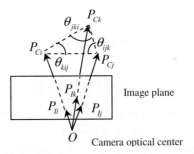

Fig. 2. The constraint based on angle. θ_{ijk}, θ_{jki}, θ_{kij} respectively represent the angles $\angle P_{Ci}P_{Cj}P_{Ck}$, $\angle P_{Cj}P_{Ck}P_{Ci}$, $\angle P_{Ck}P_{Ci}P_{Cj}$ which are invariables. P_{Ci}, P_{Cj}, P_{Ck} can be denoted as $\rho_i \bar{p}_i$, $\rho_j \bar{p}_j$, $\rho_k \bar{p}_k$ respectively.

3.2.2 The Angle Constraint

The angle constraint is shown in Figure 2. Points i, j and k are arbitrary three points from n points. They can be denoted as P_{Ci}, P_{Cj}, P_{Ck} in camera coordinate and P_{Wi}, P_{Wj}, P_{Wk} in world coordinate respectively. The angles connected by the three points are invariable so we can obtain three equations as follows:

$$\angle P_{Wi}P_{Wj}P_{Wk} = \angle P_{Ci}P_{Cj}P_{Ck}$$
$$\angle P_{Wk}P_{Wi}P_{Wj} = \angle P_{Ck}P_{Ci}P_{Cj}$$
$$\angle P_{Wj}P_{Wk}P_{Wi} = \angle P_{Cj}P_{Ck}P_{Ci}$$

They can also be expressed by the invariable angles between the vectors $\overline{ij}$, $\overline{jk}$, $\overline{ki}$ in the camera coordinate and the world coordinate. The constraint equations then become

$$(\rho_k \bar{p}_k - \rho_i \bar{p}_i)^T (\rho_i \bar{p}_i - \rho_j \bar{p}_j) = (P_{Wk} - P_{Wi})^T (P_{Wi} - P_{Wj})$$
$$(\rho_i \bar{p}_i - \rho_j \bar{p}_j)^T (\rho_j \bar{p}_j - \rho_k \bar{p}_k) = (P_{Wi} - P_{Wj})^T (P_{Wj} - P_{Wk}) \qquad (2)$$
$$(\rho_j \bar{p}_j - \rho_k \bar{p}_k)^T (\rho_k \bar{p}_k - \rho_i \bar{p}_i) = (P_{Wj} - P_{Wk})^T (P_{Wk} - P_{Wi})$$

Take the angle θ_{ijk} for instance, and let $C_{ijk} = (P_{Wk} - P_{Wj})^T (P_{Wi} - P_{Wj})$, which can be calculated based on known 3D world coordinates of these points. Then deduce the constraint equations of $\rho_i \rho_j (i = 1...n, j = 1...n)$ as follows:

$$P_{ik}\rho_{ik} - P_{ij}\rho_{ij} - P_{jk}\rho_{jk} + P_{jj}\rho_{jj} - C_{ijk} = 0 \qquad (3)$$

For the n points, if we find all the constraint equations, we can attain $3 \times C_n^3 = n(n-1)(n-2)/2$ equations. But these equations are not all independent. So we give a strategy to choose the special $n(n-1)/2$ constraint equations which can be used to figure out the camera pose. The strategy is combined by two categories of constraint equations. In the first category, the complete graph connected by n-1 points $(V_1, V_2 ... V_{n-1})$ have $C_{n-1}^2 = (n-1)(n-2)/2$ connections. The vertices of each connection and the nth point V_n can create an angle constraint equation and take V_n as the vertex of the angle. In the second category, choose an arbitrary Hamiltonian circuit from the graph of n-1 points $(V_1, V_2 ... V_{n-1})$. Each two connected edges of the Hamiltonian circuit can deduce an angle constraint equation and the number of this kind of constraint is $n-1$. Then, the sum of the constraint equation is $C_{n-1}^2 + n - 1 = n(n-1)/2$.

Take $n = 5$ for example, the set of points is $\{V_1, V_2, V_3, V_4, V_5\}$. The Figure 3 shows the chosen strategy of the constraint equations. In Figure 3 (a), the dotted lines represent the complete graph connected by the $n-1$ points $\{V_1, V_2, V_3, V_4\}$. Then we can choose the angles $\{\angle V_1 V_5 V_2, \angle V_1 V_5 V_3, \angle V_1 V_5 V_4, \angle V_2 V_5 V_3, \angle V_2 V_5 V_4, \angle V_3 V_5 V_4\}$.

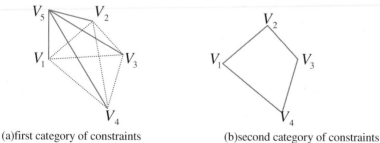

(a)first category of constraints   (b)second category of constraints

Fig. 3. The strategy of choosing the angle constraint, when n=5

In the figure 3 (b), we can easily choose the Hamiltonian circuit $V_1 \rightarrow V_2 \rightarrow V_3 \rightarrow V_4 \rightarrow V_1$ and the angles $\{\angle V_1 V_2 V_3, \angle V_2 V_3 V_4, \angle V_3 V_4 V_1, \angle V_4 V_1 V_2\}$. The amount of the angles is $n(n-1)/2 = 10$.

We can produce a system of $n(n-1)/2$ quadratic equations in n variables $\rho_i (i = 1...n)$ as Ansar's NPL method. The system can be denote as

$$M_{\frac{n(n-1)}{2} \times \left(\frac{n(n+1)}{2}+1\right)} \times L_{\left(\frac{n(n+1)}{2}+1\right) \times 1} = 0 \tag{4}$$

Where $M_{\frac{n(n-1)}{2} \times \left(\frac{n(n+1)}{2}+1\right)}$ is the coefficients matrix of the quadratic equation system, and $L_{\left(\frac{n(n+1)}{2}+1\right) \times 1}$ is the quadratic term variables which can be represented as

$$L = \begin{pmatrix} \rho_{12} & \rho_{13} & \cdots & \rho_{n-1,n} & \rho_{11} & \rho_{22} & \cdots & \rho_{n,n} & \rho \end{pmatrix}^T$$

Kipnis and Shamir proposed an algorithm named "relinearization" in CRYPTO' 99 which can solve this kind of overdetermined quadratic equations, in which the number of the equations is εn^2, where $0 < \varepsilon < 1/2$. The principle of the algorithm is considering the quadratic terms like $\rho_i \rho_j (i = 1...n, j = 1...n)$ in our approach as independent variables. The number of such variables is $n(n+1)/2$ in our system. Consider $\rho = 1$ as a variable then Equation (4) is a homogenous linear system in $n(n+1)/2+1$ variables. The number of independent equations is $n(n-1)/2$, which is less than $n(n+1)/2+1$. Then the solution of the linear system is weighted sum of $(n(n+1)/2+1) - n(n-1)/2 = n+1$ basic solutions which can be denoted as

$$L = \sum_{i=1}^{n+1} \lambda_i v_i \tag{5}$$

where $v_i (i = 1 \square n+1)$ are basic solutions of the linear system, $\lambda_i (i = 1 \square n+1)$ are the weight of each basic solution. Actually, the variables from the set of $\rho_i \rho_j (i = 1...n, j = 1...n)$ are correlated. We can produce the constraint equations of $\lambda_i (i = 1 \square n+1)$ according to the correlations of these variables. For example, we know $\rho_{ab} \rho_{cd} = \rho_{ac} \rho_{bd}$, and then the constraint equation can be expressed as

$$\sum_{i=1}^{n+1} \lambda_{ii} \left(v_i^{ab} v_i^{cd} - v_i^{ac} v_i^{bd} \right) + \sum_{i=1}^{n+1} \left(\sum_{j=i+1}^{n+1} \lambda_{ij} \left(v_i^{ab} v_j^{cd} + v_j^{ab} v_i^{cd} - v_i^{ac} v_j^{bd} - v_j^{ac} v_i^{bd} \right) \right) = 0 \tag{6}$$

where $\lambda_{ij} = \lambda_i \lambda_j (i = 1...n+1, j = 1...n+1)$. And we can produce a set of equations because of the correlations of $\rho_i \rho_j (i = 1...n, j = 1...n)$. The linear equations system combined by the equation like Equation (6) has $(n+1)(n+2)/2$ variables $\lambda_{ij} (i = 1...n+1, j = 1...n+1)$. In Table 1, we demonstrate the number of equations from different constraint forms of $\rho_i \rho_j (i = 1...n, j = 1...n)$. When n>=4, $(C_n^2 + nC_{n-1}^2 + 2C_n^4) > (n+1)(n+2)/2$.

We can solve the liner equation system directly by using the SVD method and derive the weights λ_i .Substitute λ_i into equation (5), we attain the value of ρ_{ij} and finally deduce the scale factor $\rho_i (i = 1...n)$. Using the equation $P_{Ci} = \rho_i \bar{p}_i$, we can recover 3D camera points and estimate the rotation matrix R and translation vector T from correspondence of two set of $P_{Wi} (i = 1...n)$ and $P_{Ii} (i = 1...n)$.

Table 1. The number of equations in variables $\lambda_{ij} (i = 1...n+1, j = 1...n+1)$

(a, b, c, d are different numbers from 1 to n)

Constraint form	The number of constraint equations
$\rho_{aa}\rho_{bb} = \rho_{ab}\rho_{ab}$	$C_n^2 = \dfrac{n(n-1)}{2}$
$\rho_{aa}\rho_{bc} = \rho_{ab}\rho_{ac}$	$nC_{n-1}^2 = \dfrac{n(n-1)(n-2)}{2}$
$\rho_{ab}\rho_{cd} = \rho_{ac}\rho_{bd} = \rho_{ad}\rho_{bc}$	$2C_n^4 = \dfrac{n(n-1)(n-2)(n-3)}{12}$

3.3 Parameter Modification Method

Actually, the detected 2D image points are noisy and the noise is propagated to $P_{ij} = \bar{p}_i^T \bar{p}_j$. Then, the coefficients of the quadratic equation in (4) and (6) contain noise. We can just obtain the approximated solution of ρ_{ij} and λ_{ij} . There are different choices which can deduce $\rho_i (i = 1...n)$ from ρ_{ij} or deduce $\lambda_i (i = 1...n+1)$ from λ_{ij} . For instance, we can obtain ρ_1 from $\pm\sqrt{\rho_{11}}$ or from $\pm\sqrt{\rho_{12}\rho_{13}/\rho_{23}}$. But $\sqrt{\rho_{11}}$ is not exactly equal to $\sqrt{\rho_{12}\rho_{13}/\rho_{23}}$, we can just write the approximately equation $\sqrt{\rho_{11}} \approx \sqrt{\rho_{12}\rho_{13}/\rho_{23}}$. Then ρ_1 is slightly different depend on the different deduced formula. In this paper, we present a method which can reduce the noise to derive ρ_i and λ_i .

We produce the matrix M_ρ , which denote as

$$M_\rho = \begin{bmatrix} \rho_{11} & \rho_{12} & \cdots & \rho_{1n} \\ \rho_{12} & \rho_{22} & \cdots & \rho_{2n} \\ \vdots & \vdots & \ddots & \vdots \\ \rho_{1n} & \rho_{2n} & \cdots & \rho_{nn} \end{bmatrix} \tag{7}$$

Ideally due to the correlation between the element $\rho_{ij}(i=1...n, j=1...n)$ in M_ρ, we can easily know the rank of M_ρ is $rank(M_\rho)=1$. But actually approximated solution of ρ_{ij} destroys the correlation, and then $rank(M_\rho) \neq 1$. Decompose the matrix M_ρ by the SVD method, and then we have $M_\rho = UDV^T$, where the matrix D is diagonal matrix with nonnegative real numbers on the diagonal. The diagonal entries of D are known as singular values of M_ρ and D has the property that $rank(M_\rho) = rank(D)$. To reduce the noise of M_ρ, we can retain the biggest singular value and set the other smaller singular values to zeros. As follows we deduce the new singular value matrix D^* from D, where

$$D = \begin{bmatrix} d_1 & 0 & \cdots & 0 \\ 0 & d_2 & \cdots & 0 \\ \vdots & \vdots & \ddots & \vdots \\ 0 & 0 & \cdots & d_n \end{bmatrix}, \quad D^* = \begin{bmatrix} d_1 & 0 & \cdots & 0 \\ 0 & 0 & \cdots & 0 \\ \vdots & \vdots & \ddots & \vdots \\ 0 & 0 & \cdots & 0 \end{bmatrix} \tag{8}$$

Then, we can derive the noise reduced matrix of M_ρ as

$$M_\rho^* = UD^*V^T \tag{9}$$

The elements of M_ρ^* are $\rho_{ij}^*(i=1...n, j=1...n)$ which can be consider as the modification of $\rho_{ij}(i=1...n, j=1...n)$. Then ρ_1^* can be represented as $\sqrt{\rho_{11}^*}$ or $\sqrt{\rho_{12}^*\rho_{13}^*/\rho_{23}^*}$, where $\sqrt{\rho_{11}^*} = \sqrt{\rho_{12}^*\rho_{13}^*/\rho_{23}^*}$. We can apply the same method to deduce $\lambda_i^*(i=1...n+1)$ from λ_{ij}.

4 Experiment Results

4.1 Synthetic Experiments

In the synthetic experiments, we produce the synthetic data as follows: randomly produce n points $P_{Ci}(i=1...n)$ in the camera coordinate and create translation vector T and rotation matrix R from random Euler angles. We can calculate the 3D world co-ordinates $P_{Wi}(i=1...n)$ of the points. For the calculated camera, the projected 2D image points $P_{Ii}(i=1...n)$ can also be determined. We add Gaussian noise to the ac-tual image points $P_{Ii}(i=1...n)$ and produce the 2D image points $P_{Ii}'(i=1...n)$. Then, we can test the pose estimation algorithm using the input 3D world points $P_{Wi}(i=1...n)$ and their corresponding 2D image points with noise.

In our experiments, we assume the camera effective focal lengths are $f_u = 800$ pixel, $f_v = 800$ pixel, the image resolution is 640×480 and the principal point is $(u_0, v_0) = (320, 240)$. We produce $n = 100$ points in the camera coordinate, in which the ranges of these points are $x_c \in [-3.2, 3.2]$, $y_c \in [-2.4, 2.4]$, $z_c \in [4,8]$. Four algorithm are compared, including Ansar's NPL, Morenno-Noguer's Epnp, our approach named NPLBA (N Points Linear Based on Angle Constraints) and the algorithm NPLBA+SVD (using the algorithm NPLBA and the SVD method to reduce the parameter noise).

We compare three kinds of error, re-projection error, rotation error and translation error. The re-projection error is defined as

$$RMS = \sqrt{\left(\sum_{i=1}^{100} \left| P_{li} - P_{li}^* \right|\right)/100}$$

The rotation error is defined as

$$E_R = \frac{\left\| R - R^* \right\|}{\left\| R \right\|}$$

The translation error is defined as

$$E_T = \frac{\left\| T - T^* \right\|}{\left\| T \right\|}$$

where $P_{li}^* (i = 1...100)$, R^*, T^* respectively represent the re-projection points, the rotation matrix and the translation vector estimated by different pose estimation algorithm. $P_{li} (i = 1\square 100)$, R and T are true values of 2D image points, the rotation matrix and the translation vector. $\| \ \|$ represents the calculation of the matrix norm.

Experiment 1(dependence on the number of points): We vary the number of the points from 6 to 12 for all four algorithms and add 1×1 pixels Gaussian noise to all the synthetic image points. We do 100 trials for each algorithm and the mean re-projection, rotation and translation errors are demonstrated in figure 4. The result shows that NPLBA+SVD has the least re-projection and rotation errors, its translation error is slight greater than Epnp's but less than NPL's.

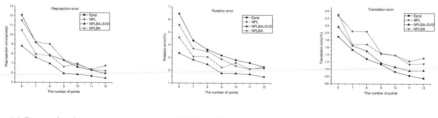

(a) Re-projection error   (b) Rotation error   (c)Translation error

Fig. 4. Mean re-projection, rotation and translation errors using different number of points

Experiment 2(dependence on noise level): We add different image noise with variances from 0.5 to 3.5 in pixel for all four algorithms and the number of point is 7 here. We also do 100 trials for each circumstance for each algorithm and obtain the mean re-projection, rotation and translation errors which are shown in figure 5. The experiment in the figure 5 demonstrated that NPLBA+SVD has the least re-projection and rotation errors and its translation error is nearly the same level with Epnp's, but less than NPL's. We can also find that the algorithm NPLBA is more accurate than NPL, and the error of NPLBA+SVD is less than NPLBA which demonstrated the effectiveness of the SVD method to reduce the parameter noise.

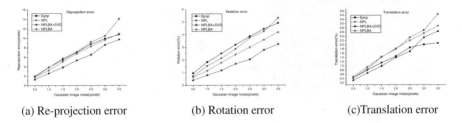

(a) Re-projection error   (b) Rotation error   (c)Translation error

Fig. 5. Mean re-projection, rotation and translation errors using different image noise levels

4.2 Real Image

The real image data was obtained from Professor Janne Heikkilä's Camera Calibration toolbox for matlab. We can load these data on the website: http://www.ee.oulu.fi/~jth/calibr/. The real image shown in figure 6(a) was captured by an CCD camera (Sony SSC-M370CE) whose resolution was 768×576 pixels and the actual physical size was $6.2031 \text{mm} \times 4.6515 \text{mm}$. According to the camera parameters provided in the literature [11], we have the principal point at $(u_0, v_0) = (367.3353, 305.9960)$ and the focal length $f = 8.3459 \text{mm}$ and the scale factor $s_u = 1.00377723$. We can deduce the effective focal $f_u = 1037.3$ pixels, $f_v = 1033.5$ pixels. Using the calibration toolbox we obtain the accurate pose as follows

$$R = \begin{bmatrix} 0.7151 & 0.0318 & -0.6983 \\ -0.2082 & 0.9633 & -0.1694 \\ 0.6673 & 0.2665 & 0.6954 \end{bmatrix} \quad T = \begin{bmatrix} -0.6378 & -100.2466 & 316.7065 \end{bmatrix}^T$$

We consider these estimated pose as real camera pose.

In this experiment, the 3D world coordinates of the circular center points and their corresponding 2D image points are know as a prior. Randomly choose six points from the point set as the input of the algorithm. The Figure 6(b) shows the re-projection point using the algorithm NPLBA+SVD, where the stars represent the real image points and the circles represent the re-projection point on the image plane. These two categories points are closed to each other, which demonstrate the algorithm's feasibility.

Table 2 shows the compared experiment results using real image. The results are the average error of 1000 trials for each algorithm. It is shown that the algorithm NPLBA+SVD has the minimum re-projection error and rotation error. And its translation error is slightly greater than Epnp's but less than NPL's. We can also find that NPLBA has less re-projection error, rotation error and translation error than NPL's.

So we can obtain the same conclusion as the synthetic experiments, in which we can conclude that the SVD method to reduce noise is effective and the NPLBA is more accurate than NPL.

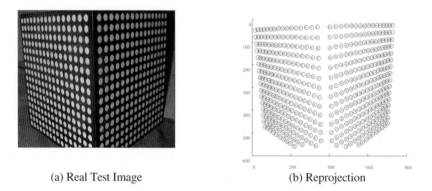

(a) Real Test Image   (b) Reprojection

Fig. 6. Real image experiments. (a) the real model using in the experiment, in which the center points of the white circles are know as the 3D points in the world coordinate; (b) the stars represent the real image points and the circles represent the re-projection points in the image plane using NPLBA+SVD.

Table 2. The pose estimation errors using 6 points for different algorithm

	Epnp	NPL	NPLBA	NPLBA+SVD
RMS(pixels)	5.2929	7.1052	5.3151	4.4522
$E_R(\%)$	2.81	3.23	2.60	1.90
$E_T(\%)$	2.00	4.48	4.26	4.06

5   Conclusion

This paper presents a linear camera pose estimation algorithm based on the angle constraints of vectors connected by two points. These angle constraint equations are deduced by the invariables of the angle connected by arbitrary three points in 3D point set. We can choose the equations applying some strategies to reduce the computational complexity. Compared with Ansar's NPL which based on the distance constraints between 3D points, our algorithm uses more strict geometric constraints which lead to more accurate estimated pose. In order to reduce the parameter noise when we deduce the parameters from their quadratic terms, an effective SVD method is proposed. Synthetic data and real image experiments demonstrate the feasibility and accuracy of our algorithm.

Acknowledgements

This research is supported by National Nature Science Foundation of China, project No.90920301.

References

1. Ansar, A., Daniilidis, K.: Linear pose estimation from points or lines. IEEE Trans. Pattern Analysis Machine Intell. 25(5), 578–589 (2003)
2. Kipnis, A., Shamir, A.: Cryptanalysis of the HFE public key cryptosystem by relinearization. In: Wiener, M. (ed.) CRYPTO 1999. LNCS, vol. 1666, pp. 19–30. Springer, Heidelberg (1999)
3. Quan, L., Lan, Z.D.: Linear N-point camera pose determination. IEEE Trans. Pattern Analysis Machine Intell. 21(8), 774–780 (1999)
4. Moreno-Noguer, F., Lepetit, V., Fua, P.: Accurate non-iterative on solution to the PnP problem. In: Proc. IEEE Conf. on Computer Vision (2007)
5. Fiore, P.D.: Efficient linear solution of exterior orientation. IEEE Trans. Pattern Analysis Machine Intell. 23(2), 140–148 (2001)
6. Lowe, D.G.: Fitting parameterized three-Dimensional models to images. IEEE Trans. Pattern Analysis Machine Intell. 13(5), 441–450 (1991)
7. Haralick, R.M.: Pose estimation from corresponding point data. IEEE Trans. Systems, Man, and Cybernetics 19(6), 1426–1446 (1989)
8. Liu, Y., Huang, T.S., Faugeras, O.D.: Determination of camera location from 2-d to 3-d line and point correspondences. IEEE Trans. Pattern Analysis Machine Intell. 12(1), 28–37 (1990)
9. Hartley, R.I.: Minimizing algebraic error in geometric estimation problems. In: Proc. IEEE Conf. on Computer Vision, pp. 469–476 (1998)
10. Umeyama, S.: Least-Squares estimation of transformation parameters between two point patterns. IEEE Trans. Pattern Analysis Machine Intell. 13(4), 376–380 (1991)
11. Heikkilä, J.: Geometric camera calibration using circular control points. IEEE Trans. Pattern Analysis Machine Intell. 22(10), 1066–1077 (2000)

Feature-Preserving 3D Thumbnail Creation with Voxel-Based Two-Phase Decomposition

Pei-Ying Chiang, May-Chen Kuo, and C.-C. Jay Kuo

Ming Hsieh Department of Electrical Engineering
University of Southern California, Los Angeles, CA 90089-2546
peiyingc@usc.edu

Abstract. We present a feature-preserving 3D thumbnail system for efficient 3D models database browsing. The 3D thumbnail is simplified from the original model so it requires much less hardware resource and transferring time. With topology-preserved 3D thumbnails, the user can browse multiple 3D models at once, and view each model from different angles interactively. To well preserve the topology of the original model, we propose an innovative voxel-based shape decomposition approach, which identifies meaningful parts of a 3D object, the 3D thumbnail is then created by approximating each individual part with fitting primitives. Experimental results demonstrates that the proposed approach can decompose a 3D model well to create a feature-preserving 3D thumbnail.

1 Introduction

The number of 3D models grows rapidly due to the popularity of various applications. Conventional 3D search engines display static 2D thumbnails on the search page for easier browsing. For capturing the best shot of a 3D object, these 2D thumbnails are pre-captured manually, which is very time-consuming. Some researchers attempted to develop systems that can take the best snapshot for a 3D object automatically [1]. This fixed selection rule does not work well for all objects and the result can just as easily capture the wrong features. Yet it is usually impossible to display every important features from a fixed angle.

In this work, we present an innovative feature-preserving 3D thumbnail creation system that helps users browse 3D models efficiently. That is, users can browse multiple 3D models with pre-generated 3D thumbnails, and view each thumbnail from different viewpoints interactively. Usually, to render complicated 3D models simultaneously demands much hardware resource and transfer time that degrades the system performance significantly. The system performance can be improved by rendering the simplified models, i.e. the 3D thumbnails, which requires much less hardware resource and transfer time.

We have developed two approaches to generate a 3D thumbnail; namely, mesh-based and voxel-based approaches. The mesh-based thumbnail creation approach was examined in our previous work [2]. Although many techniques for the mesh-based simplification of 3D models were proposed before, most of them did not address the issue of feature-preserving simplification. For example, the limbs and

G. Bebis et al. (Eds.): ISVC 2010, Part I, LNCS 6453, pp. 108–119, 2010.

the body are important features in a 3D human model, but the existing mesh simplification techniques tend to meld them together. Thus, we proposed a new mesh decomposition technique to preserve the topology of the original model. We first identify significant parts of a 3D model and then simplify each part. The framework in [2] was built upon the surface-based mesh decomposition method [3]. However, the technique in [3] are limited in some scenarios. For example, while an animal mesh can be decomposed into a main body and several protruding limbs successfully, some meaningful parts, such as the head and the neck, cannot be decomposed furthermore. In addition, a decomposed mesh may become fragmented if its adjacent parts were taken apart, which resulted in inaccurate size measurement.

In this work, we develop a voxel-based approach to improve the decomposition result. We convert a 3D polygonal model into a 3D binary voxel grid, then extract its rough skeleton with the thinning operation. The main challenge in the voxel-based approach is that the skeleton obtained using 3D thinning operations or other skeletonization algorithms contains defects that tend to result in a messy 3D thumbnail. To address this problem, we develop the two-phase decomposition process to refine the skeleton. In detail, the skeleton is first decomposed into multiple groups roughly and the voxelized model is decomposed into parts accordingly in the first phase. In the second phase, the skeleton and the voxelized model is re-decomposed more precisely based on the previous result. Finally, the 3D thumbnail will be created by approximating each part with fitting primitives. The significance of this work lies in the two-phases decomposition procedure, where a volumetric model is used in shape decomposition and simplification to overcome several issues associated with the mesh-based approach. The proposed voxel-based approach can decompose a model and extract its skeleton more accurately, and the decomposed model is not fragmented as compared to surface-based approaches. Generally, the resulting 3D thumbnails can well represent the features of the 3D objects.

The rest of this paper is organized as follows. Related works are reviewed in Sec. 2. The framework of the whole system is presented in Sec. 3. The first-phase and the second-phase shape decomposition processes are described in Sec. 4 and 5, respectively. Performance evaluation of obtained results is conducted in Sec. 6. Finally, concluding remarks and future research directions are given in Sec. 7.

2   Review of Previous Work

Various mesh/shape decomposition algorithms were developed to decompose the mesh into small parts based on properties of interest. Shamir [4] provided a thorough survey on mesh decomposition techniques, which can be classified into two major categories depending on how objects are segmented; namely, segmentation is done based on 1) semantic features or 2) geometric primitives.

Methods in the first category is to mimic human perception in psychology or shape recognition to retrieve meaningful volumetric components. Katz and Tal [5] introduced a fuzzy-based cutting method to decompose meshes into meaningful components, where over-segmentation and jaggy boundaries between sub-objects can be avoided. Further performance improvement can be achieved using multi-dimensional scaling, prominent feature point representation and core extraction. Lin *et al.* [3] proposed another decomposition scheme built upon cognitive psychology, where the protrusion, the boundary strength and the relative size of a part were taken into account. Liu *et al.* [6] developed a part-aware surface metric for encoding part information at a point on a shape. These methods focus on human perception and shape analysis in stead of math formulations. Theories arising from psychology, *e.g.* the minimal rule, separate theory and visual salience are used to analyze crucial components for shape decomposition.

Methods in the second category conduct decomposition based on geometric properties of meshes such as planarity or curvature to create surface patches. Cohen-Steiner *et al.* [7] proposed an error-driven optimization algorithm for geometric approximation of surfaces. They used an idea similar to the Lloyd algorithm and reduced the approximation error by clustering faces into best-fitting regions repeatedly. Other types of proxies were employed to replace mesh patches by Wu and Kobbelt [8] so that their scheme can be more effective for spherical, cylindrical and rolling ball blends. Attene *et al.* [9] proposed a hierarchical face-clustering algorithm for triangle meshes with various fitting primitives in an arbitrary set.

In the practical implementation, we may involve one or more segmentation methods as described above. No matter which category a method belongs to, each method has its own strength and weakness. It is not proper to claim the superiority of an algorithm just because it works well for a certain type of models since it could be deprecated when being applied to others.

3   System Overview

An overview of the proposed thumbnail creation system is depicted in Fig. 1. The main difference between this work and our previous work [2] is that a voxel-based shape decomposition scheme is proposed instead of the mesh-based approach. In this work, the polygonal model is first rasterized into a binary 3D voxel grid and the coarse skeleton is extracted from the volumetric model by the thinning operation, using the tools [10] and [11] respectively. For the rest of the paper, we use *object voxels* to represent the volumetric model and *skeleton voxels* to denote the thinning skeleton as shown in Fig. 2(a). After voxelization and thinning, the first-phase shape decomposition process is used to decompose the skeleton and object voxels. The first phase also provides information to the second phase to adjust the skeleton by eliminating detected defects in the thinning result. The decomposition results can then be fine-tuned in the second phase. The remaining modules include:

1. Parts extraction and pose normalization
   We decompose the original model into several significant parts, and the PCA transformation is applied to each part for pose normalization.

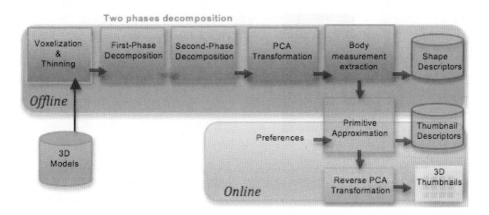

Fig. 1. Overview of the proposed system

2. Body measurement

The body measurement (*i.e.* the radius of the surrounding surface) of each part is taken along its principal axes.

3. Primitive approximation

The 3D thumbnail is created by approximating each individual part with fitting primitives and all parts can be put together with the reverse PCA transformation.

The shape descriptor and thumbnail descriptor are generated offline for each model, so the thumbnail can be downloaded and rendered online efficiently. The above steps are built upon [2]. In the following two sections, we will focus on the two-phase decomposition process, which is the main contribution of this work.

4   First-Phase Decomposition

The first-phase decomposition divides the skeleton voxels into multiple groups and then partitions the object voxels accordingly. The decomposition processes of both the skeleton voxels and object voxels are as described below.

4.1   Decomposition of Skeleton Voxels

The skeleton voxels obtained from the thinning operation are a bunch of discrete voxels in the 3D grid. To extract a meaningful skeleton from them to represent a 3D model, we define a rule to link discrete voxels. First, we classify the skeleton voxels into three categories:

- End_{SK}: the voxel which has only one neighbor;
- $Joint_{SK}$: the voxel which have more than two neighbors;
- $Normal_{SK}$: the voxel which has exactly two neighbors.

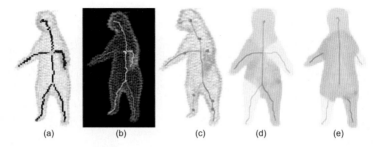

Fig. 2. An example to illustrate skeleton classification and decomposition: (a) object voxels representing a volumetric model are shown in light blue, and skeleton voxels classified as End_{SK}, $Joint_{SK}$, and $Normal_{SK}$ are shown in red, yellow, and black, respectively; (b) skeleton voxels are decomposed into multiple groups and shown in different colors; (c) turning points ($Peak_{SK}$) representing local peaks are shown in purple dot; (d) the first-phase shape decomposition result; (e) the ideal shape decomposition for the base part is shown in red

The classification task is easy. That is, we can simply check the 26-adjacent voxels in a $3 \times 3 \times 3$ grid. Once skeleton voxels are classified, we link adjacent skeleton voxels that belong to the same group. We start to create a group with one of the End_{SK} or $Joint_{SK}$ voxels and link it to the next adjacent voxel until another $End_{SK}/Joint_{SK}$ is met. As a result, all skeleton voxels can be linked and divided into different groups such as the example shown in Fig. 2(b).

After all skeleton voxels are linked, we extract the turning point so that the model can be decomposed more precisely. The turning point of a skeleton, denoted by $Peak_{SK}$, represents the local peak of the skeleton and can be extracted by analyzing the curve of the skeleton. Fig. 2(c) shows an example of extracted turning points. It is worthwhile to point out that we do not partition the skeleton with turning points in the first-phase decompostion since this will complicate the following process. The usage of the turning points will be discussed in Sec. 5.2.

4.2   Decomposition of Object Voxels

After the decomposition of skeleton voxels, object voxels which represent the volumetric model can be decomposed accordingly. Although each object voxel could be assigned to its nearest skeleton voxel by computing their Euclidean distance, this do not work well in general. For example, the distance from a finger tip to anther finger tip is not equal to their Euclidean distance since there is no straight path between them. Thus, we define the distance between object voxel V_j to skeleton voxel K_i as

$$d(V_j, K_i) = \|V_j, K_i\| \text{ , if a straight path } \overline{V_j K_i} \text{ exists}$$
$$= \text{ infinity , otherwise.}$$

To check if a straight path exists between V_j and K_i, we create a line segment $\overline{V_j K_i}$. Voxels along this line segment can be derived by interpolation. If there is an empty voxel lies within $\overline{V_j K_i}$, the distance between them are set to infinity. Accordingly, object voxels can be assigned to different parts according to their belonging skeleton voxels.

The object voxels are classified into two types based on the relationship with their neighbors; namely, *surface* and *interior* voxels. The object voxel that does not have 26 neighbors in a $3 \times 3 \times 3$ grid is called a surface voxel. Otherwise, it is an interior voxel. Since our objective is to estimate the radius of the surrounding surface along the skeleton, our interest lies in the decomposition of surface voxels. Furthermore, we would like to reduce the computational complexity. Based on the above two reasons, we focus only on the decomposition of surface voxels and ignore interior voxels. Finally, an example of the first-phase decomposition result is shown in Fig. 2(d), where each surface voxel is assigned to its nearest skeleton voxel to form a group of the same color.

After the first-phase decomposition, each skeleton voxel, K_i, has a associated list, $List(K_i)$, recording its associated surface voxels. This list is used to calculate the average radius, $Radius(K_i)$, of its surrounding surface. The information of the associated surface voxels and the average radius will be used in the second-phase decomposition.

5   Second-Phase Decomposition

The skeleton of a muti-tubular objects, such as animals, normally has its protruding skeletons intersecting with the base skeleton. There are two kinds of mis-decomposition in the first-phase: Object voxels which belong to the base part are mistakenly assign to adjacent protruding parts, and vice versa. Comparing Fig. 2(d) to the ideal decomposition in Fig. 2(e), a portion of the main body is mistakenly classified to the arm while the main body is mistakenly divided at the middle. The protruding skeleton that cuts across the boundary makes the surrounding area ambiguous and difficult to decompose. To address the problem, the base part and the protruding parts of a 3D model will be identified in the second-phase decomposition. The decomposed skeleton voxels are re-adjusted by invalidating the redundant segment and extending the base skeleton which was mistakenly divided at the intersection. Afterwards, the object voxels will be re-decomposed accordingly.

5.1   Identification of Base Part

To identify the base part, we define weighted accumulated distance for each part P_i as

$$\rho(P_i) = d_E(cen(P_i), cen(P)) \times s(P_i) + \Sigma_{\forall j \neq i}(d_P(mid(P_i), mid(P_j)) \times s(P_j)), \quad (1)$$

where $cen(P_i)$ and $cen(P)$ are the center of mass of P_i and the model P, respectively, $d_E()$ is the Euclidean distance, $s(P_i)$ is the number of object voxels belonging to P_i, and $mid(P_i)$ is the skeleton voxel of P_i that is closest to $cen(P_i)$.

Fig. 3. The base part identification procedure: (a) the path along the skeleton from one part to another, where the green dots represent centers of different parts; (b) identified base part shown in red

Since the distance between two parts is not equal to their Euclidean distance, we define the the the distance from P_i to P_j as

$$d_P(mid(P_i), mid(P_j)) = \|path(mid(P_i), mid(P_j))\|, \qquad (2)$$

where $path(mid(P_i), mid(P_j))$ contains all the skeleton voxels along the path from the $mid(P_i)$ to $mid(P_j)$, and $\|.\|$ is the path length. The idea is illustrated in Fig 3(a).

Thus, the base part which is closer to all other parts can be identified by finding the part with the minimal accumulated distance. The experimental results of this algorithm are shown in Fig. 3(b).

5.2   Adjustment of Decomposed Skeleton Voxels

To adjust the skeleton, we first constrain the segment of the protruding skeleton to its own region. Second, we extend the base skeleton by merging the protruding skeleton selectively. Finally, the protruding part is further fine-tuned by the usage of turning points. An example is illustrated in Fig. 4.

Adjusting Protruding Skeleton. For a protruding skeleton whose orientation is different from the base skeleton (*i.e. the angle between their orientations is larger than a threshold*), we aim to find its boundary and invalidate its segment that goes beyond the boundary. The boundary can be detected using the property that the surrounding surface radius along the protruding skeleton toward to the base part increases dramatically such as the surrounding surface along the arm as shown in Fig. 2(c). The corresponding skeleton voxel is called the cutting point. Then, skeleton voxels from the cutting point to the intersection with the base part should belong to the base part rather than the protruding part.

In practice, the surrounding surface radius can vary and it is difficult to define a normal range of the surface radius throughout the entire protruding part. To address this problem, we narrow down the search region for the boundary by

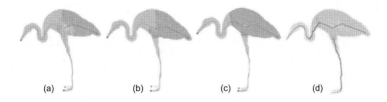

Fig. 4. The adjustment of decomposed skeleton voxels: (a) the result from the first-phase decomposition; (b) invalidating the segment of the protruding skeleton that goes beyond its boundary; (c) extending the base skeleton by merging a segment of a protruding skeleton, (d) dividing groups into sub-groups by Peak_{SK}. As shown in the example, the protruding parts such as the legs and the neck of the crane model are decomposed into subgroups with the selected turning point.

considering the surrounding surface radii along the base skeleton. That is, the average radius δ_1 along the base skeleton is the minimum threshold and the largest base radius δ_2 is the maximum threshold of the search region. Now, consider a protruding part that has successive skeleton voxels $\{ SK_1, SK_2, \ldots, SK_n \}$, and SK_n is the end voxel that is connected to the base part. We calculate the Euclidean distance from each skeleton voxel to SK_n. Only the skeleton voxel whose distance to the base part is within the search range $[\delta_1, \delta_2]$ will be examined to see if it is the cutting point. That is, if the successive skeleton voxels $\{ SK_i, \ldots, SK_j \}$ has the distance to the base within $[\delta_1, \delta_2]$, we search for the cutting point among SK_i to SK_j. If no cutting point is found in this region, we cut at SK_j instead.

After the cutting point is selected, all skeleton voxels from the cutting point to the joint of the base skeleton are removed from the protruding part. The surface voxels previously assigned to these invalidated skeletons are re-assigned to one of the closest base skeletons. Fig. 4(b) is an example of improved decomposition result by adjusting protruding skeletons (*i.e.* the skeleton of the leg).

Extending Base Skeleton. For each protruding group whose orientation is similar to the base group, we aim to merge the base skeletons which was partitioned inaccurately, such as the front body in Fig. 4(a). The decision of mergence/disjoint is made based on two criteria. One is to find the protruding boundary, where the surrounding surface radius increases dramatically. The other is to determine the turning point along the skeleton. Even if the radius deviation along the protruding skeleton is small, the base part should not be extended to the position where the protruding part starts to bend, such as the boundary between the crane's neck and the body as shown in Fig. 4(c). As a result, the search region for the cutting point is from the joint to the closet turning point. If there is no cutting point found in the region, we cut at the turning point.

Fine-Tuning of Protruding Parts with Turning Points. As the final step, we divide each protruding group into subgroups according to the curvature of

Fig. 5. Comparison of skeletons obtained with the thinning operation and the refined skeleton

its skeleton. If the joint between two subgroups is bended, we can detect the turning point and divide them accordingly. For example, the leg part in Fig. 4(c) will be further divided into a leg and a foot as the result shown in Fig.4(d).

5.3   Re-decomposition of Object Voxels

Finally, a re-decomposition process is carried out for object voxels. Here, we do not need to re-calculate the closest skeleton voxel for each object voxel. Object voxels that belonged to invalidated skeleton voxels in the first phase have already re-assigned to the closest base skeleton in the second phase, and the assignment of all other object voxels is not changed since they still belong to the same closest skeleton voxel. Instead, we only need to assign the new group id to each object voxel according to the skeleton re-decomposition result. Each object voxel is assigned to the same group as its associated skeleton voxel.

6   Experimental Results

In the experiments, we applied the proposed two-phase decomposition approach to a collection of 3D models. All of the models were pre-converted into the same obj format, normalized to the same range, and voxelized to the corresponding volumetric models. In the voxelization process, the smaller size the voxel is chosen (higher resolution), more features can be captured. However, it demands larger memory and its thinning result tends to contain more noise. We decided to use a grid of size $80 \times 80 \times 80$ since it captures the most important features with a reasonable amount of memory while unnecessary details could be reasonably discarded. In addition, the thinning process may take very long time

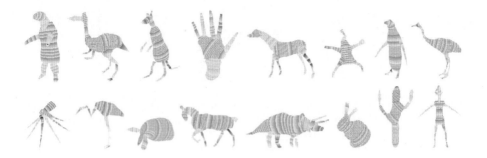

Fig. 6. Thumbnail results obtained by the proposed voxel-based approach, where the original 3D models are decomposed into multiple parts and each part is approximated by a fitting primitive

to conduct for higher resolution models. For example, it took about 27 mins to finish thinning a bunny model which was voxelized into $500 \times 500 \times 500$ grid. On the other hand, it took less than 1 sec for thinning of the same model which was voxelized into $80 \times 80 \times 80$ grid.

The experiment was run on a desktop computer with an Intel Core 2 Duo 2.53 GHz CPU, and 4G RAM. The first phase decomposition took 0.83 sec and the second phase decomposition took 0.35 sec on the average. Assigning object voxels to different parts in the second phase took much less time than the first phase since we only re-assigned the object voxels whose associated skeleton voxels were invalidated. The average processing time for create a thumbnail descriptors with our voxel-based approach was about 3.6 second.

The two-phase decomposition process refines the skeleton obtained from the thinning algorithm. The protruding skeleton that goes beyond the boundary is removed, the base skeleton is extended and each part is segmented more accurately. Fig. 5 shows the improvement of the skeleton re-decomposition result, both the original skeleton decomposition and the improved skeleton decomposition result are illustrated. All identified base parts are shown in red color and protruding parts are shown in different colors. Each model is decomposed into its significant parts more accurately with the two-phase decomposition procedure. Except for the decomposition process, the other modules described in Sec. 3 remain the same as those in [2]. The significant parts of a model were approximated by fitting primitives individually.

Several created thumbnails, each of which consists of 90 primitives, are shown in Fig. 6. The thumbnail results generated by the proposed voxel-based approach preserve more details of the original models. Some subgroups that could not be separated by the surface-based approach can be separated now. Fig. 7 shows several examples whose significant parts are still well preserved even when the models are greatly simplified.

We compare the surface-based approach in [2] and the voxel-based approach in this work in Fig. 8. Although the surface-based approach can identify the main body and several protruding parts as shown in Fig. 8(a_1). It has two main

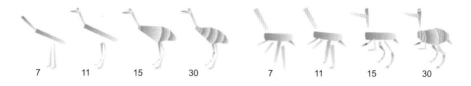

Fig. 7. Thumbnails approximated by a different number of primitives by the voxel-based approach, where subparts are not as easily breakable as those obtained by the surface-based approach as presented in [2]

shortcomings. First, the surface-based decomposition is less precise; namely, some meaningful parts which do not directly protrude from a base part cannot be further decomposed. For example, the foot and the head cannot be further separated from the leg and the neck, respectively. Second, the decomposed mesh may not represent the shape well for measurement since it become fragmented while its adjacent parts were taken apart. For example, there are missing pieces in the simplified crane model as shown in Fig. 8(a_2). Its skeleton and the body measurement extracted from the base part leans toward the upper body due to the missing pieces on the bottom as shown in Fig. 8(a_3). These shortcomings are resolved by the proposed voxel-based approach. Figs. 8 (b_1)-(b_3) show the improvement of the skeleton and the body measurement using the voxel-based approach. Besides, the file size of a created thumbnail is similar as we described in [2] since the 3D thumbnail is constructed with the same primitive in both approaches; The size of a thumbnail composed of 90 primitives is about 8KB.

Finally, it is worthwhile to point out one shortcoming of the proposed voxel-based approach. The decomposition result is highly affected by the qualify of the skeleton obtained in the beginning stage. If the skeleton does not represent the correct structure of an 3D model well, the proposed approach fails to decompose the model. Please see the last two thumbnails in Fig. 5. First, the skeleton in the middle part of the cactus model is skewed, and it does not have the same direction as the upper part and the lower part. Thus, the base part of the cactus fails to extend to the right direction. Second, the alien model has two big holes

Fig. 8. The comparison of the surface based approach (a) and the voxel-based approach (b). (a_1) The mesh decomposed by the surface-based technique [3]; (a_2) the bottom view of the decomposed base part with several pieces missing; (a_3) the extracted skeleton and the body measurement; (a_4) the resultant thumbnail shows the base part is leaning upward, since the bottom pieces are missing; (b_1) the volumetric shape decomposed by the voxel-based approach; (b_2) the improved skeleton and body measurement; and (b_3) the improved resultant thumbnail.

in the face and the skeleton obtained with thinning cannot capture this feature correctly. As a result, one half of the face is missing in the created thumbnail. To improve these thumbnail results, we need a better skeletonization technique for 3D models, which is a future research item.

7 Conclusion and Future Work

An innovative voxel-based two-phase decomposition approach was presented to resolve limitations encountered in [2]. In this work, the skeleton is decomposed into multiple groups and then the volumetric model is decomposed into significant parts accordingly. The significant parts of a 3D model can be well-preserved by its thumbnail representation even the original model is greatly simplified. As compared with the surface-based approach in [2], the voxel-based approach can represent the shape of each part better and decompose the model more accurately.

There are three possible future extensions of this work. First, in the case that the skeleton of a 3D model extracted by the thinning process fail to represent the structure of the object, the decomposition result can be severely affected. More advanced skeletonization techniques are needed to extract a finer skeleton. Second, different primitive types can be used in primitive approximation for performance comparison. Last, the textures of 3D models can also be considered as possible enhancement.

References

1. Mortara, M., Spagnuolo, M.: Semantics-driven best view of 3d shapes. In: IEEE International Conference on Shape Modelling and Applications 2009, vol. 33, pp. 280–290 (2009)
2. Chiang, P.Y., Kuo, M.C., Silva, T., Rosenberg, M., Kuo, C.C.J.: Feature-preserving 3D thumbnail creation via mesh decomposition and approximation. In: Qiu, G., Lam, K.M., Kiya, H., Xue, X.-Y., Kuo, C.-C.J., Lew, M.S. (eds.) PCM 2010. LNCS, vol. 6298, Springer, Heidelberg (2010)
3. Lin, H.Y.S., Liao, H.Y.M., Lin, J.C.: Visual salience-guided mesh decomposition. IEEE Transactions on Multimedia 9, 45–57 (2007)
4. Shamir, A.: A survey on mesh segmentation techniques. Computer Graphics Forum 27, 1539–1556 (2008)
5. Katz, S., Tal, A.: Hierarchical mesh decomposition using fuzzy clustering and cuts. In: SIGGRAPH 2003, pp. 954–961. ACM, New York (2003)
6. Liu, R.F., Zhang, H., Shamir, A., Cohen-Or, D.: A part-aware surface metric for shape analysis. Computer Graphics Forum (Eurographics 2009) 28 (2009)
7. Cohen-Steiner, D., Alliez, P., Desbrun, M.: Variational shape approximation. In: SIGGRAPH 2004, pp. 905–914. ACM, New York (2004)
8. Wu, J., Kobbelt, L.: Structure recovery via hybrid variational surface approximation. Computer Graphics Forum 24, 277–284(8) (2005)
9. Attene, M., Falcidieno, B., Spagnuolo, M.: Hierarchical mesh segmentation based on fitting primitives. Vis. Comput. 22, 181–193 (2006)
10. Binvox. 3D mesh voxelizer, http://www.cs.princeton.edu/~min/binvox/
11. Thinvox. 3D thinning tool, http://www.cs.princeton.edu/~min/thinvox/

Learning Scene Entries and Exits Using Coherent Motion Regions

Matthew Nedrich and James W. Davis

Dept. of Computer Science and Engineering
Ohio State University, Columbus, OH, USA
{nedrich,jwdavis}@cse.ohio-state.edu

Abstract. We present a novel framework to reliably learn scene entry and exit locations using coherent motion regions formed by weak tracking data. We construct "entities" from weak tracking data at a frame level and then track the entities through time, producing a set of consistent spatio-temporal paths. Resultant entity entry and exit observations of the paths are then clustered and a reliability metric is used to score the behavior of each entry and exit zone. We present experimental results from various scenes and compare against other approaches.

1 Introduction

Scene modeling is an active area of research in video surveillance. An important task of scene modeling is learning scene entry and exit locations (also referred to as sources and sinks [1]). Understanding where objects enter/exit a scene can be useful for many video surveillance tasks, such as tracker initialization, tracking failure recovery (if an object disappears but is not near an exit, it is likely due to tracker failure), camera coverage optimization, anomaly detection, etc.

In this paper we offer a novel approach for learning scene entries and exits using only weak tracking data (multiple short/broken tracks per object). Most existing approaches for learning scene entries and exits rely on strong tracking data. Such data consists of a set of reliable long-duration single-object trajectories. When an object enters the scene, it needs to be detected and tracked until it leaves the scene resulting in a single trajectory capturing the path of the object. Here, the beginning and end of the trajectory correspond to a scene entry and exit observation, respectively. Given enough of these trajectories, the set of corresponding entry and exit observations could be clustered into a set of entry and exit locations. However, collecting such a set of reliable trajectories can be cumbersome, slow, and expensive, especially for complex and crowded urban environments where tracking failures are common. Further, for many strong trackers, real-time multi-object tracking in highly populated scenes is not feasible. Alternatively, randomly selecting one of many objects in the scene to track and accumulating these trajectories over time could be unreliable as the sampled trajectories may not be representative of the true scene action (missing the more infrequent entries/exits). Such an approach also requires a long duration of time to accumulate enough trajectories.

G. Bebis et al. (Eds.): ISVC 2010, Part I, LNCS 6453, pp. 120–131, 2010.

To compensate for these issues, we instead employ data produced from a weak tracker, which provides multiple and frequently broken "tracklets". We use a modified version of the Kanade-Lucas-Tomasi (KLT) tracker [2] for our work in this paper. Such trackers are capable of locally tracking multiple targets in real-time, and are thus well suited for busy urban environments. Using a weak tracker provides a simple way to detect and track all motion in the scene, though the produced tracklets are more challenging to analyze than reliable trajectories produced by a strong tracker. As an object moves through the scene it may produce multiple tracklets that start and stop along its path. Thus, instead of one reliable trajectory representing the motion of an object, there is a set of multiple and frequently broken tracklets. The goal of our approach is to build a mid-level representation of action occurring in a scene from multiple low-level tracklets. We attempt to cohere the tracklets and construct "coherent motion regions" and use these regions to more reliably reason about the scene entry/exit locations. A coherent motion region is a spatio-temporal pathway of motion produced by multiple tracklets of one underlying entity. Here, we define an entity as a stable object or group of objects. Thus, an entity may be a person, group of people, bicycle, car, etc.

Our approach first detects entities at a frame level. Next, we track the detected entities across time via a frame-to-frame matching technique that leverages the underlying weak tracking data from which the entities are constructed. Thus, entry and exit observations are derived from entities, not individual tracks, entering and exiting the scene. We then cluster the resulting entry/exit observations and employ a convex hull area-reduction technique to obtain more spatially accurate clusters. Finally, we present a novel scoring metric to capture the reliability of each entry/exit region. We evaluate our method on four scenes of varying complexity, and compare our results with alternative approaches.

2   Related Work

Most existing work on learning entry and exit locations relies on strong tracking data. In [3], trajectory start and end points are assumed to be entry and exit observations, respectively. They cluster these points via Expectation-Maximization (EM) to obtain the entry and exit locations. Using a cluster density metric, they label clusters with low density as noise and discard them, leaving a final set of entry and exit locations. In [1] an EM approach is also used to cluster trajectory start and stop points, and they also attempt to perform trajectory stitching to alleviate tracking failures. Such approaches, however, are not applicable to weak tracking data. In [4] a scene modeling framework is described in which they cluster trajectories to learn semantic regions. Only entry and exit observations that exist near the borders of semantic regions are considered when learning entry and exit locations. A similar constraint is also employed in [5] where they only consider states (which they define as a grid area of the scene and motion direction) near the borders of an activity mask to be eligible to be entry or exit states, though their state space may be constructed from weak tracking data.

Weak tracking data has been used for many applications in computer vision. Some applications employ a low-level per-frame feature clustering as we do to detect entities. In [6] weak tracking is used to perform person counting. They first condition the short and broken trajectories to smooth and extend them. Then, for each frame they perform trajectory clustering on the conditioned trajectories to associate trajectory observations existing through that frame. Using clustering constraints such as a defined person size, they leverage the number of clusters to obtain a final person count. A similar approach is used in [7], though the clustering problem is formulated in a Bayesian framework.

Both [8] and [9] attempt to leverage the idea of a "coherent motion region" from trajectories, though in [8] their regions are constructed using a user-defined bounding box that represents the size of a person. In [9], they define a coherent motion region as a consistent flow of motion which they learn via trajectory clustering, however the trajectories they cluster are collected over a long time window and may be the result of motion occurring at very different times.

3   Framework

Our framework first involves entity detection in each frame using weak tracking data, followed by entity tracking, where we associate entities frame-to-frame, producing a set of hypothesized entity entry and exit observations.

3.1   Entity Detection

For a given frame, there exists a set of track observations (assuming there is object motion in the frame). The first step in our approach is to cluster these track observations, and thus learn a set of entities for each frame. Trajectory clustering approaches are used in both [6] and [7], but given the complexity of such trajectory clustering techniques, and their sensitivity to parameter selection, we instead use a modified version of mean-shift clustering [10].

Given a set of tracklet observations in a frame of interest, we perform a modified mean-shift clustering as follows. For an observation point $p^* = (x, y)$, we wish to find the nearest density mode from other observation points p_i. Each iteration of mean-shift clustering will move p^* closer to the nearest mode, and is computed as

$$p^*_{new} = \frac{\sum_{i=1}^{n} p_i \cdot w_{vel} \cdot K(\frac{|p_i - p^*|}{h})}{\sum_{i=1}^{n} w_{vel} \cdot K(\frac{|p_i - p^*|}{h})} \tag{1}$$

where K is a kernel (we use the Gaussian kernel), h is the kernel bandwidth, and w_{vel} is a velocity weight (to ensure that only points moving in similar directions as p^* are clustered together) computed as

$$w_{vel} = \begin{cases} \frac{1}{1+exp(-\frac{\cos(\phi)}{\sigma})} & \text{if } |\phi| < \frac{\pi}{2} \\ 0 & \text{otherwise} \end{cases} \tag{2}$$

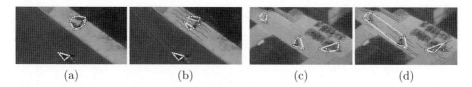

(a)   (b)   (c)   (d)

Fig. 1. Our mean-shift clustering approach (a) and (c) compared to the trajectory clustering approach in [4] (b) and (d)

Here, ϕ is the velocity angle between p^* and p_i (velocity is computed using observations in the previous two frames), and σ defines the rate of weight transition (we use $\sigma = 0.07$ for our experiments). Thus, only points that are spatially close and traveling in similar directions will seek to the same mode.

We also introduce a blend parameter to the velocity weight computation in Eqn. (2). Rather than using the initial velocity of p^* for each iteration (as it seeks to the nearest mode), we start with the initial velocity of p^* and slowly blend it with the velocity of valid surrounding points as it converges toward a mode. Doing so ensures tighter convergence among points, as computing velocities over such a short window is subject to noise. To compute the velocity of p^* at iteration k, we use $dx_{p^*}^k = (1 - \alpha) \cdot dx_{p^*} + \alpha \cdot dx_{avg}$ and $dy_{p^*}^k = (1 - \alpha) \cdot dy_{p^*} + \alpha \cdot dy_{avg}$. Here, dx_{avg} and dy_{avg} are weighted averages of the velocities of nearby points computed as

$$dx_{avg} = \frac{\sum_{i=1}^{n} dx_{p_i} \cdot w_{vel} \cdot K(\frac{|p_i - p^*|}{\frac{h}{2}})}{\sum_{i=1}^{n} w_{vel} \cdot K(\frac{|p_i - p^*|}{\frac{h}{2}})} \tag{3}$$

where dy_{avg} is computed in the same manner. The blend parameter α is a linear function of the mean-shift iteration number (increasing from 0 to 1). Points that converge to the same mode are defined as belonging to the same entity. The outcome is a set of point clusters (entities), which we then track. Two examples are shown in Fig. 1 (a) and (c).

We compared our modified frame-based mean-shift clustering technique to detect entities to the trajectory clustering technique used in [4], which uses Spectral Clustering on a trajectory similarity matrix built using a modified Hausdorff distance metric to obtain a set of trajectory clusters. The trajectory clustering approach occasionally produced unexpected results, as seen in Fig. 1 (b) and (d). In Fig. 1 (b), it over-clusters the object on the sidewalk into three entities, while our mean-shift approach clusters it as one entity (see Fig. 1 (a)). In Fig. 1 (d), [4] under-clusters the two objects on the left, while over clustering the object on the right. Each approach would sometimes over-cluster and under-cluster, though our approach performed in a more reliable manner overall. Further, our mean-shift clustering approach only has one main parameter - the kernel bandwidth. The trajectory clustering approach is very sensitive to its parameter settings.

3.2   Entity Tracking

After detecting entities, we next associate them frame to frame. This problem can be formulated as an ad-hoc blob tracking problem, though it can leverage the underlying weak tracking data that formed the entities. We use the following graph-based assignment to associate entities. Let V_a be the set of entities that exist in frame a, and V_b be the set of entities that exist in the subsequent frame b. We construct a bipartite graph $G(V, E)$ where V is the vertex set, E is the edge set, and $V = V_a \cup V_b$. We connect v_{a_i} to v_{b_j} with an edge e_{ij}, if the vertices (entities) are connected by at least one shared trajectory from the underlying weak tracking data. Unlike [11], we do not perform any adjustments to these entity interactions. Thus, entities are free to interact with splits, merges, or any combination of the two from frame to frame. We define an entity exit event as an entity from V_a that has no edge connecting it to V_b. Likewise, we define an entity entry event as an entity from V_b that is not connected to an entity in V_a. If an entity from one set (V_a or V_b) shares a tracklet with multiple entities from the other set, we consider this to be an "interaction" (e.g. split, merge). Entity tracks between entries, exits, and interactions correspond to the coherent motion regions. The set of coherent motion regions that begin due to an entity entering the scene correspond to likely entry observations. Likewise, the set of coherent motion regions that end due to an entity exiting the scene correspond to likely exit observations. The set of coherent motion regions that begin with an entry observation and end with an exit observation are especially useful, as they correspond to entities that were able to be tracked reliably from the time they entered the scene until they exited. For such cases, we are able to strongly associate their entry and exit observations, however, we do not require such entity tracks. Figure 2 (a) shows the original weak tracking entry observations, and (b) shows the resultant entity entry observations using our framework.

4   Entries and Exits

From our entity detection and tracking framework described above, we accumulate a set of entity entry/exit location observations. We now explain how we learn entry/exit "regions" from these observations, and how we score each region.

We first perform standard mean-shift clustering on our set of entry locations (and then exits) using the same kernel bandwidth as was used in Sect. 3.1. The result is a set of entry (and exit) clusters. We choose mean-shift clustering over an EM approach (as in [3]) for a few reasons. Mean-shift clustering is able to localize on cluster modes automatically, without knowledge of the number of clusters, as would be required with an EM approach. Model selection techniques such as Bayesian Information Criterion (BIC) [12], for an EM approach, may still sometimes suffer from over fitting (as explained in [13]). Further, the mean-shift clusters better represent the shape of non-Gaussian regions.

After clustering the data we attempt to remove outliers in each cluster, and localize on the area of highest density within each cluster. To accomplish this we first determine the convex hull of each cluster. We then area-reduce the convex hull by removing unreliable observations. We compute a density score for the points in each cluster via kernel density estimation [14] using only the points in the cluster. We then remove points one at a time in ascending order of density score, and compute the change in area of the new convex hull after each point is removed (the removal of outlier points will cause a large reduction of convex hull area). Thus, we have a distribution of convex hull area changes (from removing each point). We compute the variance of this distribution (assuming a zero mean), and select observations greater than $\sigma = 1.5$ standard deviations away. Of the cluster points that produced these outlier convex hull area changes, we choose the point with the highest density score, and discard all cluster points with lower density scores. Thus, we have a new set of points which better represent the true mass of the cluster. To generalize the shape of this new region, we compute a kernel density surface using the new set of remaining points, determine the point who is lowest on the density surface, and slice the surface at that density value. The perimeter of the slice outlines the final entry or exit region. Thus, unlike [3], our entry and exit clusters reflect the true spatial density and distribution of their underlying observations (which may not be Gaussian).

4.1   Entry/Exit Zone Reliability

We now describe how we validate our entry and exit regions to distinguish reliable entries and exits from those that are the result of noise or partial scene occlusions. In [3], they compute an entry/exit region density and then label regions with density below an arbitrary threshold as noise. Such an approach will not work well if scene traffic is imbalanced, as entries/exits with low popularity (and thus low density), may be regarded as noise. Also, if a scene is very noisy, this method may also classify noise as being a good entry/exit region. We define a good entry region as one with entity tracks emanating out of it, and a good exit region as one with entity tracks flowing into it. Entry regions whose entry-only tracks (or exit regions whose exit-only tracks) exhibit bidirectional activity are unreliable regions, and may be the result of areas with a high rate of tracking failure, partial scene occlusions, or scene noise (trees or other such movement that the tracker may pick up). Further, for entry regions, other tracks in the scene should not intersect the region in the same emanating direction that defines the region (i.e., the entry region should not be a "through" state). Such a scenario would indicate that another entry region exists behind the current one. The same idea holds for exit regions. Thus, we attempt to capture both the consistency of the entry/exit entity tracks that define each region, as well as the consistency of the interaction between other entity tracks and each entry/exit region.

Using the entry/exit entity tracks that define each region, we learn the distribution of directions that these tracks leave (for entries), and enter (for exits), the region by quantizing the angle that each track intersects the region into one of b bins (we use $b = 8$ in our experiments). This histogram is normalized to provide a probabilistic measure for the directions that entry/exit tracks leave/enter each region. From this distribution r, we compute a directional consistency function $\hat{r}$, which accounts for any symmetry of the entity track distribution for each entry and exit region in the following manner. For a bin i with probability $r(\theta_i)$, every other bin probability $r(\theta_j)$ is subtracted from $r(\theta_i)$, in a weighted manner such that bin angles that are directly opposite of i receive high weight (as they correspond to bi-directional behavior), and bin angles close to i receive lower weight. For a region k,

$$\hat{r}_k(\theta_i) = \frac{\max\left[0, \sum_{j=1}^{b} w_{ij} \cdot (r_k(\theta_i) - r_k(\theta_j))\right]}{\sum_{j=1}^{b} w_{ij} \cdot r_k(\theta_j)} \tag{4}$$

where w_{ij} is an angle similarity weight that give more emphasis to angles corresponding to bidirectional behavior with respect to θ_i, and is computed as

$$w_{ij} = \begin{cases} \exp(1 + \cos(|\theta_i - \theta_j|)) & \text{if } \cos(|\theta_i - \theta_j|) < 0 \\ 0 & \text{otherwise} \end{cases} \tag{5}$$

Thus, θ_j is ignored if it is within 90 degrees of θ_i, and most heavily weighted when it is exactly opposite of θ_i. For a region k, $\sum_{i=1}^{b} \hat{r}_k(\theta_i)$ is 0 when the tracks leaving an entry (or entering an exit) are completely symmetric (and thus strongly bi-directional), reflecting that the region is unreliable, and $\sum_{i=1}^{b} \hat{r}_k(\theta_i)$ is 1 when a region is completely non-symmetric, and thus reliable.

In addition to this directional consistency measure, we also need to capture the way in which other tracks interact with each entry or exit region. We combine the previous directional consistency measure (Eqn. (4)) with an interaction component to compute a total reliability score, ψ, for each region k

$$\psi_k = \left(\sum_{i=1}^{b} \hat{r}_k(\theta_i)\right) \cdot \left(1 - \min\left[1, \frac{\sum_{i=1}^{b} \hat{r}_k(\theta_i) \cdot M_k(\theta_i)}{\sum_{i=1}^{b} \hat{r}_k(\theta_i) \cdot N_k(\theta_i)}\right]\right) \tag{6}$$

Here, N_k is the number of entity tracks that define an entry/exit region k, and $N_k(\theta_i)$ is the number of entity tracks that leave an entry (or enter an exit) region k at angle i. Similarly, M_k is the number of other entity tracks in the scene that intersect an entry/exit region, and $M_k(\theta_i)$ is the number of those entity tracks that intersect and entry/exit region k at angle i. As described earlier, for a reliable region these tracks should not intersect an entry region in the same direction as the entry tracks that leave it (or intersect an exit region in the same direction as its tracks exit). The first term $\sum_{i=1}^{b} \hat{r}_k(\theta_i)$ is the directional consistency term for the entry/exit region and acts as a prior (Eqn. (4)). This score will dominate the total score if $M_k = 0$ (no tracks intersect the region k). If $M_k > 0$, and the intersecting tracks support the region as being reliable, then $\sum_{i=1}^{b} \hat{r}_k(\theta_i) \cdot M_k(\theta_i) \approx 0$. If however there are tracks that intersect the region and contribute evidence that the region may be unreliable (due to tracking failures, the region being partially occluded, etc.), $\frac{\sum_{i=1}^{b} \hat{r}_k(\theta_i) \cdot M_k(\theta_i)}{\sum_{i=1}^{b} \hat{r}_k(\theta_i) \cdot N_k(\theta_i)}$ will approach 1 as the number of discrediting intersecting tracks approaches N_k, and grow large

(a)   (b)   (c)   (d)

Fig. 2. (a) Weak tracking entries, (b) entity entries, (c) potential entry regions, and
(d) entries with reliability score $\Psi > 0.75$. [Best viewed in color]

if the number of tracks $> N_k$, penalizing the region score. If N_k is 0, then we
define the region as unreliable. Lastly, we compute a final region score as

$$\Psi_k = \frac{1}{1 + exp(-\frac{\psi_k - \mu}{\sigma})} \qquad (7)$$

where μ and σ should be determined based on the scene noise, tracking reliability,
etc. For our experiments we used $\mu = 0.5$ and $\sigma = 0.15$. Formulating the final
score as such allows for noise tolerance, and makes the model able to adapt to
scenes with different noise levels. Figure 2 (c) shows entry regions detected for
a scene, and (d) shows the resultant reliable entry regions ($\Psi > 0.75$).

In addition, we also compute a popularity score, $Popularity = N_k$, for an
entry/exit region k (number of entry/exit tracks that define the region).

5 Experiments

We performed experiments to compare our learned entry/exit regions to those
using other approaches. We employ data from four scenes of varying complexity
captured from cameras mounted on four and eight story buildings, at a resolution
of 640×480. The duration of the datasets used were 1 hour (Scene 4), 2 hours
(Scene 1), and 3 hours (Scenes 2 and 3). We also show how our framework may
be used to learn relationships between entries and exits in the scene.

5.1 Entry/Exit Region Evaluation

We compared our results to the methods described in [3] and [5] (as [5] also uses
weak tracking data to determine entries/exits). In [5], they partition the scene
into a grid of states, where each state is defined by a grid cell location and motion
direction. They then map each tracklet to a set of states. They define an entry
and exit weight metric to score each state s_i as $W_E = C_{start} \cdot max(0, 1 - \frac{C_{in}}{C_{start}})$
(entry weight), and $W_X = C_{stop} \cdot max(0, 1 - \frac{C_{out}}{C_{stop}})$ (exit weight). Here, C_{start}
is the number of tracklets that start in state s_i, C_{in} is the number of tracklets
that transition into s_i, C_{stop} is the number of tracklets that stop in state s_i,
and C_{out} is the number of tracklets that transition out of state s_i. Low weight

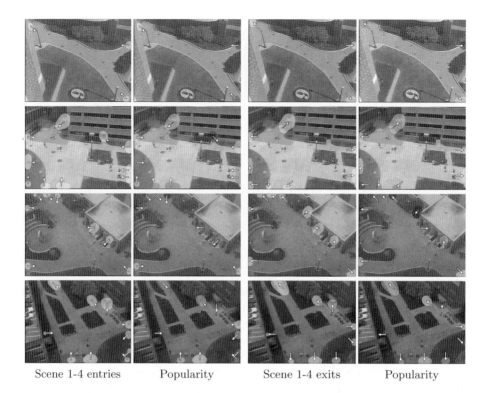

Scene 1-4 entries   Popularity   Scene 1-4 exits   Popularity

Fig. 3. Entry and exit regions (using our proposed method) with a reliability score $\Psi > 0.75$ and popularity > 10 tracks. [Best viewed in color]

entry and exit states can then be removed to obtain a final set of entry and exit states. They also use a binary activity mask to force entry/exit states to be on the border of the scene activity, though we do not employ this technique to allow for a fair comparison (neither our method, nor the method in [3] have such constraints). In [3], they use EM to cluster entry and exit track points. They then use a density metric to determine which clusters to keep, computed as W/E where W is the ratio of points belonging to a cluster, and E is the area of the Gaussian ellipse, calculated as $E = \pi \cdot I_1 \cdot I_2$ where I_1 and I_2 are the eigenvalues of the covariance matrix. We used our *entity* track entry/exit observations as input for this method, as it is intractable and noisy (see Fig. 2 (a)) to learn entries/exits using weak tracking (tracklet) start/stop observations.

For our method, we chose a reliability score threshold of $\Psi > 0.75$ to select a set of final entry/exit states. We also remove any remaining regions with a popularity score less than 10. Results from our method can be seen in Fig. 3. We display those regions along with the strongest track entry/exit angle for each region. We also display entry/exit region popularity by the strength of the entry/exit region color. It is worth noting that we experimented with various kernel bandwidth (h) values (from 10 to 20), and found the results did not vary significantly. For our experiments we used $h = 13$. For the method described in [5], we display states with an entry/exit weight at least 0.75 of the max entry/exit weight. The states are displayed by arrows over a scene grid denoting the state location and direction. For the method described in [3], we clustered the entity entry and exit observations using EM, with a relatively large number of clusters (as they do). We display entry/exit regions with density scores > 0.01 of the of the maximum density score (produced the best results). Results for these two alternative methods can be seen in Fig. 4.

For Scene 1, our method was able to learn all expected entry and exit regions. The methods in [5] and [3] both learn incorrect entries/exits due the streetlight occlusion in this scene. Our method is able to recognize this as a partial scene occlusion due to the motion symmetry and entity tracks intersecting the region. Further, [5] fails to learn the entry/exit on the top right of the scene, and [3] learns various noise clusters which result from the KLT tracker picking up bushes in the scene that move due to wind. The motion of such action is inconsistent, and thus is flagged by our method as being an unreliable region.

For Scene 2 our method was able to learn the desired entry/exit regions with the exception of the region in the top left of the scene. This is due to the region being difficult to track, and as such this region exhibits symmetric behavior. The method in [5] also does not learn this region. Both [5] and [3] again learn entry/exit regions that result from the streetlight occlusion in this scene. Also, [3] also tends to over-cluster many valid regions, and both methods learn noisy regions in the middle of the scene.

In Scene 3, all of the regions that we learn are valid entry/exit regions, though we also learn one exit region corresponding to activity at a picnic table. For this scene, [3] fails to learn the ramp entry and exit on the top left of the scene (due to low traffic) and learns a few noisy entry regions near a bush. The method in [5] also fails to learn the ramp entry and exit on the top left, as well as the exit on the top of the scene and one of the building exits. Further, [5] learns a noisy entry and exit region near the streetlight occlusion.

For Scene 4 our method learns all expected regions with the exception of the top left sidewalk. Again, this is due to this region being difficult to track. The method from [5] also fails to learn this region as an entry, and [3] only learns it as an exit. Further, [3] fails to learn the parking garage entry/exit regions as well as the building entry region, and it also over-clusters the entry near the top of the scene. In addition, [5] fails to learn the building and parking garage entries and exits, as well as the entry near the top of the scene. It also learns various noisy regions in the middle of the scene.

Choosing a robust density threshold proved to be a difficult challenge for [3]. A threshold that works well for one scene may not generalize for other scenes, as seen in the results. Also, neither method enforces entry/exit directional consistency, and thus may learn very noisy entries/exits (method [3] in Scene 1), or regions that result from partial scene occlusion (e.g., streetlight examples in Scenes 1, 2, and 3). Overall, the proposed approach produced the best results.

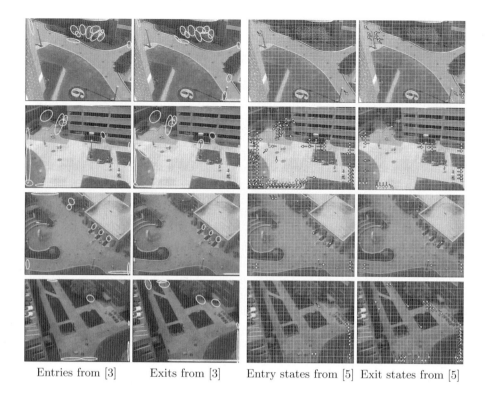

Entries from [3]   Exits from [3]   Entry states from [5]   Exit states from [5]

Fig. 4. Entry and exit regions using the method in [3] (cols 1 and 2). Entry and exit states using the method in [5] (cols 3 and 4). [Best viewed in color]

5.2   Semantic Scene Actions

In addition to using our method to learn scene entries and exits, we can also learn relationships between entries and exits. This is possible using the set of coherent motion regions that begin with an entry and end with an exit observation. This information could be used to see where people go without caring about the actual path they take. The top entry/exit relationships for Scenes 2-4 can be seen in

Scene 2   Scene 3   Scene 4

Fig. 5. Strongest activities from various scenes displayed as arrows from entry regions to exit regions. [Best viewed in color]

Fig. 5. We were able to learn the relationships via entities constructed from weak tracking data using our framework. Such analysis usually requires large amounts of strong tracking data.

6   Conclusion

We proposed a novel framework to learn scene entries and exits using weak tracking data. We described a method that forms weak tracking data into coherent motion regions by detecting entities at a frame level and tracking the entities through time, resulting in a set of reliable spatio-temporal paths. We also introduced a novel scoring metric that uses the activity in the scene to reason about the reliability of each entry/exit region. Results from multiple scenes showed that our proposed method was able to learn a more reliable set of entry/exit regions as compared to existing approaches. This research was supported in part by the US AFRL HE Directorate (WPAFB) under contract No. FA8650-07-D-1220.

References

1. Stauffer, C.: Estimating tracking sources and sinks. In: Proc. Second IEEE Event Mining Workshop (2003)
2. Shi, J., Tomasi, C.: Good features to track. In: CVPR (1994)
3. Makris, D., Ellis, T.: Automatic learning of an activity-based semantic scene model. In: AVSS (2003)
4. Wang, X., Tieu, K., Grimson, E.: Learning semantic scene models by trajectory analysis. In: Leonardis, A., Bischof, H., Pinz, A. (eds.) ECCV 2006. LNCS, vol. 3953, pp. 110–123. Springer, Heidelberg (2006)
5. Streib, K., Davis, J.: Extracting pathlets from weak tracking data. In: AVSS (2010)
6. Rabaud, V., Belongie, S.: Counting crowded moving objects. In: CVPR (2006)
7. Browstow, G., Cipolla, R.: Unsupervised Bayesian detection of independent motion in crowds. In: CVPR (2006)
8. Cheriyadat, A., Bhaduri, B., Radke, R.: Detecting multiple moving objects in crowded environments with coherent motion regions. In: POCV (2008)
9. Cheriyadat, A., Radke, R.: Automatically determining dominant motions in crowded scenes by clustering partial feature trajectories. In: Proc. Int. Conf. on Distributed Smart Cameras (2007)
10. Cheng, Y.: Mean shift, mode seeking, and clustering. PAMI, 17 (1995)
11. Masoud, O., Papanikolopoulos, N.: A novel method for tracking and counting pedestrians in real-time using a single camera. IEEE Trans. on Vehicular Tech. 50 (2001)
12. Schwarz, G.: Estimating the dimension of a model. Ann. of Statist. 6 (1978)
13. Cruz-Ramirez, N., et al.: How good are the Bayesian information criterion and the minimum description length principle for model selection? A Bayesian network analysis. In: Gelbukh, A., Reyes-Garcia, C.A. (eds.) MICAI 2006. LNCS (LNAI), vol. 4293, pp. 494–504. Springer, Heidelberg (2006)
14. Parzen, E.: On estimation of a probability density function and mode. Ann. Math. Statist. 33 (1962)

Adding Facial Actions into 3D Model Search to Analyse Behaviour in an Unconstrained Environment

Angela Caunce, Chris Taylor, and Tim Cootes

Imaging Science and Biomedical Engineering, The University of Manchester, UK

Abstract. We investigate several methods of integrating facial actions into a 3D head model for 2D image search. The model on which the investigation is based has a neutral expression with eyes open, and our modifications enable the model to change expression and close the eyes. We show that the novel approach of using separate identity and action models during search gives better results than a combined-model strategy. This enables monitoring of head and feature movements in difficult real-world video sequences, which show large pose variation, occlusion, and variable lighting within and between frames. This should enable the identification of critical situations such as tiredness and inattention and we demonstrate the potential of our system by linking model parameters to states such as eyes closed and mouth open. We also present evidence that restricting the model parameters to a subspace close to the identity of the subject improves results.

1 Introduction

Face tracking is an important area in car safety but is notoriously difficult in such an unconstrained environment because of the continuously changing lighting, harsh shadows, pose variation, and occlusion, particularly if the camera is mounted behind the steering wheel (Fig. 1). Monitoring the driver's emotional state and attentiveness adds a further level of complexity. We present here investigations into extending a successful face tracking system [1] to include facial actions with a view to analysing behaviour.

In [1], a 3D matching approach was presented that performed at least as well as an established 2D AAM based system [2] on large datasets of near frontal images [3, 4], and surpassed it on large rotations. The sparse 3D shape model, based on the active shape model [5], had a neutral expression with eyes open and mouth closed. Using examples of independent facial actions in the eye and mouth areas, we have constructed a model of facial actions and we show that by integrating this with the method in [1] the model parameters can be linked to states such as the openness of the eyes and mouth. This gives a single system via which pose, identity, behaviour, and attentiveness can be monitored.

Our approach to integrating facial actions varies from that of other authors in several ways. In [6] the authors include expression examples in the training set for a single model capable of facial actions, whereas we have treated them separately. In [7], Amberg et al. used a combined model strategy where the ID was built using PCA on neutral faces and the expression modes were concatenated onto this model having

G. Bebis et al. (Eds.): ISVC 2010, Part I, LNCS 6453, pp. 132–142, 2010.

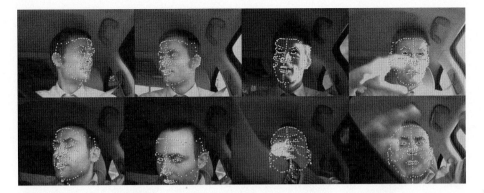

Fig. 1. Some example search results on real in-car videos. These frames illustrate the feature actions which can be handled as well as occlusion, variable lighting, and extreme pose.

been generated by performing PCA on the observed expression deformations. Similarly in [8] authors used the same principle but with a modeling method equivalent to PCA. This assumes that the deformations are transferrable between individuals. We also make this assumption but create a completely independent action model and show that improved results are obtained when this model is kept separate from the ID model. The same notion of transferrable actions is applied in [9] where the model shape is fixed and only parameters controlling pose and 'animation units' are estimated in the search. The 'animation units' are similar to the independent facial actions used in the building of our model except that we build a second model of the same form as the identity model and allow the identity parameters to vary also during search. This means that the model can be applied, without prior knowledge, to any set of still images or video and can be modified, as we will show, to learn the identity of the subject over time.

2   Search Method

The basic search system uses a sparse 3D shape model based on [5] for pose invariance, and continuously updated view-based local patches for illumination compensation. The model has a neutral expression eyes open and mouth closed and was built from a subset of 238 vertices from 923 head meshes (Fig. 2). Each mesh was created from a manual markup of photographs of an individual (Fig. 2). A generic mesh was warped [10] to fit each markup and therefore the vertices are automatically corresponded across the set, unlike other systems where the models are built from range data and correspondence must be approximated [7, 8, 11, 12].

The co-ordinates of the vertices for each example were concatenated into a single vector and Principal Component Analysis was applied to all the point sets to generate a statistical shape model representation of the data. A shape example $\mathbf{x}$ can be represented by the mean shape $\overline{\mathbf{x}}_{\mathbf{ID}}$ plus a linear combination of the principle modes of the data concatenated into a matrix $\mathbf{P}_{\mathbf{ID}}$:

$$\mathbf{x} = \overline{\mathbf{x}}_{ID} + \mathbf{P}_{ID}\mathbf{b}_{ID} \qquad (1)$$

where the coefficients $\mathbf{b}_{ID}$ are the model parameters for shape $\mathbf{x}$. We use the subscript **ID** for identity since this model has a neutral expression.

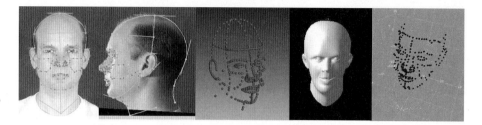

Fig. 2. The model is built using manual markups. Each subject is marked in front and profile to create a 3D point set *(3 left)*. A mesh is then warped to match each subject and a subset of vertices on features of interest is extracted from each mesh *(right)*.

The search is initialized using the Viola-Jones (V-J) face detector [13] on each still image or at the start of a sequence. The 3D shape model is placed within the box adopting its default (mean) shape and either 0° rotation or a fixed angle based on prior knowledge, for example if the camera is located below the head in a car. On each frame after the first in a sequence, the search is initialised with the shape and pose from the previous frame, unless the search fails in which case the model is reinitialised on the same frame using the V-J detector and default attributes.

Fig. 3. The patches are generated from the mean texture *(left)*. The appearance of the patches changes to match the current pose of the model *(right)*.

Each model point has a view-based local patch similar to [14] but continuously updated to reflect the exact model pose at each iteration. The patches are sampled from an average texture generated from 913 subjects (Fig. 3). Variation in the texture is not modelled. The target location for most points is found by comparing the normalised patch to a neighbourhood around each point. However, points currently at approximately 90 degrees to the viewing angle search along a profile for the highest edge strength as their new target.

The search is performed iteratively over several resolutions, low to high, completed at each before moving on to the next. Once each point has a new 2D target location the current z co-ordinate of the model is used to estimate the depth of the target point. This assumes an orthogonal projection. Finally, the shape model is fitted in 3D in a 2 stage process extended from the 2D case [5].

3   Facial Actions Model

It is difficult to obtain corresponded 3D head data for a spectrum of emotions, so we chose to build a more versatile model of facial actions. Only 8 examples were used to build the model: Mouth open; Eyes closed; Smile; Mouth turned down; Brow raised; Brow lowered; Grimace; and neutral (Fig. 4). These were all created by modifying the same neutral head mesh used to generate the examples in the identity model. This meant that they were automatically corresponded with those examples as well as each other. Using a single example of each facial action assumes that they are transferrable between individuals. After construction utilising 99.5% of the variance in the data the model had 5 modes of variation which, by observation, had the primary functions: mouth open; brow raise/lower; smile; grimace; and eyes close (Figure 4).

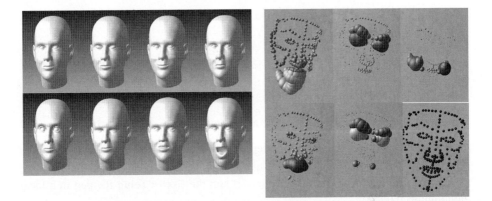

Fig. 4. *(Left)*: The neutral mesh (*top left*) was modified to create facial actions. *(Right)*: The 5 action model modes corresponded broadly to: mouth open; brow raise; smile; grimace; and eyes closed. The size of the point indicates the relative amount of movement the shade is the direction – dark and light points move oppositely. A reference point set is shown *bottom right.*

4   Integration Methods

We investigated two ways to combine these models: either as a single model incorporating identity and facial actions; or as two separate models.

4.1   Combined-Model

To build a single model of the form (1) which combines both aspects into one matrix $\mathbf{P_C}$ it is essential that the columns of $\mathbf{P_C}$ be orthogonal so that the model can be fitted as in [5] and so that any action components in the identity matrix $\mathbf{P_{ID}}$ are removed. To do this the columns of the identity model are orthogonalised ($\mathbf{P_{IDO}}$) with respect to those of the actions model ($\mathbf{P_A}$) using the Gramm-Schmidt method.

$$\mathbf{x} = \overline{\mathbf{x}}_{ID} + \mathbf{P}_C \mathbf{b}_C; \qquad\qquad \mathbf{P}_C = \left[\mathbf{P}_A | \mathbf{P}_{IDo} \right] \qquad\qquad (2)$$

The parameter vector $\mathbf{b}_C$ contains parameters for both action and identity. The mean of the combined model is taken from that for identity.

4.2   Alternating ID and Action

An alternative novel method was devised which uses the two models separately by alternating between them. At each iteration two model fits are made but only one search. First the ID model is used to search the image and then to fit to the target, next the action model is fitted to the same target. However, the mean of the actions model is replaced in the action fit by the current result from the identity fit and then the result from the action fit is incorporated into the identity model mean for the next identity fit (at the next iteration). This is summarised in the following model equations showing the forms of the models used at each iteration K:

$$\mathbf{x}^{K(1)} = \overline{\mathbf{x}}^{K(1)} + \mathbf{P}_{ID} \mathbf{b}_{ID}^{K}; \qquad\qquad \overline{\mathbf{x}}^{K(1)} = \overline{\mathbf{x}}_{ID} + \mathbf{P}_A \mathbf{b}_A^{K-1} \qquad (3)$$

$$\mathbf{x}^{K(2)} = \overline{\mathbf{x}}^{K(2)} + \mathbf{P}_A \mathbf{b}_A^{K}; \qquad\qquad \overline{\mathbf{x}}^{K(2)} = \overline{\mathbf{x}}_{ID} + \mathbf{P}_{ID} \mathbf{b}_{ID}^{K} \qquad (4)$$

Where *K(1)* and *K(2)* refer to the 1st and 2nd fit at each iteration, $\overline{\mathbf{x}}_{ID} + \mathbf{P}_{ID} \mathbf{b}_{ID}^{K}$ is the current identity result and is used as the action model mean, $\overline{\mathbf{x}}_{ID} + \mathbf{P}_A \mathbf{b}_A^{K-1}$ is devised from the current action result and is used as the identity model mean, and $\mathbf{b}_A^0$ is the zero vector. By alternating the models in this way the expression component does not have to be removed from the identity model and the risk of being trapped in a local minimum is reduced.

4.3   Search Regimes

Apart from using the combined model we examined several variations for the 2-step approach. In [1] the authors used only a subset of 155 vertices for search although all 238 were considered during fitting. We examined this case and the case where all points were used to search. We also considered alternative schemes where the actions model was included at all resolutions or only introduced at the highest (last) resolution. Table 1 shows the full list of methods.

5   Driver Monitoring and ID Smoothing

Model search was performed on difficult real-world data displaying extreme poses, occlusion, and variable lighting – both within and between frames (Fig. 1). Since each video was of a single subject we also investigated the possibility that restricting the model parameters around the observed identity of the driver may improve the search.

To do this we used a running mean and variance on the identity parameters **b**:

$$\overline{\mathbf{b}}^t = \frac{\mathbf{b}_r^t}{w_r^t}; \qquad \mathrm{var}^t(b_j) = \frac{v_{jr}^t}{w_r^t} - \overline{b}_j^2 \qquad j = 1\square\ m; \qquad w_r^t = aw_r^{t-1} + (1-a)w^t \qquad (5)$$

$$\mathbf{b}_r^t = a\mathbf{b}_r^{t-1} + (1-a)\mathbf{b}^t w^t; \qquad v_{jr}^t = av_{jr}^{t-1} + (1-a)b_j^{t2}w^t \qquad j = 1\square\ m \qquad (6)$$

where m is the number of identity modes, r indicates running totals and a is the proportion carried forward at each frame, in this case 0.9. The weights w^t are based on the match value of the model at time t, and v_{jr}^t is an intermediate variance calculation. The model parameters were restricted, by truncation, to lie within an ellipsoid about the estimated mean. Table 1 shows a full list of two-step search methods.

Table 1. List of two-step methods with the labels from Figure 8

Method			Actions model at:
1 (ActNIDS)	No ID smoothing	155 points	All resolutions
2 (ActHiNIDS)			Highest resolution
3 (AllPtsNIDS)		238 points	All resolutions
4 (AllPtsActHiNIDS)			Highest resolution
5 (Act)	ID smoothing	155 points	All resolutions
6 (ActHi)			Highest resolution
7 (AllPts)		238 points	All resolutions
8 (AllPtsActHi)			Highest resolution

6   Video Search Results

A qualitative examination of the action model parameters vs the action in the videos indicates that there is a strong link between some parameters and different states of the driver. Figure 5 shows the link between the 'brow' and 'blink' modes and the eye state of the driver.

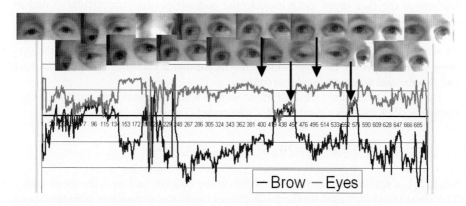

Fig. 5. The action parameters for brow and eyes are closely linked to the eye state in a sequence

In a similar way Fig. 6 shows the relationship between the parameters for mouth open and grimace and the video sequence. As with the eyes open parameter in Fig. 5 the mouth state is not described by the mouth open parameter alone, broadly speaking both parameters have to be low valued for the mouth to be open.

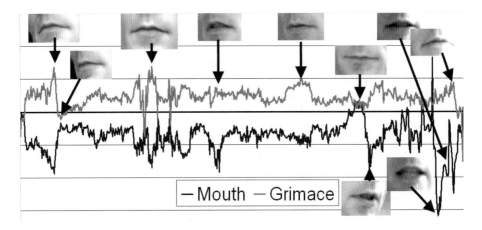

Fig. 6. The relationship between action parameters and mouth state

Figures 5 and 6 indicate that by devising appropriate classifiers the various states of the driver could be quantified. The next sections describe the development of a classifier and a quantitative examination of driver search results.

7   Training the Eye-State Classifier

We used the systems in 4.1 and 4.2 to search a data set of images where the subjects were posed, some with eyes closed, in front of a uniform background. A qualitative examination of the results for the two-step method indicated that it worked better on some individuals than others. In almost every case the model had correctly located the features of the face (eyes, nose, mouth) but had not always successfully closed its eyes. In order to build and test a discriminant vector we used only results where the eye-state was correct. The combined model (section 4.1) only displayed closed eyes on 4 of the relevant images, in general the upper eyelid was placed close to the eye-brow instead, opening the eyes wide. This meant that a classifier could not be built and this method was not considered further.

The model parameters from half of the sample were used to devise a simple 2-class linear classifier for eyes closed. Fisher's formula (7) was used.

$$\hat{w} = S_W^{-1}(m_1 - m_2); \qquad S_W = S_1 + S_2 \tag{7}$$

where S_i is a scatter matrix, S_W is the within groups scatter, and m_i is the mean vector. The classification is based on the projection of the model parameters onto this vector and the centre point was adopted as the threshold:

$$c = \hat{\mathbf{w}} \bullet \left(\frac{\mathbf{m}_1 + \mathbf{m}_2}{2} \right) \qquad (8)$$

The vector from search method 4 was used to vary the parameters of the action model. Fig. 7 shows a representation of this discriminant mode, which illustrates movement consistent with blinking.

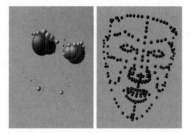

Fig. 7. *(Left)* The discriminant mode for blinking. The size of the point indicates how much it moves relative to the other points and the light and dark colours indicate opposite directions. A reference point set is also shown *(right)*.

The discriminant vector was then used to classify the other half of the data. The results for all methods (1-4) were similar with the average percentage of correct classifications at 94.6%, false positives at 5.1% and false negatives at 0.3%. The number of false positives may be reduced by moving away from the simple centrepoint threshold but this was not investigated on this data.

From these results it was not possible to decide conclusively on the best two-step method. To do this the training and test data were pooled and used to build a discriminant vector for each method (1-4) and these were used on real world data as described in the next section.

8 Detecting Closed Eyes

A 2000 frame video, with 1753 frames where the driver was visible, was labelled so that the frames where the driver's eyes were closed were so marked. The parameter values for each frame were projected onto the discriminant vectors and the ROC curve was plotted for various methods (Fig. 8). This graph shows that, in general: methods where the actions model is introduced at the highest resolution (darker lines) perform better; using all points is better than a reduced set; and ID smoothing also gives an improvement. Method 8, which does all these things, produces the best results. A good compromise between sensitivity and false alarm rate is achieved where the true positive rate (TPR) is 80% and the false positive rate is 10%. When the false alarm frames were examined two things were noted: many of the false alarms were at extreme pose angles; and many were neighbouring the true positive frames, in other words they were an extension of the 'blink'. Preliminary investigations show that by defining contiguous frames as a blink it is possible to increase the true positive rate to 90% and reduce the number of false alarms by 69%. Authors sometimes report on blinks and sometimes on frames. Our results do not match the performance quoted in [15] on frames, although the experimental protocol is unclear which makes comparison difficult, but they surpass those in [16] on similar in-car data reporting on blinks. Other authors have reported better 'blink' results but in controlled situations such as office or simulator environments [17, 18]. Our results are also comparable to those obtained using the latest texture based discriminative approach in [19] where the 'Eyes Open' 'accuracy' was quoted as 92.52% on the PubFig database. If we define accuracy as *(True Positives + True negatives)/Total* we obtain 89% but the in-car video includes full-frame images with harsh shadows and large amounts of lighting, occlusion, and pose variation not generally found in images of public figures. The advantages of this technique over such an approach are that there is no pose normalisation step [19], or need for multiple pose-related classifiers, since our model works under rotation and can thus report at large deviations from front, whilst providing the pose information itself.

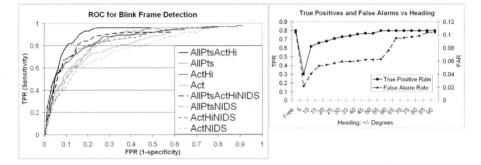

Fig. 8. (Left): ROC curves based on thresholds for blink frames. ActHi indicates the action model was introduced at the highest resolution only, otherwise at all resolutions. AllPts indicates that 238 points were used to search otherwise 155. NIDS indicates that there was no ID smoothing. (Right): Blink frame true positive and false alarm rates vs estimated headings.

We also looked at the effect of heading on sensitivity and false alarm rate. We included only the detections where the driver's heading was estimated to be between a range of angles and found that at 55 degrees the TPR for frames was at its maximum (80%) but the False Positives were reduced at around 6% (Fig. 8). Since the driver is likely to be active, although possibly distracted, at larger poses than this, it is more crucial to correctly monitor eye state at reduced extremes. These results indicate that not only is the number of false positives more acceptable in the frontal range but also that correct detections are still being made even at angles of over 50 degrees. In [20], although closed eye detection rates were higher, the authors acknowledged that no examples were rotated through more than 30 degrees.

9   Discussion

We have shown that a 3D model built from independent facial action examples can be integrated with an identity model in a novel two-step strategy to monitor the state of a subject with a view to analyzing behaviour. Our investigations have shown that a simple linear discriminant vector, based on the action model parameters, can quantify eye state, which is crucial in a driver monitoring application. Our results are comparable to those obtained using other techniques including texture based discriminative approaches. Evidence indicates that the results can be improved in a video sequence by restricting the identity model parameters to a subspace defined by the observed mean and variance for that subject. Indications are that the parameters of the action model are related to other states such as mouth open and future work will investigate the possibility of a more comprehensive analysis of the driver using this technique.

The results presented in this paper were obtained using a non-optimised debug system taking between 1 and 2 seconds to process a single image. We now have a modified and optimized version of the system which can search live webcam images at up to 30 FPS on a quad processor machine.

By combining the discriminant values and other information, such as pose and feedback from controls, it should be possible to monitor a driver, or other subject, to forestall critical situations. Future work will include developing techniques to utilize these parameters along with temporal information to identify and predict micro and macro states.

Acknowledgements

This project is funded by Toyota Motor Europe who provided the driver videos. We would like to thank Genemation Ltd. for the 3D data markups and head textures.

References

1. Caunce, A., Cristinacce, D., Taylor, C., Cootes, T.: Locating Facial Features and Pose Estimation Using a 3D Shape Model. In: International Symposium on Visual Computing, Las Vegas, pp. 750–761 (2009)
2. Cristinacce, D., Cootes, T.: Automatic feature localisation with constrained local models. Pattern Recognition 41, 3054–3067 (2007)
3. Jesorsky, O., Kirchberg, K.J., Frischholz, R.W.: Robust Face Detection Using the Hausdorff Distance. In: International Conference on Audio- and Video-based Biometric Authentication, Halmstaad, Sweden, pp. 90–95 (2001)
4. Messer, K., Matas, J., Kittler, J., Jonsson, K.: XM2VTSDB: The Extended M2VTS Database. In: International Conference on Audio- and Video-based Biometric Person Authentication, Washington DC, USA (1999)
5. Cootes, T.F., Cooper, D.H., Taylor, C.J., Graham, J.: Active Shape Models - Their Training and Application. Computer Vision and Image Understanding 61, 38–59 (1995)
6. Zhang, W., Wang, Q., Tang, X.: Real Time Feature Based 3-D Deformable Face Tracking. In: European Conference on Computer Vision, vol. 2, pp. 720–732 (2008)
7. Amberg, B., Knothe, R., Vetter, T.: Expression Invariant 3D Face Recognition with a Morphable Model. In: International Conference on Automatic Face Gesture Recognition, Amsterdam, pp. 1–6 (2008)
8. Basso, C., Vetter, T.: Registration of Expressions Data Using a 3D Morphable Model. Journal of Multimedia 1, 37–45 (2006)

9. Dornaika, F., Orozco, J.: Real time 3D face and facial feature tracking. Real-Time Image Processing 2, 35–44 (2007)
10. Bookstein, F.L.: Principal Warps: Thin-Plate Splines and the Decomposition of Deformations. IEEE Transactions on Pattern Analysis and Machine Intelligence 11, 567–585 (1989)
11. Blanz, V., Vetter, T.: A Morphable Model for the Synthesis of 3D Faces. In: SIGGRAPH, pp. 187–194 (1999)
12. Ishiyama, R., Sakamoto, S.: Fast and Accurate Facial Pose Estimation by Aligning a 3D Appearance Model. In: International Conference on Pattern Recognition, vol. 4, pp. 388–391 (2004)
13. Viola, P., Jones, M.J.: Robust Real-Time Face Detection. International Journal of Computer Vision 57, 137–154 (2004)
14. Gu, L., Kanade, T.: 3D Alignment of Face in a Single Image. In: International Conference on Computer Vision and Pattern Recognition, New York, vol. 1, pp. 1305–1312 (2006)
15. Flores, M.J., Armingol, J.M., Escalera, A.d.l.: Real-Time Drowsiness Detection System for an Intelligent Vehicle. In: IEEE Intelligent Vehicles Symposium, Netherlands, pp. 637–642 (2008)
16. Wu, J., Trivedi, M.M.: Simultaneous Eye Tracking and Blink Detection with Interactive Particle Filters. In: Advances in Signal Processing, pp. 1–17 (2008)
17. Chau, M., Betke, M.: Real Time Eye Tracking and Blink Detection with USB Cameras. Boston University - Computer Science Technical Report 2005-12, 2005-2012 (2005)
18. Lalonde, M., Byrns, D., Gagnon, L., Teasdale, N., Laurendeau, D.: Real-time eye blink detection with GPU-bases SIFT tracking. In: IEEE Conference on Computer and Robot Vision, pp. 481–487 (2007)
19. Kumar, N., Berg, A.C., Belhumeur, P.N., Nayar, S.K.: Attribute and Simile Classifiers for Face Verification. In: IEEE International Conference on Computer Vision (2009)
20. Bartlett, M.S., Littlewort, G.C., Frank, M.G., Lainscsek, C., Fasel, I.R., Movellan, J.R.: Automatic Recognition of Facial Actions in Spontaneous Expressions. Journal of Multimedia 1, 22–35 (2006)

Aggregating Low-Level Features for Human Action Recognition

Kyle Parrigan and Richard Souvenir

Department of Computer Science, University of North Carolina at Charlotte
{kparriga,souvenir}@uncc.edu

Abstract. Recent methods for human action recognition have been effective using increasingly complex, computationally-intensive models and algorithms. There has been growing interest in automated video analysis techniques which can be deployed onto resource-constrained distributed smart camera networks. In this paper, we introduce a multi-stage method for recognizing human actions (e.g., kicking, sitting, waving) that uses the motion patterns of easy-to-compute, low-level image features. Our method is designed for use on resource-constrained devices and can be optimized for real-time performance. In single-view and multi-view experiments, our method achieves 78% and 84% accuracy, respectively, on a publicly available data set.

1   Introduction

The problem of human action recognition from video is fundamental to many domains, including automated surveillance and semantic searching of video collections. As computers have become increasingly powerful, proposed solutions have grown equally in sophistication by introducing complex underlying models. For a traditional system with a camera (or set of cameras) connected to a powerful server, the complexity of the model and processing required for model building and analysis is, generally speaking, no longer an issue. However, in the emerging area of distributed smart cameras, where individual nodes have limited computing resources, it is important to use models which are efficient in computing time and space.

In this paper, we present a multi-stage method for human action recognition based on aggregating low-level image features. Our approach was motivated by observing output from marker-based motion capture. Even without a skeletal overlay, it is usually apparent what action is being performed just by observing the movement of the markers. Figure 1 shows four keyframes of a video obtained by calculating Harris corners in each frame. Compared to the output from motion capture, there are two noticeable differences. First, not all of the points discovered are related to the action of interest. Second, the points are not stable; a feature in one frame may disappear in subsequent frames. In spite of these issues, our goal is to explore whether or not these sets of low-level features can be used as the basis for discriminative human action recognition.

G. Bebis et al. (Eds.): ISVC 2010, Part I, LNCS 6453, pp. 143–152, 2010.

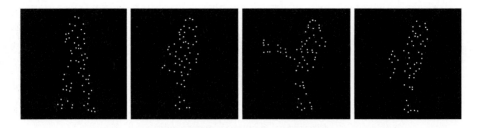

Fig. 1. Keyframes of low-level feature points of someone kicking

The primary contribution of this work is the development of a multi-stage method using low-level features for human action recognition. For each stage, we aim to make design decisions which minimize processing and storage, so that this method can be deployed on networks of resource-constrained distributed smart cameras. We show that this approach provides good performance on the publicly-available IXMAS data set [1].

2   Related Work

The literature on human motion analysis and action recognition is vast; see [2,3] for recent surveys. There is a large diversity of approaches – from the well-known temporal templates of Bobick and Davis [4] which model actions as images that encode the spatial and temporal extent of visual flow in a scene to approaches which try to fit high dimensional human body models to video frames (e.g., [5]).

Most related to our approach are methods based on using sets of local features (e.g., [6]). Some approaches use optical flow vectors as the basic feature for action recognition [7,8]. Gilbert et al. [9] introduce a method which learns compound features using spatio-temporal corner points. Laptev and Lindeberg [10] use optic flow and Gaussian derivatives to create several different local descriptors. The work of Mori and Malik [11] is similar to our hierarchical approach in that higher-level features are built from edge points. However, this method uses computationally-expensive methods for aggregation and matching.

3   Method

Our hierarchical approach to action recognition starts with low-level features computed directly from the video frames. We classify each feature based on the low-level detector used, and we extend recent methods in 2D image classification to our work in action recognition.

3.1   Low-Level Image Features

Our choice of feature representation is based on the premise that the motion of some set of informative points (even in the presence of noise) is sufficient for human action recognition. Finding points of interest in images and video is a fundamental problem in image processing and computer vision; there is much work on interest point detection [12,13].

- Algebraic Corners [14]: Corner points are defined by the intersection of two linear contours detected by fitting edge points to hyperbolas.
- Canny Edges [15]: These features are edge points that remain after a multiphase filtering and contour extension process.
- Fast Radial Interest Points [16]: Interest points are obtained using image gradients to find areas of high radial symmetry.
- Harris Corners [17]: Corner points are defined as edge points with large variations in two orthogonal directions.
- Scale-Space Extremal Points: This is the interest operator of the well-known SIFT [18] feature descriptor. These are points of locally extreme values after a difference of Gaussians filter in scale space.

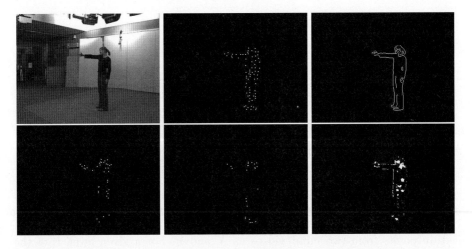

Fig. 2. Comparison of interest point operators. (a) shows a sample frame from a video of a person pointing. The next five images show the feature points selected by: (b) Harris corner detection, (c) Canny edge detection, (d) the scale-space extremal point interest operator of SIFT, (e) Algebraic corner detection, and (f) Fast Radial detection.

Figure 2 shows a sample frame from a video of a person pointing and the output of the 5 interest point operators (in order from left-to-right): Harris corner detection, Canny edges, scale-space extremal points used by the SIFT keypoint detector, Fast Radial interest points, and Algebraic Corners. The set of points differed based on the interest operator used. Empirically, we observed that, for most actions, it was generally possible to determine the action taking place, regardless of the interest operator. In Section 4, we show action recognition results for various combinations of these features.

3.2   Modeling Feature Point Sets

For each input video, we now have a 3D point cloud of features. To compactly model these feature sets, we take the approach of recent methods from 2D image classification (e.g., [19,20]). We extend the approach of Lazebnik et al. [20] for use with 3D (x, y, t) data in action recognition.

The method in [20] models an image, represented as a set of quantized image features, as a hierarchical pyramid of histograms. At each level, l, where $l = 0, \dots, L$ for some user-provided number of levels, L, an image is successively divided (in half in each dimension) into a series of increasingly finer grid boxes. For each grid box, a histogram of the distribution of features is maintained. Comparing two images then becomes a series of histogram intersection operations for corresponding bins in the pyramid. Intuitively, images which are closely related should have more matches at the smaller-sized bins of the pyramid. We extend this process for 3D data of video.

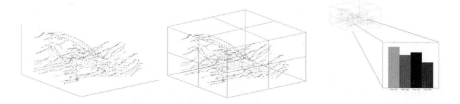

Fig. 3. Illustration of computing spatial pyramids. This figure follows the progression from raw point cloud, (a) to bin partitioning (b) and histogram construction (c).

Figure 3 illustrates the process for converting a 3D point cloud of quantized features into a histogram pyramid. Figure 3(a) shows the input to the process. Using the notation of [20], we represent the set of 3D quantized (with C classes) feature points from two input videos as X and Y. For a given level, l, of the pyramid, the video volume is partitioned into $N = 2^{3l}$ bins. Figure 3(b) illustrates this partition at the first level. The number of feature points from X and Y located in any given bin i is represented by the histograms $H_X^l(i)$ and $H_Y^l(i)$ respectively, as shown in Figure 3(c) for a single bin. The *histogram intersection function* [21] defines the number of matches between X and Y at a level l as:

$$\mathcal{I}(H_X^l, H_Y^l) = \sum_{i=1}^{2^{3l}} \min\left(H_X^l(i), H_Y^l(i)\right). \tag{1}$$

Levels of finer resolution are weighted more heavily, so the result of the histogram intersection is scaled by $\frac{1}{2^{L-l}}$. This leads to the *pyramid match kernel* [19]:

$$\kappa^L(X, Y) = \frac{1}{2^L}\mathcal{I}^0 + \sum_{l=1}^{L} \frac{1}{2^{L-l+1}}\mathcal{I}^l \tag{2}$$

where the histogram intersection function (Equation 1) at level l is written as $\mathcal{I}^l$ for convenience. Finally, we perform this operation separately, for each of the C feature types. The sub-vectors of X and Y for points in class c (where $c = 1, \ldots, C$) are denoted by as X_c and Y_c. The final kernel is then the sum of the pyramid match kernels for each class:

$$K^L(X, Y) = \sum_{c=1}^{C} \kappa^L(X_c, Y_c). \tag{3}$$

Figure 4 illustrates the histogram pyramid matching process for two input videos after they are transformed to the pyramid representations.

This approach of modeling feature sets using spatial pyramids has benefits both in terms of efficiency and for use in discriminative action classification. First, the pyramids can be efficiently constructed, in a manner similar to building a kd-tree. Second, it has been shown [19] that the method of computing pyramid similarity using histogram intersection (Equation 3) results in a Mercer kernel. This allows us to use this similarity method with powerful (yet computationally

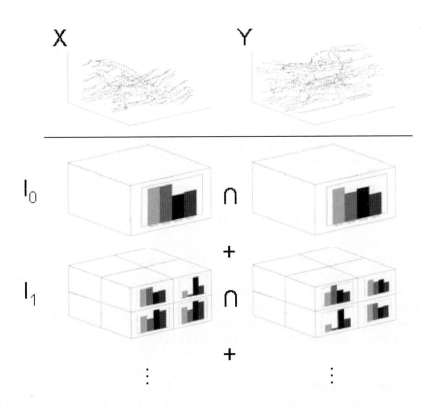

Fig. 4. Histogram Pyramid Matching. This figure illustrates the histogram pyramid matching process for two input videos after they are transformed to the pyramid representations.

efficient in the evaluation phase) kernel-based classification methods, such as support vector machines [22].

4   Experiments

For the results in this section, we used the Inria XMAS Motion Acquisition Sequences (IXMAS) dataset [1] of multiple actors performing a variety of common actions (e.g., waving, punching, sitting). This data was collected by 5 calibrated, synchronized cameras. We tested our approach by performing both single- and multi-view experiments.

For each video in our training set, we calculate feature points using the various feature transforms described in   3.1. Table 1 lists the implementation details and parameter settings used for each method. To remove noise points from the training data, we filter out any points outside the silhouette of the actor. Finally, we create spatial pyramids (Section 3.2) experimenting with various numbers of levels for the pyramids. We account for scale- and duration-invariance by limiting the spatial extent for the spatial pyramids to a bounding box enclosing the calculated regions of motion, rather than original size of the video.

Table 1. Implementation details for each feature detection method

Feature	Implementation	Parameter Settings
Algebraic Corners	Authors'[14] code	Curvature:1E-4, Fit Err:2, Window Sz:6 Edge Dist.:0.3, Corner Angle:0.05
Canny Edges	Matlab edge function	method:'canny'
Fast Radials	Peter Kovesi [23]	Radii:[1 3 5], Alpha:2
Harris Corners	Peter Kovesi [23]	Sigma:1, Thresh:10, Radius:2
SIFT	VL_Feat Library [24]	Default Settings

4.1   Parameter Estimation and Feature Selection

The two major free parameters to our system are the size of the histogram pyramids and the combination of low-level features used. The number of levels, L, in the histogram pyramids affects the dimensionality of the pyramids constructed and has an impact on the balance of efficiency and accuracy. The combination of features had a slight impact on efficiency, but mainly affects the overall accuracy as some combinations outperform others. We empirically tested various values of L and combinations of features using the IXMAS data.

Number of pyramid levels. The number of histogram bins grows exponentially with the number of the number of levels L. To determine the number of levels that provide the best trade-off between efficiency and accuracy, we ran experiments for increasing values of L using a variety of low-level features and action examples from the IXMAS data set. For, $L = 1, \ldots, 4$, the classification accuracies were 8.7%, 49.5%, 74.5%, and 76% for an experiment using Canny edges,

Harris corners, and Fast Radials. The diminishing returns on accuracy suggests that using $L > 4$ would not give us enough improvement to warrant the significant degradation in time efficiency.

Feature Selection. To determine the optimal combination of features, we performed several experiments using various combinations of features. Using 10 of the actions in the data set and 7 of the actors as training data, we measured the classification accuracy on the data from 5 different actors. We tested each feature individually. As shown in Table 2, Harris Corners and Canny Edges achieved the highest accuracies. We then tested a variety of combinations of features. The combination of Canny Edges, Harris Corners, and Fast Radials was the most accurate in our preliminary experiments, and is used in the full classification task.

Table 2. Classification accuracy for individual and combinations of features

Feature	Accuracy	Feature(s)	Accuracy
Algebraic Corners(AC)	70%	AC/CE/HC	73.5%
Canny Edges(CE)	71.5%	CE/HC/FR	75%
Fast Radials(FR)	68.5%	AC/CE/FR/HC	74.5%
Harris Corners(HC)	72.5%	CE/FR/HC/SIFT	73.5%
SIFT	54%	CE/FR/HC/SIFT	73%

4.2 IXMAS Data Results

To test our approach, we used the Inria IXMAS data set. We selected 10 of the actions (checking watch, crossing arms, scratching head, sitting, standing, waving, punching, kicking, walking, and picking up) and split the 36 actors into training and testing sets of 18 actors each.

For classification, we use the Matlab support vector implementation and the histogram pyramid match kernel (Equation 3). We employ a *one-versus-one* scheme for multi-class classification where, for k action classes, $\binom{k}{2}$ pairwise binary classifiers are constructed. To classify a new action, each classifier makes a prediction, and the class with the highest total is selected.

For the videos in our testing set, we did not assume that robust foreground-background segmentation would be available. However, using an online motion localization method, similar to [25], we could filter noise points distant from the action of interest. Once we found the region of motion in the space-time volume, we use the maximal extents in the x-, y-, and t- directions to delineate a bounding box similar to the method for the training set. No additional filtering of the distractor points that were located within these bounds was performed.

Our first experiment tests the single-view classification accuracy of our approach. Figure 5 shows the resulting confusion matrix. Each row represents the instances of the predicted class and each column represents the instances of the actual class. For each class, the true positive rate is shown along the diagonal of the matrix with the color of each cell also representing the value (warmer colors represent higher numbers.) For each action, our method predicted the correct class a majority of the time and for certain actions, the accuracy was above 90%.

Next, to assess the use of this method in a distributed smart camera network, we performed a multi-camera simulation. Each action was captured by 4 cameras placed around an actor. Similar to the previous experiment, single-view action

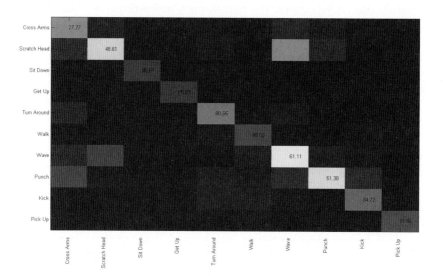

Fig. 5. Confusion Matrix for single-view experiment

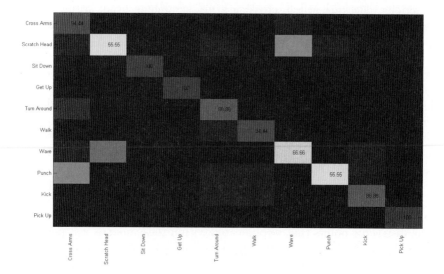

Fig. 6. Confusion Matrix for multi-camera simulation

recognition was performed on video from each camera. Then, the results of the 4 cameras were aggregated by simply selecting the mode (i.e., each camera "voted" for an action and the majority action was selected). The results are shown in Figure 6. We achieved 100% accuracy on 3 of the 10 actions, and the accuracy of each class increased compared to the single-view version, sometimes quite significantly. The overall accuracy using this simple aggregation scheme was 84%, which is competitive with the 93% and 95% accuracy rates on this data reported in [1] and [26], respectively. However, those multi-view methods both use computationally-intensive 3D integration models.

5   Conclusion and Future Work

In this paper, we presented a method for efficiently aggregating low-level features for discriminative human action recognition. Our long term plans are to deploy this method onto a network of resource-constrained distributed smart cameras. In order to achieve this, we need to address issues of robust motion localization, view-independence, and more robust multi-view aggregation. The results of this preliminary work are promising in that very high recognition rates can be achieved using these low-level features and a simple aggregation approach.

References

1. Weinland, D., Ronfard, R., Boyer, E.: Free viewpoint action recognition using motion history volumes. Comput. Vis. Image Underst. 104, 249–257 (2006)
2. Moeslund, T.B., Hilton, A., Krüger, V.: A survey of advances in vision-based human motion capture and analysis. Comput. Vis. Image Underst. 104, 90–126 (2006)
3. Wang, L., Hu, W., Tan, T.: Recent developments in human motion analysis. Pattern Recognition 36, 585–601 (2003)
4. Davis, J.W., Bobick, A.F.: The representation and recognition of human movement using temporal templates. In: Proc. IEEE Conference on Computer Vision and Pattern Recognition, pp. 928–934 (1997)
5. Ramanan, D., Forsyth, D., Zisserman, A.: Strike a pose: tracking people by finding stylized poses. In: IEEE Computer Society Conference on Computer Vision and Pattern Recognition, CVPR 2005, vol. 1, pp. 271–278 (2005)
6. Oikonomopoulos, A., Patras, I., Pantic, M.: Spatiotemporal salient points for visual recognition of human actions. IEEE Transactions on Systems, Man, and Cybernetics, Part B: Cybernetics 36, 710–719 (2005)
7. Fathi, A., Mori, G.: Action recognition by learning mid-level motion features. In: IEEE Conference on Computer Vision and Pattern Recognition, CVPR 2008, pp. 1–8 (June 2008)
8. Efros, A.A., Berg, A.C., Mori, G., Malik, J.: Recognizing action at a distance. In: Proceedings of the Ninth IEEE International Conference on Computer Vision, ICCV 2003, Washington, DC, USA, p. 726. IEEE Computer Society, Los Alamitos (2003)
9. Gilbert, A., Illingworth, J., Bowden, R.: Scale invariant action recognition using compound features mined from dense spatio-temporal corners. In: Forsyth, D., Torr, P., Zisserman, A. (eds.) ECCV 2008, Part I. LNCS, vol. 5302, pp. 222–233. Springer, Heidelberg (2008)

10. Laptev, I., Lindeberg, T.: Local descriptors for spatio-temporal recognition. In: MacLean, W.J. (ed.) SCVMA 2004. LNCS, vol. 3667, pp. 91–103. Springer, Heidelberg (2006)
11. Mori, G., Malik, J.: Recovering 3d human body configurations using shape contexts. IEEE Transactions on Pattern Analysis and Machine Intelligence 28, 1052–1062 (2006)
12. Schmid, C., Mohr, R., Bauckhage, C.: Evaluation of interest point detectors. International Journal of Computer Vision 37, 151–172 (2000)
13. Moreels, P., Perona, P.: Evaluation of features detectors and descriptors based on 3d objects. International Journal of Computer Vision 73, 263–284 (2007)
14. Willis, A., Sui, Y.: An algebraic model for fast corner detection. In: 2009, IEEE 12th International Conference on Computer Vision, pp. 2296–2302 (2009)
15. Canny, J.: A computational approach to edge detection. IEEE Trans. Pattern Anal. Mach. Intell. 8, 679–698 (1986)
16. Loy, G., Zelinsky, A.: Fast radial symmetry for detecting points of interest. IEEE Transactions on Pattern Analysis and Machine Intelligence 25, 959–973 (2003)
17. Harris, C., Stephens, M.: A combined corner and edge detection. In: Proceedings of the Fourth Alvey Vision Conference, pp. 147–151 (1988)
18. Lowe, D.G.: Distinctive image features from scale-invariant keypoints. Int. J. Comput. Vision 60, 91–110 (2004)
19. Grauman, K., Darrell, T.: The Pyramid Match Kernel: Efficient Learning with Sets of Features. Journal of Machine Learning Research (JMLR) 8, 725–760 (2007)
20. Lazebnik, S., Schmid, C., Ponce, J.: Beyond bags of features: Spatial pyramid matching for recognizing natural scene categories. In: 2006 IEEE Computer Society Conference on Computer Vision and Pattern Recognition, vol. 2, pp. 2169–2178 (2006)
21. Swain, M.J., Ballard, D.H.: Color indexing. Int. J. Comput. Vision 7, 11–32 (1991)
22. Cortes, C., Vapnik, V.: Support-vector networks. In: Machine Learning, pp. 273–297 (1995)
23. Kovesi, P.D.: MATLAB and Octave functions for computer vision and image processing. School of Computer Science & Software Engineering, The University of Western Australia (2000),
http://www.csse.uwa.edu.au/~pk/research/matlabfns/
24. Vedaldi, A., Fulkerson, B.: VLFeat: An open and portable library of computer vision algorithms (2008), http://www.vlfeat.org
25. Kulkarni, K., Cherla, S., Kale, A., Ramasubramanian, V.: A framework for indexing human actions in video. In: The 1st International Workshop on Machine Learning for Vision-based Motion Analysis, MLVMA 2008 (2008)
26. Souvenir, R., Parrigan, K.: Viewpoint manifolds for action recognition. EURASIP Journal on Image and Video Processing 2009, 13 pages (2009)

Incorporating Social Entropy for Crowd Behavior Detection Using SVM

Saira Saleem Pathan, Ayoub Al-Hamadi, and Bernd Michaelis

Institute for Electronics, Signal Processing and Communications (IESK)
Otto-von-Guericke-University Magdeburg, Germany
{Saira.Pathan,Ayoub.Al-Hamadi}@ovgu.de

Abstract. Crowd behavior analysis is a challenging task for computer vision. In this paper, we present a novel approach for crowd behavior analysis and anomaly detection in coherent and incoherent crowded scenes. Two main aspects describe the novelty of the proposed approach: first, modeling the observed flow field in each non-overlapping block through social entropy to measure the concerning uncertainty of underlying field. Each block serves as an independent social system and social entropy determine the optimality criteria. The resulted in distributions of the flow field in respective blocks are accumulated statistically and the flow feature vectors are computed. Second, Support Vector Machines are used to train and classify the flow feature vectors as normal and abnormal. Experiments are conducted on two benchmark datasets PETS 2009 and University of Minnesota to characterize the specific and overall behaviors of crowded scenes. Our experiments show promising results with 95.6% recognition rate for both the normal and abnormal behavior in coherent and incoherent crowded scenes. Additionally, the similar method is tested using flow feature vectors without incorporating social entropy for comparative analysis and the detection results indicate the dominating performance of the proposed approach.

1 Introduction

Due to the high growth in population, places such as trains stations, airports, and subways became highly congested, whereas in social, religious and political events, mass gathering of people are observed. Consequently, the main challenge is human safety and to avoid catastrophic situations [1]. For computer vision community, crowd behavior analysis is an attractive research area with challenging issues due to high object densities, self-evolving mechanism and non-uniform dynamics of the crowd.

The conventional approaches in computer vision seem to work for low density scenes aiming to capture subject-based activities, for example, abandon baggage detection [2] which are likely to fail in crowded scenes. The objective of crowd behavior analysis is very diversified (i.e. sparse [3] crowd density analysis to coarse level analysis [4] [5]). A general overview and an insight from the perspective of different commodities in crowd analysis is given by Zhan et al. [6]. Various

G. Bebis et al. (Eds.): ISVC 2010, Part I, LNCS 6453, pp. 153–162, 2010.

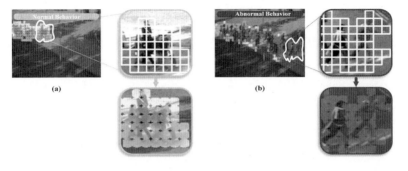

Fig. 1. (a). An example scene showing normal behavior whereas the detection results(bottom) are marked with yellow circles. (b). Shows a sample frame for abnormal event and the red circles depict the abnormal behavior.

approaches are proposed for modeling the recurrent characteristic of crowd dynamics, which can be categorized into two types namely model based approaches and event based approaches.

In the model based approaches, an earlier work is proposed by Andrade et al. [7] to model the eccentric states of individual's interactions using ergodic HMM for constrained environments. Mehran et al. [8] estimate the social force on motion flow based advected particles and simulate the normal interaction forces on these particles, implicitly. Anomalous situations are detected when the particle interaction forces are deviated from the pre-trained model. Kratz and Nishino [9] used Hidden Markov Model (HMM) to detect abnormal events in crowded scenes, but useful information about the motion patterns is lost in their proposed cuboid based windowing strategy because the coherent meaningful features are separated in different cuboids. One notable point is that development of such models require a list of the behavior patterns to indicate most likely situations. In the event based approaches, some earlier works [10][11] attempted to detect and track the objects across the intervals to determine the crowd activities. Albio et al. [12] addresses the crowd density and the abnormality detection problem by computing the statistical characteristics of corner points. Similar problem is addressed by [13] in which a probabilistic model is computed using orientation and density of flow patterns. Recently, Wu et al. [3] capture the dynamics of subjects in high density crowded scenes that are taken from a distance. The captured flow field is used to advect and track the particles overlaid onto frame. However, the resulting tracks are highly inconsistent and discriminating between usual and unusual events is extremely difficult.

Throughout the literature, the anomaly detection is treated as a context-sensitive term which merely rely on instantaneous motion features. Consequently, capturing the certain motion properties in situations where the crowded scene contains any number of concurrent and sparse human activities are extremely difficult. Therefore observed flow field tends to result in uncertain information and lead to the plausible outcome. For example, in coherent crowds (i.e. marathon) the object may move with common dynamics which is relatively easy to model.

But many scenes (i.e. shopping centers) contain completely random motion of objects resulting in a complicated dynamics and difficult to model the overall dynamics. However, it is found that for an appropriate modeling of scenes, the reliable flow patterns can play an important basis for supporting effective detection of anomalies. Unfortunately, none of these approaches [9] [8] [3] considered uncertainty in the observed optical flow and overlooked the limitations of optical flow techniques [14].

In this paper, we construct a novel top-to-down framework for analyzing the crowd behavior by incorporating social entropy. The effectiveness of the proposed approch is demonstrated by identifying the specific and overall behaviors using Support Vector Machines (SVM) as shown in Fig. 1. The main contributions are: first we obtain the reliable flow field through social entropy and compute the flow feature vectors by statistically aggregating the underlying reliable flow field in corresponding blocks; second, we are able to localize the specific and overall behaviors with SVM in coherent and incoherent crowded scenes. The paper is structured as: section 2 presents the proposed approach. Experimental results and discussions are presented in section 3. Finally, the concluding remarks are sketched in section 4.

2   Proposed Approach

In our proposed approach, we give a generic formulation and its concept to deplete the flow field uncertainties which faithfully reveals the characteristics of crowd dynamics by incorporating social entropy. The concept social entropy is originated from the field of social sciences and used in Social Entropy Theory [15]. Social entropy empirically determines a quantitative metric to emulate the system's information with a high degree of accuracy. The reliable flow field inside blocks are modeled statically resulting in unique feature vectors, whereas SVM is used to characterize the governing crowd dynamics as normal or abnormal behavior. Fig. 2 shows the various stages followed in the design of the proposed framework.

2.1   Pre-processing

In pre-processing, we build the initial background model which is generated by using Gaussian Mixture Model (GMM). The foreground is extracted robustly with background subtraction, whereas the background model is updated through MDI [16] for each time step. However, currently, we are not handling the problem of shadows.

2.2   Block Formation

Abrupt and independent activities in crowded scenes result in incoherency, which collectively defines the self-organizing nature of crowd. Each frame is sectioned into N by M blocks of size (i.e. $size = 16$), which is selected after conducting

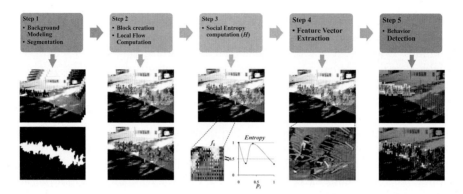

Fig. 2. Overview of the proposed approach for behavior detection in crowds

empirical studies over the dataset (i.e. PETS 2009). In parallel, a grid of two by two is placed over the detected ROI which we refer as points of interest (POI). Optical flow[17] is measured on each POI instead of computing holistically for example in [8] (note. ideally each block of $size = 16$, contains $POI = 64$ on which the optical flow is computed). This supports a fast mechanism of computation and avoids information redundancy; however, Andrade et al.[7] performed a median filtering and use PCA to reduce the feature space. By doing so, each region represents an independent system, and we can capture the local activities without discriminating overall frames. Next, for each block, a two dimension level of information is computed containing the optical flow vector and the flow density:

$$F_t = [B_1, ..., B_n], \quad B_{(i,t)} = \{f_1, ..., f_k\}, \quad f_k = (o, \rho)$$

where F_t contains n blocks $B_{(i,t)}$, $i = \{0, 1, ...n\}$ at time t. Each block contains k POIs which we consider as a 2D distribution of flow vector $o = (u, v)$, and flow magnitude $\rho = \sqrt{u^2 + v^2}$ in the rest of our paper.

Our main contention in obtaining the local motion pattern is that by using global information of motion flow field, it is difficult to reveal the required level of detail, which can differentiate the coherent and incoherent dynamics. However, it is found that due to the limitation of optical flow technique, flow field is not reliable and uncertain [14]. For this purpose, we formalize social entropy on each block, which is explained in the following section.

2.3   Social Entropy

To address this issue, we begin by describing each of our block $B_{(i,t)}$ as an independent social system and use social entropy as an optimization strategy to handle the uncertain [14] flow field distribution. Social entropy provides the context of crowd behaviors allowing us to define quantitative metric for flow field uncertainty in both aspects (i.e. the normal and abnormal behavior). Later, the distribution of homogeneity (i.e. certain in terms of normal and abnormal) and

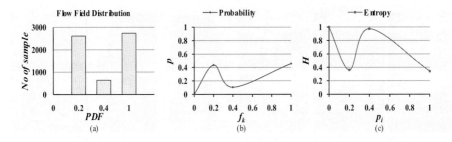

Fig. 3. (a). Shows the normalized distribution (PDF) of flow field based on training samples. (b). Green curve indicates the probability p of the flow field f_k whereas in (c). Red curve shows the corresponding Shannon's entropy H of respective probabilities p_i.

heterogeneity (i.e. uncertain in terms of normal and abnormal) of behaviors are evaluated on the training samples as shown in Fig. 3(a) while social entropy is applied to determine the certain and uncertain ranges of flow fields thus lead to reliable measure of system behavior.

In the following, a mathematical basis for calculating the social entropy of normal and abnormal flow characteristics for our social system during training process is described. We argue that measured uncertainty in the underlying social system reflects the uncertain behavior of the computed flow field which results in the incorrect characterization of crowd behavior. The uncertainty metric is computed using Shannon [18] information entropy H as:

$$H = -\sum_{i=1}^{C} p_i ln p_i \tag{1}$$

where 2D distribution f_k in each block (i.e. $B_{(i,t)}$) is categorized into normal and abnormal motion flows (i.e. Behavior category $C = 2$) and p_i determines the probability of certainty and uncertainty as shown in Fig. 3(b). As a result, low entropy H is observed in the distributions p_i that are sharply peaked around few values, whereas those that are spread more evenly across many values will have higher entropy H. On this basis, we obtained a lookup table in Fig. 3(c) substantiating the reliable and unreliable instances of f_k in each block during the testing phase.

Categorization of Certain f_k : when entropy H is minimum, it depicts 100% probability of certainty. In Fig. 3(c), we are able to select our potential flow field f_k which falls in the entropy range from $(0 - 0.6)$.

Categorization of Uncertain f_k : Unlike above postulate, if H is above 0.6, the flow field is recognized as an uncertain outcome f_k due to noise and limitation of the optical flow technique.

Fig. 4. Test sequences of PETS 2009 and UMN datasets: Normal (top) and abnormal(bottom)

2.4 Feature Vector Computation

From the above section, we obtain reliable flow field inside each block $B_{(i,t)}$. A discrete value provides weak evidence of crowd behaviors, so instead, we apply the statistical measure on the computed reliable flow field (i.e. o_k and ρ_k) on $POIs$ in each block and aggregate it into a flow feature vector $V_{B_{(i,t)}}$. Feature extraction for crowd behavior analysis is described as follows.

$$V_{B_{(i,t)}} = \overline{f_k} = (\overline{o_k}, \overline{\rho_k}) \qquad (2)$$

2.5 Behavior Detection

In computer vision, event classification and identification using discriminative techniques through learning from experimental data (i.e. feature space) have considerably been studied. Among many conventional parametric classifier, SVM offers an effective way to optimally model data. SVM learner defines the hyperplanes for the data and maximum margin is found between these hyper-planes. Radial Basis Function (RBF) Gaussian kernel $(k(x, y) = e^{-\|x-y\|^2/2\sigma^2})$ is used in the proposed approach which has performed robustly with the given number of features and obtained better results as compared to other kernels [19].

Using SVM as a classifier with flow feature vectors, we are able to distinguish specific and overall normal and abnormal dynamics in crowded scenes unlike in [8]. As discussed earlier, the computed flow feature vectors reveal the reliable characteristics in the scene which are localized as normal and abnormal behaviors by SVM.

3   Experiments and Discussion

The proposed approach is tested on publicly available datasets from PETS 2009 [20] and University of Minnesota (UMN) [21]. Fig. 4 shows the samples of test sequences. Ideally, the normal situation is represented by the usual walk of the large number of people, whereas the abnormal situations (i.e. running,

scenario	training set	frames
S1.L1	13-57	220
S1.L1	13-59	240
S1.L2	14-06	200
S1.L3	14-17	90
S1.L3	14-33	343

Fig. 5. On the left, tables presents the training process, and on the right figure shows the detection results on PETS 2009. The left frames indicate absolute normal (yellow) behavior, the middle frames show the transition (green) and the right frames depict absolute abnormal (red) behaviors along the dotted time-line in each row.

panic and dispersion) are observed when individuals or group of individuals deviate from the normal behavior. There is a major distinction between these two datasets, for example, in PETS 2009, the abnormality begins gradually unlike UMN dataset, which makes PETS more challenging due to the transitions from normal to abnormal situations.

The PETS S3 which comprise our test set contains different crowd activities such as walking, running and dispersion. For learning, table given in Fig. 5 describes the scenarios of the training sets used in the training process. A qualitative presentation, indicating the ground truth (GT) and the detection result (DR) in each row for normal, transition (i.e. the crowd behavior is neither absolute normal nor absolute abnormal) and abnormal situations in the sequences, is shown in Fig. 5. The color-bars define the respective crowd behaviors and timings of the occurrences whereas the incorrect localizations of the crowd behaviors are marked with respective colors of false detections in Fig. 5. The detection results of crowd behaviors are demonstrated in Fig. 6 where the normal behaviors are marked as yellow and abnormal behaviors are highlighted with red squares in frames. The transition situation is inferred from the percentage of normal and abnormal

Table 1. Confusion Matrix: Proposed Approach

Event	Normal	Abnormal
Normal	95.8	4.2
Abnormal	0.8	99.2

Table 2. Confusion Matrix: Unrefined Flow Features

Event	Normal	Abnormal
Normal	82.0	18.0
Abnormal	14.6	85.4

Fig. 6. Shows the detection results on PETS 2009. Left (yellow) frames shows the normal behavior detection, the middle (green) frames shows the certain ratio of normal and abnormal behavior and right (red) frame depicts the abnormal behaviors.

behaviors. The results show that the proposed approach is capable of locating the specific and overall abnormalities in the regions which are occupied by the crowd.

To validate the performance of our proposed approach for the detection of the crowd dynamics, we conduct the similar experiments using the flow feature vectors without incorporating the social entropy called unrefined flow feature vectors keeping all the parameters of the classifier same. Table. 1 shows the confusion matrix of the probability of normal and abnormal behavior analysis for each class. The calculated results indicate that the proposed method outperforms the similar method based on unrefined flow feature vectors as shown in Table. 2. The diagonal elements in the confusion matrices represent the percentage probability of recognition for each class in the group. Misclassification between the classes is shown by the non-diagonal elements which are observed due to a prominent motion field at object leg parts as compared to body and head of the objects.

We evaluate the diversity of proposed approach by conducting experiments on another dataset by UMN, which contains 11 different videos of different scenarios showing normal and escape cases. The confusion matrix in Table. 3 shows the detection probability of crowd behaviors using the proposed method. Besides, we conduct experiments using unrefined flow feature vectors for comparative analysis as shown in Table. 4 which clearly indicates the performance of the proposed work. The results in Fig.7 demonstrates the detection of crowd behaviors for normal and panic situations in the crowd.

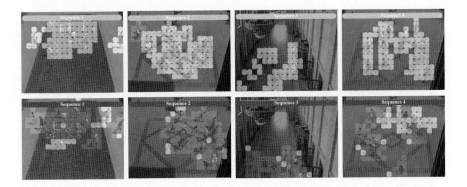

Fig. 7. Shows the detection results on UMN dataset. Top frames show the normal behavior detection indicated by yellow squares and bottom frames depict the abnormal behaviors marked with red squares. It is observed that in escape sequences due to sudden dissipation the transition situation does not occur.

Table 3. Confusion Matrix: Proposed Approach

Event	Normal	Abnormal
Normal	90.3	9.7
Abnormal	2.8	97.2

Table 4. Confusion Matrix: Unrefined Flow Features

Event	Normal	Abnormal
Normal	80.2	19.8
Abnormal	21.6	78.4

4   Conclusion

In this paper, we introduce a method to analysis the crowd behaviors incorporating the social entropy using SVM. A top to down methodology is adapted which capture the crowd characteristics based on flow features whereas the uncertainties in flow data inside each block are dealt with social entropy. Later, the flow feature vectors are classified with SVM to detect and localize the specific and overall behavior in the crowd. The results show that the proposed approach is capable of detecting the governing dynamics of the crowd and captures the transition's period successfully with 95.6% detection rate and outperform when compared to similiar approach using unrefined flow feature vectors.

Acknowledgement

This work is supported by BMBF-Frderung and Transregional Collaborative Research Centre SFB/TRR 62 funded by German Research Foundation (DFG).

References

1. Helbing, D., Johansson, A., Al-Abideen, H.Z.: The dynamics of crowd disasters: An empirical study. An Empirical Study. Phys. Rev. 75, 046109 (2007)

2. Ferryman, J.: (PETS 2007), http://www.cvg.rdg.ac.uk/PETS2007
3. Wu, S., Moore, B.E., Shah, M.: Chaotic invariants of lagrangian particle trajectories for anomaly detection in crowded scenes. In: IEEE Conference on Computer Vision and Pattern Recognition (2010)
4. Ge, W., Collins, R.: Marked point processes for crowd counting. In: IEEE Computer Society Conference on Computer Vision and Pattern Recognition, pp. 2913–2920 (2009)
5. Ali, S., Shah, M.: Floor fields for tracking in high density crowd scenes. In: Forsyth, D., Torr, P., Zisserman, A. (eds.) ECCV 2008, Part II. LNCS, vol. 5303, pp. 1–14. Springer, Heidelberg (2008)
6. Zhan, B., Monekosso, D., Remagnino, P., Velastin, S., Xu, L.: Crowd analysis: A survey. Machine Vision Application 19, 345–357 (2008)
7. Andrade, E.L., Scott, B., Fisher, R.B.: Hidden markov models for optical flow analysis in crowds. In: Proceedings of the 18th International Conference on Pattern Recognition, pp. 460–463. IEEE Computer Society, Los Alamitos (2006)
8. Mehran, R., Oyama, A., Shah, M.: Abnormal crowd behavior detection using social force model. In: IEEE Computer Society Conference on Computer Vision and Pattern Recognition, pp. 935–942 (2009)
9. Kratz, L., Nishino, K.: Anomaly detection in extremely crowded scenes using spatio-temporal motion pattern models. In: IEEE Conference on Computer Vision and Pattern Recognition (2009)
10. Boghossian, A., Velastin, A.: Motion-based machine vision techniques for the management of large crowds. In: Proceedings of 6th IEEE International Conference on ICECS Electronics, Circuits and Systems, vol. 2, pp. 961–964 (2002)
11. Maurin, B., Masoud, O., Papanikolopoulos, N.: Monitoring crowded traffic scenes. In: 5th IEEE International Conference on Intelligent Transportation Systems, pp. 19–24 (2002)
12. Albiol, A., Silla, M., Albiol, A., Mossi, J.: Video analysis using corner motion statistics. In: Performance Evaluation of Tracking and Surveillance Workshop at CVPR 2009, pp. 31–37 (2009)
13. Benabbas, Y., Ihaddadene, N., Djeraba, C.: Global analysis of motion vectors for event detection in crowd scenes. In: Performance Evaluation of Tracking and Surveillance Workshop at CVPR 2009, pp. 109–116 (2009)
14. Bruhn, A., Weickert, J., Schnörr, C.: Combining the advantages of local and global optic flow methods. In: Van Gool, L. (ed.) DAGM 2002. LNCS, vol. 2449, pp. 454–462. Springer, Heidelberg (2002)
15. Bailey, K.D.: Social entropy theory: An overview. Journal Systemic Practice and Action Research 3, 365–382 (1990)
16. Al-Hamadi, A., Michaelis, B.: An intelligent paradigm for multi-objects tracking in crowded environment. Journal of Digital Information Management 4, 183–190 (2006)
17. Bouguet, J.Y.: Pyramidal implementation of the lucas kanade feature tracker: Description of the algorithm (2002)
18. Shannon, C.E.: A Mathematical Theory of Communication. University of Illinois Press, US (1949)
19. Cristianini, N., Shawe-Taylor, J.: An Introduction to Support Vector Machines: and Other Kernel-based Learning Methods, 1st edn. Cambridge University Press, Cambridge (2000)
20. Ferryman, J., Shahrokni, A.: PETS 2009 (2009), http://www.cvg.rdg.ac.uk/PETS2009
21. UMN: (Detection of unusual crowd activity), http://mha.cs.umn.edu

Introducing a Statistical Behavior Model into Camera-Based Fall Detection

Andreas Zweng, Sebastian Zambanini, and Martin Kampel

Computer Vision Lab, Vienna University of Technology
Favoritenstr. 9/183, A-1040 Vienna, Austria

Abstract. Camera based fall detection represents a solution to the problem of people falling down and being not able to stand up on their own again. For elderly people who live alone, such a fall is a major risk. In this paper we present an approach for fall detection based on multiple cameras supported by a statistical behavior model. The model describes the spatio-temporal unexpectedness of objects in a scene and is used to verify a fall detected by a semantic driven fall detection. In our work a fall is detected using multiple cameras where each of the camera inputs results in a separate fall confidence. These confidences are then combined into an overall decision and verified with the help of the statistical behavior model. This paper describes the fall detection approach as well as the verification step and shows results on 73 video sequences.

1   Introduction

Nowadays, the future aging population and demographic change calls for new, innovative systems for health care [1]. In this paper we address the problem of elderly falling at home. Elderly have an increased risk of falling and studies have shown that an immediate alarming and helping heavily reduces the rate of morbidity and mortality [2]. Widely used systems are based on mobile devices to be worn by the users. This "push-the-button" solution has the drawback that it requires the user's capability and willingness to raise the alarm. A camera-based system is a less intrusive and more flexible solution, able to detect different kinds of events simultaneously.

In the past, a number of approaches have been proposed for camera-based fall detection. One group of methods is based on the detection of the characteristic temporal sequence of fall actions. This is accomplished by training parametric models like Hidden Markov Models using simple features, e.g. projection histograms [3] or the aspect ratio of the bounding box surrounding the detected human [4,5]. However, such systems need a lot of labeled training data and their applicability is further limited due to the high diversity of fall actions and the high number of different negative actions which the system should not classify as fall. Another group of methods explicitly measures the human posture and motion speed between consecutive frames. The underlying assumption is that a fall is characterized by a transition from a vertical to a horizontal posture with an unusually increased speed, in order to discern falls from normal actions like

G. Bebis et al. (Eds.): ISVC 2010, Part I, LNCS 6453, pp. 163–172, 2010.

sitting on a chair or lying on a bed. In this manner, in the past various features have been used for camera-based fall detection, including the aspect ratio of the bounding box [6] or orientation of a fitted ellipse [7] for posture recognition and head tracking [8] or change rate of the human's centroid [9] for motion speed. In order to derive a decision for a fall or no fall from the extracted features, parametric classifiers like Neural Networks [10] or empirically determined rules are applied [8,9,11]. A more general approach for abnormal event detection has also shown to be able to implicitly detect fall events in home environments [12]. However, approaches which do not explicitly detect a fall event have the drawback, although being adaptive, of classifying every unknown event (i.e. not learned from the training data) as suspicious. In general, representing and handling the vast amount of possible different actions at home is more complex than explicitly detecting a specific event like a fall.

From our point of view, a more efficient way for introducing learned behavior into the system is to learn zones of usual activity. In [13] this idea was raised for the monitoring of elderly in home environments. Their system uses overhead tracking to automatically learn zones of usually little motion (e.g. bed, chair). Such a learned model gives a strong clue for fall detection: for instance, lying on the floor is much more suspicious than lying on the bed.

In this paper, we extend this idea by combining such a statistical behavior model with individual fall detection in a multiple camera network. In the first stage of the system, features describing human posture and motion speed are extracted to derive confidence values for a fall event for each camera. In the second stage, the outputs from the first stage are combined with a statistical behavior model (called *accumulated hitmap*) that represents the likelihood that activity occurs in the specific scene area. In the final step, individual camera confidence values are fused to a final decision. Thus, we gain robustness by both using information from multiple cameras and by including a model describing the usual areas of activity in the scene. All steps of the overall method are basically focused on simplicity, low computational effort and therefore fast processing without the need of high-end hardware.

The remainder of this paper is structured as follows. Section 2 describes our methodology in detail. Experiments on a dataset of 73 video sequences are reported and discussed in Section 3. Section 4 gives a conclusion.

2   Methodology

Fig. 1 shows the method's workflow. For each camera the moving objects are extracted for feature computation. The features are then used to detect three different states (*"standing"*, *"in between"* and *"lying"*). By using a feature called "motion speed" computed in previous frames, a confidence value is computed for each camera input for fall detection. A voting algorithm combines the four confidences to a final decision. The detection of such a fall is verified with a statistical behavior model called accumulated hitmap which describes the local frequency of foreground objects in the scene. The accumulated hitmap is also computed from each camera input and combined for the verification step.

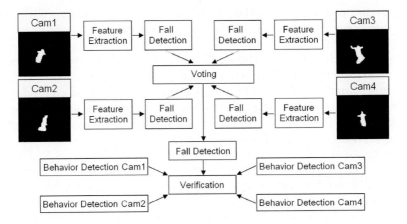

Fig. 1. The workflow of the presented fall detection method

In the following, the individual steps of our fall detection approach are described in detail.

2.1   Human Silhouette Extraction

Segmentation of the person from the background is the first step in our fall detection process. The *Colour Mean and Variance* approach has been used to model the background [14]. Motion detection has been applied with the *NRGB colour space* in order to remove shadows from the segmentation result. The detection has been done using the *RGB+NRGB* combination described in [15]. The resulting blobs (connected components) are then filtered by size to remove noise and small objects from the labeled foreground image due to the fact, that we are only interested in moving persons. The humans in the videos are tracked by a simple assignment of blobs in frame t and frame $t + 1$ using the distance of the blobs computed from the center of mass. The result of human silhouette extraction is a mask with a rough silhouette of the human in each camera frame, i.e. a set of silhouette pixels $\mathcal{P}_{c,i}$, where c is the camera index and i is the frame index.

2.2   Posture Estimation

The goal of the posture estimation step is to decide if the person is in an upright position. The goal of low computational effort renders, for instance, sophisticated model-based approaches for posture recognition infeasible. Therefore, posture estimation is kept simple and estimates basically the general orientation of the human body, i.e. standing/vertical or lying/horizontal. Based on empirical experiments, we have chosen a set of features for posture recognition which needs low computational effort for extraction but offers a comparatively strong discriminative power. In particular, the following features are extracted at every frame with index i from the set of pixels $\mathcal{P}_i$ representing the person:

- **Bounding Box Aspect Ratio** (B_i): The height of the bounding box surrounding the person divided by its width.
- **Orientation** (O_i): The orientation of the major axis of the ellipse fitted to the person, specified as the angle between the major axis and the x-axis.
- **Axis Ratio** (A_i): The ratio between the lengths of the major axis and the minor axis of the ellipse fitted to the person.

Please not that these features are not view-invariant: for instance, a high value for B_i could arise from a person standing but also from a person lying on the floor in direction of the camera's optical axis. Therefore, we propose a multiple camera system that is able to overcome the influence of weak features extracted from non-optimal views through a final fusion step. This step is described in Section 2.5. Inspired by Anderson et al. [11], we define three posture states in which the person may reside: *"standing"*, *"in between"* and *"lying"*. Sets of primarily empirically determined fuzzy thresholds in the form of trapezoidal functions [16] are assembled to interpret the intra-frame features and relate them to the postures. Thus, each feature value results in a confidence value in the range $[0, 1]$ on each posture, where the confidences of one feature sum up to 1 for all postures. These are then combined to assign a confidence value for each posture which is determined by a weighted sum of all feature confidences. The membership functions for the orientation O_i are exemplarily shown in Fig. 2.

2.3　Fuzzy-Based Estimation of Fall Confidence Values

From the computed confidence values for the different postures, for every frame a confidence value for a fall event is computed. The motion speed of the person can be exploited for fall detection since unintentional falling causes a certain motion speed of the body. Therefore, we compute a motion speed feature M_i at every frame i which expresses the amount of change that happens between consecutive frames,

$$M_i = |\mathcal{P}_i \backslash (\mathcal{P}_i \cap \mathcal{P}_{i-1})| / |\mathcal{P}_i| \tag{1}$$

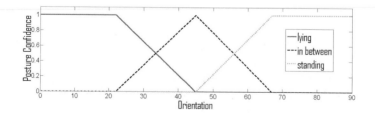

Fig. 2. Membership functions for the three postures and the intra-frame feature O_i

We combine this feature with the estimated posture confidences under the assumption that a fall is defined by a relatively high motion speed, followed by a period with a *"lying"* posture. Thus, the confidence for a fall event at frame i is computed as the motion speed M_i multiplied by the confidence values for the posture *"lying"* for the next k frames.

2.4   Combination with Learned Accumulated Hitmap

A accumulated hitmap is a matrix which counts consecutive foreground pixels. Each foreground pixel from a binary image increases the corresponding point in the hitmap by 1. The idea of computing a hitmap stems from [17] where each pixels' foreground frequency is modeled. The hitmap at a given pixel position is increased as long as the pixel in the input image is classified as a foreground pixel. The values in the hitmap therefore specify the duration of stay of objects in the scene at a given position. The pixels in the hitmap have to be decreased if the corresponding pixel in the input image is no foreground pixel anymore. Due to noise and misclassifications, the pixel value at a given position in the hitmap is decreased after a waiting period of n frames of consecutive background pixels in the input image. In Fig. 3 one frame of an image sequence and the corresponding trained hitmaps are illustrated. Lighter pixels correspond to higher frequented areas.

In the training stage the hitmap is continuously computed and the maximum value at each pixel position is stored. In the classification stage, the trained hitmap is used as a reference for normality in the scene and compared with the current computed hitmap in order to detect unexpected situations. The attribute of the hitmap is that it detects an unexpectedly long abidance of foreground objects according to the trained map. Due to the reason, that frequenting an area where the trained hitmap is almost zero, should be detected instantly compared to a longer stay of a person in an area where the trained hitmap has higher values,

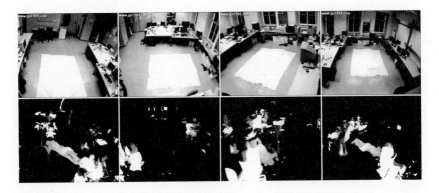

Fig. 3. Four camera views (top row) and the corresponding trained hitmaps (bottom row)

the pixels are weighted using the following equations. At first, the difference of the non-weighted hitmaps is computed:

$$H_{diff_{x,y}} = \left| H_{t_{x,y}} - H_{o_{x,y}} \right| \tag{2}$$

where $H_{t_{x,y}}$ is the trained hitmap and $H_{o_{x,y}}$ is the current computed hitmap at position (x, y). The difference is then weighted with the information of the trained hitmap:

$$H_{unexp_{x,y}} = (1 + \frac{\alpha}{max(H_t) - H_{o_{x,y}}})^{(\frac{H_{diff_{x,y}}}{2})} \tag{3}$$

if $H_{o_{x,y}} > H_{t_{x,y}}$ and 0 otherwise, where $H_{o_{x,y}} = max(H_t) - 1$ if $H_{o_{x,y}} \geq max(H_t)$. $max(H_t)$ is the maximum value in the trained hitmap and α is a regulating parameter set to the value of 50 for out dataset of fall detection sequences. The basis of the equation has its limits at 1 and at $(1 + \alpha)$. This means that higher values of the current hitmap $H_{o_{x,y}}$ raise the value of the basis and the exponent is additionally higher for small values of $H_{t_{x,y}}$. The exponent has been found empirically and has the property to quickly find unexpected behavior in areas of low frequency.

The computation of the bounding box aspect ratio, orientation and axis ratio sometimes lacks of accuracy due to an incorrect human silhouette extraction. Therefore the values of the features can cause false positives which also happens when a person is occluded by desks, chairs or other objects. The accumulated hitmap is used to verify a fall, detected by the previously mentioned fall detection. Therefore the sum of $H_{unexp_{x,y}}$ in the area of the bounding box which encloses the detected person is computed. A fall is affirmatively verified if the fall confidence surpasses a defined threshold and the hitmap sum surpasses a defined threshold in one of the frames before or after the current frame. In our setup we defined this verification window to be as long as four seconds (two seconds before and two seconds after the frame which has to be verified). The fall confidence threshold is chosen at a lower level (e.g. 0.5) for the combination with the hitmap compared to the initial fall detection approach in order to get more false alarms which will then be verified by the hitmap. The threshold for the verification using the hitmap depends on the size of the blobs in the image and was found empirically. The idea of the verification window is illustrated in Fig. 4.

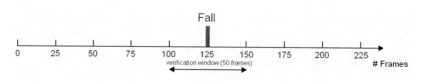

Fig. 4. Verification window in case of a detected fall

2.5   Multiple Camera Fusion

The detection of a fall with the described features requires a viewing angle where the feature values of a standing person differ from the feature values of a lying person. The second view from left in Fig. 5 shows a view where the orientation of a lying person and a standing person are similar. Therefore we use a multiple camera setup with four cameras. The combination of the results from the cameras is done as follows. The detection is computed for each camera and an alarm is fired if one of the results is higher than a defined threshold (i.e. fall confidence=0.9). This fall is then verified by the average of the hitmap values in the surrounding frames as described in Section 2.4. The average is computed for each frame and compared to a defined threshold.

3   Experiments

In order to thoroughly evaluate our fall detection method, test sequences were acquired that follow the scenarios described by Noury et al. [18]. Hence, a test set consisting of various types of falls as well as various types of normal actions was created. Four cameras with a resolution of 288×352 and frame rate of 25 fps were placed in a room at a height of approx. 2.5 meters. The four camera views are shown in Fig. 3. Five different actors simulated the scenarios resulting in a total of 49 positive (falls) and 24 negative sequences (no falls).

For the given test set, the parameter k defining the considered time period of the *"lying"* posture for fall detection (see Section 2.3) was set to 10 seconds. Please note that in a real scenario this parameter has to be set to a higher value. In our simulated falls the lying periods are considerably shorter than they would be in case of a real fall event, for obvious reasons.

Since our method delivers confidence values for a fall event as well as a hitmap value in every tested frame, we report its sensitivity and specificity in the form of ROC curves by varying the two thresholds. For generation of the ROC curve, true positives and false positives were counted as the number of positive and negative sequences, respectively, where a fall confidence and a hitmap value above the thresholds could be found. The performance of fall detection is compared to the combination with the accumulated hitmap and to the hitmap itself. The results are shown in Fig. 6.

In Fig. 6 and 7, *FD HM* shows the performance of the accumulated hitmap for detecting falls, *FD* represents the performance of fall detection without the

Fig. 5. Four camera views of a video sequence

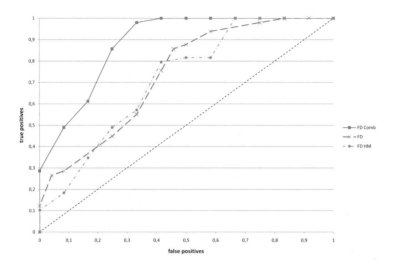

Fig. 6. ROC curves from the evaluation of 73 video sequences using the multiple camera fusion

verification of the accumulated hitmap, and *FD comb* shows the performance with verification of the fall detection using the accumulated hitmap. The evaluation is done using multiple camera fusion as described in Section 2.5. We compared our results using the multiple camera fusion with the results of the algorithms using a single camera (see Fig. 7). We observe a similar performance for the fall detection approach using the verification step compared to the multi camera fusion because the false positives from the fall detection are negatively verified from the hitmap for moving persons, since the hitmap is zero for moving persons. False positives from fall detection where a person is standing or sitting are positively verified from the hitmap since the hitmap values increase in such cases and therefore result in false positives. Our results approve the idea of verifying the detected fall since the fall detection and the hitmap approach perform worse when only one camera is used but the combined approach performs similar compared to the multiple camera setup. The bad performance of fall detection using only the hitmap on the single camera setup (e.g. 35% false positives at a true positive rate of 10%, see Fig. 7) can be explained as follows. The hitmap is evaluated by a simple threshold. High values of the hitmap represents if a person is sitting, standing or lying for 10 seconds or more. If the threshold is taken below these cases and above falls where persons are only lying for a few seconds, most of the falls are not detected but the sitting and standing persons are detected as falls, which results in this form of the ROC curve in Fig. 7.

The results show that the combination of the fall detection and the hitmap outperforms the individual approaches using multiple camera fusion as well as for a single camera setup. However, a problem of the accumulated hitmap is, that it is high for lying persons as well as for persons standing or sitting on the same position for a unexpectedly long time. A long time in this case is when the hitmap values exceed the trained hitmap values in the area of the persons location.

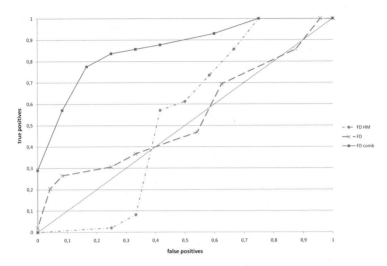

Fig. 7. ROC curves from the evaluation of 73 video sequences using a single camera

Therefore the hitmap approach should not be used for fall detection without a combination of other features. Fall detection without verification as described in this work lacks of accuracy due to problems in the segmentation process. The viewing angle of the camera is also important for the mentioned features to achieve a sufficient discriminative power of the chosen features to correctly classify different postures. However, a person can fall in different directions, therefore a multi camera system is needed to find falls for every orientation of the body. The evaluated ROC curves shown in Fig. 6 and 7 demonstrate that the performance of proposed fall detection approach is similar for a single camera system and a multi camera system.

4   Conclusions and Outlook

This paper showed a new methodology for fall detection with a multiple camera setup and verification using a statistical behavior model. The verification step improves the classification rate of falls for the multiple camera setup as well as for a single camera setup. The results on the multiple camera setup shows a huge improvement of the fall detection and the hitmap alone but not on the combination using the verification. In future work we intend to compute three dimensional volumes of the hitmap for the verification of falls for a further improvement in robustness and classification rate. Alternatively, more sophisticated fusion strategies will be investigated in order to better exploit the power of using multiple cameras and having multiple decisions.

Acknowledgement

This work was supported by the Austrian Research Promotion Agency (FFG) under grant 819862.

References

1. Grundy, E., Tomassini, C., Festy, P.: Demographic change and the care of older people: introduction. European Journal of Population 22, 215–218 (2006)
2. Wild, D., Nayak, U., Isaacs, B.: How dangerous are falls in old people at home? British Medical Journal 282, 266–268 (1981)
3. Cucchiara, R., Grana, C., Prati, A., Vezzani, R.: Probabilistic posture classification for human-behavior analysis. SMC-A: Systems and Humans 35, 42–54 (2005)
4. Anderson, D., Keller, J., Skubic, M., Chen, X., He, Z.: Recognizing falls from silhouettes. In: International Conference of the Engineering in Medicine and Biology Society, pp. 6388–6391 (2006)
5. Toreyin, B., Dedeoglu, Y., Çetin, A.: HMM based falling person detection using both audio and video. In: Sebe, N., Lew, M., Huang, T.S. (eds.) HCI/ICCV 2005. LNCS, vol. 3766, pp. 211–220. Springer, Heidelberg (2005)
6. Tao, J., Turjo, M., Wong, M., Wang, M., Tan, Y.: Fall incidents detection for intelligent video surveillance. In: Fifth International Conference on Information, Communications and Signal Processing, pp. 1590–1594 (2005)
7. Thome, N., Miguet, S., Ambellouis, S.: A Real-Time, Multiview Fall Detection System: A LHMM-Based Approach. IEEE TCSVT 18, 1522–1532 (2008)
8. Rougier, C., Meunier, J., St-Arnaud, A., Rousseau, J.: Monocular 3D head tracking to detect falls of elderly people. In: International Conference of the Engineering in Medicine and Biology Society, pp. 6384–6387 (2006)
9. Lin, C., Ling, Z., Chang, Y., Kuo, C.: Compressed-domain Fall Incident Detection for Intelligent Homecare. Journal of VLSI Signal Processing 49, 393–408 (2007)
10. Huang, C., Chen, E., Chung, P.: Fall detection using modular neural networks with back-projected optical flow. Biomedical Engineering: Applications, Basis and Communications 19, 415–424 (2007)
11. Anderson, D., Luke, R., Keller, J., Skubic, M., Rantz, M., Aud, M.: Linguistic summarization of video for fall detection using voxel person and fuzzy logic. Computer Vision and Image Understanding 113, 80–89 (2009)
12. Nater, F., Grabner, H., Van Gool, L.: Exploiting simple hierarchies for unsupervised human behavior analysis. In: International Conference on Computer Vision and Pattern Recognition (2010) (to appear)
13. Nait-Charif, H., McKenna, S.: Activity summarisation and fall detection in a supportive home environment. In: International Conference on Proceedings of the Pattern Recognition, vol. 4, pp. 323–326 (2004)
14. Wren, C., Azarbayejani, A., Darrell, T., Pentland, A.: Pfinder: Real-time tracking of the human body. PAMI 19, 780–785 (1997)
15. Kampel, M., Hanbury, A., Blauensteiner, P., Wildenauer, H.: Improved motion segmentation based on shadow detection 6, 1–12 (2007)
16. Zadeh, L.: Fuzzy sets. Information and Control 8, 338–353 (1965)
17. Ermis, E., Saligrama, V., Jodoin, P.M., Konrad, J.: Abnormal behavior detection and behavior matching for networked cameras. In: Second ACM/IEEE International Conference on Distributed Smart Cameras, pp. 1–10 (2008)
18. Noury, N., Fleury, A., Rumeau, P., Bourke, A., Laighin, G., Rialle, V., Lundy, J.: Fall detection–Principles and methods. In: International Conference of the Engineering in Medicine and Biology Society, pp. 1663–1666 (2007)

On Contrast-Preserving Visualisation of Multispectral Datasets

Valeriy Sokolov[1], Dmitry Nikolaev[1], Simon Karpenko[1], and Gerald Schaefer[2]

[1] Institute for Information Transmission Problems (Kharkevich Institute),
Russian Academy of Sciences, Moscow, Russia
[2] Department of Computer Science, Loughborough University, Loughborough, U.K.

Abstract. Multispectral datasets are becoming increasingly common. Consequently, effective techniques to deal with this kind of data are highly sought after. In this paper, we consider the problem of joint visualisation of multispectral datasets. Several improvements to existing methods are suggested leading to a new visualisation algorithm. The proposed algorithm also produces colour images, compared to grayscale images obtained through previous methods.

1 Introduction

Following the increased need of applications such as remote sensing, multispectral datasets are becoming increasingly common. Consequently, effective techniques to deal with this kind of data are highly sought after, and the development of segmentation and classification techniques based on specific properties of reflection spectra of different objects has been a very active research area in recent years. There is a natural necessity for manual verification of correctness of the results obtained by such algorithms, as well as a necessity for manual choice of areas to be processed.

For example, in the context of remote sensing, a multispectral dataset can be seen as a set of grayscale images representing the same area. Each image of this set corresponds to a small band of the registering spectra. The number of bands varies but may be greater than 200. However, the dimensionality of our human visual system is only 3 [1], and we are therefore not well equipped to perceive such datasets directly. This problem also arises for joint visualisation of images of different origins or modalities, for example visualising physical sensing of micro-samples obtained using fluorescence microscopy [2].

In this paper, we address the fundamental problem of compound representation of multispectral datasets, i.e. sets of N grayscale images $\boldsymbol{I}(x,y) = (I_n(x,y))$, $n = 1, \ldots, N$ in a form suitable for human perception. That is, we present an effective method for multispectral visualisation. Importantly, such a visualisation is required to preserve local contrast. Moreover, our proposed algorithm produces colour images, compared to grayscale images obtained through previous methods.

G. Bebis et al. (Eds.): ISVC 2010, Part I, LNCS 6453, pp. 173–180, 2010.

2   Related Work

The simplest method for visualising multispectral data is to let the user choose three images from the dataset and unite them into one falsecolour image. However, it is clear that this approach is far from optimal, and that especially when the number of images in the set is large, the number of channels the user is unable to visualise simultaneously is still large.

Another popular method uses averaging to obtain resulting grayscale images. For this, the user specifies a vector of averaging weights $\boldsymbol{\lambda} = (\lambda_1, \ldots, \lambda_N)$. The resulting grayscale image $G(x, y)$ is then obtained by $G(x, y) = \boldsymbol{I}(x, y) \cdot \boldsymbol{\lambda}$. This procedure corresponds to projecting an image vector onto the vector $\boldsymbol{\lambda}$. The disadvantage of this method however is the possibility of losing contrast. If the vectors $\boldsymbol{I}(x_1, y_1)$ and $\boldsymbol{I}(x_2, y_2)$ for neighbour points satisfy the condition $\boldsymbol{I}(x_1, y_1) - \boldsymbol{I}(x_2, y_2) \perp \boldsymbol{\lambda}$ than after averaging we have $G(x_1, y_1) = G(x_2, y_2)$ even if the norm of the difference $\|\boldsymbol{I}(x_1, y_1) - \boldsymbol{I}(x_2, y_2)\|$ is large. Of course, one can choose some other weight vector, but it seems impossible to effectively choose such a vector for all points in a general case.

One can attempt to reduce that disadvantage by choosing the weight vector depending on statistical analysis of the source dataset. One way to achieve this is using principal component analysis (PCA). Considering the source dataset as the set of vectors in some N-dimensional vector space, the weight vector is then chosen as an eigenvector corresponding to the largest eigenvalue of the covariance matrix of that set of vectors [3].

PCA-based techniques are among the most popular method for processing multispectral datasets. However, in PCA-based visualisations, there is also the possibility of losing local contrast. Let us consider an example with a model dataset of 9 images as input. In every image the intensity of pixels equals 0 for the whole image, except for a small square where the intensity equals 1. The areas of squares are completely disjoint. Hence, one should expect to see 9 squares in the visualisation. However, applying PCA-based visualisation results in an image, shown in Fig. 1, where one of the square is "eliminated" (the bottom right square in Fig. 1).

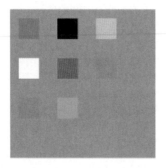

Fig. 1. PCA-based visualisation of a dataset of 9 images each containing a square

Moreover, PCA performs a projection on some diagonal in N-dimensional vector space. The cosine of the angle between this diagonal and any axis has the asymptotics $\cos\theta \sim 1/\sqrt{N}$ while $N \to \infty$. So if N is big enough, the diagonal is almost orthogonal to the choosen axis and the contribution of the corresponding image to the resulting grayscale image is very low [4].

3   Local Contrast Concept

The critique of visualisation methods for multispectral datasets presented in Section 2 is based on the observation of inconsistency between the source dataset and the resulting grayscale image. To construct a visualisation method which ensures to preserve local contrast, some authors use a formal definition of the local contrast concept.

The definition of local contrast for grayscale images is based on the gradient of a scalar function. A similar definition for a set of grayscale images, i.e. a vector function, can be derived [5]. Given a set of images $\boldsymbol{I}(x,y)$, for every point there is a square form defined by a matrix

$$C(x,y) = \begin{pmatrix} C_{xx}(x,y) & C_{xy}(x,y) \\ C_{xy}(x,y) & C_{yy}(x,y) \end{pmatrix}, \tag{1}$$

where $C_{\xi\eta}(x,y) = \sum_{i=1}^{N} \frac{\partial I_i}{\partial \xi} \frac{\partial I_i}{\partial \eta}$. The value of this form at point (x_0, y_0) on the vector (u,v) is the variation of the vector function $\boldsymbol{I}$ in direction (u,v) at point (x_0, y_0). The largest variation is achieved in the direction of the eigenvector corresponding to the largest eigenvalue λ_{max} of C. Such an eigenvector with length equal to $\sqrt{\lambda_{max}}$ may be interpreted as an analogy to the gradient of a scalar function. However, in contrast to the gradient there is no way to define the increasing direction of the vector function. Thus, there exists a problem with regards to the ambiguity of direction, i.e. a problem of signs.

The above concept of local contrast for a set of grayscale images was used by some authors to develop a template algorithm for multispectral visualisation [6, 7]. This approach consists in general of the following steps:-

1. Using the source dataset, compute the field of square forms $C(x,y)$ and compute the corresponding vector field $\boldsymbol{P}(x,y)$ (the so-called pseudo-gradient field). $\boldsymbol{P}(x,y)$ are principal eigenvectors of the form $C(x,y)$ with length equal to the square root of the corresponding eigenvalues.
2. Solve the problem of signs.
3. Construct a grayscale image $G(x,y)$ such that ∇G is similar to $\boldsymbol{P}$ in some sense.

The parameters of this template method are the chosen way to solve the sign problem and the employed measure of similarity between two vector fields to obtain the resulting grayscale image. Usually, the measure of similarity is introduced as an error function $E(x,y)$ which expresses the difference between vector fields $\boldsymbol{P}$ and ∇G at point (x,y). The total error is then computed as

$E_\Sigma = \iint\limits_\Omega E(x,y)dxdy$, where Ω is the domain on which the source dataset is defined. Hence, the third step of the visualisation is reduced to the minimisation of E_Σ in the space of scalar functions of two variables where G is from.

As for the problem of signs, a common approach is to solve it with the help of some other vector field $\boldsymbol{R}(x,y)$ called the reference field. To provide an appropriate direction to the vectors of $\boldsymbol{P}$ one should multiply $\boldsymbol{P}(x,y)$ with $sign\left(\boldsymbol{P}(x,y)\cdot\boldsymbol{R}(x,y)\right)$ at every point (x,y).

In [6] the authors use the gradient of spectral intensity of the source dataset as the reference vector field to solve the problem of signs. As error function they define

$$E_s(x,y) = |\boldsymbol{G}(x,y) - \boldsymbol{P}(x,y)|^2 . \tag{2}$$

For simplicity, we refer to this measure of similarity as linear measure since the corresponding Euler-Lagrange equation is the Poisson equation which is linear.

In [7] a similar method was proposed, and a solution to the problem of signs introduced for the case $N = 3$. First, an orthonormal basis in the N-dimension vector space is chosen, and the vectors in this basis are ordered in a manner that reflects their significance. Then one projects the source set of images onto the vectors of this basis, and a set of images $\{J_n(x,y)\}, n = 1,\ldots,N$ is thus obtained. The directions of the vectors from $\boldsymbol{P}$ with respect to the gradient of J_1 are then obtained at the points where the magnitude of that gradient is greater than some threshold. At the points where the vector $\boldsymbol{\nabla}J_1$ is small, a similar procedure is employed with the help of J_2 and so on. If the vector of the pseudo-gradient field is not related to any of $\boldsymbol{\nabla}J_i, i = 1,\ldots,N$, then the contrast at this point is small and any random direction of that vector is acceptable. The measure of vector field similarity is the same as in [6]. The authors noted the appearance of a smooth gradient in areas where the source dataset images were uniform. As a possible way to eliminate this, the authors proposed to use some non-linear measure. In a new variant of such a measure they introduced the error function

$$E_g(x,y) = -exp\left\{-E_s(x,y)/2\sigma^2\right\}, \tag{3}$$

with some σ.

4   On the Problem of Signs

As noted above, the problem of signs arises while obtaining the pseudo-gradient field $\boldsymbol{P}$. Let us consider the manner of solving this problem, which was proposed in [6]. At every point the spectral intensity $S(x,y) = \sqrt{\boldsymbol{I}^2(x,y)}$ is computed. Then, they relate the field $\boldsymbol{P}$ with the field $\boldsymbol{\nabla}S$. If the structure of the source dataset is such that there are two areas with a common boundary which have equal spectral intensities but different distributions of that intensity over the bands, then $|\boldsymbol{\nabla}S|$ at the common boundary of these areas approximately equals 0. This leads to a strong dependence of the pseudo-gradient vector direction at the common boundary on noise. The final pseudo-gradient field constructed in such a manner may demonstrate gaps along that boundary, which in turn leads to a significant warp of the resulting grayscale image.

Let us consider the test dataset introduced in [7], i.e. a Mondrian-like dataset with uniform areas of different colour but of equal intensity and triples of regions such that there is a boundary between every two regions in a triple. An example image is shown in Fig. 2.

Fig. 2. Sample Mondrian-like image

Fig. 3 demonstrates the resulting images obtained using different approaches to solve the problem of signs. The left image (a) illustrates the use of spectral intensity ordering, while the right image (b) gives the result of [7]. It is easy to see that the latter is of higher quality. Hence, the approach presented in [7] proves to be more robust for solving the problem of signs.

(a) Spectral intensity ordering   (b) Sequential ordering

Fig. 3. Visualisation results illustrating the problem of signs

We propose the following method. We use principal component analysis to obtain an orthonormal eigenbasis in the colour space of the source dataset. Then, we use the obtained basis in the manner proposed in [7] to solve the problem of signs. Consequently, this approach leads to a more accurate pseudo-gradient field.

5   Non-linear Measures

Let us consider the non-linear measure $E_g(x, y)$ proposed in [7]. We implement this measure using the gradient descent method. The grayscale image obtained using the linear measure (Fig. 3(b)) was used as initial approximation. In Fig. 4(a), the result of the minimisation with $\sigma = 0.01 \cdot I_{max}$, where I_{max} is the dynamic band of the source image is presented. We can see that many artefact boundaries appear in the resulting image in areas of uniformity. Obviously, this points to a shortcoming in the structure of this measure.

(a) Gaussian error   (b) Relative error

Fig. 4. Non-linear measure results

A good measure can be characterised as one that (1) reaches a maximum if the source boundaries and obtained boundaries are the same, (2) prevents an appearance of strong boundaries in places where they are weak, and (3) prevents disappearance of boundaries in place they are strong.

Let us consider a measure of relative error as relative difference of two vector fields $E_r(x, y)$ defined by

$$E_r(x, y) = \frac{|\boldsymbol{\nabla} G(x, y) - \boldsymbol{P}(x, y)|^2}{|\boldsymbol{\nabla} G(x, y)|^2 + |\boldsymbol{P}(x, y)|^2 + \alpha^2}. \tag{4}$$

Clearly, the first requirement is satisfied: if $|\boldsymbol{\nabla} G(x, y) - \boldsymbol{P}(x, y)|^2$ is small, the error is small. The two other requirements are satisfied too as in these cases the value of error is near to 1. This method was implemented using gradient descent, and the grayscale image obtained with the linear measure, i.e. the image of Fig. 3(b), was used as the initial approximation.

In Fig. 4(b) the result of minimisation with $\alpha = 0.007 \cdot I_{max}$ is shown. We can notice a significant improvement in comparison with the linear method of Fig. 3(b)).

6   Colour Visualisation

A visualisation which produces a grayscale image as output, although sufficient in certain cases as those given in Figs. 2 to 4, does not use the human visual system to its full ability. We therefore propose a method whose output is a colour image. Our approach proceeds in the following steps:-

1. Using one of the visualisation methods which preserve local contrast, obtain the grayscale image $Y(x, y)$.
2. Obtain a normalised vector of averaging weights such that the averaging of the source set with these weights produces a grayscale image which is similar to $Y(x, y)$. To obtain such a vector we use the minimisation of the functional

$$F(\boldsymbol{\lambda}) = \iint_{\Sigma} D(x, y) dx dy, \qquad (5)$$

where

$$D(x, y) = \left| \sum_{i=1}^{N} \frac{\lambda_i}{|\boldsymbol{\lambda}|} \boldsymbol{\nabla} I_i(x, y) - \boldsymbol{\nabla} Y(x, y) \right|^2. \qquad (6)$$

The minimisation is implemented using gradient descent, and the initial approximation is taken as $\boldsymbol{\lambda} = (1, \ldots, 1)$.
3. Project the source dataset onto the subspace which is orthogonal to the obtained vector of weights. As result, we have a set of $(N - 1)$ images. For that purpose, $(N - 1)$ vectors which form the orthonormal basis of $< \boldsymbol{\lambda} >^{\perp}$ are constructed. The corresponding $(N-1)$ images are obtained by projecting the source dataset on these vectors.
4. Project the obtained set onto the first and second eigenvectors of the PCA decomposition and construct a colour image defined by three planes in $L\alpha\beta$ colour space. The plane L is the image $Y(x, y)$ and planes α and β are two other obtained images. The $L\alpha\beta$ colour space is related to the RGB colour space by [7]

$$L = \frac{R + G + B}{3}, \quad \alpha = \frac{R - G}{\sqrt{2}}, \quad \beta = \frac{R + G - 2B}{\sqrt{6}}. \qquad (7)$$

The result of this method is presented in Fig. 5 where it is apparent that all nine squares of the test dataset are clearly visible in the resulting image. Moreover, these squares are of different colours. This corresponds to the fact that all these squares are in different planes of the source dataset.

In theory, the planes α and β may be constructed using some method preserving local contrast, but this is not rational with respect to the properties of human colour vision. The chrominance of object surfaces depends more on the properties of the object than on lighting [8], but the observable luminance depends only on the conditions of lighting and observation. As the result of adaptation to these physical laws, the vision subsystem performs relatively poorly in the estimation of absolute values of luminance but demonstrates robust differentiating of chrominance components. Therefore, the presented scheme provides a "similar" view of the areas with the same spectral composition as the human visual system.

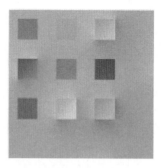

Fig. 5. Proposed visualisation of the dataset as in Fig. 1

7   Conclusions

In this paper, we considered an improved variant of a multispectral visualisation algorithm preserving local contrast. We furthermore proposed a visualisation algorithm which produces a colour image as output. The resulting algorithm allows the user to take advantage of their tri-chromatic vision system in a more adequate way.

References

[1] Hunt, R.: Measuring colour. Fountain Press (1998)
[2] Chukalina, M., Nikolaev, D., Somogyi, A., Schaefer, G.: Multi-technique data treatment for multi-spectral image visualization. In: 22nd European Conference on Modeling and Simulation, pp. 234–236 (2008)
[3] Scheunders, P.: Local mapping for multispectral image visualization. Image and Vision Computing 19, 971–978 (2001)
[4] Landgrebe, D.: On information extraction principles for hyperspectral data - a white paper (1997)
[5] Zenzo, S.D.: A note on the gradient of multi-image. Comput. Vision Graphics Image Process. 33, 116–125 (1986)
[6] Socolinsky, D., Wolff, L.: A new visualization paradigm for multispectral imagery and data fusion. In: IEEE Conf. Comp. Vis. Pat. Rec., pp. I:319–I:324(1999)
[7] Nikolaev, D., Karpenko, S.: Color-to-grayscale image transformation preserving the gradient structure. In: 20th European Conf. Modeling & Simul., pp. 427–430 (2006)
[8] Nikolaev, D., Nikolayev, P.: Linear color segmentation and its implementation. Color Vision and Image Understanding 94, 115–139 (2004)

Color Gamut Extension by Projector-Camera System

Takahiko Horiuchi, Makoto Uno, and Shoji Tominaga

Graduate School of Advanced Integration Science, Chiba University, Japan

Abstract. The color gamut of printing devices is generally smaller than that of imaging devices. Therefore, vivid color images cannot be reproduced on printed materials. This paper proposes a color gamut extension system for printed images by using a projector-camera system. The proposed system can capture a printed image using a video camera and extend the color gamut of the printed image by super-imposing a compensation image obtained from a projector device on the printed image. The compensation image is produced in the hue, saturation, and value (HSV) color space for iteratively adjusting the saturation and brightness values toward the boundary of the color gamut in the projector-camera system at each pixel. The feasibility of the proposed system is verified by experiments performed using real printed images.

1 Introduction

The range of colors produced in a given device is defined by the chromaticity and maximum luminance of the primaries; a display or a projector is typically associated with three colors [red, green, and blue (RGB)], and a printer with four colors (cyan, magenta, yellow, and black). This range of colors produced by a device is known as the "device color gamut." In recent years, new display technologies such as plasma display panels (PDPs) and liquid crystal displays (LCDs) have emerged; these technologies are capable of producing vivid colors that exceed the color gamut of standard technologies. On the other hand, most currently available printer devices including multi-primary color printing device have a smaller color gamut than light-emitting devices. Although several gamut extension algorithms, which map colors from a smaller source gamut to a larger destination gamut, have been developed [1],[2],[3],[4], it is impossible to represent colors that exceed the destination color gamut.

Recently, adaptive projection systems or the so-called "projector-camera systems" have been developed and applied used in various applications [5],[6],[7],[8], [9],[10],[11]. Tsukada et al. proposed a projector color reproduction method that enabled the true reproduction of colors when images were projected onto a colored wall rather than a white screen [5]. Grossberg proposed a method for controlling an object such that it resembles another object [6]. From the viewpoint of geometrical compensation, Bimber proposed a general correcting method for geometrical distortion when a non planar screen is used [7]. Renani et al. evaluated the performance of the screen compensation algorithms [8]. Fujii [9] constructed

G. Bebis et al. (Eds.): ISVC 2010, Part I, LNCS 6453, pp. 181–189, 2010.

a coaxial projector camera system whose geometrical distortion is independent of the screen shape. For realizing geometric compensation from the viewpoint of the user, Johnson proposed a real-time modification method for the projection image that involved feature tracking of the image [10]. These methods aim to cancel the projection distortion and color non uniformity for the projection of images on a textured, non planar screen. However, not only compensation method but also opposite techniques that enhance the non uniformity can aid human visual perception. Amano proposed an interesting technique that enhanced the appearance of color images [11]. However, this technique could not treat the color space with sufficient precision.

This paper proposes a gamut extension system for a printed image by dynamic feedback using a projector-camera system. The proposed system can extend the color gamut of a captured printed image by super-imposing a compensation color image, which is produced in hue, saturation, and value (HSV) color space, obtained from the projector device on the printed image.

2   Projector-Camera System

Figure 1 shows a system setup of the proposed gamut extension system. A compensation image is displayed on a target printed image using a Panasonic TH-AE300 projector (RGB 8bit, 1024 × 768 resolution) and the target image is captured using a Panasonic NV-GS200 camera (RGB 8bit, 720 × 480 resolution). The calibration and compensation algorithms are run on a Windows OS computer with 2GHz Intel Celeron CPU. A target printed image is irradiated by the standard D65.

In the projector-camera system, we need to know the geometric mapping between points in the projector and the image planes. We note that a projector-camera system can be designed such that the geometric mapping between the displayed and acquired images is fixed and is unaffected by the location or the shape of the scene. This is achieved by making the optics of the projection and the imaging systems coaxial. Although our system is not a coaxial one, the camera is placed on the projector as shown in Fig. 1. Hence, the geometric mapping between the camera and projector must be modeled with a mapping. We also need to know the color mapping between the projector and the camera. In this section, we explain calibration algorithms.

2.1   Geometric Calibration

To determine the geometric mapping, we use a calibration algorithm proposed by Zhang in Ref.[12]. Zhang's calibration method requires a planar checkerboard grid to be placed at different orientations in front of the camera. In our system, four planar checkerboard patterns (Black, Red, Green, Blue colors) are projected on a screen. Figure 2 shows projected patterns used in our calibration. Four colors will be also used in the color calibration in Sec.2.2. The algorithm extracts corner points of the checkerboard pattern for computing a projective

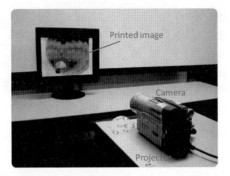

Fig. 1. A projector-camera system for the gamut extension

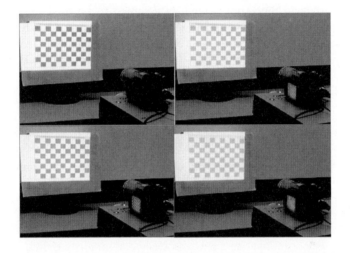

Fig. 2. Projected four calibration patterns (Black, Blue, Red, Green)

transformation (Fig. 3(a)). Afterwards, the camera interior and exterior parameters are recovered using a closed-form solution, while the third- and fifth-order radial distortion terms are recovered within a linear least-squares solution. A final nonlinear minimization of the reprojection error, solved using a Levenberg-Marquardt method, refines all the recovered parameters. After recovering the camera parameters, the camera lens distortions are corrected as shown in Fig. 3(b). Finally, geometric transformation is performed by the transparent transformation (Fig. 3(c)).

2.2   Color Calibration

Color calibration is usually performed by using an absolute standard such as a color chart. However, the absolute compensation of color is not necessary in the projector-camera feedback system. The color only has to match it between

(a) Extracted corner points (b) Corrected camera distortion   (c) Corrected geometry

Fig. 3. Process of the geometric calibration

projector and camera devices. In the case that an accurate color reproduction is required, we have to consider a lot of nonlinear characteristics of devices. However, such a calibration method by considering a lot of nonlinearities takes much computational costs. Since our system is assumed to be used in various environments, it is preferable to complete the calibration quickly. In this study, the calculation cost has priority more than strict accuracy. Therefore, color mapping is modeled with a linear mapping between the projector output RGB values and the camera input RGB values. Let us assume that the projector and the camera each have three color channels with 8 bit intensities P and C as follows:

$$
P = \begin{bmatrix} R_p \\ G_p \\ B_p \end{bmatrix}, C = \begin{bmatrix} R_c \\ G_c \\ B_c \end{bmatrix}. \tag{1}
$$

Let γ_p and γ_c be the gamma function of the projector and the camera devices. Then the linear RGB values of both devices can be expressed as

$$
\hat{P} = \left(\hat{R}_p, \hat{G}_p, \hat{B}_p \right)^T = ((R_p/255)^{\gamma_p}, (G_p/255)^{\gamma_p}, (B_p/255)^{\gamma_p},)^T \in [0,1]^3, \tag{2}
$$

$$
\hat{C} = \left(\hat{R}_c, \hat{G}_c, \hat{B}_c \right)^T = ((R_c/255)^{\gamma_c}, (G_c/255)^{\gamma_c}, (B_c/255)^{\gamma_c},)^T \in [0,1]^3. \tag{3}
$$

Then we assume that the relationship between both RGB values is characterized by a 3x4 color mixing matrix A as

$$
\hat{C} = A \begin{bmatrix} 1 \\ \hat{P} \end{bmatrix}, \qquad \hat{A} = \begin{bmatrix} r_K & r_R & r_G & r_B \\ g_K & g_R & g_G & g_B \\ b_K & b_R & b_G & b_B \end{bmatrix}. \tag{4}
$$

The matrix A becomes homogeneous form for representing the natural illuminant. In order to estimate the elements of the matrix A, we use captured four checkerboard colors $\hat{P}_K = (0,0,0)^T$, $\hat{P}_R = (1,0,0)^T$, $\hat{P}_G = (0,1,0)^T$, $\hat{P}_B = (0,0,1)^T$ in Fig. 2. Here, we can calculate the elements $\{r_K, g_K, b_K\}$ in the matrix A which represents an ambient light by projecting $\hat{P}_K = (0,0,0)^T$ and capturing the checkerboard color $\hat{C}_K$ from Eq.(4). Through the same procedure, we can obtain all elements in the matrix A by projecting $\hat{P}_R$, $\hat{P}_G$ and $\hat{P}_B$.

3   Gamut Extension Algorithm

In this section, we propose a color gamut extension algorithm using the projector-camera system. Let $\boldsymbol{P}^{*(t)}$ and $\hat{\boldsymbol{P}}^{*(t)}$ be an 8bit RGB vector projected to a printed material at time t and its linear RGB vector, respectively. As the initial projection, we set a gray image in the linear RGB space as $\hat{\boldsymbol{P}}^{*(0)} = (0.5, 0.5, 0.5)$. When the RGB intensities $\boldsymbol{P}^{*(t)}$ are projected on the printed material, the color camera captures the reflected intensities as RGB vector $\boldsymbol{C}^{(t+1)}$ at each pixel. The RGB vector is transformed into the linear RGB vector $\hat{\boldsymbol{C}}^{(t+1)}$ in the camera color space by Eq.(3). The RGB vector $\hat{\boldsymbol{C}}^{(t+1)}$ is further mapped into the projector-color space as follows:

$$\begin{bmatrix} 1 \\ \hat{\boldsymbol{P}}^{(t+1)} \end{bmatrix} = \boldsymbol{A}^{+}\hat{\boldsymbol{C}}^{(t+1)}, \tag{5}$$

where $\boldsymbol{A}^{+}$ represents the Moore-Penrose pseudoinverse of $\boldsymbol{A}$. In the projector color space, we perform the gamut extension for obtaining more vivid color. The color vector $\hat{\boldsymbol{P}}^{(t+1)}$ in the projector RGB space is transformed into HSV color space as follows:

$$\begin{bmatrix} H \\ S \\ V \end{bmatrix} = F\left(\hat{\boldsymbol{P}}^{(t+1)}\right), \tag{6}$$

where F means Smith's nonlinear conversion function from RGB to HSV [13]. The HSV stands for hue (H), saturation (S), and brightness value (V). Though there are many hue-based spaces, the authors adopt the HSV color space which is the most common cylindrical-coordinate representation of points in an RGB color model. The model is based more upon how colors are organized and conceptualized in human vision in terms of other color-making attributes, such as hue, lightness, and chroma; as well as upon traditional color mixing methods. In the HSV color space, we extend the saturation and brightness with preserving the hue angle, and obtain the extended vectors (H', S', V') as follows:

$$H' = H, \quad S' = \alpha S, \quad V' = \beta V, \tag{7}$$

where α and β mean extending factors for saturation and brightness, respectively. Then the extended colors are returned back to RGB projector space as

$$\hat{\boldsymbol{P}}^{'(t+1)} = F^{-1}\left(\begin{bmatrix} H' \\ S' \\ V' \end{bmatrix}\right), \tag{8}$$

where F^{-1} means the conversion function from HSV to RGB. Then a feedback gain matrix $\boldsymbol{G}$ is designed as

$$\boldsymbol{G} = diag\hat{\boldsymbol{P}}^{'(t+1)} diag\ \hat{\boldsymbol{P}}^{(t+1)^{-1}} \tag{9}$$

The gain factors are designed as the ratio between before and after gamut extensions and represented by the diagonal elements of the matrix $\boldsymbol{G}$. Then the linear RGB vector $\hat{\boldsymbol{P}}^{*(t+1)}$ for the projection at time t+1 is calculated as

$$\hat{\boldsymbol{P}}^{*(t+1)} = \boldsymbol{G}\hat{\boldsymbol{P}}^{*(t)}. \tag{10}$$

As the result, the RGB vector $\boldsymbol{P}^{*(t+1)}$ with 8 bit intensities for the projection at time $t + 1$ is calculated as

$$\boldsymbol{P}^{*(t+1)} = 255 \begin{bmatrix} \hat{R}_p^{*\,1/\gamma_p\,(t+1)} \\ \hat{G}_p^{*\,1/\gamma_p\,(t+1)} \\ \hat{B}_p^{*\,1/\gamma_p\,(t+1)} \end{bmatrix}. \tag{11}$$

If any elements of $\boldsymbol{P}^{*(t+1)}$ exceeds 8 bit range, the extended color is out of gamut and we keep the previous output of the projector as $\boldsymbol{P}^{*(t+1)} = \boldsymbol{P}^{*(t)}$. The dataflow of the proposed algorithm is summarized in Fig. 4. The right side of the dotted line shows processing in the camera color space, and the left side shows processing in the projector color space.

4   Experiments

The feasibility of the proposed system was evaluated by experiments which were performed in a seminar room under the Iluuminant D65. The distance between the projector-camera device and a screen was 700 mm and the size of the projected checkerboard patterns for the calibration was A4 size. The proposed algorithm was implemented on the computer by C++ language. Test samples were

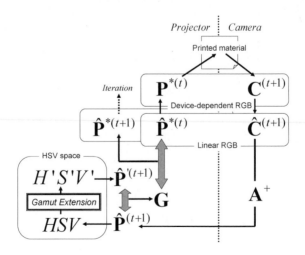

Fig. 4. The dataflow pipeline for the proposed algorithm

prepared by decreasing saturation and brightness of original pictures, and those samples were printed out on plain white papers using an inkjet printer (EPSON PM-A820).

At first, system calibration was performed by projecting four checkerboard patterns as shown in Fig.2. We investigated the error of the camera and geometric calibration by super-imposing the printed checkerboard pattern and a projected pattern. As the result, the error of the geometric calibration was less

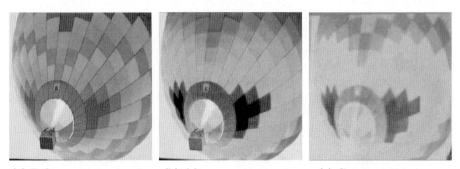

(a) Before gamut extension   (b) After gamut extension   (c) Compensation image

Fig. 5. Result of color gamut extension for "Balloon"

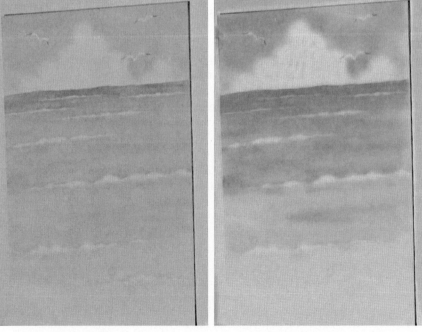

(a) Before gamut extension   (b) After gamut extension

Fig. 6. Result of color gamut extension for "Beach"

than 2 pixels. It took about 20 seconds for the calibration. After performing the calibration, we set a printed test sample on the screen. The proposed gamut extension algorithm has two parameters α and β as extending factors defined in Eq.(7). In our experiment, we determined those parameters as $\alpha = 1.35$ and $\beta = 1.1$ empirically.

Figure 5 shows a result of the gamut extension for an image "Balloon". All images were captured by a digital camera. Figure 5(a) shows the results of projecting flat-gray uncompensated initial image $\hat{P}^{*(0)}$ on the original printed image, and Fig. 5(b) shows the gamut extended image by projecting a compensation image. By comparing Fig. 5(b) with Fig. 5(a), we can confirm that a vivid color image which exceeds the printer color gamut was obtained by super-imposing the generated compensation image onto the printed image. The compensation image is shown in Fig. 5(c). The compensation image with $P^{*(t)}$ at each pixel was automatically generated by the proposed algorithm iteratively in Fig. 4. In this sample, Fig. 5(b) was obtained through 5 iterative processing. In this test sample, the mean saturation S increased from 0.30 to 0.65 and the mean brightness V increased from 0.40 to 0.55.

Figure 6 shows another test sample "Beach", which is a postcard. We confirmed that each color was appropriately extended beyond the printer gamut.

5   Conclusions

This paper has proposed a color gamut extension system for printed images by using a projector-camera system. The proposed system captured a printed image using a video camera and extended the color gamut of the printed image by super-imposing a compensation image obtained from a projector device on the printed image. The compensation image was produced in the HSV color space for iteratively adjusting the saturation and brightness values toward the boundary of the color gamut in the projector-camera system at each pixel. The feasibility of the proposed system was verified by experiments performed using real printed images.

The proposed algorithm works independently for each pixel. Therefore, the color contrast decreases when the pixel of similar hue is adjoined. Spatial gamut mapping might be required for obtaining more appropriate results and their solution remains as a future work.

References

1. Kotera, H., Mita, T., Hung-Shing, C., Saito, R.: Image-dependent gamut compression extension. In: Image Processing, Image Quality, Image Capture Systems Conferences, vol. 4, pp. 288–292 (2001)
2. Kang, B., Morovic, J., Luo, M., Cho, M.: Gamut compression and extension algorithms based on observer experimental data. ETRI J. 25, 156–170 (2003)
3. Kim, M., Shin, Y., Song, Y.L., Kim, I.: Wide gamut multi-primary display for HDTV. In: European Conference on Colour in Graphics, Imaging, and Vision, pp. 248–253 (2004)

4. Laird, J., Muijs, R., Kuang, J.: Development and evaluation of gamut extension algorithms. Color Research and Applications 34, 443–451 (2009)
5. Tsukada, M., Tajima, J.: Projector color reproduction adapted to the colored wall projection. In: European Conference on Colour in Graphics, Imaging, and Vision, pp. 449–453 (2004)
6. Grossberg, M., Peri, H., Nayar, S., Belhumeur, P.: Making one object look like another: Controlling appearance using a projector-camera system. In: IEEE Conference on Computer Vision and Pattern Recognition, vol. 1, pp. 452–459 (2004)
7. Bimber, O., Wetzstein, G., Emmerling, A., Nitschke, C., Grundhofer, A.: Enabling view-dependent stereoscopic projection in real environments. In: IEEE International Symposium on Mixed and Augmented Reality, pp. 14–23 (2005)
8. Renani, S., Tsukada, M., Hardeberg, J.: Compensating for non-uniform screens in projection display systems. In: SPIE Electronic Imaging, Color Imaging XIV: Displaying, Hardcopy, Processing, and Applications, vol. 7241 (2009)
9. Fujii, K., Grossberg, M., Nayar, S.: Projector-camera system with real-time photometric adaptation for dynamic environment. In: IEEE Conference on Computer Vision and Pattern Recognition, vol. 1, pp. 814–821 (2005)
10. Johnson, T., Fuchs, H.: Real-time projector tracking on complex geometry using ordinary imagery. In: IEEE Conference on Computer Vision and Pattern Recognition Workshop, pp. 1–8 (2007)
11. Amano, T., Kato, H.: Appearance enhancement using a projector-camera feedback system. In: IEEE International Conference on Pattern Recognition, pp. 1–4 (2008)
12. Zhang, Z.: A flexible new technique for camera calibration. IEEE Trans. Pattern Anal. and Machine Intell. 22, 1330–1334 (2000)
13. Smith, A.: Color gamut transform pairs. ACM SIGGRAPH Computer Graphics 12, 12–19 (1978)

Shading Attenuation in Human Skin Color Images

Pablo G. Cavalcanti, Jacob Scharcanski, and Carlos B.O. Lopes

Instituto de Informática, Universidade Federal do Rio Grande do Sul, Brazil

Abstract. This paper presents a new automatic method to significantly attenuate the color degradation due to shading in color images of the human skin. Shading is caused by illumination variation across the scene due to changes in local surface orientation, lighting conditions, and other factors. Our approach is to estimate the illumination variation by modeling it with a quadric function, and then relight the skin pixels with a simple operation. Therefore, the subsequent color skin image processing and analysis is simplified in several applications. We illustrate our approach in two typical color imaging problems involving human skin, namely: (a) pigmented skin lesion segmentation, and (b) face detection. Our preliminary experimental results show that our shading attenuation approach helps reducing the complexity of the color image analysis problem in these applications.

1 Introduction

Interpret the shading of objects is a important task in computer vision. This is specially true when dealing with human skin images, because the color of structures can be significantly distorted by shading effects. The occurrence of shading depends mainly on the color of the object and the light illuminating them. However, roughness of the surface, the angles between the surface and both the light sources and the camera, and the distance of the surface from both the light sources and the camera, can also significantly influence the way the scene is processed [1]. Specifically, human skin images are impacted by these factors and the analysis of these images can become difficult if the uneven illumination is not correctly understood and corrected.

In teledermatology, for example, often a standard camera color image containing a skin lesion is transmitted to a specialist, or analyzed by a pre-diagnosis system, without special attention to the illumination conditions [2][3]. However, these conditions can affect the quality of the visualization, and impact on the physician diagnosis, or limit the efficiency of the pre-screening system. Pigmented skin lesions typically have low diagnosis accuracy if the illumination condition is insufficient. These lesions usually are darker than healthy skin, and automatic approaches to segment such lesions tend to confuse shading areas with lesion areas. As a consequence, the early detection of malignant cases is more difficult

G. Bebis et al. (Eds.): ISVC 2010, Part I, LNCS 6453, pp. 190–198, 2010.

without removing shading effects from the images. Considering that melanoma is the most dangerous type of pigmented skin lesion, and that this disease results in about 10000 deaths in 40000 to 50000 diagnosed cases per year (only considering the United States of America [4]), any contribution to improve the quality of these images can be an important step to increase the efficiency of pre-diagnosis systems, and to help to detect cases in their early-stages.

Another important human skin color imaging application that is severely affected by shading effects is face detection. In this case, color images containing human skin are used in head pose estimation or in face recognition systems, and shading effects may occlude some important features of the face (e.g., eyes, nose, head geometry). Usually, it is not feasible to control the illumination condition during image acquisition, and an automatic preprocessing step to mitigate these effects is an important contribution to these systems efficiency, as will be illustrated later.

In this paper, we propose an new automatic approach to attenuate the shading effects in human skin images. In Section 2, we describe the algorithm that executes this operation. In Section 3, some preliminary experimental results of our method are shown, focusing on the benefits of this operation for the color image analysis of pigmented skin lesions and face images. Finally, in Section 4 we present our conclusions.

2   Our Proposed Shading Attenuation Method

Our method for shading effect attenuation improves on the approach proposed by Soille [5]. He proposed to correct the uneven illumination in monochromatic images with a simple operation:

$$R(x, y) = I(x, y) \ / \ M(x, y), \tag{1}$$

where, R is the resultant image, I is the original image, $M = I \bullet s$ is the morphological closing of I by the structuring element s, and (x, y) represents a pixel in these images. The main idea behind Soille method is to use the closing operator to estimate the local illumination, and then correct the illumination variation by normalizing the original image I by the local illumination estimate M. The division in Eq. 1 relights unevenly illuminated areas, without affecting the original image characteristics. Unfortunately, it is often difficult to determine an efficient structuring element for a given image, specially for human skin images that have so many distinct features, such as hair, freckles, face structures, etc. In this way, the results tends to be unsatisfactory for this type of images, as can be seen in Figs. 1(b)-(c).

Our method modifies the Soille approach by providing a better local illumination estimate M. In order to provide this local illumination estimate, we start by converting the input image from the original RGB color space to the HSV color space, and retain the Value channel V. This channel presents a higher visibility of the shading effects, as proposed originally by Soille.

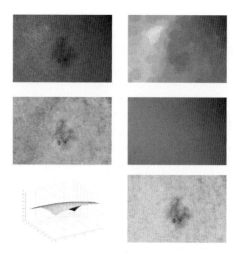

Fig. 1. Shading attenuation in a pigmented skin lesion image : (a) Input image; (b) Morphological closing of Value channel by a disk (radius = 30 pixels); (c) Unsatisfactory shading attenuation after replacing the Value channel by $R(x,y)$, as suggested by Soille [5]; (d) Local illumination based on the obtained quadric function; (e) 3D plot of the obtained quadric function; (f) Shading attenuation by using our approach.

We propose an approach inspired on the computation of shape from shading [1]. The human body is assumed to be constituted by curved surfaces (e.g. arms, back, faces, etc.) and, in the same way humans see, digital images present a smoothly darkening surface as one that is turning away from the view direction. However, instead of using this illumination variation to model the surface shape, we use this information to relight the image itself.

Let S be a set of known skin pixels (more details in Section 3). We use this pixel set to adjust the following quadric function $z(x,y)$:

$$z(x,y) = P_1 x^2 + P_2 y^2 + P_3 xy + P_4 x + P_5 y + P_6, \qquad (2)$$

where the six quadric function parameters P_i ($i = 1, ..., 6$) are chosen to minimize the error ϵ:

$$\epsilon = \sum_{j=1}^{N} [V(S_{j,x}, S_{j,y}) - z(S_{j,x}, S_{j,y})]^2, \qquad (3)$$

where, N is the number of pixels in the set S, and $S_{j,x}$ and $S_{j,y}$ are the x and y coordinates of the jth element of the set S, respectively.

Calculating the quadric function $z(x,y)$ for each image spatial location (x,y), we have an estimate z of the local illumination intensity in the image V. Replacing $M(x,y)$ by $z(x,y)$, and $I(x,y)$ by $V(x,y)$ in Eq. 1, we obtain the image $R(x,y)$ normalized with respect to the local illumination estimate $z(x,y)$. The final step is to replace the original Value channel by this new Value channel, and convert the image from the HSV color space to the original RGB color space.

As a consequence of this image relighting, the shading effects are significantly attenuated in the color image. Figs. 1(d)-(e) illustrate the results obtained with our shading attenuation method.

3   Experimental Results and Discussion

As mentioned before, our method is initialized by a set of pixels S known to be associated with skin areas. In this section, we discuss how to select this set of pixels S in two typical applications of human skin color image analysis, namely, the segmentation of pigmented skin lesions and of faces in color images. Our goal is to show that our shading attenuation approach helps in the analysis of these images, making the processing steps simpler.

3.1   Pigmented Skin Lesion Segmentation in Color Images

In this application, the focus is in the image skin area that contains the lesion. As consequence, during image acquisition, the lesion is captured in the central portion of the image, and is surrounded by healthy skin. Therefore, we assume the four image corners to contain healthy skin. This assumption is common in dermatological imaging, and also has been made by other researchers in this field [6] [7]. Therefore, we use 20×20 pixel sets around each image corner, and determine S as the union of these 1600 pixels (i.e. the four pixel sets).

Many methods have been proposed for analyzing pigmented skin lesions in dermoscopic images [8]. However, dermoscopes are tools used by experts, and there are practical situations where a non-specialist wishes to have a qualified opinion about a suspect pigmented skin lesion, but only standard camera imaging is available on site (i.e., telemedicine applications). In the following discussion we focus in this situation that justifies the use of telemedicine and standard camera imaging. To illustrate the effectiveness of our method, we compare the segmentation results in pigmented skin lesions with and without the application of our method. Usually, pigmented skin lesions correspond to local darker skin discolorations. The segmentation method used is a well known thresholding procedure based on Otsu's method [9]. This algorithm assumes two pixel classes, usually background and foreground pixels (specifically in our case, healthy and unhealthy skin pixels), and searches exhaustively for the threshold th that maximizes the inter-class variance $\sigma_b^2(th)$:

$$\sigma_b^2(th) = \omega_1(th)\omega_2(th)\left[\mu_1(th) - \mu_2(th)\right]^2, \tag{4}$$

where, ω_i are the a priori probabilities of the two classes separated by the threshold th, and μ_i are the class means. To segment the input color images, we determine a threshold th for each one of the RGB channels, and establish a pixel as lesion if at least two of its RGB values are lower than the computed thresholds. At the end, we eliminate possible small segmented regions filtering the thresholding result with a 15×15 median filter.

In Fig. 2, we present some pigmented skin image segmentation examples. These pigmented skin lesions images are publicly available in the Dermnet dataset [10]. Although our segmentation method is very simple, the application of the shading attenuation method increases its efficacy. In this way, the feature extraction and the classification procedure (typically the next steps in pre-diagnosis systems) have higher probability to produce accurate results.

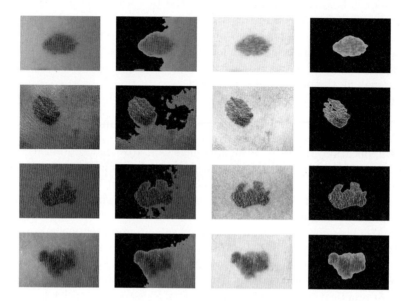

Fig. 2. Examples of pigmented skin lesion segmentation. In the first and second columns, the original images and their respective segmentation results. The third and fourth columns show the resulting images after the application of our shading attenuation method, and the respective segmentation results.

Our method may fail in some situations, as illustrated in Fig. 3. The situations illustrated in Fig. 3 are: (a) our method is adequate to model and attenuate the global illumination variation (which changes slowly), but tends to have limited effect on local cast shadows; and (b) our approach tends to fail on surface shapes that are not locally smooth, since the quadric function is not able to capture the local illumination variation in this case. In such cases, the segmentation method may confuse healthy and unhealthy skin areas. Possibly, better results could be achieved in such cases by acquiring the images in a way that surface shapes are smoother and illumination varies slowly across the scene.

3.2   Face Segmentation in Color Images

A face can be found in virtually any image location. In this case, the selection of the initialization pixel set S it is not as trivial as in the pigmented skin lesion

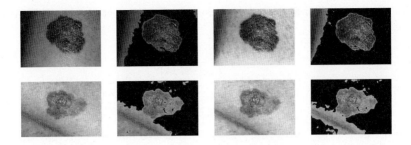

Fig. 3. Illustrations of cases where our shading attenuation method tends to fail, such as cast shadows (first line) and surface shapes not well modeled by quadric functions (second line). The first and second columns show the original images and their respective segmentation results. The third and fourth columns show the resulting images after the application of our shading attenuation method, and the respective segmentation results.

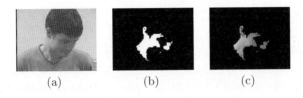

(a)   (b)   (c)

Fig. 4. Illustration of skin pixels localization using Eq. 5 : (a) Input image; (b) Binary mask; and (c) adjacent pixels identified as human skin.

segmentation problem. Therefore, we obtain the initialization pixel set S based on previously known color information [11]. A pixel is considered to be associated to a skin region in an RGB face image if :

$$R > 95 \ \wedge \ G > 40 \ \wedge \ B > 20 \ \wedge \tag{5}$$
$$\max(R, G, B) - \min(R, G, B) > 15 \ \wedge$$
$$|R - G| > 15 \ \wedge \ R > G \ \wedge \ R > B,$$

where, $\wedge$ denotes the logical operator *and*.

In Fig. 4, we present an example of the initialization skin pixels set S obtained with Eq. 5. Although this criterion to determine pixels associated to skin color is used often [11], it can be very imprecise in practical situations, specially when there is image shading. However, its use here is justifiable since all we need is a set of adjacent image pixels with skin color (i.e. likely to be located in skin regions) to initialize our error minimization operation (see Eqs. 2 and 3), and erroneous pixels should not influence significantly the final result.

Once S has been determined, the shading effects in the face image can be attenuated. To demonstrate the efficacy of our method in this application, we show the face segmentations with, and without, shading attenuation using a known Bayes Classifier for the pixels based on their corrected colors [11]. A pixel is considered skin if:

$$\frac{P(c|skin)}{P(c|\neg skin)} > \theta, \tag{6}$$

$$where \quad \theta = \kappa \times \frac{1 - P(skin)}{P(skin)}. \tag{7}$$

In Eq. 6, the a priori probability $P(skin)$ is set to 0.5, since we use the same number of samples for each class (i.e. 12800 skin pixels and 12800 non-skin pixels). The constant κ also is set to 0.5, increasing the chance of a pixel be classified as skin, and $P(c|skin)$ and $P(c|\neg skin)$ are modeled by Gaussian joint probability density functions, defined as:

$$P = \frac{1}{2\pi|\sum|^{1/2}} \times e^{-\frac{1}{2}(c-\mu)^T \sum^{-1}(c-\mu)}, \tag{8}$$

where, c is the color vector of the tested pixel, and μ and $\sum$ are the distribution parameters (i.e., the mean vector and covariance matrix, respectively) estimated based on the training set of each class (skin and non-skin).

The constant κ also is set to 0.5, increasing the chance of a pixel be classified as skin, and $P(c|skin)$ and $P(c|\neg skin)$ are modeled by Gaussian joint probability density functions, defined as:

$$P = \frac{1}{2\pi|\sum|^{1/2}} \times e^{-\frac{1}{2}(c-\mu)^T \sum^{-1}(c-\mu)}, \tag{9}$$

where, c is the color vector of the tested pixel, and μ and $\sum$ are the distribution parameters (i.e., the mean vector and covariance matrix, respectively) estimated based on the training set of each class (skin and non-skin).

Figs. 5 and 6 illustrate some face segmentation examples. These face images are publicly available in the Pointing'04 dataset [12]. The images in Fig. 5 show four different persons, with different physical characteristics and different poses (i.e. angles between their view direction and the light source), resulting in different shading effects. Clearly, the skin pixels, and consequently the face, is better segmented after we apply our shading attenuation method in all these different situations. In Fig. 6, we present four examples of the same person, just varying her head pose (the angle between her view direction and the light source). It shall be observed that even when the face is evenly illuminated, the face is better segmented after using our shading attenuation method. However, inaccuracies may occur near facial features partially occluded by cast shadows (e.g. near the nose and the chin). Based on these results, it should be expected that algorithms that extract facial features (e.g., eyes, mouth and nose) would perform their tasks more effectively, which helps in typical color image analysis problems such as head pose estimation or face recognition.

Fig. 5. Face segmentation examples. In the first and second columns are shown the original images, and their respective segmentation results. In the third and fourth columns, are shown images after the application of our shading attenuation method, and their respective segmentation results.

Fig. 6. Face segmentation examples for the same person varying its head pose. In the first and second columns are shown the original images, and their respective segmentation results. In the third and fourth columns, are shown images after the application of our shading attenuation method, and their respective segmentation results.

4   Conclusions

This paper presented a method for attenuating the shading effects in human skin images. Our preliminary experimental results indicate that the proposed method is applicable in at least two typical color image analysis problems where human skin imaging is of central importance. In the case of pigmented skin lesion segmentation, our shading attenuation method helps improving the lesion detection, and, hopefully, contributes for the early identification of skin cancer cases. We also studied the application of our shading attenuation method as a tool to increase the robustness of face segmentation, and our experiments suggest that potentially it can contribute to improve the efficiency of head pose estimation and facial recognition systems. We plan to further develop our approach using more complex quadric functions, and do a more extensive testing of our shading attenuation method in typical color imaging applications.

Acknowledgments

The authors would like to thank CNPq (Brazilian National Council for Scientific and Technological Development) for funding this project.

References

1. Shapiro, L., Stockman, G.: Computer Vision. Prentice Hall, Englewood Cliffs (2001)
2. Whited, J.D.: Teledermatology research review. Int. J. Dermatol. 45, 220–229 (2006)
3. Massone, C., Wurm, E.M.T., Hofmann-Wellenhof, R., Soyer, H.P.: Teledermatology: an update. Semin. Cutan. Med. Surg. 27, 101–105 (2008)
4. Melanoma Research Project, http://www.melresproj.com
5. Soille, P.: Morphological operators. In: Jähne, B., Haußecker, H., Geißler, P. (eds.) Handbook of Computer Vision and Applications, vol. 2, pp. 627–682. Academic Press, San Diego (1999)
6. Celebi, M.E., Kingravi, H.A., Iyatomi, H., Aslandogan, Y.A., Stoecker, W.V., Moss, R.H., Malters, J.M., Grichnik, J.M., Marghoob, A.A., Rabinovitz, H.S., Menzies, S.W.: Border detection in dermoscopy images using statistical region merging. Skin Res. Technol. 14, 347–353 (2008)
7. Melli, R., Grana, C., Cucchiara, R.: Comparison of color clustering algorithms for segmentation of dermatological images. In: Reinhardt, J.M., Pluim, J.P.W. (eds.) Medical Imaging 2006: Image Processing, SPIE, vol. 6144, p. 61443S (2006)
8. Maglogiannis, I., Doukas, C.: Overview of advanced computer vision systems for skin lesions characterization. IEEE Transactions on Information Technology in Biomedicine 13, 721–733 (2009)
9. Otsu, N.: A threshold selection method from gray-level histograms. IEEE Transactions on Systems, Man and Cybernetics 9, 62–66 (1979)
10. Dermnet Skin Disease Image Atlas, http://www.dermnet.com
11. Vassili, V.V., Sazonov, V., Andreeva, A.: A survey on pixel-based skin color detection techniques. In: Proc. Graphicon 2003, pp. 85–92 (2003)
12. Gourier, N., Hall, D., Crowley, J.L.: Estimating face orientation from robust detection of salient facial features. In: Proceedings of Pointing 2004, ICPR, International Workshop on Visual Observation of Deictic Gestures, Cambridge, UK (2004)

Color Constancy Algorithms
for Object and Face Recognition

Christopher Kanan, Arturo Flores, and Garrison W. Cottrell

University of California San Diego, Department of Computer Science and Engineering
9500 Gilman Drive, La Jolla, CA 92093-0404
{ckanan, aflores, gary}@cs.ucsd.edu

Abstract. Brightness and color constancy is a fundamental problem faced in computer vision and by our own visual system. We easily recognize objects despite changes in illumination, but without a mechanism to cope with this, many object and face recognition systems perform poorly. In this paper we compare approaches in computer vision and computational neuroscience for inducing brightness and color constancy based on their ability to improve recognition. We analyze the relative performance of the algorithms on the AR face and ALOI datasets using both a SIFT-based recognition system and a simple pixel-based approach. Quantitative results demonstrate that color constancy methods can significantly improve classification accuracy. We also evaluate the approaches on the Caltech-101 dataset to determine how these algorithms affect performance under relatively normal illumination conditions.

1 Introduction

Perceptual constancy is the tendency to perceive objects in an invariant manner despite changes in illumination, orientation, distance, position, and other factors. It greatly enhances our ability to visually discriminate items in our environment and has been studied extensively by psychologists and neuroscientists. Many algorithms have been developed to make the features in an image more invariant. Here, we focus on color and brightness constancy for discrimination.

Algorithms for color constancy have been developed for multiple reasons. The majority of algorithms, such as Retinex [1], have been designed primarily to enhance photographs taken under various illuminants. Their ability to improve object and face recognition has been infrequently investigated, despite the fact that this is a very important trait for an image recognition system to have. A deployed facial identification system that can only identify people under ideal lighting conditions would be of little practical use. Some methods attempt to attain a degree of invariance by using training on labeled images with varying color and illumination. However, this is an ideal situation and labeled training data of the same object/person under different lighting is not always available.

In many cases, color constancy algorithms are evaluated for their qualitative performance, i.e., how well an image is restored to normal lighting conditions. However, a color constancy method may have poor qualitative performance, yet still preserve important differences while weakening superficial ones. Even a qualitatively poor method may aid in image classification.

G. Bebis et al. (Eds.): ISVC 2010, Part I, LNCS 6453, pp. 199–210, 2010.

In this paper we review a variety of simple color constancy and contrast normalization techniques that could be readily used in an object recognition system. We compare both algorithms from both computer vision and computational neuroscience, which has not been previously examined. We then conduct experiments using a classifier applied directly to the image pixel channels immediately after applying color constancy algorithms. We also investigate how the approaches perform in a state-of-the-art descriptor-based framework by extracting SIFT descriptors from images and using the Naive Bayes Nearest Neighbor framework [2] for classification.

2   Methods

We have focused on color constancy algorithms that are simple, fast, and easy to implement so that they can be widely adopted by the image recognition community. A sophisticated and computationally intensive approach, like Local Space Average Color [3], is unlikely to be adopted by the object recognition community as a pre-processing step. We briefly review the methods we evaluate since some of them are not well known in the computer vision community. We also propose the Retina Transform.

2.1   Cone Transform

For humans and most vertebrates, color constancy begins with the retina where photoreceptors transduce light into neural signals. Our visual system copes with a 10-billion fold change in ambient light intensity daily, yet the activity of retinal ganglion cells, the output neurons of the retina, varies only 100-fold [4]. The photoreceptors modulate their responses using a logarithm-like function, which plays a role in luminance adaptation. These transformations are used by many computational neuroscientists in models of vision [5,6,7] to compensate for luminance and color changes across multiple orders of magnitude in natural images.

We adopt a simple model of the photoreceptors found in computational neuroscience [5,6,7]. First, we convert an image from RGB color space to LMS color space [8], approximating the responses of the long-, medium-, and short-, wavelength cone photoreceptors, which is done using the transformation

$$\begin{pmatrix} L \\ M \\ S \end{pmatrix} = \begin{pmatrix} 0.41 & 0.57 & 0.03 \\ 0.06 & 0.95 & -0.01 \\ 0.02 & 0.12 & 0.86 \end{pmatrix} \begin{pmatrix} R \\ G \\ B \end{pmatrix}, \tag{1}$$

where R, G, and B are the red, green and blue dimensions in RGB color space. Negative values are set to 0 and values greater than 1 are set to 1. This is followed by the nonlinear cone activation function applied each channel

$$I_{cone}(z) = \frac{\log(\epsilon) - \log(I_{LMS}(z) + \epsilon)}{\log(\epsilon) - \log(1 + \epsilon)}, \tag{2}$$

where $\epsilon > 0$ is a suitably small value and $I_{LMS}(z)$ is the image in the LMS colorspace at a particular location z. We use $\epsilon = 0.0005$ in our experiments. Essentially equation 2 is a logarithm normalized to be between 0 and 1. Although this method has been used by many computational neuroscientists, this is the first time that its discriminative effects have been evaluated.

2.2   Retina Transform

The Cone Transform discussed in Section 2.1 only captures the functionality of the retina's photoreceptors. The other cells in the retina transform this representation into a variety of opponent channels and a luminance channel, which are transmitted to the brain via retinal ganglion cells (RGCs) [4]. We developed a simple retina model to investigate its effect on discrimination. We are only modeling the retina's foveal RGCs, which encode the highest resolution information. We first apply the Cone Transform from Section 2.1 to the image. Then, the following linear transform is applied to the three Cone Transform channels

$$
\begin{pmatrix} O_1 \\ O_2 \\ O_3 \\ O_4 \\ O_5 \end{pmatrix} = \begin{pmatrix} \frac{1}{\sqrt{2}} & \frac{1}{\sqrt{2}} & 0 \\ \frac{1}{\sqrt{4\gamma^2+2}} & \frac{1}{\sqrt{4\gamma^2+2}} & -\frac{2\gamma}{\sqrt{4\gamma^2+2}} \\ -\frac{\gamma}{\sqrt{2\gamma^2+4}} & -\frac{\gamma}{\sqrt{2\gamma^2+4}} & \frac{2}{\sqrt{2\gamma^2+4}} \\ \frac{1}{\sqrt{\gamma^2+1}} & -\frac{\gamma}{\sqrt{\gamma^2+1}} & 0 \\ -\frac{\gamma}{\sqrt{\gamma^2+1}} & \frac{1}{\sqrt{\gamma^2+1}} & 0 \end{pmatrix} \begin{pmatrix} L_{\text{cone}} \\ M_{\text{cone}} \\ S_{\text{cone}} \end{pmatrix}, \tag{3}
$$

where $\gamma > 0$ is the inhibition weight. The RGC opponent inhibition weight is generally less than excitation [9], so we use $\gamma = 0.5$ throughout our experiments. O_1 contains ambient luminance information, similar to magnocellular RGCs [10]. O_2 and O_3 contain blue/yellow opponent information similar to koniocellular RGCs [10], and O_4 and O_5 contain red/green opponent information similar to parvocellular RGCs [10]. After applying Equation 3, the RGC output is computed by applying an elementwise logistic sigmoid, a common neural network activation function, to each of the channels.

2.3   Retinex

Perhaps the most well known color constancy algorithm is Retinex [1], which attempts to mimic the human sensory response in psychophysics experiments. Retinex is based on the assumption that a given pixel's lightness depends on its own reflectance and the lightness of neighboring pixels. We use the Frankle-McCann version of Retinex [1]. Unlike the other methods, Retinex is a local approach; it performs a series of local operations for each pixel and its neighbors. Refer to [1] for details. We test Retinex in both LMS and RGB color space.

2.4   Histogram Normalization

Histogram normalization, or equalization, adjusts the contrast of an image using the image's histogram allowing the values to be more evenly distributed [11], i.e., it spreads out the most frequent intensity values. While this approach works well for greyscale images, applying it to each channel of a color image could greatly corrupt the color balance in the image. We instead first convert the image to Hue, Saturation, and Value (HSV) color space and then apply histogram normalization only to the value (brightness) channel. This avoids corruption of the image's color balance.

2.5 Homomorphic Filtering

Homomorphic Filtering [12,13] improves brightness and contrast by modifying an image's illumination based on the assumption than an image's illumination varies slowly and that reflectance varies rapidly. It computes the Fast Fourier Transform (FFT) of an image's element-wise logarithm and then does high pass filtering to preserve reflectance while normalizing low-frequency brightness information. The image is restored using an inverse FFT and exponentiating. Like histogram normalization, we apply homomorphic filtering to the Value channel in HSV color space and then convert back to RGB color space.

2.6 Gaussian Color Model

Geusebroek et al. [14] developed the Gaussian Color Model derived using Kubelka-Munk theory, which models the reflected spectrum of colored bodies. Abdel-Hakim and Farag [15] advocated its use as a preprocessing step prior to running SIFT on each channel and [16] also found it worked well. While the model's derivation is sophisticated, it is implemented using a linear transform

$$\begin{pmatrix} E_1 \\ E_2 \\ E_3 \end{pmatrix} = \begin{pmatrix} 0.06 & 0.63 & 0.27 \\ 0.30 & 0.04 & -0.35 \\ 0.34 & -0.6 & 0.17 \end{pmatrix} \begin{pmatrix} R \\ G \\ B \end{pmatrix}, \tag{4}$$

where R, G, and B are the red, green and blue dimensions in RGB color space.

2.7 Grey World

Grey World [17] assumes that the average red, green, and blue values in an image are grey. A typical way of enforcing the grey world assumption is to find the average red, green, and blue values of the image $\mathbf{v} = (\mu_R, \mu_G, \mu_B)^T$. The average of these three values determines an overall grey value for the image $g = (\mu_R + \mu_G + \mu_B)/3$. Each color component is then scaled by how much it deviates from the average, the scale factors are $\mathbf{s} = (\frac{g}{\mu_R}, \frac{g}{\mu_G}, \frac{g}{\mu_B})^T$. Qualitatively, this method only works well if there are a sufficiently large number of different colors exhibited in a scene [3]. Despite this, it may still be beneficial for discrimination.

2.8 Opponent Color Space

Van de Sande et al. [16] examined several algorithms to improve color descriptors. Two of the approaches seemed to work fairly well, the Gaussian Color Model discussed in Section 2.6, and an approach based on opponent channels similar to the algorithm we proposed in Section 2.1. Changing to Opponent Color Space is done using the linear transform

$$\begin{pmatrix} O_1 \\ O_2 \\ O_3 \end{pmatrix} = \begin{pmatrix} \frac{1}{\sqrt{2}} & -\frac{1}{\sqrt{2}} & 0 \\ \frac{1}{\sqrt{6}} & \frac{1}{\sqrt{6}} & -\frac{2}{\sqrt{6}} \\ \frac{1}{\sqrt{3}} & \frac{1}{\sqrt{3}} & \frac{1}{\sqrt{3}} \end{pmatrix} \begin{pmatrix} R \\ G \\ B \end{pmatrix}, \tag{5}$$

where R, G, and B are the red, green, and blue components of the image. O_1, the red-green opponency channel, and O_2, the blue-yellow opponency channel, contain representations that are shift-invariant with respect to light intensity [16]. O_3 contains the intensity information, which lacks the invariance properties of the other channels.

3 Experiments: Simple System

Before testing each of the preprocessing techniques discussed in Section 2 in a descriptor-based object recognition system, we evaluated them using simple 1-nearest neighbor and linear support vector machine[1] (LSVM) [18] classifiers. We applied these classifiers directly to the preprocessed pixels by treating the images as high dimensional vectors. We use the RGB and LMS images as control methods. All images are resized to make the smallest dimension of the image 128, with the other dimension resized accordingly to preserve the image's aspect ratio. The pixel values of all images are normalized to be between 0 and 1 by dividing by 255. The default standard RGB color space is converted to linear RGB space by undoing the gamma correction if necessary.

3.1 ALOI

The Amsterdam Library of Object Images (ALOI) dataset [19] is a color dataset with 1,000 different small objects and over 48,000 images (768×576). Illumination direction and color is systematically varied in this dataset, with 24 illumination direction configurations and 12 illumination color configurations. We train on one image per category with uniform illumination conditions and test on seven images per category with altered lighting conditions (see Fig. 1).

Our results on ALOI are given in Figure 2. Almost all preprocessing methods had the effect of increasing classification accuracy, sometimes as much as by almost 20% in the case of Cone Transform over LMS alone. For this dataset, Grey World decreased performance, presumably due to the lack of sufficient color variation. A significant increase in classification accuracy is seen when using Retinex, Cone Transform, Homomorphic Filtering, and Histogram Normalization.

3.2 AR Face Dataset

The Aleix and Robert (AR) dataset[2] [20] is a large face dataset containing over 4,000 color face images (768×576) under varying lighting, expression, and dress conditions. We use images from 120 people. For each person, we train on a single image

---

[1] We used the LIBLINEAR SVM implementation, available at:
   http://www.csie.ntu.edu.tw/~cjlin/liblinear/
[2] Available at: http://cobweb.ecn.purdue.edu/~aleix/aleix_face_DB.html

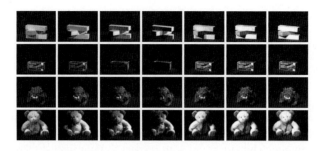

Fig. 1. Sample images from ALOI dataset [19]. Leftmost column is the training image (neutral lighting), the other columns correspond to test images.

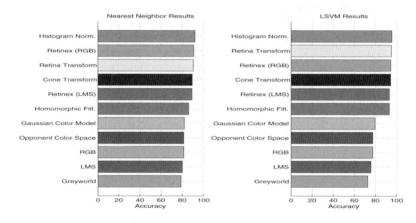

Fig. 2. Simple system results on the ALOI dataset. Histogram normalization, Retinex, and the Cone Transform all improve over the baseline RGB approach.

with normal lighting conditions and test on six images per individual that have irregular illumination and no occlusions (see Fig. 3). Our results are given in Figure 4. For both nearest neighbor and LSVM classifiers, Retinex, Homomorphic Filtering, and Grey World improve over the RGB baseline.

4   Experiments: Descriptor-Based System

Our results in Section 3 help us determine which of the methods we evaluated are superior, but object recognition systems generally operate on descriptors instead of the raw pixels. To test these approaches in a state-of-the-art recognition system we adopt the Naive Bayes Nearest Neighbor (NBNN) framework [2]. NBNN is simple and has been shown to be better than using more complicated approaches, such as bag-of-features with a SVM and a histogram-based kernel [2]. For example, using only SIFT descriptors NBNN achieves 65.0% accuracy on the Caltech-101 dataset [21] using 15 training instances per category compared to 56.4% in the popular Spatial Pyramid Matching

Fig. 3. Sample images from the AR dataset [20]. The leftmost column is the training image (neutral lighting) and the other columns correspond to test images.

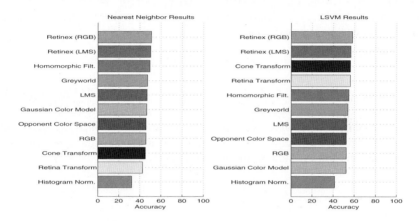

Fig. 4. Simple system results on AR dataset

(SPM) approach [22]. NBNN relies solely on the discriminative ability of the individual descriptors making it an excellent choice for evaluating color normalization algorithms.

NBNN assumes that each descriptor is statistically independent (i.e., the Naive Bayes assumption). Given a new query image Q with descriptors $\mathbf{d}_1, \ldots, \mathbf{d}_n$, the distance to each descriptor's nearest neighbor is computed for each category C. These distances are then summed for each category and the category with the smallest total is chosen. The algorithm can be summarized as:

The NBNN Algorithm
1. Compute descriptors $\mathbf{d}_1, \ldots, \mathbf{d}_n$ for an image Q.
2. For each C, compute the nearest neighbor of every $\mathbf{d}_i$ in C: $NN_C(i)$.
3. $\hat{C} = \arg\min_C \sum_{i=1}^{n} \mathrm{Dist}\,(i, \mathrm{NN}_C\,(i))$.

As in [2], $\mathrm{Dist}\,(x, y) = \|\mathbf{d}_x - \mathbf{d}_y\|^2 + \alpha \|\ell_x - \ell_y\|^2$, where ℓ_x is the normalized location of descriptor $\mathbf{d}_x$, ℓ_y is the normalized location of descriptor $\mathbf{d}_y$, and α modulates the influence of descriptor location. We use $\alpha = 0$ for all of our experiments except Caltech-101, where we use $\alpha = 1$ to replicate [2].

All images are preprocessed to remove gamma correction as in Section 3. We then apply a color constancy algorithm, followed by extracting SIFT [23] descriptors from each channel of the new representation and concatenating the descriptors from each channel. For example, for the Cone Transformation, 128-dimensional SIFT descriptor is extracted from each of the three channels resulting in a 384 dimensional descriptor. We use same SIFT implementation, VLFeat[3] [24], throughout our experiments. After all descriptors are extracted from the training dataset, the dimensionality of the descriptors is reduced to 80 using principal component analysis (PCA). This greatly speeds up NBNN, since it stores all descriptors from all training images. The remaining components are whitened, and the post-PCA descriptors are then made unit length.

As in Section 3, we also extract SIFT descriptors from images without applying a color constancy algorithm as a control method, using both RGB and LMS color spaces. We conduct experiments on ALOI [19] and AR Faces [20]. We also look at performance on Caltech-101 [21] to determine if any of the approaches are detrimental under relatively normal lighting conditions.

4.1   ALOI

We use the same subset of ALOI [19] that was used in Section 3.1. The results on ALOI are provided in Fig. 5. Retinex, the Cone Transform, and Homomorphic Filtering all improve over the RGB baseline. Note that the type of classifier interacts significantly with the color constancy method used, suggesting that the classifier type needs to be taken into account when evaluating these algorithms.

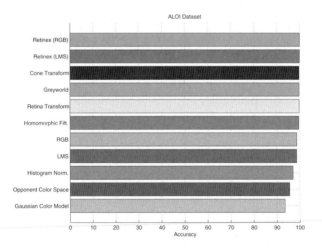

Fig. 5. Results on the ALOI dataset using each of the approaches discussed with NBNN and SIFT descriptors

For direct comparison with the experiment in [16], we also test the accuracy of each approach as a function of lighting arrangement (between one and three lights surrounding each object are on). For this experiment, we use the "l8c1" image in ALOI as the

---

[3] Code available at: http://www.vlfeat.org/

training instance, and test on the other seven lighting arrangements. Results can be seen in Fig. 6. Note the high performance achieved by Retinex and the Cone Transform. Our results are comparable or better to those in [16] for the three approaches in common: RGB, Opponent Color Space, and the Gaussian Color Model. This is likely because we used the NBNN framework, while [16] used the SPM framework [22]. While, [16] found that RGB was inferior to Opponent and the Gaussian Color models, we found the opposite. This is probably because the portion of the image that is illuminated normally will be matched very well in the NBNN framework, whereas the SPM framework will pool that region unless a sufficiently fine spatial pyramid is used; however, in [16] a $1 \times 1, 2 \times 2$, and 1×3 pyramid was used. Fig. 6 shows why Retinex, the Cone Transform, and Greyworld perform well, showing that they are resistant to lighting changes.

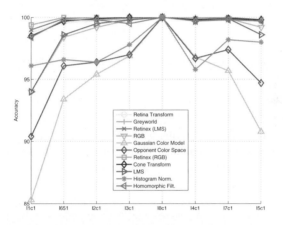

Fig. 6. Results for ALOI as a function of lighting arrangement at increasingly oblique angles, which introduces self-shadowing for up to half of the object

4.2   AR Face Dataset

The same subset of AR [20] used earlier in Section 3.2 is used with NBNN. The results are shown in Fig. 7. Retinex, the Cone Transform, and the Retina Transform all improve classification accuracy compared to the RGB baseline.

4.3   Degradation of Performance under Normal Conditions

Sometimes making a descriptor more invariant to certain transformations can decrease its discriminative performance in situations where this invariance is unneeded. To test this we evaluate the performance of each of the preprocessing algorithms on the Caltech-101 dataset[4] [21] using NBNN. We adopt the standard evaluation paradigm [2,22]. We train on 15 randomly selected images per category and test on 20 images per category,

---

[4] Available at:
   http://www.vision.caltech.edu/Image_Datasets/Caltech101/

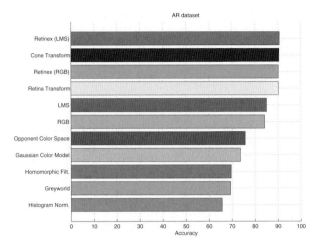

Fig. 7. Results on the AR dataset using each of the approaches discussed with SIFT-based descriptors and NBNN

unless fewer than 20 images are available in which case all of the ones not used for training are used. We then calculate the mean accuracy per class, i.e., the mean of the normalized confusion matrix's diagonal. We perform 3-fold cross validation and report the mean. Our results are given in Fig. 8. Retinex in RGB color space performed best, achieving $66.7 \pm 0.4\%$ accuracy. For comparison, using greyscale SIFT descriptors [2] achieved 65% accuracy in the NBNN framework. The Gaussian Color Model, Opponent Color Space, and Histogram Normalization all degrade performance over the RGB baseline.

5   Discussion

Our results demonstrate that color constancy algorithms can greatly improve discrimination of color images, even under relatively normal illumination conditions such as in Caltech-101 [21]. We found that both the Opponent and Gaussian Color representations were not effective when combined with the NBNN framework. This conflicts with the findings of van de Sande et al. [16]. This is probably due to differences in our approaches. They only used descriptors detected using a Harris-Laplace point detector (a relatively small fraction) and used the SPM framework, whereas we used all descriptors and the NBNN framework. Using all of the descriptors could have caused more descriptor aliasing at regions that were not interest points.

We generally found that the only method to take the local statistics of an image into account, Retinex [1], was one of the best approaches across datasets for both the simple and descriptor-based systems. However, this superiority comes with additional computational time. In our MATLAB implementation we found that Retinex with two iterations required 20 ms more than just converting to a different color space. This is a relatively small cost for the benefit it provides.

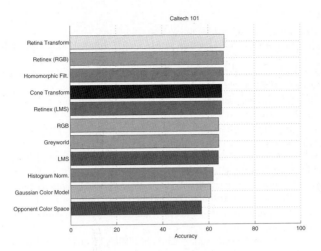

Fig. 8. Results on the Caltech-101 dataset using each of the approaches discussed

SIFT features [23] exhibit similar properties to neurons in inferior temporal cortex, one of the brain's core object recognition centers and it uses difference-of-Gaussian filters, which are found early in the primate visual system. Likewise, for the SIFT-based NBNN framework we found that the biologically-inspired color constancy algorithms worked well across datasets, unlike non-biologically motivated approaches like histogram normalization which performed well on some, but not all datasets.

6   Conclusion

We demonstrated that color constancy algorithms developed in computer vision and computational neuroscience can improve performance dramatically when lighting is non-uniform and can even be helpful for datasets like Caltech-101 [21]. This is the first time many of these methods, like the Cone Transform, have been evaluated and we proposed the Retina Transform. Among the methods we used, Retinex [1] consistently performed well, but it requires additional computational time. The Cone Transform also works well for many datasets. In future work we will evaluate how well these results generalize to other descriptors and object recognition frameworks, use statistical tests (e.g. ANOVAs) to determine how well the models compare instead of only looking at means, investigate when color is helpful compared to greyscale, and examine how each of the color constancy methods remaps an image's descriptors, perhaps using a cluster analysis of the images to see how the methods alter their separability.

Acknowledgements. This work was supported by the James S. McDonnell Foundation (Perceptual Expertise Network, I. Gauthier, PI), and the NSF (grant #SBE-0542013 to the Temporal Dynamics of Learning Center, G.W. Cottrell, PI and IGERT Grant #DGE-0333451 to G.W. Cottrell/V.R. de Sa.).

References

1. Funt, B., Ciurea, F., McCann, J.: Retinex in MATLAB. Journal of Electronic Imaging 13, 48 (2004)
2. Boiman, O., Shechtman, E., Irani, M.: In defense of Nearest-Neighbor based image classification. In: CVPR 2008 (2008)
3. Ebner, M.: Color Constancy Based on Local Space Average Color. Machine Vision and Applications 11, 283–301 (2009)
4. Dowling, J.: The Retina: An Approachable Part of the Brain. Harvard University Press, Cambridge (1987)
5. Field, D.: What is the goal of sensory coding? Neural Computation 6, 559–601 (1994)
6. van Hateren, J., van Der Schaaf, A.: Independent component filters of natural images compared with simple cells in primary visual cortex. Proc. R Soc. London B 265, 359–366 (1998)
7. Caywood, M., Willmore, B., Tolhurst, D.: Independent components of color natural scenes resemble V1 neurons in their spatial and color tuning. Journal of Neurophysiology 91, 2859–2873 (2004)
8. Fairchild, M.: Color appearance models, 2nd edn. Wiley Interscience, Hoboken (2005)
9. Lee, B., Kremers, J., Yeh, T.: Receptive fields of primate retinal ganglion cells studied with a novel technique. Visual Neuroscience 15, 161–175 (1998)
10. Field, D., Chichilnisky, E.: Information Processing in the Primate Retina: Circuitry and Coding. Annual Review of Neuroscience 30, 1–30 (2007)
11. Pizer, S., Amburn, E., Austin, J., Cromartie, R., Geselowitz, A., Romeny, B., Zimmermann, J., Zuiderveld, K.: Adaptive histogram equalization and its variations. Computer Vision, Graphics, and Image Processing 39, 355–368 (1987)
12. Oppenheim, A., Schafer, R., Stockham Jr., T.: Nonlinear filtering of multiplied and convolved signals. Proceedings of the IEEE 56, 1264–1291 (1968)
13. Adelmann, H.: Butterworth equations for homomorphic filtering of images. Computers in Biology and Medicine 28, 169–181 (1998)
14. Geusebroek, J., van Den Boomgaard, R., Smeulders, A., Geerts, H.: Color Invariance. IEEE Transactions on Pattern Analysis and Machine Intelligence 23, 1338–1350 (2001)
15. Abdel-Hakim, A., Farag, A.: CSIFT: A SIFT Descriptor with Color Invariant Characteristics. In: CVPR 2006 (2006)
16. van De Sande, K., Gevers, T., Snoek, C.: Evaluating Color Descriptors for Object and Scene Recognition. Transactions on Pattern Analysis and Machine Intelligence 32, 1582–1596 (2010)
17. Buchsbaum, G.: A spatial processor model for object colour perception. Journal of the Franklin Institute 310, 337–350 (1980)
18. Fan, R.E., Chang, K.W., Hsieh, C.J., Wang, X.R., Lin, C.J.: LIBLINEAR: A library for large linear classification. The Journal of Machine Learning Research 9, 1871–1874 (2008)
19. Geusebroek, J., Burghouts, G., Smeulders, A.: The Amsterdam library of object images. International Journal of Computer Vision 61, 103–112 (2005)
20. Martinez, A., Benavente, R.: The AR Face Database. CVC Technical Report #24 (1998)
21. Fei-fei, L., Fergus, R., Perona, P.: Learning generative visual models from few training examples: an incremental Bayesian approach tested on 101 object categories. In: CVPR 2004 (2004)
22. Lazebnik, S., Schmid, C., Ponce, J.: Beyond Bags of Features: Spatial Pyramid Matching for Recognizing Natural Scene Categories. In: CVPR 2006 (2006)
23. Lowe, D.: Distinctive image features from scale-invariant keypoints. International Journal of Computer Vision 60, 91–110 (2004)
24. Vedaldi, A., Fulkerson, B.: VLFeat: An Open and Portable Library of Computer Vision Algorithms (2008)

Chromatic Sensitivity of Illumination Change Compensation Techniques

M. Ryan Bales, Dana Forsthoefel, D. Scott Wills, and Linda M. Wills

Electrical and Computer Engineering, Georgia Institute of Technology, Atlanta, GA

Abstract. Illumination changes and their effects on scene appearance pose serious problems to many computer vision algorithms. In this paper, we present the benefits that a chromaticity-based approach can provide to illumination compensation. We consider three computationally inexpensive illumination models, and demonstrate that customizing these models for chromatically dissimilar regions reduces mean absolute difference (MAD) error by 70% to 80% over computing the models globally for the entire image. We demonstrate that models computed for a given color are somewhat effective for different colors with similar hues (increasing MAD error by a factor of 6), but are ineffective for colors with dissimilar hues (increasing MAD error by a factor of 15). Finally, we find that model choice is less important if the model is customized for chromatically dissimilar regions. Effects of webcamera drivers are considered.

1 Introduction

The wide proliferation and relatively low cost of USB webcameras make them attractive sensors for inexpensive computer vision platforms. Such platforms are useful for many applications including video surveillance, tracking, and recognition. Algorithms in these applications often rely on a degree of perceptual constancy to function properly. They observe trends in color pixel values to learn the appearance of background, and to identify features of interest. Illumination change is a common problem that such vision algorithms must face. Changes in lighting intensity, spectrum, or physical position alter the appearance of otherwise unchanged pixels, and can affect how a scene is perceived. It is desirable to compensate for illumination changes to improve the robustness of vision algorithms. Before illumination compensation can be performed, however, it is necessary to quantify the effects of lighting changes on images.

The purpose of this work is to determine the sensitivity of illumination change models to chromaticity, with the aim of improving illumination compensation techniques. Prior work in the field has discussed the problems of modeling illumination change globally—using one set of model parameters across the entire image—and has described the benefits of dividing images into arbitrary tiles. By considering each tile individually, the effects of spatially varying illumination and surface reflectance can be accommodated. In this paper, we propose that the effects of illumination changes have a significant dependence on surface color, and that the effectiveness of illumination models can be improved by segmenting the image into chromatically dissimilar

G. Bebis et al. (Eds.): ISVC 2010, Part I, LNCS 6453, pp. 211–220, 2010.

regions and separately computing compensations for each region. Spatial regionalization combined with chromatic regionalization will likely lead to additional benefits, but is not a requirement and we do not examine the compound effects here.

A set of color targets is illuminated by a controllable light source, and sets of images are taken with a webcamera under varying intensity levels and spectra. Choice of illumination model, chromatic regionalization of model parameters, and the webcamera's driver settings are examined for their effects on the effectiveness of illumination compensation. These controlled illumination experiments show that the choice of illumination model becomes less important when such chromatic regionalization is used. Computing illumination compensation models for each chromatic region reduces error by 70% to 80% on average as compared to applying a global compensation model across the entire image. Applying a model customized for one color to a color of different hue results in 15 times the error of that color's custom model. These trends can guide the development of computationally efficient illumination compensation techniques for webcamera-based vision platforms.

The rest of this paper is organized as follows. Section 2 discusses related work in the field of color and illumination change. Section 3 describes the experimental setup and evaluates several illumination models, driver settings and chromatic regionalization. Section 4 summarizes conclusions and describes future experiments.

2 Related Work

Illumination changes are generally categorized into two types: internal changes involve changes to the intensity or spectrum of the light source, while external changes result from the physical movement of the source with respect to the scene. Several studies have provided insight on the nature of scene response to internal illumination change. In particular, the choice of illumination model has received considerable attention as a tradeoff between computational complexity and accuracy.

Finlayson et al. [1] show that under certain conditions, several color constancy theories can be achieved by a diagonal matrix transform if an appropriate change of basis is applied to the sensor response function. Mindru et al. [2] test diagonal and affine transformations on images of a scene that are subjected to changes in lighting and viewpoint. They propose that the affine model best reduces error most consistently, and is worth the additional computational complexity. Gros [3] tests eight illumination models on images of a static scene in RGB space. A least median squares algorithm is used to find the optimum global parameter values for each model, and the error remaining between image pairs after compensation is calculated. For the case of intensity change, multiplication of the pixel triple by a single coefficient are found to be sufficient to account for most of the change, with more complicated models reducing the error marginally further. Spectral changes in the illumination source require models that adjust each color channel independently of the others.

Several approaches have been taken toward illumination compensation, motivated by a variety of applications. Photometric stereo [4] and physical models tend to require several training images of each scene of interest under specific known lighting conditions [5] or camera settings [6]. Human interaction with the system can also aid in color recognition to refine a model [7]. Spatially local statistics [8], [9] are computationally efficient but can fail in the presence of occlusions and interlaced textures. Features less sensitive to illumination such as special color spaces [10], [11] and edges [12], [13] can still vary, and are most often used as cues in conjunction with other approaches. Many illumination compensation techniques have been designed recently for skin tones and facial recognition [14], [15], but applications in image retrieval [16], object tracking [17], and video compression [9] have also motivated work in the field. The work presented in this paper provides a foundation for chromatically-oriented illumination compensation in color video processing.

3   Experimental Setup

This section describes the physical setup of the experiments, followed by analyses of the resulting data. The experiments used a fixed-focus Logitech USB webcamera on a stationary mount, as would be used in a surveillance application. A target was constructed consisting of twenty color chips (five hues with four saturations each) placed on a sheet of dark foamboard (Fig. 1a). The color target was oriented perpendicularly to the camera's optical axis, with the camera positioned 1.5 meters away. A light source was constructed using nine standard, independently controllable light fixtures (Fig. 1b). A plastic diffuser was used to diminish shadows, and the experiment was conducted in a dark room without external light sources. Lights were turned on three at a time to provide three consistent, discrete intensity levels. In addition, three bulb types were used (two incandescent, one fluorescent) to produce changes in spectrum. Bulb information is given in Table 1. The light source was located coaxially with and 3 meters behind the camera, and elevated 0.5 meters to reduce glare. A light meter was used to ensure that the light distribution across the target was uniform.

During the first stage of data collection, the webcamera's driver is set to automatic until the gain, exposure, contrast, and white balance settings stabilize. This step is performed on a mid-intensity scene. The driver is then switched to static operation, so the same settings are used for all subsequent image captures. Ten images are captured for each intensity level and bulb type, after allowing 10 minutes between each transition to allow the light source to reach steady state. The second stage of data collection repeats this process with the driver left on automatic, allowing the webcamera to adjust for each scene. The 10-image sequences are captured after the driver stabilizes.

3.1   Target Color Consistency

Sensors based on CMOS and CCD technologies are subject to many noise sources: temperature fluctuations, support electronics and the digital conversion process can all introduce noise into the final image. These effects can cause minor differences between images taken of an otherwise static scene. Thus, temporal and spatial averaging steps are used to minimize noise effects in our analysis of illumination changes.

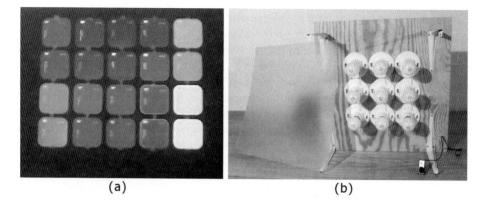

Fig. 1. (a) Color chips arranged on target. (b) Light board featuring a 3x3 arrangement of light fixtures, and detached diffuser. The Logitech camera is visible in the bottom right.

Table 1. Information about the light bulbs used in controlled illumination experiments

	Type	Power (W)	Output (lum)	Note
A	Incandescent	60	630	Full Spectrum
B	Incandescent	52	710	Soft White
C	Fluorescent	13	825	Soft White

The image sets are first tested for color consistency in the absence of illumination changes. Each set of 10 images is averaged together, and the average standard deviation if the images in each set is found to be less than 1.6% in RGB space. One average image is generated for each illumination condition, and is used in subsequent experiments. The color chips shown in Fig. 1a are chosen to provide manageable regions of reasonably consistent colors. Computations are performed on 40 x 40 pixel windows within each chip, chosen to exclude the text and labels visible on the chips.

3.2   Model Effectiveness and Chromatic Locality

Three mathematical models are tested for their effectiveness at illumination change compensation. We limit the set of evaluated models to those lowest in computational cost. These models are shown in Equations 1-3, where P is the 3x1 RGB pixel being transformed, D is a 3x3 diagonal matrix, and T is a 3x1 translation vector. Models are computed in RGB space. For each lighting transition, a least mean square algorithm is used to compute the optimum parameters (α, β, γ, x, y, and z) for each model. First the parameters of each model are tuned for global application across the image. The image is compensated by transforming each pixel by the model being tested. Then the mean absolute difference (MAD) is computed between the compensated image and the original. The calculation for MAD is shown in Equation 4, where N is the number of pixels in the regions being compared; R, G, and B represent the pixel components, and subscripts 1 and 2 denote the regions being compared. We accept MAD as a suitable metric for evaluating model performance, as our interests lie primarily in compensation to assist downstream object detection and tracking algorithms. Next, the models are optimized for and applied to each of the 20 color chips in the target, and the MAD is computed between each pair of compensated and original chips.

$$D * P = (\alpha R, \beta G, \gamma B) . \tag{1}$$

$$P+T=(R+x, G+y, B+z) . \tag{2}$$

$$D*P+T=(\alpha R+x, \beta G+y, \gamma B+z) . \tag{3}$$

$$MAD = \frac{1}{N} \sum_{i=1}^{N} \left| R_{i,1} - R_{i,2} \right| + \left| G_{i,1} - G_{i,2} \right| + \left| B_{i,1} - B_{i,2} \right| . \tag{4}$$

Table 2 shows the average MADs for the cases of no model applied, models applied globally, and models customized for individual color chips. Three lighting transitions are tested in which three, six, and nine identical lights are turned on—for example, transitioning from three lights to six lights (3-6) of the same type. Data is presented for the webcamera driver set to static operation, and for the driver set to automatic adjustment. Table 3 is organized in the same fashion and shows data for changes in light spectrum where A (full spectrum incandescent), B (soft white incandescent), and C (soft white fluorescent) denote light bulb types with different spectra.

Table 2 shows that in the case of globally calculated models, model selection has a significant impact on goodness-of-fit. The D*P+T model consistently results in the lowest error, and all three global models noticeably degrade as the magnitude of the intensity change increases. However, computing model parameters separately for each color not only achieves 70% to 80% lower error than the globally applied models, but also achieves a more consistent error rate regardless of the model used or the magnitude of the intensity change. Enabling the automatic driver measurably improves globally applied models, and does not significantly affect the performance of color-wise models, which still achieve 40% to 50% lower error than the globally computed models. This indicates that it is no more difficult to compensate for simultaneous changes in intensity and driver settings than it is to compensate for intensity change alone. Thus, from a steady-state point of view, it is reasonable to leave the automatic driver enabled to improve the camera's dynamic range.

To obtain the data in Table 3, we compare images taken under illumination from different light bulb types but similar intensities (e.g., 6 type A bulbs versus 6 type B bulbs). We see from the Uncompensated MAD row that the raw image differences caused by spectrum changes are lower in magnitude than those caused by the intensity changes. However, the automatic webcamera driver does little to mitigate the effects of the spectral changes. For changes in spectrum, optimizing the models based on chromatic region achieves 70% to 80% lower MAD error than the globally optimized models. Also, the global models gain little benefit from the automatic driver.

Table 2. Effectiveness of illumination models and color regionalization on reducing MAD error caused by intensity changes. Number pairs 3-6, 6-9, and 3-9 denote the magnitude of the intensity transition (ex: transitioning from 3 bulbs to 6 bulbs of the same type).

		Driver Static			Driver Automatic		
		3-6	6-9	3-9	3-6	6-9	3-9
Uncompensated MAD		145.2	101.7	246.9	24.4	18.4	10.4
Global	D*P	28.8	25.6	41.9	10.1	7.3	8.1
	P+T	21.2	19.9	32.2	9.3	6.5	8.1
	D*P+T	12.4	18.8	22.8	7.8	5.9	7.5
Chromatic Regions	D*P	4.3	4.3	4.5	4.7	3.9	4.6
	P+T	4.8	4.9	5.3	4.5	4.0	4.6
	D*P+T	4.2	4.3	4.6	3.7	3.5	3.7
Avg Error Reduction		76%	79%	84%	52%	42%	45%

Table 3. Effectiveness of illumination models and color regionalization on reducing MAD error caused by spectrum changes. Letter pairs A-B, B-C, and A-C denote the bulb type transition (ex: transitioning from type A bulbs to the same number of type B bulbs).

		Driver Static			Driver Automatic		
		A-B	B-C	A-C	A-B	B-C	A-C
Uncompensated MAD		34.7	39.6	42.6	38.5	42.0	41.7
Global	D*P	24.5	30.7	33.9	26.9	24.3	27.5
	P+T	18.6	26.3	27.5	19.8	19.3	19.1
	D*P+T	17.4	23.8	25.4	18.3	17.2	17.7
Chromatic Regions	D*P	9.8	8.9	6.7	10.0	8.2	8.6
	P+T	5.0	5.0	5.1	5.1	4.6	4.4
	D*P+T	4.1	4.3	4.5	4.4	3.9	3.8
Avg Error Reduction		70%	78%	81%	71%	73%	75%

3.3 Specificity of Model Parameters to Color

We observe that color-specific models achieve much better results than applying a model tuned to an entire image. Presumably, this effect is because a globally computed model is a compromise between the many colors and surfaces present in a scene. Next we wish to verify how color-specific the parameters of the illumination models are. We conduct this experiment by again applying a least mean square algorithm to compute the optimum parameters for each of three models for each color chip in the target. This time, the optimum models for each chip are applied to each of the other chips, and the MAD error is calculated between each compensated chip and its instance in the original image. This demonstrates how well the model parameters for each color work for each of the other colors. For compactness, we only show the chipwise results for the D*P model as a representative sample (Table 4), followed by the average results over all chips for each model (Table 5).

Table 4 shows the chip color for which the models are computed (G, B, P, R, Y indicate green, blue, purple, red, and yellow respectively, while the number indicates the saturation). The Self-Correct column shows the MAD error for the model applied to the chip for which it was optimized. The Similar Hue column shows the average

MAD error resulting from the model being applied to other chips of similar hue (i.e., the model for G1 applied to G2, G3 and G4). The Dissimilar Hue column shows the average MAD error resulting from the model being applied to the remaining chips.

Table 4. Effectiveness of color-specific illumination models on reducing the MAD error of various surface colors. This table represents an average over all intensity and spectrum changes.

Chip	Self-Correct	Similar Hue	Dissimilar Hue
G1	4.4	19.0	53.7
B1	4.2	12.8	53.1
P1	3.6	17.5	46.9
R1	4.2	18.3	54.3
Y1	4.0	22.7	68.9
G2	4.2	17.1	54.8
B2	3.9	10.7	44.4
P2	4.4	11.2	40.2
R2	3.5	15.7	53.8
Y2	4.5	24.9	62.2
G3	4.7	18.4	62.3
B3	3.6	13.1	47.7
P3	3.6	13.4	36.5
R3	4.1	14.3	65.2
Y3	6.4	21.1	77.9
G4	4.7	18.9	44.0
B4	3.6	10.9	41.9
P4	3.9	10.4	38.5
R4	5.3	23.8	83.0
Y4	5.2	20.0	74.2

The data supports our initial hypothesis—that optimum illumination models depend heavily on surface color, and that even with spatially uniform lighting, globally-tuned models are insufficient to compensate all of the colors and surfaces in a scene. Models achieve low error rates when applied to the color for which they are tuned (similar to error in Table 3). The models work better for surfaces of similar hue than for surfaces of different hues. Table 5 shows the data for 3 illumination models, and for 4 separate lighting transitions: dim to medium intensity (3-6), dim to bright intensity (3-9), dim spectrum changes (transitions between 3 bulbs of each type), and bright spectrum changes (transitions between 9 bulbs of each type). Regardless of the illumination change type, corrections do well when applied to the chromatic regions for which they are calculated. As intensity changes increase in severity, compensation effectiveness decreases for Similar Hue and Dissimilar Hue regions. Error increases by an average factor of 6 when chromatically optimized models are applied to colors of similar hue, and by an average factor of 15 when models are applied to colors of dissimilar hue. The diagonal transformation D*P proves most effective for Similar Hue regions during intensity changes, while the translation P+T is most effective for Similar Hue regions during spectrum changes.

Table 5. The average MAD error of chromatically-optimized illumination models applied to identical, similar and dissimilar colors. Data is shown for small and large intensity changes, and for changes in spectrum with low intensity (3 bulbs) and with high intensity (9 bulbs).

		Intensity (3-6)		Intensity (3-9)		Spectrum (Dim)		Spectrum (Bright)	
		AVG MAD	STD MAD	AVG MAD	STD MAD	AVG MAD	STD MAD	AVG MAD	STD MAD
D*P	Self Correct	4.4	0.9	4.5	1.0	8.5	6.2	6.7	3.9
	Similar Hue	22.5	14.7	27.9	17.0	41.4	41.4	41.6	37.9
	Dissimilar Hue	58.1	17.7	75.6	20.4	140.4	198.5	99.8	120.0
P+T	Self Correct	5.0	1.2	5.3	1.5	5.0	0.9	5.1	1.2
	Similar Hue	24.3	20.5	30.4	26.1	22.1	16.1	26.7	22.3
	Dissimilar Hue	37.8	12.6	48.7	13.6	39.7	10.3	45.4	11.6
D*P+T	Self Correct	4.3	0.9	4.6	1.1	4.3	0.7	4.5	1.0
	Similar Hue	38.0	20.7	43.5	23.2	32.4	17.5	36.1	18.9
	Dissimilar Hue	90.5	41.5	104.3	48.7	82.1	32.2	89.8	36.1

3.4   Realistic Scene Application

To test these observations on a more realistic scene with a wider diversity of surfaces, we capture new sets of images of a scene populated with various objects positioned at various angles (Fig. 2). A square region was selected from each of 12 objects in the scene (3 each of blue, red, green, and yellow). The objects differ in saturation and surface reflectance. Table 6 shows the results of applying each chromatic region's illumination change model to similar and dissimilar chromatic regions, formatted similarly to Table 5. Observations drawn from previous data hold for this scene as well. Models applied to surfaces with Similar Hue in the scene of Fig. 2 are slightly less effective than for the controlled surfaces of Fig. 1b due to differences between chromatic regions in surface reflectance and orientation.

Fig. 2. Realistic scene featuring a diversity of colors, surface reflectances, and orientations

Table 6. Average MAD error from applying three illumination models to the chromatic regions for which they were computed, chromatic regions with similar hue, and chromatic regions with dissimilar hue. Here, the models were applied to various surfaces in a realistic scene (Fig. 2).

		Intensity (3-6)		Intensity (3-9)	
		AVG MAD	STD MAD	AVG MAD	STD MAD
D*P	Self Correct	4.0	1.8	4.1	1.8
	Similar Hue	17.8	8.2	23.8	9.5
	Dissimilar Hue	68.4	14.1	83.7	12.1
P+T	Self Correct	6.0	2.4	7.3	3.0
	Similar Hue	40.1	11.8	53.8	16.9
	Dissimilar Hue	68.0	12.2	93.8	16.7
D*P+T	Self Correct	5.5	2.3	6.7	2.9
	Similar Hue	48.9	21.9	57.9	24.7
	Dissimilar Hue	101.8	25.9	121.5	30.5

4   Conclusions and Future Work

In this paper we show the significance of color to illumination changes in images captured by low-cost webcameras. Three illumination models are evaluated for their effectiveness in accounting for changes in lighting intensity and spectrum. The more complicated D*P+T model results in the smallest error out of the three models tested. However, by computing model parameters independently for each chromatically distinct region (without necessarily dividing the image into arbitrary spatial tiles), MAD error is reduced by an average of 70% to 80% compared with that achieved by globally calculated models. Furthermore, chromatic regionalization drastically reduces the variation in error due to model choice. This suggests that the least computationally expensive model (P+T) could be chosen in some applications to improve runtime performance in exchange for an acceptable penalty to accuracy. We have demonstrated that colors with different hues have significantly different illumination change responses, and that applying a chromatically optimized model to a color of dissimilar hue increases MAD error by an average factor of 15. Finally, we have presented evidence that a webcamera's automatic driver does not generally increase the complexity of illumination corrections. The driver does not reduce illumination compensation effectiveness for spectrum changes, and helps stabilize images after intensity changes.

This work explicitly tests the color dependency of illumination change in a way we have not seen in prior literature, and provides a compelling argument for using color regionalization in illumination modeling. Work is currently underway to exploit these relationships in a comprehensive compensation algorithm using automatic color regionalization.

References

1. Finlayson, G.D., Drew, M.S., Funt, B.V.: Diagonal Transforms Suffice for Color Constancy. In: Proc. of 4th International Conference on Computer Vision, pp. 164–171 (1993)
2. Mindru, F., Van Gool, L., Moons, T.: Model estimation for photometric changes of outdoor planar color surfaces caused by changes in illumination and viewpoint. In: Proc. International Conference on Pattern Recognition, vol. 1, pp. 620–623 (2002)
3. Gros, P.: Color illumination models for image matching and indexing. In: Proc. International Conference on Pattern Recognition, pp. 576–579 (2000)
4. Horn, B.K.P.: Robot Vision, pp. 185–216. McGraw-Hill, New York (1986)
5. Hager, G.D., Belhumeur, P.N.: Real-time tracking of image regions with changes in geometry and illumination. In: Proc. of IEEE Conference on Computer Vision and Pattern Recognition, pp. 403–410 (1996)
6. Wu, H., Jiang, P., Zhu, J.: An Illumination Compensation Method for Images under Variable Lighting Condition. In: Proc. of the IEEE Conference on Robotics, Automation and Mechatronics, pp. 1022–1026 (2008)
7. Makihara, Y., Shirai, Y., Shimada, N.: Online learning of color transformation for interactive object recognition under various lighting conditions. In: Proc. of the International Conference on Pattern Recognition, pp. 161–164 (2004)
8. Young, S., Forshaw, M., Hodgetts, M.: Image comparison methods for perimeter surveillance. In: Proc. of the International Conference on Image Processing and Its Applications, vol. 2, pp. 799–802 (1999)
9. Kamikura, K., Watanabe, H., Jozawa, H., Kotera, H., Ichinose, S.: Global brightness-variation compensation for video coding. IEEE Transactions on Circuits and Systems for Video Technology 8(8), 988–1000 (1998)
10. Drew, M.S., Jie, W., Ze-Nian, L.: Illumination-invariant color object recognition via compressed chromaticity histograms of color-channel-normalized images. In: Proc. 6th International Conference on Computer Vision, pp. 533–540 (1998)
11. Ming, Z., Bu, J., Chen, C.: Robust background subtraction in HSV color space. In: Proc. of SPIE, vol. 4861, pp. 325–332 (2002)
12. Hossain, M.J., Lee, J., Chae, O.: An Adaptive Video Surveillance Approach for Dynamic Environment. In: Proc. International Symposium on Intelligent Signal Processing and Communication Systems, pp. 84–89 (2004)
13. Wang, Y., Tan, T., Loe, K.: A probabilistic method for foreground and shadow segmentation. In: Proc. International Conference on Image Processing, vol. 3, pp. 937–940 (2003)
14. Xie, X., Lam, K.: An Efficient Illumination Compensation Scheme for Face Recognition. In: Proc. of the International Control, Automation, Robotics and Vision Conference, vol. 2, pp. 1240–1243 (2004)
15. Wong, K., Lam, K., Siu, W.: An efficient color compensation scheme for skin color segmentation. In: Proc. of the International Symposium on Circuits and Systems, vol. 2, pp. 676–679 (2003)
16. Tauber, Z., Ze-Nian, L., Drew, M.S.: Locale-based visual object retrieval under illumination change. In: Proc. of the International Conference on Pattern Recognition, vol. 4, pp. 43–46 (2000)
17. Yelal, M.R., Sasi, S.: Human tracking in real-time video for varying illumination. In: Proc. IEEE International Workshop on Intelligent Signal Processing, pp. 364–369 (2005)

Study on Image Color Stealing in Log-Polar Space

Hiroaki Kotera

Kotera Imaging Laboratory

Abstract. This paper proposes a cluster-to-cluster image color transform algorithm. Suntory Flowers announced the development of world's first blue rose "APPLAUSE". Since roses lack the blue pigment, it was long believed to be impossible. The key to success lies in the introduction of blue gene from pansy into rose. In the previous paper, PCA matching model was successfully applied to a seasonal color change in flowers, though it's not real but virtual. However, the tonal color transitions between the different color hues such as red rose and blue pansy were not so smooth but unnatural because of spatially independent color blending. In addition, the clear separation of blue or purple petal colors from greenish backgrounds is not always easy too. The paper improves the color transform algorithm in the two points, firstly, the clear color separation by introducing a "complex log" color space and secondly, the smoothed tonal color transition by introducing a "time-variant" matrix for PCA matching. The proposed algorithm is applied to ROI (Region Of Interest) image color transform, for example, a blue rose creation from red rose by continuous color stealing of pansy blue.

1 Introduction

Image segmentation plays an important role in many applications. Color clustering is a low-level task in the first stage of color image segmentation. The color of nature changes with passing time. Natural images are composed of clustered color objects with similarity to be shared each other. A concept of color transfer between two images was introduced by Kotera's PCA matching model [1] and advanced by Reinhard [2] as "scene color transfer" model. "Color stealing" by Barnsley [3] was a new concept of Fractal-based color sharing and used for synthesizing a new image by picking up a region color in one image and moving it into another image. Mochizuki [4] applied this idea to CG as "stealing autumn color". Our previous papers [5][6] extended the PCA matching model to a time-variant color transform and applied to imitate a seasonal color change in flowers. The model worked well for transferring a petal color in the source image into a different petal color in the target image, provided that the hue change between the source and target petal clusters is gentle.

Recently, Suntory Flowers succeeded in the development of world's first blue rose "APPLAUSE". Since roses lack the blue pigment, it was long believed to be impossible. The key to success lies in the introduction of blue gene from pansy into a source rose. Hearing this exciting news, we tested the PCA matching model to create a bluish rose from reddish roses by stealing a pansy blue. Of course, it's not a real but a virtual flower. Though, the color transitions in hue and tone were not so smooth but unnatural when the source and target images have a large difference in their color tones. This

G. Bebis et al. (Eds.): ISVC 2010, Part I, LNCS 6453, pp. 221–230, 2010.
© Springer-Verlag Berlin Heidelberg 2010

unnaturalness comes from the spatially-independent color blending between source and target clusters. In addition, it's not always easy to separate bluish or purple petals clearly distinguishing from the greenish background.

This paper improves the color stealing algorithm by introducing the new ideas of

a) *"log-polar mapping"* for clear segmentation of petal area by k-means clustering.
b) *"time-variant PCA matching matrix"* for smoothed color transitions from source to target.

Fig.1 overviews the proposed color stealing model and its application to creating a unique bluish rose from red, pink, orange, or yellow roses.

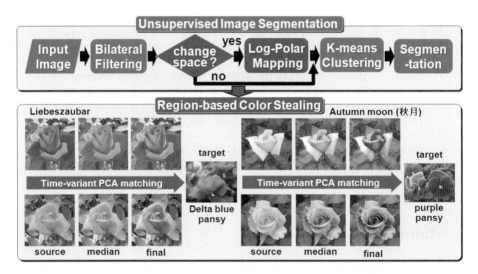

Fig. 1. Overview of segmentation-based image color stealing model

2   Pre-processing Filtering for K-Means Clustering

In practice, *k-means* clustering algorithm has been conveniently used for unsupervised image segmentation. Since *k-means* has a drawback in nonuse of spatial information, *JSEG* [7] introduced an excellent post-processing of region growing and region merging to avoid over segmentations. Instead of post-processing, this paper introduces a joint spatial-range bilateral filter to the *k-means* clustering as a pre-processing to make smooth the textural regions.

2.1   Joint LAB-Range Bilateral Filter

Before segmentation, $\{L^*, a^*, b^*\}$ images are pre-processed by a *bilateral filter* to make the *"texture"* area smooth without degrading the edge sharpness. The filtered pixel value $I_F(q)$ at central position q is given as a weighted sum of its surround pixels at p, where the spatial filter G_S works active or inactive if the range filter G_R has a high value for the low-gradient areas or a low value for the high-gradient edges.

The normal joint LAB-range bilateral filter is given by

$$I_F(q) = \frac{1}{K(q)} \sum_{p \subset \Omega} G_S(\|p-q\|) G_R(|I(p)-I(q)|) I(p)$$

$$\text{where, } K(q) = \sum_{p \subset \Omega} G_S(\|p-q\|) G_R(|I(p)-I(q)|) \qquad (1)$$

$$\text{and } G_S(x) = exp\left\{-\frac{(x)^2}{2\sigma_S^2}\right\}, \ G_R(x) = exp\left\{-\frac{(x)^2}{2\sigma_R^2}\right\}$$

Here, CIELAB values of $I(p)=L*(p)$, $a*(p)$, $b*(p)$ at pixel position p are assigned.

2.2　Joint LAB-Range vs. Joint Hue-Range Bilateral Filter

Since the each object in the image is considered to have its own surface color-hue, the range-filter G_R may be better replaced by the hue functions $H(p)$ and $H(q)$ as

$$I_{Filt}(q) = \frac{1}{K(q)} \sum_{p \subset \Omega} G_S(\|p-q\|) G_R(|H(p)-H(q)|) I(p)$$

$$\text{where, } K(q) = \sum_{p \subset \Omega} G_S(\|p-q\|) G_R(|H(p)-H(q)|) \qquad (2)$$

$$H(p) = \begin{cases} 0 \ (achromatic \ zone) \ for \ |a*(p)| \leq \delta \cap |b*(p)| \leq \delta \\ (2\pi)^{-1} tan^{-1}[b*(p)/a*(p)] \qquad for \ a*(p) > \delta > 0 \\ 0.5 + (2\pi)^{-1} tan^{-1}[b*(p)/a*(p)] \ for \ a*(p) < -\delta \\ 1.0 - (2\pi)^{-1} tan^{-1}[b*(p)/a*(p)] \ for \ a*(p) > \delta \cap b*(p) < -\delta \end{cases} \qquad (3)$$

Now the hue range is normalized to $0 \leq H(p) \leq 1$ and an achromatic zone is set around 0 < $|a*|$, $|b*|$< δ for small positive value δ to avoid any artifacts in gray area. In comparison with normal k-means, the pre-processing bilateral filter clearly improves the segmentation accuracy as shown in Fig.2. Though the joint hue-range bilateral filter seems to be a little bit better, the joint LAB-range bilateral filter is easy to use from a point of simplicity and computation costs.

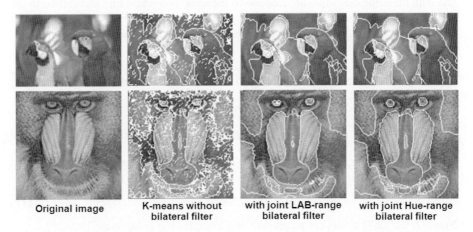

| Original image | K-means without bilateral filter | with joint LAB-range bilateral filter | with joint Hue-range bilateral filter |

Fig. 2. Effect in bilateral filter for image segmentation by k-means clustering

3   Color Mapping to Complex Log-Polar Space

3.1   Motivation for Using Log-Polar Transform

Schwartz's complex-logarithmic model [8] is known as a topographic Log-Polar Transform (LPT) of visual field onto the cortex. Though LPT is a space-variant image encoding scheme used in computer vision, here it's applied to the better separation of image color clusters from a point of mapping characteristics of LPT.

Two-dimensional log mapping function maps a complex number z to another complex number $log(z)$ as

$$z = x + jy = \rho e^{j\theta}, \text{ where, } \rho = |z| = \sqrt{x^2 + y^2}, \ \theta = tan^{-1}\left(\frac{y}{x}\right) \quad (4)$$

$$lo\, g(z) = u + jv = log(\rho) + j\theta, \ j = \sqrt{-1} \quad (5)$$

It maps the Cartesian coordinates (x, y) to the *log-polar* space notated as (u, v).

Since the origin itself is a singularity, the CBS (Central Blind Spot) model is introduced not to have the negative radii by setting the blind spot size $\rho_0 \leq \rho$ as

$$(u, v) \triangleq \left(log_a\left(\frac{\rho}{\rho_0}\right), \ \theta\right) \quad (6)$$

Considering the discrete *log-polar* space with R rings and S sectors for integer numbers of $u = 1, 2, \cdots, R$ and $v = 1, 2, \cdots, S$, we get the following relations [9] as

$$a = exp[log(\rho_R/\rho_0)/R] \quad (7)$$

$$\Delta\rho(u) = \rho_u - \rho_{u-1} = \rho_0 a^{u-1}(a-1) \quad (8)$$

$$\Delta\theta = \theta_v - \theta_{v-1} = 2\pi/S, \quad \vartheta_v = v\Delta\theta \quad (9)$$

Since the source data points (x, y) in a sector area $(\Delta\rho(u), \ \Delta\theta)$ are mapped to the new coordinates (u, v), points on the circles around the origin with equal radii are placed at the parallel vertical lines. While, points on the lines outward from the origin separated by equal angle are mapped onto the parallel horizontal lines.

Now assigning the (x, y) coordinates to the (a^*, b^*) values, the colors on the linear line with the same hue angle are mapped to the same horizontal line and shifted to the vertical directions for the different hue angles. As well, the colors on the circle with the same chroma (radius) are mapped to the same vertical line and shifted to the horizontal directions for the different radii.

3.2   Complex Log-Polar Transform in CIELAB Color Space

Applying this method to CIELAB color space, Eq. (4) is replaced by

$$z = a^* + jb^* = \rho e^{j\theta}, \text{ where, } \rho = |z| = \sqrt{a^{*2} + b^{*2}}, \ \theta = tan^{-1}\left(\frac{b^*}{a^*}\right) \quad (10)$$

Where, angle θ denotes the color hue in uniform perceptual color space CIELAB.

L^* value is also converted to the same *logarithmic* scale with the offset bias μ as

$$w = log(L^* + \mu) \quad (11)$$

Now, $\{L^*, a^*, b^*\}$ colors are mapped to new complex log-polar space $\{w, u, v\}$.

Fig.3 illustrates how the Munsell 1600 color chips are mapped onto log-polar space. It's shown that the chips with similar color hues tend to be mapped onto the horizontal lines separated vertically depending on their hue angles and the chips with similar radii from the center onto the vertical lines separated horizontally depending on their chroma values. Hence, the purplish/bluish petal colors may be better separated from the greenish back in the LPT space rather than normal CIELAB space.

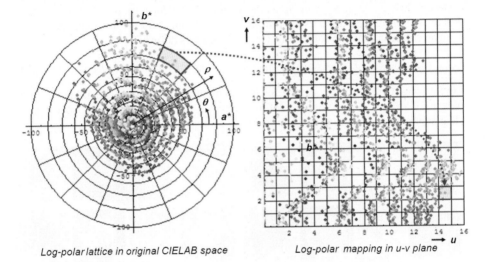

Log-polar lattice in original CIELAB space   Log-polar mapping in u-v plane

Fig. 3. Complex log-polar mapping of Munsell colors (1600 chips)

3.3   Cluster Separability in Complex Log-Polar Space

Fig.4 shows the segmentation results for typical roses by k-means with bilateral filter. In the case of segmentation into $K=2$ classes, most of vivid color petals are well separated from the background without LPT. While some of the dull-hued purplish or bluish petals were hard to separate clearly in normal CIELAB space. When the image colors are mapped onto complex LPT space, their clusters are occasionally relocated to be easier for segmentation. Of course, the complex LPT is not always superior to normal CIELAB but may be selected as the occasion demands. We need any criterion to judge which space has the better separability for the given color clusters.

As a measure of goodness in clustering, the invariant criterion function is estimated based on the scatter matrices [10] as follows.

The scatter matrix for the *k-th* cluster in subset D_k is described as

$$S_k = \sum_{x \in D_k} (x - m_k)(x - m_k)^t \; ; m_k = \frac{1}{N_k} \sum_{x \in D_k} x \tag{12}$$

Where, N_k denotes the number of pixels in class k.

The within-cluster scatter matrix S_W is given by the sum of S_k as

$$S_W = \sum_{k=1}^{K} S_k \tag{13}$$

While, the between–cluster scatter matrix S_B is defined by

$$S_B = \sum_{k=1}^{K} N_k (\boldsymbol{m}_k - \boldsymbol{m})(\boldsymbol{m}_k - \boldsymbol{m})^t ; \boldsymbol{m} = \frac{1}{N} \sum_{x \in D} \boldsymbol{x} = \frac{1}{N} \sum_{k=1}^{K} N_k \boldsymbol{m}_k \qquad (14)$$

The total scatter matrix $\boldsymbol{S}_T$ is the sum of $\boldsymbol{S}_W$ and $\boldsymbol{S}_B$ as given by

$$\boldsymbol{S}_T = \sum_{x \in D} (\boldsymbol{x} - \boldsymbol{m})(\boldsymbol{x} - \boldsymbol{m})^t = \boldsymbol{S}_W + \boldsymbol{S}_B \qquad (15)$$

Note that $\boldsymbol{S}_T$ doesn't depend on how the set of samples is partitioned into clusters. The between-cluster scatter $\boldsymbol{S}_B$ goes up as the within-cluster $\boldsymbol{S}_W$ goes down. Now, we can define an optimum partition as the criterion that minimizes $\boldsymbol{S}_W$ or maximizes $\boldsymbol{S}_B$.

The criterion function for the cluster separability is defined by

$$\boldsymbol{J}_{B/W} = T_{race}\left[\boldsymbol{S}_W^{-1}\boldsymbol{S}_B\right] = \sum_{i=1}^{3} \lambda_i ; \ \lambda_i \ is \ the \ i-th. \ eigen \ value \qquad (16)$$

Fig.5 shows a comparative sample for the segmentation in normal CIELAB vs. complex LPT spaces. The complex LPT outperforms the normal CIELAB space in its clear separation of petals with the higher $J_{B/W}$ scores.

Fig. 4. Segmentation results for typical roses by k-means with bilateral filter

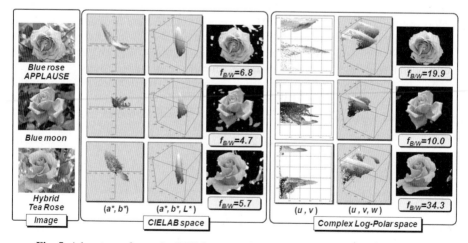

Fig. 5. Advantage of complex LPT for separating dull-hued bluish or purplish petals

4   Segmentation-Based Color Stealing

4.1   Cluster-to-Cluster Principal Component Matching

The key to color transfer between two different objects in source and target images is based on "cluster-to-cluster" PCA color matching algorithm [11] as follows.

First, the source color vector $_sX$ in image S and the target color vector $_DX$ in image T are projected onto the vectors $_sY$ and $_TY$ in the common PCA space by Hotelling Transform as

$$_sY =\ _sA(_sX -\ _s\mu), \quad _TY =\ _TA(_TX -\ _T\mu)$$
$$where, \ _s\mu = E\{_sX\} \ and \ _T\mu = E\{_TX\}: mean \ vectors \tag{17}$$

$_sA$ and $_TA$ are the eigen vectors of covariance matrices $_sC_X$ and $_TC_X$ for $_sX$ and $_TX$.

Thus the covariance matrices $_sC_Y$ and $_TC_Y$ for $_sY$ and $_TY$ are diagonalized as given by

$$_sC_Y =\ _sA(_sC_X)_s A' = diag\{_s\lambda_1,\ _s\lambda_2,\ _s\lambda_3\}$$
$$_TC_Y =\ _TA(_TC_X)_T A' = diag\{_T\lambda_1,\ _T\lambda_2,\ _T\lambda_3\} \tag{18}$$

Where, $\{_s\lambda_i\}$ and $\{_T\lambda_i\}$ are the eigen values of $_sY$ and $_TY$.

Second, the source color vector $_sY$ and the target color vector $_TY$ are mapped onto the same PCA axes and $_sY$ is transformed to match $_TY$ by the scaling matrix $_sS_T$ as follows.

$$_TY = (_sS_T) \cdot (_sY)$$
$$where, \ _sS_T = diag\left\{\sqrt{_T\lambda_1/_s\lambda_1},\sqrt{_T\lambda_2/_s\lambda_2},\sqrt{_T\lambda_3/_s\lambda_3}\right\} \tag{19}$$

Connecting Eq. (17) to Eq. (19), the colors $\{_sX\}$ in the source cluster S is transformed to the set of destination colors $\{_DX\}$ that is approximately matched to the colors $\{_TX\}$ in the target cluster T by the matrix M_C. M_C matches the color hue by cluster rotation and the variance by scaling as

$$_DX = M_C(_sX -\ _s\mu) +\ _T\mu \cong\ _T X$$
$$where, \ M_C = (_TA^{-1})(_sS_T)(_sA) \tag{20}$$

4.2   PCA Matching to Time-Variant Blended Cluster

In the previous paper [5][6], the PCA matching model was applied to a time-variant sequential color transfer to imitate a seasonal color change in flowers. A time-varying median image is created by matching the source cluster S to that of blended median cluster R just as the same method as cross dissolving [12]. The median cluster R is given by stealing the pixels from the segmented target image by the ratio of α_n and blending them with the source S by the ratio of $(1-\alpha_n)$ as follows.

$$R(n) = (1-\alpha_n)S + \alpha_nT \ \ for \ \alpha_n = n/N \ ; \ n = 0,1,\cdots,N \tag{21}$$

The conventional cross dissolving method causes a double exposure artifact due to the mixture of independent pixels between the source and target images, while the PCA matching algorithm gives a better time-variant color change from S to T by just substituting an every blended cluster R for the target T. Indeed, since the color transfer is limited to the segmented areas in S and T, the colors in source cluster S changes gradually approaching to those in the target cluster T with suppression of double exposure artifact. However this extension model didn't always create smooth tonal color changes between the images different color hues such as reddish and bluish roses, though, of course, it worked well between the similar color hues.

4.3   Color Stealing by Time-Variant PCA Matching Matrix

An improved algorithm is proposed to make the smoother color transitions between any two images. Instead of time-variant color blending function in Eq. (21), the time-variant PCA color matching matrix M_C is modified to be variable according to the time-variant ratio of α_n as

$$M_C(n) = (1-\alpha_n)I + \alpha_n M_C \;\; for \; \alpha_n = n/N \,; \; n = 0,1,\cdots,N$$

(22)

$M_C(n)$ changes from I to M_C according to the time sequence $n=0 \sim 1$, where I denotes 3 x 3 identity matrix.

Substituting $M_C(n)$ for M_C in Eq. (20), the colors $\{_SX\}$ in the source cluster S is transformed to the destination colors $\{_DX\}$ with time sequence n, finally matching to the colors $\{_TX\}$ in the target cluster T like as

$$_DX = \{(1-\alpha_n)I + \alpha_n M_C\}(_SX - _S\mu) + _T\mu$$
$$where, \; M_C = (_T A^{-1})(_S S_T)(_S A)$$

(23)

Fig.5 illustrates the basic concept of time-variant PCA matching matrix model.

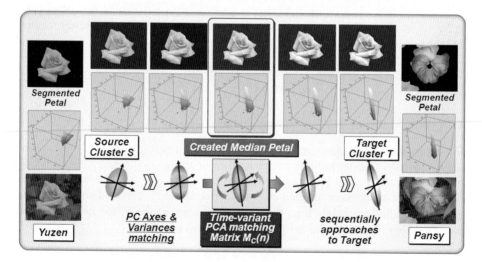

Fig. 6. Time-variant PCA matching matrix model for color stealing

5   Experiments on Blue Rose Creation by Stealing Pansy Blue

In 2004, Suntory Flowers Limited announced the successful development of the world's first blue rose "APPLAUSE", with nearly 100% blue pigment in the petals. Because roses lack blue pigment, their biotechnology research since 1990, introduced a blue gene from pansies into roses.

Now the proposed time-variant PC matching matrix model is applied to create a blue rose from a red rose by stealing the bluish colors from pansy as same as Suntory, though this is, of course, not a real but a virtual effect in computer color imaging.

The results are compared with the conventional cross resolving method and our previous time-variant blending model. The proposed time-variant PCA matching matrix model clearly resulted in the smoother color transitions in the hue and gradation of intermediate (median) images as shown in Fig.7.

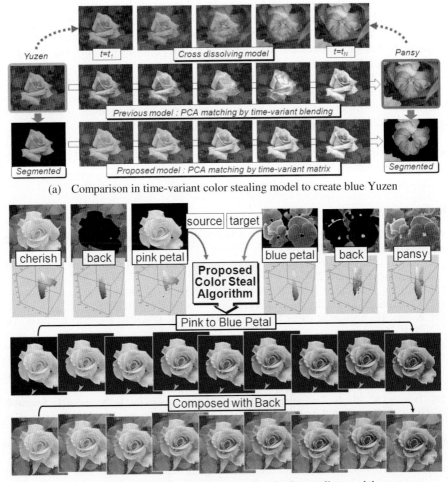

(a)   Comparison in time-variant color stealing model to create blue Yuzen

(b)   Creation of blue Cherish by proposed color stealing model

Fig. 7. Creation of Blue Rose by stealing Pansy Blue

6   Conclusions

The paper proposed a novel approach to a segmentation-based ROI image color transformations. Image segmentation is the basis for Computer Vision, but there is no royal road to the unsupervised clustering. K-means is a most popular algorithm for separating the color object with unique surface-hue without any learning samples.

Firstly, the pre-processing by joint LAB-range bilateral filtering proved to be very effective for separating the petal area from the complex background.

Secondly, paying our attention to the hue-oriented image color distributions, the color separability in clusters is newly discussed from a point of Schwartz's complex LPT mapping. Since the pixel colors on the same linear hue lines with different hue angles are mapped onto the same horizontal lines but separated vertically, such color clusters are remapped to be more separable. Although the LPT is a well-known space-variant image encoding scheme useful for computer vision but hasn't any meaningful relation to the geometrical design in color space, it resulted in the distinct cluster separability for the segmentation of dull-hued cold color roses better than the normal CIELAB space. The invariant criterion on the cluster separability is estimated by using the scatter matrices and the complex LPT outperforms normal CIELAB space with the higher $J_{B/W}$ scores.

However, the k-means clustering in LPT space is not always superior to normal CIELAB but is image-dependent. In order to switch on/off the LPT according to the image color distribution, a quick pre-estimation tool for the cluster separability is necessary and is left behind as a future work.

Lastly, the segmentation-based cluster-to-cluster PCA color matching algorithm has advanced by introducing a time-variant matching matrix. The proposed color stealing model is successfully applied to a time-variant virtual blue rose creation from usual reddish or warm color roses, resulting in the smoothed color transitions.

References

[1] Kotera, H., et al.: Object-oriented Color Matching by Image Clustering. In: Proc. CIC6, pp. 154–158 (1998)
[2] Reinhard, E., et al.: Color transfer between images. Proc. IEEE Comp. Graph. Appl., 34–40 (September/October 2001)
[3] Barnsley, M.: Super Fractals. Cambridge Press, New York (2006)
[4] Mochizuki, S., et al.: Stealing Autumn Color. In: SIGGRAPH (2005)
[5] Kotera, H.: Sequential Color Stealing of SOUVENIR D'ANNE FRANK. In: Proc. AIC 2009 (2009)
[6] Kotera, H.: Morphological Color Change in Morning Glory. In: Proc. CIC17, pp. 79–84 (2009)
[7] Deng, Y., Manjunath, B.S.: IEEE Trans. Patt. Anal. Machine Intell. 23(9), 800–810 (2001)
[8] Schwartz, E.L.: Spatial Mapping in the Primate Sensory Projection: Analytic Structure and Relevance to Perception. Biological Cybernetics 25, 181–194 (1977)
[9] Traver, J.V., Pla, F.: Designing the lattice for log-polar images. In: Nyström, I., Sanniti di Baja, G., Svensson, S. (eds.) DGCI 2003. LNCS, vol. 2886, pp. 164–173. Springer, Heidelberg (2003)
[10] Duda, R.O., Hart, P.E., Stork, D.G.: Pattern Classification. John-Wiley & Sons, West Sussex (2001)
[11] Kotera, H., Horiuchi, T.: Automatic Interchange in Scene Colors by Image Segmentation. In: Proc. CIC12, pp. 93–99 (2004)
[12] Grundland, M., et al.: Cross Dissolve without Cross Fade. In: Eurographics 2006, vol. 25(3) (2006)

How to Overcome Perceptual Aliasing in ASIFT?

Nicolas Noury, Frédéric Sur, and Marie-Odile Berger

Magrit Project-Team, UHP / INPL / INRIA, Nancy, France

Abstract. SIFT is one of the most popular algorithms to extract points of interest from images. It is a scale+rotation invariant method. As a consequence, if one compares points of interest between two images subject to a large viewpoint change, then only a few, if any, common points will be retrieved. This may lead subsequent algorithms to failure, especially when considering structure and motion or object recognition problems. Reaching at least affine invariance is crucial for reliable point correspondences. Successful approaches have been recently proposed by several authors to strengthen scale+rotation invariance into affine invariance, using viewpoint simulation (*e.g.* the ASIFT algorithm). However, almost all resulting algorithms fail in presence of repeated patterns, which are common in man-made environments, because of the so-called perceptual aliasing. Focusing on ASIFT, we show how to overcome the perceptual aliasing problem. To the best of our knowledge, the resulting algorithm performs better than any existing generic point matching procedure.

1 Introduction and Related Works

One of the first steps in many computer vision applications is to find correspondences between points of interest from several images. Applications are *e.g.* photography stitching [1], object recognition [2], structure from motion [3], robot localization and mapping [4], etc. Points of interest belong to "objects" viewed from different camera positions. Thus, their definition ought to be insensitive to the aspect of the underlying object. Besides, it is desirable to attach vectors to these points which describe a surrounding patch of image, in order to find correspondences more easily. Ideally, these vectors should not change across the views. In the pinhole camera model, 3D objects are transformed via projective mappings. However, the underlying object is generally unknown. With the additional assumption that points of interest lie on planar structures, points and descriptors should be invariant to homographies. Since affine mappings are first-order approximations of homographies, this weaker invariance is often considered sufficient.

In his groundbreaking work [2], D. Lowe explains how to extract scale+rotation invariant keypoints, the so-called SIFT features. Some authors have tried to reach affine invariance (see *e.g.* MSER [5], Harris / Hessian Affine [6] and the survey [7], or [8] for semi-local descriptors). Although these latter methods have been proved to enable matching with a stronger viewpoint change, all of them are prone to

G. Bebis et al. (Eds.): ISVC 2010, Part I, LNCS 6453, pp. 231–242, 2010.

fail at a certain point. A more successful approach has been recently proposed
by several authors (*e.g.* [9–11]), in which viewpoint simulation is used to increase
scale+rotation to affine invariance. These papers demonstrate that this dramatically improves the number of matches between two views compared to MSER or
Harris/Hessian Affine, even with a strong viewpoint change.

Let us explain viewpoint simulation, and especially Morel and Yu's ASIFT [10]
which we aim at improving. In ASIFT, affine invariance of image descriptors
is attained by remarking from Singular Value Decomposition that any affine
mapping A (with positive determinant) can be decomposed as

$$A = \lambda R_\psi \begin{pmatrix} t & 0 \\ 0 & 1 \end{pmatrix} R_\phi \tag{1}$$

where $\lambda > 0$, R_ψ and R_ϕ are rotation matrices, $\phi \in [0, 180°)$, $t \geqslant 1$.

Since SIFT is scale+rotation invariant, a collection of affine invariant (ASIFT)
descriptors of an image I is obtained by extracting SIFT features from the
simulated images $I_{t,\phi}$ with

$$I_{t,\phi} = \begin{pmatrix} t & 0 \\ 0 & 1 \end{pmatrix} R_\phi(I). \tag{2}$$

Indeed, the location of the SIFT keypoints is (nearly) covariant with any scale
and rotation change λR_ψ applied to $I_{t,\phi}$, and the associated descriptor does
(almost) not change. From [10], it is sufficient to discretize t and ϕ as: $t \in$
$\{1, \sqrt{2}, 2, 2\sqrt{2}, 4\}$ and $\phi = \{0, b/t, \ldots, kb/t\}$ with $b = 72°$ and $k = \lfloor t/b \cdot 180° \rfloor$.

The next step is to match ASIFT features between two images I and I'. A
two-scale approach is proposed in [10]. First, the $I_{t,\phi}$ and $I'_{t',\phi'}$ are generated
from downsampled images (factor 3), then SIFT features extracted from each
pair $(I_{t,\phi}, I'_{t',\phi'})$ are matched via the standard algorithm from [2], namely that
nearest neighbours are selected provided the ratio of the Euclidean distance between the nearest and the second nearest is below some threshold (0.6 in ASIFT).
The deformations corresponding to the M pairs (M typically set to 5) that yield
the largest number of matches are used on the full-resolution I and I', giving
new SIFT features that are matched by the same above-mentioned criterion. The
obtained correspondences are then projected back to I and I', provided already-
placed correspondences are at a distance larger than $\sqrt{3}$. This strategy is used
to limit the computational burden and also prevents redundancy between SIFT
features from different deformations. A subsequent step consists in eliminating
spurious correspondences with RANSAC by imposing epipolar constraints.

Lepetit and Fua [9] use the same decomposition as in eq. (1). Since their points
of interest are invariant neither to scale nor to rotation, they have to discretize
or randomly sample the whole set of parameters λ, t, ψ, ϕ.

Let us also mention Molton *et al.* work [11]. Small planar image patches are
rectified through homographies in a monocular SLAM application. In this framework, an on-the-fly estimation of the camera motion and of the 3D normal of the
patch is available. Thus there is no need to generate every possible rectification,

making it effective in a real-time application. This provides a richer description of the 3D scene than with standard point features.

Aim and organization of the article. As noted in [10], ASIFT fails when confronted to repeated patterns. In this work we propose to secure ASIFT against it. Section 2 explains why repeated patterns are important and call for a special treatment in *every* image matching applications. Section 3 describes the proposed algorithm. We also improve the selection of the relevant simulated images and the back-projection step, while enabling non nearest neighbour matches. Experiments are presented in Section 4. The proposed algorithm has also an increased robustness to large viewpoint changes.

2   Perceptual Aliasing and Point Matching

Perceptual aliasing is a term coined by Whitehead and Ballard in 1991 [12]. It designates a situation where *"a state in the world, depending upon the configuration of the sensory-motor subsystem, may map to several internal states; [and] conversely, a single internal state may represent multiple world states"*. In computer vision applications and especially point of interest matching, invariant features make it possible to overcome the first part of the perceptual aliasing. A viewpoint invariant feature such as ASIFT is indeed supposed to give a unique representation of the underlying 3D point, whatever the camera pose. However, repeated patterns are also uniquely represented although they do not correspond to the same 3D point. This makes almost all point matching algorithms fail when confronted to repeated patterns, except when explicitly taking them into account in an *ad hoc* application (*e.g.* [13]). Some authors even get rid of them at an early stage (*e.g.* in [14], patterns occurring more than five times are *a priori* discarded). The problem is of primary importance since repeated patterns are common in man-made environments. Just think of two views of a building: correctly matching the windows is simply impossible when considering only invariant descriptors. Additional geometric information is needed.

The problem is all the more relevant as in most applications, matching (or sometimes tracking) points of interest usually consists in two independent steps: 1) point of interest matching by keeping the "best" correspondence with respect to the distance between the associated descriptors, then 2) correspondence pruning by keeping those that are consistent with a viewpoint change. A popular choice for step 1) is nearest neighbour matching, which yet gives false correspondences, partly because of perceptual aliasing. The nearest neighbour has indeed no reason to be a correct match in case of repeated patterns. Step 2) is often a RANSAC scheme, which keeps only the correspondences consistent with the epipolar geometry (fundamental or essential matrix) or with a global homography (for planarly distributed points or for a camera rotating around its optical center). Since ASIFT uses this two-step scheme to match simulated images, it is not able to retrieve from perceptual aliasing. If the images mostly show repeated patterns, ASIFT even simply fails as in Section 4, Figures 5 and 6.

We have recently proposed [15] a new one-step method to replace both above-mentioned steps 1) and 2). It is a general algorithm to match SIFT features between two views, and it is proved to be robust to repeated patterns. The present contribution is to incorporate it into the ASIFT algorithm. Let us briefly describe the method (which is a generalization of [16]). Considering N_1 points of interest x_i with the associated SIFT descriptor D_i from image I, and N_2 points of interest x'_j with descriptor D'_j from image I', one aims at building a set of correspondences $(x_i, x'_j)_{(i,j) \in S}$ where S is a subset of $[1 \ldots N_1] \times [1 \ldots N_2]$, which is the "most consistent set with respect to a homography" among all possible sets of correspondences. Let us note that the model in [15] also copes with general epipolar geometry; we will see in Section 3 why we focus on homographies. The consistency of S is measured in [15] as a *Number of False Alarms* (NFA) derived from an *a contrario* model (see the books [17, 18] and references therein):

$$\text{NFA}(S, H) = (\min\{N_1, N_2\} - 4)\, k! \binom{N_1}{k} \binom{N_2}{k} \binom{k}{4} f_D(\delta_D)^k f_G(\delta_G)^{k-4} \quad (3)$$

where:
- the homography H from I to I' is estimated from four pairs from S,
- k is the cardinality of S,
- $\delta_D = \max_{(i,j) \in S} \mathsf{dist}(D_i, D'_j)$ where dist is a metric over SIFT descriptors,
- f_D is the cumulative distribution function of δ_D and is empirically estimated from I and I', yielding an adaptive measure of resemblance,
- $\delta_G = \max_{(i,j) \in S} \max\{d(x'_j, H x_i), d(x_i, H^{-1} x'_j)\}$ where d is the Euclidean distance between two points,
- f_G is the cumulative distribution function of δ_G.

Several possibilities for dist are investigated in [15]. We choose here to use the CEMD-SUM metric introduced in [19], based on an adaptation of the Earth's Mover Distance for SIFT descriptors. In particular, it is proved to behave better with respect to the quantization effects than the standard Euclidean distance.

For the sake of brevity, we elaborate here neither on the statistical model giving f_G and f_D nor on the definition of the NFA and kindly refer the reader to [15]. Let us simply say that $f_D(\delta_D)^k f_G(\delta_G)^{k-4}$ is the probability that *all* points in S are mapped to one another through H (with precision δ_G), while *simultaneously* the associated descriptors are similar enough (with precision δ_D), *assuming that points are independent.* If this probability is very low, then the independence assumption is rejected following the standard hypothesis testing framework. There must be a better explanation than independence, and each pair of points probably corresponds to the same 3D point. The advantage of this framework is that it automatically balances the resemblance between descriptors and the geometric constraint. Considering a group with a very low probability, all of its descriptors are close to one another (photometric constraint) and each of its points projects close to the corresponding point in the other image via H

(geometric constraint, which is not covered at all by nearest neighbour matching). Mixing both constraints makes it possible to correctly associate repeated patterns. Additionally, it is permitted to match non-nearest neighbours, provided they satisfy the geometry. As we will see in Section 4, this provides a number of correspondences that are never considered in standard SIFT matching.

Now, instead of measuring the probability $f_D(\delta_D)^k f_G(\delta_G)^{k-4}$ (of a false positive in hypothesis testing) which naturally decreases as k grows, the NFA is introduced in the *a contrario* literature. One can prove (see [15, 17–19] for further information) that a group such that NFA $\leq \varepsilon$ is expected to appear less than ε times under independence hypothesis (hence the term *Number of False Alarms*). Thus, comparing groups of different sizes via the NFA is sound. As noted in [15], small groups can win over large ones if they are very accurate (that is, descriptors are very similar and points are nearly perfectly related by a homography).

Since the combinatorial complexity of the problem is very large, a heuristic-driven search based on random sampling is given in [15], in order to reach the group S with the (hopefully) lowest NFA. Let us also mention that this method does not need application-specific parameters.

Remark that Hsiao *et al.* [20] have very recently proposed to use ASIFT for 3D object recognition, improving pose estimation when facing strong viewpoint changes, thanks to the numerous point correspondences. As in [15] for 2D/2D matching, they solve correspondences between 3D points and 2D points by simultaneously taking account of photometric resemblance and pose consistency. Their algorithm is thus robust to repeated patterns.

3   Improving ASIFT

We explain here how we modify the ASIFT algorithm by incorporating the NFA criterion (yielding the Improved ASIFT algorithm, I-ASIFT in the sequel). The basic idea is to replace the nearest neighbour matching between generated images with the matching algorithm of [15], *i.e.* seek the group of correspondences consistent with a homography, with the lowest NFA. The back-projection of the matching features to the original images is also improved. The algorithm for both Improved ASIFT and Standard ASIFT is explained in Figure 1, where the proposed modifications are highlighted. A running-example is provided on Figure 2. Figure 3 compares with standard SIFT matching and ASIFT.

Let us discuss the modifications. First, we replace in step 3 the nearest neighbour criterion by the above-mentioned method. The reason to use homography constraint is that when simulating affine transformations, one expects that some of them will correctly approximate homographies related to planar parts of the scene (possibly related to *virtual* planes, in the sense that points may be distributed on a plane which has no physical meaning). Then, each group of correspondences between simulated images should correspond to points lying over a planar structure, and consequently be associated via a homography. In standard ASIFT, the number of groups (*i.e.* of considered pairs of generated images) is

Data: two images I and I'.

1. For both images, generate the new collection of images $I_{t,\phi}$ and $I'_{t',\phi'}$ (eq. (2)):
I-ASIFT - use $t, t' \in \{1, \sqrt{2}, 2\}$ and ϕ, ϕ' as in ASIFT (the range of t is the same as in [20], sufficient if the viewpoint change is not too extreme)
ASIFT - first low resolution, then full resolution simulation only for a limited number of $(t, \phi), (t'\phi')$, as explained in Section 2.

2. Extract the SIFT features from the generated images.

3. Match the SIFT features between the pairs of generated images:
I-ASIFT - for each pair $(I_{t,\phi}, I'_{t',\phi'})$ extract the group of point correspondences with the lowest NFA (eq. (3), see discussion).
ASIFT - for each pair from the limited set of step 1, match each feature from $I_{t,\phi}$ to its nearest neighbour in $I'_{t',\phi'}$, provided the ratio between the distances to the nearest and to the second nearest neighbour is below 0.6

4. Back-project the matched SIFT keypoints from the $I_{t,\phi}$'s and $I'_{t',\phi'}$'s to I and I':
I-ASIFT - keep groups with $\log(\mathbf{NFA}) < -50$, then sort them increasingly along their NFA. Starting from the first group, back-project a pair of matching features only if each feature do not fall in the vicinity of any already-placed feature. The vicinity is defined as the back-projection in I (resp. I') of the circle around the feature extracted from the simulated images, with radius equal to the SIFT scale (minimum $\simeq 2$ pixels).
ASIFT - back-project the matching features only if there is no already-placed feature at a distance less than $\sqrt{3}$ pixels.

5. Discard possible false correspondences:
I-ASIFT - use *a contrario* RANSAC [16] to check consistency with epipolar geometry or to homography, depending on the case of interest.
ASIFT - use a contrario RANSAC to check consistency with epipolar geometry only (not mentioned in [10], but mandatory in the implementation from [21]).

Output: a set of corresponding points of interest between I and I'.

Fig. 1. Improved ASIFT (**I-ASIFT**) and Standard ASIFT (*ASIFT*)

limited *a priori* to five. In our framework, it would lead to select correspondences from a fixed number of planar pieces. On the contrary, we keep groups with log(NFA) below -50. This amounts generally to keeping between 5 (fully planar scene) and $\simeq 70$ (multi-planar scene) groups. There is no need to keep a larger number of groups since groups with the largest NFA would be made of redundant points or would be made of a few inconsistent points filtered by the final RANSAC.

To improve the back-projection of step 4, we propose to use the NFA as a goodness-of-fit criterion. As remarked by Hsiao *et al.* [20], viewpoint simulation methods give anyway a large number of correspondences, some of them being concentrated in the same small area. The NFA criterion balances the size of a group and its accuracy as explained earlier. It seems to us that favouring groups with the lowest NFA is sounder than systematically favouring large groups. In addition, when back-projecting points we thoroughly select correspondences from their scale in order to prevent accumulations in small areas (note that our criterion is stricter than the one in ASIFT). Getting correspondences uniformly and densely distributed across the 3D scene is important for structure and motion applications (as in [20]).

Let us remark that repeated patterns bring specific problems that RANSAC cannot manage. As remarked in [15], if the repeated patterns are distributed along the epipolar lines, then it is simply impossible to disambiguate them from two views (as in Figure 6, ACM+F). Theoretically, I-ASIFT could also suffer from it. However, it would require that: 1) one of the group consists in a bunch of shifted patterns consistent with a homography (as in group 51 on figure 2), 2) this group is large enough and has a very low NFA (otherwise most points are redundant with already-placed points), and 3) points are along the associated epipolar lines (otherwise they are discarded by the final RANSAC). Thus I-ASIFT is more robust to this phenomenon.

4   Experiments

We compare the proposed I-ASIFT, noted I-ASIFT+F (resp. I-ASIFT+H) when the final RANSAC is based on fundamental matrix (resp. homography), with:

– standard SIFT matching (that is, nearest neighbour + distance ratio condition, between 0.6 and 0.8 to give the best possible results), followed by the RANSAC from [16]. We note this algorithm NNR+F if RANSAC imposes epipolar geometry (fundamental matrix), or NNR+H for homography constraint;

– the *a contrario* matching algorithm from [15], which permits SIFT matching with repeated patterns, noted ACM+F or ACM+H;

– ASIFT, whose implementation is kindly provided by Morel and Yu [21].

We use Vedaldi and Fulkerson's code for SIFT [22]. The reader is kindly asked to zoom in the pdf file.

Figure 4 is an assessment on a pair of images with a very strong viewpoint change. NNR and ACM simply fail here, Harris/Hessian Affine and MSER give less than 2 matches (see [10]). Viewpoint simulation is thus needed. I-ASIFT provides more correspondences than ASIFT, which are distributed in a dense fashion while ASIFT accumulates them in small areas. This is mainly caused by the distance ratio threshold set to 0.6 in ASIFT, which discards too many correspondences in some generated image pairs. However, using a higher value leads to a larger rate of outliers, especially when considering large perspective deformations. The more sophisticated matching in I-ASIFT automatically adapts the resemblance metric between descriptors from each pair of simulated images.

Figure 5 shows an experiment with almost only repeated patterns which can still be disambiguated after a careful examination. In this case, ASIFT fails. More precisely, it gives some good correspondences, but they are buried in a large amount of false matches, and cannot be retrieved with the final RANSAC. NNR does not give any correspondence. Since the viewpoint change is not too

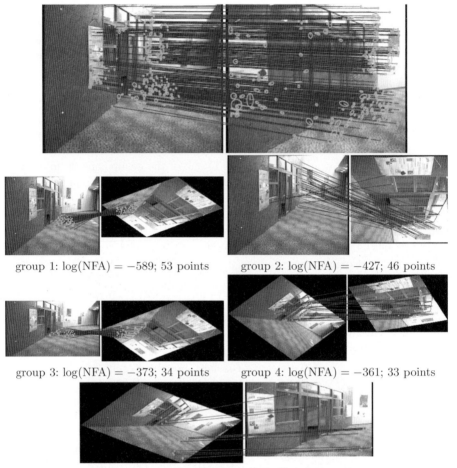

group 1: log(NFA) = −589; 53 points   group 2: log(NFA) = −427; 46 points

group 3: log(NFA) = −373; 34 points   group 4: log(NFA) = −361; 33 points

group 51: log(NFA) = −319; 13 points

Fig. 2. *Running example.* **Top**: 210 correspondences found with I-ASIFT. Each green ellipse is the backprojection in the original images of the circle with a radius equal to the SIFT scale in the simulated image. **Below (groups 1 to 4)**: correspondences from the four pairs $(I_{t,\phi}, I'_{t',\phi'})$ corresponding to the groups with the lowest NFA. One can see that these groups actually correspond to points over a quite small piece of plane. In this experiment, 65 such groups are kept (with log(NFA) < −50). The last 10 groups yield only 14 correspondences. As a comparison, the four groups shown here yield 116 correspondences. With our scheme, points from group 3 (resp. 4) redundant with those of group 1 (resp. 2) are not back-projected to I and I'. Note that points on the wall or on the carpet are scattered among several groups. Indeed, the strong induced homographies need to be approximated with several affine mappings. In **group 51**, the matching algorithm is trapped by perceptual aliasing: descriptors are alike but the correspondences are consistent with a homography "by chance". 10 points from this group are back-projected, but all of them are discarded by the final RANSAC imposing consistency to epipolar geometry.

Fig. 3. *Running example.* **Left**: SIFT matching (nearest neighbour + ratio set to 0.8), cleaned by the same RANSAC as in ASIFT [16]. **Right**: ASIFT. 97 matches are found for SIFT, 153 for ASIFT. For a fair comparison, ASIFT was run with the same resolution as I-ASIFT (no downsampling). Some points on the carpet are not correctly matched. A bunch of wrong correspondences can indeed be seen on the foreground, because of repeated patterns falling by chance near the associated epipolar lines.

Fig. 4. *The Creation of Adam* (from [10, 21]). **Left**: ASIFT. 100 matches. **Right**: I-ASIFT+H. 124 matches are retrieved with t in the range $\{1, \sqrt{2}, 2, 2\sqrt{2}, 4\}$ as in ASIFT. 49 groups are kept. The range $\{1, \sqrt{2}, 2\}$ (which we use for all other experiments of this article with I-ASIFT) still gives 29 matches (19 groups), not shown here.

Fig. 5. *Flatiron Building.* **Left**: ACM+H finds 21 matches, 13 of which are nearest neighbours, 5 are second nearest neighbours, the 3 remaining matches are between 3rd and 7th nearest neighbours. **Right**: I-ASIFT+H finds 44 matches, 29 of which are nearest neighbours, 7 are 2nd nearest, the 8 remaining matches are between 3rd and 6th nearest neighbours. 15 groups are kept.

strong, ACM+H still finds 21 correspondences, all correct. I-ASIFT+H finds 44 correspondences, all correct. One can see that the NFA criterion (eq. (3)) permits us to match features which are not nearest neighbours (30% in I-ASIFT+H). Of course, such correspondences are never considered in standard SIFT matching.

Figure 6 is another experiment with repeated patterns. To the best of our knowledge, I-ASIFT is the only generic point matching algorithm able to retrieve

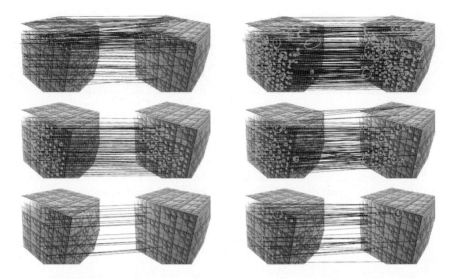

Fig. 6. *Synthetic Cube.* **Top left**: ASIFT. 101 correspondences, almost half of them are not correct. Many matching patterns are actually shifted. **Top right**: I-ASIFT+F. 192 correspondences. A careful examination proves that almost all are correct. Only 102 among them are nearest neighbours, the others match between 2nd and 8th nearest neighbours. 49 groups are kept. **Middle left**: ACM+H. 83 matches (only 40% of them are nearest neighbours), patterns of the "dominant" plane are correctly retrieved (homography constraint). **Middle right**: ACM+F. 102 matches (55% are nearest neighbours). False correspondences can be seen. This is simply unavoidable with two-view geometry, since in this experiment many wrongly associated repeated patterns (correct for the photometric constraint) lie along the corresponding epipolar lines (thus correct for the geometric constraint). **Bottom left**: NNR+H. 19 matches, corresponding to shifted patterns. **Bottom right**: NNR+F. 42 matches, many of them are not correct.

a large number of correct correspondences over the three visible sides of the cube. This is a highly desirable feature for structure and motion applications. One can also see that the ACM+H method (used in step 3 of I-ASIFT) is able to cope with repeated patterns. Let us remark that in this experiment we get *one* of the consistent solutions. However, we cannot eliminate the hypothesis that the cube has been rotated by 90°. In some cases, a certain amount of perceptual aliasing cannot be reduced from the information contained in images.

Figure 7 shows that I-ASIFT is also more robust to strong viewpoint changes than ASIFT. It is due to the proposed strategy consisting in back-projecting features from simulated images, which automatically selects a large number of groups of correspondences consistent with a local homography, contrary to ASIFT where most correspondences actually come from the same pair of simulated images. NNR+F and ACM+F do not give any set of correspondences.

Fig. 7. *Leuven Castle*: two distant images from M. Pollefeys' sequence. **Left**: ASIFT. 94 matches, only among points from the same façade. Note that repeated windows yield false correspondences with the fourth window in the second image (which is not present in the first one.) **Right**: I-ASIFT+F. 118 matches (24% are not nearest neighbours), distributed over the whole building. Except for two, all of them are correct. The fundamental matrix is then estimated over the retrieved set of correspondences. The epipolar lines (in yellow in the right images) corresponding to some handpicked points (from the left images) prove that I-ASIFT permits to reliably estimate the camera motion. The points associated to the handpicked ones are indeed less than 1 pixel away from the corresponding epipolar line. In contrast, the camera motion cannot be retrieved from ASIFT. As a comparison, MSER gives 5 matches, and Harris/Hessian Affine 20-30 matches mainly between wrongly associated repeated patterns. (code from Mikolajczyk et al.'s `www.featurespace.org`).

5 Conclusion

The main contribution of this article is to change the matching paradigm of ASIFT (namely nearest neighbour matching) to a more sophisticated one which aggregates sets of correspondences consistent with a local homography. It is not limited to nearest neighbour and yields dramatic results when confronted to repeated patterns. The resulting algorithm is also more robust than ASIFT, MSER, or Harris/Hessian Affine to large viewpoint changes, showing promising capacities for Structure From Motion applications.

References

1. Brown, M., Lowe, D.: Automatic panoramic image stitching using invariant features. International Journal of Computer Vision 74, 59–73 (2007)
2. Lowe, D.: Distinctive image features from scale-invariant keypoints. International Journal of Computer Vision 60, 91–110 (2004)
3. Gordon, I., Lowe, D.: Scene modelling, recognition and tracking with invariant image features. In: Proc. International Symposium on Mixed and Augmented Reality (ISMAR), pp. 110–119 (2004)
4. Se, S., Lowe, D., Little, J.: Vision-based global localization and mapping for mobile robots. IEEE Transactions on Robotics 21, 364–375 (2005)
5. Matas, J., Chum, O., Urban, M., Pajdla, T.: Robust wide-baseline stereo from maximally stable extremal regions. Image and Vision Computing 22, 761–767 (2004)
6. Mikolajczyk, K., Schmid, C.: Scale & affine invariant interest point detectors. International Journal of Computer Vision 60, 63–86 (2004)

7. Mikolajczyk, K., Tuytelaars, T., Schmid, C., Zisserman, A., Matas, J., Schaffalitzky, F., Kadir, T., Gool, L.V.: A comparison of affine region detectors. International Journal of Computer Vision 65, 43–72 (2006)

8. Musé, P., Sur, F., Cao, F., Gousseau, Y., Morel, J.M.: An a contrario decision method for shape element recognition. International Journal of Computer Vision 69, 295–315 (2006)

9. Lepetit, V., Fua, P.: Keypoint recognition using randomized trees. IEEE Transactions on Pattern Analysis and Machine Intelligence 28, 1465–1479 (2006)

10. Morel, J.M., Yu, G.: ASIFT: A new framework for fully affine invariant image comparison. SIAM Journal on Imaging Sciences 2, 438–469 (2009)

11. Molton, N.D., Davison, A.J., Reid, I.D.: Locally planar patch features for real-time structure from motion. In: Proc. British Machine Vision Conference, BMVC (2004)

12. Whitehead, S., Ballard, D.: Learning to perceive and act by trial and error. Machine Learning 7, 45–83 (1991)

13. Schaffalitzky, F., Zisserman, A.: Planar grouping for automatic detection of vanishing lines and points. Image and Vision Computing 18, 647–658 (2000)

14. Schaffalitzky, F., Zisserman, A.: Automated location matching in movies. Computer Vision and Image Understanding 92, 236–264 (2003)

15. Noury, N., Sur, F., Berger, M.O.: Determining point correspondences between two views under geometric constraint and photometric consistency. Research Report 7246, INRIA (2010)

16. Moisan, L., Stival, B.: A probabilistic criterion to detect rigid point matches between two images and estimate the fundamental matrix. International Journal of Computer Vision 57, 201–218 (2004)

17. Cao, F., Lisani, J., Morel, J.M., Musé, P., Sur, F.: A theory of shape identification. Lecture Notes in Mathematics, vol. 1948. Springer, Heidelberg (2008)

18. Desolneux, A., Moisan, L., Morel, J.M.: From Gestalt theory to image analysis: a probabilistic approach. Interdisciplinary applied mathematics. Springer, Heidelberg (2008)

19. Rabin, J., Delon, J., Gousseau, Y.: A statistical approach to the matching of local features. SIAM Journal on Imaging Sciences 2, 931–958 (2009)

20. Hsiao, E., Collet, A., Hebert, M.: Making specific features less discriminative to improve point-based 3D object recognition. In: Proc. Conference on Computer Vision and Pattern Recognition, CVPR (2010)

21. Morel, J.M., Yu, G.: ASIFT. In: IPOL Workshop (2009), http://www.ipol.im/pub/algo/my_affine_sift (Consulted 6.30.2010)

22. Vedaldi, A., Fulkerson, B.: VLFeat: An open and portable library of computer vision algorithms (2008), http://www.vlfeat.org/ (Consulted 6.30.2010)

Speeding Up HOG and LBP Features for Pedestrian Detection by Multiresolution Techniques

Philip Geismann and Alois Knoll

Robotics and Embedded Systems
Technische Universität München
Boltzmannstrasse 3, 85748 Garching, Germany
{geismann,knoll}@cs.tum.edu

Abstract. In this article, we present a fast pedestrian detection system for driving assistance. We use current state-of-the-art HOG and LBP features and combine them into a set of powerful classifiers. We propose an encoding scheme that enables LBP to be used efficiently with the integral image approach. This way, HOG and LBP block features can be computed in constant time, regardless of block position or scale. To further speed up the detection process, a coarse-to-fine scanning strategy based on input resolution is employed. The original camera resolution is consecutively downsampled and fed to different stage classifiers. Early stages in low resolutions reject most of the negative candidate regions, while few samples are passed through all stages and are evaluated by more complex features. Results presented on the INRIA set show competetive accuracy performance, while both processing and training time of our system outperforms current state-of-the-art work.

1 Introduction

The ability of driver assistance and automotive safety systems to handle dangerous traffic situations highly depends on the quality of their environmental perception. The detection of pedestrians is of special interest in this context, because they are the most vulnerable road users. From a detection point of view, finding persons in images is among the most challenging tasks in object recognition due to the high variability of human appearance. Furthermore, ressource issues are important, because a car safety system should be able to detect relevant pedestrians as soon as possible.

Most image-based recognition approaches today use sliding window techniques, thus solving the problem of object localiziation by image patch classification. In a moving camera environment, the window sizes and positions depend on the real world distances of objects. Scanning procedures considering only relevant positions are commonly referred to as "smart sliding windows". Current

G. Bebis et al. (Eds.): ISVC 2010, Part I, LNCS 6453, pp. 243–252, 2010.

research in pedestrian classification is mainly focused on improving accuracy and speed.

The complexity of the pedestrian class is handled via strong machine learning methods by computing discriminative features inside an image region. The first approach that showed excellent results was the work of Dalal/Triggs [1] who used local histograms of oriented gradients (HOG) as features and a linear support vector machine classifier. While HOGs (and similar gradient based features like edge orientation histograms (EOH) [2]) proved to be be very accurate for pedestrian recognition [3], [4], they also have weaknesses: Due to the nature of the histogram, image regions with different content can lead to a similar gradient histogram. Therefore, more features have been used in addition to HOGs to overcome these disadvantages, the most noteable being the local binary pattern (LBP) [5], [6]. Authors succeeded in utilizing more discriminant features, but usually coming at high computation costs [7],[8].

To deal with speed issues, several concepts have been introduced in the last years. Most commonly, boosted classifiers in a rejecting cascade structure can speed up sliding window techniques significantly, since the detection of one or more pedestrians in an image is a rather rare event [9],[10]. Additionally, boosting as feature selection provides an efficient way to choose discriminant features from a large feature pool [11]. However, training time is often very high, since the boosting process can be seen as a a search for the optimal features/classifiers to reach a certain training goal. For most boosting algorithms, the majority of computed features and weak classifiers trained is never used in the final classifier. Another promising speedup mechanism is the use of different resolutions in a coarse-to-fine strategy. In works like [12], [13], classification of subregions in an image is performed as a multiple stage process. By computing simple to complex features in coarse-to-fine resolutions, the full sliding window approach can be thinned out and thus accelerated. The outcomes of these approaches have similarities to rejecting cascades, but use the training data in a more efficient and time saving way.

In this paper, we propose a multiresolution system that detects pedestrians in different sizes using HOG and LBP features, which currently appear to be the most powerful feature types for pedestrian recognition. A general system overview is given in Figure 1 (a). We show that the performance is competetive while the execution and training time is reduced significantly compared to current state-of-the-art works.

2   Approach

Previous studies have shown that using different types of features for pedestrian classification can achieve a great improvement. Ideally, complementary features can mutually compensate for their particular flaws and make the whole process more robust. In the following section, we give an overview of the two feature types used and some detail on efficient implementation.

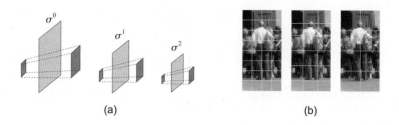

Fig. 1. (a) Multiscale detection in multiple resolutions. Each resolution r is obtained by downsampling factor α^r. The detection process starts in the lowest resolution, regions classified as positives are passed to the next resolution stage. (b) Block grids for three descriptors in different sizes (image patches are upscaled).

2.1 Feature Extraction in Multiple Scales

In order to detect pedestrians in various distances from the camera (near to far), features must be extracted at several sizes, called *scales* throughout this article. This could be achieved by resizing the input image and shifting a detector window with constant size. However, since the number of scaling steps is usually quite high (covered distances range from 0m to 40m), it is expensive to compute all features for every step, even if the images get smaller. To reduce the number of feature computations, we desire a feature representation that allows us to scale block features rather than the input image itself. Therefore, we make use of the concept of the integral image or integral histogram ([14]) for both types of features. This mechanism allows for rapid block sum computation in constant time for any scale size. It is especially helpful for sliding windows, since evaluated image regions usually have strong overlap. In practice, grids of block features are resized and shifted across the image in different scales. Note that although the overall system uses multiple resolutions, different scales still must be evaluated at each resolution. As a result, actual feature computation is done only for a few downscaling resolutions, whereas the scales in between are obtained by direct scaling of features.

Usually, integral images are computed globally for the whole image. However, in a multiresolution approach it makes sense to provide each resolution stage with a mask image of the previous outcoming regions. This way, the integral image feature computation is ideally performed only on small image areas. The following subsections show how integral images can be utilized for HOG and LBP features.

Histograms of Oriented Gradients (HOG). HOG features originate from the work of Lowe [15] and have been successfully used in pedestrian classification by Dalal/Triggs [1]. It seems to be one of the best features for capturing edge and shape information, while being sensitive to noisy background edges and clutter. In the original approach, pixel gradients are extracted from a spatial grid of

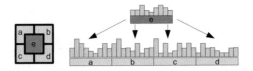

Fig. 2. An additional center block is used for approximation of center pixel weighting

$7 \times 15 = 105$ overlapping blocks. Each block consists of four 8×8 pixel cells, for which orientation histograms with 9 bins are computed. Four cell vectors are concatenated to one block vector and normalized at feature level, all block vectors together form the final feature vector for one instance. As stated before, we make use of integral images in our approach similar to the one proposed in [10]. An important parameter in the original approach is the spatial pixel weighting, which is difficult in the integral image approach since spatial information on pixel level is usually lost. To overcome this limitation, we use an additional cell in the center of the original 16x16 pixel blocks. Its histogram is distributed among the four covered cells as depicted in Figure 2. Due to the integral image, the additional center cell computation consumes only little extra time. Note that the feature dimension remains the same, as the additional center cell only influences the histogram values of the four cell vectors, but otherwise is not included in the feature vector.

We trained a detector as in [1] with the additional center cell weighting. As Figure 4(a) shows, this procedure performs better than the original approach on the INRIA set. Thus we were able to transform the original HOG descriptor into a more efficient structure. To use the detector in different scales, we are now able to resize the inner HOG blocks and positions rather than the image.

Local Binary Patterns (LBP). LBP features originate from texture analysis and were introduced by Ojala et al. in [16]. In its basic form, the LBP is a simple and efficient texture descriptor that encaptures intensity statistics in a local neighborhood.

LBP features can be computed efficiently and are invariant to monotonic graylevel changes, which makes them a good complementary feature for a HOG detector. LBPs describe relations between a center pixel value and pixels surrounding it in a circle. A LBP is usually denoted $LBP_{P,R}$, where R is the radius of the circle to be sampled and P is the number of sampling points. The bit pattern is transformed into a decimal code by binomial weighting:

$$LBP_{P,R}(x,y) = \sum_{i=0}^{P-1} b_i 2^i \tag{1}$$

Following the suggestions of Wang et al. [6], the best LBP variant in combination with blocks of 16x16 pixels turned out to be $LBP_{8,1}$. An illustration of $LBP_{8,1}$ is given in Figure 3 (a). One problem with LBP is that although it

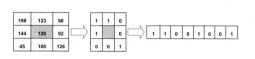 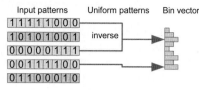

Fig. 3. (a) Basic LBP feature processing. (b) Only uniform patterns are used, and every pattern falls into the same bin as its inverse pattern.

can be computed efficiently, the actual feature is formulated as a histogram of LBP pattern frequencies inside a block region. Thus, memory consumption can become quite large since there are 256 possible values for an 8 bit pattern of $LBP_{8,1}$. As stated in [16], most patterns occuring in natural images can be formulated as *uniform patterns*: They form the subset of LBPs with at most two transitions between 0 and 1. Uniform patterns also appear to be most suitable for classification because they are robust and insensitive to noisy image data. For $LBP_{8,1}$, there are 58 uniform LBPs in total. In [5], the authors present encoding schemes that describe the patterns in terms of angle and length to further reduce dimension space. However, [6] outperformed their approach at the cost of using all 58 dimensional data.

In order to reduce the complexity and still maintain the good accuracy of HOG-LBP, we found out that the dimension of 58 can be reduced by treating every uniform bit pattern equally to its bit inverse representation without loss of discriminative power. Exceptions are the two bit patterns 00000000 and 11111111 which are highly discriminative and also very common in real world images since the latter describes completely homogenous areas. Starting from originally 256 different values, the use of uniform patterns and the inversion operation on 56 of 58 patterns leaves us with $(58 - 2)/2 + 2 = 30$ values to distinguish. The feature extraction process is illustrated in Figure 3 (b). To construct a block LBP feature, all non-uniform patterns are neglected, and the remaining ones are sorted into bins. For each bin we use an integral image for quick calculation. Although it may seem to be ressource consuming, the practical combination of multiresolution and masking techniques ensure that the LBP integral images must only be computed globally on the low resolution images. After L2 normalization for each separate block, the block vectors are concatenated into one LBP feature vector.

We implemented the HOG-LBP detector of [6] (without occlusion handling) with the reduced dimensionality, while all other parameters were the same as in the original approach. Figure 4(b) shows that the performance of our HOG-LBP detector with only 30 different LBPs is comparable to the results of [6]. The final detectors we build using LBP features are designed as grids of non-overlapping blocks. Block sizes differ depending on the training resolution, but we make sure that the size for one cell block is not below 12 pixels (see Table 1 for details). Also, the blocks are arranged in a non-overlapping manner, as it was shown that overlapping achieves no better discriminance.

2.2   Multiresolution Framework

The multiresolution approach is based on the idea that most of the candidate regions in an image do not contain interesting structures, and that many of them can be sorted out even in a low resolution. The detection process starts by consecutively downsampling the input image (see Figure 1 (a)), starting from resolution $r = 0$ to the lowest resolution. In each resolution r, only regions classified as positive samples are passed on to the next stage. Note that this multiresolution approach also includes a multiscale approach, meaning that multiple scales of regions are evaluated in each resolution. The idea of this is quite similar to the idea behind cascaded classifiers, where usually few simple features in the first stages reject most of the input regions. This information must be derived from the complete training data during boosting. The training process of a multiresolution classifier set is much simpler and faster, since the input training data of early stages consists only of low resolution images and simple detectors.

We construct our multiresolution HOG-LBP detector following the training algorithm of [13]. In short it can be summarized as follows (assume resolutions from $r = 0..n$), with slight modifications:

- Train a classifier for the lowest resolution $r = n$, using the downsampling ratio α^n for both positive and negative training images
- For each resolution:
  - Adapt the stage threshold of the previously trained classifier until a user defined target quality criterion is met
  - Create the training set for the current classifier by using false positive detections of all classifiers trained so far plus a fixed negative training set
  - Train a classifier in the current resolution set, using the downsampling ratio α^n for both positive and negative training images
  - Inner Bootstrapping: Run the current classifier on the training set and retrain b times, using false positives as negative training input.

In this algorithm, some parameters must be chosen manually beforehand, for instance the target rates of each stage or the total number of resolutions used. In our case, we did not set target rates but trained each classifier with a fixed number of two rounds of bootstrapping. The training led to a multiresolution classifier set of only three classifiers. It should be mentioned that the sizes and resolutions chosen were influenced by the fact that training and testing was done using the INRIA test set. For further application areas, different descriptor sizes and resolutions might be more appropriate. Table 1 gives some details about the descriptor parameters and size ratios. The detector grids of the three different training sizes are visualized in Figure 1 (b). The images are upscaled for illustration, but actually have the sizes given in Table 1. As downsampling factor we used $\alpha = 0.625$. Note that for each classifier (except the very first one), a bootstrapping routine is performed several times. In contrast to the original algorithm, we performed bootstrapping for classifiers 1 and 2 also on the whole training set, not only on the misclassified samples of previous stages. This may seem redundant as the later stages might never receive these additional samples by the early stages. However, as in practice the training set size is limited, classifiers using strong features like HOG and LBP can run into an overlearning state quickly, which is why additional samples help the generalization performance.

Table 1. Multiresolution HOG-LBP

Resolution	patch size	LBP cell size	LBP block size	LBP detector size	HOG cell size	HOG block size	HOG detector size
1	24x48	(12,12)	(12,12)	(2, 4)	(8,8)	(16,16)	(2,5)
2	48x96	(16,16)	(16,16)	(3, 6)	(8,8)	(16,16)	(4, 9)
3	64x128	(16,16)	(16,16)	(4, 8)	(8,8)	(16,16)	(7,15)

3   Experimental Results

All experiments were conducted on the INRIA database. It consists of 1208 pedestrian images for training and 563 for testing plus their left-right reflections, all scaled to 64x128 pixel size. In addition, there are negative images containing no persons for both training and testing. Training was performed by downsampling the training images in the way listed in Table 1. All classifiers are linear support vector machines (SVM), a maximum-margin classifier which is well suited for high-dimensional data. Also, the negative examples went through the same downsampling process (linear sampling). We tested the performance on the INRIA set in two basic experiments: First, we used the integral image approach on the negative test images, meaning that we resized the detector windows and left the images unchanged. However, since this procedure differs from the way INRIA data is usually tested, we performed a second experiment to ensure comparability. In the second setting, we used the same testing methodology as in [1], where a fixed detector window is shifted across the resized negative images. The resulting DET curves were basically identical, which shows that the proposed feature scaling is generally applicable and unbiased towards larger negative image samples.

3.1   Accuracy

The DET curve plot in Figure 4(b) shows the detection-error-tradeoff for our multiresolution system and selected state-of-the-art systems. The different points of the multiresolution classifier curve were created by stepwise linear modification of the threshold of each classifier stage. The curve shows that the combination of HOG and LBP features in a multiresolution framework can compete with the current state-of-the-art HOG-LBP. Figure 5 shows the output regions of each classifier stage on INRIA test images. The last stage classifier (in the highest resolution) is basically the one proposed by [6] (except for the mentioned modifications), which means that its accuracy is not harmed by the early rejecting

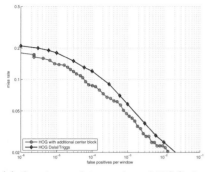 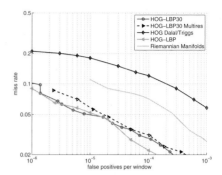

(a) Our integral image based HOG descriptor with additional center blocks.

(b) Our integral image based HOG-LBP detector (HOG-LBP 30) and the multiresolution detectors (HOG-LBP30 Multires) in comparison to current state-of-the-art detectors ([6],[8]).

Fig. 4. Classification performance on the INRIA test set

classifiers. As stated before, interestingly this behavior was achieved without tuning of threshold parameters. Further optimization of learning parameters such as stage thresholds could result in better performance. The results show that the combination of HOG and LBP feature descriptors also work in lower resolutions. The gap between multiresolution HOG-LBP and the original HOG-LBP can partly be explained by the fact that we employed a L2 normalization scheme for both LBP and HOG, since experiments showed that the original L1 square norm performed weaker than L2 normalization when using LBP30.

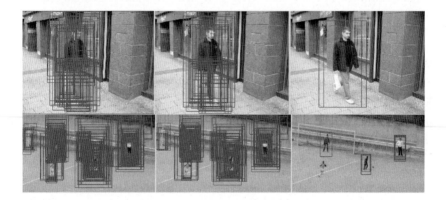

Fig. 5. Screenshots of the detection results of three stages on INRIA test set

3.2   Training/Testing Speed

The training of the proposed system took less than an hour for all three classifiers, including bootstrapping. This is faster than reported training times for recent cascade methods in the dimension of several days [10]. Although training and bootstrapping has to be done three times instead of one time in a monolithic detector, the negative bootstrapping sets are much smaller, which again leads to fewer memory consumption and less training time. Table 2 shows some insights on the detection time of the system measured on a 3GHz DualCore PC with 3GB RAM. All times are given in ms and were obtained on 320x240 images using a sparse scan with a stride of 8 pixels. The first two columns give an idea of the complexity of the feature extraction times if each stage had to evaluate every incoming region without rejection. It can be seen that especially the high resolution detector would be time consuming, as it uses the most complex detector grid. The rightmost columns resemble the system behavior in practice, where little time is spent evaluating the most interesting regions. In other words, a complete fullscan in the lowest resolution rejects about 94.46% of the search windows on the INRIA set while achieving a detection rate of 99.28%. In practice, the system runs roughly at 15Hz.

Table 2. Detection times [ms] for a sparse scan of an 320x240 image

Resolution	Baseline time		Hierarchical classification time	
	HOG	LBP	HOG	LBP
24x48	32	21	32	21
48x96	55	42	3	6.5
64x128	144	87	5	17

4   Conclusions

We proposed a fast system for pedestrian detection with very good accuracy. To achieve good classification performance, we extracted strong local HOG and LBP block features from grid descriptors in different resolutions. Utilizing the integral image method, feature extraction of block features inside a region was performed in constant time. We trained multiple classifiers in a coarse-to-fine rejecting strategy, which showed similar speedup effect like rejecting cascades. The approach was tested on the well-known challenging INRIA person set and showed competetive results regarding accuracy. Although we use the system for the detection of pedestrians, the general idea can be applied to the detection of other object classes as well.

References

1. Dalal, N., Triggs, B.: Histograms of oriented gradients for human detection. In: IEEE Computer Society Conference on Computer Vision and Pattern Recognition, CVPR 2005, vol. 1, pp. 893–886 (2005)

2. Geronimo, D., Lopez, A., Ponsa, D., Sappa, A.: Haar wavelets and edge orientation histograms for On Board pedestrian detection. In: Pattern Recognition and Image Analysis, pp. 418–425 (2007)
3. Enzweiler, M., Gavrila, D.M.: Monocular pedestrian detection: Survey and experiments. IEEE Transactions on Pattern Analysis and Machine Intelligence 31, 2179–2195 (2009)
4. Dollar, P., Wojek, C., Schiele, B., Perona, P.: Pedestrian detection: A benchmark. In: IEEE Conference on Computer Vision and Pattern Recognition, CVPR 2009, pp. 304–311 (2009)
5. Mu, Y., Yan, S., Liu, Y., Huang, T., Zhou, B.: Discriminative local binary patterns for human detection in personal album. In: IEEE Conference on Computer Vision and Pattern Recognition, CVPR 2008, pp. 1–8 (2008)
6. Wang, X., Han, T.X., Yan, S.: An HOG-LBP human detector with partial occlusion handling. In: IEEE International Conference on Computer Vision, ICCV 2009 (2009)
7. Agarwal, A., Triggs, W.: Hyperfeatures - multilevel local coding for visual recognition. In: Leonardis, A., Bischof, H., Pinz, A. (eds.) ECCV 2006. LNCS, vol. 3951, pp. 30–43. Springer, Heidelberg (2006)
8. Tuzel, O., Porikli, F., Meer, P.: Pedestrian detection via classification on riemannian manifolds. IEEE Transactions on Pattern Analysis and Machine Intelligence 30, 1713–1727 (2008)
9. Viola, P., Jones, M., Snow, D.: Detecting pedestrians using patterns of motion and appearance. In: Ninth IEEE International Conference on Computer Vision, Proceedings, vol. 2, pp. 734–741 (2003)
10. Zhu, Q., Avidan, S., chen Yeh, M., ting Cheng, K.: Fast human detection using a cascade of histograms of oriented gradients. In: CVPR 2006, pp. 1491–1498 (2006)
11. Chen, Y., Chen, C.: Fast human detection using a novel boosted cascading structure with meta stages. IEEE Transactions on Image Processing 17, 1452–1464 (2008)
12. Geismann, P., Schneider, G.: A two-staged approach to vision-based pedestrian recognition using haar and HOG features. In: 2008 IEEE Intelligent Vehicles Symposium, Eindhoven, Netherlands, pp. 554–559 (2008)
13. Zhang, W., Zelinsky, G., Samaras, D.: Real-time accurate object detection using multiple resolutions. In: IEEE 11th International Conference on Computer Vision, ICCV 2007, pp. 1–8 (2007)
14. Porikli, F.: Integral histogram: A fast way to extract histograms in cartesian spaces. In: Proc. IEEE Conf. on Computer Vision and Pattern Recognition, vol. 1, pp. 829–836 (2005)
15. Lowe, D.G.: Distinctive image features from Scale-Invariant keypoints. International Journal of Computer Vision 60, 91–110 (2004)
16. Ojala, T., Pietikinen, M., Harwood, D.: A comparative study of texture measures with classification based on featured distributions. Pattern Recognition 29, 51–59 (1996)

Utilizing Invariant Descriptors for Finger Spelling American Sign Language Using SVM

Omer Rashid, Ayoub Al-Hamadi, and Bernd Michaelis

Institute for Electronics, Signal Processing and Communications (IESK)
Otto-von-Guericke-University Magdeburg, Germany
{Omer.Ahmad,Ayoub.Al-Hamadi}@ovgu.de

Abstract. For an effective vision-based HCI system, inference from natural means of sources (i.e. hand) is a crucial challenge in unconstrained environment. In this paper, we have aimed to build an interaction system through hand posture recognition for static finger spelling American Sign Language (ASL) alphabets and numbers. Unlike the interaction system based on speech, the coarticulation due to hand shape, position and movement influences the different aspects of sign language recognition. Due to this, we have computed the features which are invariant to translation, rotation and scaling. Considering these aspects as the main objectives of this research, we have proposed a three-step approach: first, features vector are computed using two moment based approaches namely Hu-Moment along with geometrical features and Zernike moment. Second, the categorization of symbols according to the fingertip is performed to avoid mis-classification among the symbols. Third, the extracted set of two features vectors (i.e. Hu-Moment with geometrical features and Zernike moment) are trained by Support Vector Machines (SVM) for the classification of the symbols. Experimental results of the proposed approaches achieve recognition rate of 98.5% using Hu-Moment with geometrical features and 96.2% recognition rate using Zernike moment for ASL alphabets and numbers demonstrating the dominating performance of Hu-Moment with geometrical features over Zernike moments.

1 Introduction

Human Computer Interaction (HCI) is emerged as a new field which aims to bridge the communication gap between humans and computers. An intensive research has been done in computer vision to assist HCI particularly using gesture and posture recognition [1][2]. Many pioneering techniques have been proposed to solve the research issues however a natural mean of interaction still remains and yet to address.

Sign language recognition is an application area for HCI to communicate with computers and is categorized into three main groups namely finger spelling, word level sign and non-manual features [3]. In sign language, Hussain [4] used Adaptive Neuro-Fuzzy Inference Systems (ANFIS) model for the recognition of Arabic Sign Language. In his approach, gloves are used for detection of fingertip and

G. Bebis et al. (Eds.): ISVC 2010, Part I, LNCS 6453, pp. 253–263, 2010.

Fig. 1. Shows the framework of posture recognition system

wrist location with six different colors. Similarly, another approach is proposed by Handouyahia et al. [5] which presents a recognition system based on shape description using size functions for International Sign Language. Neural Network is used to train alphabets from the features computed for sign languages. However, the computed features in their proposed approach are not rotation invariant.

Other approach includes the Elliptic Fourier Descriptor (EFD) used by Malassiotis and Strintzis [6] for 3D hand posture recognition. In their system, orientation and silhouettes from the hand are used to recognize 3D hand postures. Similarly, Licsar and Sziranyi [7] used Fourier coefficients from modified Fourier descriptor approach to model hand shapes for the recognition of hand gestures. Freeman and Roth [8] suggested a method by employing orientation histogram to compute features for the classification of gesture symbols, but huge training data is used to resolve the rotation problem. Through out the literature, it is observed that the coarticulation such as hand shape, position and remains a fundamental research objective in vision based hand gesture and posture recognition systems.

The main contribution of the paper can be elaborated in two aspects: two set of invariant feature vectors are extracted to handle the coarticulation issues and the performance analysis of these feature vectors are demonstrated; and the symbols are categorized according to the detected fingertip by computing the curvature analysis to avoid the mis-classification among the posture signs. The remainder of the paper is organized as follows. Section 2 demonstrates the posture recognition system for ASL. Experimental results are presented in section 3 to show the performance of proposed approaches. Finally, the concluding remarks are sketched in section 4.

2   Posture Recognition System

In this section, components of proposed posture recognition systems are presented as shown in Fig. 1.

2.1   Pre-processing

The image acquisition is done by Bumblebee2 camera which gives 2D images and depth images. The depth image sequences are exploited to select region of interest for segmentation of objects (i.e. hands and face) where the depth lies in range from 30 cm to 200 cm (i.e. in our experiments) as shown in Fig. 2(a). In this region, we extract the objects (i.e. hands and face) from skin color distribution

Fig. 2. (a) Original Image with selected depth region (b) Results of Normal Gaussian distribution using the depth Information (c) Detected hands and face

and are modeled by normal Gaussian distribution characterized by mean and variance as shown in Fig. 2(b). We have used YC_bC_r color space because skin color lies in a small region of chrominance components where as the effect of brightness variation is reduced by ignoring the luminance channel. After that, skin color image is binarized and the contours are extracted by computing chain code representation for detection of hands and face as shown in Fig. 2(c).

2.2 Feature Extraction

Two different approaches are employed and analyzed in the proposed approach for the extraction of posture features and are described in the following section.

Hu Moments and Geometrical Feature Vectors: In the first approach, Hu-Moment (i.e. statistical feature vector (i.e. FV)) F_{Hu} and geometrical feature vectors F_{Geo} are computed for posture recognition and is formulated as:

$$F_{Hu,Geo} = F_{Hu} \wedge F_{Geo} \tag{1}$$

Statistical FV: Hu-Moments [9] are derived from basic moments which describe the properties of objects shape statistically (i.e. area, mean, variance, covariance and skewness etc). Hu [9] derived a set of seven moments which are translation, orientation and scale invariant. Hu invariants are extended by Maitra [10] to be invariant under image contrast. Later, Flusser and Suk [11] derived the moments which are invariant under general affine transformation. The equations of Hu-Moments are defined as:

$$\phi_1 = \eta_{20} + \eta_{02} \tag{2}$$

$$\phi_2 = (\eta_{20} - \eta_{02})^2 + 4\eta_{11}^2 \tag{3}$$

$$\phi_3 = (\eta_{30} - 3\eta_{12})^2 + (3\eta_{21} - \eta_{03})^2 \tag{4}$$

$$\phi_4 = (\eta_{30} + \eta_{12})^2 + (\eta_{21} + \eta_{03})^2 \tag{5}$$

$$\phi_5 = (\eta_{30} - 3\eta_{12})(\eta_{30} + \eta_{12})[(\eta_{30} + \eta_{12})^2 - 3(\eta_{21} + \eta_{03})^2] + (3\eta_{21} - \eta_{03})(\eta_{21} + \eta_{03})[3(\eta_{30} + \eta_{12})^2 - (\eta_{21} + \eta_{03})^2] \tag{6}$$

$$\phi_6 = (\eta_{20} - \eta_{02})[3(\eta_{30} + \eta_{12})^2 - (\eta_{21} + \eta_{03})^2] + 4\eta_{11}(\eta_{30} + \eta_{12})(\eta_{21} + \eta_{03}) \tag{7}$$

$$\phi_7 = (3\eta_{12} - \eta_{03})(\eta_{30} + \eta_{12})[(\eta_{30} + \eta_{12})^2 - 3(\eta_{21} + \eta_{03})^2] +$$
$$(3\eta_{12} - \eta_{30})(\eta_{21} + \eta_{03})[3(\eta_{30} + \eta_{12})^2 - (\eta_{21} + \eta_{03})^2] \qquad (8)$$

These seven moments are derived from second and third order moments. However, zero and first order moments are not used in this process. The first six Hu-Moments are invariant to reflection [12], however the seventh moment changes the sign. Statistical FV contain the following set:

$$F_{Hu} = (\phi_1, \phi_2, \phi_3, \phi_4, \phi_5, \phi_6, \phi_7)^T \qquad (9)$$

where ϕ_1 is the first Hu-Moment. Similar is the notation for all other features in this set.

Geometrical FV: Geometrical FV consist of circularity and rectangularity, and are computed to exploit hand shapes with the standard shapes like circle and rectangle. This feature set varies from symbol to symbol and is useful to recognize ASL signs. Geometrical FV is stated as:

$$F_{Geo} = (Cir, Rect)^T \qquad (10)$$

Circularity Cir and rectangularity $Rect$ are measures of shape that how much object's shape is closer to circle and rectangle respectively. These are defined as:

$$Cir = \frac{Perimeter^2}{4\pi \times Area} \ , \ \ Rect = \frac{Area}{l \times w} \qquad (11)$$

Length l and width w are calculated by the difference of largest and smallest orientation in the rotation. Orientation of object is calculated by computing the angle of all contour points using central moments. Statistical and geometrical FV set are combined together to form a feature vector set and is defined as:

$$F_{Hu,Geo} = (\phi_1, \phi_2, \phi_3, \phi_4, \phi_5, \phi_6, \phi_7, Cir, Rect)^T \qquad (12)$$

Zernike Moments: Teague [13] examines that Cartesian moment can be replaced by orthogonal basic set (i.e. Zernike polynomial), resulting in an orthogonal moment set. The magnitudes of Zernike moments are invariant to rotation and reflection [14]. However, translation and scaling invariance can easily be achieved like central moments.

$$A_{pq} = \frac{(p+1)}{\Pi} \sum_x \sum_y I(x,y)[V_{pq}(x,y)], x^2 + y^2 \le 1 \qquad (13)$$

where $I(x,y)$ is image pixel and p and q defines the moment-order. Zernike polynomials $V_{pq}(x,y)$ are defined in polar form $V_{pq}(r,\theta)$ as:

$$V_{pq}(r,\theta) = R_{pq}(r)e^{-jq\theta} \qquad (14)$$

where R_{pq} is a radial polynomial and is defined as:

$$R_{pq}(r) = \sum_{s=0}^{\frac{p-|q|}{2}} \frac{(-1)^s (p-s)! r^{p-2s}}{s! \left(\frac{p+|q|}{2} - s\right)! \left(\frac{p-|q|}{2} - s\right)!} \qquad (15)$$

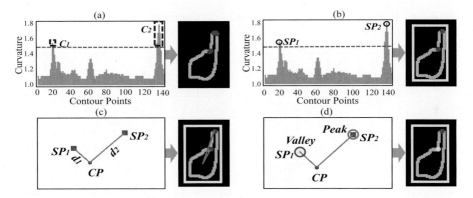

Fig. 3. (a) It shows clusters (i.e. C_1 and C_2) whose threshold is above $\sqrt{2}$. (b) Maximum local extreme selected contour point (i.e. SP_1 and SP_2) from these clusters. Red points show values above threshold $\sqrt{2}$ (i.e. candidates for fingertips).(c) Red points show the selected contour points (i.e. SP_1 and SP_2). Distance is calculated from center point (CP) and normalization is done. (d) Normalized values greater than 0.5 are detected as fingertip (i.e. peak) marked by red point. Yellow marks represent values less than 0.5.

We have used Zernike moments upto $4th$ order moment. The feature vector set for Zernike moment is as under:

$$F_{Zernike} = (Z_1, Z_2, Z_3, Z_4, Z_5, Z_6, Z_7, Z_8, Z_9)^T \tag{16}$$

Normalization: The normalization is done for features to keep them in a particular range and is defined as:

$$c_{min} = \mu - 2\sigma \ , \ c_{max} = \mu + 2\sigma \tag{17}$$

$$nF_i = (F_i - c_{min}) \ / \ (c_{max} - c_{min}) \tag{18}$$

$nF_{Hu,Geo}$ is the normalized feature for Hu-Moment with geometrical FV . c_{max} and c_{min} are the respective maximum and minimum values used for the normalization of these features. Similar is the case of Zernike moments (i.e.$nF_{Zernike}$).

2.3   Fingertip Detection for Categorization

Given the contour of detected hand, curvature is estimated by considering the neighbor contour points to detect the fingertip [15]. Mathematically, curvature gives the ratio of length (i.e. sum of distances that a curve has) and displacement measures the distance from the first to last point if curve covers a straight line. It is computed from the following equation:

$$curv\,(k) = length/displacement \tag{19}$$

$$length = \sum_{i=(k-n/2)}^{i=(k+n/2)} \|(P_i - P_{i+1})\| \tag{20}$$

Table 1. Confusion Matrix of one Detected Fingertip

Sign	A	B	D	I	H/U
A	99.8 / 98.98	0.0 / 0.02	0.0 / 0.0	0.0 / 0.8	0.2 / 0.2
B	0.0 / 0.1	98.18 / 96.3	1.0 / 1.6	0.0 / 0.0	0.82 / 2.0
D	0.0 / 0.0	0.0 / 2.0	98.67 / 96.1	1.33 / 1.9	0.0 / 0.0
I	0.58 / 2.0	0.0 / 0.0	0.8 / 2.1	98.62 / 94.62	0.0 / 0.38
H/U	0.0 / 0.0	3.08 / 2.6	0.0 / 0.4	0.24 / 0.8	96.68 / 96.2

Table 2. Confusion Matrix of Two Detected Fingertips

Symbol	C	L	P	Q	V	Y
C	98.65 / 96.95	0.25 / 0.3	0.0 / 0.0	0.75 / 0.1	0.0 / 0.2	0.35 / 0.65
L	0.38 / 1.2	98.5 / 96.2	0.0 / 0.0	0.76 / 1.6	0.0 / 0.2	0.36 / 0.8
P	0.0 / 0.0	0.0 / 0.1	98.74 / 95.5	1.26 / 3.2	0.0 / 0.0	0.0 / 1.2
Q	0.0 / 0.0	0.0 / 0.5	3.78 / 4.2	96.22 / 94.4	0.0 / 0.0	0.0 / 0.9
V	0.2 / 0.0	0.0 / 0.0	0.0 / 0.0	0.0 / 1.3	99.35 / 97.4	0.45 / 1.3
Y	0.0 / 0.0	0.0 / 0.1	0.0 / 0.25	0.0 / 1.05	0.7 / 1.3	99.3 / 96.3

$$displacement = \left\| \left(P_{k-n/2} - P_{k+n/2} \right) \right\| \tag{21}$$

where k is the boundary point of object at which curvature $curv(k)$ is estimated, n is total number of pixels used for curvature estimation, and P_i and $P_{(i+1)}$ are the objects boundary points.

The principle objective is to find high curvature values from contour points which results in detection of peaks from hands contour and tends to represent the fingertip. We have adaptively determine number of contour points (i.e. after conducting empirical experiments) by exploiting the depth information which provides information about the distance of object from the camera. In this way, a candidate for the fingertip is selected when curvature value is greater than $\sqrt{2}$.

Experimental results show contour of left hand with threshold greater than $\sqrt{2}$. In Fig. 3(a), there are two clusters named C_1 and C_2 and maximum value from these clusters is selected using maximum local extreme value. The resulted points are marked as a fingertip (i.e. SP_1 and SP_2) as shown in Fig. 3(b). It is observed that both the peaks (i.e. ∩) and valleys (i.e. ∪) can be inferred as a fingertip. Therefore, the next step is to remove valleys from being detected as a fingertip. For this purpose, selected contour points (i.e. SP_1 and SP_2) are taken and their distances are computed from the center point CP of object as shown in Fig. 3(c). The normalization is done and these points are scaled ranging from 0 to 1. We pick the points whose values are greater than 0.5 for fingertip detection. In Fig. 3(d), red mark represents the fingertip whereas the yellow mark points to a valley (not a fingertip). In this way, fingertips are successfully detected for the categorization of symbols using above defined criteria.

Four groups are formed for ASL alphabets according to number of detected fingertips (i.e.Group I with no fingertip,Group II with one fingertip,Group III with two fingertips and Group IV with three fingertips). Numbers are not grouped with alphabets because some numbers are similar to alphabets, so it is hard to classify them together (i.e. 'D' and '1' are same with small change of thumb).

Table 3. Confusion Matrix of Fingertips (i.e. all)

FT/ Gr.	1	2	3	4
0	99.8	0.2	0.0	0.0
1	1.2	96.8	2.0	0.0
2	0.0	0.0	95.1	4.9
3	0.0	0.0	0.9	99.1

Table 4. Confusion Matrix of ASL Numbers

Nr.	0	1	2	3	4	5	6	7	8	9
0	99.8 / 99.1	0.2 / 0.9	0.0 / 0.0	0.0 / 0.0	0.0 / 0.0	0.0 / 0.0	0.0 / 0.0	0.0 / 0.0	0.0 / 0.0	0.0 / 0.0
1	0.3 / 1.1	99.4 / 97.7	0.3 / 1.2	0.0 / 0.0	0.0 / 0.0	0.0 / 0.0	0.0 / 0.0	0.0 / 0.0	0.0 / 0.0	0.0 / 0.0
2	0.0 / 0.0	0.0 / 0.98	98.3 / 96.52	0.4 / 0.6	0.0 / 0.0	0.0 / 0.0	1.3 / 1.9	0.0 / 0.0	0.0 / 0.0	0.0 / 0.0
3	0.0 / 0.0	0.0 / 0.0	0.4 / 1.0	98.2 / 96.6	0.9 / 1.3	0.0 / 0.0	0.5 / 0.6	0.0 / 0.0	0.0 / 0.5	0.0 / 0.0
4	0.0 / 0.0	0.0 / 0.0	0.0 / 0.8	0.2 / 1.2	98.2 / 94.6	1.6 / 3.4	0.0 / 0.0	0.0 / 0.0	0.0 / 0.0	0.0 / 0.0
5	0.0 / 0.0	0.0 / 0.0	0.0 / 0.0	0.0 / 1.0	2.4 / 3.8	97.6 / 95.2	0.0 / 0.0	0.0 / 0.0	0.0 / 0.0	0.0 / 0.0
6	0.0 / 0.0	0.0 / 0.0	0.8 / 1.6	0.6 / 1.2	0.0 / 0.0	0.0 / 0.0	98.6 / 97.2	0.0 / 0.0	0.0 / 0.0	0.0 / 0.0
7	0.0 / 0.0	0.0 / 0.0	0.0 / 0.0	0.0 / 0.0	0.0 / 0.0	0.0 / 0.0	0.7 / 1.3	98.3 / 96.5	0.6 / 1.6	0.4 / 0.6
8	0.0 / 0.0	0.0 / 0.0	0.0 / 0.0	0.4 / 0.3	0.0 / 0.0	0.0 / 0.0	0.2 / 0.8	0.4 / 0.8	98.4 / 97.6	0.6 / 0.8
9	0.0 / 0.0	0.0 / 0.0	0.0 / 0.0	0.5 / 0.8	0.0 / 0.0	0.0 / 0.0	0.0 / 0.0	0.4 / 1.75	0.5 / 1.35	98.6 / 96.1

2.4   Classification

Classification is the last step in posture recognition where a posture symbol is assigned to one of the predefined classes. In proposed approach, a set of thirteen ASL alphabets and ten ASL numbers are classified using Support Vector Machines (SVM). SVM [16][17] is a supervised learning technique for optimal modeling of data. It learns decision function and separates data class to maximum width. SVM learner defines the hyper-planes for data and maximum margin is found between these hyper-planes. Because of the maximum separation of hyper-planes, it is also considered as a margin classifier. We have used Radial Basis Function (RBF) Gaussian kernel which has performed robustly with given number of features and provided optimum results as compared to other kernels.

3   Experimental Results

A database is built to train posture signs which contain 3600 samples taken from eight persons on a set of thirteen ASL alphabets and ten numbers. Classification results are based on 2400 test samples from five persons and test data used is entirely different from training data. As Zernike moments $nF_{Zernike}$ and Hu-moments with geometrical FV $nF_{Hu,Geo}$ are invariant to translation, rotation and scaling, therefore posture signs are tested for these properties.

3.1   ASL Alphabets

Thirteen alphabets are categorized into four groups according to fingertips and are presented as follows:

Group I: There is no misclassification between 'A' and 'B' because these two signs are very different. Precisely, statistical FV show correlation in Hu-Moments for

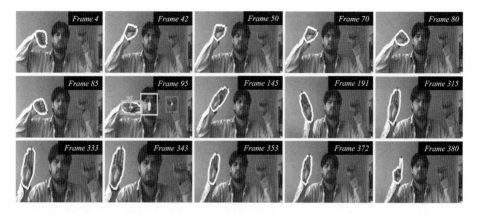

Fig. 4. Sequence starts with recognition of posture signs 'A' till frame 90 with different orientations and back and forth movement. Similarly, posture sign 'B' is recognized from frame 91 to frame 372. 'D' is the last recognized sign in sequence.

features (i.e. $\phi_3, \phi_4, \phi_5, \phi_6, \phi_7$) whereas in Zernike moments, features (i.e. Z_4, Z_8, Z_9) are very different for these posture symbols. Discrimination power of SVM can be seen from this behavior and it classifies posture signs 'A' and 'B' correctly.

Group II: Table 1 shows confusion matrix of classes with one fingertip detected for Hu-Moment with geometrical FV (i.e. above in fraction $\frac{99.8}{98.98}$) and Zernike moments (i.e.below in fraction $\frac{99.8}{98.98}$). It can be seen that from first approach, alphabet 'A' results in least misclassification whereas 'H'/'U' has the maximum misclassification with other posture alphabets. Misclassifications between 'B' and 'H/U' are observed in Hu-Moment features during back and forth movement.

Group III: Confusion matrix of group with two detected fingertips is shown in Table 2. During experiments, it is seen that highest misclassification exists between 'P' and 'Q' signs because of their shape and geometry. Besides, statistical FV in this group lies in similar range and thus possesses strong correlation which leads to misclassification between them. Similar is the case of Zernike moments.

Group IV: The posture sign 'W' is always classified because it is the only element in this group.

3.2   ASL Numbers

The classes for ASL numbers are tested for translation, orientation and scaling, and are shown in Table 4. It is found that the least misclassification of number '0' with other classes because its geometrical features are entirely different from others. Highest misclassification exists between numbers '4' and '5' as the similarity between these signs (i.e. thumb in number '5' is open) is very high. Zernike moments also give similar results as the first approach. Average recognition rate for ASL numbers are 98.6% and 96.4% from Hu-Moment with geometrical FV and Zernike moments respectively.

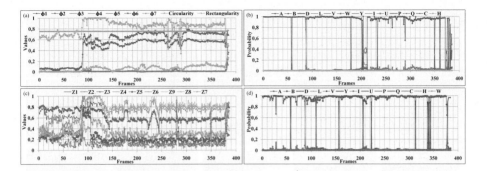

Fig. 5. (a) and (b) Shows the feature vectors values and classification probability for Hu-Moment with circularity and rectangularity. (c) and (d) shows the feature vectors and classification results for Zernike moment.

3.3   Misclassification of Fingertip Groups

The fingertip probability classification of a group with other groups is shown in Table 3. It is observed that the misclassification exists in neighboring groups and is due to the reason that only +/- 1 detected fingertip is wrong or wrongly detected for posture signs. A strong misclassification between the detected fingertips of group III and group IV can be seen in Table 3. Wrong fingertip detection from group III leads to the detection of three fingertips and thus 'W' is recognized from posture sign. Similarly, if two fingertips are detected in group IV, it leads to the classification from group with two fingertips detected class signs.

3.4   Test Sequences

Fig. 4 presents the sequence in which detected posture signs are 'A', 'B' and 'D'. In this sequence, first posture sign changes from 'A' to 'B' is occurred in frame 90 followed by another symbol change at frame 377 from 'B' to 'D'. The symbols in this sequence are tested for rotation and scaling, and it can be seen that Hu-Moment with geometrical FV doesn't affect much under these conditions. However, in our case, Zernike moments are affected under rotation which leads to misclassification. Moreover, it is observed that Zernike moments are most robust to noise than Hu-Moment with geometrical FV. Fig. 5(a) and (b) presents feature values of Hu-Moment with geometrical FV and classification probabilities of this approach for test sequence samples. Moreover, Zernike moments features and classification probabilities are presented in Fig. 5(c) and (d).

Fig. 6(a) presents the sequence in which classified posture symbols are 'W', 'A', 'Y', 'D' and 'V'. Classification probabilities of Hu-Moment with geometrical features and Zernike moments are shown in Fig. 6(b) and (c) respectively. It can be seen from the graphs that classification errors are occured due to wrong fingertip detection and thus results in classification of symbol in a different group.

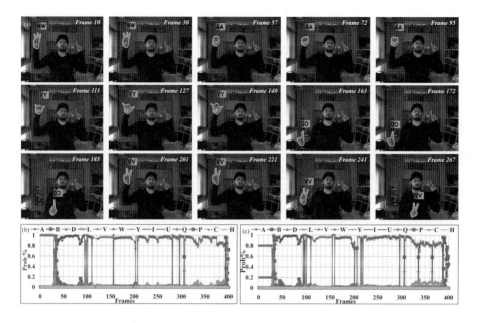

Fig. 6. (a) Presents the sequence in which recognized posture symbols are 'W', 'A','Y','D' and 'V'. (b) shows classification probabilities of Hu-Moment with geometrical features and (c) shows classification probabilities of Zernike moments.

Results show that recognition rate of feature vector computed from Hu-Moments with the geometrical features (i.e. 98.5%) dominate the performance over recognition outcome of feature vectors measured from Zernike moments (i.e. 96.2%).

4   Conclusion

In this paper, we have proposed a three-step approach for recognition of ASL posture signs. Two moment based approaches are analyzed and their features are extracted which are invariant to translation, orientation and scaling. Besides, the fingertip is detected for ASL alphabets and used as a measure to categorize thus avoids the misclassifications of posture signs. SVM is applied for recognition of ASL signs and recognition results of both approaches are computed and analysed. The recognition results based on Hu-Moment with geometrical FV outperforms the other approach using feature vectors computed with Zernike moments.

Acknowledgement

This work is supported by Transregional Collaborative Research Centre SFB/TRR 62 Companion-Technology for Cognitive Technical Systems funded by DFG and Forschungspraemie (BMBF-Froederung, FKZ: 03FPB00213).

References

1. Jaimes, A., Sebe, N.: Multimodal human-computer interaction: A survey. Computer Vision and Image Understanding, 116–134 (2007)
2. Karray, F., Alemzadeh, M., Saleh, J., Arab, M.: Human-computer interaction: Overview on state of the art. International Journal on Smart Sensing and Intelligent Systems 1, 137–159 (2008)
3. Bowden, R., Zisserman, A., Kadir, T., Brady, M.: Vision based interpretation of natural sign languages. In: Proc. 3rd ICCVS, pp. 391–401 (2003)
4. Hussain, M.: Automatic recognition of sign language gestures. Master thesis, Jordan University of Science and Technology (1999)
5. Handouyahia, M., Ziou, D., Wang, S.: Sign language recognition using moment-based size functions. In: Int. Conf. of Vision Interface, pp. 210–216 (1999)
6. Malassiotis, S., Strintzis, M.: Real-time hand posture recognition using range data. Image and Vision Computing 26, 1027–1037 (2008)
7. Licsar, A., Sziranyi, T.: Supervised training based hand gesture recognition system. In: ICPR, pp. 999–1002 (2002)
8. Freeman, W., Roth, M.: Orientation histograms for hand gesture recognition. In: Int. Workshop on Automatic FG Recognition, pp. 296–301 (1994)
9. Hu, M.: Visual pattern recognition by moment invariants. IRE Trans. on Information Theory 8, 179–187 (1962)
10. Maitra, S.: Moment invariants. In: IEEE Conf. on CVPR., vol. 67, pp. 697–699 (1979)
11. Flusser, J., Suk, T.: Pattern recognition by affine moment invariants. Journal of Pattern Recognition 26, 164–174 (1993)
12. Davis, J., Bradski, G.: Real-time motion template gradients using intel cvlib. In: IEEE ICCV Workshop on Framerate Vision (1999)
13. Teague, M.: Image analysis via the general theory of moments. Journal of the Optical Society of America, 920–930 (1980)
14. Bailey, R., Srinath, M.: Orthogonal moment features for use with parametric and non-parametric classifiers. IEEE Trans. on PAMI 18, 389–399 (1996)
15. Kim, D.H., Kim, M.J.: A curvature estimation for pen input segmentation in sketch-based modeling. Computer-Aided Design 38, 238–248 (2006)
16. Cristianini, N., Taylor, J.: An Introduction to Support Vector Machines and other kernel based learning methods. Cambridge University Press, Cambridge (2001)
17. Lin, C., Weng, R.: Simple probabilistic predictions for support vector regression. Technical report, National Taiwan University (2004)

Bivariate Feature Localization for SIFT Assuming a Gaussian Feature Shape

Kai Cordes, Oliver Müller, Bodo Rosenhahn, and Jörn Ostermann

Institut für Informationsverarbeitung (TNT)
Leibniz Universität Hannover
{cordes,omueller,rosenhahn,ostermann}@tnt.uni-hannover.de

Abstract. In this paper, the well-known SIFT detector is extended with a bivariate feature localization. This is done by using function models that assume a Gaussian feature shape for the detected features. As function models we propose (a) a bivariate Gaussian and (b) a Difference of Gaussians. The proposed detector has all properties of SIFT, but provides invariance to affine transformations and blurring. It shows superior performance for strong viewpoint changes compared to the original SIFT. Compared to the most accurate affine invariant detectors, it provides competitive results for the standard test scenarios while performing superior in case of motion blur in video sequences.

1 Introduction

The precise detection of feature points in an image is a requirement for many applications in image processing and computer vision. In most cases, feature points are used to establish correspondences between different images containing the same scene captured by a camera. These corresponding points can be used to perform measurements to reconstruct the geometry of the observed scene [1] or to detect objects. To obtain reasonable and comparable feature points, the image signal surrounding the feature position is analyzed and distinctive characteristics of this region are extracted. These characteristics are usually assembled to a vector which describes the feature. This vector, called descriptor [2], is used to establish correspondences by calculating a similarity measure (L_2 distance) between the current descriptor and the feature descriptors of a second image. Feature detection and correspondence analysis is more challenging if the baseline between the cameras capturing the images is large. The region description of corresponding features before and after the viewpoint change should be the same. Several methods have been proposed in the literature to model invariants, such as brightness change and rotation [3, 4], scale change [2, 5, 6], affine transformation [7–12], and perspective transformation [13].

For viewpoint changes, the mapping between the local regions of two images can be modeled by an affine transformation. This is motivated by the assumption that the scene surface is locally planar and perspective effects are small on local

G. Bebis et al. (Eds.): ISVC 2010, Part I, LNCS 6453, pp. 264–275, 2010.

regions [11]. Based on affine normalization, the Harris-Affine [8] and Hessian-Affine [9] detectors determine the elliptical shape with the second moment matrix of the intensity gradient. In [14], maximally stable extremal regions (MSER) are constructed using a segmentation process. Then, an ellipse is fit to each of these regions. Although these detectors initially do not select features in an affine scale space, their evaluation shows excellent performance for significant viewpoint changes [11, 15].

Extensive work has been done on evaluating feature point detectors and descriptors [9, 11, 16–18]. Commonly, the most important criterion for affine invariant detectors is the repeatability criterion [9, 18]. For evaluation, the reference test set of images offered in [11] is widely used. The data set contains sequences of still images with changes in illumination, rotation, perspective, and scale. Using this data set, the most accurate localization of affine invariant features is performed by the Harris-Affine, Hessian-Affine, and MSER detector [15]. However, their performances depend on the image transformation emphasized throughout the specific test sequence and the type of texture present in the images.

Since the results of different detectors lead to differently structured features, the detectors are categorized by the type of features they detect. In [15], these categories for the affine invariant feature detectors are *corner detectors* (Harris-Affine), *blob detectors* (Hessian-Affine), and *region detectors* (MSER). For computer vision challenges like object detection, the combination of complementary features can be used to increase the performance [11, 19, 20].

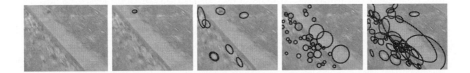

Fig. 1. Examples of motion blurred image content. From left to right: input image, Hessian-Affine, MSER, SIFT, and our approach DoG-Affine. Harris-Affine provides no feature for this image (part of the sequence shown in Figure 5).

The SIFT detector performs very well under moderate affine distortions. It has been shown [12] that in this case the repeatability rates are better than those from affine invariant detectors. However, the performance of SIFT decreases under substantial viewpoint changes. Also, elliptically shaped features are not localized accurately. Hence, our motivation is to enhance the affine invariant properties of SIFT while preserving its performance for moderate affine distortions. This is done without building an affine scale space pyramid, but by using a modified localization scheme that assumes (a) a bivariate model and (b) a Gaussian shape for the features located in the scale space. These two characteristics can also increase the stability of the detection in images with motion blur as shown in Figure 1. In this example, an elliptical description of the feature is preferable to represent the region. The detection of Harris-Affine, Hessian-Affine, and MSER lead to only a few features. In video sequences taken by a moving camera, motion blur is prevalent even if it is not visible to the human eye [21].

This work shows that our approach for determining the feature localization increases the performance of the SIFT detector in case of strong viewpoint changes as well as motion blur. The **contributions** of this paper are

- the extension of the SIFT detector with a bivariate localization procedure,
- the comparison of the extended detector to the most accurate state of the art affine invariant feature detectors using synthetic and natural images, and
- the evaluation of the detectors using a structure from motion approach with naturally motion blurred images.

In the following Section, the SIFT detector is briefly presented. In Section 3, the proposed affine invariant extension of SIFT is derived. Section 4 shows experimental results using synthetically constructed and real image data. In Section 5, the paper is concluded.

2   The SIFT Detector

The **S**cale **I**nvariant **F**eature **T**ransform provides distinctive feature points which can be used to establish correspondences between images with the same scene content. The resulting features are invariant to illumination, rotation, and scale changes between the analyzed images. The SIFT detector demonstrates superior performance in this field. However, affine invariance that is commonly used to approximate the perspective distortion resulting from a viewpoint change is not modeled.

The workflow of SIFT using an image as input data is shown in Figure 2. First, feature points are characterized as extrema in the Difference of Gaussians pyramid of the image. The Difference of Gaussians pyramid is used as an approximation of the Laplacian pyramid, which has been proven to provide stable scale invariant features. In the next step, the localization is refined by an interpolation of the 26 surrounding grid points with x-, y-, and scale - coordinates. The interpolation is done by fitting a second order function to the Difference of Gaussians output. It provides subpel and subscale accuracy of the localization of a feature point. After the localization, features are rejected that do not fulfill the critera of (a) minimal contrast, and (b) maximal ratio of principal curvatures. This guaranties the repeatability stability of the accepted features. In order to apply an orientation parameter to a feature, the main orientation of the surrounding image gradients is estimated. If the orientation estimation procedure leads to ambiguous results, multiple features with the same coordinates, but different orientation are constructed. Finally, a 128 dimensional vector is computed using the surrounding gradient orientations. This vector is called descriptor. It is used to establish the correspondence to a feature in the other image. Correspondences between two images are found by associating feature points with a minimal distance between their descriptors.

Fig. 2. Workflow diagram of the detection of image features with the SIFT detector. The part modified in our approach is marked with a dotted box border.

3   Bivariate Feature Localization

In order to estimate the localization parameters of a feature, a parabolic interpolation of the Difference of Gaussians signal is used by SIFT based feature detectors [2, 6, 10, 16]. If the neighborhood of a feature in the input image has Gaussian shape, the output of the Difference of Gaussians filter has Difference of Gaussians shape [22]. Therefore, the approach of interpolating with a parabolic curve is suboptimal and can lead to localization errors. This motivates the approximation of a scale space extremum using a Gaussian (Section 3.1) or a Difference of Gaussians (Section 3.2). Thereby, we can expect a better approximation for features with a Gaussian shape. An example result compared to the other affine invariant detectors for a Gaussian feature is shown in Figure 3.

3.1   Gaussian Function Model

The Gaussian function model we propose with the covariance matrix $\Sigma = \begin{pmatrix} a^2 & b \\ b & c^2 \end{pmatrix}$ and its determinant $|\Sigma| = det(\Sigma)$ is the following

$$G_{\mathbf{x}_0,\Sigma}(\mathbf{x}) = \frac{r_G}{\sqrt{|\Sigma|}} \cdot e^{-\frac{1}{2}((\mathbf{x}-\mathbf{x}_0)^\top \Sigma^{-1}(\mathbf{x}-\mathbf{x}_0))} \tag{1}$$

Here, $\mathbf{x}_0 = (x_0, y_0) \in [-1; 1]$ is the subpel position of the feature point. The parameters a, b, c define the surrounding elliptical region and r_G is a peak value parameter. The parameter vector $\mathbf{p_G} = (x_0, y_0, a, b, c, r_G)$ determines a member of the function model.

3.2   Difference of Gaussians Function Model

If we assume a Gaussian image signal $G_{\mathbf{x}_0,\Sigma}$ surrounding a feature point as in Equation (1), the correct response of the Difference of Gaussians filter is a Difference of Gaussians function with modified standard deviations σ. This motivates the following Difference of Gaussians regression function

$$\begin{aligned} D_{\mathbf{x}_0,\sigma} &= r_D(G_{\mathbf{x}_0,\Sigma_\sigma} - G_{\mathbf{x}_0,\Sigma_{k\sigma}}) * G_{\mathbf{x}_0,\Sigma} \\ &= r_D(G_{\mathbf{x}_0,\Sigma_\sigma+\Sigma} - G_{\mathbf{x}_0,\Sigma_{k\sigma}+\Sigma}) \end{aligned} \tag{2}$$

with $\Sigma_\sigma = \begin{pmatrix} \sigma^2 & 0 \\ 0 & \sigma^2 \end{pmatrix}$ and the standard deviation σ of the detected scale of the feature. Like in equation (1), r_D is a peak value parameter.

Fig. 3. Different results of feature localization of an elliptical Gaussian blob obtained by the Harris-Affine, Hessian-Affine, MSER, and our approaches Gaussian-Affine and DoG-Affine (from left to right). The right three images show desirable results.

Like the Gaussian function (Section 3.1), the Difference of Gaussians function can be described by a six-dimensional parameter vector $\mathbf{p_D} = (x_0, y_0, a, b, c, r_D)$.

3.3   Optimization and Computational Complexity of the Approaches

The bivariate Gaussian (Section 3.1) and the Difference of Gaussians function models (Section 3.2) evaluate 9 and 27 pixel values, respectively. A member of the function model is determined by a parameter vector $\mathbf{p} = (x_0, y_0, a, b, c, r)$. The parameter vector $\mathbf{p}$ is identified through a regression analysis which is initialized with the fullpel position, a circular shape $\Sigma = \mathbf{E}$, and a peak level parameter equal to 1. To obtain the optimal parameter vector, the distance between the pixel neighborhood and the model function is minimized using the Levenberg-Marquardt algorithm.

Convergence. As only the feature localization scheme of SIFT is modified, the original SIFT and the proposed localization approaches use the same fullpel features selected in the scale space pyramid as input. Nevertheless, some features are rejected after the localization by SIFT that are considered valid using our approaches and vice versa. The overall numbers of features per image N_f are approximately equal as shown in Table 1 for the *Graffiti* sequence. The number of features that are localized with our methods but discarded by SIFT, are 173 and 210 for Gaussian-Affine and DoG-Affine, respectively.

Computation Time. The computation time of the two approaches using the same SIFT implementation basis[1] is shown in Table 2. Here, The *Graffiti* (Section 4) sequence is processed. The overall time needed is divided by the number of detected features. Compared to the original SIFT (*o-sift*), the computational overhead of Gaussian-Affine (*gauaff*) and DoG-Affine (*dogaff*) is a factor of 2.7 and 7.1, respectively. While the Difference of Gaussians function model appears to be mathematically correct, the Gaussian-Affine provides a good approximation.

---

[1] http://web.engr.oregonstate.edu/~hess/

Table 1. Detected number of features of the original SIFT (*o-sift*) and our approaches Gaussian-Affine (*gauaff*) and DoG-Affine (*dogaff*). About 200 of the features localized by our methods are rejected by the original SIFT - detector as shown in the columns gauaff \ o-sift and dogaff \ o-sift, respectively.

	o-sift	gauaff	gauaff \ o-sift	dogaff	dogaff \ o-sift
N_f	1148	1234	173	1270	210

Table 2. The computation time used per feature for the original SIFT (*o-sift*) and our approaches Gaussian-Affine (*gauaff*) and DoG-Affine (*dogaff*)

o-sift	gauaff	dogaff
$0.97ms$	$2.67ms$	$6.91ms$

4   Experimental Results

For the evaluation of our approaches, we use the following input data:

- synthetic test images to prove the accuracy of our approaches for Gaussian shaped features (Figure 4)
- natural image set with viewpoint changes to evaluate the competitiveness compared to the most accurate affine invariant detectors [11]
- natural motion blur images to show the improvement in a structure from motion approach (Figure 5)

As pointed out in [15], the most accurate affine invariant feature point detectors are the Hessian-Affine [10] (*hesaff*), the Harris-Affine [9] (*haraff*), and Maximally Stable Extremal Regions (MSER) [14] (*mseraf*). Hence, these detectors will serve as benchmarks for the comparison with our detectors Gaussian-Affine (Section 3.1) and DoG-Affine (Section 3.2), denoted as *gauaff* and *dogaff*. The input data description used for the evaluation follows in Section 4.1. The results are presented in Section 4.2.

4.1   Data Construction

Synthetic Image Data. The synthetic test images are constructed using elliptical Gaussian blobs with varying size and orientation determining the covariance matrix Σ_0. The image size is 64×64. Some examples are shown in Figure 4. The elliptical region is defined by the center position (x_0, y_0) and the ground truth covariance matrix $\Sigma_0 = \begin{pmatrix} a_0^2 & b_0 \\ b_0 & c_0^2 \end{pmatrix}$. Gaussian noise is added to the images. All resulting covariance matrices estimated in the results section will be invariant to a global scale factor. Therefore, a normalization is applied to the covariance matrix results.

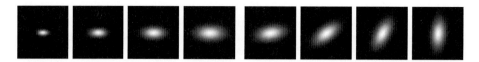

Fig. 4. Some examples of the synthetic test images with varying size (left) and angle (right) determined by the covariance matrix Σ_0

Reference Image Data Set. For the evaluation with natural images, the well-known Mikolajczyk data set[2] is used. The data set provides image pairs with a camera performing a rotational movement, or observing a planar scene. Thus, all detected feature ellipses in one image can be mapped to the corresponding feature ellipses in the other image by a homography. The estimated homography matrices are also provided within the data set. These images are widely used by the computer vision community, especially for evaluating feature descriptors.

Natural Motion Blur Images. To obtain natural motion blured images, a sequence is captured from a moving vehicle. Radial distortion is compensated. The scene content consists of rigid geometry. The level of motion blur as well as the viewpoint angle increases with a decreasing distance of camera and scene structure. Some images of the sequence (94 frames) are shown in Figure 5.

Fig. 5. Some example frames $f_0, f_{30}, f_{60}, f_{90}$ of the natural motion blur sequence (94 frames). The motion blur level on the sidewalk is higher than on the wall.

4.2   Results

Synthetic Images. The methods lead to different results regarding the localization and the amount of detected features as shown in Figure 3. The subjectively best results are obtained by MSER and our approaches Gaussian-Affine and DoG-Affine, which are very similar (see Figure 3). For an objective evaluation of the synthetic images, the Surface Error [9] is calculated. The surface error is a percentage value that is minimal if a detected ellipse area is exactly matching the ellipse determined by the ground truth values. For the evaluation, the detected covariance matrices are normalized. This avoids the results being dependent on a global scale of the features. If multiple features are detected, the best of them

---

[2] www.robots.ox.ac.uk/~vgg/research/affine/index.html

is chosen. For the evaluation, Gaussian noise of 30 dB is added to the images. The results are shown in Figure 6. Our detector DoG-Affine provides the best results for this type of features with added noise. The Gaussian-Affine detector yields nearly the same results with slightly increased error. Harris-Affine, a *corner detector*, results in the highest Surface Errors of the four detectors. Ellipses with small sizes of ≤ 2.5 are not detected by Hessian-Affine.

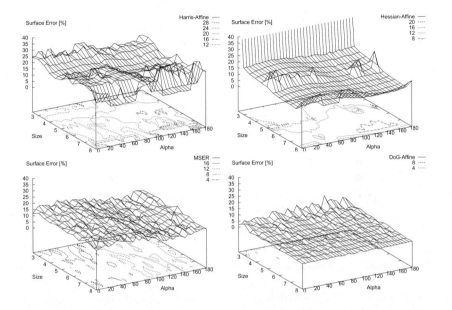

Fig. 6. Surface Error for the test signals shown in Figure 4 with varying angle alpha and varying size. Gaussian noise of 30 dB is added to the images. From left top to bottom right: Harris-Affine, Hessian-Affine, MSER, and our approach DoG-Affine. Contour lines are displayed in the ground plane. Together with the legend at the top right of each diagram, they indicate the error level.

Natural Image Pairs with Viewpoint Change. The image data sets provide image pairs with associated homography matrices that map all features of one image to the corresponding features in the other image. For the evaluation, the repeatability criterion with default parameters is used [9]. A correspondence is deemed correct, if the surface error between the ellipses of the features is below 40 %. The repeatability value is the ratio of correct correspondences and the minimal number of feature points detected in the first and the second image. The detectors SIFT, Gaussian-Affine, and DoG-Affine use the same default SIFT threshold values [6]. The evaluation results are shown in Figure 7. On top, the repeatability values between the first and the other images are displayed. The bottom row shows the number of correct matches. It can be seen that our

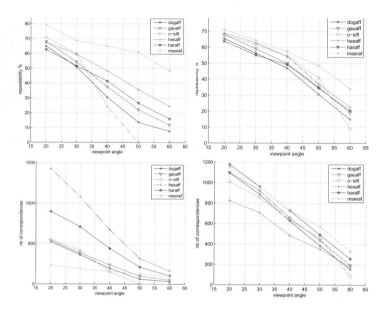

Fig. 7. Results of the *Graffiti* sequence (left) and the *Wall* sequence (right) demonstrating perspective change. On top, the repeatability curves are shown, the bottom diagrams show the number of correctly detected feature pairs.

detectors Gaussian-Affine and DoG-Affine provide better results for strong viewpoint change than the original SIFT approach. MSER performs best for these still images with well-bounded textures. For most natural images in this setup, the Gaussian-Affine provides slightly better repeatability results than the DoG-Affine approach. Overall, our approaches provide competitive accuracy, MSER and Hessian-Affine provide slightly better repeatability rates. This result is also valid for the other test scenarios [11], which are not shown here.

Motion Blur Sequence. The test scenario for the motion blur images evaluates a structure from motion technique like in [1] to demonstrate the usability of the feature detectors in the field of multiview scene reconstruction. Resulting from the camera motion, blurring occurs as well as a moderate viewpoint angle between consecutive frames. The motion blur level as well as the viewpoint angle increase with a decreasing distance between scene content and camera position. For the comparison, feature point sets are generated for each of the tested detectors. For the reconstruction of the scene, feature correspondences are established using the SIFT descriptor [6]. Then, the outliers are removed using the fundamental matrix and the RANSAC algorithm [23]. Finally, the cameras A_k and 3D object points $\mathbf{P}_j$ are estimated and refined using incremental bundle adjustment [1, 24]. The incremental bundle adjustment minimizes the reprojection error, which is defined by the distances between an estimated 3D object point $\mathbf{P}_j$

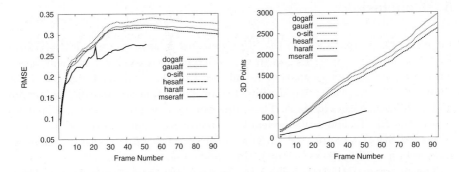

Fig. 8. Structure from motion results using different detectors for the natural motion blur sequence for each frame k. Harris-Affine, Hessian-Affine, MSER, Gaussian-Affine, DoG-Affine. Left: mean reprojection error [*pel*], right: Number of valid 3D object points until frame k. haraff and hesaff fail in frame $k_s = 2$.

projected by the estimated camera matrix $\mathbf{A}_k$ and the detected positions of corresponding 2D feature point $\mathbf{p}_{j,k}$ in the images k (see e.g. [24]).

$$RMSE = \sqrt{\frac{1}{JK} \sum_{j=1}^{J} \sum_{k=1}^{K} d(\mathbf{p}_{j,k}, \mathbf{A}_k \mathbf{P}_j)^2} \tag{3}$$

The $RMSE$ is used for the evaluation which is shown for each frame k in Figure 8, left. The number of valid reconstructed object points J for each k is shown in Figure 8, right. It can be seen that the proposed detectors Gaussian-Affine and DoG-Affine provide a more accurate reconstruction of the scene than the original SIFT detector. The reconstruction with DoG-Affine provides a smaller RMSE than with Gaussian-Affine, but about 11% less object points. The estimation using Harris-Affine, Hessian-Affine, or MSER fails after an early frame k_s in the bundle adjustment step. In Table 3 the failure frame number k_s, the overall RMSE and the number of object points in the sequence for each detector are shown. The main reason for the failure is that the motion blurred sideway region is not represented by sufficiant accurate features as shown in Figure 1.

Table 3. Detection results of the natural motion blur sequence (see Figure 5, 94 frames). After frame number k_s the bundle adjustment diverges if $k_s < 94$. The overall resulting number of valid object points is denoted with J.

	haraff	hesaff	mseraf	o-sift	gauaff	dogaff
k_s	2	2	52	94	94	94
J	30	31	641	2760	2962	2627
$RMSE$	–	–	0.278	0.325	0.309	0.301

5   Conclusion

In this work, an extension to the SIFT detector is proposed which enables the detection of elliptical structures in the Difference of Gaussians gradient signal. Therefore, the subpel estimation technique of the original SIFT detector is exchanged by a regression analysis using two proposed function models, a Gaussian and a Difference of Gaussians. These function models provide features that are rejected by SIFT. The resulting feature detector has all properties of the widely respected SIFT detector, but provides invariance to affine transformation and motion blur while preserving the stable convergence behavior.

The accuracy of the approach is proved using synthetically constructed image data with added Gaussian noise. For the viewpoint change scenario, our approach shows superior performance compared to SIFT, while being competitive to state of the art affine invariant feature detectors. The natural motion blur images show superior structure from motion reconstruction results for our approach compared to all other detectors regarding accuracy and stability.

References

1. Pollefeys, M., Gool, L.V.V., Vergauwen, M., Verbiest, F., Cornelis, K., Tops, J., Koch, R.: Visual modeling with a hand-held camera. International Journal of Computer Vision (IJCV) 59, 207–232 (2004)
2. Brown, M., Lowe, D.G.: Invariant features from interest point groups. In: British Machine Vision Conference (BMVC), pp. 656–665 (2002)
3. Canny, J.: A computational approach to edge detection. IEEE Transactions on Pattern Analysis and Machine Intelligence (PAMI) 8, 679–698 (1986)
4. Harris, C., Stephens, M.: A combined corner and edge detector. In: Alvey Vision Conference, pp. 147–151 (1988)
5. Lindeberg, T.: Feature detection with automatic scale selection. International Journal of Computer Vision (IJCV) 30, 79–116 (1998)
6. Lowe, D.G.: Distinctive image features from scale-invariant keypoints. International Journal of Computer Vision (IJCV) 60, 91–110 (2004)
7. Lindeberg, T., Garding, J.: Shape-adapted smoothing in estimation of 3-d shape cues from affine deformations of local 2-d brightness structure. Image and Vision Computing 15, 415–434 (1997)
8. Mikolajczyk, K., Schmid, C.: An affine invariant interest point detector. In: Heyden, A., Sparr, G., Nielsen, M., Johansen, P. (eds.) ECCV 2002. LNCS, vol. 2350, pp. 128–142. Springer, Heidelberg (2002)
9. Mikolajczyk, K., Schmid, C.: Scale & affine invariant interest point detectors. International Journal of Computer Vision (IJCV) 60, 63–86 (2004)
10. Mikolajczyk, K., Schmid, C.: A performance evaluation of local descriptors. IEEE Transactions on Pattern Analysis and Machine Intelligence (PAMI) 27, 1615–1630 (2005)
11. Mikolajczyk, K., Tuytelaars, T., Schmid, C., Zisserman, A., Matas, J., Schaffalitzky, F., Kadir, T., Gool, L.V.: A comparison of affine region detectors. International Journal of Computer Vision (IJCV) 65, 43–72 (2005)

12. Yu, G., Morel, J.M.: A fully affine invariant image comparison method. In: IEEE International Conference on Acoustics, Speech and Signal Processing (ICASSP), Washington, DC, USA, pp. 1597–1600. IEEE Computer Society, Los Alamitos (2009)
13. Köser, K., Koch, R.: Perspectively invariant normal features. In: IEEE International Conference on Computer Vision (ICCV), pp. 1–8 (2007)
14. Matas, J., Chum, O., Urban, M., Pajdla, T.: Robust wide baseline stereo from maximally stable extremal regions. In: British Machine Vision Conference (BMVC), vol. 1, pp. 384–393 (2002)
15. Tuytelaars, T., Mikolajczyk, K.: Local invariant feature detectors: a survey. In: Foundations and Trends in Computer Graphics and Vision, vol. 3 (2008)
16. Ke, Y., Sukthankar, R.: Pca-sift: A more distinctive representation for local image descriptors. In: International Conference on Computer Vision and Pattern Recognition (ICCV), pp. 506–513 (2004)
17. Schmid, C., Mohr, R., Bauckhage, C.: Comparing and evaluating interest points. In: IEEE International Conference on Computer Vision and Pattern Recognition (ICCV), pp. 230–235 (1998)
18. Schmid, C., Mohr, R., Bauckhage, C.: Evaluation of interest point detectors. International Journal of Computer Vision (IJCV) 37, 151–172 (2000)
19. Mikolajczyk, K., Leibe, B., Schiele, B.: Multiple object class detection with a generative model. In: IEEE Conference on Computer Vision and Pattern Recognition (CVPR), vol. 1, pp. 26–36 (2006)
20. Sivic, J., Russell, B.C., Efros, A.A., Zisserman, A., Freeman, W.T.: Discovering objects and their location in images. In: IEEE International Conference on Computer Vision, vol. 1, pp. 370–377 (2005)
21. Vatis, Y., Ostermann, J.: Adaptive interpolation filter for h.264/avc. IEEE Transactions on Circuits and Systems for Video Technology 19, 179–192 (2009)
22. Cordes, K., Müller, O., Rosenhahn, B., Ostermann, J.: Half-sift: High-accurate localized features for sift. In: IEEE Conference on Computer Vision and Pattern Recognition (CVPR) Workshop, Miami Beach, USA, pp. 31–38 (2009)
23. Fischler, R.M.A., Bolles, C.: Random sample consensus: A paradigm for model fitting with application to image analysis and automated cartography. Communications of the ACM 24, 381–395 (1981)
24. Thormählen, T., Hasler, H., Wand, M., Seidel, H.P.: Merging of feature tracks for camera motion estimation from video. In: 5th European Conference on Visual Media Production (CVMP), pp. 1–8 (2008)

Linear Dimensionality Reduction through Eigenvector Selection for Object Recognition

F. Dornaika[1,2] and A. Assoum[3]

[1] University of the Basque Country, San Sebastian, Spain
[2] IKERBASQUE, Basque Foundation for Science, Bilbao, Spain
[3] LaMA Laboratory, Lebanese University, Tripoli, Lebanon

Abstract. Past work on Linear Dimensionality Reduction (LDR) has emphasized the issues of classification and dimension estimation. However, relatively less attention has been given to the critical issue of eigenvector selection. The main trend in feature extraction has been representing the data in a lower dimensional space, for example, using principal component analysis (PCA) without using an effective scheme to select an appropriate set of features/eigenvectors in this space. This paper addresses Linear Dimensionality Reduction through Eigenvector selection for object recognition. It has two main contributions. First, we propose a unified framework for one transform based LDR. Second, we propose a framework for two transform based DLR. As a case study, we consider PCA and Linear Discriminant Analysis (LDA) for the linear transforms. We have tested our proposed frameworks on several public benchmark data sets. Experiments on ORL, UMIST, and YALE Face Databases and MNIST Handwritten Digit Database show significant performance improvements in recognition that are based on eigenvector selection.

1 Introduction

In most computer vision and pattern recognition problems, the large number of sensory inputs, such as images and videos, are computationally challenging to analyze. In such cases it is desirable to reduce the dimensionality of the data while preserving the original information in the data distribution, allowing for more efficient learning and inference. In order to understand and analyze the multivariate data, we need to reduce the dimensionality. If the variance of the multivariate data is faithfully represented as a set of parameters, the data can be considered as a set of geometrically related points lying on a smooth low-dimensional manifold. The fundamental issue in dimensionality reduction is how to model the geometry structure of the manifold and produce a faithful embedding for data projection. During the last few years, a large number of approaches have been proposed for constructing and computing the embedding. We categorize these methods by their linearity. The linear methods, such as principal component analysis (PCA) [1], multidimensional scaling (MDS) [2] are evidently effective in observing the Euclidean structure. The nonlinear methods such as locally linear embedding (LLE) [3], Laplacian eigenmaps [4], Isomap [5] focus on preserving the geodesic distances.

G. Bebis et al. (Eds.): ISVC 2010, Part I, LNCS 6453, pp. 276–285, 2010.

Linear dimensionality reduction (LDR) techniques have been increasingly important in pattern recognition [6] since they permit a relatively simple mapping of data onto a lower-dimensional subspace, leading to simple and computationally efficient classification strategies. Although the field has been well developed for computing the dimension of mapping for the supervised and unsupervised cases, the issue of selecting the new components in the new-subspace has not received much attention. In most practical cases, relevant features in the new embedded space are not known a priori. Finding out what features to use in a classification task is referred to as feature selection. Although there has been a great deal of work in machine learning and related areas to address this issue these results have not been fully explored or exploited in emerging computer vision applications. Only recently there has been an increased interest in deploying feature selection in applications such as face detection, vehicle detection, and pedestrian detection.

The main trend in feature extraction has been to represent the data in a lower dimensional space computed through a linear or non-linear transformation satisfying certain properties. The goal is to find a new set of features that represents the target concept in a more compact and robust way but also to provide more discriminative information. Most efforts in LDR have largely ignored the feature selection problem and have focused mainly on developing effective mappings (i.e., feature extraction methods).

In this paper, we propose a feasible approach to the problem of linear dimensionality reduction: how many and which eigenvectors to retain. Our proposed method can be seen as an extension of [7]. In [7], the authors use PCA and feature selection for the task of object detection. They demonstrate the importance of feature selection (eigenvector selection) in the context of two object detection problems: vehicle detection and face detection. However, our proposed work differs from [7] in three aspects. First, we address Linear Dimensionality Reduction and feature selection for object recognition—a more generic problem than object detection. Second, our study is not limited to the use of the classical PCA. Third, we propose a novel scheme for selecting eigenvectors associated with a concatenation of two linear embedding techniques.

The remainder of the paper is organized as follows. Section 2 briefly describes some backgrounds on Linear Dimensionality Reduction schemes as well as on feature selection. Section 3 presents the proposed approach for the estimation of embedding through feature selection. Section 4 presents some experimental results obtained with several benchmark data sets.

2 Backgrounds

2.1 Linear Dimensionality Reduction

All Linear Dimensionality Reduction schemes (including the most recent ones) compute a matrix transform that maps the original data samples to a new subspace. In general, this computation is carried out by solving an eigen decomposition problem. Let $\mathbf{W}$ be the matrix formed by concatenating the relevant obtained eigenvectors. The linear transform is given by $\mathbf{W}^T$. The dimension of each eigenvector is equal to that of the original data samples. Estimating the number of the relevant eigenvectors (the columns of matrix $\mathbf{W}$) is usually carried out using some classical schemes (e.g., using the percentage of retained variability) or by performing a series of experiments on a given validation dataset (e.g., plotting the recognition rate versus the number of retained eigenvectors). However, how many and which eigenvectors to retain from the global set of computed eigenvectors is still an open problem. It has been argued that many discarded eigenvectors by classical schemes can have much discriminative information. In this study, we propose a unified framework that selects the most appropriate eigenvectors for a given classification problem. Among all the subspace learning methods, PCA and LDA are two most famous ones and have become the most popular techniques for biometrics applications such as face recognition and palmprint recognition. PCA is a popular method for unsupervised linear dimensionality reduction [8]. It projects the original data into a low-dimensional space, which is spanned by the eigenvectors associated with the largest eigenvalues of the covariance matrix of all data samples. Turk and Pentland [1] proposed the first application of PCA to face recognition. Since the basis vectors constructed by PCA had the same dimension as the input face images, they were named "Eigenfaces". Unlike PCA, LDA is a supervised method which takes full consideration of the class labels. While PCA seeks directions that are efficient for representation, LDA seeks directions that are efficient for discrimination. LDA is a supervised technique which has been shown to be more effective than PCA in many applications. It aims to maximize the between-class scatter and simultaneously minimize the within-class scatter. In many cases, the concatenation of PCA and LDA has proven to improve the recognition performance as well as to solve the possible singularities of LDA (Small Sample Size problem). Without loss of generality, we apply our proposed frameworks for LDR with eigenvector selection on the linear embedding techniques: PCA, LDA, and PCA+LDA.

2.2 Feature Selection

In a great variety of fields, including pattern recognition and machine learning, the input data are represented by a very large number of features, but only few of them are relevant for predicting the label. Many algorithms become computationally intractable when the dimension is high. On the other hand, once a good small set of features has been chosen, even the most basic classifiers (e.g. 1-nearest neighbor, 1NN) can achieve desirable performance. Therefore, feature selection, i.e. the task of choosing a small subset of features which is sufficient to predict the target labels well, is critical to minimize the classification error. At the same time, feature selection also reduces training and inference time and leads to a better data visualization as well as to a reduction of measurement and storage requirements. Roughly speaking, feature selection algorithms have two key problems: search strategy and evaluation criterion. According to the criterion, feature selection algorithms can be categorized into filter model and wrapper model. In the wrapper model, the feature selection method tries to directly optimize the performance of a specific predictor (classification or clustering algorithm). The main drawback of this method is its computational deficiency. In the filter model, the feature selection is done as a preprocessing, without trying to optimize the performance of any specific predictor directly [9]. This is usually achieved through an evaluation function and a search strategy is used to select a feature subset that maximizes this evaluation function. A comprehensive discussion of feature selection methodologies can be found in [10,11].

3   Eigenvector Selection for Object Recognition

We propose two frameworks for LDR: i) the first framework addresses the linear dimensionality reduction given by one matrix transform (e.g., the PCA transform), and ii) the second framework addresses the linear dimensionality reduction given by the concatenation of two matrix transforms (e.g., PCA+LDA or PCA+Locality Preserving Projection (LPP) [12]).

3.1   One Transform Based LDR

In this section, we propose a generic scheme for eigenvector selection. The proposed scheme can be used by any LDR technique. In our study, this scheme will be used with PCA and LDA. We will adopt a wrapper technique for eigenvector selection. The evaluation strategy will be directly encoded by the recognition accuracy over validation sets. Without any loss of generality, the classifier used after Linear Dimension Reduction and eigenvector selection will be the KNN classifier. This classifier is one of the oldest and simplest methods for pattern classification and it is one of the top 10 algorithms in data mining [13]. The adopted search strategy will be carried out by a Genetic Algorithm (GA). We use a simple encoding scheme in the form of a bit string (a chromosome) whose length is determined by the total number of eigenvectors. Each eigenvector is associated with one bit in the string. If the ith bit is 1, then the ith eigenvector is selected, otherwise, that eigenvector is ignored. Each string thus represents a different subset of eigenvectors.

Evaluation Criterion. The goal of feature selection is to use less features to achieve the same or better performance. Therefore, the fitness evaluation contains two terms: (1) accuracy and (2) the number of features selected. The performance of the KNN classifier is estimated using a validation data set which guides the GA search. Each feature subset contains a certain number of eigenvectors. If two subsets achieve the same performance, while containing different number of eigenvectors, the subset with fewer eigenvectors is preferred. Between accuracy and feature subset size, accuracy is our major concern. We used the fitness function shown below to combine the two terms:

$$Fitness = c_1 \, Accuracy + c_2 \, Zeros \tag{1}$$

where *Accuracy* corresponds to the classification accuracy on a validation set for a particular subset of eigenvectors, and *Zeros* corresponds to the number of eigenvectors not selected (i.e., zeros in the individual). c_1 and c_2 are two positive coefficients with $c_2 << c_1$. In our implementation, c_1 was set to 10^4 and c_2 to 1.

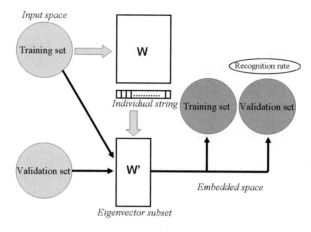

Fig. 1. Evaluating the fitness of a given individual string in the Genetic Algorithm. The matrix **W** denotes the whole set of eigenvectors. The matrix **W**′ denotes a putative subset of eigenvectors. The corresponding recognition rate will be used for computing the fitness of the string (Eq. (1)).

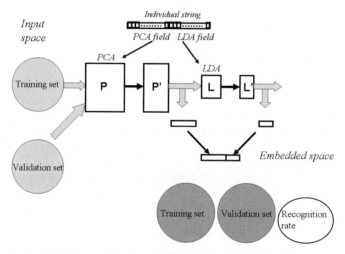

Fig. 2. Evaluating the fitness of a given individual string in the Genetic Algorithm for the two transform based LDR

Search Strategy: A Genetic Algorithm. Genetic Algorithms (GAs) are biologically motivated adaptive systems based on natural selection and genetic recombination [14]. In the standard GA, candidate solutions are encoded as fixed length vectors–strings. The initial population of solutions is chosen randomly. These candidate solutions are allowed to evolve over a certain number of generations. At each generation, the fitness of each string is calculated; this is a measure of how well the string optimizes the objective function. Subsequent generations are created through a process of selection, recombination, and mutation. In all of our experiments, we used a population size of 100. The maximum number of generations is set to 50 generations.

Figure 1 illustrates how the fitness of every individual string is evaluated using training and validation sets. The matrix $\mathbf{W}$ denotes the whole set of eigenvectors. The matrix $\mathbf{W}'$ denotes the selected subset of eigenvectors. The corresponding recognition rate will be used as the first term in Eq. (1) in order to compute the fitness of the string.

3.2   Two Transform Based LDR

In the previous section, we have described the linear dimensionality reduction through feature selection when the linear embedding is given by one single matrix transform. In this section, we describe our proposed framework for eigenvector selection whenever the linear embedding is given by the concatenation of two matrix transforms. Our proposed framework is generic in the sense that it can be used with any pair of linear dimensionality approaches. However, for the sake of clarity and without loss of generality, we assume that the first linear embedding is given by PCA and the second linear embedding is given by LDA. Note that when face images are mapped using this concatenation we get the embedding of Fisherface. One possible solution is to perform classical PCA and LDA in sequence. The feature selection scheme is then invoked in order to select the most appropriate eigenvectors for both linear transforms using the concatenated selected features. However, this scheme does not take into account the fact that the second embedding (in our case it is LDA) is already built on the total set of eigenvectors of the first embedding (in our case it is PCA). Thus, we propose a novel flexible selection scheme that is summarized in Figure 2. In this case, we design a chromosome with two fields. The first field is associated with the PCA eigenvectors (its size is set in the same manner of the one transform based DLR) and the second field is associated with the LDA eigenvectors (the size is fixed to the number of classes minus one). As can be seen, for every subset of selected PCA eigenvectors there will be a different global set of LDA eigenvectors. In other words, for a given training set, the global PCA matrix is fixed but the LDA matrices (global and subset) will depend on the solution encoded by the current chromosome. Once again, the best chromosome is searched for by maximizing the fitness measure (1) over a validation set. To this end, we use the same Genetic Algorithm.

4   Experimental Results

4.1   Benchmark Data Sets

The data sets used are ORL face data set, UMIST face data set, YALE face data set and MNIST handwritten digit data set[1]. The ORL data set contains 400 face

---

[1] ORL, UMIST, YALE, and MNIST data sets can be retrieved respectively from
http://www.cl.cam.ac.uk/research/dtg/attarchive/facedatabase.html
http://www.shef.ac.uk/eee/research/vie/research/face.html
http://see.xidian.edu.cn/vipsl/database_Face.html
http://yann.lecun.com/exdb/mnist/

images of 40 individuals. Each individual has 10 gray images with size 92×112. The UMIST data set contains 575 gray images of 20 different people. The size of each image is 112×92 pixels. The YALE face data set contains 165 images of 15 persons. Each individual has 11 images of size 320×243 pixels. The MNIST Handwritten Digit data set is composed of 70000 sample digits. Each digit has 6000 examples for training and 1000 examples for testing. In our experiment, only one tenth of MNIST data set is used. The details of these data sets are described in Table 1. Some samples of these data sets are shown in Figure 3.

Table 1. Details of benchmark data sets

Data set	Data set size (N)	Sample dimension (D)	Nb. classes (C)
ORL	400	10304	40
UMIST	575	10304	20
YALE	165	77760	15
MNIST	7000	784	10

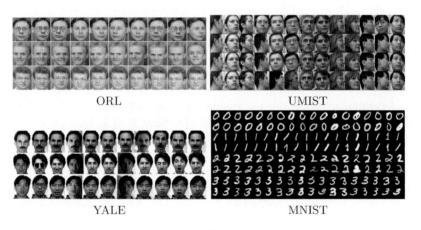

ORL   UMIST

YALE   MNIST

Fig. 3. Some samples in ORL, UMIST, YALE, and MNIST data sets

4.2   Method Comparison

We empirically evaluate the improvement obtained by our proposed frameworks on the above data sets. We have performed a number of experiments and comparisons to demonstrate the importance of eigenvector selection for object recognition based on linear embedding. First, recognition experiments are conducted on the data sets using the classical Linear Dimensionality Reduction schemes, i.e. using the top eigenvectors, followed by the KNN classifier (K=3). Second, recognition experiments are conducted on the same data sets using the Linear Reduction schemes with feature selection. In this scheme, we use ten fold cross-validation scheme. For each fold, 70 % of the samples are used for training and the remaining samples are used for testing. For all data sets used, we retained the top 150 eigenvectors obtained by the classical PCA. The top 150 eigenvectors

Table 2. Average recognition rates obtained with four data sets

	PCA	PCA with FS	PCA+LDA	PCA+LDA with FS
ORL	91.7%	97.8%	95.7%	100%
UMIST	95.7%	99.4%	98.7%	100%
YALE	78.4%	97.3%	97.5%	99.8%
MINST	85.0%	93.7%	77.2%	93.6%

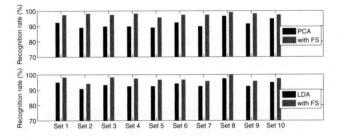

Fig. 4. Recognition rates using linear dimensionality reduction schemes applied on ORL data set. The blue bars correspond to the classical dimensionality reduction schemes, and the red bars to their use with feature selection (eigenvector selection) paradigm (Section 3.1).

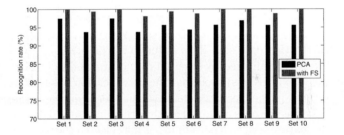

Fig. 5. Recognition rates using linear dimensionality reduction and Feature Selection applied on UMIST dataset. The blue bars correspond to the PCA dimensionality reduction scheme. The red bars correspond to its use with the feature selection paradigm (Section 3.1).

correspond to an average variability (over ten training sets) of 96.5% , 97.7%, 99.9%, and 97.6% for ORL, UMIST, YALE, and MNIST data sets, respectively.

Figure 4 shows the recognition rate for ORL data set for all 10 partitions. The two plots correspond to the PCA and LDA schemes, respectively. The blue bars correspond to the classical linear mapping (150 dimensions for PCA, 39 dimensions for LDA). The red bars correspond to the mapping obtained by the same linear methods with Feature Selection (Section 3.1). As can been seen, the use of feature selection paradigm has improved the recognition rate. Figures 5 and 6 illustrate the recognition rate obtained for UMIST and YALE data sets, respectively.

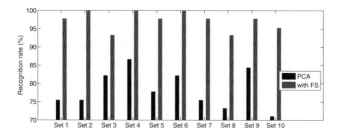

Fig. 6. Recognition rates using linear dimensionality reduction and Feature Selection applied on YALE dataset. The blue bars correspond to the PCA dimensionality reduction scheme. The red bars correspond to its use with feature selection paradigm (Section 3.1).

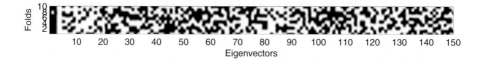

Fig. 7. The selected eigenvectors for YALE data set and for 10 folds. In this array, a white dot means that the corresponding eigenvector has been selected. As can be seen, in almost all folds the first three eigenvectors have not been selected.

We observe that the improvement obtained with YALE data set is larger than the one obtained with other data sets. This is due to the fact that, in the YALE data set, the face images are affected by several types of variability (e.g., non uniform lighting and facial expressions) such that the classification based on the classical mapping with the top eigenvectors did not show a performance having the same order of magnitude as the ones obtained for the other data sets. This was proven by the fact that the feature selection paradigm has not selected the top three eigenvectors which correspond to illumination variations (Figure 7). Table 2 summarizes the average recognition rate over the 10 partitions for the four data sets. The last column corresponds to the proposed two transform based LDR (Section 3.2). As can be seen, the second framework has provided high recognition rates for face datasets ($C \geqslant 15$).

5   Conclusion

In this paper, we presented two unified frameworks for identifying relevant eigenvectors for LDR based multi-class recognition problems. The first framework uses one matrix transform. The second framework uses the concatenation of two matrix transforms. The procedure of eigenvector identification is based on maximizing the recognition rate with a genetic algorithm. Experiments are conducted on four benchmark data sets using the linear embedding given by PCA and LDA. These experiments have shown that the recognition rate obtained by the two proposed frameworks has increased and that the compression of data is improved. We have found that this holds true for LDA embedding whenever the number of classes is greater than 10.

References

1. Turk, M., Pentland, A.: Eigenfaces for recognition. Journal of Cognitive Neuroscience 3, 71–86 (1991)
2. Borg, I., Groenen, P.: Modern Multidimensional Scaling: theory and applications. Springer, New York (2005)
3. Roweis, S.T., Saul, L.K.: Nonlinear dimensionality reduction by locally linear embedding. Science 290, 2319–2327 (2000)
4. Belkin, M., Niyogi, P.: Laplacian eigenmaps for dimensionality reduction and data representation. Neural Computation 15(6), 1373–1396 (2003)
5. Tenenbaum, J.B., de Silva, V., Langford, J.C.: A global geometric framework for nonlinear dimensionality reduction. Science 290, 2319–2323 (2000)
6. Martinez, A.M., Zhu, M.: Where are linear feature extraction methods applicable? IEEE Trans. Pattern Analysis and Machine Intelligence 27(12), 1934–1944 (2005)
7. Suna, Z., Bebisa, G., Miller, R.: Object detection using feature subset selection. Pattern Recognition 37, 2165–2176 (2004)
8. Jolliffe, I.: Principal Component Analysis. Springer, New York (2002)
9. Mitra, P., Murthy, C., Pal, S.: Unsupervised feature selection using feature similarity. IEEE Trans. Pattern Analysis and Machine Intelligence 24, 301–312 (2002)
10. Guyon, I., Elisseeff, A.: An introduction to variable and feature selection. Journal of Machine Learning Research 3, 1157–1182 (2003)
11. Liu, H., Yu, L.: Toward integrating feature selection algorithms for classification and clustering. IEEE Trans. Knowledge Data Engineering 17, 494–502 (2005)
12. He, X., Yan, S., Hu, Y., Niyogi, P., Zhang, H.Z.: Face recognition using laplacianfaces. IEEE Trans. Pattern Analysis and Machine Intelligence 27(3), 328–340 (2005)
13. Wu, X., Kumar, V., et al.: Top 10 algorithms in data mining. Knowledge Information Systems 14(1), 1–37 (2008)
14. Srinivas, M., Patnaik, L.: Genetic algorithms: a survey. IEEE Computer 27(6), 17–26 (1994)

Symmetry Enhanced Adaboost

Florian Baumann, Katharina Ernst, Arne Ehlers, and Bodo Rosenhahn

Institut für Informationsverarbeitung
Leibniz Universität Hannover
Appelstraße 9a, 30167 Hannover, Germany

Abstract. This paper describes a method to minimize the immense training time of the conventional Adaboost learning algorithm in object detection by reducing the sampling area. A new algorithm with respect to the geometric and accordingly the symmetric relations of the analyzed object is presented. Symmetry enhanced Adaboost (SEAdaboost) can limit the scanning area enormously, depending on the degree of the objects symmetry, while it maintains the detection rate. SEAdaboost allows to take advantage of the symmetric characteristics of an object by concentrating on corresponding symmetry features during the detection of weak classifiers. In our experiments we gain 39% reduced training time (in average) with slightly increasing detection rates (up to 2.4% and up to 6% depending on the object class) compared to the conventional Adaboost algorithm.

1 Introduction

This paper introduces an extension of the object detection framework proposed by Viola and Jones [1] in which the well established Adaboost algorithm [2, 3] is utilized to form a strong classifier.

Adaboost in one of its various forms is widely used in many tasks of object detection. Applied by the framework of Viola and Jones it allows high detection rates in real-time classification. Because of its economic consumption of processing power it is a good choice for applications, such as face detection, on mobile devices like digital cameras and camera phones. One drawback of the algorithm is the extensive training phase, due to the huge number of weak classifiers and the iterative training.

Our contribution is to reduce this problem by exploiting symmetric object characteristics, since symmetries can be observed all over nature [4], see also Figure 1. Not only constructed objects (e.g. buildings, cars, etc.) have axes of symmetry but also humans, animals and plants. We want to take advantage of this phenomenon by exploiting this fact in the learning and detection phase of the conventional Adaboost algorithm. In the experiments we demonstrate the applicability of our method on two example scenarios. The first scenario is the optical inspection of SMD-components while our second scenario approaches face detection.

G. Bebis et al. (Eds.): ISVC 2010, Part I, LNCS 6453, pp. 286–295, 2010.
© Springer-Verlag Berlin Heidelberg 2010

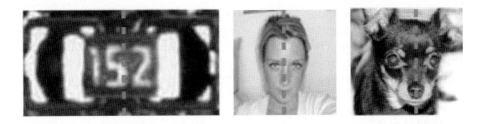

Fig. 1. Objects with symmetric characteristics: left: SMD-component symmetry, middle: face symmetry, right: animal symmetry

This paper is structured as follows: In the next section all required tools for boosting are briefly described. In section 3 the modified symmetry-enhanced Adaboost algorithm will be defined in detail. The final section presents the experimental results on SMD-components and face detection and a comparison to the conventional Adaboost algorithm for object detection as proposed by Viola and Jones [1].

2   Adaboost Algorithm

Adaptive Boosting (short: Adaboost), a machine learning algorithm proposed by Freund and Schapire [2, 3] is the foundation for our work. This algorithm is based upon the boosting theory of Kearns developed in 1988 [5]. In collaboration with Vazirani [6] the theory was enhanced and finally elaborated by Freund and Schapire in 1995 to the popular Adaboost algorithm.

2.1   Adaboost by Freund and Schapire

Adaboost combines multiple weak classifiers to one strong classifier in the fashion of a weighted sum. These weights are determined during a round-based training phase. In each round a given weak classifier is evaluated on the training set resulting in higher weights for better performing classifiers. An important aspect of Adaboost is that also weights are assigned to the training examples. These weights are adapted after each round to increase the influence of incorrectly classified examples. The only required inputs besides the number of rounds in the training phase are a collection of positive and negative image examples as well as the set of weak classifiers.

Adaboost yields good results in various object detection applications but has the shortcoming of an extensive training phase.

Several variants of boosting algorithms have been developed in the last years, e.g. LPBoost, SoftBoost, MILBoost, FloatBoost or S-Adaboost [7–11]. LPBoost focuses on minimizing the generalization error by maximizing the soft margin. Its drawback is to determine a good upper bound for the number of iterations.

SoftBoost [8] overcomes this problem by applying a relative entropy regularization yielding in an upper bound that is logarithmic in the number of examples. But the generalization error of SoftBoost decreases only slowly at the beginning of the training. Entropy Regularized LPBoost [7] adds the same relative entropy regularization to LPBoost and thus combines its high performance with a good upper bound. In contrast to Adaboost MILBoost [9] introduces a cost function which is combined with the feature selection. A backtrack mechanism is provided by FloatBoost [10] that allows to remove weak classifiers having low performance after each iteration step of Adaboost. In S-Adaboost [11] the Divide and Conquer Principle is utilized to allow Adaboost to divide the input space into subspaces of e.g different classification difficulty and to classify them using dedicated strong classifiers. A non-linear combination of these classifiers joins the advantages of simpler but more general and stronger but more specific classifiers.

Whereas these methods concentrate on the weighting functions iteration [7, 8] and feature analysis [9, 10], we propose to integrate geometric prior information in the feature selection to speed up the training phase. Assuming a symmetry property in the training of object classes is a simple but effective way to reduce the enormous sampling size and to exploit properties of natural objects. Therefore, our method is completely different to earlier approached modifications. But our contribution can easily be applied to the above mentioned adaptions of Adaboost to further enhance the performance.

2.2   Framework by Viola and Jones

Based on the algorithm of Freund and Schapire [2], Viola and Jones introduced Adaboost into the field of face detection [1]. They boosted weak classifiers using Haar-like features and the representation by integral images. Viola and Jones achieved excellent results, furthermore their framework is working in real-time.

The name Haar-like feature comes from their similarity to Wavelet transformations with Haar basis functions. Haar-like features are simple rectangle features, which are a good choice to represent weak classifiers in boosting. Each feature is described by a position (offset) and a +1/-1-profile. Figure 2 shows some typical features for object detection.

The integral image offers a fast computation of rectangle feature values. For each image location (x,y) the sum of pixels above and to the left of (x,y) is pre-calculated and stored. Using integral images the Haar-like features can be computed in an very efficient way.

To deal with the huge amount of weak classifiers in the training phase they proposed an algorithm that searches in each training round over the set of possible weak classifiers and selects the one with the lowest classification error by passing only a single time through a sorted list. For more information about the feature selection mechanism see [1].

3   Symmetry Enhanced Adaboost

3.1   Comprising of Symmetric Object Characteristics by Inserting Symmetry Features

Many natural objects have symmetric characteristics. Faces can be divided into two halves by inserting one symmetry axis. Other objects from nature for example flowers, butterflies or snowflakes can be divided into several segments. Even viruses are showing a very high symmetry in several axes. Regarding SMD-Components we can divide each picture into four quadrants by inserting two symmetry axes. We want to benefit from this symmetry to reduce the training time and to stabilize the detection rate of Adaboost. We make the assumption that each selected classifier in an image section implicates at least one symmetric classifier. Reflected along one of the symmetry axes this symmetric classifier should hold a similar low classification error. In that way the best classifier selected in each round is used to detect a symmetric classifier in a reduced, mirrored search area. Furthermore to refuse classifiers providing higher error rates we apply a threshold to assure that the symmetric classifier produces a classification error nearby the original feature. By setting the threshold in our experiments to 0.025 the classification error of the symmetric classifier is only allowed to be 0.025 higher than the error of the original classifier. The classification error is computed as the sum of the weights of the misclassified examples images. As these weights are normalized, the classification error is within the range $[0, 1]$. It must be taken into consideration that the position of the object differs from picture to picture in little variabilities. Thus a variance in x- and y-direction is inserted. This variance extends the sampling area of the symmetric object area to a few pixel in both directions.

Fig. 2. Examples of a detected feature and its corresponding symmetry feature

Hence a possible deviation in the position of the object can be adjusted, and the algorithm has more flexibility in choosing the best corresponding classifier. By this idea the symmetric characteristics of objects are utilized to reduce the sampling area. The sampling of only half or quarter of the image suffices in our application, while features are placed in the other image area by computing the symmetries.

The reflected features are based on features found in the reduced sampling area. In our application we use one symmetric axis in the vertical position for SMD- and face detection. In order to get the corresponding symmetry features we must reflect only the vertical features. Line- and horizontal features can stay in their configuration. But in the case of two symmetric axes we need to apply a reflection to vertical and horizontal features. Figure 2 shows four examples of a feature and its corresponding symmetry feature detected by SEAdaboost in faces and SMD-Components.

3.2   Modified Algorithm

Based on the notation in [1] the SEAdaboost algorithm can be summarized as follows:

- **Given:** Labeled images $(x_1, y_1), ..., (x_m, y_m)$, with $y_j \in \{1, 0\}$
- **Select** optimal symmetry axis of images $(x_1, y_1), ..., (x_m, y_m)$
- **Initialize** weights $w_{1,j} = \frac{1}{2n}, \frac{1}{2l}$ if $y_j = 0,1$ respectively, with n denoting the number of negative and l of positive images,
- For round number $r = 1, ..., R$:
  1. **Normalize** weights $w_{r,j}$:
  $$w_{r,j} = \frac{w_{r,j}}{\sum_{k=1}^{m} w_r, k} \tag{1}$$

  2. **Select** best weak classifier in the image region limited by the symmetric axis with respect to the error:
  $$\kappa_r = \min \sum_k w_k \, |h(x_k, f, p, \tau) - y_k| \tag{2}$$

  3. **Define** $h_r(x)$ with minimal error κ_r:
  $$h_r(x) = h(x, f_r, p_r, \tau_r) \tag{3}$$

  4. **Select** best symmetric weak classifier with respect to the error and the variance ψ in x- and y-direction:
  $$\kappa_{r,sym} = \min \sum_k w_k \, |h(x_k, f, p, \tau, \psi) - y_k| \tag{4}$$

  5. **Define** $h_{r,sym}(x)$ with minimal error $\kappa_{r,sym}$ and respect to a threshold ρ:
  $$h_{r,sym}(x) = h(x, f_{r,sym}, p_{r,sym}, \tau_{r,sym}, \psi, \rho) \tag{5}$$

  6. **Update** the weights:
  $$w_{r+1,j} = w_{r,j} \cdot \chi_r^{1-\lambda_j} \cdot \chi_{r,sym}^{(1-\lambda_{j,sym})\beta_r}, \tag{6}$$
  $$\chi_r = \frac{\kappa_r}{1 - \kappa_r} \quad \text{and} \quad \chi_{r,sym} = \frac{\kappa_{r,sym}}{1 - \kappa_{r,sym}}$$

     with

  $$\lambda_j = \begin{cases} 0 & \text{, if } x_j \text{ classified correctly by the weak classifier} \\ 1 & \text{, else.} \end{cases}$$

and with

$$\lambda_{j,sym} = \begin{cases} 0 & \text{, if } x_j \text{ classified correctly by the symmetric weak classifier} \\ 1 & \text{, else.} \end{cases}$$

and with

$$\beta_r = \begin{cases} 1 & \text{, if } \kappa_{r,sym} < \kappa_r + \rho \\ 0 & \text{, else.} \end{cases}$$

- The strong classifier is:

$$Z(x) = \begin{cases} 1 & \text{, if } \sum_{r=1}^R \delta_r h_r(x) + \delta_{r,sym} h_{r,sym}(x) \geq \frac{1}{2} \sum_{r=1}^R \delta_r + \delta_{r,sym} \\ 0 & \text{, else.} \end{cases}$$

(7)

with $\delta_r = log\frac{1}{\chi_r}$, $\quad \delta_{r,sym} = \beta_r log\frac{1}{\chi_{r,sym}}$

After preprocessing and selecting the optimal symmetry axes the labeled image examples are initialized, i.e. the weights of each positive and negative image example are set with respect to the respective training set size. In the beginning of each round the image weights are normalized (1).

Afterwards the algorithm searches in every round for the best weak classifier, in an area limited by at least one symmetric axis (2). Searching means to select the best classifier that splits the images in two labeled groups and thereby produces the lowest error. One great advantage of this mechanism is that we only need to pass a single time through a sorted list by calculating an error based on the image weights. More information about the feature selection mechanism is given in [1].

Once a weak classifier is detected and defined (3), the regarding feature is reflected and moved to the symmetric area (4). In this area we search with respect to a variance for the corresponding symmetric classifier. As shown in Figure 2 the symmetric classifiers can be moved from the reflected position by several pixels.

Optimally, the algorithm selects two weak classifiers each round, whereas the selection of the symmetry classifier is subject to restrictions. This classifier should only be selected and defined (4,5), if the produced classification error is at maximum given by the threshold ρ added to the error of the already selected classifier. If the classifier cannot fulfill this condition, it is rejected. In that way the optimal weak classifiers with respect to the weights of the example images are selected.

In the next step the weights of the example images are updated regarding its classification success by the newly selected weak classifiers. This provides the opportunity to prefer incorrect classified images in the following rounds. After the step of updating the weights the next round starts.

Finally after the expiration of round R the algorithm forms a strong classifier by a linear combination of the selected (conventional and symmetric) weak classifiers. Each weak classifier is weighted by its own produced error in (2) and (4). The strong classifier can be applied to unseen images.

4 Experimental Results

We evaluate SEAdaboost in two fields of application, face detection and the optical inspection of SMD-components. Whereas for faces, the training data are obviously faces vs. non-faces, for SMD-components it is different, since we want to detect wrong positioned or missing components, as well as components with defect solder joints, see Figure 3.

Fig. 3. Detection of different faces (credits to AT&T Laboratories, Cambridge) and non-faces / SMD-components and SMD-components with different kinds of failures

One challenge in the classification of SMD-components is given by image examples, which contain reflections of other components nearby. The difficulty in the training of these examples lies in classifying defects of solder joints correctly, while detecting reflections as those. With regard to the face detection we need to locate faces in different poses or with different expressions. Furthermore our method should be robust to changing backgrounds or illumination/lighting disparities. For the face detection training we use the AT&T database of faces. Therefore we give credits to AT&T Laboratories Cambridge. The database contains ten different images each of 40 people varying in the lighting, facial expressions (open / closed eyes, smiling / not smiling) and facial details (glasses / no glasses). The images were taken against a dark homogeneous background with the people in an upright, frontal position.

We divided our image bases into a training and validation set using a 67/33 ratio. Crowther and Cox [12] illustrated that especially for small bases like our set of SMD-components a split containing only a small part for validation is not recommendable. They suggested to select a ratio between 50/50 and 70/30.

We demonstrate the difference between the classification results of the conventional Adaboost and SEAdaboost, i.e. we show the minimization of training time in Table 1.

Table 1. Experimental results of face- and SMD-detection in comparison. The SEAdaboost achieves the highest detection rate while the training time reduces significantly.

Application	Algorithm	Training time	Hit-rate	Rounds
Face	Adaboost	34.0 hours	89.56%	18
Face	SEAdaboost	21.0 hours	90.72%	18
SMD	Adaboost	36.0 hours	91.6%	10
SMD	SEAdaboost	23.0 hours	97.5%	10

By incorporating the idea of symmetry in the conventional Adaboost algorithm and the thereby reduced sampling area, Symmetry enhanced Adaboost effectively stabilizes learning results. In the scenario of face detection our method increases the detection rate slightly between 0.2% and 2.5% and in application on SMD-components between 2.4% and 6.0% depending on the number of training rounds.

The dependency of the detection rate from the number of training rounds respectively training time is shown for both object classes in Figure 4 and 5. The test results indicate significantly, that the detection rate increases with a higher number of training rounds for all implemented algorithms.

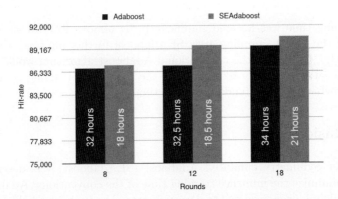

Fig. 4. Rounds/training time vs. detection rate applied to faces. With a higher number of rounds, the detection rate increases.

The sets of corresponding symmetry classifiers allow the detection of objects by smaller features which can be fitted tighter to the significant regions in images. Figure 6 demonstrates that the region around the eyes is represented by the symmetric features more precisely than by a conventional three-rectangle feature. Also small rotations of the object can be better compensated.

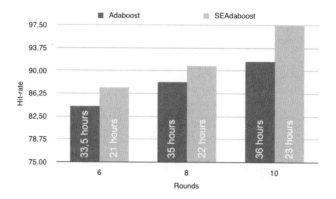

Fig. 5. Rounds/training time vs. detection rate applied to SMD-components. With a higher number of rounds, the detection rate increases.

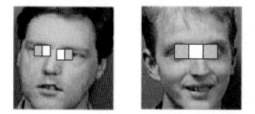

Fig. 6. Regions around the eye detected by SEAdaboost exploiting a pair of symmetric two-rectangle features and by Adaboost using a conventional three-rectangle feature

The training time was lowered significantly by almost 40%. In spite of the reduction of the object sampling area, SEAdaboost further improves the classification results. Obviously the conventional Adaboost algorithm would achieve the same detection accuracy by increasing the number of training rounds, but with the drawback of an extended training time.

5   Conclusions

In this paper Symmetry enhanced Adaboost (SEAdaboost) was described, a method to minimize the immense training time of the conventional Adaboost algorithm, while minor increasing the detection rate. The basic idea behind SEAdaboost is to exploit symmetries of objects, which are often present in certain object classes. We applied our algorithm to the field of face detection and classification of SMD-components. Our experimental results show that SEAdaboost reduces the learning phase drastically. The training time is reduced by almost 40%, while the detection rate could be slightly improved. Thus SEAdaboost qualifies to detect objects with symmetrical characteristics (e.g. components, faces). An application of SEAdaboost to objects with asymmetric characteristics would lead to a higher training time and high error rates of the symmetry features. However this effect could be used to make a statement concerning the object symmetry exploiting the weights of the symmetry features. Thus an application of SEAdaboost would also be of great convenience to estimate an object symmetry.

Our approach can easily be combined with other extensions, such as LPBoost, SoftBoost, MILBoost, FloatBoost or S-Adaboost [7–11].

References

1. Viola, P., Jones, M.J.: Robust real-time face detection. International Journal of Computer Vision 57, 137–154 (2004)
2. Freund, Y., Schapire, R.E.: A short introduction to boosting. Journal of Japanese Society for Artificial Intelligence 14, 771–780 (1999)
3. Freund, Y., Schapire, R.E.: Experiments with a new boosting algorithm. In: Proceedings of the Thirteenth International Conference, pp. 148–156 (1996)
4. Loy, G., Eklundh, J.O.: Detecting symmetry and symmetric constellations of features. In: Leonardis, A., Bischof, H., Pinz, A. (eds.) ECCV 2006. LNCS, vol. 3952, pp. 508–521. Springer, Heidelberg (2006)
5. Kearns, M.: Thoughts on hypothesis boosting. Unpublished manuscript (1988)
6. Kearns, M., Vazirani, U.V.: An introduction to computational learning theory. MIT Press, Cambridge (1994)
7. Warmuth, M.K., Glocer, K.A., Vishwanathan, S.: Entropy regularized lpboost. In: Freund, Y., Györfi, L., Turán, G., Zeugmann, T. (eds.) ALT 2008. LNCS (LNAI), vol. 5254, pp. 256–271. Springer, Heidelberg (2008)
8. Warmuth, M.K., Glocer, K., Raetsch, G.: Boosting algorithms for maximizing the soft margin. In: Advances in Neural Information Processing Systems, vol. 20, pp. 1585–1592. MIT Press, Cambridge (2008)
9. Viola, P., Platt, J.C., Zhang, C.: Multiple instance boosting for object detection. In: Advances in Neural Information Processing, vol. 18, pp. 1417–1426 (2007)
10. Li, S.Z., Zhang, Z., Shum, H.Y., Zhang, H.: Floatboost learning for classification. In: Advances in Neural Information Processing Systems. Microsoft Research Asia (2002)
11. Jiang, J.L., Loe, K.F.: S-adaboost and pattern detection in complex environment. In: Proceedings of the 2003 IEEE Computer Society Conference on Computer Vision and Pattern Recognition (CVPR 2003), pp. 413–418 (2003)
12. Crowther, P.S., Cox, R.J.: A method for optimal division of data sets for use in neural networks. In: Khosla, R., Howlett, R.J., Jain, L.C. (eds.) KES 2005. LNCS (LNAI), vol. 3684, pp. 1–7. Springer, Heidelberg (2005)

Object Category Classification
Using Occluding Contours

Jin Sun[1], Christopher Thorpe[2], Nianhua Xie[1], Jingyi Yu[2], and Haibin Ling[1]

[1] Computer and Information Science Department, Temple University, Philadelphia, PA, USA
[2] Department of Computer and Information Science, University of Delaware,
Newark, DE, USA

Abstract. *Occluding contour* (OC) plays important roles in many computer vision tasks. The study of using OC for visual inference tasks is however limited, partially due to the lack of robust OC acquisition technologies. In this work, benefit from a novel OC computation system, we propose applying OC information to category classification tasks. Specifically, given an image and its estimated occluding contours, we first compute a distance map with regard to the OCs. This map is then used to filter out distracting information in the image. The results are combined with standard recognition methods, bag-of-visual-words in our experiments, for category classification. In addition to the approach, we also present two OC datasets, which to the best of our knowledge are the first publicly available ones. The proposed method is evaluated on both datasets for category classification tasks. In all experiments, the proposed method significantly improves classification performances by about 10 percent.

1 Introduction

Occluding contour (OC) is well known to play important roles in many vision tasks [1, 2]. Unlike regular photograph, an occluding contour image removes the effects of illumination, texture, and appearance while maintaining important edge and silhouette information. In computer vision, researchers have been seeking to develop new contour-based visual inference algorithms for many years [4]. In many visual inference tasks, a big challenge is to locate foreground object boundaries from the sea of all kinds of edge contours. Despite the known importance of OC, acquiring high quality OC in complex environment has been a long-standing challenging task [2].

In this paper we study the method and efficacy of using OC information for visual category classification, which is among the most important vision tasks [17]. We first use a novel multi-flash based OC acquisition device to get the initial OC estimation. This step provides us occluding contours that are more accurate than those from other existing methods. Once the OC data is ready, they can be used to improve visual inference tasks such as category classification.

The basic idea is to use occluding contours for feature filtering. Similar strategy appeared in [3]. Regions that are close to an OC are more likely to contain valuable shape related information and less pruning to distracting texture noises. Therefore, OC can be used to trim local visual features and then prepare a "purified" shape-related feature set for high level vision tasks such as visual recognition, detection, tracking,

G. Bebis et al. (Eds.): ISVC 2010, Part I, LNCS 6453, pp. 296–305, 2010.

etc. Specifically, for an image and its estimated OC image, a distance map is generated from the occluding contours. The distance map is used to filter out distracting local features in the image. The improved feature set is then combined with standard bag-of-visual-words model for visual category classification.

Another contribution of this paper is the benchmark datasets, which are the first such datasets to the best of our knowledge. We designed two datasets with both color images and OC images. The first dataset simulates the ideal occluding contours by manually picking OCs from normal edges maps (Canny edges [5]). In contrast, the occluding contours in the second dataset are automatically computed from the OC-Cam introduced in Section 2. We conducted category classification experiments on both datasets. In all experiments, the OCs information in the proposed method could significantly help improve classification performances.

The rest of the paper is organized as follows. Section 1.1 summarizes related work. Then, we introduce the OC acquisition device in Section 2. After that, the proposed OC-based category classification method is described in Section 3. Section 4 presents the experiments. Finally, Section 5 concludes the paper.

1.1   Related Work

As mentioned above, OC image could eliminate many distracting effects while maintaining important edge and silhouette information. To acquiring OCs, traditional passive image processing methods are often not robust enough to classify scene edges (e.g., occluding contours vs. material or texture edges) [8]. In particular, when foreground objects are surrounded by complicated background, it would be highly difficult to identify the occlusion boundaries.

Recent advances in computational photography have suggested that active illumination techniques can utilize shadows to effectively extract occlusion boundaries. For example, aerial imagery techniques can first detect shadows in a single intensity image and then infer building heights by assuming the ground geometry and surface reflectance models [11–13]. It is also possible to strategically cast shadows onto scene objects to recover their geometry [19]. In our work, shadows of objects are produced by multi-flash camera, which we refer readers to [18] for a complete review of this device.

In computer vision, researchers have been seeking to develop contour-based visual inference algorithms for many years. For example, contour information has been widely used in object recognition and localization tasks [4, 7, 15, 16, 20]. Most previous studies either assume that shapes of target objects are known, or work directly on the contours obtained from low- or middle-level edge extraction processes. Our work is different in that we explicitly use occluding contours achieved through the hardware directly.

Among many visual inference tasks, we choose visual category classification to demonstrate the effectiveness of using OCs. Category classification is an important research topic and has been attracting a large amount of research attention recently [17]. Our method is closely related to the bag-of-visual-words model [10, 21], which have been demonstrated excellent performance on several benchmark datasets [22, 23]. The proposed method can be viewed as an extension of these methods in the aspect of feature selection.

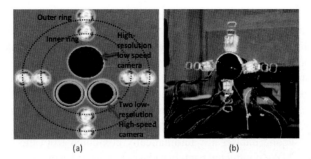

Fig. 1. The Occluding Contour Camera (OC-Cam) system design. (a) An OC-Cam uses one high resolution infrared camera and a pair of high speed visible light cameras. They are surrounded by multiple rings of controllable infrared LED lights. (b) The infrared camera and the LEDs form a multiple-flash camera.

Fig. 2. Examples of occluding contours achieved by the OC-Cam system. Left: original normal images. Middle: corresponding OC images. Right: result of Canny edges [5] for comparison.

2   Extracting Occluding Contours

In this section we briefly introduce the occluding contour acquisition devices used in this study and the postprocessing steps. It is worth noting that both acquisition and postprocessing are fully automatic.

2.1   Occluding Contour Camera

Our solution for acquiring the OC data is to construct a novel *Occluding Contour Camera* (OC-Cam). The OC-Cam extends the previously proposed multi-flash camera that composes of a single image sensor with four flashes evenly distributed about the camera's center of projection, as shown in Figure 1. To acquire the contour data, the multi-flash camera takes successive photos of a scene, each with a different flash turned on. The location of the shadows generally abuts depth discontinuities and changes along with the flash position. All the depth edge pixels hence can be detected by analyzing shadow variations. For example, turning on the left flash will result in the shadows to the right of the depth edge. We can then traverse the image horizontally and identify the pixels that transition from the non-shadow region to the shadow region.

The major limitation of the multi-flash camera is that it is difficult to determine the proper camera-flash baseline, i.e., the distance the flash lies from the center of projection of the camera. For example, the shadows may appear detached from the boundary when

Fig. 3. Left: original normal images. Middle: original occluding contours. Right: results after postprocessing.

the baseline is too large or may disappear when the baseline is too small. Since our goal is to acquire the occluding contours for various types of objects, it is important that we dynamically adjust the flash baseline, e.g., for acquiring both the internal and the external occluding contours of the object.

To achieve this goal, our OC-Cam mounts multiple rings of flashes around the central camera to support dynamic flash-camera baselines. In our implementation, we synchronized the LED flashes and the central viewing camera using the APIs provided by the PointGrey Research. The APIs allows the programmer to configure the camera, trigger image capture and also gives the programmer access to general purpose registers on the camera that can act as input/output ports depending on the configuration. Specifically, we use these registers as a four bit output port to send signals to the flash hardware. During the capture process each shot will take one picture at the given frame rate while at that time one of the flashes is illuminating the scene. For moderate frame rates such as twenty frames per second, synchronizing the flash sequence with the shutter is extremely important.

In order to control multiple rings of flashes with the same four bit port architecture, we further modify the control hardware: instead of using each bit to directly control a given flash, all flashes will be triggered sequentially from one bit. To do this, the pulse from one bit on the port will increment a counter. The output of the counter is fed into a decoder that indexes each flash. Therefore, the binary output from the counter can select an output line on the decoder which triggers the appropriate flash. We then use another bit to control which ring of flashes is used. Several images from the system are shown in Figure 2.

2.2 Postprocessing

The original OC image contains noises and irrelevant broken lines, hence a postprocessing is needed for further usage. First, an image filter with certain threshold (100/255 in gray level in the experiments setting) is convolved with the original image to produce binary image where black pixels indicate edge and white pixels indicate irrelevant background. Then the morphologic operators, closing and opening, are conducted on the image successively. Intuitively, closing joins the broken lines into connected line components (2x2 neighborhood pixels in the experiment setting) while opening removes those connected components less than certain amount of pixels in area (100 pixels in the experiments setting). Figure 3 shows some results of the postprocessing.

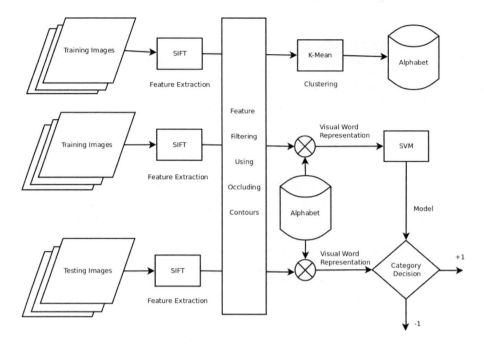

Fig. 4. Flow chart of OC guided category classification

3   Category Classification Using Occluding Contours

In this section we explore how to bring the rich shape information carried in OCs into category classification tasks. The overview of the process is shown in Figure 4.

3.1   Feature Filtering Using Occluding Contours

We propose to advance the state-of-the-art visual recognition algorithms by exploring the role of OCs as a *feature filter*. Specifically, OCs can help high-level vision tasks to get "purified" shape related features. This "purified" feature set in turn leads to improved object representation.

Let an input image be $I : \Lambda \to [0, 1]$, where $\Lambda \subset \mathbb{R}^2$ is the grid I defined on. The feature extraction of I is represented by a process $\mathcal{F}(I)$, which results in a set of local features. Without loss of generality, we denote the feature set as

$$\mathcal{F}(I) = \{(\mathbf{x}_i, \mathbf{f}_i)\}, \tag{1}$$

where $\mathbf{x}_i \in \mathbb{R}^2$ indicates the position of the ith feature and $\mathbf{f}_i \in \mathbb{R}^{n_f}$ indicates the n_f-dimensional feature descriptor. Specifically, in our experiment SIFT [14] is used, such that n_f=128.

The OC image of I is then denoted as $I_{oc} : \Lambda \to [0, 1]$. Our task is to use I_{oc} to trim $\mathcal{F}(I)$. A natural strategy is to use I_{oc} directly by eliminating any feature $(\mathbf{x}_i, \mathbf{f}_i)$ such

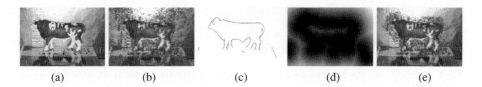

| (a) | (b) | (c) | (d) | (e) |

Fig. 5. Use occluding contours for feature filtering. (a) The original image. (b) The original image with local features. (c) The OC image. (d) The distance map. (e) Original image with filtered local features.

that $\mathbf{x}_i$ is within a distance of an OC pixel. Precisely, the new feature set $\mathcal{G}$ is defined by:

$$\mathcal{G}(\mathcal{F}(I), I_{mask}) = \{(\mathbf{x}_i, \mathbf{f}_i) \in \mathcal{F}(I) : I_{mask}(\mathbf{x}_i) < \tau\}, \tag{2}$$

where $I_{mask}(\mathbf{x})$ is the distance transform map of $I_{oc}(\mathbf{x})$ and τ is the distance threshold. In particular our experiment uses Euclidean distance. In the new feature set $\mathcal{G}$, feature descriptors will stay close to occluding contour therefore provide better description of the target objects. An example of the filtering process is shown in Figure 5.

3.2   Category Classification Using Bag-of-Visual-Word Model

We follow the idea of Bag-of-Visual-Word approach [21] to represent the images as histogram of visual words. The independent features are generated by SIFT 128-dimensional feature descriptor. After that the alphabet of visual words, i.e. codewords dictionary, is formed by k-means clustering. The new image thus could be represented by histogram of visual words in the alphabet. The main difference here is that we apply the OC information to filter out irrelevant features whenever possible, as shown in Figure 4.

The discriminative method Support Vector Machine (SVM) is used in our approach as classifier. For implementation, we choose the LibSVM package [6] and a Gaussian kernel defined by

$$K(\mathbf{s}_i, \mathbf{x}) = \exp\left(-\frac{||\mathbf{s}_i - \mathbf{x}||^2}{2\sigma^2}\right), \tag{3}$$

where $\mathbf{s}_i$ denotes support vectors, $\mathbf{x}$ represents the feature representation (histogram of visual words) of the input image and σ is the covariance parameter for the Gaussian kernel.

4   Experiments

To evaluate the proposed method, two datasets are created containing both color images and corresponding OC images. The proposed method is conducted on both datasets in comparison with the original bag-of-visual-word method. In the following, we use BOW as abbreviation for the bag-of-visual-word and BOW+OC for the proposed method.

Fig. 6. Example images misclassified by the standard BOW but correctly classified using our proposed BOW+OC method

Table 1. Category classification experiments on the Category-16 dataset

Method	BOW	BOW+OC
Classification rate (%)	43.57±1.97	49.82±2.27

4.1 Synthetic Occluding Contours

We first create a dataset containing 16 categories selected from the Caltech 256 dataset [9]. Each of the 16 categories contains 20 images. For each image in the dataset, we generate a "simulated" occluding contour image by first using Canny edge detector and then manually removing non-occluding contours. In other words, for each image I in the dataset, there is a corresponding OC mask I_{oc}. In the rest of the paper we call this dataset *Category-16*.

To demonstrate how OCs can help with category classification tasks, both BOW and BOW+OC frameworks are conducted on the Category-16 dataset. For the experiment, we randomly divide the dataset into training and testing sets, with 10 of total 20 images per category for training and the rest for testing. The experimental result is summarized over 5 random splits. The average classification rate is listed in Table 1. It shows that OC, even when used in a very simple way, can substantially improve recognition rate. Fig. 6 shows several examples that are misclassified by BOW but correctly classified by BOW+OC.

4.2 Real Dataset with Complicated Background

Another dataset we build contains five categories: car, cow, cup, dog and horse. Each of the five categories contains 24 images (accompanied with OC images) taken from six different objects. For every object, images are shot from four poses: 0°, 90°, 180° and 270° horizontal rotating from the default pose. For each image in the dataset, we generate the occluding contour image from the device introduced in Section 2. Therefore every image I in the dataset has a corresponding gray level OC image I_{oc}. In the rest of the paper we call this dataset *Category-5*. Figure 7 shows example images of this dataset. We will make this dataset publicly available after publishing this paper.

Similar to the experiment on the synthetic dataset, both BOW and BOW+OC frameworks are conducted on the Category-5 dataset to demonstrate how OCs can help with the category classification tasks. To make experiment result consistent, we still randomly divide the dataset into training and testing sets, with 12 of total 24 images per category for training and the rest for testing. The experimental result is summarized over 100 random splits. The average classification rate is listed in Table 2. The result has shown that OC method has significant improvement in recognition rate: around

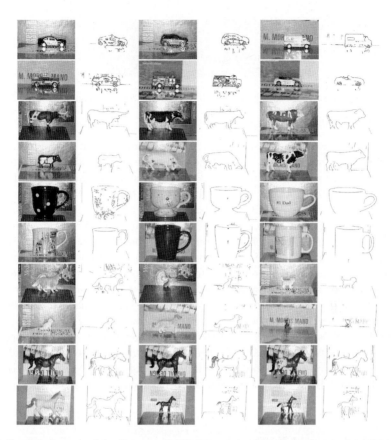

Fig. 7. Example images of the Category-5 OC dataset, six objects per class and one image pair (color image and OC image) per object

Fig. 8. Example images misclassified by the standard BOW but correctly classified using our proposed BOW+OC method

Table 2. Category classification experiments on the Category-5 dataset

Method	BOW	BOW+OC
Classification rate (%)	56.17±7.78	66.50±8.22

10%. Examples in Figure 8 show some examples that are misclassified by BOW but successfully classified by BOW+OC.

Fig. 9. Example images misclassified by the BOW+OC but correctly classified using standard BOW method

The above two experiments show clearly that OC information can be used to improve the performance of visual classification. It is also worth studying when the proposed method "hurts" the performance. In Figure 9 we show some examples which are misclassified by BOW+OC but successfully classified by BOW. One possible reason of misclassification by BOW+OC could be that sometimes it over-eliminates feature descriptors from the image. The setting of τ, i.e. the distance threshold, currently is determined empirically. This problem becomes noticeable when the input images are of certain types: background is solo-colored, background is uniformly textured, etc. These types share one property in common: feature descriptors are aggregated close to the object and are little scattered around the background in the OC image. Receiving those images as input, the BOW+OC method might eliminate some valuable feature descriptors around the object hence reduce the recognition rate while on the other hand standard BOW method will benefit from keeping all features. Though the limitation it appears in these scenarios, BOW+OC actually satisfied our expectation: when the standard BOW method can handle object classification in uniform background but suffer from complicated background, BOW+OC performs much better recognition rate according to the experimental results above.

5   Conclusions and Future Work

This paper investigates using the shape information from Occluding Contour (OC) to improve visual inference tasks, with focus on category classification. To this end, a new method is proposed that uses occluding contours as a feature filter to improve the image representation used in category classification. The improved representation is then combined with the bag-of-visual-words model for classification tasks. The proposed method clearly improves the performance on two datasets.

The applications of occluding contours are by no means limited to category classification. In fact, we expect the study in this paper to motivate rich future work toward different fields in computer vision, such as object localization. The datasets presented in this paper can therefore serve as benchmarks for future study as well. Aside from application of OC information in visual inference, we are also interested in improving the process of OC acquisition.

Acknowledgment. Ling is supported by NSF under Grant IIS-1049032. Thorpe and Yu are supported under NSF Grant IIS-CAREER-0845268 and an Air Force Young Investigator Award.

References

1. Maire, M., Arbelaez, P., Fowlkes, C., Malik, J.: Using contours to detect and localize junctions in natural images. In: CVPR (2008)
2. Martin, D., Fowlkes, C., Malik, J.: Learning to detect natural image boundaries using local brightness, color, and texture cues. IEEE Trans. Pattern Anal. Mach. Intell. (2004)
3. Lee, Y.J., Grauman, K.: Shape discovery from unlabeled image collections. In: CVPR (2009)
4. Biederman, I., Ju, G.: Surface vs. edge-based determinants of visual recognition. Cognitive Psychology 20, 38–64 (1988)
5. Canny, J.: A computational approach to edge detection. IEEE Trans. Pattern Anal. Mach. Intell. 8, 679–714 (1986)
6. Chang, C., Lin, C.: LIBSVM: a library for support vector machines (2001), http://www.csie.ntu.edu.tw/~cjlin/libsvm
7. Ferrari, V., Fevrier, L., Jurie, F., Schmid, C.: Groups of adjacent contour segments for object detection. IEEE Trans. Pattern Anal. Mach. Intell. 30(1), 36–51 (2008)
8. Forsyth, D.: Computer Vision - A Modern Approach. Prentice Hall, Englewood Cliffs (2002)
9. Griffin, G., Holub, A., Perona, P.: Caltech-256 object category dataset. Technical Report 7694, California Institute of Technology (2007)
10. Fei-Fei, L., Perona, P.: A bayesian hierarchical model for learning natural scene categories. In: CVPR (2005)
11. Huertas, A., Nevatia, R.: Detecting buildings in aerial images. Comput. Vision Graph. Image Process. 41(2), 131–152 (1988)
12. Irvin, R., Mckeown, D.: Methods for exploiting the relationship between buildings and their shadows in aerial imagery, vol. 19, pp. 1564–1575 (1989)
13. Lin, C., Nevatia, R.: Building detection and description from a single intensity image. Comput. Vis. Image Underst. 72(2), 101–121 (1998)
14. Lowe, D.G.: Distinctive image features from scale-invariant keypoints. Int. J. Comput. Vision 60(2), 91–110 (2004)
15. Lu, C., Latecki, L.J., Adluru, N., Yang, X., Ling, H.: Shape guided contour grouping with particle filters. In: Proceedings of the Tenth IEEE International Conference on Computer Vision, ICCV (2009)
16. Maji, S., Malik, J.: Object detection using a max-margin hough transform. In: IEEE Computer Society Conference on Computer Vision and Pattern Recognition, CVPR (2009)
17. Ponce, J., Hebert, M., Schmid, C., Zisserman, A. (eds.): Toward Category-Level Object Recognition. LNCS, vol. 4170. Springer, Heidelberg (2006)
18. Raskar, R., han Tan, K., Feris, R., Yu, J., Turk, M.: Non-photorealistic camera: depth edge detection and stylized rendering using multi-flash imaging. ACM Transactions on Graphics 23, 679–688 (2004)
19. Savarese, S., Rushmeier, H., Bernardini, F., Perona, P.: Shadow carving. In: IEEE International Conference on Computer Vision, vol. 1, p. 190 (2001)
20. Shotton, J., Blake, A., Cipolla, R.: Contour-based learning for object detection. In: ICCV, pp. 503–510 (2005)
21. Willamowski, J., Arregui, D., Csurka, G., Dance, C., Fan, L.: Categorizing nine visual classes using local appearance descriptors
22. Xie, N., Ling, H., Hu, W., Zhang, X.: Use Bin-Ratio Information for Category and Scene Classification. In: CVPR (2010)
23. Zhang, J., Marszalek, M., Lazebnik, S., Schmid, C.: Local features and kernels for classification of texture and object categories: A comprehensive study. International Journal of Computer Vision 73(2), 213–238 (2007)

Fractal Map: Fractal-Based 2D Expansion Method for Multi-scale High-Dimensional Data Visualization

Takanori Fujiwara[1], Ryo Matsushita[1], Masaki Iwamaru[2], Manabu Tange[3], Satoshi Someya[1], and Koji Okamoto[1]

[1] The Graduate School of Frontier Sciences, The University of Tokyo
[2] SGI Japan, Ltd.
[3] Department of Technical Engineering, Shibaura Institute of Technology

Abstract. Visualization of high-dimensional data is difficult to realize and manipulate with 2D display. For example, visualizing time-varying volume data (4D) with volume rendering and animation has spatial and temporal shielding, and data of 5 or more dimensions cannot be visualized on 2D display with existing methods. In this paper, we propose a method that expands high-dimensional data onto a 2D image plane. The proposed method uses the self-similarity of the fractal shape and achieves multi-scale high-dimensional data visualization on 2D display. With this method, we can visualize the entire domain of high-dimensional data without occlusions. Also, one-to-one correspondence in the elements of high-dimensional data and its 2D expansion enables us to manipulate high-dimensional data with 2D expanded result as an interface.

1 Introduction

Volume data visualization is one of the most important issues in the field of scientific visualization. Volume rendering and animation are often used in visualizing volume data (3D) or time-varying volume data (4D) on 2D display. However, with these methods, the visualization results are view and/or time dependent images, which have spatial and temporal shielding. And, in case of higher (5D, 6D, ⋯) dimensional data, we can't visualize the data with existing scientific visualization methods. But, in the field of information visualization, there are various 2D and 3D-based methods [1] for visualizing complex hierarchical and/or large-scale data. It is useful to apply their features in the subject of scientific visualization.

In this paper, we propose a method that expands high-dimensional data into a 2D fractal shape. The proposed method provides view and/or time independent visualization and enables the display of data of 5 or more dimensions. It also enables multi-scale visualization of data with self-similarity of fractal. Therefore, we can get the result images clearly and the images show where to focus, thus the proposed method reduces the cognitive load. Furthermore, we can combine the proposed method with existing scientific visualization methods, and grasp data from various perspectives.

G. Bebis et al. (Eds.): ISVC 2010, Part I, LNCS 6453, pp. 306–315, 2010.

2   Previous Studies

2.1   Scientific Visualization Methods

One of the major 3D-based methods for scalar volume data is volume rendering. The basic techniques of volume rendering were developed in 1988 (Levoy [2]; Drebin [3] et al.; Sabella [4]; Upson and Keeler [5]) and have since been modified and extended [6]. This method directly visualizes the entire volume onto an image plane using a ray-casting algorithm. Volume rendering is suitable for grasping the spatial distribution and continuity of values or for recognizing shapes because the 3D structure is directly projected onto the image plane.

On the other hand, one limitation of the volume rendering method is that it is view-dependent in that it is based on light transport models and the human visual system. The visualization result changes according to the position and direction of the viewpoint, and must be re-calculated whenever the viewpoint is translated or rotated. Volume rendering may be referred to as a 3D-to-2D method, that is, the visualization target is a 3D volume and the result is a 2D image. This reduction of dimension is carried out by ray casting. However, the combination of view-dependency and casting causes a shielding problem. Furthermore, it should be noted that in volume rendering, the contributions of each sampling point on a casting ray are totaled, which means that the full uniqueness of the structure cannot be reproduced by volume rendering because each voxel is not visualized individually. By rendering volume data from various viewpoints, we can interpret its properties more easily. The rendering parameters must also be adaptable. Therefore, a manipulation system for rotating, translating, and scaling the viewpoint, and parameter adjustment functions are essential for the application of volume rendering.

2.2   Information Visualization Methods

In the information visualization field, there is also a range of 2D- or 3D-based visualization methods. However, these methods are not intended for scientific visualization. There are numerous information visualization techniques, and a number of bold and novel concepts have xbeen reported, some of which might be applied to the field of scientific visualization. The advantages and disadvantages of 2D- and 3D-based methods discussed above actually more strongly affect information visualization methods.

The tree map [7] is a 2D-based method for representing hierarchical data structures. Hierarchy is indicated by a nested set of rectangles. Two types of node attributes can be visualized concurrently by assigning them to rectangle area and color. This method provides an overview of the node structure and data attributes and is free from occlusion. The Heiankyo-view [8] is a modification of a tree map that optimizes the alignment and packing of icon nodes.

There have been few attempts to introduce fractal concepts to information visualization. In fractal views [9], a multi-resolution visualization method maintains the total amount of displayed information nearly constant by considering the Hausdorff dimension of the data structure.

2D- and 3D-based methods have also been combined in the field of information visualization. Tory et al. examined the effectiveness of 2D/3D combination displays for orientation and relative positioning tasks [10]. They reported that 3D displays with appropriate cues (e.g., shadows) could be most effective for appropriate navigation and relative positioning. On the other hand, combination 2D/3D displays are useful for precise navigation and positioning. Harald et al. [11] combined 2D and 3D scatter plots with interactive focus + context features to overcome the problems of 3D point cloud visualization. The interaction of 2D and 3D scatter plots is accomplished by brush stroke user inputs.

The Value and Relation (VaR) display [12] is a 2D map visualization method for large data sets with hundreds of dimensions. Using this method, as its name suggests, not only the value of multi-dimensional large scale data but the relationships among the dimensions can be visualized, and so users grasp the associations among dimensions easily. In the VaR display, data values are represented as glyphs, which are produced by pixel-oriented [13] or alternative techniques such as the X-ray scatter plots. Dimensional relationships are reflected in the positions of the glyphs, and the positions are decided using an MDS algorithm [14] or Jigsaw map [15]. Using the VaR display, users can interactively detect patterns within large data sets.

3   2D Expansion Visualization Method of Volume Data

In this section, we showe the existing expansion method inspired by information visualization techniques [16]. This method visualizes volume data with an octree structure onto 2D image by expanding voxels into a fractal shape, which is called the Sierpinski carpet (SC) . The 2D visualization solves the occlusion problems in 3D visualization.

3.1   Resolution and Elements of Volume Data and Sierpinski Carpet

The volume data for resolution n $(R = n)$ consists of 8^n elements (voxels) . We define that $R = n$ indicates the volume data is divided into 2^n components on each dimension. Figure 1 shows the elements of the volume data for $R = 1, 2, 3$.

And the SC for $R = n$ also consists of 8^n elements (squares) . We define that the SC of 8 elements corresponds to $R = 1$, and by applying the subdivision scheme, the resolution of the SC increases by one. Figure 2 shows the elements of the SC for $R = 1, 2, 3$. Then we utilize the correspondence of the number of elements in volume data and in the SC for $R = n$.

3.2   2D Expansion of Volume Data

The SC has the same number of elements as that of the original volume data for the same resolution, we can achieve 2D expansion of volume data by defining a mapping rule for each element in the volume data and the SC. Figure 3 shows the mapping rule for $R = 1$. For higher resolution volume data, the expansion can be done by applying the mapping rule for $R = 1$ recursively. In addition, the mapping rule for $R = 1$ can be applied to volume data and SC locally, and we can obtain multi-resolution visualization result. Thus, we can display whole octree structure of volume data using the SC. And, by using multi-resolution method, we can cull unnecessary information and it's easier to focus to where should be noted, so that the cognitive load would be reduced. Furthermore, the expansion method can cooperate with the existing methods like volume rendering, and since each sub-square in SC is a one-to-one corresponded with each voxel in volume data, by pointing to a sub-square, we can assign any voxel in the volume data and apply numerical functions like subtraction of the pointed value from values of a whole area. And we convert scalar value of each element of volume data to color in RGBA. Then, we decide each element of SC has the same RGB value to the corresponding element of volume data and an alpha value of 1.0 (each element of volume data has an alpha value depending on its scalar value). By setting colors this way, we can compare expanding and volume rendering result.

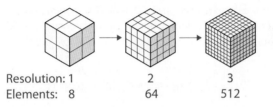

Resolution: 1   2   3
Elements: 8   64   512

Fig. 1. Octree structure of voxels when $R = 1, 2, 3$

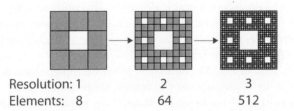

Resolution: 1   2   3
Elements: 8   64   512

Fig. 2. Sierpinski carpets when resolutions $R = 1, 2, 3$

4   2D Expansion Visualization Method of High-Dimensional Data

In the previous section, we explained the method that can expand volume data (3D) to 2D image. In this section, we explain the extension method that can expand higher dimensional data. The proposed method is realized via using combinational fractal shapes which is made by combining from two basic fractal shapes.

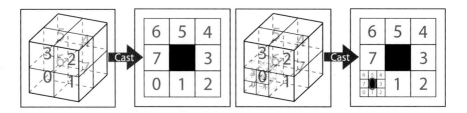

Fig. 3. Left: An example of expansion rule for volume data with R=1, Right: An example of multi-resolution expansion

4.1 Dimensions, Resolution and Elements of High-Dimensional Data

m ($m \geq 2$) dimensional data for $R = n$ consists of 2^{mn} elements. We define R=n for m dimensional data that indicates the m dimensional data is divided into 2^n on each dimension. For example, $m = 2, 3, 4, 5, 6$ dimensional data have 4^n, 8^n, 16^n, 32^n, 64^n elements. 2^{mn} elements can be transformed as Eq. 1 according to whether dimension is even ($m = 2k$, $k \geq 1$) or odd ($m = 2k + 1$) .

$$2^{mn} = \begin{cases} 4^{kn} & (m = 2k) \\ (8 \cdot 4^{k-1})^n & (m = 2k + 1) \end{cases} \tag{1}$$

4.2 Combinational Fractal Shapes, Resolution and Elements

The fractal shape divided evenly into four squares recursively (F4, Fig. 4) and the SC for $R = 1$ consists of 4 and 8 elements. By combining together these two fractal shapes for R=1, fractal shapes which of a new count of elements can be created. For example, when we replace each square of the SC with F4 for $R = 1$, we obtain the fractal shape which has 32 elements for $R = 1$ and 32^n elements for $R = n$ (Left part of Fig. 5) . Similarly, when we replace each square of the SC with F4 for $R = k + 1$, we obtain the fractal shape which has $8 \cdot 4^{k-1}$ elements for $R = 1$ and $(8 \cdot 4^{k-1})^n$ elements for R=n. And, in the case of the fractal shape for $R = 1$ which is same as F4 for $R = k$, the fractal shape has 4^k elements for $R = 1$, $(4^k)^n$ elements for $R = n$ (Right part of Fig. 5 shows an example with $k = 2$) . Note that these numbers for $R = n$ ($(8 \cdot 4^{k-1})^n$ and $(4^k)^n$) correspond to the numbers in Eq. 1.

4.3 2D Expansion of High-Dimensional Data

We obtained the fractal shapes which have same counts of elements with high-dimensional data. Then, high-dimensional data can be expanded onto a 2D image by giving the mapping rule for $R = 1$ and applying the rule recursively. Figure 6 shows an example of the mapping rule for 4D data. Moreover, as with the 3D to 2D mapping rule showed in section 3, the mapping rule can be applied locally, and we can also obtain multi-resolution visualization result of high-dimensional

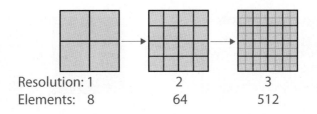

Resolution: 1   2   3
Elements: 8   64   512

Fig. 4. 4 split fractal shapes (F4) for $R = 1, 2, 3$

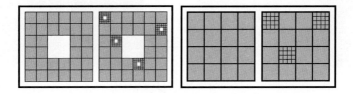

Fig. 5. 32 and 16 split fractal shapes for $R = 1$ and subdivided locally

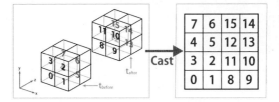

Fig. 6. An example of mapping rule for time-varying volume data (4D)

data. Other useful features (cooperating with the existing methods, retrieving positions of original high-dimensional data and applying numerical functions) are also available.

5 Application Examples of 2D Expansion Visualization

In this section, we showe the application examples, visualization of 3D pressure data obtained in numerical simulation of airflow such as that behind a model car [17], the pressure data with time variation (4D) and 5D lattice random walk.

5.1 Visualization of 3D Pressure Data

Figure 7 shows the visualization example of 3D Pressure Data. In this example, we obtained non-spatial shielding result and with the cooperation of volume rendering, we can grasp and manipulate the data with an understanding of 3D figuration. In this application, we got multi-resolutional result according to the variance assigned with "Split Ratio Bar", controlled color appearance with "Min Max Bars", subtracted the value of the selected region on 2D result. And this

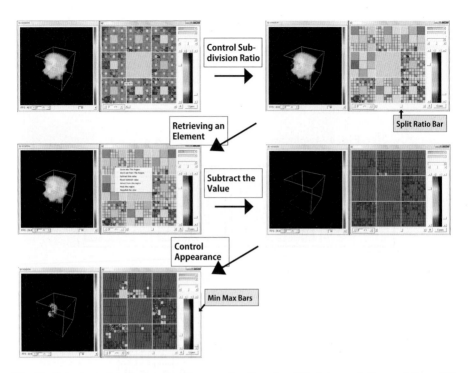

Fig. 7. An example of visualization application for 3D data and its workflow. We subtracted the value of the selected region and highlight the regions of higher values with "Min Max Bars".

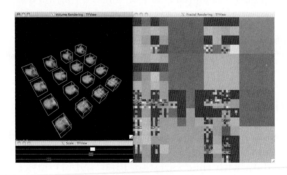

Fig. 8. An example of visualization application for 4D data

14	22	25	29	30	31
18	10	26	28	27	15
12	20			7	23
8	24			11	19
16	4	3	5	17	13
0	1	2	6	9	21

Fig. 9. 5D mapping rule

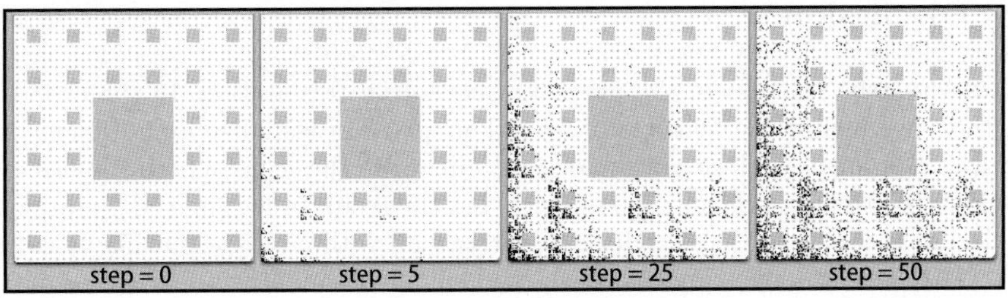

Fig. 10. 5D random walk visualization. Black points are walking elements.

application uses blank spaces of SC for visualizing the average value. We can recognize easily where is the high pressure part using 2D expansion without view dependence. Also, by subtracting the value of selected region, we can find the regions where of bigger values. For example, when we select the average value of all the areas, we can find the regions where of bigger value than the average value. With the 2D expansion result and use of it as an interface, we can manipulate and grasp 3D figuration and its value of the data without overlooking and misunderstanding of data values.

5.2   Visualization of Time-Varying 3D Pressure Data

Figure 8 shows the visualization result of time-varying 3D pressure data. We obtained the result which has no spatial shielding and temporal shielding by applying the proposed method for 4D data. As well as the previous subsection, we can get multi-resolutional result and control the color appearance.

5.3   Visualization of 5D Lattice Random Walk

As an example of application for high dimensional data, figure 10 shows the visualization result of 5D lattice random walk. 5D random walk is the expansion version of 1D, 2D and 3D random walk. 5D version has directions of movement in 5D space. We set 3000 elements at coordinate $(0, 0, 0, 0, 0)$ which is the initial position and the elements will walk to the positions of which the coordinate is composed by integers from 0 to 7. In this application, we set a 5D mapping rule as following. Since 5D data for $R = 1$ has 32 elements, when some coordinate of element is $(d_1, d_2, d_3, d_4, d_5)(d_1, d_2, d_3, d_4, d_5 = \{0, 1\})$, we assign a number$(d_1 + 2d_2 + 2^2 d_3 + 2^3 d_4 + 2^4 d_5)$ as ID to the element. And these elements are mapped into a 5D fractal shape for $R = 1$ as figure 9. According to this rule, an element from original 5D data which is far away from $(0, 0, 0, 0, 0)$ is set to a position in 5D fractal shape which is far from a position of ID=0. Since the rule is applied recursively, in a case of a bigger R, similarly to the case for $R = 1$, an element of a big offset from $(0, 0, 0, 0, 0)$ will be also mapped to a position far away from the position of ID=0 on the SC in the overall view. We observed the diffusion of elements at figure 10.

6   Discussion

We discuss the features of the proposed method in this section. In the proposed method, we utilize the features of fractal shapes and visualize high-dimensional data as 2D image. And we can achieve visualizing all the areas of high-dimensional data without view and time dependence, get the result reflecting the hierarchy of high-dimensional data, and reduce cognitive load with multi-resolution visualization. Furthermore, by using the result of 2D image as an interface, direct pointing to a region of high-dimensional data and manipulating the regional values are realized. Additionally, the proposed method can cooperate with the existing methods, and we can grasp the data multilaterally. With recursive application of the mapping rule for $R = 1$, the result for $R = n$ has same properties with $R = 1$ in an overall view. So we can obtain effective visualization result with a proper determination of the mapping rule for $R = 1$. However, the proposed method has also its limitation. Continuity of data positions is lost in 2D expanding, thus grasping the figuration of high-dimensional data is difficult. Therefore, the proposed method is considered more useful in a task of grasping values distribution according to positions than in a task of grasping the figuration.

7   Conclusion

In this paper, we proposed a fractal-based 2D expansion method for high-dimensional data visualization. Utilizing the features of fractal shapes enables us to visualize hierarchical and multi-resolution data. The proposed method can be used in visualization of 3D and 4D data without view and time dependence and enables the visualization of data of 5 or more dimensions on 2D display. But, the proposed method has also its limitation. It is difficult to grasp the figuration of data. But, this limitation can be supplemented with cooperation of the existing methods.

In case of visualizing the data which has spatial and time dimensions, such as the time-varying volume data, it is difficult to track the temporal changes, because the proposed method expands spatial and temporal dimensions together. In the future, we will develop a dimension reduction visualization method for high-dimensional data which expands dimensions separately.

References

1. Teyseyre, A.R., Campo, M.R.: An overview of 3d software visualization. IEEE Transactions on Visualization and Computer Graphics 15, 87–105 (2009)
2. Levoy, M.: Display of surfaces from volume data. IEEE Computer Graphics and Applications 8, 29–37 (1988)
3. Drebin, R., Carpenter, L., Hanrahan, P.: Volume rendering. ACM SIGGRAPH Computer Graphics 22, 74 (1988)
4. Sabella, P.: A rendering algorithm for visualizing 3D scalar fields. In: Proceedings of the 15th Annual Conference on Computer Graphics and Interactive Techniques, p. 58. ACM, New York (1988)

5. Upson, C., Keeler, M.: V-buffer: visible volume rendering. In: Proceedings of the 15th Annual Conference on Computer Graphics and Interactive Techniques, p. 64. ACM, New York (1988)

6. Kaufman, A., Mueller, K.: Overview of volume rendering. In: The Visualization Handbook, pp. 127–174 (2005)

7. Johnson, B., Shneiderman, B.: Tree-maps: a space-filling approach to the visualization of hierarchical information structures. In: Proceedings of the 2nd Conference on Visualization 1991, VIS 1991, pp. 284–291. IEEE Computer Society Press, Los Alamitos (1991)

8. Itoh, T., Yamaguchi, Y., Ikehata, Y., Kajinaga, Y.: Hierarchical data visualization using a fast rectangle-packing algorithm. IEEE Transactions on Visualization and Computer Graphics, 302–313 (2004)

9. Koike, H.: Fractal views: a fractal-based method for controlling information display. ACM Trans. Inf. Syst. 13, 305–323 (1995)

10. Tory, M., Moller, T., Atkins, M., Kirkpatrick, A.: Combining 2D and 3D views for orientation and relative position tasks. In: Proceedings of the SIGCHI Conference on Human Factors in Computing Systems, pp. 73–80. ACM, New York (2004)

11. Piringer, H., Kosara, R., Hauser, H.: Interactive focus+context visualization with linked 2d/3d scatterplots. In: International Conference on Coordinated and Multiple Views in Exploratory Visualization, pp. 49–60 (2004)

12. Yang, J., Hubball, D., Ward, M.O., Rundensteiner, E.A., Ribarsky, W.: Value and relation display: Interactive visual exploration of large data sets with hundreds of dimensions. IEEE Transactions on Visualization and Computer Graphics 13, 494–507 (2007)

13. Keim, D.: Designing pixel-oriented visualization techniques: theory and applications. IEEE Transactions on Visualization and Computer Graphics 6, 59–78 (2000)

14. Bentley, C., Ward, M.: Animating multidimensional scaling to visualize N-dimensional data sets. In: Proceedings of the 1996 IEEE Symposium on Information Visualization (INFOVIS 1996), p. 72. IEEE Computer Society, Los Alamitos (1996)

15. Wattenberg, M.: A note on space-filling visualizations and space-filling curves (2005)

16. Iwamaru, M., Okamoto, K.: Multiresolutional volume expansion and visualization with fractal diagram. Journal of the Visualization Society of Japan 28, 75–78 (2008)

17. Tsubokura, M., Kobayashi, T., Nakashima, T., Nouzawa, T., Nakamura, T., Zhang, H., Onishi, K., Oshima, N.: Computational visualization of unsteady flow around vehicles using high performance computing. Computers & Fluids 38, 981–990 (2009)

Visual Network Analysis of Dynamic Metabolic Pathways

Markus Rohrschneider[1], Alexander Ullrich[1], Andreas Kerren[2],
Peter F. Stadler[1], and Gerik Scheuermann[1]

[1] Leipzig University, Department of Computer Science, Germany
[2] Linnaeus University, School of Computer Science, Physics and Mathematics (DFM), Sweden

Abstract. We extend our previous work on the exploration of static metabolic networks to evolving, and therefore dynamic, pathways. We apply our visualization software to data from a simulation of early metabolism. Thereby, we show that our technique allows us to test and argue for or against different scenarios for the evolution of metabolic pathways. This supports a profound and efficient analysis of the structure and properties of the generated metabolic networks and its underlying components, while giving the user a vivid impression of the dynamics of the system. The analysis process is inspired by Ben Shneiderman's mantra of information visualization. For the overview, user-defined diagrams give insight into topological changes of the graph as well as changes in the attribute set associated with the participating enzymes, substances and reactions. This way, "interesting features" in time as well as in space can be recognized. A linked view implementation enables the navigation into more detailed layers of perspective for in-depth analysis of individual network configurations.

1 Introduction

Metabolic networks, the set of chemical compounds and their interactions that constitute life in the most basic sense, are the best studied biological networks. With the plethora of genomic, proteomic and metabolomic data available it becomes possible to study cell behavior. However, to understand the underlying principles of life and gaining further insights about the metabolism of cells for the use in biotechnological applications, e.g., pharmaceutical target prediction or metabolic engineering, we need tools to model and analyze the metabolic processes, pathways, and networks. There exist successful means for the reconstruction of metabolic networks from annotated genomes [1], the analysis of these networks in terms of elementary pathways [2], and description of their behavior with the help of ODE models [3]. Further insight into the development of kinetic models of metabolic networks addressing rate laws of the involved enzymes is provided in [4]. The situation becomes more difficult when we want to explain the evolutionary mechanisms of these systems, i.e., the formation of metabolic pathways or the emergence of complex network properties. Although, several scenarios exist that provide some insight into the evolution of metabolic pathways [5], only few aspects are well understood. Especially, the first steps in early metabolism evade observation

G. Bebis et al. (Eds.): ISVC 2010, Part I, LNCS 6453, pp. 316–327, 2010.

by conventional approaches. To this end, Ullrich et al. [6] developed a multi-level computational model to study the transition to life: the evolution of metabolic pathways
from catalyzed chemical reactions. The simulation approach implements components
on different scales in a more realistic manner than has been done so far.

In this work we introduce a plug-in for exploring dynamic graphs extending the
existing graph visualization software previously described in [7]. The implementation
of the extension was primarily driven by the given data and the requirements stated by
the scientists providing it. These include

1. Overview of the complete series of evolving metabolic networks, i.e. involvement
   of metabolites, reactions and enzymes, and evaluation of key properties, e.g. quantity (concentration) and activity (participation in pathways)
2. Analysis of dynamics in the network's topology and attribute set. Compare networks of different time steps and analyze topology dynamics in more detail.
3. Elementary pathway analysis of selected network generations. Time series analysis
   of attributes associated with selected node.

For the analysis of the simulation results, an efficient visualization system tailored
to suit our needs is of utmost importance. The main function of the software introduced in this article lies in the analysis of metabolic networks in general and studying
the evolution and dynamic behavior of metabolism in particular. This is achieved by
providing an insightful overview on different scales (e.g., on the metabolite-, pathway-,
or network-level) and different angles (e.g., dynamics in topology vs. attribute dynamics) of the vast amount of extracted information. Being able to observe all components
(individually or together) for the entire simulation time in one representation gives us
a much deeper understanding of the system's dynamics than any statistical analysis or
static view can provide. By means of one sample simulation, we show the possibilities
of the tool and which potential general insights we can gain.

2   Related Work

To deal with large biochemical networks several methods and tools have been developed. Simple approaches try to visualize the complete network on the screen and use
zooming and panning for navigation. Examples are common graph drawing or network
analysis tools [9,10]. Other approaches, such as KGML-ED [11], improve the navigation
between single pathways by providing an hierarchical overview and functions to zoom
into the top nodes of the hierarchy, or by extending the pathway by connected pathways
within the same frame. Our own recent development [7] realized a grid-based visualization approach for metabolic networks supported by a focus&context view. This view
is based on a Table Lens method [12], which provides multiple foci and together with
the grid-layout the preservation of the user's mental map, see below. A good overview
on open problems and challenges in biological network visualization is provided by the
papers [7,13]. They provide a comprehensive list of related work, however not focused
on the visualization of dynamic biochemical pathways. Oldiges et al. [14] address the

specific problem of metabolic network model visualization. However, their article is particularly related to the numerical analysis of dynamic biochemical systems with less emphasis on the visual analysis of the dynamics of the network topology. To the best of our knowledge, there is no other visualization tool that focuses on this specific task.

In general, the visualization of dynamic graphs is a well-known area in the graph drawing community [15]. Dynamic graph drawing addresses the problem to layout graphs, which evolve over time by adding and deleting edges and nodes. This results in an additional esthetic criterion known as *preserving the mental map* [16]. Ad-hoc approaches compute a new layout for the entire graph after each time step using algorithms developed for static graph layout, see for example those presented in the book [9]. In most cases this approach produces layouts which violate the mental map. In our own work, we follow the basic idea of the so-called *Foresighted Layout* (FL) of dynamic graphs [17]. Given a sequence of n graphs, a global layout is computed, which induces a layout for each of the n graphs. The FL-algorithm is generic in the sense that it takes a static graph drawing algorithm as a parameter. It optimally preserves the mental map. An algorithm for drawing a sequence of graphs online, i.e., where the graph sequence to be laid out is not known in advance, was presented by Frishman and Tal [18].

The general design of our plug-in is based on standard coordinated and multiple view visualization techniques. An excellent starting point for related work of this kind of visualization techniques is the annual conference series on Coordinated & Multiple Views in Exploratory Visualization (CMV) or the work of Roberts [19]. In our case, the coordination between the different views is mainly done by brushing techniques. The work of Moody et al. [20] focuses on the visualization of dynamic networks in general and the evolution of social networks in particular. The authors state two common approaches: plotting network summary statistics as line graphs over time and examining separate images of the network at each point in time. Our work has been inspired by these two techniques.

3   The Model

In this section we introduce a computational model of early metabolism for studying the emergence and evolution of catalyzed chemical reaction networks. The model consists of a graph-based artificial chemistry allowing for realistic kinetic behavior and a proto cell-like entity that inhabits the artificial chemistry and that is exposed to changes (e.g., mutations, source) and selection against other proto cells.

The artificial chemistry of this model is motivated by the chemist's intuition of molecules and chemical reactions. Consequently, molecules are modeled as labeled graphs, with atoms as nodes and bonds as edges. Given this representation, it is easy to see that chemical reactions can be understood as graph transformations, or in computer science terms, as simple graph rewriting rules. Metabolic networks are expanded using a stochastic network generator inspired by Faulon [21]. For simplicity, reaction rates were computed here based on topological indices (Wiener number [22]) of the educt and product molecules of the reactions. The simulation takes two molecules as steady input, namely, the sequential and cyclic form of glucose.

The proto cell contains a simple cyclic genome with several RNA- genes encoding for a particular reaction type (graph rewriting rule) through a sophisticated genotype-phenotype mapping [23]. The genome is subject to mutation, deletion, duplication and horizontal gene transfer events. Therefore, reactions can occur, change and disappear from the proto cell or even get copied to a neighbor. In each generation, only half of all proto cells is selected and generates an identical copy. There is steady influx of metabolites from the environment and out flux of produced metabolites in way of biomass production. The constitution of either may change during the course of the simulation.

The metabolism of a proto cell is evaluated by analyzing the stoichiometric matrix and fulfilling steady-state and inequality constraints to compute the set of elementary pathways [1].

4   The Data

The simulation is run as an adaptive walk over 100 generations. In the analysis of the simulation results, several types of information on different levels are processed. Most importantly, the structure of the metabolic network in form of a bipartite labeled graph is stored in a GraphML [24] file. Metabolites and reactions are the nodes of the graph. An edge leading from a metabolite to a reaction indicates that the respective metabolite is an educt in the reaction, an edge from a reaction to the metabolite node marks the metabolite as product. The labels for reactions are unique identifiers giving insight to their function. The metabolite label is its canonical SMILES string [25], a unique structural representation that is easily readable for chemists. Further, the concentration (number of molecules) for each metabolite is included in the network information.

In addition to the network information, flux information—the set of elementary pathways through the network—is made available to our visualization tool in a simple text file. Extremal nodes are listed. These represent the metabolites transferred into the cell and those that are used as biomass or excreted into the environment, respectively. For each reaction it is noted whether it is present in a particular elementary pathway or not (0 or 1).

All types of information are generated for each generation. Since the simulation has several parameters and input options, the data can be very diverse in size and number of files as well as complexity. Here lies also one important merit of this visualization tool. Choosing an "interesting" simulation run for further analysis from the range of possible simulations. The visualization of all levels and generations combined allows an efficient decision process that is of particular importance in a development and testing stage.

5   The Visualization Framework

In this chapter we focus on different visualization techniques implemented to support the data analysis process. In this context, mental map preservation is a key require-ment for analyzing dynamic networks [16]. Changes in the graph drawing from one network generation to the next should be minimal if the topological changes are small. We achieve the requirement of mental map preservation by following the idea of [17] and create a foresighted layout by constructing the *Set Union graph* $\hat{G} = (\hat{V}, \hat{E})$ with $\hat{V} = \bigcup_{i=1}^{n} V_i$ and $\hat{E} = \bigcup_{i=1}^{n} E_i$ where $(V_i, E_i) = G_i$ are the networks after i generations (see Figure 1). After the preceding cycle removal, we lay out $\hat{G}$ using Sugiyama's method for directed acyclic graphs [26,27]. This layout method is suitable for our visualization, because the constructed graph contains very few cycles, and the general direction of fluxes through the network is suggested by the graph drawing, i.e., from top (source) to bottom (sink). To emphasize the importance of extremal nodes—metabolites existing in the cell with no reaction producing them (source metabolites) and metabolites with no reaction consuming them (biomass production)—we connect them to a global source or sink node, i.e, the resulting acyclic graph becomes a so-called *st-graph* [9]. The set union graph contains the elements of all time steps. Layouting this graph ensures the nodes' positions to remain constant when changing to a different network generation.

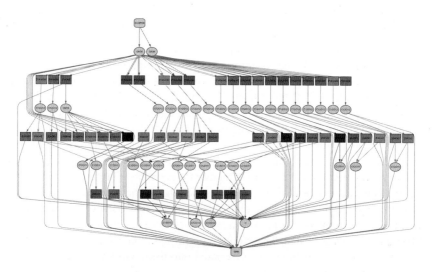

Fig. 1. Set Union Graph laid out using Sugiyama layout algorithm. The reaction nodes (rectangles) are colored according to their first appearance (red: earlier, blue: later).

The three requirements stated in the introduction meet Ben Shneiderman's mantra of information visualization [28]. In the following we describe the visual analysis process based on the scheme *"Overview first, zoom and filter, details on demand"*.

5.1 Overview

After construction of the Set Union Graph and associating the flux information with the graph elements, the primary objective of the overview visualization is to give the user a general idea of the network elements—metabolites, reactions and enzymes—involved, their life time, and the development of fundamental attributes associated with the network elements over time (see Figure 2). When presenting the Set Union Graph (a), a given node coloring scheme distinguishes between older and newer nodes. The time of first occurrence of a node in the network determines its color. The node appearing first is red, the node appearing last is blue. Node colors in between are interpolated using the color scale depicted in Figure 7(c), third from left. The user may choose, whether this scheme is applied to reaction nodes, to molecule nodes, or both.

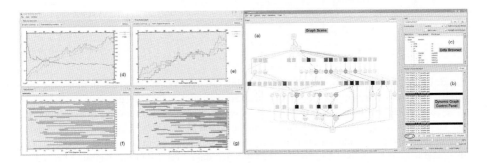

Fig. 2. Graphical User Interface of the Dynamic Graph Analysis plug-in. *Overview visualization*: Time Series Charts of selected attributes (d,e) display attribute dynamics over time. Interval Charts (f,g) represent the dynamic topology of the graph in terms of life times of metabolites, enzymes, and reactions. In (g), horizontal bars depicting the nodes' life time have been overlaid with the attribute *Fluxes through node*. The Graph Scene (a) shows the Set Union Graph with the applied node coloring scheme. As for *Zoom and Filter*, the user may select different network generations (b) to apply the set operators for filtering elements.

Further insight into the life times of metabolites and reactions give the interval diagrams depicted in Figures 2(e,f). Except for the artificially inserted environment nodes (global source and sink), each row represents a node in the graph. Horizontal bars depict the life time and may be overlaid with additional information, e.g. node degree, fluxes through that node, and concentration for metabolite nodes. In addition to interval diagrams, time series charts (d,e) summarize selected attributes and display their dynamics over time. The user can again choose the subset of nodes to be taken into account (metabolites, enzymes, or reactions) and the attribute set (node number, node degree, number of elementary fluxes through a node, and concentration values), and combine these time series in any way for comparison.

5.2   Zoom and Filter

In this analysis step, the user wants to detect "interesting features" in the overview and select individual networks for further inspection. Interesting in an evolving metabolic network may be periods of stabilities or instabilities in a topological sense—appearance of new reactions or metabolites—as well as in terms of flux behavior—changes of associated attributes.

The straightforward approach is to simply browse the time line. For that purpose, we have implemented a linked view connecting the diagrams of the overview visualization with the dynamic graph in the Graph Scene. The screen shot of the software given in

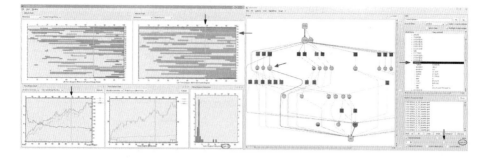

Fig. 3. Linked View realization facilitates browsing different graph snapshots in time. The blue arrows indicate the current position in time, red arrows indicate the selected node in the current generation. These components of the graphical user interface are also sensitive to user input and can be used for navigation. Selecting a node in the Graph Scene (r.h.s.) highlights the associated row in the appropriate interval chart as well as the associated point in time in all charts. The five diagrams given on the l.h.s. display the following data. Top: Life time diagram of reactions overlaid with the number of pathways through each reaction node. Life time diagram of metabolites overlaid with each node's degree. Bottom: Time series chart giving number of nodes, edges, and nodes-to-edges-ratio. Time series chart of summarized node degree (minimum, maximum, average) over all metabolites. Node degree histogram of the currently displayed graph generation.

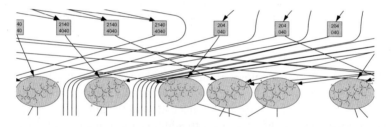

Fig. 4. Semantic Zoom: Below a certain level-of-detail threshold, the chemical structure of the molecule is shown instead of the totals formula

Figure 3 gives an impression on that type of navigation. The user may jump directly to that time point of interest by clicking into any of the displayed diagrams to further inspect the associated network. For each point in time, the current attributes are visualized in the nodes. For metabolites, the concentration values are depicted by the "filling level" of the node. Additionally, the node sizes and edge widths represent the number of elementary pathways these elements participate in.

For comparing different network generations from a topological point of view, the user may select a number of time steps and apply operators on the node and edge sets of the chosen graph to filter certain elements of the super graph. Set operators include *AND, OR,* and *DIFF* for the symmetrical difference between different network snapshots. This is used for detecting subset relations and selecting appearing or disappearing elements (see Figure 2b).

Finally, we have extended the semantic zoom capability in our visualization tool. As the user increases the level of detail, the chemical structures of metabolites is rendered within the associated nodes (see Figure 4).

5.3   Detailed View

In this section, the user takes a closer look on the emergence of individual elementary pathways (fluxes) in a single network evolution step. The aim is to further investigate elements being more or less likely to participate in pathways through the metabolic network and to identify individual elementary pathways. As described in the previous section, the user has identified reactions and metabolites preferred to form pathways as well as key enzymes with high activity. Again, interactivity plays a crucial role in this analysis step. There are two methods of operation: First, the user can select any number of elementary pathways to be highlighted in the Graph Scene displaying the current network generation. Second, the previously identified key elements can be selected in the Graph Scene for highlighting all associated elementary pathways. See the screen shot given in Figure 5. We again implemented set operators on the selected nodes applied for the flux visualization. We found that this is a highly flexible and intuitive way to detect pathways running through all the selected elements – *AND* operator, at least one of the selected elements – *OR*, or none of the selected elements – *NOT*.

Concerning the attribute dynamics associated with an enzyme, reaction, or metabolite, we take advantage of the linked view implementation depicted in Figure 3 to display the attribute development of the selected node over time. Selecting a different node instantly updates the displayed time series of the chosen attribute.

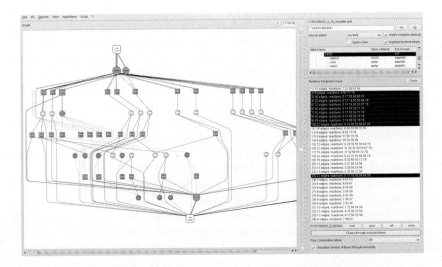

Fig. 5. Details on Demand: Interactive flux analysis for one chosen time step (here: t=99). Individual elementary pathways can be selected for visualization. All pathways through molecule *C6O5* are highlighted.

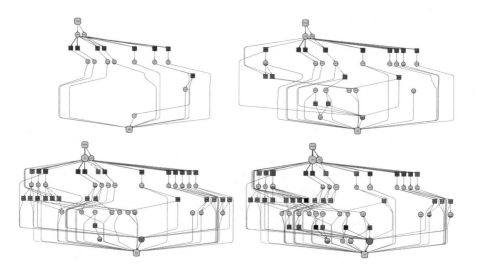

Fig. 6. A series of simulated metabolic networks after 11, 31, 67, and 100 generations. The squares represent reactions, circles represent metabolites. Node size and edge width encode for the number of minimal pathways in which the respective object is involved. Note the dark gray "filling level" of the metabolite nodes depicting the current concentration value.

6   Results

In this analysis we wanted to investigate the early steps in the formation and evolution of metabolic pathways and interpret our findings in terms of existing evolutionary scenarios. We will focus on three popular theories, that can be compared nicely to our results. One of the first theories proposed on this matter is backward or retrograde evolution [29], stating that pathways evolved upwards, in the need of finding beneficial substrates due to depletion of metabolites. Contrary, forward evolution [30] suggests the opposite direction of pathways evolution. Due to ever further processing of molecules for energy production, pathways evolve in such a way that ancient enzymes are upstream along the pathway, while younger enzymes are further downstream. The third scenario is the patchwork model [31], which explains the formation of new metabolic pathways through recruiting of enzymes from already existing pathways.

The four snapshots in Figure 6 showing the metabolic network at different points in time are aligned to the union graph over all generations. Thus, we can see that in the first steps the reactions upstream in the network are added. The pathways are formed further in this forward direction. Looking at the last generation, basically all pathways from source to sink follow the forward evolution scenario, with older (red) enzymes being at the top (upstream) and younger (blue) enzymes more at the bottom (downstream). This observation is further established through the interval graph for all chemical reactions in Figure 7. The reactions are here ordered according to their position in the graph. There is a clear trend of older reactions being on the top and younger ones following

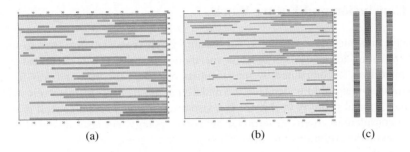

(a)   (b)   (c)

Fig. 7. Life time diagram of metabolites (a) and reactions (b). Their position in the diagram (y-axis) reflects the associated nodes' positions in the graph layout. Reactions close to the source metabolites are in upper positions, reactions close to the sink metabolites are placed at the bottom. In (c), our scenario (first bar) can be compared to the three evolution models: retrograde evolution, forward evolution, patchwork model.

more downstream. If we compare the colored bars (Figure 7c) showing the enzyme age distribution for our results and the three scenarios mentioned above, the pathway evolution again seems to explain our results best. Therefore, it appears that in the early phase of metabolic evolution, forward evolution is dominant.

We turn now to the evolution of general properties of the metabolic networks from our simulation. The numbers of metabolites and chemical reactions (see Figure 8a) develop with almost the same rate. This indicates that most metabolites are only involved in exactly one reaction. Combining this reasoning with the observation that the maximal node degree of metabolites increases significantly more than their average node degree (see Figure 8b), we can conclude that our metabolic networks evolved one or only a few highly connected metabolites, so called hub-metabolites, and probably has a scale-free node-degree distribution, typical for real-world metabolic networks. Another observation is the steady increase of the average enzyme connectivity while the average metabolite connectivity converges. The explanation for the latter is the high number of metabolites involved in only one reaction. A similar trend will likely arise in more complex stages for enzyme connectivity as well.

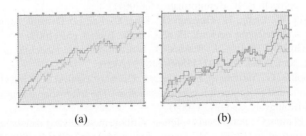

(a)   (b)

Fig. 8. Tracking selected attributes over time. (a) Number of metabolites (green) and reaction nodes (red). (b) Node degree (maximum and average) of metabolites (green) and enzyme nodes (red).

7   Conclusion and Future Work

We have presented an extension to our existing graph visualization system to support the exploration and analysis of dynamic metabolic networks. The development process was intensively accompanied by the scientists providing the data and was found to be extremely helpful to understand the underlying mechanisms of metabolic network and biochemical pathway evolution. The visualization could reveal general properties of the considered systems in terms of network topology, but also answered specific questions on the evolution of metabolic networks and the emergence of pathways within the network.

We found that interactivity plays a crucial role in the analysis process. It was successfully implemented using the linked view method for intuitive navigation in time as well as within a selected network configuration. We intend to examine more simulation runs with different parameter configurations to compare the results and to gain a deeper understanding of metabolic network evolution.

For laying out the constructed set union graph, the Sugiyama method has proven to produce the best results. The layout algorithm was a suitable choice due to the fact that the considered network contained only a few number of cycles, and therefore, the observed elementary pathways followed the general direction from top (source nodes) to bottom (sink nodes). The major disadvantage of this layout method is the amount of space required for the drawing. The number of graph elements in the super graph was small enough for a feasible application of this layout algorithm. Datasets with more generations can become very large and too complex for using the applied graph layout. However, there is room for improvement, since many elements in the super graph do not overlap in time and may therefore occupy the same position reducing the total space for the layout.

References

1. Palsson, B.O.: Systems Biology: Properties of Reconstructed Networks. Cambridge University Press, New York (2006)
2. Gagneur, J., Klamt, S.: Computation of elementary modes: a unifying framework and the new binary approach. BMC Bioinformatics 5 (2004)
3. Yang, K., Ma, W., Liang, H., Ouyang, Q., Tang, C., Lai, L.: Dynamic simulations on the arachidonic acid metabolic network. PLoS Comput. Biol. 3, e55 (2007)
4. Steuer, R., Gross, T., Selbig, J., Blasius, B.: Structural kinetic modeling of metabolic networks. Proc. Natl. Acad. Sci. 103, 11868–11873 (2006)
5. Caetano-Anollés, G., Yafremava, L.S., Gee, H., Caetano-Anollés, D., Kim, H.S., Mittenthal, J.E.: The origin and evolution of modern metabolism. The International Journal of Biochemistry & Cell Biology 41, 285–297 (2009)
6. Ullrich, A., Flamm, C.: Functional evolution of ribozyme-catalyzed metabolisms in a graph-based toy-universe. In: Heiner, M., Uhrmacher, A.M. (eds.) CMSB 2008. LNCS (LNBI), vol. 5307, pp. 28–43. Springer, Heidelberg (2008)
7. Rohrschneider, M., Heine, C., Reichenbach, A., Kerren, A., Scheuermann, G.: A novel grid-based visualization approach for metabolic networks with advanced focus & context view. In: Eppstein, D., Gansner, E.R. (eds.) GD 2009. LNCS, vol. 5849, pp. 268–279. Springer, Heidelberg (2010)

8. Kerren, A., Ebert, A., Meyer, J. (eds.): Human-Centered Visualization Environments. LNCS, vol. 4417. Springer, Heidelberg (2007)
9. Di Battista, G., Eades, P., Tamassia, R., Tollis, I.G.: Graph Drawing: Algorithms for the Visualization of Graphs. Prentice Hall, New Jersey (1999)
10. Görg, C., Pohl, M., Qeli, E., Xu, K.: Visual Representations. In: [8], pp. 163–230
11. Klukas, C., Schreiber, F.: Dynamic exploration and editing of KEGG pathway diagrams. Bioinformatics 23, 344–350 (2007)
12. Rao, R., Card, S.K.: The table lens: merging graphical and symbolic representations in an interactive focus+context visualization for tabular information. In: Conference Companion on Human Factors in Computing Systems, CHI 1994, p. 222. ACM, New York (1994)
13. Albrecht, M., Kerren, A., Klein, K., Kohlbacher, O., Mutzel, P., Paul, W., Schreiber, F., Wybrow, M.: On open problems in biological network visualization. In: Eppstein, D., Gansner, E.R. (eds.) GD 2009. LNCS, vol. 5849, pp. 256–267. Springer, Heidelberg (2010)
14. Oldiges, M., Noack, S., Wahl, A., Qeli, E., Freisleben, B., Wiechert, W.: From enzyme kinetics to metabolic network modeling - visualization tool for enhanced kinetic analysis of biochemical network models. Eng. Life Sci. 6 (2006)
15. Branke, J.: Dynamic graph drawing. In: Drawing Graphs: Methods and Models, London, UK, pp. 228–246. Springer, Heidelberg (2001)
16. Misue, K., Eades, P., Lai, W., Sugiyama, K.: Layout adjustment and the mental map. Journal of Visual Languages and Computing 6, 183–210 (1995)
17. Diehl, S., Görg, C., Kerren, A.: Preserving the mental map using foresighted layout. In: Proc. of Joint Eurographics-IEEE TVCG Symp. on Vis., VisSym 2001, pp. 175–184. Springer, Heidelberg (2001)
18. Frishman, Y., Tal, A.: Online dynamic graph drawing. IEEE TVCG 14, 727–740 (2008)
19. Roberts, J.C.: Exploratory visualization with multiple linked views. In: MacEachren, A., Kraak, M.J., Dykes, J. (eds.) Exploring Geovisualization. Elseviers, Amsterdam (2004)
20. Moody, J., McFarland, D., Bender-deMoll, S.: Dynamic network visualization. American Journal of Sociology 110 (2005)
21. Faulon, J.L., Sault, A.G.: Stochastic generator of chemical structure. 3. Reaction network generation. J. Chem. Inf. Comp. Sci. 41, 894–908 (2001)
22. Wiener, H.: Structural determination of paraffin boiling points. Journal of the American Chemical Society 69, 17–20 (1947)
23. Ullrich, A., Flamm, C.: A sequence-to-function map for ribozyme-catalyzed metabolisms. In: ECAL. LNCS, vol. 5777. Springer, Heidelberg (2009)
24. Brandes, U., Eiglsperger, M., Herman, I., Himsolt, M., Marshall, M.S.: Graphml progress report: Structural layer proposal. In: Mutzel, P., Jünger, M., Leipert, S. (eds.) GD 2001. LNCS, vol. 2265, pp. 501–512. Springer, Heidelberg (2002)
25. Weininger, D.: SMILES, a chemical language and information system. 1. introduction to methodology and encoding rules. J. Chem. Inf. Comput. Sci. 28, 31–36 (1988)
26. Sugiyama, K., Tagawa, S., Toda, M.: Methods for visual understanding of hierarchical systems. IEEE Trans. Systems, Man, and Cybernetics 11, 109–125 (1981)
27. Gansner, E.R., Koutsofios, E., North, S.C., Vo, K.P.: A technique for drawing directed graphs. IEEE Trans. Software Eng. 19, 214–230 (1993)
28. Shneiderman, B.: The eyes have it: A task by data type taxonomy for information visualizations. In: VL, pp. 336–343 (1996)
29. Horowitz, N.H.: On the evolution of biochemical syntheses. Proc. Natl. Acad. Sci. USA 31, 153–157 (1945)
30. Cordon, F.: Tratado evolucionista de biologa. Aguilar Ediciones, Madrid (1990)
31. Jensen, R.A.: Enzyme recruitment in evolution of new function. Annu. Rev. Microbiol. 30, 409–425 (1976)

Interpolating 3D Diffusion Tensors in 2D Planar Domain by Locating Degenerate Lines

Chongke Bi[1], Shigeo Takahashi[1], and Issei Fujishiro[2]

[1] Graduate School of Frontier Sciences, The University of Tokyo, Japan
[2] Department of Information and Computer Science, Keio University, Japan

Abstract. Interpolating diffusion tensor fields is a key technique to visualize the continuous behaviors of biological tissues such as nerves and muscle fibers. However, this has been still a challenging task due to the difficulty to handle possible degeneracy, which means the rotational inconsistency caused by degenerate points. This paper presents an approach to interpolating 3D diffusion tensors in 2D planar domains by aggressively locating the possible degeneracy while fully respecting the underlying transition of tensor anisotropy. The primary idea behind this approach is to identify the degeneracy using minimum spanning tree-based clustering algorithm, and resolve the degeneracy by optimizing the associated rotational transformations. Degenerate lines are generated in this process to retain the smooth transitions of anisotropic features. Comparisons with existing interpolation schemes will be also provided to demonstrate the technical advantages of the proposed approach.

1 Introduction

Recent development of visualization techniques for tensor fields has provided an effective means of understanding biological tissues especially in medical applications. Diffusion tensor magnetic resonance imaging (DT-MRI) is such an example where the associated tensor fields are obtained by measuring the motion of water molecules.

In general, a tensor field is obtained as a grid of tensor samples, and thus requires appropriate interpolation of such discrete samples to explore the structures of underlying features. For interpolating diffusion tensor fields, it is important to retain the smooth transition of anisotropic features inherent in the given tensor fields, especially around degenerate points, where at least two of three eigenvalues are equivalent [1]. Zheng et al. [2] proved that degenerate line, which connects degenerate points (the detail will be introduced in Section 4), is the most stable topological structure for 3D tensors, while degenerate points are unstable. However, existing interpolating schemes cannot obtain these degenerate lines to retain the smooth transition of anisotropic features. Figure 1 describes such limitations. Figure 1(a) presents a diffusion tensor field containing two degenerate points, and a degenerate line is obtained using our scheme, as shown in Figure 1(e). However, in Figures 1(b) and (c), we cannot observe such degenerate line while discontinuities appear in Figure 1(d). Note that the color of each ellipsoid indicates the anisotropy of the corresponding tensor value, which is represented by the FA value (Eq. (4)) of the tensor.

G. Bebis et al. (Eds.): ISVC 2010, Part I, LNCS 6453, pp. 328–337, 2010.

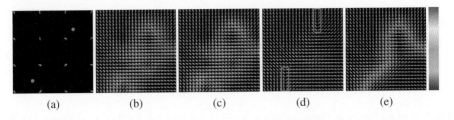

Fig. 1. Interpolating a diffusion tensor field containing two degenerate points. (a) Original tensor samples. Results with the (b) component-wise, (c) Log-Euclidean, (d) geodesic-loxodrome, and (e) proposed interpolation scheme.

This paper presents an approach to interpolating diffusion tensor fields by locating degenerate points and generating degenerate lines. The main idea is to cluster discrete tensor samples with similar anisotropy and orientation using a minimum spanning tree strategy, in order to locate degenerate points, which are connected by degenerate lines. Figure 1(e) presents the result that a degenerate line is obtained between the two degenerate points, which is the primary advantage of the proposed method over the existing interpolation schemes. In this paper, we introduce our method to interpolate 3D tensors in 2D planar domain, which is an initial step of our research in 3D tensor fields.

The remainder of this paper is organized as follows: Section 2 introduces several mathematical prerequisites for diffusion tensors, and then provides a brief survey on related work. Our approach for interpolating tensor fields is detailed first for 1D domain in Section 3, and then for 2D cases even with tensor degeneracy in Section 4. The effectiveness of the proposed approach is presented through the comparison with existing interpolation schemes in Section 5, followed by the conclusion of this paper in Section 6.

2   Related Work

A 3D diffusion tensor can be represented by three real eigenvalues $\lambda_1 \geq \lambda_2 \geq \lambda_3 > 0$, together with the corresponding eigenvectors e_1, e_2, and e_3 that form an orthonormal basis, which can be visualized as an ellipsoid as shown in Figure 2(a). The shape of the ellipsoid depends on the eigenvalues, which are defined as tensor anisotropy. Several metrics for evaluating such anisotropy have been proposed [3], which include linearity (C_l), planarity (C_p), sphericity (C_S), and Fractional Anisotropy (*FA*) as follows:

$$C_l = (\lambda_1 - \lambda_2)/(\lambda_1 + \lambda_2 + \lambda_3), \tag{1}$$

$$C_p = 2(\lambda_2 - \lambda_3)/(\lambda_1 + \lambda_2 + \lambda_3), \tag{2}$$

$$C_s = 3\lambda_3/(\lambda_1 + \lambda_2 + \lambda_3), \tag{3}$$

$$FA = \frac{\sqrt{3\left((\lambda_1 - \bar{\lambda})^2 + (\lambda_2 - \bar{\lambda})^2 + (\lambda_3 - \bar{\lambda})^2\right)}}{\sqrt{2\left(\lambda_1^2 + \lambda_2^2 + \lambda_3^2\right)}},$$

$$\text{where} \quad \bar{\lambda} = (\lambda_1 + \lambda_2 + \lambda_3)/3. \tag{4}$$

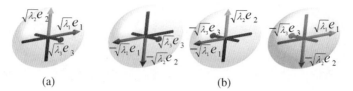

(a)   (b)

Fig. 2. Sign ambiguity of eigenvectors. (a) A tensor can be represented by an ellipsoid. The directions of the coordinate axes are represented by the arrows. Different colors are assigned to the axes for representing the magnitudes of eigenvalues. (b) Three other possible definitions of the tensor in (a).

Note that $C_l + C_p + C_s = 1$ and $0 \leq C_l, C_p, C_s, FA \leq 1$.

As described in the literature [4], the history of tensor interpolation methods started with naïve schemes as usual, including component-wise interpolation of tensor matrices. These methods, however, incur undesirable change in the tensor anisotropy, and cannot generate degenerate lines between degenerate points, as seen in Figure 1(b). In addition, positive-definiteness of the tensor matrix may not be preserved since the linear interpolation has been applied to each component of the tensor matrix in this scheme.

To alleviate this problem, Batchelor et al. [5] defined a distance function so that we can interpolate the tensors by tracking the corresponding geodesic path on a nonlinearly curved space. Their approach still incurs undesirable transition of the anisotropic features along the interpolated tensors when the associated rotational angle is relatively large. Furthermore, Fletcher et al. [6] modeled the space of diffusion tensors as a Riemannian symmetric manifold and introduced a framework for the statistical analysis of that space. However, their methods suffer from high computational costs because the geodesic path invokes long iterative numerical computations.

Recently, Arsigny et al. [7] developed a Riemannian metric called *Log-Euclidean* to provide a faster computational algorithm. This has been accomplished by transforming tensor samples into their matrix logarithms so that we can perform the tensor interpolation using Euclidean operations. However, it still incurs unnecessary change in the anisotropy of the interpolated tensors, as shown in Figure 1(c). Kindlmann et al. [8] presented a novel tensor interpolation method called the *geodesic-loxodrome*, which discriminates between the isotropic and anisotropic components of the tensors first and then interpolates each component individually. This accomplishes high quality interpolating results, however, at the cost of longer computation times again. The method may also incur undesirable discontinuity over the domain when its boundary has redundant rotation of the tensor orientation, as shown in Figure 1(d).

Any of the aforementioned approaches tried to transform tensor matrices to some specific nonlinear space and perform the interpolation by finding an optimal transition path between the tensors in that space. However, less attention has been paid to the eigenstructures of the tensor matrices. Merino-Caviedes et al. [9] developed a method for interpolating 2D diffusion tensors defined over the 2D planar domain, where they constitute a 3D Euclidean space spanned by the two eigenvalues and the angle between the primary eigenvector and the *x*-axis of the 2D domain. Hotz et al. [10] presented a sophisticated model based on the eigenstructures of the 2D diffusion tensors. In this

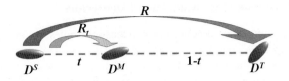

Fig. 3. Interpolating a 1D diffusion tensor field

work, they linearly interpolated between each pair of the eigenvalues and the corresponding pair of eigenvectors component-wise separately. They also located degenerate points over 2D triangulated domain. Readers can refer to a more complete survey in [11].

3   Interpolating 1D Tensor Fields

3.1   Eigenstructure-Based Tensor Representation

In this paper, an approach to interpolating diffusion tensors by employing an eigenstructure-based representation is proposed. Nonetheless, such interpolation scheme has not been fully tackled so far, because it cannot provide a unique description of a tensor.

This comes from the fact that each tensor has sign ambiguity in its eigenvector directions since both $Ae_i = \lambda_i e_i$ and $A(-e_i) = \lambda_i(-e_i)(i = 1,2,3)$ hold simultaneously, where A represents the matrix representation of the tensor. Thus, even when we suppose that the three eigenvalues suffice the condition $\lambda_1 \geq \lambda_2 \geq \lambda_3 > 0$, and the associated eigenvectors are all normalized to unit vectors to form a right-handed coordinate system, we still have four different representations for a single tensor, as shown in Figure 2.

For interpolating 1D tensor fields, we first establish the correspondence between each adjacent tensor samples, and then individually interpolate between each pair for seeking the smooth transition of tensor anisotropy values and its associated orientations.

3.2   Optimizing Correspondence between Tensors

Suppose that we have two tensor samples D^S and D^T, while their normalized eigenvectors are represented as $\{e_1^S, e_2^S, e_3^S\}$, and $\{e_1^T, e_2^T, e_3^T\}$, respectively. The rotation matrix R that transforms between D^S and D^T can be formulated as:

$$R = (p_1 e_1^T, p_2 e_2^T, p_3 e_3^T)(e_1^S, e_2^S, e_3^S)^{-1}, \tag{5}$$

where $p_i (i = 1,2,3)$ is defined to be the sign of each eigenvector e_i, in such a way that $p_i = \pm 1 \ (i = 1,2,3)$ and $\prod_{i=1}^3 p_i = 1$. The rotation angle θ between D^S and D^T is given by:

$$\theta = \arccos |(\mathrm{tr}\, R - 1)/2|, \tag{6}$$

where $\mathrm{tr}\, R$ is the trace of R. We assume $\theta \in [0, \pi/2]$, to remove redundant rotation.

3.3 Interpolation Using Eigenvalues and Eigenvectors

Having fixed the eigenvector directions of two tensor samples, we interpolate their corresponding eigenvalues and eigenvectors individually. Suppose that we calculate the interpolated tensor D^M at the ratio of $t : (1-t)$ in the range $[0, 1]$ between D^S and D^T, as shown in Figure 3. We calculate the three eigenvalues λ_i^M ($i = 1,2,3$) of D^M by linearly interpolating between the eigenvalues of D^S and D^T, and three eigenvectors e_i^M ($i = 1,2,3$) by linearly interpolating the associated rotation angle between them as:

$$\lambda_i^M = (1-t)\lambda_i^S + t\lambda_i^T, \quad \text{and} \tag{7}$$
$$(e_1^M, e_2^M, e_3^M) = R^t(e_1^S, e_2^S, e_3^S). \tag{8}$$

4   Interpolating 2D Tensor Fields

In order to extend the previous formulation to 2D planar domains, we need to handle the following two important technical issues:

1. In 2D cases, the rotational transformation depends on two parameters that define the parameterization of the 2D planar domain. We have to take care of the order of applying the rotation matrices since the rotations do not commute with each other.
2. We need to remove the rotational inconsistency around degenerate points.

4.1   Combination of Rotations for 2D Cases

For the noncommutative multiplication of rotation matrices, we alleviate the problem by employing Alexa's formulation on linear combination of transformations [12], which enables us to handle the multiplication of rotation matrices as their linear sum. For example, in the square region confined by the discrete tensor samples D_{00}, D_{01}, D_{11}, and D_{10}, where the region is defined as a 2D parametric domain $(s,t) \in [0,1] \times [0,1]$ (Figure 4). If we denote Alexa's commutative multiplication operator by $\oplus$, we can define the tensor D at parametric coordinates (s,t), using the bilinear interpolation as:

$$R = R_x^{s(1-t)} \oplus R_y^{(1-s)t} \oplus R_{xy}^{st}, \tag{9}$$

where, R_x, R_y, and R_{xy} represent the rotation matrices between D_{00} and D_{01}, D_{00} and D_{10}, and D_{00} and D_{11}, respectively. Now the eigenvectors of D can be obtained by applying R to those of D_{00}.

4.2   Locating Tensor Degeneracy

For 3D diffusion tensors, a tensor is defined as *a degenerate point* if at least two of three eigenvalues are equivalent [1]. Zheng et al. [2] proved that the most stable topological structure for 3D diffusion tensors is *a degenerate line* which connects degenerate points in tensor fields. We define the square containing a degenerate point as *a degenerate cell*.

   We will introduce how to locate the position of degenerate points. Since each degenerate point is contained in some degenerate cell, we can locate degenerate cells

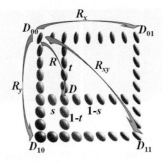

Fig. 4. Interpolating a 2D diffusion tensor field

instead of degenerate points. A minimum spanning tree (MST)-based clustering algorithm is employed so that we can group tensor samples (or clusters) that share similar anisotropic values and their associated orientations. This is accomplished by introducing the following dissimilarity metric that evaluates the proximity between the neighbor tensor samples:

$$d(D^S, D^T) = \alpha |C_l^S - C_l^T| + \beta |C_p^S - C_p^T| + \gamma(|\theta_{S,T}|/(\pi/2)), \qquad (10)$$

where, C_l^S and C_l^T represent the C_l values of the two tensor samples D_S and D_T, and C_p^S and C_p^T are the corresponding C_p values. In addition, $\theta_{S,T}$ is the minimal rotation angle between the right-handed coordinate systems defined by the two sets of eigenvector directions. This is calculated by selecting one representation for each tensor (Figure 2). This metric satisfies the fundamental axioms for metric spaces, and tries to evaluate both the differences in the anisotropy and the rotational angle between two tensors. Our experiments suggest that the parameter setting $\alpha = 4$, $\beta = 2$ and $\gamma = 1$ is reasonable for this purpose because we are more likely to group high anisotropic tensor samples in earlier stages of this clustering process. Figure 5(a) shows such an example, where we find the most similar pair of samples among the candidate adjacent pairs and connect the pair with MST-based clustering. We continue this process until all the tensor samples fall into a single cluster as shown in Figure 5(b).

After finishing this process, we can identify the pair of adjacent tensor samples as *a degenerate pair* if their rotation angle is more than $\pi/2$. By counting the number of degenerate pairs, we can locate degenerate cells. This is because a degenerate cell contains an odd number of degenerate pairs (Figure 6). Figure 5(c) shows such an example, where two degenerate points are located. However, from the interpolated result in Figure 5(d), we find that the rotational inconsistency exists just on the degenerate pairs between the two degenerate points. Therefore, we should introduce degenerate lines to remove the rotational inconsistency and connect the degenerate points.

4.3   Rotational Inconsistency Around Degenerate Points

The main idea of resolving the rotational inconsistency is to transform aforementioned degenerate pairs into non-degenerate ones. For this purpose, we optimize the rotational

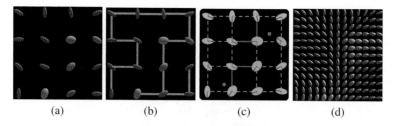

Fig. 5. MST-based clustering in tensor field: (a) Original tensor samples (b) The final MST that covers all the tensor samples. (c) Degenerate points (green points) and degenerate pairs (blue lines) obtained through the MST-based clustering algorithm. (d) Interpolated result obtained through the MST-based clustering algorithm only.

Fig. 6. Configurations of degenerate pairs in a unit square. A pair of tensor samples is drawn in blue if it is degenerate, and the square is shaded in red if it contains a degenerate point.

transformation between the two end tensor samples. This is achieved by selecting one of the two tensors, and changing the order of its eigenvectors to minimize the rotational angle in between.

Figure 7 illustrates this process. We focus on a degenerate pair indicated by the red segment in Figure 7(a), and try to transform it into non-degenerate one. We basically select one of the two tensors (circled by a broken circle in red in Figure 7(a)) as the one that has not been visited yet, while the tensor with low anisotropy is more likely to be selected if both are unvisited. The order of its eigenvectors is then rearranged in order to minimize the rotational angle between these two tensors where the first and second eigenvectors are exchanged in this case. Finally, we label the adjusted tensor as visited, and check the incident pairs as represented by the yellow segments in Figure 7(b), because the change in the tensor representation may transform the neighbor pairs into degenerate ones. Now we select another degenerate pair and repeat this process, as shown in Figure 7(c), until all the degenerate pairs are resolved into non-degenerate ones.

Actually, our method tries to introduce isotropic tensors in the region between the neighbor degenerate pairs, and these isotropic tensors constitute of degenerate lines. This has been finished by optimizing the rotational angle between the two end tensors on degenerate pairs. Therefore, our method is able to remove the rotational inconsistency by generating degenerate lines, as shown in Figure 7(d).

Furthermore, degenerate lines do not affect the anisotropy of the region without degenerate pairs. The reason is that the orders of the eigenvalues of all the tensors in such region change at the same time. For example, there is no degenerate pairs in the rightmost and bottommost cell in Figure 7(a). Before we transform the degenerate pairs into

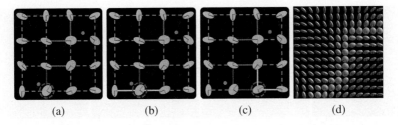

(a)   (b)   (c)   (d)

Fig. 7. Generating degenerate lines by resolving degenerate pairs of tensor samples: (a) Find one of the remaining degenerate pairs in red, and transform it into non-degenerate one. (b) Check if the pairs around the selected tensor sample have redundant rotations. (c) If they exist, select them and transform them into non-degenerate ones. We continue this process until all the remaining degenerate pairs are resolved. (d) Result after rotational inconsistency has been resolved, where a degenerate line is generated between the two degenerate points.

non-degenerate ones, all the orders of their eigenvalues are $\{\lambda_1, \lambda_2, \lambda_3\}$. After we finish transforming the degenerate pairs into non-degenerate ones, all the orders of their eigenvalues are changed into $\{\lambda_2, \lambda_1, \lambda_3\}$. Therefore, the anisotropy of interpolated tensors in such region will not be affected, as shown in Figure 7(d).

5   Results and Discussions

In this section, we demonstrate the effectiveness of our approach in the 2D tensor fields without degenerate points and with degenerate points, respectively. We also compare the results of our approach with those obtained by other existing schemes.

Figure 8(a) presents a 2D case where 6×6 discrete tensor samples guide the underlying "X"-like shape. No degenerate points are included in this dataset. Figures 8(b), (c), and (d) show interpolated tensor samples obtained using the component-wise, Log-Euclidean and our schemes, respectively. The two conventional schemes unexpectedly incur low anisotropic features around the crossing of the two anisotropic line features, while our method can still maximally preserve the underlying anisotropic structures.

Figure 9 shows results of interpolating a real human brain DT–MRI dataset ($256 \times 256 \times 30$). Figure 9(a) is the 17th axial slice of the original dataset, which is down-sampled into 128×128, and Figure 9(b) is the zoom-up view of the region boxed in a square in Figure 9(a) where two fibers intersect with each other. To interpolate the tensor samples in this region, a degenerate line should be obtained to separate these two fibers. Figures 9(c), (d), (e), and (f) show the interpolation results with the component-wise, Log-Euclidean, geodesic-loxodrome, and our schemes, respectively. The results show that our scheme can produce a degenerate line composed by lower anisotropic tensors, which separates these two fibers, and our scheme can also respect the anisotropic features of the two fibers. However, neither of the component-wise or Log-Euclidean schemes can respect the underlying anisotropic features of the left fiber appropriately. The geodesic-loxodrome scheme can fully respect the anisotropy of the two fibers. Unfortunately, all these three schemes cannot generate the degenerate line to separate the two fibers.

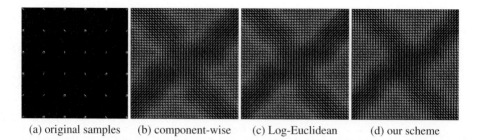

(a) original samples   (b) component-wise   (c) Log-Euclidean   (d) our scheme

Fig. 8. Interpolating a 2D diffusion tensor field. (a) Original tensor samples. Results with the (b) component-wise, (c) Log-Euclidean, and (d) our interpolation schemes.

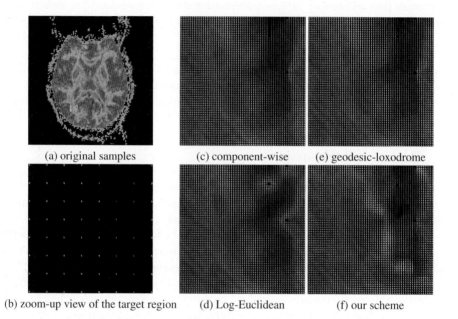

(a) original samples   (c) component-wise   (e) geodesic-loxodrome

(b) zoom-up view of the target region   (d) Log-Euclidean   (f) our scheme

Fig. 9. Human brain DT–MRI dataset: (a) Original data samples. (b) The zoom-up view of the region where two anisotropic features intersect. Interpolation results obtained using (c) component-wise, (d) Log-Euclidean, (e) geodesic-loxodrome, and (f) our interpolation schemes.

6   Conclusion

An approach to interpolating diffusion tensor fields through the analysis of the associated eigenvalues and eigenvectors has been presented in this paper. Compared with other existing interpolation schemes, the present approach can maximally respect the underlying anisotropy of the given dataset, especially in the tensor fields containing degenerate points. In our method, degenerate points can be connected by degenerate lines, which are the most stable topological structure for 3D tensors, by employing

MST-based algorithm. We also solve the non-commutative property of matrix composition by taking advantage of Alexa's linear combination of transformations [12].

However, the present approach may not be able to effectively handle noisy datasets, where such anisotropic features are rather scattered over the data domain. We are currently working on extending our 2D scheme to 3D so as to enable ones to perform detailed analysis of complex fiber structures.

Acknowledgements. We thank Haruhisa Ishida helped us implement an early version of the prototype system. This work has been partially supported by Japan Society of the Promotion of Science under Grants-in-Aid for Scientific Research (B) No. 22300037, and Challenging Exploratory Researches No. 21650019.

References

1. Hesselink, L., Levy, Y., Lavin, Y.: The topology of symmetric, second-order 3D tensor fields. IEEE Transactions on Visualization and Computer Graphics 3, 1–11 (1997)
2. Zheng, X., Pang, A.: Topological lines in 3D tensor fields. In: Proceedings of the Conference on Visualization 2004, pp. 313–320. IEEE Computer Society, Los Alamitos (2004)
3. Westin, C., Peled, S., Gudbjartsson, H., Kikinis, R., Jolesz, F.: Geometrical diffusion measures for MRI from tensor basis analysis. In: Proceedings of ISMRM 1997, p. 1742 (1997)
4. Westin, C.F., Maier, S.E., Mamata, H., Nabavi, A., Jolesz, F.A., Kikinis, R.: Processing and visualization for diffusion tensor MRI. Medical Image Analysis 6, 93–108 (2002)
5. Batchelor, P.G., Moakher, M., Atkinson, D., Calamante, F., Connelly, A.: A rigorous framework for diffusion tensor calculus. Magnetic Resonance in Medicine 53, 221–225 (2005)
6. Fletcher, P.T., Joshi, S.: Riemannian geometry for the statistical analysis of diffusion tensor data. Signal Processing 87, 250–262 (2007)
7. Arsigny, V., Fillard, P., Pennec, X., Ayache, N.: Log-Euclidean metrics for fast and simple calculus on diffusion tensors. Magnetic Resonance in Medicine 56, 411–421 (2006)
8. Kindlmann, G., Estepar, R., Niethammer, M., Haker, S., Westin, C.: Geodesic-Loxodromes for diffusion tensor interpolation and difference measurement. In: Ayache, N., Ourselin, S., Maeder, A. (eds.) MICCAI 2007, Part I. LNCS, vol. 4791, pp. 1–9. Springer, Heidelberg (2007)
9. Merino-Caviedes, S., Martin-Fernandez, M.: A general interpolation method for symmetric second-rank tensors in two dimensions. In: Proceeding of IEEE International Symposium on Biomedical Imaging, pp. 931–934 (2008)
10. Hotz, I., Sreevalsan-Nair, J., Hamann, B.: Tensor field reconstruction based on eigenvector and eigenvalue interpolation. In: Scientific Visualization: Advanced Concepts, vol. 1, pp. 110–123 (2010)
11. Laidlaw, D., Weickert, J. (eds.): Visualization and Processing of Tensor Fields. Springer, Heidelberg (2009)
12. Alexa, M.: Linear combination of transformations. ACM Transactions on Graphics 21, 380–387 (2002)

Indented Pixel Tree Plots

Michael Burch, Michael Raschke, and Daniel Weiskopf

VISUS, University of Stuttgart

Abstract. We introduce Indented Pixel Tree Plots (IPTPs): a novel pixel-based visualization technique for depicting large hierarchies. It is inspired by the visual metaphor of indented outlines, omnipresent in graphical file browsers and pretty printing of source code. Inner vertices are represented as vertically arranged lines and leaf groups as horizontally arranged lines. A recursive layout algorithm places parent nodes to the left side of their underlying tree structure and leaves of each subtree grouped to the rightmost position. Edges are represented only implicitly by the vertically and horizontally aligned structure of the plot, leading to a sparse and redundant-free visual representation. We conducted a user study with 30 subjects in that we compared IPTPs and node-link diagrams as a within-subjects variable. The study indicates that working with IPTPs can be learned in less than 10 minutes. Moreover, IPTPs are as effective as node-link diagrams for accuracy and completion time for three typical tasks; participants generally preferred IPTPs. We demonstrate the usefulness of IPTPs by understanding hierarchical features of huge trees such as the NCBI taxonomy with more than 300,000 nodes.

1 Introduction

Designing tree visualizations that scale up to millions of elements and that still give a clear impression of hierarchical structures and substructures, their sizes, as well as their depths globally and locally will speed up exploration tasks for hierarchical data in various application areas. For instance, the field of software visualization is a rich source for hierarchical data since software artifacts are hierarchically structured and can additionally be associated with attributes such as the developer who did the latest changes or the size of each software artifact. As another example, taxonomies in biology are used to classify various species into a hierarchical order. These may be enriched by an additional attribute such as the family or order name of the species.

There are many tree visualization techniques, including node-link diagrams, layered icicle plots, tree-maps, or hybrid representations consisting of at least two of them. In this paper, we propose a different kind of tree representation that makes use of indented outlines. The IPTPs can be classified as a one-and-a-half dimensional visualization approach since it is sufficient to read such plots from left to right to understand the hierarchical semantics similar to reading a text fragment.

IPTPs focus on clearly displaying the structure, the size, and the depth of a tree in a sparse and redundant-free way. IPTPs draw inner vertices as vertically arranged lines and leaf groups as horizontally arranged lines. Using this metaphor, edges are represented only implicitly by the vertically and horizontally aligned structure of the plot.

G. Bebis et al. (Eds.): ISVC 2010, Part I, LNCS 6453, pp. 338–349, 2010.

Adding explicit links to the plot would be a possible extension. However, this would not convey extra information and hence, could be considered as superfluous (chart junk) according to Tufte [1]—information in a graphical display that can be left away without affecting the understandability and readability.

The hierarchy is recursively encoded into an IPTP by placing parent nodes to the left of their underlying tree structure and by grouping leaf nodes together to the rightmost position of the plot. Consequently, each node in the hierarchy can be mapped to a representative position on the horizontal axis. This allows us to attach an attribute to each node in an aligned way and hence, allows for better comparisons among these attributes. This task would be difficult to impossible for tree-maps, layered icicle plots, or node-link diagrams.

To evaluate the readability of IPTPs, we conducted a user study that compares IPTPs to node-link diagrams by measuring completion time and accuracy for three different tasks in a within-subjects experiment. Finally, to illustrate a typical application, we apply the IPTP technique to the NCBI taxonomy that contains several hundred thousand nodes.

2   Related Work

IPTPs are inspired by the indented outline visual metaphor for displaying hierarchical data structures. Indented outline approaches are popular in file browsers such as Microsoft Explorer. An interactive approach was proposed by Engelbart and English [2] as early as forty years ago. Software developers use the principle of indentation in their everyday programming work—in an automatic and unconscious way.

In general, there is a huge body of previous research on hierarchy visualization [3]. Node-link diagrams are the most widely used approach. For example, Battista et al. [4], Herman et al. [5], and Reingold and Tilford [6] use conventional node-link diagrams to depict relationships between hierarchically ordered elements. Several variations exist for node-link diagrams that use different orientations of these rooted node-link diagrams. Attaching an attribute to all of the nodes—for example a text label—using node-link diagrams may lead to overlaps in the display. Moreover, a simultaneous comparison of all attributes is problematic since these are not aligned in the same way.

Radial node-link approaches place nodes on concentric circles, where the radii of the circles depend on the depths of the corresponding nodes in the tree [4,5,7]. Although this technique leads to a more efficient use of space, it is more difficult to judge if a set of nodes belongs to the same hierarchy level. This apparent drawback of radial diagrams can be explained by the fact that the human visual system can judge positions along a common scale with a lower error rate than positions along identical but non-aligned scales, as demonstrated in graphical perception studies by Cleveland and McGill [8]. IPTPs align nodes on the same depth on a horizontal line, which makes this judgment easier. Balloon layouts are another strategy to display hierarchies in the form of node-link diagrams: they show the hierarchical structure clearly but suffer from space-inefficiency for large and deep hierarchies [5,9] and make these judgments more difficult due to the smaller radii for deeper subtrees. Moreover, it is difficult to attach an attribute to each node for comparisons between hierarchy levels.

Tree-maps [10] are a space-filling alternative for displaying hierarchies. One drawback of tree-maps is the fact that hierarchical relationships between parent and child nodes are hardly perceived in deeply nested hierarchical structures. Nesting can be indicated by borders or lines of varying thickness—at the cost of additionally needed screen space. Tree-maps are an excellent choice when encoding quantitative data attached to hierarchy levels but IPTPs are more useful when comparing hierarchical data.

Layered icicle plots require substantial amounts of image space: they use as much area for parent nodes as the sum of all their related child nodes together. A benefit of this representation is that the structure of the displayed hierarchy can be grasped easily and moreover, this type of diagram scales to very large and deep trees. Variations of this idea are known as Information Slices [11], Sunburst [12], and InterRing [13]. These diagrams make use of polar coordinates, which may lead to misinterpretations of nodes that all have the same depth in the hierarchy. As another drawback, all icicle-oriented techniques require separation lines between adjacent elements allowing differences in hierarchy levels and nodes to be perceived. In contrast, IPTPs do not need such separation lines, resulting in a more compact representation, and attaching labels to the nodes is easier with our technique.

3   Visualization Technique

IPTPs only map vertices to graphical primitives; edges are present only indirectly— visible through the arrangement of the primitives for the vertices. Inner vertices are mapped to vertically aligned lines, whereas leaf vertices are mapped to single square objects. This asymmetric handling of inner and leaf vertices leads to a better separation of both types of visual hierarchy elements. Parent–child relationships are expressed by indentation of the corresponding geometric shapes with respect to the hierarchy levels of the respective parent and child vertices.

On the technical level, we use graph terminology to describe an information hierarchy. Then, the hierarchy can be considered as a special case of a directed graph $T = (V, E)$, where V denotes the set of vertices (i.e., nodes) and $E \subseteq V \times V$ denotes the set of edges (i.e., links) that express parent–child relationships directed from a parent to a child node.

3.1   Plotting Algorithm

Algorithm 1 provides the pseudo code of the plotting algorithm. Rendering is initiated by calling *Indented*$(r, 0, 0)$, with the root vertex $r \in V$. The algorithm plots the hierarchy $T = (V, E)$ recursively as follows. The recursion is used to traverse and render all children of the current vertex: $\{v \in V \mid (r, v) \in E\}$. The depth parameter d is incremented by one each time the recursion enters a deeper hierarchy level. Simultaneously, the horizontal position n is updated by the size of the underlying sub-hierarchy. The recursion stops at the level of leaf vertices.

The following functions implement the actual rendering process:

– **drawLeaf(n, d):** Leaf vertices are rendered as squared boxes. The parameter n is responsible for the horizontal placement of a box representing a leaf vertex, whereas

---

Algorithm 1. IPTPs

Indented(r, n, d):

    // r: vertex where plotting starts

    // n: current horizontal position (integer)

    // d: current level (depth) of hierarchy (integer)

    $C := \{v \in V \mid (r,v) \in E\};$ // set of child vertices of r

    if $\mid C \mid > 0$ **then**

        // inner vertex

        drawInnerVertex(n, d);

        for all $v \in C$ **do**

            // traverse children

            $n :=$ Indented(v, $n+1$, $d+1$);

        end for

    else

        // leaf vertex

        drawLeaf(n, d);

        $n := n+1;$ // advance horizontal position

    end if

    return n;   // horizontal end position

---

the parameter d is used to place it vertically. In general, we paint the upper left corner of the box at coordinates $(n \cdot size_h, d \cdot size_v)$.

- **drawInnerVertex(n, d):** Inner vertices are rendered as vertical rectangles to make their visual encoding significantly different from the visual encoding of leaf vertices. The upper left corner of the vertical rectangle is rendered at coordinates $(n \cdot size_h, d \cdot size_v)$ and the lower left corner at coordinates $(n \cdot size_h, (d+1) \cdot size_v)$.

The above rendering routines assume a uniform grid of graphical primitives of squared shape. The primitives may be associated with pixels, leading to a pixel-based plot. For smaller hierarchies, "fat pixels" can be used: each graphical primitive covers a square of pixels so that the elements of the plot are enlarged. We use $size_v$ as a parameter that describes the vertical size of the representation of inner vertices; $size_h$ describes the horizontal size of each graphical object (inner and leaf vertices alike) and, additionally, the height of the square representation of leaf vertices.

The parameter $size_v$ may be adjusted depending on the depth of the hierarchy, as to make good use of vertical image space. The same holds for parameter $size_h$ in the horizontal direction. Optionally, the initially squared objects for leaf nodes can be vertically flattened, see Figure 1.

The geometric arrangement assumes that the plot is rendered from left to right for the vertices (i.e., the root vertex is placed leftmost) and from top to bottom for increasing depth level of the hierarchy. If needed, the plot could be rotated by 90 degrees, leading to a vertical arrangement of vertices and horizontal mapping of depth levels.

Figure 1 (b) illustrates the IPTP for a hierarchy example. The same hierarchy is also shown as a traditional node-link diagram in Figure 1 (a) to compare both approaches side by side. For example, the root node is located at the top of the node-link diagram, whereas, in the IPTP, it is visually encoded by the leftmost and topmost vertical rectangle.

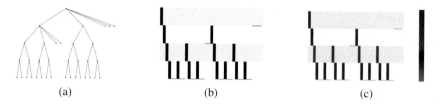

Fig. 1. A hierarchy visualized as (a) node-link diagram; (b) IPTP; (c) IPTP with color gradient

We can see from Figure 1 (b) that an IPTP is interpreted from right to left. Inspecting a number of leaf nodes always means to first orientate to the left to examine their parent node, then looking upwards and again to the left to detect the grandparent node. Inner nodes at the same depth of the hierarchy can be detected by inspecting the whole horizontal line of a specific hierarchy level. A single subtree can be explored by detecting its root node and searching for the next element on the same vertical position in the plot that is located to the right. All elements in between belong to this subtree.

3.2   Recursive Hierarchical Ordering

Transforming an unordered hierarchy into an IPTP may lead to many leaf vertices spread all across the plot. This may be problematic since the pixel-based encoding would result in dispersed single pixels that are hard to perceive. To increase the visual coherence of the plot, we therefore group leaf vertices within the same hierarchy level and place them to the right of all inner vertices within this hierarchy level. This layout principle leads to pixel-based horizontal lines when many leaf nodes belong to the same hierarchy level. The length of those horizontal lines can be judged easily and hence, levels with many leaf vertices can be readily detected.

3.3   Additional Visual Features

The basic plotting method can be enriched by further visual components for better readability or scalability. We discuss the use of color gradients, image scale-down, and high-resolution sampling for hierarchies beyond pixel resolution.

Color Coding.   Color can be applied to group hierarchy levels of the same depth in the IPTPs, leading to a clearer visual separation of deeper levels in the hierarchy. In our implementation, we use a discrete color gradient to support a user in classifying hierarchical elements by their depth.

In addition, neighboring depth levels may be indicated by horizontal background stripes of two cyclic shades of gray, chosen according to $d \bmod 2$.

Figure 1 (c) shows a visual variant of an IPTP by applying a color gradient and adding horizontal background stripes to the black colored plot in Figure 1 (b). As a benefit, leaf nodes can be detected between subsequent inner nodes in the plot even if it is scaled down beyond pixel size.

Horizontal and Vertical Scaling. The hierarchical structure remains clearly visible even if the plot is scaled to a much higher resolution. This holds for both the horizontal and vertical direction. Huge hierarchies can be explored by just inspecting static images without the need for interactive features such as scrolling vertically or horizontally, typically required in other visualization techniques.

Interaction. Our implementation supports interactive features such as expanding and collapsing of hierarchy levels, filtering and querying, exploration support for hierarchy relationships, details on demand, or applying different color scales.

4 User Study

In a user experiment, we firstly investigated the readability of IPTPs compared to node-link diagrams (NLDs), both without color gradient and interactive features. The user study was within-subjects. We chose node-link diagrams as source of comparison because they are the most widely used, well established, de-facto standard for hierarchy visualization [14]. Each test of the visualizations included three dataset sizes and three tasks. Questions were designed so that subjects had to answer in forced-choice fashion.

4.1 Stimuli and Tasks

A stochastic algorithm generated all datasets synthetically. The dataset construction was parameterized by the size of the hierarchy in terms of the number of vertices and the maximal depth. One constraint was that the presentation space was identical for both techniques. The experiment included three tasks:

- **T1:** Finding the least common ancestor of two leaf vertices
- **T2:** Checking the existence of an identical sub-hierarchy elsewhere in the plot
- **T3:** Estimating which of two sub-hierarchies was the larger one

All tasks are important when exploring attributes that are attached to all hierarchy levels since patterns in the set of attributes may be caused by corresponding hierarchy levels that may show similar hierarchical patterns. We chose the tasks as a result of a pilot study.

4.2 Study Method

We chose a within-subjects user study design with 30 participants. They had to answer questions that where recorded by an operator. Subjects could additionally mark preferences and provide comments by filling in questionnaires.

Environment Conditions and Technical Setup. The user experiment was conducted in a laboratory that was insulated from outside distractions. All visualizations were presented on a 24 inch Dell 2408 wfp ultrasharp TFT screen at a resolution of 1920×1080 pixels with 32 bit color depth. To avoid wrong results the subjects' responses were recorded by an operator pushing two specially marked keys on a PC keyboard. We assume that the delay associated with the operator for every task execution and recording loop was approximately the same amount of time (fault tolerance < 100 ms).

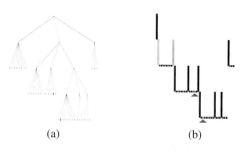

(a)   (b)

Fig. 2. Example stimuli from the study: (a) node-link diagram with color-coded target elements; (b) corresponding IPTP

Subjects. Thirty (23 male, 7 female) subjects were recruited. Sex was not considered a confounding factor for this study. Twenty-seven participants were undergraduate students of our university and three were graduate students. Twenty-three subjects were computer scientists and seven were engineers. The average age was 27 years (minimum 22, maximum 53). Subjects were paid € 10 for participating in the user experiment. Twelve stated that they were familiar with visualization techniques or had attended a lecture with this topic. Eighteen stated that they were not familiar with visualization techniques. All participants had normal or corrected-to-normal color vision, which we confirmed by an Ishihara test and a Snellen chart to estimate visual acuity.

Study Procedure. First, subjects had to fill out a short questionnaire about age, field of study, and prior knowledge in visualization techniques. Then, they read a two-page instruction manual on IPTPs and node-link diagrams. After the participants were given time to read this tutorial, we did a practice run-through of the user tasks. The time duration of the complete training was 10 minutes. During this practice test, subjects could ask questions about the visualization technique and clarify potential problems or misinterpretations. We also used the practice test to confirm that the subjects understood both IPTPs and node-link diagrams.

Then, we continued with the main evaluation that took between 15 and 20 minutes depending on the fitness of the subject. There was a "Give Up" option, but it was not used by the subjects. Tasks, tree sizes, and visualization types were randomized and balanced to compensate for learning effects. Each participant had to perform T1, T2, and T3 for each tree size and visualization type. One task consisted of seven trials per tree size. The time limit for every trial was 20 seconds.

In T1, the child nodes were marked by red colored circles in case of node-link diagrams, and by red colored triangles in case of IPTPs. Two possible ancestors were colored in green and blue, see Figure 2. In T2, one sub-hierarchy was marked with a red starting node. In T3, the starting nodes of two sub-hierarchies were marked green and blue. Subjects had to respond with "blue" or "green" (T1 and T3) and "yes" or "no" (T2). Completion times and correctness of answers were recorded for the seven trials. This procedure resulted in a total of 126 measurements for each participant since we showed the combination of two visualization techniques, three tasks, three dataset sizes, and seven trials.

After the main evaluation, subjects were given a second questionnaire in which they marked their preferences in using one of the two visualization techniques. Finally, participants were given the opportunity to provide open, unconstraint comments.

4.3 Study Results

Task Completion Time. To compare the completion times concerning their significance we analyzed the data task-pairwise with the paired t-test. Table 1 shows the averaged completion times over all 30 subjects with standard deviation and t-test values t. Also, the best technique for every series of measurements is identified. There were 5 timeouts over all measurements and all subjects. We rated each timeout with 20 seconds in the data analysis.

When comparing IPTPs to node-link diagrams over all dataset sizes, the average and median of all time measurements is approximately equal in T1 and T2. None of the t-tests indicated significant differences.

For T3, both techniques differ: For the small and large datasets, node-link diagrams are significantly faster to read. For the medium dataset, IPTPs are superior close to significance level ($t = 0.07$). It is interesting that the relative performance of the two visualization techniques varies in such a non-monotonic way with dataset size, indicating that a clearly superior technique cannot be identified.

We discuss the results of our study for the completion times when participants solved task T1 in more detail since we found there the most interesting behavior. Figure 3 shows a parallel coordinates plot for the completion times for T1. The axes are ordered from left to right by the dataset sizes and annotated by the seven trials for small, medium, and large sized tree data. Green squares indicate that the task was answered correctly by this participant whereas red squares show incorrect answers.

One can easily see that the tasks for the yellow colored IPTPs are frequently answered correctly in less than 1,000 milliseconds and the fastest answer was given after 200 milliseconds. A few participants needed more than 5,000 milliseconds to answer. The blue colored polylines for the node-link diagrams show a different behavior. The

Table 1. Comparison of task completion times. Each table element shows best technique, t-test values, and average completion times (in milliseconds) with standard deviation in parentheses.

	Task 1	Task 2	Task 3
Small	NLDs	IPTPs	NLDs
t	0.45	0.17	0.002
NLDs	1995 (545)	5328 (1634)	3576 (1215)
IPTPs	2016 (934)	5090 (1293)	4514 (1473)
Medium	IPTPs	NLDs	IPTPs
t	0.37	0.21	0.07
NLDs	2110 (541)	5930 (1723)	5107 (2183)
IPTPs	2059 (987)	6156 (2018)	4500 (1166)
Large	IPTPs/NLDs	IPTPs	NLDs
t	0.49	0.41	< 0.001
NLDs	2388 (562)	8685 (3001)	3325 (926)
IPTPs	2386 (1315)	8603 (2181)	3976 (1257)

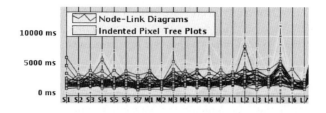

Fig. 3. Parallel coordinates plot showing the completion times and the correctness of answers for task T1 and for seven trials of each dataset size

shortest time for an answer is 980 milliseconds and there are not as many outliers with more than 5,000 milliseconds. The standard deviation is much smaller here. The parallel coordinates plot shows that task T1 was solved consistently in less than 1,000 milliseconds for IPTPs by a subgroup of the participants. This indicates that these participants used IPTPs more efficiently than node-link diagrams. We conjecture that these participants have well adopted IPTPs.

Accuracy. When performing T1 and T3, subjects had a higher accuracy with node-link diagrams. The difference was significant for the large dataset (for both T1 and T3) and the small dataset (only for T3). The pattern recognition in T2 had a higher accuracy in the IPTPs regarding to the average accuracy. However, those differences were not statistically significant.

Subjective Preferences. In a questionnaire, subjects could give their preferences when using the IPTPs or the node-link diagrams for exploring small and large datasets (see Table 2). Additionally, more general questions are documented in Table 3.

Subjects tended to prefer node-link diagrams for small datasets. When exploring large datasets, they significantly found IPTPs more useful. Table 3 shows the results

Table 2. Subject preferences: average rating in Likert scale (1 = IPTPs, 2 = NLDs) including standard deviation in parentheses

Question	Rating	Preference
Q1: Which visualization do you prefer for small datasets?	1.76 (0.43)	NLDs
Q2: Which visualization do you prefer for large datasets?	1.26 (0.45)	IPTPs

Table 3. General questions. Average rating in Likert scale (1 = I agree, 5 = I disagree) for each visualization technique.

Question	IPTPs	NLDs	t
Q3: The visualization was helpful to me.	2 (1.14)	1.9 (0.91)	0.410
Q4: I enjoyed this visualization.	2.3 (1.10)	1.9 (0.92)	0.083
Q5: I found the visualization motivating.	1.8 (0.95)	2.46 (0.90)	0.005

of questions Q3 to Q5. For Q3, both visualization techniques are rated head-to-head. Participants enjoyed using node-link diagrams (Q4) more than IPTPs (non-significant), but found IPTPs more motivating to use (Q5, significant).

In the last part of the user study, subjects could provide open comments. Thirteen subjects stated that they expected their results in reading IPTPs to further increase with more training. They also expressed their belief that learning the IPTPs is possible in a 10-minutes tutorial. Thirteen subjects said that the node-link diagrams were already very well-known from their academic studies and are intuitive because of their natural shape. Seven felt that size approximation could more easily be done in IPTPs. Seven stated that ancestor identification is comparable in both techniques.

Our last question asked what subjects associated when they saw IPTPs for the first time during the tutorial. Most of the subjects responded with associations to bar code labels, brackets in computer languages, and "something that must be high-tech".

4.4 Discussion

The comparison of the task completion times of IPTPs and node-link diagrams shows that the readability of the IPTPs is approximately the same as that of the node-link approach. The t-test results showed significant differences for only few cases.

The accuracy in the task execution is more accurate for node-link diagrams than for IPTPs with respect to T1 and T3. IPTPs reach a higher accuracy rate when subjects ascertain the existence of an identical sub-hierarchy elsewhere in the diagram. Therefore, we believe that correct pattern identification is one of the big advantages of IPTPs.

As subjects were used to node-link diagrams, they understood the basic rules of IPTPs within a short time of introduction, and could use this new approach afterwards to perform the given tasks with a comparable speed with respect to that of node-link diagrams. Subjects stated that their speed and accuracy would increase with more training.

Another very interesting aspect of IPTPs concerns the cognitive and perceptual processes when performing task T1. To get more insight into the strategical solution of this task with more than two target objects, we will extend the user experiment by using eye tracking techniques.

5   Application

The National Center for Biotechnology Information (NCBI) uses a taxonomy that contains the names of all organisms that are represented in the NCBI genetic databases with at least one nucleotide or protein sequence [15]. We demonstrate the usefulness of IPTPs by understanding hierarchical features of huge trees such as the NCBI taxonomy with some 324,000 nodes in a static diagram, which may be difficult when the same dataset is visualized in a tree-map, a layered icicle plot, or a node-link diagram. Furthermore, the plot can be scaled down immensely and the hierarchical structure remains still visible.

Figure 4 shows an IPTP of this dataset with a maximum depth of 40. The plot is color-coded by a blue-to-red color gradient and horizontally arranged gray stripes are added to enhance the perception of hierarchy levels. Since this plot is scaled down

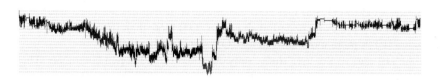

Fig. 4. An IPTP for the NCBI taxonomy in a blue to red color gradient

vertically to a high extent—more than 200 elements per pixel—a horizontal pixel-based line indicates that there are very many leaf nodes in the corresponding hierarchy level.

Several horizontally arranged lines in Figure 4 can be detected at the right side in the plot. A detail-on-demand request shows that they belong to the *Bacteria* branch and are labeled by *environmental samples* and *unclassified bacteria miscellaneous*.

Another apparent anomaly in this visualization can be detected near the middle of the horizontal axis where the hierarchy is very deeply structured. A detail-on-demand request shows that there are many elements in a subhierarchy of depth at least 36. All other subhierarchies in this taxonomy are not as deep as this part. We apply a filtering function for all elements of that depth and select the least common ancestor of all of those remaining subhierarchies. We obtain the inner node that belongs to the *Perciformes* species.

6   Conclusions

We have introduced IPTPs—a pixel-based visualization technique suitable for representing large hierarchies. Its key elements are the visualization of inner vertices as vertical lines and leaf vertex groups as horizontal lines. Edges are represented only implicitly through the layout of the plot, leading to a compact visual representation. In particular, there is no need for additional graphical elements such as separator lines in layered icicle plots, the nesting indicators in tree-maps, or the geometric link elements in node-link diagrams. Another advantage is that both inner and leaf vertices are mapped to unique horizontal positions. Therefore, annotations or other visual information may be aligned below the plot and can easily be associated with inner and leaf vertices. Moreover, attributes may be compared for all hierarchy levels simultaneously.

We evaluated IPTPs by a controlled user study that compared our plotting approach with node-link diagrams. There is no clear picture whether IPTPs or node-link diagrams are superior and almost all timing and accuracy results are in very similar ranges. However, the participants preferred the IPTPs. An interesting finding is that a basic and working understanding of the new visual representation can be learned in less than 10 minutes. We believe that further exposure and experience with IPTPs may improve the users' proficiency. This conjecture needs to be tested by a future long-term study that is beyond the scope of this paper. Similarly, further application examples in addition to biological hierarchies could be investigated in more depth.

Acknowledgements

We would like to thank Dr. Kay Nieselt, University of Tübingen, for providing the NCBI taxonomy dataset and the many participants for taking part in our user study.

References

1. Tufte, E.R.: The Visual Display of Quantitative Information, 1st edn. Graphics Press, Cheshire (1983)
2. Engelbart, D.C., English, W.K.: A Research Center for Augmenting Human Intellect. Video of Public Demonstration of NLS (1968), http://www.1968demo.org/
3. McGuffin, M.J., Robert, J.M.: Quantifying the space-efficiency of 2D graphical representations of trees. Information Visualization (2009), doi:10.1057/ivs.2009.4
4. Battista, G.D., Eades, P., Tamassia, R., Tollis, I.G.: Graph Drawing: Algorithms for the Visualization of Graphs. Prentice Hall, Upper Saddle River (1999)
5. Herman, I., Melançon, G., Marshall, M.S.: Graph visualization and navigation in information visualization: A survey. IEEE Transaction on Visualization and Computer Graphics 6, 24–43 (2000)
6. Reingold, E.M., Tilford, J.S.: Tidier drawings of trees. IEEE Transactions on Software Engineering 7, 223–228 (1981)
7. Eades, P.: Drawing free trees. Bulletin of the Institute for Combinatorics and its Applications 5, 10–36 (1992)
8. Cleveland, W.S., McGill, R.: An experiment in graphical perception. International Journal of Man-Machine Studies 25, 491–501 (1986)
9. Grivet, S., Auber, D., Domenger, J.P., Melançon, G.: Bubble tree drawing algorithm. In: Wojciechowski, K., Smolka, B., Palus, H., Kozera, R.S., Skarbek, W., Noakes, L. (eds.) Computer Vision and Graphics, Dordrecht, The Netherlands, pp. 633–641. Springer, Heidelberg (2006)
10. Shneiderman, B.: Tree visualization with tree-maps: 2-d space-filling approach. ACM Transactions on Graphics 11, 92–99 (1992)
11. Andrews, K., Heidegger, H.: Information slices: Visualising and exploring large hierarchies using cascading, semi-circular discs. In: Proceedings of the IEEE Information Visualization Symposium, Late Breaking Hot Topics, pp. 9–12 (1998)
12. Stasko, J.T., Zhang, E.: Focus+context display and navigation techniques for enhancing radial, space-filling hierarchy visualizations. In: Proceedings of the IEEE Symposium on Information Visualization, pp. 57–66 (2000)
13. Yang, J., Ward, M.O., Rundensteiner, E.A., Patro, A.: InterRing: a visual interface for navigating and manipulating hierarchies. Information Visualization 2, 16–30 (2003)
14. Andrews, K., Kasanicka, J.: A comparative study of four hierarchy browsers using the hierarchical visualisation testing environment. In: Proceedings of Information Visualization, pp. 81–86 (2007)
15. Benson, D.A., Karsch-Mizrachi, I., Lipman, D.J., Ostell, J., Sayers, E.W.: Genbank. Nucleic Acids Research 37, 26–31 (2009)

Visualizing Multivariate Hierarchic Data Using Enhanced Radial Space-Filling Layout

Ming Jia[1], Ling Li[2], Erin Boggess[1], Eve Syrkin Wurtele[2], and Julie A. Dickerson[1]

[1] Dept. of Electrical and Computer Engineering, Iowa State University, Ames, IA, USA
[2] Dept. of Genetics, Development and Cell Biology, Iowa State University, Ames, IA, USA
jiaming@iastate.edu, liling@iastate.edu, eboggess@iastate.edu,
mash@iastate.edu, julied@iastate.edu

Abstract. Currently, visualization tools for large ontologies (e.g., pathway and gene ontologies) result in a very flat wide tree that is difficult to fit on a single display. This paper develops the concept of using an enhanced radial space-filling (ERSF) layout to show biological ontologies efficiently. The ERSF technique represents ontology terms as circular regions in 3D. Orbital connections in a third dimension correspond to non-tree edges in the ontology that exist when an ontology term belongs to multiple categories. Biologists can use the ERSF layout to identify highly activated pathway or gene ontology categories by mapping experimental statistics such as coefficient of variation and overrepresentation values onto the visualization. This paper illustrates the use of the ERSF layout to explore pathway and gene ontologies using a gene expression dataset from *E. coli*.

1 Introduction

Linking high-throughput experimental data with hierarchical ontologies that relate biological concepts is a key step for understanding complex biological systems. Biologists need an overview of broader functional categories and their performance under different experimental conditions to be able to ask questions such as whether degradation pathways have many highly expressed genes, or which biological process categories are overrepresented in the data. These needs pose many unique requirements on the visualization of biological ontologies, such as being able to visualize an overview of an ontology mapped with experimental data and clearly show the non-tree connections in ontology.

Current tools which visualize biological ontologies normally employ the traditional Windows[TM] Explorer-like indented list, as are found in EcoCyc[1] and AmiGO[2], or node-link based layouts, e.g., OBOEdit[3] and BinGO[4]. These kinds of layouts are well suited for tens of nodes, however quickly become cluttered if hundreds of nodes are shown simultaneously. As a result, users often collapse the ontology to reduce its visual complexity, and only expand small portions when needed. The tradeoff of this abstraction is the loss of context of the overall ontology structure. Moreover, biological ontologies are not pure tree structures, but are directed acyclic graphs (DAG), i.e.,

G. Bebis et al. (Eds.): ISVC 2010, Part I, LNCS 6453, pp. 350–360, 2010.

they contain non-tree edges where many child nodes having multiple parents. Current software tools are not suitable for tracing such connections.

To address these problems, we propose the enhanced radial space-filling (ERSF) algorithm that uses an intuitive orbit metaphor to explicitly visualize non-tree edges, and make them appear differently than the major hierarchic structure. The ERSF is implemented in a software package based on the Google Earth API. To the best of our knowledge, this is the first application to use 3D RSF in biology and the first algorithm to visualize non-tree edges on a RSF plot.

Some preliminary results that demonstrate the use of ERSF for a single ontology dataset have been published in a workshop proceeding [5]. This work focuses on the visualization benefits of the ERSF layout on multiple datasets in terms of user requirements. The platform is also extended to visualize general ontology structures and multivariate data. Moreover, we here conducted an initial user test and summarize this feedback in Section 4.

The contributions of our ERSF-based software to information visualization area are:

- Applying the radial space-filling layout to a common but challenging visualization task in biology field.
- Enhancing the radial space-filling technique with orbits metaphor for visualizing non-tree edges in hierarchic dataset.
- Mapping key summary statistics from experimental high-throughput data on the hierarchical visualization and links with traditional parallel coordinate views.

This paper is organized as follows: Section 2 describes the properties of the biological datasets, lists the requirements for the visualization and assesses the related work. Section 3 d the ERSF layout. Section 4 reveals some interesting findings from the initial user testing.

2 Background

2.1 Ontology Data Description

An ontology is a formal explicit description of concepts, or classes in a domain of discourse [6]. Biologists use ontologies to organize biological concepts. The Gene Ontology (GO) [2] is a controlled vocabulary of gene and gene products across all species. The Pathway Ontology (PO) [7] is a recent concept that provides a controlled vocabulary for biological pathways and their functions. PO, like many other ontologies, is hierarchical data, but it is not a pure tree structure because several pathways may have multiple parents. Both ontologies are actually directed acyclic graphs. To facilitate the visualization, we first construct a spanning tree in the ontology, and then define the connections in the spanning tree as tree edges and all remaining edges as non-tree edges or cross links. The non-tree edges are of particular interest since they represent pathways that perform multiple functions.

We illustrate this application with the *E. coli* Pathway Ontology from EcoCyc [1]. The EcoCyc PO contains 442 nodes, where 289 of them are pathways or leaves. It also contains 508 edges, where 67 (13.2%) are non-tree edges. PO's for other species are slightly different, however, they are of similar scale. Another feature typical of a

PO is that the height of the hierarchy is normally low, e.g., 6 for *E. coli*, which results in a very large width/height ratio (289/6=48.1).

Another dataset we focus on is the Gene Ontology (GO) Slim [8], which are important subsets of GO that contain around 100 terms.

Besides studying PO structure, in day-to-day research, biologists need to make sense of system-wide experimental data and wish to understand how the experimental conditions affect the underlying biology. One typical type of experimental data is transcriptomics (often referred to as gene expression data), which describes the abundance of gene transcripts during an experiment. Other experimental data types include metabolomics and proteomics. For gene expression data, the original data is typically a data matrix where each row describes a gene, and each column records the expression level of genes under a certain condition, e.g., a point, treatment, or replicate.

2.2 Visualization Requirements

Based on the data described above and tasks biologists perform, the basic requirements for visualization of PO are to:

- View the whole ontology on a single screen to gain a global feeling for the data and the main hierarchical structure.
- View ontology details by navigation and/or interaction (zoom, pan, rotation).
- Map attributes on the ontology so that they are easily visible.
- Clearly show non-tree connections.

2.3 Related Work

The most widely-used representation for ontology structure is the Windows[TM] Explorer-like tree view, or indented list. One implementation of indented list (Class Browser) is evaluated in [9] with three other methods (Zoomable interface, Focus + Context, and Node-link/tree). The indented list lacks the ability to show non-tree edges. Users presented with indented list naturally think the underlining data is a pure tree structure.

Node-link graph and treemaps [10] are also widely used to visualize ontology. OBO-Edit [3] combines an indented tree browser (Tree Editor) and a graphical tree drawing (Graph Editor) which uses the node-link based layout from GraphViz[11]. BinGO[4], a Cytoscape plug-in for analyzing Gene Ontology, uses the default 2D hierarchic layout from Cytoscape. The node-link based layout is very good at showing simple hierarchical structures (e.g. contain less than 50 nodes). However when the number of entities increases, those layouts become very cluttered and incomprehensible. Fig. 1 shows the result of our PO dataset using these layout methods. We can see that the whole hierarchic structure and non-tree edges are not obvious in these views. Due to the cluttered layout for a large number of entities, researchers normally confine their view to a limited subset of the whole structure, and are thus unable to gain the global knowledge.

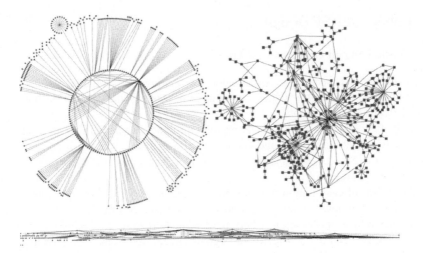

Fig. 1. The Pathway Ontology of *E.coli* from EcoCyc is shown in Cytoscape using circular layout (top left), organic layout (top right), and hierarchic layout (bottom). The ontology contains 442 nodes and 508 edges. The hierarchical structure can hardly be seen.

Treemap based systems [10] are able to visualize the whole GO with mapped data in one screen, and are suitable for identifying regions of interest. However, the hierarchical structure is hard to see in a treemap since it is a nesting-based layout which overplots the parent nodes with their children nodes [12]. Another limitation of treemap is that it lacks a meaningful representation of non-tree edges, a key requirement. As observed in [6], treemaps and other space-filling layouts normally duplicate nodes which have multiple parents. If the node being duplicated is a non-leaf node, the whole substructure rooted at this node will be duplicated as well. Thus duplicating nodes in hierarchic dataset may greatly increase a graph's visual complexity. Duplication also causes confusion for the user. For example, when user finds two regions have similar visual patterns in a treemap, they may think that they have discovered two groups of genes functioning similarly. Unfortunately, they often turn out to be the identical GO terms being drawn twice.

Besides the visualization methods mentioned above, Katifori et al. [6] have also presented many tools and layout algorithms to visualize ontologies and graphs in general. For example, a hyperbolic tree [13] can handle thousands of nodes. However, in a hyperbolic tree visualization, it is difficult to distinguish between tree and non-tree edges among hundreds of edges since they are all represented as links. Another disadvantage is that hyperbolic trees are not space efficient, and normally only a couple of pixels are used for each node. Therefore attributes (like gene expression data) mapped on nodes become hard to distinguish and interpret.

Space-filling methods are considered very space-efficient and are good for mapping attributes on node regions. Despite the disadvantages of rectangular space-filling (such as treemap), evaluations [14] find that radial space-filling (RSF) methods [15] are quite effective at preserving hierarchical relations.

3 Enhanced RSF Design

3.1 Visualizing an Ontology as a Tree

For explanatory purpose, we first assume the ontology as a pure tree structure that does not have any non-tree edges, and explain how the traditional RSF technique can visualize this simplified data.

In the RSF drawing, each circular region represents one node in the tree. The node represents an ontology term, and can either be a pathway (leaf) or a category (non-leaf). For the ease of explanation, we will interchangeably use the words ontology term, node, and region.

The height of each region is set as proportional to the height of the subtree rooted at that node. For color, initially we used structure-based coloring [15] to convey additional hierarchical information, where the leaf node regions are colored according to the color wheel and the non-leaf node regions are colored as the weighted average of its children's color. However, during the initial user testing, several users pointed out that the drawing is full of saturated color, and it is hard to distinguish orbits from the main drawing. To solve this problem, we propose and provide an option to use orbit-based coloring where every category regions are white and highly transparent.

Fig. 2a shows a small tree with eight leaf nodes and five non-leaf nodes, labeled as graph G1. Fig. 2b shows the result of using RSF in 3D on graph G1. Non-leaf nodes correspond to pathway categories.

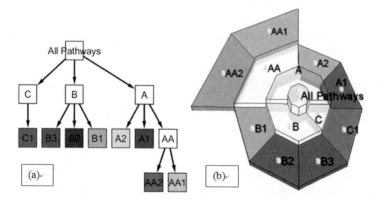

Fig. 2. Graph G1 shows hierarchical relationships among leaf nodes (pathways) and non-leaf nodes (pathway categories), drawn in hierarchic layout in Cytoscape (a) and the radial space-filling layout (b)

3.2 Visualizing Pseudo Ontology with Non-tree Edges

As noted earlier, RSF cannot support non-tree edges. To better meet the visualization requirements, we proposed the enhanced RSF layout, or ERSF, which uses orbits to represent non-tree edges. Fig. 3a shows graph G2, which adds four non-tree edges to G1. The ERSF drawing of G2 is shown in Fig. 3b where the spanning tree is drawn using traditional RSF. The metaphor of "satellite orbits" represents non-tree edges as circular links. For each child node, which has at least two parents, one orbit circle is drawn on the layer of that node. The parent that connects the node in the spanning tree is the major parent and other parents are minor parents. The region of each node is placed under the region of its major parent as in RSF. For every minor parent, a green edge from the center of its region to the orbit of the child is called the 'downlink'. The intersections between downlinks and orbits are called access points, which are represented by red dots.

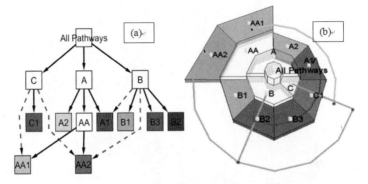

Fig. 3. Graph G2 is drawn using hierarchic layout in Cytoscape (a) and the Enhanced Radial Space-Filling (ERSF) layout using orbit-based coloring (b). In the hierarchic layout, red dashed lines represent non-tree edges. In the ERSF layout, orbits with blue and green radial links represent non-tree relations. For example, the green line extruded from B contains two red-dots: the inner one intersects with red orbit of A1 and the outer one intersects with green orbit of AA2. These orbits mean that B is the minor parent of both A1 and AA2.

To help viewers find and visually trace interesting non-tree edges, the orbits need to be distinguishable from one another. In order to do this, the orbits are first restricted to span in the middle area of each layer, thus leaving a visually apparent gap between orbits in adjacent layers. To distinguish orbits in the same layer, we make the orbit the same color as the child region originating the orbit.

3.3 Visualizing Pathway Ontology Dataset

The PO data from *E.coli* is presented with the ERSF view in Fig. 4. Compared to Fig. 1 where the same dataset is shown by node-link based layout, it is clear that ERSF can show some patterns on the overview. For example, the most orbits are concentrated on the third layer, and one category (*methylglyoxal detoxification*) contains many children in other categories because its green uplink intersects many light blue orbits.

The orbit-based coloring allows users to visually trace the orbits. For example, the category *amino acids degradation* (on left) intersects with one orange orbit. To find the child of this orbit one can visually trace the orbit along the circular curve or directly glance at the orange regions on the right side, and find the child region which originates the outer-most orbit. The red dot serves as a "shortcut" for this specific task. For instance, users can click on the red dot on the intersection of category *amino acids degradation* and the orbit and then a pop-up dialog will indicate that it connects to child *superpathway of threonine metabolism*.

It is also clear that three pathways in the category *cell structures biosynthesis* are also the children of another category *fatty acids and lipids biosynthesis*. When a user wants more information about those non-tree edges, he can rotate and zoom the view.

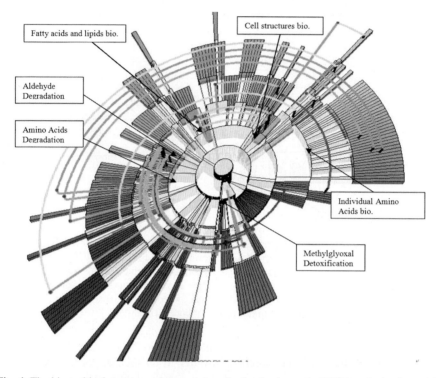

Fig. 4. The hierarchical structure of the ontology is clearly shown in ERSF method using orbit-based coloring. There are many pathways that belong to at least two categories. This kind of multiple inheritance information is difficult to see in other visualization methods.

3.4 Mapping Experimental Values on Ontology

The strategy of a biological scientist evaluating experimental data is to look for which parts of the network show significantly different measurements across different conditions. Questions such as 'Which pathways or categories are most changed under *anaerobic stress*?' can be addressed by mapping the values onto the whole Pathway Ontology.

We use animation to show the values of a series of experiments. For instance, one time-series experiment with 4 time points is presented as animation of 4 frames.

To analyze the gene expression data, we initially map the average expression value on color, and map the variation on height. The visual results of two frames are shown in Fig. 5a and Fig. 5b. The first frame shows the value of one replicate in controlled condition, while the second frame shows that of the treatment condition.

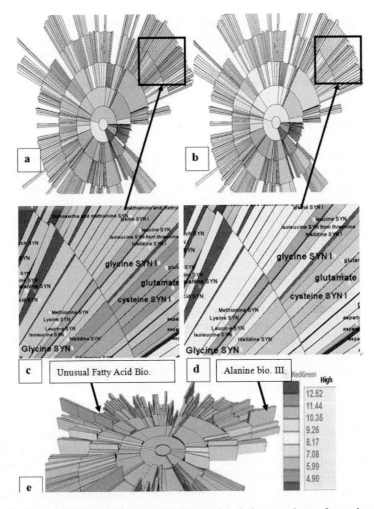

Fig. 5. Average expression values and coefficient of variation are shown for each condition. Color gradient represents values for gene expression, from green (low) to red (high). Two conditions are compared. The orange and red colors in condition 1 show that these categories have much higher expression values in condition 1 (a) than in condition 2 (b). The differences between these two conditions are more obvious when using animation. (c), (d) show the details in the *glycine biosynthesis* categories. When the view is tilted (e), the categories with high variation are shown by their higher height.

Users can tilt the view to see the height of each region (Fig. 5e). In this view, one category (*unusual fatty acid biosynthesis*) stands out, because its and its descendents have very high variation and expression values. This discovery demonstrates the benefit of using 3D to show these two attributes together. Another similar interesting discovery is pathway *alanine biosynthesis III*, which also has very high variation but very low expression values.

By switching between these two conditions, we notice that most of the pathways and categories have a greenish color under the treatment, which indicates lower expression values in the treatment condition than in the controlled condition. This is an interesting trend, since in most experiments the treatments normally have greater values. To confirm this trend, we can map the difference between these two conditions directly on the ontology. We can also map many other attributes onto the color and height of the ERSF drawing, e.g. statistical significance p-value.

4 Initial User Testing

To get some initial feedback from the users, we conducted a pilot user testing involving four users who are PhD students or postdocs in biology field. The goal is to better understand the needs of the biologist-users and to test the effectiveness of the ERSF.

Users were presented with several tasks in two categories: understand the ontology and the gene expression data. One typical user task is to find the pathways which belong to multiple categories. In order for the users to provide the most realistic and valuable feedback, they worked in a relaxed setting where the tasks were not timed.

All users who participated in the pilot user test preferred the ERSF solution to the traditional indented list and node-link based layout. They think the ability to show the whole ontology structure is an important feature, and is especially useful for the system scale experimental dataset. The reason is that knowing which parts of the whole system the experiment affected is an important goal in their research. However, this is hard to do if they are only presented with a small subset of the system. Moreover, users generally gave up on some time-consuming tasks. For example, finding the pathways that belong to at least two categories is extremely difficult using indented lists and node-link based layouts.

Another interesting phenomenon is that although ERSF provides a 3D view of the ontology, users mostly view it from the top down orientation, which is essentially a 2D ERSF layout. Therefore, when users were given the choice to map an attribute to either color or height, all of them prefer mapping the most important attribute to color. Some possible reasons include: biologists are used to traditional 2D tools, and height is hard to interpret precisely due to foreshortening [16]. Nevertheless, the 3D view provides the benefit of mapping two variables simultaneously (color and height). This ability is important for some tasks that may lead to interesting discoveries, e.g. finding pathways that both have high variation and high expression value.

5 Discussion

Fig. 4 shows that visualizing the ontology using ERSF has several advantages. First, this design clearly distinguishes between spanning tree relationships and non-tree edges. Second, compared to treemaps with a crosslink overlay [17], there are much fewer edge-crossings and the drawing is neater since orbits and links are circular and radial respectively. Third, all downlinks of a parent share only one link edge, thus the total length of those edges is the same as the length of the longest link. This property reduces the graph's visual complexity, especially when one node is the minor parent of many other child nodes.

Another benefit of using ERSF is that it does not duplicate nodes, which reduces the visual complexity compared to normal RSF. For the *EColi* PO dataset, ERSF reduced 67 duplicated nodes out of 442 nodes (15.2%). For the GO Slim dataset, since many nodes that have multiple parents are categories, RSF duplicates the whole subtree rooted at those nodes. On the contrary, ERSF reduced 38 duplicated nodes out of 112 nodes (33.9%).

When mapping the node experimental data onto regions' height, e.g. in Fig. 5, it is cumbersome to render the orbits because the orbits may be occluded by higher regions. It is also difficult to follow the orbits when the regions color is mapped by experimental values. As a result, the orbits are not shown by default when mapping attributes onto regions.

The Pathway Ontology dataset shown here contains around 500 nodes, and can be gracefully drawn in one screen. Our suggestion is to limit the data size to 1000 nodes since otherwise the peripheral regions will become as thin as one pixel in width and are difficult to be distinguished. We also noticed that in our dataset, the percentage of non-tree edges is relatively low (from 10% to 20%). The edge bundling method proposed in [18] may be helpful for dataset with high percentage (e.g. above 40%) of non-tree edges. As a result, we suggest that the ERSF method is best suited for visualizing medium-sized multivariate hierarchic data (contains 100 to 1000 nodes) and with medium-to-low percentage of multiple inheritances.

6 Conclusion and Future Work

This work focuses on effective visualization of hierarchic ontologies in biological research. To satisfy the visualization requirements, we propose the enhanced radial-space filling (ERSF) method which arranges ontology regions circularly in a 3D space and uses orbits to represent the non-tree edges. To facilitate the study of large-scale, system-level experimental data, we provide various and customizable ways to map data and statistical results on the ERSF visualization.

The proposed ERSF algorithm has two major advantages over traditional methods in biological data visualization. First, it provides easy visual identification and navigation of non-tree edges in ontology without duplicating nodes. Second, it allows large scale experimental data to be mapped and navigated on the context of the hierarchical structure of the ontology, which may lead to discoveries on a system level.

Initial testing by users has shown the tool to be preferable to their current working solutions, which have been based on indented lists and node-link layouts. A larger quantitative user study is planned in the near future.

The proposed ERSF method can also be adapted to visualize other types of hierarchic data, e.g., company hierarchy and software inheritance diagrams.

References

1. Keseler, I.M., et al.: EcoCyc: a comprehensive view of Escherichia coli biology. Nucleic Acids Res. 37(Database issue), D464-D470 (2009)
2. Carbon, S., et al.: AmiGO: online access to ontology and annotation data. Bioinformatics 25(2), 288–289 (2009)
3. Day-Richter, J., et al.: OBO-Edit–an ontology editor for biologists. Bioinformatics 23(16), 2198–2200 (2007)

4. Maere, S., Heymans, K., Kuiper, M.: BiNGO: a Cytoscape plugin to assess overrepresentation of gene ontology categories in biological networks. Bioinformatics 21(16), 3448–3449 (2005)
5. Jia, M., et al.: MetNetGE: Visualizing biological networks in hierarchical views and 3D tiered layouts. In: IEEE International Conference on Bioinformatics and Biomedicine Workshop, BIBMW 2009 (2009)
6. Katifori, A., et al.: Ontology visualization methods a survey. ACM Computing Surveys 39(4), 10 (2007)
7. Green, M.L., Karp, P.D.: The outcomes of pathway database computations depend on pathway ontology. Nucleic Acids Res. 34(13), 3687–3697 (2006)
8. Consortium, G.O.: GO Slim and Subset Guide (2009),
   `http://www.geneontology.org/GO.slims.shtml`
9. Katifori, A., et al.: Selected results of a comparative study of four ontology visualization methods for information retrieval tasks. In: Second International Conference on Research Challenges in Information Science, RCIS 2008 (2008)
10. Baehrecke, E.H., et al.: Visualization and analysis of microarray and gene ontology data with treemaps. BMC Bioinformatics 5, 84 (2004)
11. Ellson, J., Gansner, E.R., Koutsofios, E.: Graphviz and dynagraph static and dynamic graph drawing tools. Technical report, AT&T Labs - Research (2003)
12. Tekusova, T., Schreck, T.: Visualizing Time-Dependent Data in Multivariate Hierarchic Plots -Design and Evaluation of an Economic Application. In: Information Visualisation, IV 2008, Columbus, OHIO, USA (2008)
13. Munzner, T.: Exploring Large Graphs in 3D Hyperbolic Space. IEEE Computer Graphics and Applications 18(4), 18–23 (1998)
14. John, S.: An evaluation of space-filling information visualizations for depicting hierarchical structures, pp. 663–694. Academic Press, Inc., London (2000)
15. Yang, J., et al.: InterRing: a visual interface for navigating and manipulating hierarchies, pp. 16–30. Palgrave Macmillan, Basingstoke (2003)
16. Munzner, T.: Process and Pitfalls in Writing Information Visualization Research Papers. In: Information Visualization: Human-Centered Issues and Perspectives, pp. 134–153. Springer, Heidelberg (2008)
17. Fekete, J., Wang, D.: Overlaying Graph Links on Treemaps. In: Information Visualization 2003 Symposium Poster Compendium. IEEE, Los Alamitos (2003)
18. Holten, D.: Hierarchical Edge Bundles: Visualization of Adjacency Relations in Hierarchical Data. IEEE Transactions on Visualization and Computer Graphics 12(5), 741–748 (2006)

An Efficient Method for the Visualization of Spectral Images Based on a Perception-Oriented Spectrum Segmentation

Steven Le Moan[1,2], Alamin Mansouri[1], Yvon Voisin[1], and Jon Y. Hardeberg[2]

[1] Le2i, Université de Bourgogne, Auxerre, France
[2] Colorlab, Høgskolen i Gjøvik, Norway

Abstract. We propose a new method for the visualization of spectral images. It involves a perception-based spectrum segmentation using an adaptable thresholding of the stretched CIE standard observer color-matching functions. This allows for an underlying removal of irrelevant channels, and, consequently, an alleviation of the computational burden of further processings. Principal Components Analysis is then used in each of the three segments to extract the Red, Green and Blue primaries for final visualization. A comparison framework using two different datasets shows the efficiency of the proposed method.

1 Introduction

Most of today's visualization devices are based on the paradigm that a combination of three primary colors (roughly red, green and blue) is sufficient for the human eye to characterize a color [1]. However, in many applications such as remote sensing, medical or art imaging, measuring the electromagnetic behavior of a scene has to be made with more spectral precision. Analogously to the need for a high spatial resolution for an enhanced separation of the different elements of a scene, a high spectral resolution allows for a better estimation of its reflectance, and thus, a better characterization of its color properties, regardless of the conditions of acquisition (illuminant, acquisition device). Spectral imaging consists of acquiring more than three spectral components from a scene, usually dozens, each one of them representing a reduced range of wavelengths, for better spectral precision (analogously to a pixel covering a small area of the space).

However, multispectral display devices are not yet of common use and the RGB strategy is still the most widespread. Therefore, when it comes to the task of visualizing a spectral image on a traditional computer screen, only three channels can be used, which implies a dimensionality reduction.

Tri-stimulus representation of multi and hyperspectral images for visualization is an active field of research which has been thoroughly investigated over the past decades. One of the most common approaches is probably the one referred to as *true color*. It is basically achieved by use of the CMF-based band transformation: each primary (R,G and B) is the result of a distinct linear combination of spectral channels in the visible range of wavelengths (400-700nm). Even though

G. Bebis et al. (Eds.): ISVC 2010, Part I, LNCS 6453, pp. 361–370, 2010.

it generally yields a natural visual rendering, this approach does not adapt itself to the data at all, therefore, noise and redundancy are not accurately handled.

Another very common approach for dimensionality reduction is Principal Components Analysis (PCA), which has been extensively used for visualization purposes. In [2], Durand and Kerr proposed an improved decorrelation method, aiming at maximizing both the Signal-to-Noise-Ratio (SNR) and the color contrast. Later, Tyo *et al.* [3], investigated PCA for N-to-3 dimensionality reduction towards the HSV colorspace. An automatic method to find the origin of the HSV cone is also proposed in order to enhance the final color representation. Tsagaris *et al.* [4] proposed to use the fact that the red, green and blue channels, as they are interpreted by the human eye, contain some correlation, which is in contradiction to the underlying decorrelation engendered by the PCA. For that reason, the authors proposed a constrained PCA-based technique in which the eigendecomposition of the correlation matrix is forced with non-zero elements in its non-diagonal elements. Scheunders [5] proposed to achieve a spatial decorrelation by first performing a simple image segmentation before using PCA and neural-network-based transformation in each region, distinctively. In [6], Du *et al.* compared seven feature extraction techniques in terms of class separability, including PCA, Independent Components Analysis (ICA) and Linear Discriminant Analysis (LDA).

However, due to the correlation matrix computation and manipulation, PCA is known to have a poor computational efficiency. For that reason, Jia *et al.* [7] proposed to segment the image's spectrum into contiguous subgroups of bands, in order to divide the complexity of PCA. Their method takes advantage of the block-structure of the correlation matrix of the spectral image to find the different subgroups. Zhu *et al.* [8] investigated spectral segmentation techniques (equal subgroups, correlation coefficients and RGB-based) together with ICA for visualization purposes. Segmented PCA is also investigated in [9], including equal subgroups, maximum energy and spectral-signature-based partitionings. However, none of these spectrum segmentation methods accurately handles the human perception of color. Moreover, we believe that gathering only contiguous bands and not allowing a band to belong to several segments is very restrictive and does not allow for an accurate exploitation of the perceptual correlation between the three primary colors.

In this paper, we propose an efficient visualization technique for the visualization of spectral images. It involves a fast perception-oriented spectrum segmentation, based on a thresholding of the CIE standard observer CMF. Three segments are formed and each one can then be used for the extraction of the Red, Green and Blue components. Segments are allowed to overlap and to be non-contiguous, so that a band can belong to several segments. Depending on the value of the thresholding parameter, a certain amount of bands is excluded, hence an underlying removal of irrelevant channels, and, consequently, an alleviation of the computational burden of further processings. In the remainder of this paper, the spectrum segmentation technique is first presented and the dimensionality reduction problem is briefly tackled. Then, the experimental framework is detailed and results are subjectively and objectively discussed before conclusion.

2   The Proposed Method

2.1   Spectrum Segmentation

Spectrum segmentation aims at regrouping spectral channels so that bands of a same group are considered similar in some way. This allows for an alleviation of the computational burden of the feature extraction, in a divide-and-conquer fashion. In the related works mentioned in the previous section, segmentation is performed based on different criteria that we do not believe to allow for a good handling of human perception of color. Therefore, we propose to measure similarity taking such property into account. At this aim, we propose to use the CIE 1964 Supplementary Standard Colorimetric Observer Color Matching Functions (CMF) [10] which are descriptors of the chromatic response of the human eye (see figure 1).

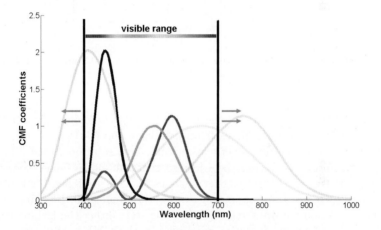

Fig. 1. The stretched CMF principle: In strong colors, the original functions. In light colors, the same ones stretched to fit a larger range of wavelengths.

The CMF are usually used to linearly combine spectral channels towards a tristimulus representation corresponding to how a human eye would see the scene [11]. In other words, each spectral channel $i \in [1..N]$ is associated with three weighting coefficients $x(i), y(i), z(i)$ roughly corresponding to its contributions to the perception of the red, green and blue. For clarity purposes, let us use the following notations $W_i^R = x(i), W_i^G = y(i), W_i^B = z(i)$.

We propose to interpret this statement as follows: the higher the weighting coefficient W_i^p, the higher the relevance for i to be a good representative of p. Consequently, we propose to cluster the CMF coefficients into two classes, by means of a binarizing threshold τ. Coefficients above τ depicts the relevant wavelengths for band selection. We note the ensemble of the corresponding channels Seg_p^τ, $p \in \{R, G, B\}$.

The spectrum segmentation is performed using normalized functions so that $max_i(W_i^P) = 1, \forall P$. Figure 2 illustrates the technique as well as the role of τ.

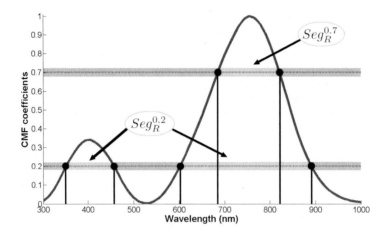

Fig. 2. The role of τ, example on the red function for $\tau = 0.2$ and $\tau = 0.7$. In both cases, the grey segments highlight the removal areas.

A problem appears when the spectral image contains channels outside the visible range of wavelengths (400-700nm). Indeed, the CMF are designed only for this part of the electromagnetic spectrum. As a solution to this, Jacobson et al. [12] proposed to stretch the CMF so that the entire image spectrum is covered, no matter what wavelengths it ranges in. This stretched CMF principle is illustrated by figure 1, for an image covering the range [300..1000] nanometers. In the case of a non-constant spectral sampling step, either the lacking channels must be replaced by interpolation methods, or the CMF coefficients must be adjusted. For computational ease, we recommend the latter solution.

Eventually, three segments are obtained, depending on the binarization threshold: Seg_R^τ, Seg_G^τ and Seg_B^τ in which the feature extractions for the red, green and blue channels will be performed, respectively.

Consequently, for a growing value of τ, the size of segments gets smaller and:

$$\tau_1 > \tau_2 \rightarrow Seg_p^{\tau_2} \in Seg_p^{\tau_1}, \forall p \in \{R, G, B\} \qquad (1)$$

According to its nature, τ allows for the moderation of the aforementioned hypothesis. If it is set to 0, the hypothesis is rejected and band selection is totally unconstrained. On the contrary, if $\tau = 1$, the hypothesis is considered perfectly relevant and there is no need to proceed with band selection since, in that case, the size of each segment is reduced to 1. As will be seen and discussed in the results section, τ also allows one to moderate the natural aspect of the representation, and hence can be manually adjusted, according to the user's need.

2.2 Dimensionality Reduction

Dimensionality reduction aims at finding a small set of bands which most accurately represents the whole spectral image. In this work, we used Principal Components Analysis (PCA), also known as Principal Components Transform (PCT) or Karhunen–Loève Transform (KLT). PCA is based on the eigendecomposition of the correlation matrix of the data, that is, in our case, the correlation matrix of a group of spectral channels. It is well-known the first PC contains generally more than 95% of the data energy. Therefore, we extract the first PC of each one of the three subgroups created by the previous step. Eventually, we obtain three highly-informative channels which are used for RGB visualization.

3 Experiments and Results

Since the spectrum segmentation part is the core of our method, we propose to use several different segmentation techniques coupled with PCA in order to focus on the partitioning aspect. The other techniques we used are the following:

- The "equal subgroups" partitioning investigated for instance by Zhu *et al.* [8] simply consists of creating three contiguous subgroups of equal size.
- The "correlation-based" segmented PCA proposed by Jia *et al.* [7] consists of taking advantage of the block structure of the correlation matrix of spectral images. It achieves partitioning by segmenting the correlation matrix into three contiguous subgroups.
- The "maximum energy" segmented PCA, proposed by Tsagaris *et al.* [9] consists of partitioning the spectrum in a way that maximizes the eigenvalue corresponding to the first PC in each contiguous subgroup.

Objective comparison of the results has been achieved by considering two aspects:

- The *natural rendering*, in order to measure how appealing it can be for human vision. The evaluation of this criterion is quite challenging since there is no consensus on what is a natural color or a natural contrast, even though there have been some attempts to define those [13]. We propose to use a pseudo *true color* representation of the image as a reference for natural rendering. Even if a *true color* image is generally obtained by neglecting the non-visible ranges of wavelengths, we used the stretched CMF to compute the latter representation, in order to make the comparison relevant. As for a comparison metric, we used the euclidean norm in the CIE L*a*b* colorspace, also known as the CIE76 ΔE_{ab}^{*}. Transformation to the L*a*b* colorspace was achieved by first converting reflectance data to XYZ vectors and then L*a*b* vectors by using a D65 standard illuminant for white point estimation. We insist on the fact that these representations are used only to assess the natural rendering, by supposing that they represent the best possible natural results. However, as stated in the introduction, using only the CMF does not allow for an accurate handling of the intrinsic properties of the image, since it is a data-independent transformation. This metric will be referred to as **NR** (Natural Rendering).

– The *perceptual class separability*, in order to measure the visual informative content of the representation, as suggested by Du *et al.* in [6]. At this aim, we manually selected 20 pixels by class, in each image. A class has been defined as a distinct visual feature (see description of the datasets). Then, each class centroid has been identified and projected in L*a*b*. The ΔE_{ab}^* distances between each couple of centroids have then been averaged. This metric will be referred to as **ICPD** (Inter-Class Perceptual Distance).

For our experiments, we used two different spectral images:

– 'Oslo' is a 160 bands remote sensing hyperspectral image, it represents a urban area in the neighborhood of Oslo (Norway). It was acquired with the HySpex VNIR-1600 sensor, developed by the Norsk Elektro Optikk (NEO) company in Norway. The sensor ranges from the early visible (400nm) to the near infrared (1000nm) with a spectral resolution of 3.7 nm. more information can be found on the constructor's website [14]. Based on a human expertise, we considered 5 classes in this image: vegetation, road, roof tops (two kinds) and cars.

– 'Flowers' is a 31 bands multispectral image from the database used in [15]. The sensor ranges only in the visible spectrum (400-720nm) with a 10nm spectral resolution. Three classes are present in this image: flower, leaves and background.

Figure 3 shows the pseudo *true color* representations of the images and results are presented on figures 4-7. Tables 1 and 2 gives the average perceptual distances with the pseudo *true color* references. Ranges of the colorspace are: $L* \in [0..100]$, $a* \in [-110.. + 110]$ and $b* \in [-110.. + 110]$.

We can see that the results from the presented technique are more naturally contrasted and thus allow for a more accurate interpretation by the human expert/enduser. As a result, the NR values are considerably better for the CMF-based segmentation than for the three other techniques. The worst natural renderings are obtained with the Maximum-ernergy-based segmented PCA. The evolution of the thresholding parameter does not yield any constant increase or decrease of the NR rate, however, we can see that the worst result is obtained for $\tau = 1$ on both images. This reveals that, by constraining too much the spectrum segmentation, the quality of the results lowers. Still, our results show that the CMF-based technique allows for the closest representations to the pseudo *true color* versions. Regarding the class-separability results, the maximum-energy-based segmented PCA gives the best results. The CMF-based method comes second and gives its best results for $\tau = 1$. However, considering the major improvement of NR brought by our approach, this latter gives an overall much

(a)   (b)

Fig. 3. Pseudo *true color* representations of the test images

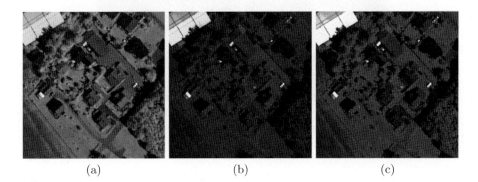

(a)   (b)   (c)

Fig. 4. RGB visualization of the 'Oslo' image according to different spectrum segmentation techniques: (a) Equal subgroups (b) Correlation-based (c) Maximum Energy

Table 1. Results for the 'Oslo' image

Segmentation technique	NR	ICPD
Equal subgroups	53.76	205.22
Correlation-based	90.87	226.72
Maximum energy	101.03	**243.21**
CMF-based, $\tau = 0.1$	**7.65**	222.01
CMF-based, $\tau = 0.5$	14.40	227.36
CMF-based, $\tau = 0.8$	11.96	224.55
CMF-based, $\tau = 1$	19.62	242.45

better tradeoff between both metrics. Subjectively, if we look at the flower in figure 6c, even if the perceptual class-separability is obviously high, the one in 7d is much less disturbing and allows for a better quicker understanding of the endmembers of the scene.

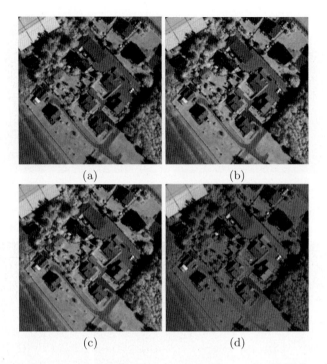

Fig. 5. RGB visualization of the 'Oslo' image according to the CMF-based segmented PCA with $\tau = 0.1, 0.5, 0.8$ and 1, respectively

Fig. 6. RGB visualization of the 'Flowers' image according to different spectrum segmentation techniques: (a) Equal subgroups (b) Correlation-based (c) Maximum Energy

The main drawback coming with an increase of the thresholding parameter is the loss of spectral channels, and thus the loss of information, which is done independently to the data itself. For that reason, one can see τ as a tradeoff parameter, allowing for the balance between natural constancy and visual informative content.

Fig. 7. RGB visualization of the 'Flowers' image according to the CMF-based segmented PCA with $\tau = 0.1$, 0.5, 0.8 and 1, respectively

Table 2. Results for the 'Flowers' image

Segmentation technique	NR	ICPD
Equal subgroups	12.37	101.34
Correlation-based	13.34	109.71
Maximum energy	22.50	**117.42**
CMF-based, $\tau = 0.1$	4.24	94.48
CMF-based, $\tau = 0.5$	**1.73**	96.33
CMF-based, $\tau = 0.8$	3.81	100.73
CMF-based, $\tau = 1$	5.70	112.30

4   Conclusion

A new method for the visualization of spectral images involving a spectrum segmentation technique based on a thresholding of color-matching functions has been introduced. Contrary to other spectral segmentation techniques, it allows for an underlying removal of irrelevant bands for visualization and thus allows for computational burden alleviation. The thresholding parameter allows for balancing the following hypothesis: the higher the color matching coefficient, the higher the relevance for the corresponding band to be a good representative of the corresponding primary. It also allows for balancing between natural constancy of the representation and the amount of bands to be removed. The results obtained by the presented technique are more naturally contrasted and thus make interpretation easier and quicker. Moreover, they contain a high class separability which reveals the presence of important visual information.

Acknowledgements

The authors would like to thank the Regional Council of Burgundy for supporting this work as well as the Norsk Elektronik Optikk company in Oslo for providing useful data.

References

1. Grassmann, H.: On the theory of compound colors. Phil. Mag. 7, 254–264 (1854)
2. Durand, J., Kerr, Y.: An improved decorrelation method for the efficient display ofmultispectral data. IEEE Trans. on Geoscience and Remote Sensing 27, 611–619 (1989)
3. Tyo, J., Konsolakis, A., Diersen, D., Olsen, R.: Principal-components-based display strategy for spectral imagery. IEEE Trans. on Geoscience and Remote Sensing 41, 708–718 (2003)
4. Tsagaris, V., Anastassopoulos, V.: Multispectral image fusion for improved rgb representation based on perceptual attributes. International Journal of Remote Sensing 26, 3241–3254 (2005)
5. Scheunders, P.: Multispectral image fusion using local mapping techniques. In: International Conference on Pattern Recognition, vol. 15, pp. 311–314 (2000)
6. Du, Q., Raksuntorn, N., Cai, S., Moorhead, R.: Color display for hyperspectral imagery. IEEE Trans. on Geoscience and Remote Sensing 46, 1858–1866 (2008)
7. Jia, X., Richards, J.: Segmented principal components transformation for efficient hyperspectral remote-sensing image display and classification. IEEE Trans. on Geoscience and Remote Sensing 37, 538–542 (1999)
8. Zhu, Y., Varshney, P., Chen, H.: Evaluation of ica based fusion of hyperspectral images for color display. In: 2007 10th International Conference on Information Fusion, pp. 1–7 (2007)
9. Tsagaris, V., Anastassopoulos, V., Lampropoulos, G.: Fusion of hyperspectral data using segmented pct for color representation and classification. IEEE Trans. on Geoscience and Remote Sensing 43, 2365–2375 (2005)
10. http://www.cie.co.at/main/freepubs.html (last check : September 11, 2010)
11. Trussell, H.: Color and multispectral image representation and display. In: Handbook of Image and Video Processing, p. 411 (2005)
12. Jacobson, N., Gupta, M.: Design goals and solutions for display of hyperspectral images. IEEE Trans. on Geoscience and Remote Sensing 43, 2684–2692 (2005)
13. Ruderman, D.: The statistics of natural images. Network: Computation in Neural Systems 5, 517–548 (1994)
14. http://www.neo.no/hyspex/ (last check : September 11, 2010)
15. Nascimento, S., Ferreira, F., Foster, D.: Statistics of spatial cone-excitation ratios in natural scenes. Journal of the Optical Society of America A 19, 1484–1490 (2002)

A New Marching Cubes Algorithm for Interactive Level Set with Application to MR Image Segmentation

David Feltell and Li Bai

School of Computer Science, University of Nottingham, UK
{dzf,bai}@cs.nott.ac.uk

Abstract. In this paper we extend the classical marching cubes algorithm in computer graphics for isosurface polygonisation to make use of new developments in the sparse field level set method, which allows localised updates to the implicit level set surface. This is then applied to an example medical image analysis and visualisation problem, using user-guided intelligent agent swarms to correct holes in the surface of a brain cortex, where level set segmentation has failed to reconstruct the local surface geometry correctly from a magnetic resonance image. The segmentation system is real-time and fully interactive.

1 Background

The level set method, introduced in [1], has shown great promise for medical image analysis [2,3,4]. However, threshold based techniques suffer when classes (e.g. tissue types) have inhomogeneous voxel intensities across the input image. In some cases the areas showing error are immediately apparent in a 3D rendering of the final segmentation. It would then be advantageous to allow for manual identification followed by autonomous correction of these areas in an efficient and fully interactive manner.

1.1 The Level Set Method

We can define an *implicit* surface as an isosurface embedded in a scalar field, $\{\mathbf{p} \mid \phi_{(\mathbf{p})} = 0\}$ - the *zero-level set* of ϕ. We further assume that $\phi < 0$ indicates inside and $\phi \geq 0$ indicates outside a volume and that the surface is advected in the normal direction. Then we arrive at the *level set equation*:

$$\frac{\partial \phi_{(\mathbf{p})}}{\partial t} = |\nabla \phi_{(\mathbf{p})}| F(\mathbf{p}) \tag{1}$$

where F is an arbitrary function controlling the speed of surface points.

Updating the values of ϕ at each point in space requires a numerical technique to evolve the zero-level set from an initial specification. In the naive case, the whole of space must be evaluated at each iteration, giving (in three dimensions)

G. Bebis et al. (Eds.): ISVC 2010, Part I, LNCS 6453, pp. 371–380, 2010.

$O(n^3)$ complexity, rather than just the area of interest - the $O(n^2)$ surface. Algorithms have been developed to overcome these issues by only performing updates on regions near the surface, rather than integrating over the entire space. The most well-known is the *narrow band* approach [5,6]. This method has since seen several developments that trade accuracy for efficiency by reducing the width of the band, along with algorithmic changes. In-particular, the *sparse field* method [7] has been widely adopted, especially for interactive-rate applications [8,9] and is used in this work.

In order to allow local and interactive modification of the level set surface, the sparse field algorithm needs to be modified. A modified sparse field algorithm that allows local modification of the level set surface is described in [10]. The level set method works particularly well with the marching cubes algorithm, as the signed distance scalar field provides the required inside/outside corner statuses. That is, each lattice node stores its distance to the object's surface, with nodes inside the object having negative (or positive, if so defined) sign. By looking for adjoining nodes with opposite sign we can find those edges that the object's surface bisects. With this information we can construct a triangle mesh representative of the isosurface, as detailed below.

2   An Extended Marching Cubes Algorithm

In order to visualise the local modification of the level set surface, we also require a method of rendering the surface in 3D as it is being updated. To reconstruct a surface from volumetric data, the original marching cubes method divides the data volume into grid cells and polygonises each cell individually. Each cell is defined as 8 corners that together form a cube, with the specific triangle configuration determined by which edges of the cell are bisected by the surface. That is, a continuous surface is approximated by finding the bisection points along each edge of each cell that the surface cuts through. To determine how the bisection points should be connected to form a triangle mesh, the marching cubes method makes use of two look-up tables. The first look-up table takes a bitwise encoded index of corner inside/outside status (8 bits) and returns a bitwise encoded list of edge vertices (12 bits), with each 'on' bit corresponding to an edge vertex that must be calculated. The second look-up table details how those vertices should be linked together to form the final triangle mesh polygonisation of the cell.

As with the level set method, the naive marching cubes implementation has $O(n^3)$ time complexity. By exploiting a narrow band data structure the complexity is reduced to $O(n^2)$. If we are only modifying a few localised areas on the surface, whilst the majority of the surface remains static, we would like to forgo the necessity of cycling through the entire band just to update these regions. Therefore, we have further modified the method to reduce the complexity to $O(1)$ for a single small region, or $O(n)$ for n such regions. We have integrated this new marching cubes algorithm with the localised sparse field algorithm in

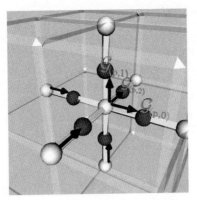

Fig. 1. Illustration of the vertex index look-up grid, $\mathcal{G}$. Each grid point $\mathbf{p}$ stores three indices $\{\mathcal{G}_{(\mathbf{p},0)}, \mathcal{G}_{(\mathbf{p},1)}, \mathcal{G}_{(\mathbf{p},2)}\}$, one for each cardinal direction. Each index acts as a reference to a vertex in the $\mathcal{V}$ vertex array.

[10], allowing for real-time visualisation of deformations by agent swarms, even for large surfaces.

2.1   Selecting Modified Mesh Fragments

We create a lattice of mesh fragments, $\mathcal{M}$, each element of which contains a polygonisation covering a set of ϕ lattice nodes. In other words, we break up the complete mesh into smaller fragments and only repolygonise those fragments that are responsible for one or more of the cells that have been modified (e.g. by agents deforming the surface), rather than repolygonising the whole surface/space.

To find and repolygonise invalidated mesh fragments, we maintain the mesh fragment lattice as well as a matching lattice of boolean flags $\mathcal{B}$ in memory, which are initially all $false$. As input we require a set of ϕ points that have been invalidated (i.e. changed in value), $\mathcal{X} \subset \mathbb{Z}^3$. We start by defining an empty array of fragment locations $\mathcal{U} \subset \mathbb{Z}^3$. We then cycle through all the points in $\mathcal{X}$ to find which fragments have been invalidated and append them to $\mathcal{U}$. Since each fragment covers multiple ϕ lattice points, the $\mathcal{B}$ lattice is required to ensure no duplicate fragment locations are added to $\mathcal{U}$. Now we have the set of invalid mesh fragments $\mathcal{U}$, we reconstruct all marked fragments, $\mathcal{M}_{(\mathbf{p} \in \mathcal{U})}$, and reset corresponding boolean flags, $\mathcal{B}_{(\mathbf{p} \in \mathcal{U})} = false$. Reconstruction of a fragment proceeds by polygonising each cell that it covers, one at a time.

Each fragment stores an array of vertices $\mathcal{V} \subset \mathbb{R}^3 \times \mathbb{R}^3$ (position and normal), which is shared amongst all cells of the fragment, allowing vertices to be reused rather than recalculated, where possible. In addition, each fragment stores a triangle array, $\mathcal{T} \subset \mathbb{Z}^3$ - a list of 3-tuple indices referencing vertices in $\mathcal{V}$. The procedure for polygonising each cell is detailed in the following subsection.

2.2   Polygonising a Cell

The first step in polygonising a cell is to calculate the index i into the look-up tables: ξ, the edge table; and Λ, the triangle table (available, for example, from [11,10]). The next step is to compute the vertices (bisection points) lying along the edges of the cell. This results in a 12 -element array, $v \subset \mathbb{Z}$ - one element for each edge of the cell being polygonised, with each element storing an index into the vertex array $\mathcal{V}$, or a $NULL$ value if a vertex is not required along that edge. The vertices are then connected to form the triangle list. Each of these steps are detailed below.

Calculating Indices into the Look-Up Tables: each corner of the cell is tested in a specific order and the index - an 8-bit value i - is modified in a bitwise fashion depending on the result. We first test each corner in turn, setting the respective bit in i to 1 if that corner is inside the volume, and 0 otherwise. When i is resolved from a bit string into an integer we get our index into the look-up tables. The 12-bit value stored in the edge look-up table at $\xi_{(i)}$ can then be used to rapidly determine which edge vertices are required. First, we must define the 12 element array v for this cell. If a bit in $\xi_{(i)}$ corresponding to a given edge is switched on, then the proper polygonisation of this cell requires a vertex that is positioned along that edge.

Fetching/Calculating an Edge Vertex: by defining a cell edge as a start point, $\mathbf{p} \in \mathbb{Z}^3$, and a direction, $d \in \{0,1,2\}$, we can define a lattice $\mathcal{G}$ that stores an index into $\mathcal{V}$ for each positive cardinal direction, as shown in Fig. 1. If the vertex has already been created and added to $\mathcal{V}$, then its index will be stored in $\mathcal{G}$ and we can simply return the value at $\mathcal{G}_{(\mathbf{p},d)}$. Otherwise we must calculate the vertex, append it to the vertex array $\mathcal{V}$, and store its index in $\mathcal{G}$ for future queries. The use of $\mathcal{G}$ allows us to reuse vertices rather than duplicate them for every triangle that shares this vertex in this and neighbouring cells. By using $\mathcal{G}$ we can ensure in $O(1)$ time that no unnecessary recalculations are performed and allow the use of index buffers in the graphics device memory, reducing the (typically large) memory footprint.

The calculation of a new vertex $\mathbf{v}$, such that $\mathbf{v}_{pos}$ gives the vertex position and $\mathbf{v}_{norm}$ gives the vertex normal, is performed by linearly interpolating the values of ϕ at the two endpoints of the edge, such that $\phi_{(\mathbf{v}_{pos})} = 0$. Similarly we can calculate the vertex normal using $\nabla\phi$ at each endpoint. Once $\mathbf{v}$ is calculated we append it to the end of the vertex array $\mathcal{V}$ and update the look-up lattice at $\mathcal{G}_{(\mathbf{p},d)}$ with the index of this vertex (which is simply the size of the array $\mathcal{V}$ minus 1).

Triangulating a Cell: each triangle is defined as a 3-tuple, $\tau \in \mathbb{Z}^3$, of indices referencing vertices stored in $\mathcal{V}$. Triangles are defined using the second look-up table, Λ, where $\Lambda_{(i)}$ gives the triangle list for this particular cell configuration and each element j of $\Lambda_{(i)}$, $\Lambda_{(i,j)} \in \{0,1,\ldots,10,11\}$, gives an index into v.

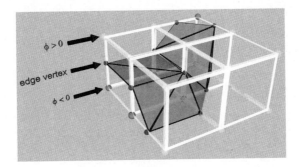

Fig. 2. A partial polygonisation of an arbitrary isosurface. A sample set of the triangle mesh configurations given in the look-up table Λ are combined to show how they fit together to create a coherent triangle mesh.

Let $(a, b, c) = (\Lambda_{(i,j)}, \Lambda_{(i,j+1)}, \Lambda_{(i,j+2)})$. The next triangle required to polygonise this cell is constructed from the indices stored in the array v, such that $\tau = (v_{(a)}, v_{(b)}, v_{(c)})$. τ is then appended to the mesh fragment's triangle array, $\mathcal{T}$. We then increment, $j \leftarrow j + 3$, to the next triangle and repeat until all triangles for this cell are fetched, as flagged by some $NULL$ value at $\Lambda_{(i,j)}$. Figure 2 shows how several triangle mesh configurations for individual cells join to create a representative triangle mesh hull of the entire isosurface.

Once all these steps are complete, $\mathcal{U}$ determines which fragments need updating in the graphics hardware, using the fragment's arrays $\mathcal{V}$ and $\mathcal{T}$ directly as vertex and index buffers, respectively.

Figure 3 shows the results of a simple experiment contracting a cube under mean curvature flow. The cumulative time taken to search for and repolygonise affected mesh fragments is plotted over 1500 time steps for different sized fragments. The largest 100x100x100 fragment covers the entire 100x100x100 lattice, so is the equivalent of performing the marching cubes algorithm on the entire lattice, creating a single mesh fragment as in the naive marching cubes method. It is clear that the smallest fragment sizes perform significantly better, tending to zero seconds per update as the zero-level set becomes smaller.

A mesh fragment size of 50x50x50 (giving eight fragments) actually performs slightly worse than simply polygonising the entire ϕ lattice as a single mesh. The target shape is centred in the lattice, so the extents reach into all eight regions and as a result all eight fragments must be repolygonised every frame. This shows an associated overhead involved in breaking the mesh into fragments, mostly as a result of repeated effort at the border between fragments where vertices are not shared and must be recalculated.

3   Level Set Image Segmentation

For our example we are using user-directed agent swarms to correct a level set segmentation of a MR brain scan. This initial segmentation result is obtained

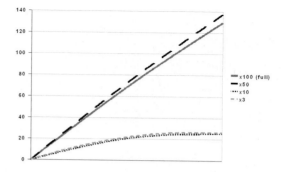

Fig. 3. Cumulative time taken in seconds, over 1500 time steps, to find and repoly-gonise affected mesh fragments during contraction under mean curvature flow. The experiment takes place in a 100x100x100-node lattice with a 50x50x50 box centred within the lattice, which then undergoes mean curvature flow. Each curve represents mesh fragments of varying size, from 100x100x100 (labelled x100) - the entire lattice as in the naive marching cubes method, down to 3x3x3 (labelled x3) - the smallest workable size.

using the standard sparse-field level set method [7]. We define a simple threshold based speed function for use in the level set equation:

$$F(\mathbf{p}) = \frac{1}{2} \left((1 - \alpha) D(\mathbf{p}) + \alpha \, \kappa(\mathbf{p}) \right) \tag{2}$$

Where: $\kappa = \nabla \bullet \frac{\nabla \phi}{|\nabla \phi|}$ is the mean curvature, implemented using the difference of normals method (covered very nicely in [15]); and:

$$D(\mathbf{p}) = 2 \left(\frac{|\mu - \mathcal{I}_{(\mathbf{p})}|^n}{\epsilon^n + |\mu - \mathcal{I}_{(\mathbf{p})}|^n} \right) - 1 \tag{3}$$

Where: $\mathcal{I}$ is the input image, assumed to be greyscale, such that $\forall \mathbf{p}, \mathcal{I}_{(\mathbf{p})} \in [0, 1]$; μ is the ideal value; ϵ is the acceptable error; and n controls the steepness of the distribution. This function gives a value in $[-1, 1)$, negative when the data point is within the threshold and positive otherwise. With our choice of signed distance representation (positive outside and negative inside the volume) a negative value for D indicates an outward expansion.

To summarise, we have a global ideal voxel intensity and a threshold. Within that threshold the surface expands, although mediated by a small curvature to smooth over noise and prevent leaks. Similarly, outside of the threshold the surface contracts, similarly mediated by the curvature term. The result of this global ideal-value-and-threshold approach gives a nice approximation, when the various free parameters are chosen well, and is relatively fast computationally. However, even after tuning the parameters until they produce the best possible segmentation, it is apparent that this approach is weak when confronted with intra-class inhomogeneity across voxel intensities. In this case the surface prematurely stops expanding in some areas. When viewed in 3D we clearly see 'holes'

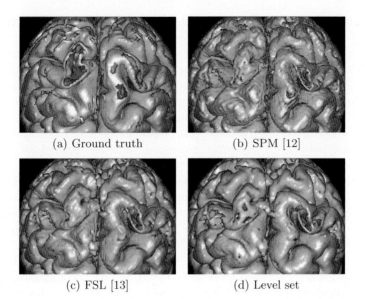

(a) Ground truth   (b) SPM [12]

(c) FSL [13]   (d) Level set

Fig. 4. A 3D rendering of the segmentation of grey matter from a BrainWeb [14] simulated MR image. Images show results using different software packages, as well as the hand-segmented ground truth. All clearly show holes in the superior end of the segmentation, including the ground truth.

in the surface. The 3D view shows these holes much more clearly than a 2D slice-by-slice comparison.

Figure 4 shows the superior end of the final segmentation result from three segmentation methods, the SPM [12] and FSL [13] software packages, our level set implementation, plus the ground truth. The ground truth, FSL and SPM volumes are stored as fuzzy objects. For these volumes, the surface is extracted as the 0.5-curve, providing the best balance between under and over classification, whereas the level set surface is extracted as the zero-curve (zero level set). We can see in all cases the segmentation suffers from holes in the final surface extraction, with the ground truth actually being the worst offender.

We would like to enable users to interactively guide the correction of these holes in real-time. Using mechanisms found in previous work on situating virtual swarms in a deformable environment [16,17,10,18], we embed multiple agents on the surface, whose movements are weighted toward areas that the user identifies with a click of the mouse. The surface in that area is corrected by the swarm, with the results of the swarm's modifications immediately visible to the user in real-time, thanks to our extensions to the marching cubes algorithm. The following section details the methods we used to embed the agents and have them interactively modify the surface about a user-selected location.

4   Populating the Zero-Level Set with Agent Swarms

Agents can be situated on the zero-level surface with an $\mathbb{R}^3$ position and velocity, moving across and modifying the surface using interpolation methods from/to the $\mathbb{Z}^3$-located ϕ lattice nodes. This is quite straightforward since the level set representation allows us to locate the surface in space and calculate the surface normal at any point very easily. Fetching the ϕ value at a given $\mathbb{R}^3$ location can use standard trilinear interpolation. This easily extends to calculating interpolated surface normals or any other numerical value as required for movement and sensory information.

Modifying the surface (that is, the zero-layer ϕ nodes) at a given $\mathbb{R}^3$ location is less straightforward. To do this, we update the values at all zero-layer points within a given range of the agent's position $\mathbf{x}$ by a value $A(\mathbf{x})$, weighted by distance from the agent using a normalised 3D Gaussian distribution. In other words, $A(\mathbf{x})$ gives the amount to raise/lower the surface by.

The decision each iteration by each agent on whether to raise/lower the surface uses population-level sampled averages of voxel intensity and standard deviation, so that decisions on surface modifications can be made on a local basis and adapt as the user directs the agents around the surface. For specifics on the A function used in this example see [18]. The ϕ locations modified by the agents are stored each frame and passed as the $\mathcal{X}$ parameter to our marching cubes algorithm as detailed in Sect. 2.1.

For movement across the surface, a *movement potential* function $P(\mathbf{x})$ is specified that represents the 'forces' pulling agents in certain directions. The function P does not in general produce a valid movement direction vector along the surface, so it must be resolved to lie along, that is, tangential to, the zero-level isosurface. Also, the agents will tend to 'drift' from the zero-curve over time and could overshoot the narrow band and end up in undefined territory, so a term must be added to compensate for this:

$$\mathbf{x} \longleftarrow \mathbf{x} + \left\| P - \nabla|\phi| \left(\nabla|\phi| \bullet P \right) \right\| - \varepsilon \nabla|\phi| \tag{4}$$

where $0 < \varepsilon \ll 1$ controls the strength of the pull back onto the zero-curve and $\nabla|\phi| = \left(\frac{\partial|\phi|}{\partial x} i + \frac{\partial|\phi|}{\partial y} j + \frac{\partial|\phi|}{\partial z} k \right)^T$.

Again, for implementation details of the P function used in this example see [18].

5   Results and Discussion

Figure 5 shows the result of running the semi-autonomous swarm surface correction algorithm on a simulated MR image taken from BrainWeb [14]. A grey matter level set segmentation created using a sub-optimally low threshold was populated by 200 agents. Lowering the threshold results in more and larger holes in the segmentation than the optimal choice (which still exhibits holes, but less obviously so). The user need only click the mouse whilst pointing in the general

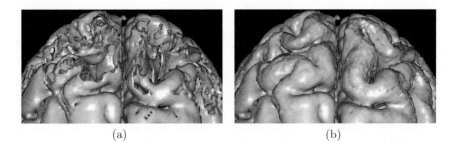

(a)   (b)

Fig. 5. Image showing level set surface with visible holes at the superior end of a grey matter segmentation. These holes are repaired by 200 agents directed by a non-expert user. (a) shows the initial surface, with multiple visible holes; (b) shows the final surface with these holes corrected.

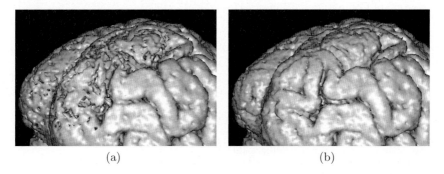

(a)   (b)

Fig. 6. A level set segmentation and swarm based correction of grey matter from a real T1 MRI scan, as opposed to a BrainWeb simulation: (a) shows a portion of the initial level set surface; and (b) shows the swarm corrected surface

area of interest and the agents very rapidly 'fill' nearby holes. To show how this methodology is not restricted to idealised BrainWeb images, a further illustration is shown in Fig. 6, where the process of level set segmentation and swarm correction is applied to a real T1 MR image of a brain.

The test platform is an Intel Xeon 3GHz CPU, with 4GB RAM and a NVIDIA Quadro FX 5800 GPU. As a straightforward marching cubes mesh visualisation tool, the system achieves between 14000 and 15000 FPS when rendering a grey matter segmentation from a 181x217x181 MR image. Once 200 agents are added and moving without modifying the surface, the system runs at between 1800 and 1900 FPS. Whilst repairing a surface this figure is reduced to between 100 and 300 FPS. More specifically, when agents are coded to raise or lower the surface randomly with a 0.1 probability, that is, on average each agent raises or lowers the surface every 10 time steps, the result runs at 110 FPS. When this probability is raised to 0.5 the result is 55 FPS. The non-linear correlation is due to agents modifying overlapping regions, meaning less level set surface points and thus associated marching cubes mesh fragments need be processed.

References

1. Osher, S., Sethian, J.A.: Fronts propagating with curvature-dependent speed: algorithms based on Hamilton-Jacobi formulations. Journal of Computational Physics 79(1), 12–49 (1988)
2. Zhukov, L., Museth, K., Breen, D., Whitaker, R., Barr, A.: Level set modeling and segmentation of DT-MRI brain data. Journal of Electronic Imaging 12(1), 125–133 (2003)
3. Cates, J.E., Lefohn, A.E., Whitaker, R.T.: GIST: an interactive, GPU based level set segmentation tool for 3D medical images. Medical Image Analysis 8(3), 217–231 (2004)
4. Shi, Y., Karl, W.: A fast level set method without solving PDEs. In: IEEE International Conference on Acoustics, Speech, and Signal Processing (ICASSP), vol. 2, pp. 97–100 (2005)
5. Adalsteinsson, D., Sethian, J.: A fast level set method for propagating interfaces. Journal of Computational Physics 118, 269–277 (1995)
6. Sethian, J.: Level Set Methods and Fast Marching Methods. Cambridge University Press, Cambridge (1999)
7. Whitaker, R.T.: A level-set approach to 3D reconstruction from range data. Computer Vision 29(3), 203–231 (1998)
8. Museth, K., Breen, D.E., Whitaker, R.T., Mauch, S., Johnson, D.: Algorithms for interactive editing of level set models. Computer Graphics Forum 24(4), 821–841 (2005)
9. Lefohn, A.E., Cates, J.E., Whitaker, R.T.: Interactive, GPU-based level sets for 3D segmentation. In: Ellis, R.E., Peters, T.M. (eds.) MICCAI 2003. LNCS, vol. 2878, pp. 564–572. Springer, Heidelberg (2003)
10. Feltell, D.: Self-organised virtual swarms in a continuous deformable environment. PhD thesis, The University of Nottingham, Nottingham, UK (2010)
11. Bourke, P.: Polygonising a scalar field (May 2009), http://local.wasp.uwa.edu.au/~pbourke/geometry/polygonise/
12. Friston, K., Ashburner, J., Kiebel, S.J., Nichols, T., Penny, W.: Statistical Parametric Mapping: The Analysis of Functional Brain Images. Academic Press, London (2007)
13. Zhang, Y., Brady, M., Smith, S.: Segmentation of brain MR images through a hidden Markov random field model and the expectation maximization algorithm. IEEE Transactions on Medical Imaging 20(1), 45–57 (2001)
14. McConnell BIC: BrainWeb: Simulated brain database, Montreal Neurological Institute (2010), http://www.bic.mni.mcgill.ca/brainweb/
15. Lefohn, A.E., Kniss, J.M., Hansen, C.D., Whitaker, R.T.: A streaming narrow-band algorithm: Interactive computation and visualization of level sets. IEEE Transactions on Visualization and Computer Graphics 10, 422–433 (2004)
16. Feltell, D., Bai, L., Jensen, H.J.: An individual approach to modelling emergent structure in termite swarm systems. International Journal of Modelling, Identification and Control 3(1), 29–40 (2008)
17. Feltell, D., Bai, L., Soar, R.: Level set brain segmentation with agent clustering for initialisation. In: International Conference on Bio-inspired Systems and Signal Processing (BIOSIGNALS), Funchal, Madeira, Portugal (2008)
18. Feltell, D., Bai, L.: 3D level set image segmentation refined by intelligent agent swarms. In: IEEE World Congress on Computational Intelligence, Barcelona, Spain (2010)

Attention-Based Target Localization Using Multiple Instance Learning

Karthik Sankaranarayanan and James W. Davis

Dept. of Computer Science and Engineering
Ohio State University
Columbus, OH, USA
{sankaran,jwdavis}@cse.ohio-state.edu

Abstract. We propose a novel Multiple Instance Learning (MIL) framework to perform target localization from image sequences. The proposed approach consists of a softmax logistic regression MIL algorithm using log covariance features to automatically learn the model of a target that persists across input frames. The approach makes no assumptions about the target's motion model and can be used to learn models for multiple targets present in the scene. The learned target models can also be updated in an online manner. We demonstrate the validity and usefulness of the proposed approach to localize targets in various scenes using commercial-grade surveillance cameras. We also demonstrate its applicability to bootstrap conventional tracking systems and show that automatic initialization using our technique helps to achieve superior performance.

1 Introduction

Object tracking is one of the most studied problems in the field of computer vision. Tracking of pedestrians is an end application in itself, or can be used for collecting trajectories for higher level behavior analysis. Traditional techniques for following particular targets in surveillance settings which use *active* PTZ cameras involve a security operator driving a joystick (to control the PTZ camera) and adjusting it to follow the target of interest. For handing off the task to an automatic tracker, the operator would have to stop and select the target (e.g., with a bounding box) for the tracker to initialize. While on the one hand, learning the target model from a single frame is not robust, requiring the operator to click on multiple successive frames is also very impractical. Moreover, due to the fast and real-time nature of these tasks, such initializations are both unreliable and require significant training. They also require the ability to control multiple input devices for driving the camera (joystick) and clicking on the video feed (mouse). In this situation, an automatic technique to learn target models that is both robust for tracking and intuitive to operate is needed.

In a typical surveillance setting, an intuitive way for an operator to initialize a tracker would be to use the joystick to loosely follow the person of interest and then have the system *automatically* learn the intended target, and then continue

G. Bebis et al. (Eds.): ISVC 2010, Part I, LNCS 6453, pp. 381–392, 2010.

tracking the target. Since the system now has a robust model that has been automatically learned using the target's persistent appearance features, it can use this model to reliably track the target in future frames. Towards this end, we propose an attention-based technique to learn target models using a machine learning approach. (Note that the attention nature of this method allows the target to be anywhere in the scene/image and therefore does not assume any particular motion model on the part of the target). A limitation of supervised learning is that it requires labeling of the individual instances, which are typically hard to obtain (this is also the case in our setting, as explained later). Multiple instance learning (MIL) is a variant of supervised learning where it relaxes the granularity at which the labels are available. Therefore, in MIL settings, the instances are grouped into "bags" which may contain any number of instances and the labeling is done at the bag level. A bag is labeled positive if it contains at least one positive instance in it. On the other hand, a bag is labeled negative if it contains all negative instances. Note that positive bags may contain negative instances.

In our problem formulation, we model images containing the target of interest as bags, and regions within the images as instances. The MIL framework is therefore well suited to this task because it is guaranteed that at least one instance in the bag/image contains the object of interest, and it is much more efficient (faster and cheaper) to label bags/images instead of individual image regions. Therefore, we take image patches from areas of motion in the image (when the camera is not moving) and create a positive bag using all of these instances. At the same time, we sample image patches from the non-moving areas and create a negative bag from these instances. (Another option is to collect all patches from the input image and build a positive bag, and use patches from a background image of the same/similar region to construct a negative bag.) By repeating this for every frame, we are guaranteed that every positive bag contains at least one patch containing the target of interest and at the same time ensures that the target is absent from all the negative bags. By training a MIL algorithm based on logistic regression, we learn a target model that can be used to classify every instance (image patch) from a new incoming frame as target or not with a certain probability. We then use this probability map over the image and threshold it to update the model for the target.

2   Related Work

While there has been much work in the areas of pedestrian detection and visual tracking separately, not much work has been done in automatic localization from the point of view of initialization of a tracker based on visual attention. Pedestrian detection approaches such as [1] are generally view specific and are therefore unsuitable for our domain since PTZ cameras overlooking a large area can have a wide range of pedestrian views (from fronto-parallel to top-down). Moreover, such approaches are not applicable for finding the most *persistent* pedestrian in the scene (the target to be tracked), which is what is required to initialize an object tracker in our operator-joystick setting. In the area of object tracking, popular approaches include appearance-based techniques such as Mean-shift [2] and Covariance tracking [3], and filtering and association-based approaches such as a Kalman filter [4] and particle filter [5]. All of these approaches require good target localization and initialization for them to work well, and assume that a good initial target model is provided.

In the area of Multiple Instance Learning, the original work of [6] proposed to learn axis-parallel rectangles for modeling target concepts. Since then there have been various algorithms proposed including Diverse Density (DD) [7], EM-DD [8], and SVM techniques [9]. More recently, in a comparison study of MIL algorithms, Multiple instance logistic regression [10] has been shown to be the state-of-the-art MIL algorithm, especially for image retrieval tasks. Our contribution is an adaptation of such a logistic regression based MIL algorithm based on the *softmax* function and a new application problem.

3   Multiple Instance Learning

In order to learn the target model from a sequence of images and localize the target of interest within a new image, we wish to build a discriminative classifier which can output the probability $p(y = 1|x)$ indicating the posterior probability that the target is present $(y = 1)$ in the image patch x. In a MIL framework, the input data is obtained in the form of positive bags (B^+) and negative bags (B^-) containing instances. More formally, the input is presented as $\{(X_1, y_1), (X_2, y_2), ..., (X_n, y_n)\}$ where $X_i = \{x_{i1}, x_{i2}, ..., x_{im}\}$ denotes bag i containing m instances and has a corresponding bag label $y_i \in \{0, 1\}$. Each instance x_{ij} is a feature vector calculated for an image patch j from bag i. The bag labels are obtained from the instances in the bags. More specifically, a bag is labeled positive if it contains *at least one* positive instance. A bag is labeled negative if it contains *all* negative instances.

Using a likelihood formulation, the correct bag classifier/labeler will maximize the log likelihood of labels over all the bags (given the MIL constraints)

$$\log \mathcal{L} = \sum_i^n \log p(y_i|X_i) \tag{1}$$

where $p(y_i|X_i)$ is the probability of the bag i (given its instances) having label y_i. As we can see, since the above likelihood formulation is expressed in terms of bag probabilities and what we want is to learn an instance-level classifier (for an instance/patch x), we will use a combining function to assemble instance-level probabilities into a bag probability. This is done using the *softmax* combining function as follows.

From the definition of positive and negative bags, we can formally express the notion of bag label in terms of its instance labels as

$$y_i = \max_j(y_{ij}) \tag{2}$$

which states that the label of a bag is the label of the instance within it which has the highest label (i.e.,$\{0, 1\}$). Notice how this formulation conforms to the definition of positive and negative bags and encodes the multiple instance assumption. Here, we incorporate a probabilistic approximation of the max operator called *softmax*, in order to combine these instance probabilities in a smoother way, so as to allow all instances to contribute to the bag label. This *softmax* function is defined as: $\text{softmax}(a_1, ..., a_m) = \sum_{j=1}^{m} (a_j \exp(\alpha a_j)) / \sum_{j=1}^{m} \exp(\alpha a_j)$. α is a constant that controls the weighting within the *softmax* function such that *softmax* calculates mean when $\alpha=0$ and max when $\alpha \to \infty$.

The bag-level probabilities for positive and negative bags are now defined as

$$p(y_i = 1|X_i) = \text{softmax}(t_{i1}, ..., t_{im}), \quad p(y_i = 0|X_i) = 1 - p(y_i = 1|X_i) \quad (3)$$

where $t_{ij} = p(y_{ij} = 1|x_{ij})$ are the instance level probabilities being combined to obtain the bag probabilities $p(y_i|X_i)$. Thus, if one of the instances is very likely to be positive, the nature of the *softmax* combining function is such that its estimate of the bag's "positive-ness" will be very high, since it gives an exponentially higher weight to such an instance, and consequently the weighted average of all the instances will also be high. Here, α controls the proportion of instances in the bag that influence the bag label. Therefore, if one has an estimate of the proportion of positive to negative instances in the positive bags (noise-level), one can appropriately tune α to reflect this, and hence learn more robust models than by simply using the max operator.

Next, to model these instance-level probabilities t_{ij}, we employ a logistic formulation given as

$$t_{ij} = p(y_{ij} = 1|x_{ij}) = \frac{1}{1 + \exp(-\mathbf{w} \cdot x_{ij})} \quad (4)$$

where the parameter vector $\mathbf{w}$ (to be learned) models the target of interest, so that the probability $p(y_{ij} = 1|x_{ij})$ calculated with Eqn. 4 would be high for an image patch x_{ij} that contains the target, and low for a patch that does not contain the target.

Now, using Eqns. 4 and 3 in Eqn. 1 along with a regularization term on $\mathbf{w}$, we can express a maximum likelihood formulation (in terms of the parameter vector $\mathbf{w}$ to be learned) as

$$\hat{\mathbf{w}} = \arg\max_{\mathbf{w}} \sum_{i \in B^+} \log \left(\frac{\sum_{j=1}^{m} t_{ij} \exp(\alpha t_{ij})}{\sum_{j=1}^{m} \exp(\alpha t_{ij})} \right) + \sum_{i \in B^-} \log \left(1 - \frac{\sum_{j=1}^{m} t_{ij} \exp(\alpha t_{ij})}{\sum_{j=1}^{m} \exp(\alpha t_{ij})} \right) - \frac{\lambda}{2} \mathbf{w}^T \mathbf{w} \quad (5)$$

where the regularization term is obtained by using the following prior on the parameter vector $\mathbf{w}$,

$$p(\mathbf{w}) \sim \mathcal{N}(0, \lambda^{-1}\mathbf{I}), \quad \lambda > 0 \quad (6)$$

to give better generalization performance [11]. To optimize Eqn. 5, we use a gradient-based optimization technique using the BFGS method [12].

The parameter vector $\hat{\mathbf{w}}$ obtained from the optimization algorithm thus represents the learned target model. Therefore, when presented with a new image, the probability that an image patch within it (with feature vector x) contains the learned target can be calculated from Eqn. 4 using $\hat{\mathbf{w}}$.

4 Image Features and Instance Model

We adopt a variation of covariance matrix features to obtain a feature vector x_{ij} for each image patch instance in our formulation. Covariance features have been shown to be robust appearance-based descriptors for modeling image regions [3]. The covariance matrix representation C_R for a given image patch R of size $W \times H$ in our framework is calculated as

$$C_R = \frac{1}{WH} \sum_{k=1}^{WH} (f_k - \mu_R)(f_k - \mu_R)^T \tag{7}$$

where $f_k = [x \; y \; r \; g \; b \; I_x \; I_y]$ is a 7 dimensional feature vector using a combination of position, color, and gradient values at each pixel location in the image patch R, and μ_R is the mean feature vector within the image patch.

We require a Euclidean distance based feature representation for Eqn. 4 whereas distances between covariance matrices are based on their eigenvalues [3]. Therefore, we use the property from [13] that eigenvalues of matrix logarithm of a covariance matrix C_R are equal to logarithms of eigenvalues of C_R. Therefore, the covariance matrix descriptor can be transformed to a feature vector representation by first calculating the matrix logarithm of C_R to obtain C_l and then stringing out the elements of the matrix C_l to obtain a vector C_v [13]. Moreover, since the matrix logarithm C_l is a symmetric matrix, it is fully specified by its bottom triangular part. Therefore, the feature vector C_v only needs to have the bottom triangular part of C_l, with the off-diagonal elements scaled by $\sqrt{2}$ to compensate for their double presence in the matrix. In our case, the 7x7 dimensional covariance matrix reduces to a 28 dimensional feature vector. We then use these log covariance-based features to model the instances x_{ij} corresponding to each image patch.

5 Target Localization Algorithm

The first step in our localization approach is to extract image patch instances from a sequence of images and use them to construct positive and negative bags. Given an input image sequence such as Fig. 1(a), we first detect regions of motion in each image by standard frame differencing (with the assumption that the target is moving). For each image, we then extract image patches from a reasonably large sample of the pixel locations marked as belonging to the motion region (the patch size can be predetermined or multiple sizes/aspect-ratios can be used). We construct a positive bag for this image using these instances since it is guaranteed to have at least one instance patch containing the desired target. Note that with this technique, instances corresponding to other parts of the scene in motion (trees, cars, noise pixels, etc.) would also be added, but that is acceptable since a positive bag can contain negative instances. At the same time, we sample a similarly large number of pixel locations from the (non-moving) background and extract image patches from these locations to construct the corresponding negative bag. Notice that this method ensures that no instance in this bag will contain the target. We similarly repeat this process for each of the input frames. This way it is guaranteed that at least one instance corresponding to the desired target is present in *each* of the positive bags and at the same time, absent from *all* the negative bags, thus satisfying our Multiple Instance assumption.

Once the positive and negative bags are constructed, we train the MIL classifier to learn the target concept by using the aforementioned optimization method. We initialize the weight parameter vector $\mathbf{w}$ uniformly at random between (-1, 1). The algorithm converges when the maximum change in the weight vector is less than a fixed small threshold ϵ.

Online Update: An important aspect of the proposed learning approach is that the learned target concept can be updated in an online manner with each new incoming frame instead of having to retrain the classifier using all the positive and negative bags collected from the beginning. We use the assumption that the target appearance does not change much with the new frame and hence would result in only a small change in the target model. Once we receive a new incoming frame, we create a positive and negative bag using the method described in Sect. 5 and run the gradient optimization algorithm, but this time with the weight vector initialized to the previously learned target model. Once the algorithm converges, we obtain the new weight vector reflecting the updated target model.

Multiple Targets: Another advantage of the proposed approach is its ability to learn multiple target concepts (if present across the input bags). Since it could be the case that more than one target is present across all input frames, there could potentially be multiple target models to be learned by the algorithm. Therefore, the proposed approach tests for multiple targets by using the first learned concept to remove all corresponding target instances from the positive bags, and then retrains to learn the next strongest concept, and so on for each remaining target. More specifically, once the algorithm converges and learns the first target model, it then uses Eqn. 4 with this model for every instance from every positive bag to calculate its probability of being the target. It then classifies each instance as positive (target) if this probability lies above a fixed threshold σ. We then update every positive bag by removing from it all the instances classified as being positive. This ensures that none of the positive bags contain even a single instance corresponding to the learned target. It is important that

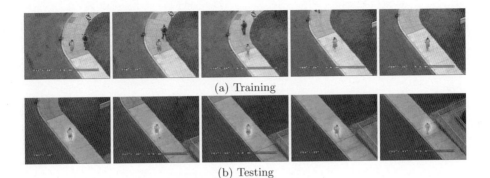

(a) Training

(b) Testing

Fig. 1. (a) Image sequence input to MIL. (b) Probability surfaces overlaid on incoming frames showing target localization (best viewed in color).

the threshold σ be picked conservatively so as to eliminate every true positive instance, even at the cost of eliminating a few false positives if necessary (we set $\sigma=0.75$). Since all instances corresponding to the first target have now been removed, re-running the optimization algorithm learns the next strongest target concept (if such a valid target is present). This process is repeated until all valid target concepts have been learned. To test whether a valid target concept has been learned or not, we calculate the log likelihood of the learned model on the input set of positive and negative bags. An extremely low value of the combined likelihood over all the bags indicates that the learned model is degenerate and there was no target concept left to learn.

6   Experiments

In this section, we present various experiments to demonstrate the proposed approach for target localization, apply it to a unique target detection and auto-locking system, demonstrate its sufficiency for learning appearance-based models to bootstrap object tracking systems, and finally, compare its performance with other manual initialization techniques.

6.1   Automatic Target Localization

Given a sequence of input images, we first constructed a set of positive and negative bags according to the technique described in Sect. 5 with patch size of 75x25 pixels. We then ran the MIL algorithm to learn a concept corresponding to a target that was common across all positive bags and absent in each of the negative bags. The parameters of the learning algorithm were set as $\alpha=3$, $\epsilon=10^{-5}$, and $\lambda=10^{-2}$. The validity of the learned target concept was checked by calculating the likelihood of the learned model across all input bags. Next, for every new incoming frame, the target model was evaluated against the input image at every possible location. This results in a probability surface corresponding to

(a) Training

(b) Testing

Fig. 2. (a) Input image sequence containing 2 valid targets. (b) Probability surfaces for incoming frames showing unique target detected at a later instant.

the new incoming frame indicating the probability of the target being present at each particular location. Figure 1(a) shows the sequence of input frames used to learn a target model. Figure 1(b) shows the probability "heatmap" overlaid on new input frames, representing the probability surface $p(y = 1|x)$ for each patch x across the image. As seen in the figure, the target (in blue) present across all input images was detected and localized by the algorithm. After this, a new pair of positive and negative bags was created for the new input frame using the on-line update technique described in Sect. 5. The target model was then updated using the MIL algorithm and the updated model was used to evaluate the next incoming frame. This process was repeated with every new frame. Figure 1(b) shows the results of target localization using model update in the new input frames. Notice also that the other person (wearing black pants) was not learned by the algorithm since that person was not present in all of the frames, thus not satisfying the MIL criterion.

6.2 Auto-locking for Unique Target Detection

A useful feature of the proposed approach with the online update is its ability to continue updating all the target models (if there are multiple targets present in the scene satisfying the MIL criterion), and continue this process until a *unique* target is detected and localized in the scene. This is possible because the likelihood evaluation from the optimization algorithm can be used to identify the number of target models learned. Thus, this feature is useful in a system which can continuously update multiple target models until only the single most persistent target remains in the scene and then use that model for active tracking. We demonstrate this ability here.

As seen in Fig. 2(a), there were 2 targets present in the scene satisfying the MIL criterion. Therefore, the algorithm learned 2 corresponding target models and these were then used for localization with the incoming frames as shown in Fig. 2(b) (note that a probability surface corresponding to each learned model

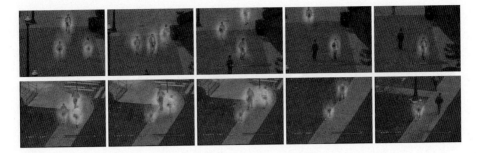

Fig. 3. Localization results from two other sequences involving multiple targets

is obtained separately, but they are shown here overlaid on the same image for compactness). Further, with each new incoming frame, the target models were updated online and eventually, when incorporating a frame where one of the targets is no longer present, the MIL criterion is violated and consequently the only single remaining target model was updated, as shown in the last two frames in Fig. 2(b). Even if the other target later re-enters the scene, they will not be localized, as the MIL criterion requires the target to be persistent across all frames from the beginning. Figure 3 shows correct localization results from 2 other sequences involving multiple targets and distractors.

6.3   Sufficiency of Learned Models for Tracking

Once a unique target is detected, the next step is to use the probability surface for localizing the target and build an appearance model that can then be used for tracking. We present experiments that demonstrate that the target models learned using the proposed approach can be used with commonly used tracking methods such as covariance and mean-shift tracking.

We use the probability surface generated by the MIL algorithm and threshold it to extract the extents of the target which can then be used to build an appearance-based model for tracking. Figure 4(a) shows the probability surface for a particular input frame. We then picked a threshold of 0.5 on the probability surface and used this to extract the area of the learned target model. Figure 4(b) shows the thresholded area of the target corresponding to Fig. 4(a), and the associated image chips in Fig. 4(c) show the results for various other input frames. Note that they all roughly correspond to the same area (target's torso and legs).

We then calculated the width and the height extents of the thresholded area and fit a bounding box around the region. This bounding box was then used to learn an appearance-based model of the target and bootstrap different trackers. This bounding box (mostly around the target's torso and legs) captured the appearance features that remained most persistent across input frames, as opposed to a larger bounding box around the entire body (including head, hands, and feet), which could potentially include several background pixels. We evaluated

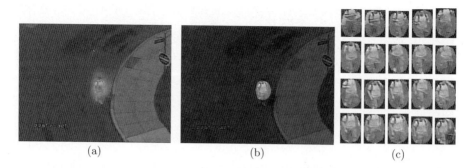

Fig. 4. (a) Probability surface. (b) Thresholded surface showing identified target area. (c) Image chips showing region used to learn appearance model for target tracking.

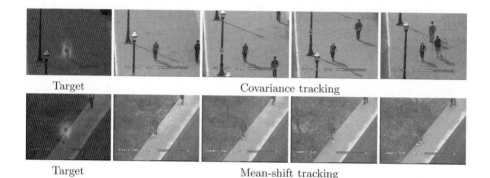

Fig. 5. Results from bootstrapping (a) covariance tracker (b) mean-shift tracker using a model learned from the proposed approach

this approach with both covariance and mean-shift trackers. Figure 5(top) shows a few frames from tracking a target using a covariance tracker, and Fig. 5(bottom) shows the results using the mean-shift tracker. In both cases, the appearance model learned using the proposed approach is reliable and sufficient to bootstrap standard object trackers.

6.4   Comparison with Manual Initialization

Using a standard test sequence, we next performed three experiments to compare the performance of our automatic localization and tracker initialization with typical manual initialization techniques to demonstrate the applicability of our approach. In the first experiment, we initialized a covariance tracker by manually specifying the location of the target and the size of the bounding box around the target in the first frame. In the second experiment, instead of manually marking the target in only the first frame, we marked its locations in each of the first 5 frames and computed a manifold mean of the covariance matrix representations

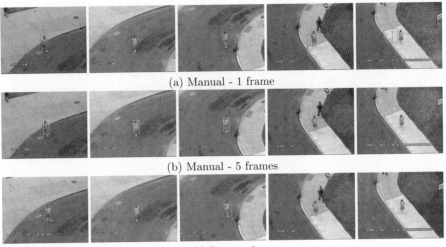

(a) Manual - 1 frame

(b) Manual - 5 frames

(c) Proposed

Fig. 6. Tracking results for (a) manual initialization with just 1 frame, (b) manual initialization using 5 frames, and (c) automatic initialization using proposed approach

from all 5 frames, and then used this model to manually initialize the covariance tracker. Even though this scenario is unrealistic in practical settings (where one cannot expect to manually select the target in every frame), we performed this experiment for a fair comparison with the proposed approach since our automatic localization uses an initial set of frames. (Note that the covariance tracker used a model update every frame). The third experiment involved tracking the target after automatically learning the target model using the proposed approach.

The results of each of the three experiments are shown in Fig. 6. As seen in Fig. 6(a), the manual initialization with a single frame performs poorly and loses the target within a few frames (as expected). The second experiment produces better results since it computes an average model across the 5 frames and consequently the model learned is less noisy. However, even in this case, we can see that the target is lost after a few frames (see Fig. 6(b)). In the third experiment (where the target is automatically localized), the target is tracked the longest (see Fig. 6(c)). The strength of employing a MIL formulation here (as opposed to a supervised approach) is that the task of identifying the best representation of the target present in the frame (and one that is also the most persistent across all frames) is ambiguous, and hence is pushed into the MIL framework. Consequently, our MIL approach outperforms the alternate initialization methods.

7   Summary and Future Work

We proposed a novel MIL framework to perform target localization from image sequences in a surveillance setting. The approach consists of a softmax logistic regression MIL algorithm using log covariance features to automatically learn the model of a target that is persistent across all input frames. The learned target model can be updated online and the approach can also be used to learn multiple targets in the scene (if present). We performed experiments to demonstrate the validity and usefulness of the proposed approach to localize targets in various scenes. We also demonstrated the applicability of the approach to bootstrap conventional tracking systems and showed that automatic initialization using our technique helps achieve better performance. In future work, we plan to explore the applicability of this approach to multi-camera systems for learning models from multiple views of targets.

Acknowledgement. We gratefully acknowledge the support of the U.S. Department of Energy through the LANL/LDRD Program under LDRD-DR project RADIUS for this work.

References

1. Leibe, B., Seemann, E., Schiele, B.: Pedestrian detection in crowded scenes. In: Proc. IEEE CVPR (2005)
2. Comaniciu, D., Ramesh, V., Meer, P.: Real-time tracking of non-rigid objects using mean-shift. In: Proc. IEEE CVPR (2000)
3. Porikli, F., Tuzel, O., Meer, P.: Covariance tracking using model update based on means on Riemannian manifolds. In: Proc. IEEE CVPR (2006)
4. Kalman, R.E.: A new approach to linear filtering and prediction problems. Transactions of the ASME–Journal of Basic Engineering 82, 35–45 (1960)
5. Arulampalam, A., Clapp, T.: A tutorial on particle filters for online nonlinear bayesian tracking. IEEE Transactions on Signal Processing 50, 174–188 (2002)
6. Dietterich, T., Lozano-Perez, T.: Solving the multiple-instance problem with axis-parallel rectangles. In: Artificial Intelligence (1997)
7. Maron, O., Lozano-Perez, T.: A framework for multiple instance learning. In: Proc. NIPS (1998)
8. Zhang, Q., Goldman, S.: Em-dd: An improved multiple instance learning technique. In: Proc. NIPS (2001)
9. Andrews, S., Hofmann, T.: Support vector machines for multiple instance learning. In: Proc. NIPS (2003)
10. Settles, B., Craven, M., Ray, S.: Multiple-instance active learning. In: Proc. NIPS (2007)
11. Nigam, K., Lafferty, J., McCallum, A.: Using maximum entropy for text classification. In: Proc. IJCAI (1999)
12. Fletcher, R.: Practical methods of optimization. In: Unconstrained Optimization, ch. 3, vol. 1 (1980)
13. Jost, J.: Riemannian geometry and geometric analysis. Springer, Berlin (2002)

Introducing Fuzzy Spatial Constraints in a Ranked Partitioned Sampling for Multi-object Tracking

Nicolas Widynski[1,2], Séverine Dubuisson[2], and Isabelle Bloch[1]

[1] Télécom ParisTech, CNRS UMR 5141 LTCI, Paris, France
[2] UPMC - LIP6, Paris, France

Abstract. Dealing with multi-object tracking in a particle filter raises several issues. A first essential point is to model possible interactions between objects. In this article, we represent these interactions using a fuzzy formalism, which allows us to easily model spatial constraints between objects, in a general and formal way. The second issue addressed in this work concerns the practical application of a multi-object tracking with a particle filter. To avoid a decrease of performances, a partitioned sampling method can be employed. However, to achieve good tracking performances, the estimation process requires to know the ordering sequence in which the objects are treated. This problem is solved by introducing, as a second contribution, a ranked partitioned sampling, which aims at estimating both the ordering sequence and the joint state of the objects. Finally, we show the benefit of our two contributions in comparison to classical approaches through two multi-object tracking experiments and the tracking of an articulated object.

1 Introduction

Sequential Monte Carlo methods, also known as particle filters, have been widely used in the vision community. Their natural dispositions for tracking purposes, their reliability to deal with non linear systems and their easiness of implementation have certainly contributed to this success. Dealing with possibly numerous interacting objects requires to represent the interactions between objects and to employ an efficient algorithm whatever the number of objects to track. We focus here on these two classical issues.

As a first contribution, we propose to introduce fuzzy spatial constraints between objects into the particle filter. This allows easily modeling potentially complex interactions between objects. To our knowledge, spatial constraints in a particle filter framework have only been used in specific ways, in a non fuzzy formalism (e.g. [1]). Fuzzy spatial constraints have shown a real interest in various domains, such as clustering [2], brain segmentation in 3D MRI images [3] or graph reasoning over fuzzy attributes [4].

The adaptation of particle filters to track several objects has been extensively addressed in the literature, in many different ways. Among these, the authors

G. Bebis et al. (Eds.): ISVC 2010, Part I, LNCS 6453, pp. 393–404, 2010.

in [5] proposed a Jump Markov System to model the number of objects, the association hypothesis between observations and objects and the indivual states. In [6], the authors use one particle filter and model interactions between objects and measures using a Joint Probabilistic Data Association Filter (JPDAF) framework. In [7], the distribution of the association hypotheses is computed using a Gibbs sampler. On another hand, as pointed out in [8], the importance sampling, used in particular in particle filters, suffers from the problem of the curse of dimensionality. This means that the particle filter requires a number of particles that exponentially increases with the number of objects. This renders the practical use of a particle filter for multiple object tracking difficult as soon as the number of objects is greater than three. Therefore, the authors in [9,10] proposed a particle filter that avoids this additional cost using a partitioned sampling strategy, based on a principle of exclusion (i.e. specifying that a measurement may be associated to at most one object). This is performed by partitioning the state space, typically considering one element of the partition per object, according to a specific ordering of the objects that we call scenario, and to select particles, using a weighted sampling, that are the most likely to fit with the real state of the object. The considered order matters since it can lead to unsuitable behaviors of the filter, such as loosing tracks, for example when the first considered object is hidden. In fact, as pointed out in [11] and as we will detail further in this article, using a specific order to estimate the state may also have many bad effects, which can be explained by a phenomenon of impoverishment of the particle set. In [11], the joint filtering distribution is represented by a mixture model, in which each mixture component describes a specific order and is estimated using a partitioned sampling technique. This idea has also been used in [12] for multi-cue fusion purposes. However, the number of particles allocated to each component is fixed, that may degrade performances of the filter when the chosen orders are not relevant [11]. In this article, we propose to jointly estimate the order in which objects are processed and states of the objects, in a so called ranked partitioned sampling strategy. This allows us to consider the whole set of possible orders and to automatically prune irrelevant scenarios.

This paper is organized as follows. Section 2 presents the fuzzy spatial constraint formalism and its introduction into a probabilistic framework. Section 3 describes the multi-object tracking based on a particle filtering modeling. Then in Section 4, we recall the principle of the partitioned sampling before introducing our ranked partitioned sampling in Section 5. We finally show results in Section 6 by considering two classical multi-object tracking experiments and the tracking of an articulated object.

2   Modeling Fuzzy Spatial Constraints

In this section, we propose to model explicitly interactions between objects via fuzzy spatial relations defined over one, two or more objects indicating to which degree the relation is satisfied. They will be considered as constraints the objects should satisfy during the tracking process, to a non-zero degree, and are therefore called fuzzy spatial constraints. Each relation is considered as a linguistic variable, taking a small number of linguistic values [13]. The granularity of this representation can be defined by the application. The semantics of each linguistic value is defined by a fuzzy set on the variable domain. Fuzzy spatial constraints may be defined by unary fuzzy operators, such as the concept of the size of an object (which may take the values *small, medium, large, ...*); by binary operators, such as the concept of relative orientation (*is to the right of, is to the left of, ...*); by ternary operators, such as the concept of local disposition (*is the first of, is in the middle of, is the last of*); and more generally *n*-any operators. In this paper, we focus on binary, ternary and quaternary operators, considering concepts of intersection, distance, angle and alignment. Fuzzy spatial constraints may be fixed during the tracking process (the topology of the configuration of the objects is fixed, although known imprecisely), may evolve during the time (fuzzy spatial concepts gradually change their values), or may be defined over time (considering imprecise fuzzy spatio-temporal constraints). Here we only consider time-independent spatial relations but we believe that the two last types of constraints may be of interest too.

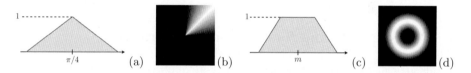

Fig. 1. Spatial fuzzy relations illustrated (a,c) in the variable domain and (b,d) in the image domain with respect to the center point. Figures (a,b) represent the value *north-east* of the concept of orientation and figures (c,d) the value *medium distance m* of the concept of distance.

Let $\mathbf{x}_t^*$ be an hypothetic state of an object and $\tilde{\mathbf{x}}_t = (\tilde{\mathbf{x}}_t^1, \ldots, \tilde{\mathbf{x}}_t^L)$ be the vector state of L objects already processed at time t. We now define a fuzzy membership function $\mu(\mathbf{x}_t^*; \tilde{\mathbf{x}}_t) \in [0,1]$ which describes to which degree an object configuration $\mathbf{x}_t^*$ satisfies the spatial constraints imposed by $\tilde{\mathbf{x}}_t$. Denoting by K the number of spatial constraints we consider, we define μ as:

$$\mu(\mathbf{x}_t^*; \tilde{\mathbf{x}}_t) = \mathop{\Xi}_{k=1}^{K} \mu^k(\mathbf{x}_t^*; \tilde{\mathbf{x}}_t) \tag{1}$$

with Ξ a fusion operator, for example a t-norm (fuzzy conjunction) [14], and $\mu^k \in [0,1]$ the membership function of the kth spatial constraint. For example, considering a binary fuzzy relation, μ^k is defined as:

$$\mu^k(\mathbf{x}_t^*; \tilde{\mathbf{x}}_t) = \mathop{\psi}_{l=1}^{L} \mu_l^k(\mathbf{x}_t^*; \tilde{\mathbf{x}}_t^l) \tag{2}$$

with ψ a fusion operator, for example a t-norm, and $\mu_l^k \in [0,1]$ the membership function of the kth spatial constraint between the current object and object l.

The shape of the function μ^k may be fixed by the application. In our experiments, we consider a triangular one to generate the orientation *north-east* and a trapezoidal one for the distance *medium*, respectively illustrated in Figures 1(a) and 1(b). In a similar way, considering ternary constraints, μ^k is defined as:

$$\mu^k(\mathbf{x}_t^*;\tilde{\mathbf{x}}_t) = \overset{L}{\underset{l_1=1}{\psi}}\ \overset{L}{\underset{l_2=l_1+1}{\psi}}\ \mu_{l_1,l_2}^k(\mathbf{x}_t^*;\tilde{\mathbf{x}}_t^{l_1},\tilde{\mathbf{x}}_t^{l_2}) \tag{3}$$

with $\mu_{l_1,l_2}^k \in [0,1]$ the membership function of the kth spatial constraint between the current object and objects l_1 and l_2. Finally, we introduce the fuzzy spatial constraints into a probabilistic framework, by defining the function $\phi(\mathbf{x}_t^*;\tilde{\mathbf{x}}_t)$ as:

$$\phi(\mathbf{x}_t^*;\tilde{\mathbf{x}}_t) \propto \mu(\mathbf{x}_t^*;\tilde{\mathbf{x}}_t)^\gamma \tag{4}$$

with $\gamma \in \mathbb{R}^+$ a fixed parameter and $\mu(\mathbf{x}_t^*;\tilde{\mathbf{x}}_t)$ the membership function of the spatial constraints defined in Equation 1. Examples of constraints will be defined in Section 6. The function ϕ allows passing from a possibilistic semantic to a probabilistic one and will be integrated in a dynamic model in a particle filtering framework (see Section 4).

3   Particle Filtering for Multi-object Tracking

In this article, we consider that the number M of objects is known. Let us consider a classical filtering problem and denote by $\mathbf{x}_t \in \mathbb{X}$ the hidden joint state of a stochastic process at time t: $\mathbf{x}_t = (\mathbf{x}_t^1,\ldots,\mathbf{x}_t^M)$ with $\mathbf{x}_t^i \in \mathbb{X}^*$ the unknown state of the ith object, and by $\mathbf{y}_t \in \mathbb{Y}$ the measurement state extracted from the image sequence. The temporal evolution of $\mathbf{x}_t$ and the measurement equation are given by $\mathbf{x}_t = f_t(\mathbf{x}_{t-1},\mathbf{v}_t)$ and $\mathbf{y}_t = h_t(\mathbf{x}_t,\mathbf{w}_t)$, where $\mathbf{v}_t$ and $\mathbf{w}_t$ are independent white noises. The non-linear Bayesian tracking consists in estimating the *posterior* filtering distribution $\Pr(\mathbf{x}_t|\mathbf{y}_{1:t})$ through a non-linear transition function f_t and a non-linear measurement function h_t. The resulting filtering density function can be expressed by $p(\mathbf{x}_t|\mathbf{y}_{1:t}) \propto p(\mathbf{y}_t|\mathbf{x}_t) \int_{\mathbb{X}} p(\mathbf{x}_t|\mathbf{x}_{t-1}) p(\mathbf{x}_{t-1}|\mathbf{y}_{1:t-1})d\mathbf{x}_{t-1}$. Particle filters are used to approximate the *posterior* density function (*pdf*) by a weighted sum of N Dirac masses $\delta_{\mathbf{x}_t^{(n)}}(d\mathbf{x}_t)$ centered on hypothetic state realizations $\{\mathbf{x}_t^{(n)}\}_{n=1}^N$ of the state $\mathbf{x}_t$, also called particles. Then, the filtering distribution $\Pr(d\mathbf{x}_t|\mathbf{y}_{1:t})$ is recursively approximated by the empiric distribution $P_N(d\mathbf{x}_t|\mathbf{y}_{1:t}) = \sum_{n=1}^N w_t^{(n)}\delta_{\mathbf{x}_t^{(n)}}(d\mathbf{x}_t)$, where $\{\mathbf{x}_t^{(n)}\}$ is the nth particle and $w_t^{(n)}$ its weight. If an approximation of $\Pr(d\mathbf{x}_{t-1}|\mathbf{y}_{1:t-1})$ is known, the process is divided into three main steps:

1. The diffusion step consists in estimating $p(\mathbf{x}_t|\mathbf{y}_{1:t-1})$ by propagating the particle swarm $\{(\mathbf{x}_{t-1}^{(n)}), w_{t-1}^{(n)}\}_{n=1}^N$ using an importance function $q(\mathbf{x}_t|\mathbf{x}_{0:t-1}^{(n)},\mathbf{y}_t)$.
2. The update step then computes new particle weights using the new observation $\mathbf{y}_t$, as: $w_t^{(n)} \propto w_{t-1}^{(n)} \dfrac{p(\mathbf{y}_t|\mathbf{x}_t^{(n)})p(\mathbf{x}_t^{(n)}|\mathbf{x}_{t-1}^{(n)})}{q(\mathbf{x}_t|\mathbf{x}_{0:t-1}^{(n)},\mathbf{y}_t)}$, such that $\sum_{i=1}^N w_t^{(n)} = 1$.

3. Resampling techniques are employed to avoid particle degeneracy problems.

When needed, the data association problem may be handled by estimating both the objects and a vector of visibility of those [9,15]. This involves a measurement process dependent on the visibility of the objects. This point will be discussed in Sections 5 and 6.1.

The presented particle filter algorithm uses, by essence, an importance sampling procedure while simulating according to a density function $q(\mathbf{x}_t|\mathbf{x}_{0:t-1}^{(n)},\mathbf{y}_t)$. In practice, this means in particular that an increase in the dimension of $\mathbf{x}_t$ systematically induces an increase in the variance of the particle weights, leading to possibly fatal impoverishment of the particle set. In the case of multi-object tracking, where $\mathbf{x}_t^i$ and $\mathbf{x}_t^j$ are two states evolving in the same space $\mathbb{X}^*$, it has been shown [10] that N^2 particles are necessary to achieve the same level of tracking performance than when tracking a single object with N particles. To tackle the dimensionality problem, the authors in [10] propose, instead of directly sampling from the joint configuration of the objects, to decompose the vector state of the objects by partitioning the state space, and then handling one object at a time. This is called partitioned sampling and presented in the next section.

4 Partitioned Sampling (PS)

The Partitioned Sampling (PS), introduced in [9,10], divides the joint state space $\mathbb{X}$ into a partition of M elements, i.e. one by object, and for each of them, applies the transition (dynamics) and performs a weighted resampling operation.

Weighted Resampling. A weighted resampling operation transforms a particle set $\{\mathbf{x}_t^{(n)},w_t^{(n)}\}_{n=1}^N$ into another one while keeping the distribution intact. Weights $\{\rho^{(n)}\}_{n=1}^N$ are called importance weights and are designed such that $\rho^{(n)} = g(\mathbf{x}_t^{(n)})/\sum_{m=1}^N g(\mathbf{x}_t^{(m)})$. The strictly positive function g is called *weighting function*, and aims at resampling the particles set according to the peaks of g. Finally the particle set $\{\tilde{\mathbf{x}}_t^{(n)}, w_t^{(n)}/\rho^{(n)}\}_{n=1}^N$ is obtained by selecting particles $\{\mathbf{x}_t^{(m)}\}_{m=1}^N$ with probabilities $\{\rho^{(m)}\}_{m=1}^N$.

Partitioned Resampling. Denoting by $\sim g_i$ the weighted resampling operation of the ith object, by $\sim$ the resampling procedure according to the particle weights $\{w_t^{(n)}\}_{n=1}^N$, and by f_i the dynamics process of the object i, possibly conditioned by objects already generated $\mathbf{x}_t^{1:i-1}$, the partitioned sampling operation is summarized in Figure 2(a).

Although any weighting function g_i should asymptotically keep the *posterior* unchanged, the objective of this step is to obtain an accurate representation of the *posterior*. Then, considering a factorization of the likelihood such that it allows us to deal with each object independently, i.e. $p(\mathbf{y}_t|\mathbf{x}_t) = \prod_{i=1}^M p_i(\mathbf{y}_t|\mathbf{x}_t^i)$, then the likelihood $h_i = p_i(\mathbf{y}_t|\mathbf{x}_t^i)$ of the object i appears to be a natural choice and leads to the diagram proposed in Figure 2(b).

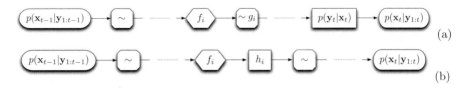

(a)

(b)

Fig. 2. (a) Diagram of the Partitioned Sampling procedure and (b) diagram of the Partitioned Sampling procedure using the likelihood as weighting function

Additionally, we propose to integrate the fuzzy spatial constraints modeled in Section 2. The simplest way to do it consists in introducing the interaction density function defined in Equation 4 into the dynamical model, leading to $f_i = p(\mathbf{x}_t^i|\mathbf{x}_{t-1}^i)\,\phi(\mathbf{x}_t^i;\mathbf{x}_t^{1:i-1})$. This model can be viewed as the pairwise Markov Random Field *prior* used in [16,1,12]. However, in a more general perspective, it is often impossible to directly generate samples from $\phi(\mathbf{x}_t^i;\mathbf{x}_t^{1:i-1})$. Then, we consider $f_i = p(\mathbf{x}_t^i|\mathbf{x}_{t-1}^i)$ whereas the likelihood integrates the interaction term, i.e., $h_i = p_i(\mathbf{y}_t|\mathbf{x}_t^i)\phi(\mathbf{x}_t^i;\mathbf{x}_t^{1:i-1})$. This procedure does not affect the *posterior* since it can be seen as an importance sampling step [1].

Discussion. The partitioned sampling is a very efficient sampling method since, by alleviating the dimension problem, it considerably reduces the computation cost. However, as discussed in [11], the order of the considered objects has a direct impact on the performance of the tracker. This is due to the M successive weighting resampling procedures performed by the algorithm. Hence, objects placed at early stages will be prone to more impoverishment effects than the others. On the other side, objects placed at the end may suffer from a lack of diversity even before being considered, which may also lead to tracking errors.

The method presents an additional difficulty. If occlusions occur, one may quite rightly handle visible objects first, and hence adopt a dynamic order strategy of the objects. The solution proposed in [9] is called *Branched Partitioned Sampling* (BPS), and consists in adding to the global estimation a vector of visibility, and then recursively grouping together particles with an identic realization of this vector, generating an hypothesis tree. This solution has however a major drawback: considering a tracking problem with M objects, the method possibly divides particles into $M!$ hypotheses, which looses the interest of the partitioned sampling since the particles no longer try to survive over a large set of N particles but over possible irrelevant sets of $N/M!$ elements.

The *Dynamic Partitioned Sampling* (DPS), proposed in [11], uses a mixture model to represent the *posterior* distribution. Each mixture component represents a specific order of processing of the objects. In their experiments, the authors used M permutation sets, deterministically defined, each one owning N/M particles. This strategy improves Partitioned Sampling results since it alleviates impoverishment effects, especially when occlusions occur. However, using a fixed small subset of possible permutations might not be robust. Moreover, splitting particles into several sets has the same drawback than the Branched Partitioned Sampling, since particles evolve into subsets.

To overcome these problems, we propose a new sampling strategy, called *Ranked Partitioned Sampling* (RPS), which jointly estimates the state of the objects and the estimation of the order of processing of the objects. Objects with highest confidence are considered in earlier stages of the scenarios. Each scenario is then confronted to each of the others, implicitly pruning unlikely branches. The adapative choice of the order of processing aims at limiting the impoverishment effect.

5   Ranked Partitioned Sampling (RPS)

Let $\mathbf{o}_t = (\mathbf{o}_t^1, \ldots, \mathbf{o}_t^M)$ be an ordered sequence of processing, i.e. a permutation over M objects. Then $\mathbf{o}_t^i$ is a random variable that indicates the position of the object i in this ordered sequence. We call a scenario the inverse permutation $\mathbf{o}_t^{-1}$ considered at a particular instant t, which we denote by $\mathbf{s}_t = (\mathbf{s}_t^1, \ldots, \mathbf{s}_t^M)$. Hence the kth component of a scenario is defined such that $\mathbf{s}_t^k \triangleq \sum_{i=1}^{M} i\, \delta_{\mathbf{o}_t^i}^k$, with δ_a^b the Kronecker function that equals 1 if $a = b$, 0 otherwise, and indicates the object processed at the step k. We first consider fixed probabilities of transition of the positions:

$$p(\mathbf{s}_t^h = i | \mathbf{s}_{t-1}^k = i) \triangleq p(\mathbf{o}_t^i = h | \mathbf{o}_{t-1}^i = k) \triangleq \alpha_{k,h} \quad \forall i \in \{1, \ldots, M\} \tag{5}$$

By first considering objects placed in the earliest stages at time $t-1$, the joint transition distribution of $\mathbf{o}_t$ is:

$$p(\mathbf{o}_t | \mathbf{o}_{t-1}) = p\left(\mathbf{o}_t^{s^1} | \mathbf{o}_{t-1}^{s^1}\right) \prod_{k=2}^{M} p\left(\mathbf{o}_t^{s^k} | \mathbf{o}_{t-1}^{s^k} \triangleq k, \mathbf{o}_t^{s^1}, \ldots, \mathbf{o}_t^{s^{k-1}}\right) \tag{6}$$

with $\mathbf{s}^k \triangleq \mathbf{s}_{t-1}^k$, the time subscript being omitted to simplify notations. The last conditional distribution in Equation 6 depends on the probabilities of transitions of positions defined in Equation 5 and inaccessible positions of already considered objects:

$$p\left(\mathbf{o}_t^{s^k} = h | \mathbf{o}_{t-1}^{s^k}, \{\mathbf{o}_t^{s^u}\}_{u=1}^{k-1}\right) = \left[1 - \sum_{j=1}^{k-1} \delta_{\mathbf{o}_t^{s^j}}^h\right] \left[\alpha_{k,h} + \frac{1}{M-k+1} \sum_{j=1}^{k-1} \alpha_{k, \mathbf{o}_t^{s^j}}\right] \tag{7}$$

The first term in the product ensures that the probability is set to 0 if the position h has already been assigned, whereas the last term uses probabilities of transition of the assigned positions to balance the distribution in a uniform way. We decompose the joint transition density so that $p(\mathbf{x}_t, \mathbf{o}_t | \mathbf{x}_{t-1}, \mathbf{o}_{t-1}) = p(\mathbf{x}_t | \mathbf{x}_{t-1}, \mathbf{o}_t) p(\mathbf{o}_t | \mathbf{o}_{t-1})$. Conditioned by the sequence order defined by $\mathbf{o}_t$, the transition density of the vector state $\mathbf{x}_t$ is decomposed considering first the objects placed in the earliest stages:

$$p(\mathbf{x}_t | \mathbf{x}_{t-1}, \mathbf{o}_t) \triangleq \prod_{k=1}^{M} p(\mathbf{x}_t^{s_t^k} | \mathbf{x}_{t-1}^{s_t^k}, \mathbf{x}_t^{s_t^1}, \ldots, \mathbf{x}_t^{s_t^{k-1}}) \tag{8}$$

However, following the choices made in Section 4, we set the dynamic process to $f_{\mathbf{s}_t^k} = p(\mathbf{x}_t^{\mathbf{s}_t^k}|\mathbf{x}_{t-1}^{\mathbf{s}_t^k})$. In the same way, the likelihood is defined as $p(\mathbf{y}_t|\mathbf{x}_t, \mathbf{o}_t) \triangleq \prod_{k=1}^{M} p_{\mathbf{s}_t^k}(\mathbf{y}_t|\mathbf{x}_t^{\mathbf{s}_t^k})$. To summarize, at a time t, for each particle, the algorithm first generates a scenario, and then, at position k of the process (scenario), it proposes a new state of the object $\mathbf{s}_t^k$ using dynamics, and computes the likelihood before resampling all the particles. The approximation of the joint filtering distribution of $\mathbf{x}_t, \mathbf{o}_t$ is obtained once the M positions have been computed. By setting $h_{\mathbf{s}_t^k} = p_{\mathbf{s}_t^k}(\mathbf{y}_t|\mathbf{x}_t^{\mathbf{s}_t^k})\phi(\mathbf{x}_t^{\mathbf{s}_t^k}; \mathbf{x}_t^{\mathbf{s}_t^{1}:\mathbf{s}_t^{k-1}})$, we obtain the diagram in Figure 3.

Fig. 3. Diagram of the Ranked Partitioned Sampling procedure using the likelihood as weighting function

There are many ways to deal with hidden objects. A current choice is to estimate a visibility vector and then proposing scenarios according to it. However, in this paper, we implicitly consider that an object at a position i is always less visible than a object at position $i - k$, with $0 < k < i$.

6   Experiments

6.1   People Tracking

We consider a public sequence [17], from which we extracted 260 frames where three pedestrians walk and occult each other. Let $\mathbf{x}_t^i = (x_t^i, y_t^i)^T$ be the unknown state of object i, with $(x_t^i, y_t^i)^T$ the $2D$ center of a person. Dimension of the rectangles surrounding persons are fixed. The dynamics is a random walk, i.e. $\mathbf{x}_t^i = f_t(\mathbf{x}_{t-1}^i, \mathbf{v}_t^i) = \mathbf{x}_{t-1}^i + \mathbf{v}_t^i$ with $\mathbf{v}_t^i \sim \mathcal{N}(\mathbf{0}_{2\times1}, \Sigma)$. For the BPS, we modeled a visibility vector, where the transition probability from state *visible* to state *hidden* was fixed to 0.2, and the transition probability from state *hidden* to state *visible* to 0.5. These probabilities have been fixed empirically, in a way that they give priority to the state *visible*, although being flexible enough to deal with sudden occlusions. For the RPS, we implicitly consider the visibility of an object by its position in the processing order, then no spatial constraint is necessary for this first experiment. For both methods, visible objects are considered first, although in the RPS the visibility vector is not modeled since the scenario vector determines the visibility of an object. The likelihood of an object is based on a distance between a color model histogram and a candidate histogram defined by the particle [18]. However, only the visible part of the object is considered to avoid penalizing hidden or partially hidden objects.

Results using $N = 500$ particles are illustrated in Figure 4. Rectangles in red correspond to the estimation of objects (i.e. the Monte Carlo expected value).

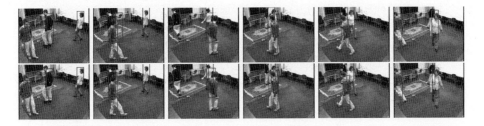

Fig. 4. People tracking results at times 5, 24, 99, 205, 212 and 259. First row: branched PS [9], second row: ranked PS (proposed approach).

As mentionned in Section 4, in the BPS, the particles may be divided into $M!$ sets, which may maintain scenarios where the visibility hypotheses are wrong. Moreover, the visibility vector is not well adapted in the case where the number of objects is greater than two, since it does not solve anymore the data association problem. These two points explain the difference of the results obtained by the BPS and the RPS (see e.g. second and last images). Figures 5(b) and 5(c) show the overall superiority of the RPS over the BPS. Figure 5(d) presents the *posterior* probabilities obtained by the RPS for a person to be considered first in the processing order induced by $\mathbf{o}_t$, where a low probability indicates that the person is likely to be partially hidden. We can appreciate the probabilities estimated for example at times 24, 99, 205 and 212 that are consistent with the sequence (Figure 4).

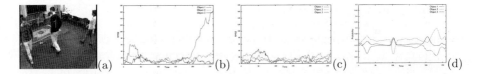

Fig. 5. (a) Indices of the pedestrians present in the scene (b) RMSE of the branched PS, (c) RMSE of the ranked PS (proposed approach) and (d) *posterior* probabilities obtained by the RPS for a person to be considered first in the processing order

6.2   Ant Tracking

This test sequence has been successfully studied in [1], using a MCMC-Based particle filtering approach. The state of the object i, $\mathbf{x}_t^i = (x_t^i, y_t^i, \theta_t^i)^T$, contains the 2D position $(x_t^i, y_t^i)^T$ and the object orientation θ_t^i. Dynamics of position and orientation are random walks. We used for this experiment an exclusion fuzzy spatial constraint: using a fuzzy semi-trapeze, two ants must not overlap more than 10% of their own areas, and from 5% the degree of satisfaction of the constraint starts to decrease. The likelihood is a simple background substraction.

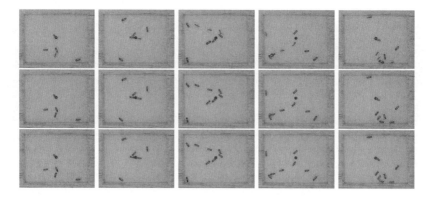

Fig. 6. Ant tracking results. First row: RPS without spatial constraints, second row: PS with spatial constraints and last row: RPS with spatial constraints.

Figure 6 shows results using a RPS with and without spatial constraints, and a PS with spatial constraints. Estimated positions of the ants are represented in red. $N = 500$ particles were used for this sequence of 750 frames. The benefit of using a simple spatial constraint is very clear here (several ants are not tracked without spatial constraints, while they are successfully tracked with such constraints). The PS and the RPS give global comparable results since all possible processing orders lead to almost identical results. Then the RPS performs as well as the PS when the order does not significantly matter and the number of objects is small enough to not suffer from an impoverishment effect.

6.3 Hand Tracking

We finally consider a problem of tracking an articulated object. The state of the object i, $\mathbf{x}_t^i = (\mathrm{x}_t^i, \mathrm{y}_t^i, \theta_t^i)^T$, contains its $2D$ center position $(\mathrm{x}_t^i, \mathrm{y}_t^i)^T$ and orientation θ_t^i. Each finger shape is fixed and expressed by vectors of 6 $2D$-control points, located on the basis of fingers, on the middle and on the fingertips. Dynamics of position and orientation are random walks. The difficulty of this application is that fingers may be partially or totally hidden. We then would like to track the hand by preserving a global consistency of the shape using fuzzy spatial constraints. Although they might be automatically learnt, we consider here fixed spatial relations between the fingers. For instance, we used four fixed fuzzy spatial constraints: angle, distance, alignment and exclusion. We defined two values of the binary constraint of angle, with an uncertainty expressed by a trapezoidal template of length support $\pi/4$: *nearly* $-\pi/8$, and *nearly* $\pi/8$; two values of the binary constraint of distance: *close* and *far*; one ternary or quaternary (it depends on the number of objets already processed) constraint of alignment: using a linear regression, fingers cannot move away from a fixed distance threshold; and one binary constraint of exclusion: no overlap between fingers is allowed. The likelihood is obtained by computing gradient values over normal lines of regular points of a finger B-spline [19].

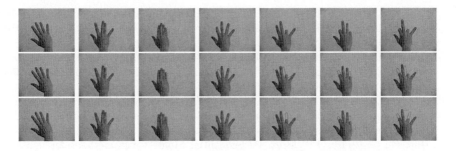

Fig. 7. Hand tracking results. First row: RPS without spatial constraints, second row: PS with spatial constraints and last row: RPS with spatial constraints.

The results are illustrated in Figure 7 and were obtained using $N = 2000$ particles, over a sequence of 800 frames. Results using a simple partitioned sampling strategy were generated using a random sequence order (in this particular sequence, fixed sequence order gave not quite as good results). RPS without spatial constraints fails as soon as a finger is hidden. The RPS obtains better results than the PS thanks to the estimation of the scenario, which allows estimating first fingers that are trusted, the visible ones, and then the other ones, which are then more constrained by the fuzzy spatial relations.

7 Conclusion

This article presents two contributions. First, we introduced fuzzy spatial constraints into a multi-object tracking based on particle filtering. This novel information allows us to easily handle constraints between objects in a unified framework. As a second contribution, the multi-object particle filter uses a ranked partitioned sampling strategy, which, like the partitioned sampling, tackles the problem of dimensionality by sequentially performing a weigthed resampling step in single object state spaces. Moreover, the simulation order proposed in the RPS is adaptive, which makes the tracking more robust and alleviates the impoverishment effect, while keeping a computation time identical to the PS one.

References

1. Khan, Z., Balch, T., Dellaert, F.: MCMC-Based particle filtering for tracking a variable number of interacting targets. IEEE Transactions Pattern Analysis and Machine Intelligence 27, 1805–1918 (2005)
2. Pham, D.: Fuzzy clustering with spatial constraints. In: IEEE International Conference on Image Processing, vol. II, pp. 65–68 (2002)
3. Colliot, O., Camara, O., Bloch, I.: Integration of fuzzy spatial relations in deformable models - Application to brain MRI segmentation. Pattern Recognition 39, 1401–1414 (2006)

4. Fouquier, G., Atif, J., Bloch, I.: Local reasoning in fuzzy attribute graphs for optimizing sequential segmentation. In: Escolano, F., Vento, M. (eds.) GbRPR 2007. LNCS, vol. 4538, pp. 138–147. Springer, Heidelberg (2007)
5. Doucet, A., Vo, B., Andrieu, C., Davy, M.: Particle filtering for multi-target tracking and sensor management. In: 5th International Conference on Information Fusion, pp. 474–481 (2002)
6. Schulz, D., Burgard, W., Fox, D., Cremers, A.: Tracking multiple moving targets with a mobile robot using particle filters and statistical data association. In: IEEE International Conference on Robotics & Automation, pp. 1665–1670 (2001)
7. Hue, C., Le Cadre, J.P., Prez, P.: Sequential Monte Carlo methods for multiple target tracking and data fusion. IEEE Transactions on Signal Processing 50, 309–325 (2002)
8. MacKay, D.J.C.: Introduction to Monte Carlo methods. In: Jordan, M.I. (ed.) Learning in Graphical Models. NATO Science Series, pp. 175–204. Kluwer Academic Press, Dordrecht (1998)
9. MacCormick, J., Blake, A.: A probabilistic exclusion principle for tracking multiple objects. International Journal of Computer Vision, 39, 57–71 (2000)
10. MacCormick, J., Isard, M.: Partitioned sampling, articulated objects, and interface-quality hand tracking. In: Vernon, D. (ed.) ECCV 2000. LNCS, vol. 1843, pp. 3–19. Springer, Heidelberg (2000)
11. Smith, K., Gatica-Perez, D.: Order matters: A distributed sampling method for multi-object tracking. In: British Maschine Vision Conference, pp. 25–32 (2004)
12. Duffner, S., Odobez, J., Ricci, E.: Dynamic partitioned sampling for tracking with discriminative features. In: British Maschine Vision Conference (2009)
13. Zadeh, L.A.: The concept of a linguistic variable and its application to approximate reasoning. Information Sciences 8, 199–249 (1975)
14. Dubois, D., Prade, H.: Fuzzy Sets and Systems: Theory and Applications. Academic Press, Inc., Orlando (1980)
15. Pérez, P., Vermaak, J.: Bayesian tracking with auxiliary discrete processes. application to detection and tracking of objects with occlusions. In: ICCV 2005 Workshop on Dynamical Vision, Beijing, China, pp. 190–202 (2005)
16. MacCormick, J., Blake, A.: Spatial dependence in the observation of visual contours. In: Burkhardt, H., Neumann, B. (eds.) ECCV 1998. LNCS, vol. 1407, pp. 765–781. Springer, Heidelberg (1998)
17. Fleuret, F., Berclaz, J., Lengagne, R., Fua, P.: Multi-camera people tracking with a probabilistic occupancy map. IEEE Transactions on Pattern Analysis and Machine Intelligence 30, 267–282 (2008)
18. Pérez, P., Hue, C., Vermaak, J., Gangnet, M.: Color-based probabilistic tracking. In: Heyden, A., Sparr, G., Nielsen, M., Johansen, P. (eds.) ECCV 2002. LNCS, vol. 2350, pp. 661–675. Springer, Heidelberg (2002)
19. Isard, M., Blake, A.: Condensation-conditional density propagation for visual tracking. International Journal of Computer Vision 29, 5–28 (1998)

Object Tracking and Segmentation in a Closed Loop

Konstantinos E. Papoutsakis and Antonis A. Argyros

Institute of Computer Science, FORTH
and
Computer Science Department, University of Crete
{papoutsa,argyros}@ics.forth.gr
http://www.ics.forth.gr/cvrl/

Abstract. We introduce a new method for integrated tracking and segmentation of a single non-rigid object in an monocular video, captured by a possibly moving camera. A closed-loop interaction between EM-like color-histogram-based tracking and Random Walker-based image segmentation is proposed, which results in reduced tracking drifts and in fine object segmentation. More specifically, pixel-wise spatial and color image cues are fused using Bayesian inference to guide object segmentation. The spatial properties and the appearance of the segmented objects are exploited to initialize the tracking algorithm in the next step, closing the loop between tracking and segmentation. As confirmed by experimental results on a variety of image sequences, the proposed approach efficiently tracks and segments previously unseen objects of varying appearance and shape, under challenging environmental conditions.

1 Introduction

The vision-based tracking and the segmentation of an object of interest in an image sequence are two challenging computer vision problems. Each of them has its own importance and challenges and can be considered as "chicken-and-egg" problems. By solving the segmentation problem, a solution to the tracking problem can easily be obtained. At the same time, tracking provides important input to segmentation.

In a recent and thorough review on the state-of-the-art tracking techniques [1], tracking methods are divided into three categories: *point tracking, silhouette tracking* and *kernel tracking*. Silhouette-based tracking methods usually evolve an initial contour to its new position in the current frame. This can be done using a state space model [2] defined in terms of shape and motion parameters [3] of the contour or by the minimization of a contour-based energy function [4, 5], providing an accurate representation of the tracked object. Point-tracking algorithms [6, 7] can also combine tracking and fine object segmentation using multiple image cues. Towards a more reliable and drift-free tracking, some point tracking algorithms utilize energy minimization techniques, such as Graph-Cuts or Belief Propagation on a Markov Random Field (MRF) [8] or on a Conditional

G. Bebis et al. (Eds.): ISVC 2010, Part I, LNCS 6453, pp. 405–416, 2010.

Random Field (CRF) [9, 10]. Most of the kernel-based tracking algorithms [11–13] provide a coarse representation of the tracked object based on a bounding box or an ellipsoid region.

Despite the many important research efforts devoted to the problem, the development of algorithms for tracking objects in unconstrained videos constitutes an open research problem. Moving cameras, appearance and shape variability of the tracked objects, varying illumination conditions and cluttered backgrounds constitute some of the challenges that a robust tracking algorithm needs to cope with. To this end, in this work we consider the combined tracking and segmentation of previously unseen objects in monocular videos captured by a possibly moving camera. No strong constraints are imposed regarding the appearance and the texture of the target object or the rigidity of its shape. All of the above may dynamically vary over time under challenging illumination conditions and changing background appearance. The basic aim of this work is to preclude tracking failures by enhancing its target localization performance through fine object segmentation that is appropriately integrated with tracking in a closed-loop algorithmic scheme. A kernel-based tracking algorithm [14], a natural extension of the popular mean-shift tracker [11, 15], is efficiently combined with Random Walker-based image segmentation [16, 17]. Explicit segmentation of the target region of interest in an image sequence enables reliable tracking and reduces drifting by exploiting static image cues and temporal coherence.

The key benefits of the proposed method are (i) close-loop interaction between tracking and segmentation (ii) enhanced tracking performance under challenging conditions (iii) fine object segmentation (iv) capability to track objects from a moving camera (v) increased tolerance to extensive changes of object's appearance and shape and, (vi) continual refinement of both the object and the background appearance models.

The rest of the paper is organized as follows. The proposed method is presented in Sec. 2. Experimental results are presented in Sec. 3. Finally, Sec. 4 summarizes the main conclusions from this work and future work perspectives.

2   Proposed Method

For each input video frame, the proposed framework encompasses a number of algorithmic steps, tightly interconnected in a closed-loop which is illustrated schematically in Fig.1. To further ease understanding, Fig.2 provides sample intermediate results of the most important algorithmic steps.

The method assumes that at a certain moment t in time, a new image frame I_t becomes available and that a fine object segmentation mask M_{t-1} is available as the result of the previous time step $t - 1$. For time $t = 0$, M_{t-1} should be provided for initialization purposes. Essentially, M_{t-1} is a binary image where foreground/background pixels have a value of 1/0, respectively (see Fig.2). The goal of the method is to produce the current object segmentation mask M_t. Towards this end, the spatial mean and covariance matrix of the foreground region of M_{t-1} is computed, thus defining an ellipsoid region coarsely representing

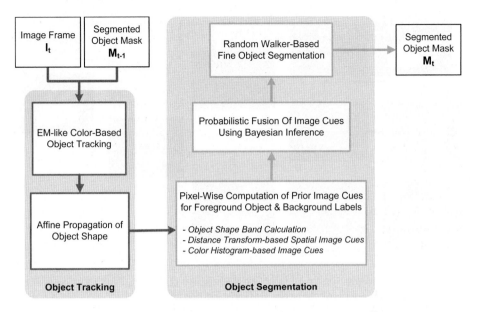

Fig. 1. Outline of the proposed method

the object at $t-1$. Additionally, a color-histogram-based appearance model of the segmented object (i.e., the one corresponding to the foreground of M_{t-1}) is computed using a Gaussian weighting kernel function. The iterative (EM-like) tracking algorithm in [14] is initialized based on the computed ellipsoid and appearance models. The tracking thus performed, results in a prediction of the position and covariance of the ellipsoid representing the tracked object. Based on the transformation parameters of the ellipsoid between $t-1$ and t, a 2D spatial affine transformation of the foreground object mask M_{t-1} is performed. The propagated object mask $M_t^{'}$ indicates the predicted position and shape of the object in the new frame I_t. The Hausdorff distance between the contour points of M_{t-1} and $M_t^{'}$ masks is then computed and a *shape band*, as in [4, 9], around the $M_t^{'}$ contour points is determined, denoted as B_t. The width of B_t is equal to the computed Hausdorff distance of the two contours. This is performed to guarantee that the shape band contains the actual contour pixels of the tracked object in the new frame. Additionally, the pixel-wise Distance Transform likelihoods for the object and background areas are computed together with the pixel-wise color likelihoods based on region-based color histograms. Pixel-wise Bayesian inference is applied to fuse spatial and color image cues, in order to compute the probability distribution for the object and the background regions. Given the estimated Probability Density Functions (PDFs) for each region, a Random Walker-based segmentation algorithm is finally employed to obtain M_t in I_t.

In the following sections, the components of the proposed method are described in more detail.

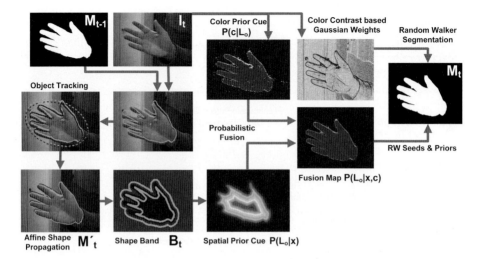

Fig. 2. Sample intermediate results of the proposed method. To avoid clutter, results related to the processing of the scene background are omitted.

2.1   Object Tracking

This section presents the tracking part of the proposed combined tracking and segmentation method (see the bottom-left part of Fig.1).

EM-Like Color Based Object Tracking: The tracking method [14] used in this work is closely related to the widely-used mean-shift tracking method [11, 15]. More specifically, this algorithm coarsely represents the objects' shape by a 2D ellipsoid region, modeled by its center $\boldsymbol{\theta}$ and covariance matrix Σ. The appearance model of the tracked object is represented by the color histogram of the image pixels under the 2D ellipsoid region corresponding to $\boldsymbol{\theta}$ and Σ, and is computed using a Gaussian weighting kernel function. Provided M_{t-1} and I_{t-1}, $\boldsymbol{\theta}_{t-1}$, Σ_{t-1} the object appearance model can be computed for time $t-1$. Given a new image frame I_t where the tracked object is to be localized, the tracking algorithm evolves the ellipsoid region in order to determine the image area in I_t that best matches the appearance of the tracked object in terms of a Bhattacharrya coefficient-based color similarity measure. This gives rise to the parameters $\boldsymbol{\theta}_t$ and Σ_t that represent the predicted object position and covariance in I_t.

Affine Propagation of Object Shape: The tracking algorithm presented above assumes that the shape of an object can be accurately represented as an ellipse. In the general case, this is a quite limiting assumption, therefore the objects' appearance model is forced to include background pixels, causing tracking to drift. The goal of this work is to prevent tracking drifts by integrating tracking with fine object segmentation.

To accomplish that, the contour C_{t-1} of the object mask in M_{t-1} is propagated to the current frame I_t based on the transformation suggested by the parameters $\boldsymbol{\theta}_{t-1}$, $\boldsymbol{\theta}_t$, Σ_{t-1} and Σ_t. A 2D spatial, affine transformation is defined between the corresponding ellipses. Exploiting the obtained Σ_{t-1} and Σ_t covariance matrices, a linear 2×2 affine transformation matrix A_t can be computed based on the square root $(\Sigma^{1/2})$ of each of these matrices. It is known that a covariance matrix is a square, symmetric and positive semidefinite matrix. The square root of any 2×2 covariance matrix Σ can be calculated by diagonalization as

$$\Sigma^{1/2} = Q\Lambda^{1/2}Q^{-1}, \tag{1}$$

where Q is the square 2×2 matrix whose i^{th} column is the eigenvector q_i of Σ and $\Lambda^{1/2}$ is the diagonal matrix whose diagonal elements are the square values of the corresponding eigenvalues. Since Σ is a covariance matrix, the inverse of its Q matrix is equal to the transposed matrix Q^T, therefore $\Sigma^{1/2} = Q\Lambda^{1/2}Q^T$. Accordingly, we compute the transformation matrix A_t by:

$$A_t = Q_t\Lambda_t^{1/2}\Lambda_{t-1}^{-1/2}Q_{t-1}^T. \tag{2}$$

Finally, C_t' is derived from C_t based on the following transformation

$$C_t' = A_t(C_t - \boldsymbol{\theta}_{t-1}) + \boldsymbol{\theta}_t. \tag{3}$$

The result indicates a propagated contour C_t', practically a propagated object mask M_t' that serves as a prediction of the position and the shape of the tracked object in the new frame I_t.

2.2   Object Segmentation

This section presents how the pixel-wise posterior values on spatial and color image cues are computed and fused using Bayesian inference in order to guide the segmentation of the tracked foreground object (see the right part of Fig.1).

Object Shape Band: An object shape band B_t is determined around the predicted object contour C_t'. Our notion of shape band is similar to those used in [4, 9]. B_t can be regarded as an area of uncertainty, where the true object contour lies in image I_t. The width of B_t is determined by the Euclidean, 2D Hausdorff distance between contours C_{t-1} and C_t' regarded as two point sets.

Spatial Image Cues: We use the Euclidean 2D Distance Transform to compute the probability of a pixel x_i in image I_t to belong to either the object L_o or the background L_b region, based on its 2D location $x_i = (x, y)$ on the image plane. As a first step, the shape band B_t of the propagated object contour C_t' is considered and its inner and outer contours are extracted. The Distance Transform is then computed starting from the outer contour of B_t towards the inner part of the object. The probability $P(L_o|x_i)$ of a pixel to belong to the object given its image location is set proportional to its normalized distance from the outer contour of the shape band. For pixels that lie outside the outer contour of B_t, it holds that $P(L_o|x_i) = \epsilon$, where ϵ is a small constant.

Similarly, we compute the Euclidean Distance Transform measure starting from the inner contour of B_t towards the exterior part of the object. The probability $P(L_b|x_i)$ of a pixel to belong to the background given its image location is set proportional to its normalized distance from the inner contour of the shape band. For pixels that lie inside the inner contour of B_t, it holds that $P(L_b|x_i) = \epsilon$.

Color Based Image Cues: Based on the segmentation M_{t-1} of the image frame I_{t-1}, we define a partition of image pixels Ω into sets Ω_o and Ω_b indicating the object and background image pixels, respectively. The appearance model of the tracked object is represented by the color histogram H_o computed on the Ω_o set of pixels. The normalized value in a histogram bin c encodes the conditional probability $P(c|L_o)$. Similarly, the appearance model of the background region is represented by the color histogram H_b, computed over pixels in Ω_b and encoding the conditional probability $P(c|L_b)$.

Probabilistic Fusion of Image Cues: Image segmentation can be considered as a pixel-wise classification problem for a number of classes/labels. Our goal is to generate the posterior probability distribution for each of the labels L_o and L_b, which will be further utilized to guide the Random Walker-based image segmentation. Using Bayesian inference, we formulate a probabilistic framework to efficiently fuse the available prior image cues, based on the pixel color and position information, as described earlier. Considering the pixel color c as the evidence and conditioning on pixel position x_i in image frame I_t, the posterior probability distribution for label L_l is given by

$$P(L_l \mid c, x_i) = \frac{P(c \mid L_l, x_i)P(L_l \mid x_i)}{\sum_{l=0}^{N} P(c \mid L_l, x_i)P(L_l \mid x_i)}, \tag{4}$$

where $N = 2$ in our case. The probability distribution $P(c \mid L_l, x_i)$ encodes the conditional probability of color c taking the pixel label L_l as the evidence and conditioning on its location x_i. We assume that knowing the pixel position x_i, does not affect our belief about its color c. Thus, the probability of color c is only conditioned on the prior knowledge of its class L_l following that $P(c \mid L_l, x_i) = P(c \mid L_l)$. Given this, Eq.(4) transforms to

$$P(L_l \mid c, x_i) = \frac{P(c \mid L_l)P(L_l \mid x_i)}{\sum_{l=0}^{N} P(c \mid L_l)P(L_l \mid x_i)}. \tag{5}$$

The conditional color probability $P(c \mid L_l)$ for the class L_l is obtained by the color histogram H_l. The conditional spatial probability $P(L_l \mid x_i)$ is obtained by the Distance-Transform measure calculation. Both of these calculations have been presented earlier in this section.

Random Walker Based Object Fine Segmentation: The resulting posterior distribution $P(L_l \mid c, x_i)$ for each of the two labels L_o and L_b on pixels x_i guides the Random Walker-based image segmentation towards an explicit and accurate segmentation of the tracked object in I_t.

Random Walks for image segmentation was introduced in [18] as a method to perform K-way graph-based image segmentation given a number of pixels with user (or automatically) defined labels, indicating the K disjoint regions in a new image that is to be segmented. The principal idea behind the method is that one can analytically determine the real-valued probability that a random walker starting at each unlabeled image pixel will first reach one of the pre-labeled pixels. The random walker-based framework bears some resemblance to the popular graph-cuts framework for image segmentation, as they are both related to the spectral clustering family of algorithms [19], but they also exhibit significant differences concerning their properties, as described in [17].

The algorithm is formulated on a discrete weighted undirected graph $G = (V, E)$, where nodes $u \in V$ represent the image pixels and the positive-weighted edges $e \in E \subseteq V \times V$ indicate their local connectivity. The solution is calculated analytically by solving $K\text{-}1$ sparse, symmetric, positive-definite linear systems of equations, for K labels. For each graph node, the resulting probabilities of the potential labels sum up to 1.

In order to represent the image structure by random walker biases, we map the edge weights to positive weighting scores computed by the Gaussian weighting function on the normalized Euclidean distance of the color intensities between two adjacent pixels, practically the color contrast. The Gaussian weighting function is

$$w_{i,j} = e^{-\frac{\beta}{\rho}(\|c_i - c_j\|)^2} + \epsilon, \tag{6}$$

where c_i stands for the vector containing the color channel values of pixel/node i, ϵ is a small constant (i.e $\epsilon = 10^{-6}$) and ρ is a normalizing scalar $\rho = \max(\|c_i - c_j\|), \forall i, j \in E$. The parameter β is user-defined and modulates the spatial random walker biases, in terms of image edgeness. The posterior probability distribution $P(L_l \mid c, x_i)$ computed over the pixels x_i of the current image I_t suggest the probability of the pixels to be assigned to the label L_l. Therefore, we consider the pixels of highest posterior probability values for the label L_l as pre-labeled/seeds nodes of that label in the formulated graph.

An alternative formulation of the Random Walker-based image segmentation method is presented in [16]. This method incorporates non-parametric probability models, that is, prior belief on label assignments. In [16], the sparse linear systems of equations that need to be solved to obtain a real-valued density-based multilabel image segmentation are also presented. The two modalities of this alternative formulation suggest for using only prior knowledge on the belief of a graph node toward each of the potential labels, or using prior knowledge in conjunction with pre-labeled/seed graph nodes. The γ scalar weight parameter is introduced in these formulations, controlling the degree of effectiveness of the prior belief values towards the belief information obtained by the random walks. This extended formulation of using both seeds and prior beliefs on graph nodes is compatible with our approach considering the obtained posterior probability distributions $P(L_l \mid c, x_i)$ for the two segmentation labels. The two Random Walker formulations that use prior models, suggest for a graph construction similar to the graph-cut algorithm [20], where the edge weights of the constructed graph can be seen as the *N-links* or *link-terms* and the prior belief values of the graph nodes for any of the potential labels can be considered as the *T-links* or the *data-terms*, in graph cuts terminology.

Regardless of the exact formulation used, the primary output of the algorithm consists of K probability maps, that is a soft image segmentation per label. By assigning each pixel to the label for which the greatest probability is calculated, a K-way segmentation is obtained. This process gives rise to object mask M_t for image frame I_t.

3   Experimental Results and Implementation Issues

The proposed method was extensively tested on a variety of image sequences. Due to space limitations, results on eight representative image sequences are presented in this paper. The objects tracked in these sequences go through extensive appearance, shape and pose changes. Additionally, these sequences differ with respect to the camera motion and to the lighting conditions during image acquisition which affects the appearance of the tracked objects.

We compare the proposed joint tracking and segmentation method with the tracking-only approach of [14]. The parameters of this algorithm were kept identical in the stand-alone run and in the run within the proposed framework. It is important to note that stand-alone tracking based on [14] is initialized with the appearance model extracted in the first frame of the sequence and that this appearance model is not updated over time. This is done because in all the challenging sequences we used as the basis of our evaluation, updating the appearance model based on the results of tracking, soon causes tracking drifts and total loss of the tracked object.

Figure 3 illustrates representative tracking results (i.e., five frames for each of the eight sequences). In the first sequence, a human hand undergoes complex articulations, whereas the lighting conditions significantly affect its skin color tone. In the second sequence, a human head is tracked despite its abrupt scale changes and the lighting variations. In the third sequence the articulations of a human hand are observed by a moving camera in the context of a continuously varying cluttered background. The green book tracked in the fourth sequence undergoes significant changes regarding its pose and shape, whereas light reflections on its glossy surface significantly affect its appearance. The fifth sequence is an example of a low quality video captured by a moving camera, illustrating the inherently deformable body of a caterpillar in motion. The sixth and seventh sequences show a human head and hand, respectively, which both go through extended pose variations in front of a complex background. Finally, the last, low resolution sequence has been captured by a medical endoscope. In this sequence, a target object is successfully tracked within a low-contrast background.

Each of the image sequences in Fig.(3) illustrating human hands or faces as well as the green book sequence consists of 400 frames of resolution 640×480 pixels, captured at a frame rate of 5-10 fps. The resolution of each frame of the image sequences illustrated in the second and the fifth row is 320×240 pixels. The last image sequence depicted in the Fig.(3), captured by a medical endoscope consists of 20 image frames of size 256×256 pixels each.

The reported experiments were generated based on a Matlab implementation, running on a PC equipped with an Intel i7 CPU and 4 GB of RAM memory. The runtime performance of the current implementation varies between 4 to 6 seconds per frame for 640×480 images. A near real-time runtime performance is feasible by optimizing both the EM-like component of the tracking method and the solution of the large sparse linear system of equations of the Random Walker formulation in the segmentation procedure.

Each frame shown in Fig.(3) is annotated with the results of the proposed algorithm and the results of the tracking method proposed in [14]. More specifically, the blue solid ellipse shows the expected position and coarse orientation of the tracked object as this results from the tracking part of the proposed methodology. The green solid object contour is the main result of the proposed algorithm which shows the fine object segmentation. Finally, the result of [14] is shown for comparison as a red dotted ellipse. Experimental results on the full video datasets are available online[1].

In all sequences, the appearance models of the tracked objects have been built based on the RGB color space. The object and background appearance models used to compute the prior color cues are color histograms with 32 bins per histogram for both the object and the background. Preserving the parameter configuration of the object tracking algorithm as described in [14], the target appearance model of the tracker is implemented by a color histogram of 8 bins per dimension.

The Random Walker segmentation method involves three different formulations to obtain the probabilities of each pixel to belong to each of the labels of the segmentation problem, as described in Sec. 2.2. The three options refer to the usage of seed pixels (pre-labeled graph nodes), prior values (probabilities/beliefs on label assignments for some graph nodes), or a combination of them. For the last option, the edge weights of the graph are computed by the Eq.(6), where the β scalar parameter controls the scale of the edgeness (color contrast) between adjacent graph nodes. The pixel-wise posterior values are computed using Bayesian inference as described in Sec. 2.2 and are exploited to guide segmentation as seed and prior values in terms of Random Walker terminology. Each pixel x_i of posterior value $P(L_l \mid x_i)$ greater or equal to 0.9 is considered as a seed pixel for the label L_l, thus as a seed node on the graph G. Any other pixel of posterior value $P(L_l \mid x_i)$ less than 0.9 is considered as a prior value for label L_l. In the case of prior values, the γ parameter is introduced to adjust the degree of authority of the prior beliefs towards the definite label-assignments expressed by

---

[1] http://www.ics.forth.gr/~argyros/research/trackingsegmentation.htm

Fig. 3. Experimental results and qualitative comparison between the proposed framework providing tracking and segmentation results (blue solid ellipse and green solid object contour, respectively) and the tracking algorithm of [14] (red dotted ellipse). See text for details.

Table 1. Quantitative assessment of segmentation accuracy. See text for details.

Segmentation option	Precision	Recall	F-measure
Priors	93,5%	92,9%	93,1%
Seeds	97,5%	99,1%	98,3%
Priors and Seeds	97,5%	99,1%	98,3%

the seed nodes of the graph. In our experiments, the β parameter was selected within the interval of $[10 - 50]$, whereas the γ ranges within $[0.05 - 0.5]$.

In order to assess quantitatively the influence of the three different options regarding the operation of the Random Walker on the quality of segmentation results, the three different variants have been tested independently on an image sequence consisting of $1,000$ video frames. For each and every of these frames ground truth information is available in the form of a manually segmented foreground object mask. Table 1 summarizes the average (per frame) precision, recall and F-measure performance of the proposed algorithm compared to the ground truth. As it can be verified, although all three options perform satisfactorily, the use of seeds improves the segmentation performance.

4   Summary

In this paper we presented a method for online, joint tracking and segmentation of an non-rigid object in a monocular video, captured by a possibly moving camera. The proposed approach aspires to relax several limiting assumptions regarding the appearance and shape of the tracked object, the motion of the camera and the lighting conditions. The key contribution of the proposed framework is the efficient combination of an appearance-based tracking with Random Walker-based segmentation that jointly enables enhanced tracking performance and fine segmentation of the target object. A 2D affine transformation is computed to propagate the segmented object shape of the previous frame to the new frame exploiting the information provided by the ellipse region capturing the segmented object and the ellipse region predicted by the tracker in the new frame. A shape-band area is computed indicating an area of uncertainty where the true object boundaries lie in the new frame. Static image cues including pixelwise color and spatial likelihoods are fused using Bayesian inference to guide the Random Walker-based object segmentation in conjunction with the color-contrast (edgeness) likelihoods between neighboring pixels. The performance of the proposed method is demonstrated in a series of challenging videos and in comparison with the results of the tracking method presented in [14].

Acknowledgments

This work was partially supported by the IST-FP7-IP-215821 project GRASP. The contributions of FORTH-ICS members I. Oikonomidis and N. Kyriazis to the development of the proposed method are gratefully acknowledged.

References

1. Yilmaz, A., Javed, O., Shah, M.: Object tracking: A survey. ACM Comput. Surv. 38, 13 (2006)
2. Isard, M., Blake, A.: Condensation: Conditional density propagation for visual tracking. International Journal of Computer Vision 29, 5–28 (1998)
3. Paragios, N., Deriche, R.: Geodesic active contours and level sets for the detection and tracking of moving objects. IEEE Transactions on PAMI 22, 266–280 (2000)
4. Yilmaz, A., Li, X., Shah, M.: Contour-based object tracking with occlusion handling in video acquired using mobile cameras. IEEE Transactions on PAMI 26, 1531–1536 (2004)
5. Bibby, C., Reid, I.: Robust real-time visual tracking using pixel-wise posteriors. In: Forsyth, D., Torr, P., Zisserman, A. (eds.) ECCV 2008, Part II. LNCS, vol. 5303, pp. 831–844. Springer, Heidelberg (2008)
6. Khan, S., Shah, M.: Object based segmentation of video using color, motion and spatial information. In: IEEE Computer Society Conference on CVPR, vol. 2, p. 746 (2001)
7. Baltzakis, H., Argyros, A.A.: Propagation of pixel hypotheses for multiple objects tracking. In: Bebis, G., Boyle, R., Parvin, B., Koracin, D., Kuno, Y., Wang, J., Pajarola, R., Lindstrom, P., Hinkenjann, A., Encarnação, M.L., Silva, C.T., Coming, D. (eds.) ISVC 2009. LNCS, vol. 5876, pp. 140–149. Springer, Heidelberg (2009)
8. Yu, T., Zhang, C., Cohen, M., Rui, Y., Wu, Y.: Monocular video foreground/background segmentation by tracking spatial-color gaussian mixture models. In: IEEE Workshop on Motion and Video Computing (2007)
9. Yin, Z., Collins, R.T.: Shape constrained figure-ground segmentation and tracking. In: IEEE Computer Society Conference on CVPR, pp. 731–738 (2009)
10. Ren, X., Malik, J.: Tracking as repeated figure/ground segmentation. In: IEEE Computer Society Conference on CVPR, pp. 1–8 (2007)
11. Comaniciu, D., Ramesh, V., Meer, P.: Kernel-based object tracking. IEEE Transactions on PAMI 25, 564–577 (2003)
12. Tao, H., Sawhney, H., Kumar, R.: Object tracking with bayesian estimation of dynamic layer representations. IEEE Transactions on PAMI 24, 75–89 (2002)
13. Jepson, A.D., Fleet, D.J., El-Maraghi, T.F.: Robust online appearance models for visual tracking. IEEE Transactions on PAMI 25, 1296–1311 (2003)
14. Zivkovic, Z., Krose, B.: An em-like algorithm for color-histogram-based object tracking. In: IEEE Computer Society Conference on CVPR, vol. 1, pp. 798–803 (2004)
15. Comaniciu, D., Ramesh, V., Meer, P.: Real-time tracking of non-rigid objects using mean shift. In: IEEE Computer Society Conference on CVPR, vol. 2, p. 2142 (2000)
16. Grady, L.: Multilabel random walker image segmentation using prior models. In: Proceedings of the 2005 IEEE Computer Society Conference on CVPR, vol. 1, pp. 763–770 (2005)
17. Grady, L.: Random walks for image segmentation. IEEE Transactions on PAMI 28, 1768–1783 (2006)
18. Grady, L., Funka-Lea, G.: Multi-label image segmentation for medical applications based on graph-theoretic electrical potentials. In: Sonka, M., Kakadiaris, I.A., Kybic, J. (eds.) CVAMIA/MMBIA 2004. LNCS, vol. 3117, pp. 230–245. Springer, Heidelberg (2004)
19. von Luxburg, U.: A tutorial on spectral clustering. Statistics and Computing 17, 395–416 (2007)
20. Boykov, Y., Funka-Lea, G.: Graph cuts and efficient n-d image segmentation. International Journal of Computer Vision 70, 109–131 (2006)

Optical Flow Estimation with Prior Models Obtained from Phase Correlation

Alfonso Alba[1], Edgar Arce-Santana[1], and Mariano Rivera[2]

[1] Facultad de Ciencias, Universidad Autónoma de San Luis Potosí,
Diagonal Sur S/N, Zona Universitaria, 78290, San Luis Potosí, SLP, México
fac@fc.uaslp.mx, arce@fciencias.uaslp.mx
[2] Centro de Investigacion en Matematicas A.C., AP 402,
Guanajuato, Gto., 36000, Mexico
mrivera@cimat.mx

Abstract. Motion estimation is one of the most important tasks in computer vision. One popular technique for computing dense motion fields consists in defining a large enough set of candidate motion vectors, and assigning one of such vectors to each pixel, so that a given cost function is minimized. In this work we propose a novel method for finding a small set of adequate candidates, making the minimization process computationally more efficient. Based on this method, we present algorithms for the estimation of dense optical flow using two minimization approaches: one based on a classic block-matching procedure, and another one based on entropy-controlled quadratic Markov measure fields which allow one to obtain smooth motion fields. Finally, we present the results obtained from the application of these algorithms to examples taken from the Middlebury database.

1 Introduction

Optical flow is the apparent velocity of a moving object in a two-dimensional image. This does not always correspond to the true velocity of the object in the 3D scene; for example, when an object approaches the viewer along the camera axis, its 3D velocity vector is parallel to this axis, but the optical flow field will correspond to a radial pattern where each pixel seems to be moving from the center outwards. Nevertheless, optical flow has numerous applications in computer and robot vision, autonomous vehicle navigation, video encoding, and video stabilization, among others. Formally, each point in the 3D moving scene is projected to a point $\mathbf{x}(t) = (x(t), y(t))$ in the 2D image plane, where t denotes time. Therefore, one can define the optical flow $\mathbf{v}(\mathbf{x}, t)$ as $\mathbf{v}(\mathbf{x}, t) = d\mathbf{x}/dt$. Discretization of t leads to the alternative definition $\mathbf{v}(\mathbf{x}, t) = \mathbf{x}(t) - \mathbf{x}(t-1)$.

Numerous methods have been proposed in the literature for the estimation of optical flow; some of these are based on gradient estimation [1], global energy minimization [2], and block matching [3]. For a comprehensive overview and comparison of the most relevant methods, please refer to [4] and [5]. The

G. Bebis et al. (Eds.): ISVC 2010, Part I, LNCS 6453, pp. 417–426, 2010.

Middlebury College also hosts an evaluation website where newly proposed algorithms can be evaluated using standard datasets [6]. In general, most methods can be classified either as local or global. Local methods try to solve the problem individually for each pixel, considering only the pixel in question and its neighborhood. In contrast, global methods typically propose a global error function of the form $U(\mathbf{v}) = M(\mathbf{v}) + R(\mathbf{v})$, where $M(\mathbf{v})$ is a matching term that penalizes differences between $f(\mathbf{x})$ and $g(\mathbf{x}+\mathbf{v}(\mathbf{x}))$ (also called the optical flow constraint), and $R(\mathbf{v})$ is a regularization term which enforces smoothness of the flow field by penalizing abrupt changes in $\mathbf{v}(\mathbf{x})$, particularly in those areas where $\mathbf{v}(\mathbf{x})$ is assumed to be homogeneous (e.g., areas which belong to the same object). Minimization of $U(\mathbf{v})$ is usually a non-trivial task which is commonly achieved by iterative techniques, such as gradient descent, Gauss-Seidel, or Markov Chain Monte Carlo (MCMC) methods, which are computationally intensive.

Most of the proposed methods, especially those which produce dense results (i.e., one motion vector per pixel), suffer from a high computational cost, which seriously limit their application to realtime systems. At the moment of writing, the top-ranking method in the Middlebury database requires 16 minutes to compute the flow field of the Urban scene [7], and only one of the reported algorithms is capable of near-realtime performance (8 fps), but it requires specialized GPU hardware [8]. Moreover, since optical flow is typically computed between two consecutive frames of a video sequence, processing at high frame rates is desirable because it increases the reliability in the detection of fast-moving objects and non-translational movements.

One popular approach for the realtime estimation of optical flow, consists in defining a large enough set of candidate motion vectors, and then applying some combinatorial optimization scheme for choosing the most adequate candidate for each pixel, so that a given cost function is minimized. For example, in a block matching approach, typical penalty functions are the sum of absolute differences (SAD), sum of squared differences (SSD), and cross-correlation. Other popular optimization methods used include: Bayesian estimation, graph cuts, and belief propagation. Under this approach, the choice of the initial candidate vector set plays an important role: if the set is too small, the algorithm may be unable to recover long displacements and the results will lack precision due to the limited number of choices; if, on the other hand, the candidate set is too large, the algorithm will be computationally more expensive and the penalty function may be more sensitive to occlusions and homogeneous regions (the aperture problem) due to the presence of spurious candidates. A common solution consists in quantizing the search space using a equally-spaced grid of vectors with integer coordinates and a maximum displacement of D along each direction; for example, motion-based video encoding typically uses D between 7 and 15. This solution, however, is far from optimal, since the number of candidate vectors increases quadratically with respect to D; for instance, with $D = 7$, the optimization algorithm would have to choose among 225 candidates per block or pixel, and it would still be unable to recover larger displacements.

This work introduces a novel method to adequately choose a small set of integer-valued candidate vectors which approximately coincide with the true displacements between two images, drastically reducing the computation time during the cost optimization process, while allowing large displacement ranges to be used. The proposed method, which is based on the computation of multiple maxima in the phase correlation function of the two images, is introduced in Section 2.1. In sections 2.2 and 2.3, full algorithms are presented using different combinatorial optimization methods which achieve varying results in terms of quality and computational cost. Section 3 presents results obtained from the application of these algorithms to scenes from the Middlebury database. Finally, some conclusions are drawn in Section 4.

2   Methodology

The phase-only correlation (POC) function $r(k)$ between two functions $f(x)$ and $g(x)$ is defined as the inverse Fourier transform of their normalized cross-spectrum; that is,

$$r(x) = \mathcal{F}^{-1} \left\{ \frac{F(k)G^*(k)}{|F(k)G^*(k)|} \right\}, \tag{1}$$

where $\mathcal{F}^{-1}$ denotes the inverse Fourier transform, F and G are the Fourier transforms of f and g, respectively, and G^* denotes the complex conjugate of G. It can be easily shown that if $g(x) = f(x - d)$, then $r(x) = \delta(x + d)$, where δ is the Dirac impulse function; this means that one can easily estimate the displacement between f and g as $\hat{d} = \arg\max\{r(x)\}$. In practice, the POC function is rarely an impulse due to the presence of noise, occlusions, border effects, and periodic textures which introduce spurious correlations; however, it is still robust enough for a variety of applications. For instance, this method has been successfully used for rigid image registration [9] [10] [11] [12], motion estimation [13], and stereo disparity estimation [14] [15]. Various interpolation schemes have also been devised to achieve subpixel accuracy [16] [17].

While most authors use the POC function to estimate a single motion vector for each pixel, here we propose using POC to obtain a small candidate set for all pixels in a large region of the image. This method is based on the hypothesis that the most significant maxima of the phase-only correlation function between two images in a video sequence are related to the displacement vectors of the objects in the scene.

2.1   Estimation of Prior Models Using Phase Correlation

Let $f(\mathbf{x})$ and $g(\mathbf{x})$ be two consecutive images in a sequence and $L = \{0, \ldots, N_x - 1\} \times \{0, \ldots, N_y - 1\}$ be a finite rectangular lattice of size $N_x \times N_y$ where the images are observed. We first divide L into smaller overlapping sub-lattices $L_1, \ldots, L_Q$ so that any two corresponding points between both images belong to at least one of the sub-lattices. For example, if the largest expected displacement magnitude is D, it makes sense that the size of the sub-lattices must be at least $(2D + 1) \times$

$(2D+1)$. In particular, we use equally-sized sub-lattices of size $W \times H$, where W and H are powers of 2, so that efficient FFT algorithms can be used to estimate the POC function; the separation between sub-lattices is approximately $W/2$ in the horizontal direction, and $H/2$ in the vertical direction; this overlap ensures that a given pixel in one image and its correspondent in the other image will always belong to at least one, but no more than four sub-lattices.

We then compute the POC function $r_q(\mathbf{x})$ between both images at each sub-lattice L_q. Specifically, we obtain sub-images $f_q(\mathbf{x}) = f(\mathbf{x} - \mathbf{x}_q)$ and $g_q(\mathbf{x}) = g(\mathbf{x} - \mathbf{x}_q)$ for $\mathbf{x} \in \{0, \ldots, W-1\} \times \{0, \ldots, H-1\}$, where $\mathbf{x}_q$ is the upper-left corner of L_q, and compute r_q as

$$r_q(\mathbf{x}) = \mathcal{F}^{-1} \left\{ \frac{F_q(\mathbf{k})G_q^*(\mathbf{k})}{|F_q(\mathbf{k})G_q^*(\mathbf{k})|} \right\}, \qquad (2)$$

where F_q and G_q are the Fourier transforms of f_q and g_q, respectively.

The next step is to obtain a set of motion vector candidates from r_q, against which all pixels in L_q will be tested. We do this simply by choosing the M most significant positive maxima from r_q. Although other selection schemes can be used, we found this simple method to perform best. One must keep in mind that $r_q(\mathbf{x})$ may be a circularly shifted version of the true POC function, so that peaks observed in the right half of the image correspond to candidate vectors in the second and third quadrants (negative x), and peaks in the bottom half correspond to candidates in the third and fourth quadrants (negative y).

With this methodology, one can obtain a candidate set $\mathcal{D}_q = \{\mathbf{v}_1, \ldots, \mathbf{v}_M\}$ for each sub-lattice L_q, or alternatively, a candidate set $\mathcal{D}_{\mathbf{x}} = \bigcup_{q:\mathbf{x} \in L_q} \mathcal{D}_q$, for each pixel $\mathbf{x}$, whose cardinality will be between M and $4M$. One can also obtain the full candidate set as $\mathcal{D} = \bigcup_q \mathcal{D}_q$.

2.2   Optical Flow Estimation Using Block-Matching

Once the candidate set is known, one must find, for each pixel $\mathbf{x}$ the best motion vector $\hat{\mathbf{v}}(\mathbf{x}) \in |\mathcal{D}_{\mathbf{x}}|$, according to some matching criteria. This can be done by minimizing a penalty function ρ which measures the differences between pixel values at $\mathbf{x}$ in f, and pixel values at $\mathbf{x} + \mathbf{v}$ in g. This minimization can thus be expressed in the following way:

$$\hat{\mathbf{v}}(\mathbf{x}) = \arg \min_{\mathbf{d} \in \mathcal{D}_{\mathbf{x}}} \{\rho(f(\mathbf{x}) - g(\mathbf{x} + \mathbf{d}))\}. \qquad (3)$$

Unfortunately, the approach described above does not introduce any regularization constraints, resulting in very noisy estimations. A better approach consists in minimizing the total penalty in a given neighborhood around pixel $\mathbf{x}$. For example, using a square-shaped window $\mathcal{W} = \{-w, \ldots, w\} \times \{-w, \ldots, w\}$ as neighborhood results in the following minimization problem:

$$\hat{\mathbf{v}}(\mathbf{x}) = \arg \min_{\mathbf{d} \in \mathcal{D}_{\mathbf{x}}} \left\{ \sum_{\mathbf{r} \in \mathcal{W}} \rho(f(\mathbf{x} - \mathbf{r}) - g(\mathbf{x} + \mathbf{d} - \mathbf{r})) \right\}, \qquad (4)$$

for all $\mathbf{x} \in L$. The matching window size w directly affects the granularity of the results. Small windows produce noisy displacement fields, mainly due to ambiguities in homogeneous regions, whereas large windows will blur object's borders and are more sensitive to projective distortions between the images (e.g., non-translational movements). Some authors suggest using multiple or adaptable windows. Note that, regardless of the window size, the sum between the brackets can be computed very efficiently using integral images [18] or aggregate cost techniques.

Typical choices for the penalty function ρ are the L^1 norm, $\rho(x) = |x|$, the L^2 norm, $\rho(x) = |x|^2$, and the total-variation approximated function $\rho(x) = \sqrt{x^2 + \epsilon^2}$ [19], which is a differentiable version of the more robust L^1 norm. However, since our combinatorial approach does not require the penalty function to be differentiable, we have chosen a slightly more robust function based on a truncated L^1 norm, given by

$$\rho(x) = \min\{|x|, \kappa R\}, \tag{5}$$

where R is the dynamic range of the image and κ is a parameter between 0 and 1.

2.3   Smooth Optical Flow Estimation Using EC-QMMFs

The block-matching approach described above assigns one of the motion vector candidates to each pixel, which results in a piecewise constant flow field. Since there are only a few candidates, this is equivalent to performing a hard segmentation of the reference image with respect to the motion of each object. Although this approach is computationally very efficient (see results below), there are various drawbacks: first, it does not provide a natural mechanism for adding more complex constraints, such as edge preserving smoothness, which are key to obtain competitive results in terms of precision. Second, the POC function can only capture integer displacements; it is possible, however, to apply interpolation methods to obtain or refine candidates with sub-pixel accuracy, but this does not prevent a block-matching scheme from producing over-quantized results. Finally, there are applications where one would like a soft-segmentation instead of a hard one; that is, instead of answering the question *what motion vector corresponds to pixel* $\mathbf{x}$*?*, we would like to answer the question *what is the probability of motion vector* $\mathbf{v}$ *being the best for pixel* $\mathbf{x}$*?* in order to make more informed decisions.

With this in mind, we propose a different estimator for the motion vector $\mathbf{v}(\mathbf{x})$ assigned to each pixel, given by a linear combination of all candidate vectors:

$$\mathbf{v}(\mathbf{x}) = \sum_{\mathbf{d} \in \mathcal{D}} b_{\mathbf{d}}(\mathbf{x})\mathbf{d}. \tag{6}$$

where $b_{\mathbf{d}}(\mathbf{x})$ is the weight of candidate $\mathbf{d}$ at pixel $\mathbf{x}$, and $\sum_{\mathbf{d} \in D} b_{\mathbf{d}}(\mathbf{x}) = 1$ for all $\mathbf{x}$. Computing these weights is a very ill-posed problem, so one must impose certain constraints to obtain an unique solution. For instance, one would like the weight field $b_{\mathbf{d}}$ to be spatially piecewise smooth, so that similar weights are observed across regions which belong to the same object. Smoothness constraints can be easily handled if we assume that b follows a Markovian model; e.g., the weights $b(\mathbf{x})$ for a given pixel $\mathbf{x}$ depend (probabilistically) on the weights of those pixels in a neighborhood of $\mathbf{x}$.

On the other hand, note that $b(\mathbf{x})$ can be seen as a probability distribution (i.e., a probability measure) over the candidate vectors for each pixel $\mathbf{x}$. If, for some pixel $\mathbf{x}$, the distribution $b(\mathbf{x})$ is nearly uniform (i.e., highly entropic), then the estimator in Eq. 6 will approach zero as the number of candidate vectors increase. This is because given a sufficiently large candidate set, each candidate will more likely be in the opposite direction to some other. For this reason, one would like the measures $b(\mathbf{x})$ to have low entropy.

One efficient model which allows one to control the degree of smoothness of a field of probability measures, while maintaining control over their entropy, is the Entropy-Controlled Quadratic Markov Measure Field model (EC-QMMF) [20]. The model is described by the energy function $U(b)$ of the field b, given by

$$U(b) = -\sum_{x \in L} \sum_{\mathbf{d} \in \mathcal{D}} b_{\mathbf{d}}^2(\mathbf{x}) \left[-\log \hat{b}(\mathbf{x}) - \mu \right] + \lambda \sum_{<\mathbf{x},\mathbf{y}>} \beta_{<\mathbf{x},\mathbf{y}>} \|b(\mathbf{x}) - b(\mathbf{y})\|^2 , \quad (7)$$

subject to

$$b_{\mathbf{d}}(\mathbf{x}) \geq 0, \ \mathbf{d} \in \mathcal{D}, \ \mathbf{x} \in L, \tag{8}$$

$$\sum_{\mathbf{d} \in \mathcal{D}} b_{\mathbf{d}}(\mathbf{x}) = 1, \ \mathbf{x} \in L, . \tag{9}$$

Here, $\hat{b}$ is the normalized likelihood measure, given by

$$\hat{b}_{\mathbf{d}}(\mathbf{x}) = \exp\{-\rho(f(\mathbf{x}) - g(\mathbf{x} + \mathbf{d}))/(\kappa R)\}/Z(\mathbf{x}), \tag{10}$$

where $Z(\mathbf{x})$ is a positive normalization constant chosen so that $\sum_{\mathbf{d}} \hat{b}_{\mathbf{d}}(\mathbf{x}) = 1$, μ is a hyperparameter which controls the entropy of the $b(\mathbf{x})$, λ is regularization parameter that controls the degree of smoothing. The sum in the second term of Eq. 7 (the regularization term) ranges over all nearest-neighbors $< \mathbf{x}, \mathbf{y} >$. Finally, the coefficients $\beta_{<\mathbf{x},\mathbf{y}>}$ are used to reduce oversmoothing at the objects' edges by locally modulating the regularization parameter λ: for a given $\epsilon < 1$, we set $\beta_{<\mathbf{x},\mathbf{y}>} = \epsilon$ if there exists an edge in I_1 between pixels $\mathbf{x}$ and $\mathbf{y}$; otherwise, we let $\beta_{<\mathbf{x},\mathbf{y}>} = (1 - \epsilon)$.

An important property of the EC-QMMF energy function (Eq. 7) is that it is quadratic on its unknowns; therefore, it can be minimized very efficiently by solving a linear system with constraints. Constraint (9) may be handled using the method of Lagrange multipliers, so that the system remains linear. Constraint (8) is handled simply by setting any negative $b_{\mathbf{d}}(\mathbf{x})$ to zero and renormalizing $b(\mathbf{x})$ after each Gauss-Seidel iteration (see [20] for details). Note that, one can set $b_{\mathbf{d}}(\mathbf{x}) = 0$ for all $\mathbf{d} \notin \mathcal{D}_{\mathbf{x}}$, and only update $b_{\mathbf{d}}(\mathbf{x})$ for $\mathbf{d} \in \mathcal{D}_{\mathbf{x}}$ during the Gauss-Seidel iterations, making the minimization process more efficient.

3   Preliminary Results

We have implemented the algorithms described above in C using the OpenCV library [21] to perform some of the computations. Our aim is not to compete with state of the art methods in terms of quality or accuracy, but to show how the computation of multiple POC peaks may provide a way to derive efficient optical flow algorithms capable of obtaining usable results for realtime or near-realtime applications. Since the well-known Middlebury evaluation system [6] focuses on accuracy, and does not provide information about the test platform of each of the reported methods (which may vary from slow Matlab code to fast GPU or FPGA implementations), we decided not to include our Middlebury results in this work, and performed instead a quantitative comparison between our method and the OpenCV implementations of the classic Lucas-Kanade (LK) [1] and Horn-Schunck (HS) [2] methods. While these are not state-of-the-art methods, they are relatively efficient and provide a good comparison baseline under the same test platform. All tests have been performed in an Intel Core2Duo workstation running at 2.4 Ghz with 2 Gb of RAM; however, the code has not been thoroughly optimized to take advantage of parallel processing (e.g., by means of SSE2 instructions and multi-threading). In the case of the EC-QMMF method, we were not aiming for realtime performance; therefore, we applied an additional median filter to the magnitude and angle of the resulting flow field.

The parameters used for the estimation of the candidate set (Section 2.1) are: sublattice size $W = H = 128$, number of candidates per sublattice $M = 6$, and maximum displacement magnitude $D = 32$. For the block matching approach we use a matching window of size $w = 7$. The penalty function used is the truncated L^1 norm (5) with $\kappa = 0.2$. For the EC-QMMF approach, we use $\epsilon = 0$ (no regularization at the edges), $\lambda = 3$ and $\mu = 1$, and perform 20 and 100 iterations of the Gauss-Seidel method. For the OpenCV algorithms, we have used a 15×15 window for the LK method, and $\lambda = 0.001$ for HS, with 1000 iterations.

Our results are summarized in Table 1, where the following methods are compared: Lucas-Kanade in OpenCV (LK), Horn-Schunck in OpenCV (HS), Block Matching with prior models (P1), and EC-QMMF with prior models and 20 Gauss-Seidel iterations (P2) or 100 iterations (P3). For each test scene from the Middlebury database [6], we compute the average end-point error (in pixels), the average angular error (in degrees), and the computation time (in seconds). Figure 1 shows the resulting flow fields obtained with the proposed approaches. Note that the Block Matching approach with prior POC models (Algorithm P1) provides a similar computational performance than OpenCV's LK implementation, but with a much improved accuracy. It is worth noting that, at the time of publication, the computation times reported in the Middlebury database for the Urban sequence (from the evaluation set) typically range from a few seconds to several minutes, with only two methods reporting times of less than one second (0.12 and 0.97 s); however, these two methods are based on GPU implementations. On the other hand, our block-matching based method can process the Urban scene in 0.275 s, and is capable of processing 320×240 sequences at

Table 1. Comparative results using the Middlebury test set as benchmark. Methods tested are Lucas-Kanade (LK), Horn-Schunck (HS), and the proposed algorithms using prior models obtained from the POC function: Block Matching (P1), EC-QMMF with 20 iterations (P2), and EC-QMMF with 100 iterations (P3).

End-point error (pixels)

	Dimetrodon	Grove2	Grove3	Hydrangea	Rubberwhale	Urban2	Urban3	Venus
LK	1.443	2.872	3.543	3.166	0.497	8.424	7.471	3.620
HS	1.521	2.815	3.444	3.111	0.469	7.922	6.899	3.369
P1	0.452	0.543	1.027	0.632	0.412	1.765	2.086	0.565
P2	0.449	0.530	0.998	0.658	0.373	1.673	2.042	0.477
P3	0.434	0.516	1.007	0.679	0.359	1.607	1.963	0.443

Angular error (degrees)

	Dimetrodon	Grove2	Grove3	Hydrangea	Rubberwhale	Urban2	Urban3	Venus
LK	34.97	61.16	51.02	50.69	14.54	54.68	61.85	56.70
HS	37.08	60.77	49.61	46.61	13.97	49.24	60.80	52.20
P1	9.60	7.09	10.83	7.14	11.79	13.15	17.29	7.24
P2	9.55	6.78	10.87	7.66	10.15	13.10	16.82	5.03
P3	9.27	6.62	10.95	7.67	9.89	12.48	16.32	4.44

Processing time (seconds)

	Dimetrodon	Grove2	Grove3	Hydrangea	Rubberwhale	Urban2	Urban3	Venus
LK	0.118	0.160	0.158	0.120	0.117	0.158	0.158	0.090
HS	2.465	4.377	5.559	3.201	3.366	5.497	6.829	3.398
P1	0.107	0.162	0.139	0.104	0.103	0.144	0.141	0.053
P2	2.200	2.751	3.680	3.421	2.075	4.922	4.727	1.218
P3	6.857	7.991	10.255	9.645	6.658	14.523	14.434	4.080

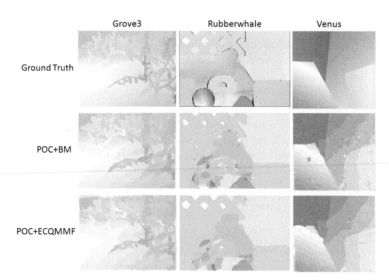

Fig. 1. Results obtained with Block Matching and EC-QMMF (100 iterations) using prior models obtained from phase correlation

25 fps in a typical PC without careful optimization. The EC-QMMF approach demands considerably more computational resources, however, it is still fairly efficient and also incorporates additional control over the smoothness and edge definition of the resulting fields, providing a viable platform for the development of efficient high quality motion estimation methods.

4   Conclusions

In this work, a technique for the estimation of prior models (motion vector candidates) for combinatorial optical flow estimation algorithms is presented. These models are obtained from the maxima of the phase-only correlation in a relatively large area of the image. Our hypothesis that the estimated models correspond to the true displacements is verified experimentally with satisfactory results. The main advantage of the proposed method is that it may provide a very reduced set of candidates (either for each pixel, region, or for the full image), making the optimization process significatively more efficient in computational terms. Although our preliminary tests were performed in a typical dual core PC, implementation on parallel architectures such as FPGAs or GPUs is expected to run at high frame rates.

Acknowledgements. This work was supported by grant PROMEP/UASLP/-10/CA06. A. Alba was partially supported by grant PROMEP/103.5/09/573.

References

1. Lucas, B.D., Kanade, T.: An Iterative Image Registration Technique with an Application to Stereo Vision. In: Proc. of Imaging Understanding Workshop, pp. 121–130 (1981)
2. Horn, B.K.P., Schunck, B.G.: Determining Optical Flow. Artificial Intelligence 17, 185–203 (1981)
3. Essannouni, F., Haj Thami, R.O., Salam, A., Aboutajdine, D.: An efficient fast full search block matching algorithm using FFT algorithms. International Journal of Computer Science and Network Security 6, 130–133 (2006)
4. Barron, J., Fleet, D., Beauchemin, S.: Performance of optical flow techniques. International Journal of Computer Vision 12, 43–77 (1994)
5. Baker, S., Scharstein, D., Lewis, J.P., Roth, S., Black, M.J., Szeliski, R.: A database and evaluation methodology for optical flow. Technical Report MSR-TR-2009-179, Microsoft Research (2009)
6. Baker, S., Scharstein, D., Lewis, J.P., Roth, S., Black, M., Szeliski, R.: Middlebury stereo vision page (2007), http://vision.middlebury.edu/flow/
7. Sun, D., Roth, S., Black, M.J.: Secrets of Optical Flow Estimation and Their Principles. In: Proc. IEEE CVPR 2010. IEEE, Los Alamitos (2010)
8. Rannacher, J.: Realtime 3D motion estimation on graphics hardware. Bachelor thesis, Heidelberg University (2009)
9. De Castro, E., Morandi, C.: Registration of Translated and Rotated Images Using Finite Fourier Transforms. IEEE Transactions on Pattern Analysis and Machine Intelligence 9, 700–703 (1987)

10. Reddy, B.S., Chatterji, B.N.: An FFT-Based Technique for Translation, Rotation, and Scale-Invariant Image Registration. IEEE Transactions on Image Processing 5, 1266–1271 (1996)
11. Keller, Y., Averbuch, A., Moshe, I.: Pseudopolar-based estimation of large translations, rotations, and scalings in images. IEEE Transactions on Image Processing 14, 12–22 (2005)
12. Keller, Y., Shkolnisky, Y., Averbuch, A.: The angular difference function and its application to image registration. IEEE Transactions on Pattern Analysis and Machine Intelligence 27, 969–976 (2005)
13. Chien, L.H., Aoki, T.: Robust Motion Estimation for Video Sequences Based on Phase-Only Correlation. In: 6th IASTED International Conference on Signal and Image Processing, pp. 441–446 (2004)
14. Takita, K., Muquit, M.A., Aoki, T., Higuchi, T.: A Sub-Pixel Correspondence Search Technique for Computer Vision Applications. IECIE Trans. Fundamentals E87-A, 1913–1923 (2004)
15. Muquit, M.A., Shibahara, T., Aoki, T.: A High-Accuracy Passive 3D Measurement System Using Phase-Based Image Matching. IEICE Trans. Fundam. Electron. Commun. Comput. Sci. E89-A, 686–697 (2006)
16. Foroosh, H., Zerubia, J.B., Berthod, M.: Extension of phase correlation to subpixel registration. IEEE Transactions on Image Processing 11, 188–200 (2002)
17. Shimizu, M., Okutomi, M.: Sub-pixel estimation error cancellation on area-based matching. International Journal of Computer Vision 63, 207–224 (2005)
18. Viola, P., Jones, M.: Robust Real-time Object Detection. International Journal of Computer Vision 57, 137–154 (2002)
19. Charbonnier, P., Blanc-Féraud, L., Aubert, G., Barlaud, M.: Deterministic Edge-Preserving Regularization in Computed Imaging. IEEE Transactions on Image Processing 6, 298–311 (1997)
20. Rivera, M., Ocegueda, O., Marroquin, J.L.: Entropy-Controlled Quadratic Markov Measure Field Models for Efficient Image Segmentation. IEEE Trans. Image Process. 16, 3047–3057 (2007)
21. Bradski, G., Kaehler, A.: Learning OpenCV: Computer Vision with the OpenCV Library. O'Reilly Media, Sebastopol (2008)

Conservative Motion Estimation from Multi-image Sequences

Wei Chen

U. S. Naval Research Laboratory, Remote Sensing Division,
Code 7233, 4555 Overlook Ave. S.W., Washington, D.C. 20375, USA

Abstract. Motion estimation in image sequences is a fundamental problem for digital video coding. In this paper, we present a new approach for conservative motion estimation from multi-image sequences. We deal with a system in which most of the motions in the scene are conservative or near-conservative in a certain temporal interval with multi-image sequences. Then a single conservative velocity field in this temporal range can across several successive frames. This system can be proved to be fully constrained or over-constrained when the number of frames is greater than two. A framework with displaced frame difference (DFD) equations, spatial velocity modeling, a nonlinear least-squares model, and Gauss-Newton and Levenberg-Marguardt algorithms for solving the nonlinear system is developed. The proposed algorithm is evaluated experimentally with two standard test image sequences. All successive frames except the last one (used for reference frame) in this conservative system can be synthesized by the motion-compensated prediction and interpolation based on the estimated motion field. This framework can estimate large scale motion field that across more than two successive frames if most of the motions in the scene in the temporal interval are conservative or near-conservative and has better performance than the block matching algorithm.

1 Introduction

Motion estimation in image sequences is a fundamental problem for digital video coding. Many methods for motion estimation have been proposed based on the conservative constraint of the optical flow intensity accompanied by some smoothness constraints or additional assumptions [1] – [14]. Almost all digital video algorithms employ motion estimation for reduction of the temporal redundancy in the video signal and increasing the compression efficiency.

Most of these works impose spatial smoothness of the velocity field on the system with alternative form of the optical flow conservative constraint equations, cost functions, and optimization strategies [1] – [14]. Recently, several new frameworks have been presented for the motion estimation in term of conservative constraint of optical flow intensity [15]-[18]. An innovative approach to motion determination based on a conservative mechanical model is proposed by Chen [17]. The motion fields are recognized as the velocity field and displacement field, and the displacement field is

G. Bebis et al. (Eds.): ISVC 2010, Part I, LNCS 6453, pp. 427–436, 2010.

proved to be equivalent to a conservative velocity field. Furthermore, the inverse problem for determining the conservative velocity in an image sequence has been proved to be fully constrained without any additional smoothness constraint. One of the numerical approaches to conservative velocity estimation proposed is based on theories and equations suggested by Chen [17].

In this paper, we consider the case in which the motion estimation is based only on multi-image sequences (more than two successive frames) for the application of video coding. To obtain higher compression ratio, the motion field estimation should across as much multiple temporal frames as possible so that all previous or latter frames can be reconstructed by the motion-compensated prediction and interpolation given start or final frame. In this case, all intermediate frames between the start and final frames are dropped for the removal of temporal redundancy. However, the dropped frame information in a short temporal range may be useful for the motion estimation. It is clear that the system for the purpose of the digital video compression can provide a condition with more than two successive frames to form a fully constrained system. We can find more than one displaced frame difference (DFD) equations on a fixed pixel point from more than two successive frames if the motion of all moving objects in the scene is conservative or near-conservative. Then these DFD equations contain a unique conservative velocity field that crosses all these frames in this conservative system.

This new approach for motion estimation with data of multi-image sequences is developed based on a framework proposed in the author's previous work [15], [16] with a new nonlinear inverse model.

This paper is organized as follows: In section 2, a set of system equations are derived. In Section 3, we apply the new technique to two test video image sequences used in computer vision. Finally, conclusions are drawn in last section.

2 Optical Flow Estimation

Efficient frameworks for the motion estimation based only on two successive image frames have been proposed in the author's previous works [15]-[18]. In this paper, we are concerned with the estimation of the projected displacement or conservative velocity field for a conservative system from more than two successive frames if most of motion in the scene is conservative or near-conservative.

2.1 The DFD Equations

Assuming $I(x, y, t)$ is the image intensity at a position coordinate (x, y) at time t, the temporal integral form of the optical flow conservative constraint can be expressed by the DFD equations for two or more image sequences [1]. The DFD equations are defined by

$$DFD_{ijs} = I(i+(t_s-t_1)u_{ij}, j+(t_s-t_1)v_{ij}, t_s) - I_{ij}(t_1) = 0 \; \forall \; (1 < s \le M) \,. \tag{1}$$

where $I_{ij}(t) = I(i, j, t)$, $u_{ij} = u_{ij}(t_1) = u(i, j, t_1)$ and $v_{ij} = v_{ij}(t_1) = v(i, j, t_1)$ are optical flow intensity, two components of conservative velocities on pixel points at time $t = t_1$,

$s \in [2, ..., M]$, and M is the number of successive frames. The number of two component velocity u_{ij} and v_{ij} at a fixed pixel point is equal to two. Since the number of the DFD equations (1) for all $s \in [2, ..., M]$ is equal to $M - 1$, the system is fully constrained or over-constrained if $M > 2$ for the conservative velocity field.

The DFD system equations on all pixels indicate that the system is nonlinear. Since there exist many local minimum solutions under lower degrees of the over-constraint [19], a spatial velocity field modeling is necessary.

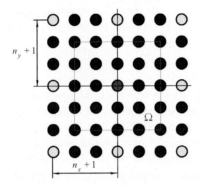

Fig. 1. Spatial interpolation points marked by blue dots and nodes marked by yellow and red with $\{p, q\}$ and $\{k, l\}$ indices in an image scene, respectively

2.2 Velocity Field Modeling

A set of the DFD equations in (1) provides sufficient number of equations to solve the conservative velocity field if $M > 2$. This is a nonlinear system problem and can be solved by a successful framework proposed by Chen [18].

Any two-dimensional function can be expressed by a Lagrange's bilinear polynomial [15]

$$f(x, y) = \sum_{\alpha=0}^{1} \sum_{\beta=0}^{1} f(p + \alpha n_x, q + \beta n_y) H_{p+\alpha n_x, q+\beta n_y}(x, y), \qquad (2)$$

where function $H_{a,b}(x, y)$ is defined by

$$H_{a,b}(x, y) = \frac{1}{n_x n_y} \begin{cases} (n_x - x + p)(n_y - y + q) & (a = p \cap b = q) \\ (x - p)(n_y - y + q) & (a = p + n_x \cap b = q), \\ (n_x - x + p)(y - q) & (a = p \cap b = q + n_y) \\ (x - p)(y - q) & (a = p + n_x \cap b = q + n_y) \end{cases}$$

and n_x and n_y are the number of interpolation points on x and y directions, and quantized indices on nodes are functions of x and y are given by

$$\{p, q\} = \{n_x \left\lfloor \frac{x}{n_x} \right\rfloor, n_y \left\lfloor \frac{y}{n_y} \right\rfloor\},$$

where $\lfloor \ \rfloor$ denotes an integer operator [15].

The two component velocity field on pixel points with horizontal and vertical co-ordinates x and y in an image can be expressed by the following bilinear polynomial functions with first order continuity that holds for all $N_x \times N_y$ image globally

$$\{u_{ij}, v_{ij}\} = \sum_{\alpha=0}^{1}\sum_{\beta=0}^{1}\{u_{p+\alpha n_x, q+\beta n_y}, v_{p+\alpha n_x, q+\beta n_y}\}H_{p+\alpha n_x, q+\beta n_y}(i, j), \qquad (3)$$

All velocity off nodes (blue color points shown in Fig. 1) can be calculated by equa-tion (3) using the velocity on node points expressed as u_{pq} and v_{pq}.

We define a parameter n which controls the degree of the over-constrained system by

$$n_x = n_y = n \geq 1 \quad (M > 2).$$

All velocity vectors v_{ij} in (3) are no longer independent variables for $n > 1$ except on node points and all independent velocities on nodes are connected with each other in whole image scene by equation (3).

We can control the number of interpolation points related to the resolution of the displacement field and the degree of the over-constrained by adjusting the parameter $n \geq 1$ to When the parameter n is equal to one, all node points and pixel positions are overlapped together and the system becomes a fully constrained ($M = 3$) or over-constrained system ($M > 3$). The degree of the over-determined DFD equations for the nonlinear least-squares can be controlled by the adjustable parameters n_x, n_y, and M.

A nonlinear least-squares model described in the next section is used to solve the over-constraint system with velocity field u_{pq} and v_{pq} on nodes as optimized parameters.

2.3 Global Optimization

A cost function that is a sum of the total errors of the DFD in (1) based on least-squares principle is given by

$$J = \sum_{i,j}\sum_{s=2}^{M}DFD_{ijs}^{2}, \qquad (4)$$

where i and j go over all pixels in $N_x \times N_y$ image ($i \in [0, N_x]$ and $j \in [0, N_y]$) and $M > 2$. Minimizing the cost function in (4) for given indices k and l for all node points in an image, we find the following independent system equations

$$\sum_{i,j}\sum_{s=2}^{M}DFD_{ijs}\{\frac{\partial DFD_{ijs}}{\partial u_{kl}}, \frac{\partial DFD_{ijs}}{\partial v_{kl}}\} = 0.$$

To solve the nonlinear least-squares problem, we employ Gauss-Newton iteration methods with Levenberg-Marguardt algorithm. This Levenberg-Marguardt method can improve converge properties greatly in practice and has become the standard of nonlinear least-squares routines.

Fig. 2. An example of the motion vector positional distribution on a Claire image (352 × 288) by the block-based model and the proposed framework with 16 × 16 block size ($n = 16$)

Furthermore, employing a principle similar to that of the Gauss-Seidel iteration algorithm, we make use of updated values of u_{kl} and v_{kl} as soon as they become available during the iteration processing. Since there exist many local minimum solutions under lower degrees of the over-constrained [17], an algorithm of progressve relaxation of pyramid constraint that adapts a variable resolution of the velocity structure during the iterations is employed in this algorithm [18].

Fig. 2 shows an example of the motion vector positional distributions on an image by the block matching algorithm (vectors on blocks) [1], [2], [6], [7] that has been utilized in video coding and the proposed framework (vectors on nodes located on the corner (node) of the blocks) with 16 × 16 block size ($n = n_x = n_y = 16$). The block matching algorithm estimates a single motion vector for each block and assumes that this vector is uniform within the block. The vector field by the block-based model is not continuous. However, the velocities on nodes estimated by the current framework are continuous and globally optimized, and all velocity vectors on pixel points can be changed and calculated by the modeled field function (3). If the block size is $n \times n$ with dimension $N_x \times N_y$ images, then the number of transmitted or stored motion vectors for both proposed and block matching estimators are equal to $(N_x / n + 1) \times (N_y / n + 1)$ and $(N_x / n) \times (N_y / n)$, respectively. Using almost the same number of the velocity vectors in a fixed block size for both approaches, the current framework (the velocity field with C^1 continuity obtained by global optimal strategies) can provide much higher accuracy performance than the block-based model (the velocity field with C^0 continuity obtained by local searching strategies).

2.4 Motion-Compensated Processing

Motion-compensated prediction (MCP) and interpolation (MCI) are a fundamental component of most proposed or adopted standards in digital video coding. A new approach for the motion-compensated processing has been proposed in the author's previous work [19]. There are three major factors that can determine the quality of interpolated images: the accuracy of motion estimation, dynamic motion modeling, and the use of appropriate MCI equations. Motion estimation from image sequences provides an initial velocity field which describes spatial variations of the motion in the scene. The dynamic motion model can determine the motion field evolution versus time in a temporal interval based on the initial velocity field. Finally, the motions in the scene observed by the optical flow intensity have to be described by an appropriate equation which is employed for synthesizing the MCI images [19].

In this paper, all computations of the MCP and MCI are based on the framework proposed in the author's previous work [19].

3   Experiments

In this section, we demonstrate experimental results to evaluate the performance of the estimator using the algorithms proposed in this paper. We apply the proposed estimator to two standard video sequences Taxi and Claire. The Peak Signal-to-Noise Ratio (PSNR) of the resulting DFD in (1) between original frame and predicted frame is used to evaluate how good the motion estimations for the video image sequences are. All estimated conservative velocity fields in this paper are based on raw data image sequences without any filter processes.

Methods of the motion estimation have been proposed with a wide range of computational complexities depending on the applications [1]-[14]. We are concerned with the applications for reduction of the temporal redundancy in the video signal and increasing the compression efficiency in this paper. For convenience, we compare the proposed estimator to the block matching estimator because both estimators utilize almost the same number of motion vectors for motion field estimation.

A Claire image sequence (196×196) with frames from 25 to 27 on left of the first row is shown in Fig. 3. The motion fields are estimated with 2×2 and 4×4 block sizes ($n = 2$ and 4) by the proposed method and block matching algorithm (using frame 25 and 27) from the image sequences. The synthesized MCP_{25} image with the PSNR = 45.0 dB and 43.8 dB by comparison with the original frame 25 is shown in Fig. 3, respectively. All interpolated images $MCI_{25+1/2}$, MCI_{26}, and $MCI_{26+1/2}$ as shown in Fig. 3 are interpolated by suggested framework [19].

The last test case uses the Taxi image sequences as shown in Fig. 4. The velocity fields are estimated from the six successive frames form 1 to 6 with different block sizes (4×4 and 8×8) by current estimator (with all frames) and block matching algorithm (with frame 1 and 6). The synthesized MCP_1 image with the PSNR = 39.2 dB and 37.0 dB by comparison with the original frame 1 is shown in Fig. 4, respectively. All interpolated images from MCI_2 to MCI_5 as shown in Fig. 4 are reconstructed by suggested framework [21].

The MCP images showed in Fig. 3 and 4 demonstrate that MCP images always have higher accuracy PSNR than the interpolated MCI intermediate images between start frame and final frame for a smaller block size. Because of the quantization errors and the motions in the scene are not exactly conservative, the MCI images have a little lower PSNR values than the MCP_1 image. These alignment errors of the moving

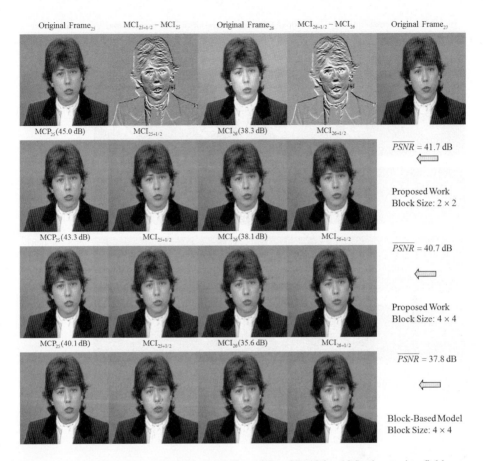

Fig. 3. A Claire image sequence with frame from 25 to 27 (196 × 196), the motion fields are estimated by current method and block matching algorithm with different block sizes (2 × 2 and 4 × 4) as the motion estimation parameters, MCP and MCI motion pictures synthesized by the proposed framework [19], and inter-frame difference images of $MCI_{25+1/2} - MCP_{25}$ and $MCI_{26+1/2} - MCI_{26}$

objects in the scene between the original and synthesized images caused by speed and path differences can be visually missed. Humans are sensitive only to the distortions of the moving objects, artifacts, and dirty window effects in the motion pictures, but are less sensitive to the alignment errors of the moving objects in the scene caused by acceleration (or deceleration) or path in a very short temporal interval.

Both synthesized images using motion fields estimated by proposed framework and block matching algorithm indicate that the first motion field is more accurate than the second one. Dirty window effects and observable distortions of the moving objects can be found in all MCI images synthesized by block matching algorithm as shown in figures 3 and 4. However, clear motion pictures have been synthesized with motion fields estimated by current work.

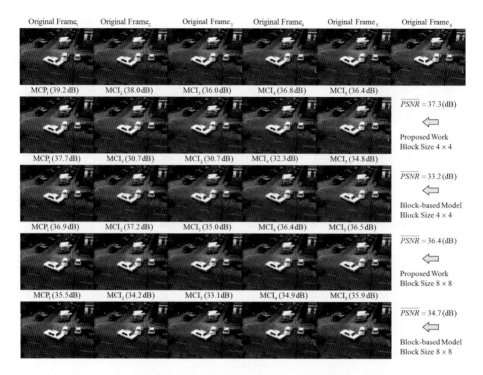

Fig. 4. Taxi image sequences ($256 \times 191 \times 6$). Motion fields are estimated by the proposed method and block matching algorithm with different block sizes (4×4 and 8×8) as the motion estimation parameters. Synthesized motion pictures are synthesized by the proposed framework [19].

4 Conclusion

Estimation of motion in image sequences is a fundamental problem for digital video coding. In this paper, we present a new approach for conservative motion estimation from multi-image sequences.

We deal with a system in which most of the motions in the scene are conservative or near-conservative in a certain temporal interval with multi-image sequences. Then a single conservative velocity field in this temporal range can across several successive frames. The number of unknown two component velocity u_{ij} and v_{ij} at a fixed pixel point is equal to two. Since the number of the DFD equations (1) is equal to $M - 1$, the system is fully constrained or over-constrained if $M > 2$ for the conservative velocity field. A framework with displaced frame difference (DFD) equations, spatial velocity modeling, a nonlinear least-squares model, and Gauss-Newton and Levenberg-Marguardt algorithms for solving the nonlinear system has been developed.

The proposed algorithm has been evaluated experimentally with two standard test image sequences. All successive frames except the last one (used for reference frame) in this conservative system can be synthesized by the motion-compensated prediction and interpolation based on the estimated motion field. This framework can estimate large scale motion field that across more than two successive frames if most of the motions in the scene in the temporal interval are conservative or near-conservative and has better performance than the block matching algorithm.

Acknowledgements

This research work was supported by the Office of Naval Research through the project WU-4279-10 at the Naval Research Laboratory. The author would like to thank the Institut für Algorithmen und Kognitive Systeme, KOGS/IAKS Universität Karlsruhe, Germany, and signal analysis and machine perception laboratory of the Ohio State University for providing image sequences.

References

1. Tekalp, A.M.: Digital Video Processing, p. 83. Prentice Hall PTR, Englewood Cliffs (1995)
2. Stiller, C., Konrad, J.: Estimating motion in image sequences: A tutorial on modeling and computation of 2D motion. IEEE Signal Processing Magazine, 70–91 (1999)
3. Horn, B., Shunck, B.: Determining optical flow. Artificial Intelligence (17), 185–203 (1981)
4. Lucas, B.D.: Generalized image matching by the method of differences, PhD thesis, Carnegie Mellon Univ. (1984)
5. Robbins, J.D., Netravali, A.N.: Recursive motion compensation: A review. In: Huang, T.S. (ed.) Image Sequence Processing and Dynamic Scene Analysis, Berlin, Germany, pp. 76–103. Springer, Heidelberg (1983)
6. Glazer, F., et al.: Scene matching by hierarchical correlation. In: Proc. IEEE Comp. Vision Pattern Recognition Conf., Washington, DC (June 1983)
7. Ghanbari, H., Mills, M.: Block matching motion estimations: New results. IEEE Trans. Circuit Syst. 37, 649–651 (1990)
8. Bigun, J., Granlund, G.H., Wiklund, J.: Multidimensional orientation estimation with applications to texture analysis and optical flow. IEEE TPAMI (1991)
9. Black, M.J., Anandan, P.: The robust estimation of multiple motions: parametric and piecewise smooth flow fields. Computer Vision and Image Understanding 63(1), 75–104 (1996)
10. Heitz, F., Bouthemy, P.: Multimodal estimation of discontinuous optical flow using Markov random fields. IEEE Transactions on Pattern Analysis and Machine Intelligence 15(12), 1217–1232 (1993)
11. Nesi, P.: Variational approach to optical flow estimation managing discontinuities. Image and Vision Computing 11(7), 419–439 (1993)
12. Weickert, J., Schnörr, C.: A theoretical framework for convex regularizers in PDE-based computation of imagemotion. International Journal of Computer Vision 45(3), 245–264 (2001)
13. Bruhn, A., Weickert, J., Schnorr, C.: Lucas/Kanade Meets Horn/Schunck: Combining Local and Global Optic Flow Methods. International Journal of Computer Vision 61(3), 211–231 (2005)
14. Brox, T.: Highly Accurate Optic Flow Computation with Theoretically Justified Warping. International Journal of Computer Vision 67(2), 141–158 (2006)

15. Chen, W., Mied, R.P., Shen, C.Y.: Near-surface ocean velocity from infrared images: Global Optimal Solution to an inverse model. J. Geophys. Res. 113, C10003 (2008), doi:10.1029/2008JC004747
16. Chen, W.: A Global Optimal Solution with Higher Order Continuity for the Estimation of Surface Velocity from Infrared Images. IEEE Trans. Geosci. Rem. Sens. 48(4), 1931–1939 (2010)
17. Chen, W.: Determination of Conservative Velocity from an Image Sequence (2010) (submitted for publication)
18. Chen, W.: Estimation of Conservative Velocity from an Image Sequence (2010) (submitted for publication)
19. Chen, W.: Motion-Compensated Processing with Conservative Motion Model (2010) (submitted for publication)

Gradient-Based Modified Census Transform for Optical Flow

Philipp Puxbaum and Kristian Ambrosch

AIT Austrian Institute of Technology,
A-1220 Vienna, Austria
philipp.puxbaum@ait.ac.at, kristian.ambrosch@ait.ac.at

Abstract. To enable the precise detection of persons walking or running on the ground using unmanned Micro Aerial Vehicles (MAVs), we present the evaluation of the MCT algorithm based on intensity as well as gradient images for optical flow, focusing on accuracy as well as low computational complexity to enable the real-time implementation in light-weight embedded systems. Therefore, we give a detailed analysis of this algorithm on four optical flow datasets from the Middlebury database and show the algorithm's performance when compared to other optical flow algorithms. Furthermore, different approaches for sub-pixel refinement and occlusion detection are discussed.

1 Introduction

For the fast and precise preparation of search and rescue missions after major chemical accidents, the information about the position of persons trying to escape the poisoned area is of essential importance. Here, we are working on a light weight computer vision system that can be deployed in unmanned Micro Aerial Vehicles (MAVs) for the detection of persons walking or running on the ground. Therefore, we analyzed a correspondence algorithm originally designed for high accuracy in real-time stereo vision, the Modified Census Transform (MCT) on gradient and intensity images (IGMCT) [1], on its performance in optical flow for the precise computation of the objects' 3D movement in two following images. In this work, we present an optimization of the algorithm's parameters for optical flow as well as an evaluation of its accuracy when compared to other popular optical flow algorithms.

For the evaluation we are using two well known optical flow algorithms that are still often considered for real-time applications, even if they cannot be considered state-of-the-art any more, proposed by Lucas and Kanade [2] as well as Horn and Schunk [3]. For the comparison with a recently published algorithm, we are also using the algorithm published by Ogale and Aloimonos [4]. Furthermore, we are presenting the results for the Sum of Absolute Differences (SAD) algorithm, implemented in the framework as the IGMCT, as well as the overall results for the MCT on intensity images only.

G. Bebis et al. (Eds.): ISVC 2010, Part I, LNCS 6453, pp. 437–448, 2010.
© Springer-Verlag Berlin Heidelberg 2010

2   Related Work

To determine the Optical Flow in real time it is often enforced to implement algorithms on dedicated hardware like Field-Programmable-Gate-Arrays (FP-GAs) or Application-Specific-Integrated-Circuits (ASICs). Because of the heavy computational load software solutions are usually less attractive for real-time implementations. To reach a fast calculation speed for either retrieving vector maps for the optical flow or the disparity map in stereo vision, region-based matching algorithms are highly preferred. Here, two very popular algorithms are the SAD and the Census Transform. An overview of correlation/distance measures for region-based matching can be found in [5].

In [6] Claus et al. proposed an FPGA implementation of the Census Transform for optical flow. They used a ternary decision for the Census Transform instead of a binary as introduced in [7]. They used an 8-neighborhood at two distances for the transformation. The resulting vector field is sparse. So the Optical Flow is calculated for a defined number of features. The whole system is implemented as a System-on-Chip (SoC) using an XC2VP30 FPGA from Xilinx with two embedded PowerPC processor cores. With an image resolution of 640x480, a search window of 15x15 and approximately 17000 features they reached a frame rate of about 45 fps.

In [8] Barron-Zambrano et al. implemented three different techniques using FPGAs. A correlation-based approach using SAD, a differential-based technique as proposed in [2] and an energy/frequency-based approach. Here, the algorithms' parameters in their simulations were: an 8x8 pixels search window and 4x4 reference window for the SAD algorithm, a 4-point central difference mask in space and time (5x5x5 window) for the differential method and 5x5 spatial mask for the Gabor filter, and 5 frames in time. The design was implemented using Xilinx tools 9.2i targeted to a Virtex-4 XC4LX160ff1513-12, one of the three FP-GAs available in the DNK8000PCI DiniGroup FPGA prototyping board. They reached a frame rate of 50 fps at an image resolution of 200x200.

The hardware implementation of the SAD algorithm of Fukuoka and Shibati in [9] reached a frame rate of about 1300 fps at an image resolution of 68x68 pixels.

Another Optical Flow implementation in [10] uses a cross-correlation algorithm. A fast calculation is achieved by parallel processing using MMX/SSE technology of modern general purpose processors. The calculations were performed on a Pentium 4 processor with 2.8 GHz and 512MB RAM, achieving a frame rate of 10 fps at an image resolution of 640x480 pixels.

3   Gradient-Based MCT for Optical Flow

The Census Transform, originally proposed by Zabih and Woodfill [7] is a non-parametric transformation of dedicated blocks from the intensity image into bit vectors. The Census Transform has various advantages like illumination invariance, when compared to other matching methods like Sum of Absolute Differences (SAD) or Sum of Squared Differences (SSD). It has a good tradeoff between quality and resource usage, especially because a hardware implementation requires only simple logic elements and allows parallelized computation.

The Census Transform is realized with a comparison function ξ (eq. 1) which transforms the intensity values into 0 or 1. This function compares the intensity value of the center pixel i_1 with the other pixels i_2 in the neighborhood.

$$\xi(i_1, i_2) = \begin{cases} 1 & | & i_1 > i_2 \\ 0 & | & i_1 \leq i_2 \end{cases} \tag{1}$$

The result of the comparison function is then concatenated to a bit vector (eq. 2). The length of the bit vector is the same as the size of the transformed block, defined as s_t.

$$T_{Census}(I, x, y, s_t) = \bigotimes_{[n,m]} \xi[I(x, y), I(x + n, y + m)] \tag{2}$$

$$n, m \epsilon [-\frac{s_t - 1}{2}, \frac{s_t - 1}{2}] \tag{3}$$

After the transformation of the intensity image the matching costs have to be estimated. Here, for two bit vectors the costs are determined by their hamming distance (eq. 4). The variable t_1 is the Census Transformed bit vector of $I_1(x, y)$ and t_2 is a bit vector in I_2 at position $(x + d_x, y + d_y)$. The matching costs have to be calculated for all disparities d_x and d_y.

$$C_{Census}(t_{1_{x,y}}, t_{2_{x+d_x, y+d_y}}) = hdist(t_{1_{x,y}}, t_{2_{x+d_x, y+d_y}}) \tag{4}$$

To improve the transformation, the incorporation of gradient images is desirable. Unfortunately, the original Census Transform is not capable to handle blocks with a saturated center value, which is considerably often the case in gradient images, especially near edges in the intensity images. Here, the Modified Census Transform (MCT) (eq. 5), originally proposed by Froeba and Ernst [11] for face recognition algorithms, leads to significant advantages as proposed in [1], since it compares the average intensity in a block instead of the center pixel. Here, we apply the MCT on x- and y-gradient images using Sobel filters. To enable a fast implementation only the absolute gradients are calculated.

$$T_{MCT}(I, x, y, s_t) = \bigotimes_{[n,m]} \xi[\bar{I}(x, y), I(x + n, y + m)] \tag{5}$$

Using gradient images and intensity image for matching procedure not only extends the information processed by the algorithm, it also extends the computational costs by a factor of 3. To reduce the complexity we decided to reduce the MCT computation to a sparse one using a sequential sparse factor of 2.

Hereafter, the calculated matching costs are aggregated (eq. 6) over a quadratic block with size s_a. The aggregation allows robust results but also smoothes the vector field.

$$a_{x,y,d} = \sum_n \sum_m c_{x+m,y+n,d} \tag{6}$$

$$n, m \epsilon [-\frac{s_a - 1}{2}, \frac{s_a - 1}{2}] \tag{7}$$

To reduce the complexity of the cost aggregation we use Integral Images [12] for constant processing time, independent of the block size s_a.

$$II(x, y) = \sum_{x'=0}^{x} \sum_{y'=0}^{y} I(x', y') \tag{8}$$

The block sum can be calculated easily by adding, respectively subtracting four values (eq. 9).

$$II(x, y, s_a) = II(x + h, y + h) - II(x + h, y - h) - \\ II(x - h, y + h) + II(x - h, y - h) \tag{9}$$

$$h = \frac{s_a - 1}{2} \tag{10}$$

After the cost aggregation the minimal matching costs have to be found and the corresponding disparity has to be identified. However, this vector field is only pixel-accurate. To improve the results and gain sub-pixel accuracy a sub-pixel refinement is implemented in the algorithm. Based on the fact that Optical Flow calculates two-dimensional disparities, it is necessary to fit a two-dimensional geometrical structure. For this reason we analyzed two approaches. The first one is to fit a paraboloid, as Krsek et al. did in [13], but for other purpose. Another approach as shown in [14] is to fit parabolas, one for the horizontal and one for the vertical refinement. For the refinement the matching costs are used. The paraboloid equation (eq. 11) is solved using the eight neighbored matching costs (eq. 12 and 13).

$$ax^2 + by^2 + cxy + dx + ey + f = z \tag{11}$$

$$\mathbf{X}\mathbf{A}^T = \mathbf{z} \tag{12}$$

$$\begin{pmatrix} x_1^2 & y_1^2 & x_1 y_1 & x_1 & y_1 & 1 \\ & & \vdots & & & \\ x_n^2 & y_n^2 & x_n y_n & x_n & y_n & 1 \end{pmatrix} \begin{pmatrix} a \\ b \\ \vdots \\ f \end{pmatrix} = \begin{pmatrix} z_1 \\ \vdots \\ z_n \end{pmatrix} \tag{13}$$

The equation system is solved with least squares method (eq. 14).

$$X^T X A = X^T z \tag{14}$$

With the solved parameters $a...f$ the coordinates where the minimum occurs, can be calculated with eq. 15.

$$x_{sub} = \frac{ce - 2bd}{-c^2 + 4ab}, \quad y_{sub} = -\frac{2ae - cd}{-c^2 + 4ab} \tag{15}$$

The parabola fitting [14] can be performed in a much easier way, leading to a highly reduced computational complexity. It is only necessary to calculate the minimum of the horizontal and vertical fitted parabola (eq. 16).

$$\hat{d}_{sub} = \frac{y_0 - y_2}{2y_0 - 4y_1 + 2y_2} \tag{16}$$

After the vector field has been refined it is necessary to detect the occlusions in the image. The Optical Flow can only be estimated when regions are visible in both images. If an object disappears from the first to the second image this region has to be marked as occluded. For this reason two methods are implemented which make an occlusion detection possible. One method, the left/right consistency check is a preferred one in Stereo Vision. For the consistency check the Optical Flow has to be calculated two times, once from first to the second image and once from the second image to the first. In case that a region is not occluded the vectors will have the same direction but different algebraic signs. Occluded regions will be matched to different regions. The calculation of the angular error and a defined threshold makes the occlusion detection possible:

$$\kappa(OF_{12}, OF_{21}) = \begin{cases} occluded, & AE(OF_{12}(x,y), -OF_{21}(x + u(OF_{12}(x,y)), \\ & y + v(OF_{12}(x,y)))) > \varepsilon \\ else & OF_{12}(x,y) \end{cases} \tag{17}$$

Another method to detect occlusions and mismatched areas is to calculate confidence values. Confidence values are measures how good a determined disparity can be trusted. It calculates the difference between the absolute minimal costs and second lowest minima. For better real-time capabilities, the minimal costs and the neighborhood, defined over a block size are simply excluded from this new minimum search.

$$\delta(c_{min_{old}}, c_{min_{new}}) = \begin{cases} occluded, & c_{min_{new}} - c_{min_{old}} < \varepsilon \\ else & d(c_{min_{old}}) \end{cases} \tag{18}$$

4   Parameter Optimization and Experimental Results

To compare the algorithms the evaluation database of Baker et al. in [15] is used. Four datasets with ground truth are used for this evaluation. The images are called Dimetrodon, RubberWhale, Grove2 and Venus dataset and are online available.[1] For comparison the average angular error (AAE) and the endpoint error (AEP) are calculated. Furthermore the good matches of certain thresholds (AE: 10°, 5°, 2.5°; EP: 1, 0.5 pixels) are estimated.

For optimization purposes the parameters of all algorithms (except from Ogale and Aloimonos [4]) have to be varied. The algorithm of Horn and Schunck [3] has two parameters, the lagrange factor and the number of iterations. The software implementation out of the openCV[2] library is used for the evaluation. The test series has shown that with 12800 iterations and a lagrange factor of 0.0001 the best results are achieved. As the test series ended at 12800 after a quadruplication of iterations the improvement was marginal. The algorithm reached for the RubberWhale dataset adequate results but failed at the others (fig. 1).

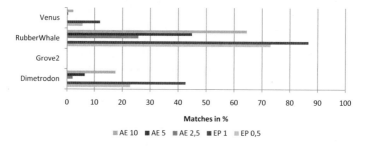

Fig. 1. Analysis of good matches with optimized parameters of the algorithm of Horn/Schunck [3]

For the Lucas/Kanade algorithm [2] we also used an openCV implementation. Here, only one parameter has to be optimized. The block size for grouping pixels can varied from 3 to 15. Bigger block sizes lead to run-time errors in the implementation. At a window size of 15, the best results were achieved. In fig. 2 the good matches are presented. The algorithm, as that of Horn/Schunck [3] delivers only for the RubberWhale dataset appropriate results and fails at the others.

The third algorithm from Ogale/Aloimonos [4] is evaluated with a precompiled code, which is available online[3]. This code only delivers pixel-accurate results, but the results are much better than the algorithms evaluated before as can be seen in fig. 3.

The SAD algorithm, a very popular matching algorithm, has no transformation and only one parameter besides the disparity range. After the absolute difference is calculated the aggregation block size can be chosen. For an aggregation block size of 15 pixels the best results are achieved.

---

[1] http://vision.middlebury.edu/flow/data/
[2] http://opencv.willowgarage.com/wiki/
[3] http://www.cs.umd.edu/~ogale/download/code.html

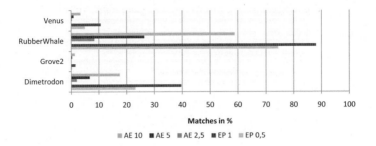

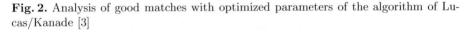

Fig. 2. Analysis of good matches with optimized parameters of the algorithm of Lucas/Kanade [3]

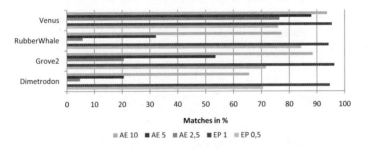

Fig. 3. Analysis of good matches of the algorithm of Ogale/Aloimonos [4]

Calculating the Optical Flow using the MCT, applied on the gradient and intensity images leads to the best results. This algorithm needs more computational resources than, i.e., the SAD algorithm, because of the required transformation step. The parameters are the Census block size, over which the bit vector is generated, and the aggregation block size. The best results are achieved with a Census block size of 3 pixels and an aggregation block size of 25 pixels. With this large block size robust results are possible but it also smoothes the vector field. The weighting factor between gradient and intensity Census is 2:1.

The evaluation of the two sub-pixel refinement methods has shown that the paraboloid fitting (2D) produced better results. The problem that paraboloid fitting brings with it, is that due to solving the equation with least squares method results bigger than 1 or smaller than -1 can appear. In this case the result is out of borders and the parabola fitting (1D) [14] is used instead. For real-time implementations this strategy would lead to a higher computational complexity. The results of the comparison are shown in Tab. 1 for all four datasets. For the first three datasets the improvement for the angular matches is good. A higher accuracy could be gained. The Venus images are modified stereo images. Therefore the improvement isn't so big according to the other sets.

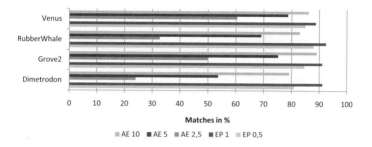

Fig. 4. Analysis of good matches of the SAD algorithm

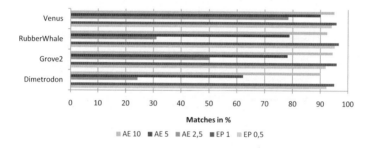

Fig. 5. Analysis of good matches of the IGMCT algorithm

Table 1. Comparison of the sub-pixel refinement "parabola fitting" and "paraboloid fitting"

	Ø[degrees,pixel]		matches [%]				
Dimetrodon	AE	EP	10°	5°	2,5°	1 pixel	0,5 pixel
1D	10,6419	0,2179	90,1192	62,1942	24,3574	95,0479	92,2981
2D & 1D	10,1427	0,2051	92,4958	66,7442	28,3889	94,9848	93,0607
Impr.	-0,4992	-0,0128	2,3766	4,55	4,0315	-0,0631	0,7626
Grove2	AE	EP	10°	5°	2,5°	1 pixel	0,5 pixel
1D	4,7915	0,3324	94,3984	78,1621	50,1979	95,8662	91,9681
2D & 1D	4,3935	0,3165	94,3851	87,3669	55,0426	95,8066	92,6038
Impr.	-0,3980	-0,0159	-0,0133	9,2048	4,8447	-0,0596	0,6357
RubberWhale	AE	EP	10°	5°	2,5°	1 pixel	0,5 pixel
1D	6,7783	0,1782	92,4424	78,8183	31,0470	96,6530	95,1949
2D & 1D	6,4642	0,1677	91,7989	80,6242	42,9291	96,7867	95,2686
Impr.	-0,3141	0,0105	-0,6435	1,8059	11,8821	0,1337	0,0737
Venus	AE	EP	10°	5°	2,5°	1 pixel	0,5 pixel
1D	4,9614	0,3429	95,0056	89,9812	78,4254	95,7863	94,0520
2D & 1D	4,8601	0,3422	95,4693	91,6241	78,6341	95,5971	93,6497
Impr.	-0,1013	-0,0007	0,4637	1,6429	0,2087	-0,1892	-0,4023

For comparison of all tested algorithms the average errors and matches over all four datasets are calculated and displayed in Tab. 2. The IGMCT delivers the best results except the AE-matches with threshold $2,5°$ where the intensity Census is a bit better. For the comparison of all tested algorithms the average errors and matches over all four datasets are calculated and displayed in Tab. 2. The IGMCT delivers the best results except the AE-matches where the intensity Census is a bit better. For visual comparison of the different algorithms the estimated Optical Flows for all tested algorithms for the RubberWhale dataset are displayed in Fig. 6 through 9 using the color coding from Baker et al. [15].

Table 2. Comparison of all tested Optical Flow algorithms (Average over all four datasets)

	Ø[degrees,pixel]		matches [%]				
Algorithm	AE	EP	10°	5°	2,5°	1 pixel	0,5 pixel
HS [3]	35,54	1,89	21,08	12,92	6,96	35,31	25,38
LK [2]	43,38	2,11	20,23	8,56	2,66	34,97	25,69
OA [4]	9,18	0,42	81,28	48,59	26,88	95,18	75,60
SAD	10,23	0,49	84,29	69,16	41,77	90,85	84,58
IMCT	7,45	0,30	90,50	76,73	49,45	94,85	91,12
IGMCT	6,79	0,27	92,99	77,29	46,00	95,84	93,38

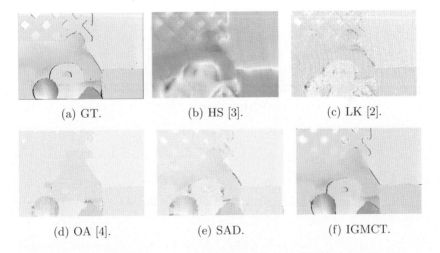

(a) GT.   (b) HS [3].   (c) LK [2].

(d) OA [4].   (e) SAD.   (f) IGMCT.

Fig. 6. Optical Flows for the RubberWhale dataset with ground truth

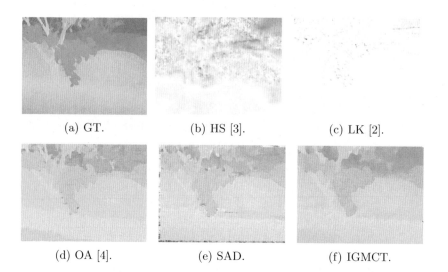

(a) GT.   (b) HS [3].   (c) LK [2].

(d) OA [4].   (e) SAD.   (f) IGMCT.

Fig. 7. Optical Flows for the Grove2 dataset with ground truth

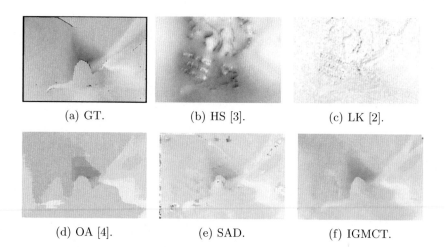
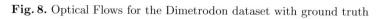

(a) GT.   (b) HS [3].   (c) LK [2].

(d) OA [4].   (e) SAD.   (f) IGMCT.

Fig. 8. Optical Flows for the Dimetrodon dataset with ground truth

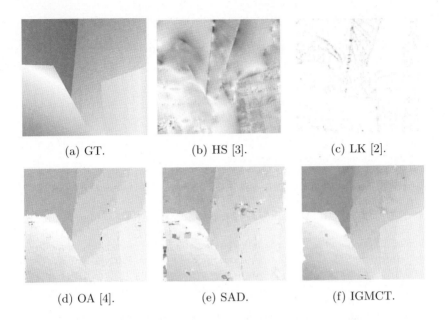

(a) GT.   (b) HS [3].   (c) LK [2].

(d) OA [4].   (e) SAD.   (f) IGMCT.

Fig. 9. Optical Flows for the Venus dataset with ground truth

5   Conclusion

Even if the MCT on the intensity images already produces very accurate results for a real-time algorithm, the use of the gradient images further reduces the average angular error by 0,66. Here, the average endpoint error is nearly equal for both algorithms. However, the correct matches within a maximum deviations within 1 and 0,5 pixels show a considerable improvement. For the maximum deviation of 1 pixel, the resulting accuracy of 95,84% is even 0,66% higher than the result of the pixel accurate Ogale and Aloimonos algorithm, a state-of-the-art non-real-time algorithm. However, the use of a sparse transform enables a resource efficient implementation of the IGMCT on real-time embedded systems at nearly the same computational costs as the original Census Transform. For the detection of occluded regions, the consistency check produced the best results. Here, the high image quality of the Middebury datasets enforces very dense results and the confidence value can be considered to be very useful in real-world images taken with industrial cameras. Taking into account that today's DSPs often incorporate a bit count operation further reduces the performance requirements for the calculation of the Hamming Distance, making the IGMCT a good choice not only for hardware-based, but also for software-based real-time implementations.

Acknowledgements

The work leading to the proposed results were funded by the KIRAS security research program of the Austrian Ministry for Transport, Innovation and Technology (www.kiras.at).

References

1. Ambrosch, K.: Mapping Stereo Matching Algorithms to Hardware. PhD thesis, Vienna University of Technology (2009)
2. Lucas, B., Kanade, T.: An iterative image registration technique with an application to stereo vision. In: Proceedings of Imaging Understanding Workshop, pp. 121–130 (1981)
3. Horn, B., Schunck, B.: Determining optical flow. MIT - Artificial Intelligence Memo No.572 (1980)
4. Ogale, A., Aloimonos, Y.: A roadmap to the integration of early visual modules. International Journal of Computer Vision: Special Issue on Early Cognitive Vision 72 (2007)
5. Giachetti, A.: Matching techniques to compute image motion. Image and Vision Computing 18 (2000)
6. Claus, C., Laika, L., Jia, L., Stechele, W.: High performance fpga based optical flow calculation using the census transformation. In: The Intelligent Vehicles Symposium (2009)
7. Zabih, R., Woodfill, J.: Non-parametric local transforms for computing visual correspondence. In: Eklundh, J.-O. (ed.) ECCV 1994. LNCS, vol. 801, pp. 151–158. Springer, Heidelberg (1994)
8. Barron-Zambrano, J., Torres-Huitzil, C., Cerda, M.: Flexible architecture for three classes of optical flow extraction algorithms. In: International Conference on Reconfigurable Computing and FPGAs, pp. 13–18 (2008)
9. Fukuoka, Y., Shibata, T.: Block-matching-based cmos optical flow sensor using only-nearest-neighbor computation. In: IEEE International Symposium on Circuits and Systems, pp. 1485–1488 (2009)
10. Sadykhov, R.K., Lamovsky, D.V.: Fast cross correlation algorithm for optical flow estimation. In: Proceedings of the 7th Nordic Signal Processing Symposium, pp. 322–325 (2006)
11. Froeba, B., Ernst, A.: Face detection with the modified census transform. In: Proceedings of the Sixth IEEE Conference on Automatic Face and Gesture Recognition (2004)
12. Viola, P., Jones, M.: Robust real-time face detection. International Journal of Computer Vision (IJCV) 57
13. Krsek, P., Pajdla, T., Hlaváč, V.: Estimation of differential structures on triangulated surfaces. In: 21st Workshop of the Austrian Association for Pattern Recognition (1997)
14. Shimizu, M., Okutomi, M.: Precise sub-pixel estimation on area-based matching. In: Proceedings of the Eight IEEE International Conference on Computer Vision (2003)
15. Baker, S., Scharstein, D., Lewis, J., Roth, S., Black, M., Szeliski, R.: A database and evaluation methodology for optical flow. In: Proceedings of the Eleventh IEEE Conference on Computer Vision (2007)

Depth Assisted Occlusion Handling in Video Object Tracking

Yingdong Ma and Qian Chen

Centre for Digital Media Computing
Shenzhen Institutes of Advanced Technology, Shenzhen, China

Abstract. We propose a depth assisted video object tracking algorithm that utilizes a stereo vision technique to detect and handle various types of occlusions. The foreground objects are detected by using a depth and motion-based segmentation method. The occlusion detection is achieved by combining the depth segmentation results with the previous occlusion status of each track. According to the occlusion analysis results, different object correspondence algorithms are employed to track objects under various occlusions. The silhouette-based local best matching method deals with severe and complete occlusions without assumptions of constant movement and limited maximum duration. Experimental results demonstrate that the proposed system can accurately track multiple objects in complex scenes and provides improvements on dealing with different partial and severe occlusion situations.

1 Introduction

Detection and tracking of moving objects is a very important research area of computer vision and has a wide range of applications. Many researchers have investigated object tracking and different approaches have been developed. These approaches can be classified into four groups: tracking by statistical method [1, 2], region-based tracking [3], feature-based tracking [4] and appearance-based tracking [5]. Many of these approaches can achieve good results in some cases, such as when the target has distinct colour distribution from the backgrounds. However, multiple objects tracking is still a difficult task due to various problems, including inaccurate motion vector estimation, variation of the non-rigid object appearance, and confusions in multiple targets' identities when their projections in the camera image are close.

Depth information has been used for object tracking in recent years. This exploits the advantages of a stereo-vision system's ability to segment objects from a cluttered background, separating objects at different depth layers under partial occlusion, and adding depth feature to enhance object appearance models. While some tracking methods focus on the usage of depth information only [6], other depth-based object tracking approaches make use of depth information on better foreground segmentation [7], plan view construction [8], object scale estimation [9], and enhanced object shape descriptors [10]. Depth information is also used as a feature to be fused in a

G. Bebis et al. (Eds.): ISVC 2010, Part I, LNCS 6453, pp. 449–460, 2010.

Bayesian network or a maximum-likelihood model to predict 3D object positions [11].

To our knowledge, depth information is not widely used in occlusion analysis and occlusion handling. Firstly, the depth map is often much noisier than the corresponding grey-level or colour image. The quality of depth map is also sensitive to textureless areas and the distance from the camera. These disadvantages of depth data make it more suitable for combining with other appearance features to enhance the object model in a tracking system. Secondly, when occluding objects have similar disparity ranges due to close interaction, they cannot be separated by depth-based segmentation methods. Moreover, the depth information can be totally lost when severe and complete occlusion occurs. In this work, we present a depth assisted multiple objects tracking algorithm which focuses on addressing three fundamental challenges: stable object segmentation from cluttered background; depth assisted occlusion analysis and depth-based occlusion handling.

2   Stereo Assisted Moving Object Detection

In this work, a pair of calibrated CCD cameras is used to get the left and right images. Disparity estimation is implemented using a dense stereo matching technique. Firstly, candidate points are found in the right image which match points in the left image. Secondly, the Sum of Absolute Differences (SAD) of the candidate point pairs is calculated within a rectangular window. Two points are regarded as corresponding points if their minimum SAD is less than a threshold. The disparity between the corresponding points is then calculated as described in [12].

Foreground Object Segmentation: In order to recover the depth information from a disparity map, a depth density image is generated by transforming the depth information on the XY plane (disparity map) to the XZ plane. Points are projected to the XZ plane according to their horizontal position and their grey level, where X is the width of the depth map and the range of Z is [0, 255]. Because an object has similar grey level in the disparity map, the influence of the 3D points of this object is cumulative on the depth density image. The depth density image will contain large values in areas with a high density of 3D points.

Foreground object regions segmentation is performed based on the depth density image using a region growing method. After a morphological opening and closing operation to remove noisy points and connect nearby points, all 8-orientation-connected points are grouped as one object. The disparity map is segmented according to the depth density image. The x-coordinates of each connected component in the depth density image shows the location and width of each foreground object. All corresponding points in the disparity map are labelled as different objects according to their grey level.

Due to the stereo matching ambiguity or depth discontinuities at textureless areas, depth maps are often much noisier than their corresponding colour images. Moreover, stable obstacles in the depth map should be removed to save computation time in the tracking step. We use the motion masks generated by change detection to refine the

Fig. 1. Depth-based segmentation. (left to right) The input left colour frame, The Disparity map, The Depth density image, Depth-based object detection.

silhouette of segmented objects. These segmented object regions are then used in the following sections for object tracking. Fig.1 shows the depth-based segmentation examples.

3 Depth Assisted Object Tracking

We analyse the occlusion situations in the current frame according to the depth segmentation results and the occlusion rate of each track in the previous frame. The tracking system uses different strategies to achieve accurate tracking according to the occlusion analysis result. The flowchart of the tracking system is shown in Fig.2.

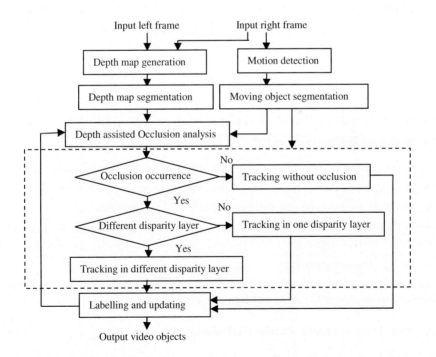

Fig. 2. Flowchart of the depth assisted tracking system

3.1 Depth Assisted Occlusion Analysis

The proposed depth assisted occlusion analysis method judges the occurrence and end of partial occlusion based on two clues. The first clue comes from the depth density image and the second one is each object's occlusion rate from the previous frame. Let R_i^k denotes the ith foreground object region in frame k. If there are no vertical over-lays between the corresponding blob of R_i^k and other foreground blobs in the depth density image, region R_i^k is labelled as a candidate non-occlusion object.

The occlusion state of region R_i^k is also checked by means of occlusion rate. An occlusion rate γ_i^k is assigned to each segmented moving object region, which records the percentage of occluded part of each object in the previous frame.

$$\gamma_i^{k-1} = \sum_{j=1}^{m} (N_{R_{ij}^{k-1}} \mid d_j > d_i)/N_{R_i^{k-1}}, R_{ij}^{k-1} = R_i^{k-1} \cap S_j^{k-1} \tag{1}$$

where N_R is the number of pixels in region R. d_i is the average disparity value of region R_i, where larger d_i means object region R_i is closer to the camera. R_{ij} is the overlaid area between R_i and the jth object's silhouette S_j. All γ_i^0 are initialized to zero for non-occluding objects. When partial occlusion occurs in the first frame, the occlusion rate of an occluded object is initialized as the ratio between its bounding box area and the bounding box area of the entire occluding region.

According to the above two conditions, various occlusions are predicted following these rules:

1. Non-occlusion: there is no overlaid part between each pair of foreground blobs in the current u-projection image and all γ_i^{k-1} have the value of zero.

2. Partial occlusion occurs: all γ_i^{k-1} have the value of zero and there is overlaid part between at least one pair of foreground blobs in the current u-projection image.

3. Serious/Full occlusion occurs: the value of any γ_i^{k-1} is larger than 0.6

4. End of partial occlusion: there is no overlaid part between each pair of foreground blobs in the current u-projection image. Each blob corresponds to one object and at least one γ_i^{k-1} has non-zero value.

5. End of full occlusion: a partial occluded foreground region matches to a track whose occlusion rate is 1.0 in the last frame

3.2 Video Tracking under Partial Occlusion

The video objects correspondence under non-occlusion is achieved through the short-est three-dimensional Euclidean distance between tracks in the previous frame and object regions in the current frame. Once a foreground objects region R_i^k finds its corresponding track, the track is updated using the object's silhouette S_i^k.

An occlusion is detected when two or more foreground object regions in the depth density image starts to overlay each other. Partial occlusion occurs when objects get close or one pass in front of others. It can be divided into two types based on the disparity ranges of the occluding objects: occlusion in different disparity layers and occlusion in one disparity layer. In the later case, occluding objects cannot be segmented as individual objects by using the depth-based segmentation. We use an Iterative Silhouette Matching algorithm (ISM) to tackle this problem. After object correspondence, tracks and their occlusion rate are updated using their corresponding object's silhouettes.

Partial Occlusion in Different Disparity Layers. When occluding objects have different disparity layers, they can be segmented in the depth map and be sorted in descending order by means of their disparity ranges. Object correspondence under different disparity layers is based on silhouette matching. The distance matrix $D^k(i, j)$ is measured using the Kullback-Leibler distance between two colour distributions. Let $O^k = \{O_1^k, O_2^k, \Box, O_n^k\}$ denotes the occluding objects and S^{k-1} denotes the existing tracks (object silhouettes in the last frame). The colour distribution of each occluding object and existing tracks are $P_i^k(u)$ and $Q_j^{k-1}(u)$, respectively.

$$D^k(i, j) = \sum_{u=1}^{N_{bin}} P_i^k(u) \log \frac{P_i^k(u)}{Q_j^{k-1}(u)} \tag{2}$$

where $i = 1, \Box, n, j = 1, \Box, m$. N_{bin} is the number of bins, $N_{bin} = N_{red} \times N_{green} \times N_{blue}$. Fig.3 shows an example of partial occlusion tracking under different disparity layers.

Fig. 3. Partial occlusion tracking under different disparity layers. (left to right) three tracks from the last frame, depth map segmentation result and matching result.

Partial Occlusion in One Disparity Layers. When occluding objects have similar disparity range, the proposed segmentation method will label these objects as one foreground region. Various algorithms, including Maximum-likelihood [13], fuzzy k-means [14], and K nearest neighbourhood classifier [15], have been developed to estimate the separate object's locations. To achieve better object separation, we propose an depth-based ISM algorithm to separate occluding objects. Assuming that there are n occluding objects O^k in the object region $R^k_{occlusion}$, their corresponding tracks in the previous frame are S^{k-1}. The algorithm has the following operations:

1. According to the depth ordering of S^{k-1}, find the corresponding component of the track with the largest depth, $S^{k-1}_{foremost}$ (i.e. the smallest distance from the camera), within $R^k_{occlusion}$ using the colour histogram-based silhouette matching algorithm. Label it as $O^k_{foremost}$.

2. Remove $O^k_{foremost}$ from $R^k_{occlusion}$. Replace the track with the one which has the next largest depth and find its corresponding component within the remaining area of $R^k_{occlusion}$.

3. Repeat step two until all tracks in O^k find their corresponding regions in $R^k_{occlusion}$.

4. Update the occlusion rate and depth order for each object. If two or more objects have similar average disparity ranges, update their depth order according to their occlusion rate: the smaller γ^k_i an object has, the closer the object is to the camera Fig.4 shows an example of partial occlusion in one disparity layer.

Fig. 4. Partial occlusion in one disparity layer. (left to right) the depth map and its segmentation result, in which two persons are labelled as one object region; two tracks from the last frame and their matching result.

3.3   New Object, Splitting and Severe Occlusion Handling

The idea of severe occlusion handling is that full occluded objects will reappear later as partial occluded objects behind their occluder. We propose a severe occlusion handling algorithm without motion restriction on the occluded objects and no limited maximum duration of occlusions but we assume that the shapes of these objects do not change sharply during severe or full occlusion.

Let $S^k_{ij} = \{S^k_{i1}, S^k_{i2}, \Box, S^k_{im}\}$ be the severe or full occluded tracks and the foremost track is S_i. If a partial occluded foreground region in a later frame, k+n, cannot find the matching track and its occluder is S^{k+n}_i, the local best match of each S_{ij} within $R^{k+n}_{occlusion}$ is acquired by:

$$T(x, y) = \arg\min_{(x,y)} \frac{1}{N} \sum_{(x',y') \in S^p_j} [I(x + x', y + y') - S^p_j(x', y')]^2 \qquad (3)$$

where $R_{occlusion}^{k+n}$ is the object region, $T(x, y)$ is the transformation parameters, N is the number of pixels in S_j^p, and I denotes the current frame. The silhouette of each S_{ij} for splitting handling is the one in frame p, which has minimum non-zero occlusion rate. If more than 50 percent of the foreground object region is within the local best match area of S_j^p, a splitting of S_j is detected. The occlusion rate of S_j is updated using (1) and remove S_j from the full occlusion list S_{ij}. Otherwise, the foreground object region indicates a new object and updates its occlusion rate accordingly.

4 Experimental Results

The proposed tracking algorithm has been tested on some video sequences. The input to the algorithm is a pair of video sequences and the output is a set of video objects that are labelled using bounding boxes with different colours over time in the left video sequence. These test videos are recorded from two fixed CCD cameras, which are set up at different locations, including an office, a laboratory, and a reception area. In these video sequences, various types of occlusions are involved: partial occlusion in different disparity layers, partial occlusion in one disparity layer, and short-term severe occlusion. Non-occlusion is also included in some video sequences for completeness.

The algorithm is implemented using C on a personal computer with 3.0GHz CPU and 2GB memory. The video resolution is 640×480 pixels (24 bits per pixel). For a typical partial occlusion scenario, an average computational time of 0.237s per frame is achieved.

4.1 Tracking without Occlusion

An example of tracking objects that are not occluded is shown in Fig.5. In this sequence a person walks in the scene, places a box, and leaves the scene. This example illustrates the ability of the proposed algorithm handling deformable objects and splitting. The splitting occurs at frame 45 and frame 46. However, due to the shadows on the floor, two foreground objects are segmented as one object in the depth map. The box is labelled as a new object at frame 51. From then on, the person and the box are tracked as two individual objects.

4.2 Tracking under Partial Occlusion in Different Disparity Layers

We demonstrate the tracking of partially occluded objects in Fig.6. In this example, partial occlusion occurs at the beginning and continues throughout the sequence. Three persons in this sequence make different movements: the person on the left makes random movement on a chair; the person in the middle dynamically changes his disparity range by moving fore-and-aft and the person on the right rotates throughout the sequence. In most of the frames, they have different disparity ranges.

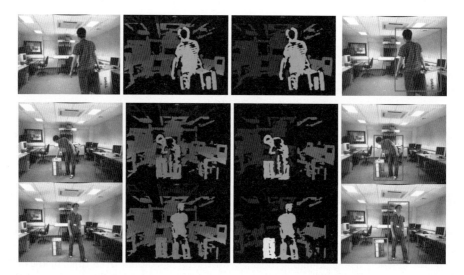

Fig. 5. Example of tracking objects that are not occluded. From top to down, the frame numbers are 24, 42, and 51, respectively. First and second column: the input left frame and the corresponding depth map. Third column: the depth map segmentation results. Fourth column: the video tracking results.

The proposed segmentation method can successfully detect them under partial occlusion. All of them are coherently labelled over time even though the visual feature of the foremost human head changes drastically.

Fig. 6. Tracking results with partial occlusion in different disparity layers. From top to down, left to right, the frame numbers are 1, 5, 15, 20, 25, 30, 35, and 40, respectively.

4.3 Tracking under Partial and Severe Occlusion

Tracking under partial and complete occlusion is illustrated in Fig.7. The partial occlusion occurs at frame 24 and finishes at frame 52 due to the close interaction between two persons in a disparity layer closer to the cameras. The occlusion rate of the occluded person exceeds 0.6 at frame 47, where the tracking system enters the severe occlusion mode. The person with red bounding box walks in a farther disparity layer and pauses for a while behind others. He is completely occluded by others from frame 31 to 36 and under partial occlusion from frame 37 to 41. From frame 42 to 52, he is under complete occlusion again. Fig.8 shows his silhouettes before, under, and after partial occlusion.

In most of the frames where partial or severe occlusion occurs, the proposed depth map segmentation method can detect foreground objects despite the cluttered background. But it groups objects, which have similar disparity ranges, into one region as shown in Fig.4. This example illustrates the robustness, and importance of using the proposed silhouette matching algorithm. With silhouette matching algorithm, partially occluded objects in one foreground region are tracked successfully.

Fig. 7. Tracking under partial and severe occlusion. The frame numbers are 20, 24, 30, 33, 37, 41, 45, and 49, from top to down, left to right, respectively.

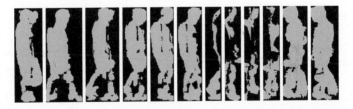

Fig. 8. Silhouette of the far most person for frame 20, 22, 24, 26, 27, 28, 30, 37, 39, 41, 54, and 56, from left to right

As an algorithm comparison, the video sequence as shown in Fig.7 is used to test other tracking techniques. In Fig.9, the results are obtained from the template matching method proposed by Nguyen and Smeulders [16] whereas Fig.10 illustrates the tracking results of Porikli and Tuzel's work [17], which is based on the mean shift algorithm. In both of the two tests, the foreground objects are segmented using the background subtraction method. Before partial occlusion occurs, the templates of each person in the scene are recorded. The template matching technique works well in the case of non-occlusion and partial occlusion due to the updating of template by an appearance filter but failed under severe occlusion. The far most person is missed from frame 41 to frame 49, where severe occlusion occurs. From frame 44 to frame 48, the person with green bounding box is also missed due to the same reason. From frame 50, the template matching method can recover from the severe occlusion and matches all three persons again.

The tracking results of the mean shift algorithm are illustrated in Fig.10. The occluded persons are lost when severe occlusion occurs. After splitting, the far most person is matched again as his location does not changed during severe occlusion. However, the mean shift algorithm combines the person with green bounding box with the foremost person, and after splitting in frame 49, labels him as a new object.

Fig. 9. Tracking results of template matching method proposed by Nguyen and Smeulders [16]. The frame numbers are 42, 45, 49, and 54, from top to down, left to right, respectively.

Fig. 10. Tracking results of the mean shift algorithm proposed by Porikli and Tuzel [17]. The frame numbers are 20, 30, 43, and 54, from left to right, respectively.

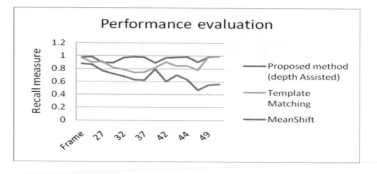

Fig. 11. Comparison of three tracking algorithms

Fig.11 shows the performance evaluation of three tracking algorithms in terms of the recall measure, which shows how much relevant object pixels the proposed algorithm has extracted. The algorithm comparison starts from frame 24, the frame before occlusion, and stops at frame 54, where the occlusion ends. The pixels of objects under full occlusion are considered as correct corresponding pixels and those belonging to wrong labelled objects are regarded as wrong corresponding pixels. From Fig.11, we can observe that the proposed depth assisted tracking algorithm achieves better tracking performance than other approaches.

4.4 Object Tracking with and without Depth Map Segmentation

Fig.12 illustrates the object segmentation results with and without depth map segmentation. When stationary foreground objects and the background have similar colours, spatial and temporal based methods cannot guarantee correct segmentation results. As shown in Fig.12 (f), the person far away from the camera is missed in several frames due to the cluttered background and the small amount of movement that he makes. However, with the proposed depth map segmentation, the correct object masks can be obtained.

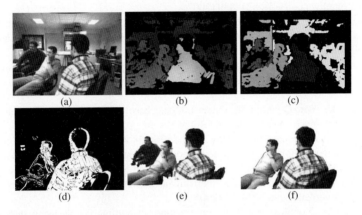

Fig. 12. Object segmentation with and without depth map. (a) Colour frame. (b) Disparity map. (c) Depth-based segmentation. (d) Motion mask. (e) Depth-motion-MRF model based segmentation. (f) Motion and colour based segmentation.

5 Conclusion

In this work, we introduce a novel depth assisted object tracking algorithm aimed at improving the robustness and accuracy to different types of occlusion. The proposed algorithm consists of four components, including depth-based object segmentation, depth assisted occlusion analysis, depth-based partial occlusion handling and depth-ordering based severe occlusion handling. The main contribution of this algorithm is to utilize depth information in these components to handle various occlusion situations effectively and robustly. The new methods developed to address different tracking problems can be summarized as follows.

Firstly, the depth density image based object detection method is introduced to extract foreground regions, which are further refined by motion detection masks. Secondly, the segmented object regions along with the occlusion rate of each tracks in the previous frame form the basis of the depth assisted occlusion analysis method. Thirdly, different object tracking strategies are employed according to the various occlusion situations. Finally, when severe and full occlusion occurs, with the help of depth ordering of each track, the local best matching method is effective and robust for splitting handling.

The experimental results presented have confirmed the performance of our proposed object tracking algorithm under different challenging occlusion situations.

References

1. Pan, J., Hu, B., Zhang, J.: Robust and accurate object tracking under various types of occlusions. IEEE Transactions on Circuits and Systems for Video Technology 18(2) (2008)
2. Zhu, J., Lao, Y., Zheng, Y.: Effective and robust object tracking in constrained environments. In: Proceedings of IEEE International Conference on Acoustics, Speech and Signal Processing, pp. 949–952 (2008)
3. Guo, Y., Hsu, S., Sawhney, H.S., Kumar, R., Shan, Y.: Robust object matching for persistent tracking with heterogeneous features. IEEE Transactions on Pattern Analysis and Machine Intelligence 29(5), 824–839 (2007)
4. Yilmaz, A., Li, X., Shah, M.: Contour-based object tracking with occlusion handling in video acquired using mobile cameras. IEEE Transactions on Pattern Analysis and Machine Intelligence 26(11), 1531–1536 (2004)
5. Jepson, A., Fleet, D., El-Maraghi, T.: Robust online appearance models for visual tracking. IEEE Transactions on Pattern Analysis and Machine Intelligence 25(10), 1296–1311 (2003)
6. Parvizi, E., Wu, Q.: Multiple object tracking base on adaptive depth segmentation. In: IEEE Conference on Computer and Robot Vision, pp. 273–277 (2008)
7. Krotosky, S.J., Trivedi, M.M.: On Color-, Infrared-, and Multimodal-Stereo Approaches to Pedestrian Detection. IEEE Transactions on Intelligent Transportation Systems 8(4) (2007)
8. Harville, M.: Stereo person tracking with short and long term plan-view appearance models of shape and color. In: IEEE International Conference on Advanced Video and Signal based Surveillance, pp. 522–527 (2005)
9. Beymer, D., Konolige, K.: Real-time tracking of multiple people using stereo. In: IEEE Frame Rate workshop (1999)
10. Tang, F., Harville, M.: Fusion of local appearance with stereo depth for object track-ing. In: IEEE Computer Vision and Pattern Recognition Workshops, pp. 1–8 (2008)
11. Okada, R., Shirai, Y., Miura, J.: Object tracking based on optical flow and depth. In: IEEE International Conference on Multisensory Fusion and Integration for Intelligent Systems, pp. 565–571 (1996)
12. Huang, Y., Fu, S., Thompson, C.: Stereovision-based object segmentation for automotive applications. EURASIP Journal on Applied Signal Processing, 2322–2329 (2005)
13. Senior, A.: Tracking people with probabilistic appearance models. In: ECCV Workshop on Performance Evaluation of Tracing and Surveillance Systems, pp.48–55 (2002)
14. Cavallaro, A., Steiger, O., Ebrahimi, T.: Tracking video objects in cluttered background. IEEE Transactions on Circuits and Systems for Video Technology 15(4), 575–584 (2005)
15. Zhu, L., Zhou, J., Song, J.: Tracking multiple objects through occlusion with online sampling and position estimation. Pattern Recognition 41(8), 2447–2460 (2008)
16. Nguyen, H.T., Smeulders, A.W.M.: Fast occluded object tracking by a robust appearance filter. IEEE Transactions on Pattern Analysis and Machine Intelligence 26(8), 1099–1104 (2004)
17. Porikli, F., Tuzel, O.: Human body tracking by adaptive background models and mean-shift analysis. In: IEEE International Workshop on Performance Evaluation of Tracking and Surveillance (2003)

Acquisition Scenario Analysis for Face Recognition at a Distance

P. Tome[1], J. Fierrez[1], M.C. Fairhurst[2], and J. Ortega-Garcia[1]

[1] Biometric Recognition Group - ATVS
Escuela Politecnica Superior - Universidad Autonoma de Madrid
Avda. Francisco Tomas y Valiente, 11 - Campus de Cantoblanco 28049 Madrid, Spain
{pedro.tome,julian.fierrez,javier.ortega}@uam.es
[2] Department of Electronics, University of Kent, Canterbury, Kent CT2 7NT, UK
{M.C.Fairhurst}@kent.ac.uk

Abstract. An experimental analysis of three acquisition scenarios for face recognition at a distance is reported, namely: close, medium, and far distance between camera and query face, the three of them considering templates enrolled in controlled conditions. These three representative scenarios are studied using data from the NIST Multiple Biometric Grand Challenge, as the first step in order to understand the main variability factors that affect face recognition at a distance based on realistic yet workable and widely available data. The scenario analysis is conducted quantitatively in two ways. First, we analyze the information content in segmented faces in the different scenarios. Second, we analyze the performance across scenarios of three matchers, one commercial, and two other standard approaches using popular features (PCA and DCT) and matchers (SVM and GMM). The results show to what extent the acquisition setup impacts on the verification performance of face recognition at a distance.[1]

Keywords: Biometrics, face recognition, at a distance, on the move.

1 Introduction

Face and iris are two of the most relevant biometrics used nowadays in many user recognition applications [1,2]. A new research line growing in popularity is focused on using these biometrics in less constrained scenarios in a non-intrusive way, including acquisition "On the Move" and "At a Distance" [3]. Imagine a scenario where the people do not have to stop in front of a sensor to acquire a picture of the face: simply, they walk through an identification bow. This kind of scenarios are still in their infancy, and much research and development is

---

[1] P. T. is supported by a FPU Fellowship from Univ. Autonoma de Madrid. Part of this work was conducted during a research stay of P.T. at Univ. of Kent funded by the European Action COST2101. This work has been partially supported by projects Bio-Challenge (TEC2009-11186), Contexts (S2009/TIC-1485), TeraSense (CSD2008-00068) and "Cátedra UAM-Telefónica".

G. Bebis et al. (Eds.): ISVC 2010, Part I, LNCS 6453, pp. 461–468, 2010.

needed in order to achieve the levels of precision and performance that certain applications require.

The new field of biometrics at a distance is enabled mainly thanks to: 1) recent advances in sensing technology [2], and 2) new algorithms and methods to deal with varying factors (e.g., illumination, movement, pose, distance to the camera), which in this case are less controlled than the ideal situations commonly considered in biometrics research.

As a result of the interest in these biometric applications at a distance, there is now a growing number of research works studying how to compensate for the main degradations found in uncontrolled scenarios [4]. Nevertheless, there is almost no experimental knowledge about the main variability factors found in specific scenarios, which may help in devising robust methods for biometrics at a distance tailored to specific applications of practical importance. The contribution of the present paper is toward this end, by analyzing quantitatively three scenarios of face recognition at a distance, namely: close, medium and far distance between subject and camera. This analysis is conducted quantitatively at two levels for the considered scenarios: 1) main data statistics such as information content, and 2) performance of recognition systems: one commercial, and two other based on popular features (PCA and DCT) and matchers (SVM and GMM).

The scenarios under study are extracted from the NIST Multiple Biometric Grand Challenge [5], which is focused on biometric recognition at a distance using iris and face. In particular, we use a subset of this benchmark dataset consisting of images of a total of 112 subjects acquired at different distances and varying conditions regarding illumination, pose/angle of head, and facial expression.

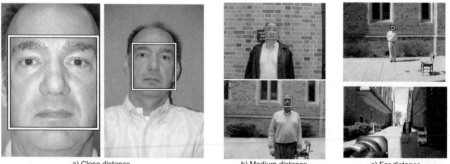

a) Close distance   b) Medium distance   c) Far distance

Fig. 1. Example images of the three scenarios: a) close distance, b) medium distance, and c) far distance

The paper is structured as follows. Sect. 2 describes the dataset and scenarios under study. Sect. 3 analyzes the main data statistics of the scenarios. Sect. 4

studies the performance of the three considered recognition systems on the different scenarios. Sect. 5 finally discusses the experimental findings and outlines future research.

2 Scenario Definition

The three scenarios considered are: 1) "close" distance, in which the shoulders may be present; 2) "medium" distance, including the upper body; and 3) "far" distance, including the full body. Using this three general definitions we marked manually all the 3482 face images from the 147 subjects present in the dataset NIST MBGC v2.0 Face Stills [5]. Some examples images are depicted in Fig. 1. A portion of the dataset was discarded (360 images from 89 subjects), because the face was occluded or the illumination completely degraded the face. Furthermore, although this information is not used in the present paper, all the images were marked as indoor or outdoor.

Finally, in order to enable verification experiments considering enrollment at close distance and testing at close, medium, and far distance scenarios, we kept only the subjects with at least 2 images in close and at least 1 image in both of the two other scenarios. The data selection process is summarized in Table 1, where we can see that the three considered scenarios result in 112 subjects and 2964 face images.

Table 1. Number of images of each scenario constructed from NIST MBGC v2.0 Face Visible Stills

Num. users	Close distance	Medium distance	Far distance	Discarded images	Total
147	1539	870	713	360	**3482**
	At least *2 images* per user	At least *1 images* per user			
112	1468	836	660		**2964**

3 Scenario Analysis: Data Statistics

3.1 Face Segmentation and Quality

We first segmented and localized the faces (square areas) in the three acquisition scenarios using the VeriLook SDK discussed in Sect. 4.1. Segmentation results are shown in Table 2, where the segmentation errors increase significantly across scenarios, from only 1.43% in close distance to 82.57% in far distance. Segmentation errors here mean that the VeriLook software could not find a face in the image due to the small size of faces and increment of variability factors. For all the faces detected by VeriLook, we conducted a visual check, where we observed 3 and 10 segmentation errors for medium and far distance, respectively.

Table 2. Segmentation results based on errors produced by face Extractor of VeriLook
SDK

	Close distance	Medium distance	Far distance	Discarded	Total
Num. Images	1468	836	660	360	**3324**
Errors	21	151	545		**848**
Errors(%)	1.43%	18.06%	**82.57%**		

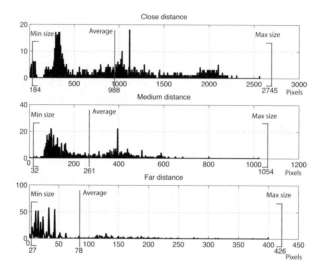

Fig. 2. Histograms of face sizes for each scenario (side of the square area in pixels)

All the segmentation errors were then manually corrected by manually marking the eyes. The face area was then estimated based on the marked distance
between eyes.

The resulting sizes of the segmented faces are shown in Fig. 2, where we
observe to what extent the face size decreases with the acquisition distance. In
particular, the average face size in pixels for each scenario is: 988×988 for close,
261×261 for medium, and 78×78 for far distance.

Another data statistic we computed for the three scenarios is the average face
quality index provided by VeriLook ($0 =$ lowest, $100 =$ highest): 73.93 for close,
68.77 for medium, and 66.50 for far distance (see Fig. 3, computed only for the
faces correctly segmented by VeriLook). As stated by VeriLook providers, this
quality index considers factors such as lightning, pose, and expression.

3.2　Information Content

The entropy of the face images in the different acquisition scenarios represents
a quantitative assessment of the information content in the gray levels of the
images. In principle, an image acquired in controlled conditions (illumination,

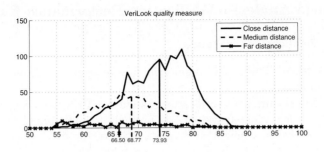

Fig. 3. Histogram of face quality measures produced by VeriLook SDK

clean background, neutral pose, ...) would have less entropy than other image acquired at a distance in uncontrolled conditions. In Fig. 4 (top), this effect is patent: the farther the distance the higher the entropy. When considering only the information within the segmented faces, as shown in Fig. 4 (down), the opposite occurs: the farther the distance the lower the entropy. These two measures (increase in entropy of the full image, and decrease in entropy of the segmented faces), can therefore be seen, respectively, as a quantitative measure of the scenario complexity increase due to background effects, and the reduction in information within the region of interest due to acquisition scenario change.

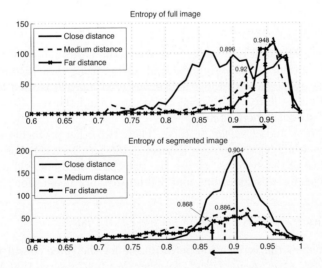

Fig. 4. Histograms of entropy for full images (top) and segmented faces (down) for the three scenarios with their corresponding average value

4 Scenario Analysis: Verification Performance Evaluation

4.1 Face Verification Systems

- **VeriLook SDK.** This is the commercial face recognition system provided by Neurotechnology[2].
- **PCA-SVM system.** This verification system uses Principal Component Analysis (PCA). The evaluated system uses normalized and cropped face images of size 64×80 (width $\times$ height), to train a PCA vector space where 96% of the variance is retained. This leads to a system where the original image space of 5120 dimensions is reduced to 249 dimensions. Similarity scores are computed in this PCA vector space using a SVM classifier with linear kernel.
- **DCT-GMM system.** This verification system also uses face images of size 64×80 divided into 8×8 blocks with horizontal and vertical overlap of 4 pixels. This process results in 285 blocks per segmented face. From each block a feature vector is obtained by applying the Discrete Cosine Transform (DCT); from which only the first 15 coefficients ($N = 15$) are retained. The blocks are used to derive a world GMM Ω_w and a client GMM Ω_c [6]. From previous experiments we obtained that using ($M = 1024$) mixture components per GMM gave the best results. The DCT feature vector from each block is matched to both Ω_w and Ω_c to produce a log-likelihood score.

4.2 Experimental Protocol

Three main experiments are defined for the verification performance assessment across scenarios:

- *Close2close.* This will give us an idea about the performance of the systems in ideal conditions (both enrollment and testing using close distance images). About half of the close distance subcorpus (754 images) is used for development (training the PCA subspace, SVM, etc.), and the rest (714 images) is used for testing the performance.
- *Close2medium*, and *close2far* protocol. These two other protocols use as training set the whole close distance dataset (1468 face images). For testing the performance of the systems, we use the two other datasets: 836 medium distance images for *close2medium*, and 660 far distance images for *close2far*.

4.3 Results

In Fig. 5 we show the verification performance for the three considered scenarios: *close2close*, *close2medium*, and *close2far*. We first observe that VeriLook is the best of the three systems in *close2close* with an EER around 7%. At the same time, this commercial system is the most degraded in uncontrolled conditions, with an EER close to 40% in *close2far*, much worse than the other two

---

[2] http://www.neurotechnology.com/

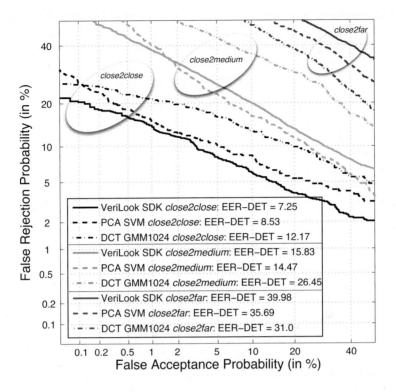

Fig. 5. Verification performance results for the three scenarios and three systems considered

much simpler systems. This result corroborates the importance of analyzing and properly dealing with variability factors arising in biometrics at a distance.

We also observe in Fig. 5 that the GMM-based system works better in far distance conditions than the other systems, although being the less accurate in *close2close* and *close2medium*. This result demonstrates the greater generalization power of this simple recognition approach, and its robustness against uncontrolled acquisition conditions.

Based on this observation, we finally conducted a last experiment simplifying the DCT-GMM complexity in order to enhance its generalization power, seeking for a maximum of performance in the challenging *close2far* scenario. The verification performance results are given in Table 3 as EER for decreasing DCT-GMM complexity (N = DCT coefficients, M = Gaussian components per GMM). The results indicate in this case that decreasing the recognition complexity (i.e., improving the generalization power) of this simple recognition method does not help in improving its robustness against uncontrolled conditions. In other words, the DCT-GMM recognition complexity initially considered ($N = 15, M = 1024$), is the most adequate for the *close2far* scenario studied here.

Table 3. Verification performance of the DCT-GMM system for different configurations

EER	M Gaussians								
N Coeff.	close2close			close2medium			close2far		
DCT	1024	128	8	1024	128	8	1024	128	8
15	12.17	14.62	20.06	26.45	29.06	36.19	31.01	32.52	38.74
10	13.22	15.97	19.62	26.09	28.72	34.90	29.80	32.83	38.58
5	17.66	19.80	22.15	31.72	34.60	35.43	33.46	37.07	39.37

5   Discussion and Future Work

An experimental approach towards understanding the variability factors in face recognition at a distance has been reported. In particular, we have conducted a data-driven analysis of three realistic acquisition scenarios at different distances (close, medium, and far), as a first step towards devising adequate recognition methods capable to work in less constrained scenarios. This data-driven analysis has been made for a subset of the benchmark dataset NIST MBGC v2.0 Face Stills.

Our analysis has been focused on: 1) data statistics (segmented face sizes, quality and entropy measures), and 2) verification performance of three systems. The results showed that the considered systems degrade significantly in the far distance scenario, being more robust to uncontrolled conditions the most simple approach.

Noteworthy, the scenarios considered in the present paper differ not only in the distance factor, but also in illumination and pose (being the illumination variability much higher in far distance than in close distance). Based on the data statistics obtained and the performance evaluation results, a study of the effects of such individual factors is source for future research.

References

1. Zhao, W., Chellappa, R., Phillips, P.J., Rosenfeld, A.: Face recognition: A literature survey. ACM Comput. Surv. 35, 399–458 (2003)
2. Matey, J., Naroditsky, O., Hanna, K., Kolczynski, R., LoIacono, D., Mangru, S., Tinker, M., Zappia, T., Zhao, W.: Iris on the move: Acquisition of images for iris recognition in less constrained environments. Proc. of the IEEE 94, 1936–1947 (2006)
3. Li, S.Z., Schouten, B., Tistarelli, M.: Biometrics at a Distance: Issues, Challenges, and Prospects. In: Handbook of Remote Biometrics for Surveillance and Security, pp. 3–21. Springer, Heidelberg (2009)
4. Robust2008: Robust biometrics: Understanding science and technology (2008), http://biometrics.cylab.cmu.edu/ROBUST2008
5. MBGC: Multiple biometric grand challenge. (NIST - National Institute of Standard and Technology)
6. Galbally, J., McCool, C., Fierrez, J., Marcel, S., Ortega-Garcia, J.: On the vulnerability of face verification systems to hill-climbing attacks. Pattern Recognition 43, 1027–1038 (2010)

Enhancing Iris Matching Using Levenshtein Distance with Alignment Constraints

Andreas Uhl and Peter Wild

Department of Computer Sciences,
University of Salzburg, A-5020 Salzburg, Austria

Abstract. Iris recognition from surveillance-type imagery is an active research topic in biometrics. However, iris identification in unconstrained conditions raises many proplems related to localization and alignment, and typically leads to degraded recognition rates. While development has mainly focused on more robust preprocessing, this work highlights the possibility to account for distortions at matching stage. We propose a constrained version of the Levenshtein Distance (LD) for matching of binary iris-codes as an alternative to the widely accepted Hamming Distance (HD) to account for iris texture distortions by e.g. segmentation errors or pupil dilation. Constrained LD will be shown to outperform HD-based matching on CASIA (third version) and ICE (2005 edition) datasets. By introducing LD alignment constraints, the matching problem can be solved in $O(n \cdot s)$ time and $O(n+s)$ space with n and s being the number of bits and shifts, respectively.

1 Introduction

Unconstrained iris recognition is a relatively new branch in iris-based identification. It is driven by the demands to push biometric image acquisition towards an extraction of biometric signals with the subject of interest moving or being at-a-distance from biometric sensors. Advantages of such systems comprise better usability, higher throughput, and the ability to acquire biometric measurements without required cooperation. While iris recognition in reasonably constrained environments provides high confidence authentication with equal error rates (EERs) of less that 1% [1], a reduction of constraints is quite challenging. First-generation prototype iris identification systems designed for stand-off video-based iris recognition, e.g Sarnoff's Iris-on-the-move [2], or General Electric's Stand-off Iris Recognition system [3], have proven the feasibility of iris recognition from surveillance-type imagery. But also the need for better segmentation techniques than usually applied in still-image iris recognition to account for distortions like motion blur, defocusing or off-axis gaze direction has been identified as a main issue. Challenges like the Iris Challenge Evaluation (ICE) and Multiple Biometric Grand Challenge (MBGC) have provided standardized datasets to aid in finding solutions to these problems.

G. Bebis et al. (Eds.): ISVC 2010, Part I, LNCS 6453, pp. 469–478, 2010.

Most publications regarding iris recognition in unconstrained environments aim at more sophisticated preprocessing techniques to successfully localize and segment images of the human eye. Proença et al. [4] identify the critical role of segmentation and observe a strong relationship between translational segmentation inaccuracies and recognition error rates. Matey et al. [5]. assess the effect of resolution, wavelength, occlusion and gaze as the most important factors for incorrect segmentation and give a survey of segmentation algorithms. While iris boundaries have been modeled as circles, ellipses and more complex shapes, still whatever model is used, the processing chain of almost all iris recognition algorithms resembles Daugman's standard approach [6] very close: After successful determination of the inner and outer pupil centers, the iris-ring texture of a person's eye is unwrapped and further processed by feature extraction modules. Refinements of this model usually refer to more sophisticated generation of noise-masks determining pixels containing eyelashes, eyelids, or other types of distortions. The majority of feature extraction approaches extracts binary output (iris-codes) from the obtained normalized textures [1], and employs the fractional Hamming Distance (HD) over different bit shifts (to account for rotational alignment) between iris-codes in order to determine a degree of similarity at matching stage. Indeed very few studies have proposed new or compared different binary similarity and distance measures, and it is common agreement, that HD is the best method for this task.

Similarity measure selection is a problem encountered in various fields. Cha et al. [7] compare several binary vector similarity measures including a new variable credit similarity measure (altering credit for zero-zero and one-one matches) for iris biometric authentication. Their proposed metric improved and generalized HD measures by introducing weights in order to give greater importance to error pixels in a neighborhood of error pixels. However, in order to determine parameters, a separate training stage is needed and the trained contributing factor was reported to vary considerably depending on the application data. A more exhaustive hierarchically clustered summary of binary similarity and distance measures can be found in Choi et al. [8].

In this paper, we propose the use of the Levenshtein Distance (LD) [9] for iris matching in order to tolerate segmentation inaccuracies and distortions caused by the linearity of Daugman's normalization model [6]: Each iris point is mapped into a doubly dimensionless coordinate system to account for elastic deformation of the iris texture. Yuan et al. [10] identified drawbacks of this iris normalization model and confirm claims in [11] that a linear stretching seems not enough to catch the complex nature of pupil dilation. The novelty of our aproach lies in considering deformation at the matching - and not at normalization - stage. We perceive a significant improvement of recognition accuracy using this method compared to traditional HD at only moderate additional time overhead. Improvements to common implementations of LD computation, i.e. alignment constraints, are suggested and evaluated for two different open iris databases with respect to recognition accuracy and time complexity. As a by-product of our evaluation, we also highlight the impact of shifts on iris recognition accuracy.

The remainder of this paper is structured as follows: An introduction to LD and its application in iris recognition is given in Sect. 2. Experimental setup, employed feature extraction algorithms and datasets are presented in Sect. 3. Evaluations are discussed in Sect. 4 and Sect. 5 concludes this work.

2   Levenshtein Distance in Iris Recognition

The Levenshtein Distance dates back to the 1960s [9] and is a well-known classical distance measure in pattern recognition for sequences of possibly different length. The idea to use LD for biometrics is not new, e.g. Schimke et al. [12] employ an adapted version of LD based on event-string modelling and a nearest neighbor classifier for online signature verification. In this work, we assess the usability of LD for iris recognition in order to cope with imperfect normalization of iris textures. We employ LD at the matching stage to enhance iris recognition accuracy, as outlined in Fig. 1.

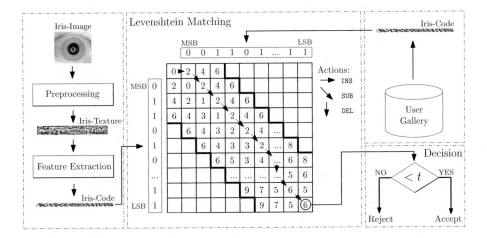

Fig. 1. System Architecture: the basic operation mode of the proposed system

The inherent idea of LD is to employ inexact matching allowing a sequence to exhibit additional or lack parts of another similar sequence. Similarity is defined by an optimal transformation of one sequence into the other by three operations: *INS* (insert), *DEL* (delete) and *SUB* (substitute). Each operation is associated with a cost, c_{INS} and c_{DEL} are scalar values, $c_{SUB}(a,b)$ is a function depending on the symbols $a, b \in \{0,1\}$ at specific positions of the two sequences to be compared (in iris recognition we consider binary sequences only). Typically, $c_{SUB}(a,b) = 0 \Leftrightarrow a = b$ for symbol b replacing a. The LD is also called Edit Distance and can be calculated by Dynamic Time Warping (DTW) [13], a dynamic programming algorithm to align the sequences. Let $A \in \{0,1\}^m$ and $B \in \{0,1\}^n$

be binary sequences of length m and n (as we apply LD to iris-codes of fixed length, typically $m = n$ holds). DTW uses a matrix D of size $(m+1) \times (n+1)$, which is incrementally computed:

$$D[0,0] := 0 \tag{1}$$

$$\forall i > 0 : D[i,0] := D[i-1,0] + c_{DEL}, \tag{2}$$

$$\forall j > 0 : D[0,j] := D[0,j-1] + c_{INS}, \tag{3}$$

$$\forall i,j > 0 : D[i,j] := \min(D[i-1,j] + c_{DEL},$$
$$D[i,j-1] + c_{INS},$$
$$D[i-1,j-1] + c_{SUB}(A[i],B[j])). \tag{4}$$

The invariant maintained throughout the algorithm is, that the subsequence $A[1..i]$ can be transformed into $B[1..j]$ using a minimum of $D[i,j]$ operations. The Levenshtein Distance of A and B is $LD(A,B) := D(m,n)$. A traceback algorithm is used to find the optimal alignment, i.e. an alignment path results from joining nodes and depending on the local direction an optimal (not necessary unique) sequence of transformation operations is derived, see Fig. 1. However, as the matching stage needs the distance measure only, we can avoid a storage of the entire matrix and reduce space complexity from $O(n^2)$ (assuming $n = m$ in our application domain) to $O(n)$ with column-wise computation. Still, time complexity stays at $O(n^2)$ with this modification, which is not useful for commercial applications. The traditional HD dissimilarity measure needs $O(n \cdot s)$ time and $O(n+s)$ space with s being the number of bit shifts, which is usually a very small constant number.

In order to further reduce LD time requirements, we define additional constraints: Computations are restricted to an evaluation region S on matrix D shaped as a stripe from top-left to bottom-right (see Fig. 1), i.e. technically we define $S := \{(x,y) : |x - y| \le s\}$ and set:

$$\forall (i,j) \notin S : D[i,j] = 0. \tag{5}$$

By this modification, we enforce a maximum local deviation of the patterns by s shifts. Like in the previously refined implementation, we keep track of the last column only. The resulting final algorithm solves the matching problem in $O(n \cdot s)$ time and $O(n + s)$ space with n and s being the number of bits and shifts, respectively. It is worth to notice, that the last optimization does no longer deliver the exact Levenshtein Distance. However, it is a very natural constraint to give an upper limit on relative shifts, since it should not be too difficult to get estimators for eye tilt (e.g. localization of eye lids).

Finally, in order to obtain a normalized distance value, we note that for $m = n$ the condition $LD(A,B) \le HD(A,B)$ holds, and we therefore divide the result by the code size n. Note, that for LD the triangle inequality holds. The advantage of LD over traditional HD lies in its ability of non-linear alignment. For this reason, it is widely accepted in computational biology, e.g. to estimate alignments of DNA.

3   Experimental Setup

In order to test the performance of the proposed iris matching technique, we employ existing iris recognition systems and replace the iris matching module by the implementation outlined in Sect. 2, see Fig. 1. The transparent application in existing iris biometric solutions (no re-enrollment of the user gallery is necessary) is a key advantage of the proposed approach. All used system components as well as employed biometric databases are described in more detail as follows.

3.1   Databases

For experimental evaluation we employ two different iris databases:

- *CASIA*: we select all 1307 left eyes out of a collection of 2655 NIR illuminated indoor images with 320×280 pixel resolution of the open CASIA-V3-Interval[1] iris database. This dataset reflects performance on high quality input.
- *ICE*: this dataset contains all 1425 right eyes out of 2953 NIR illuminated images with 640 × 480 pixel resolution of the open ICE-2005[2] iris database. We selected this test set for lower quality input, as some images have noticeable blur and occlusions.

In case of the *CASIA* dataset, 4028 genuine (intra-class) and 15576 imposter (inter-class) comparisons were executed, for the *ICE* database results refer to 12214 genuine and 7626 imposter comparisons, i.e. we matched all genuine pairs, but only the first template between users.

3.2   Normalization

Depending on the type of input data, we employ two different iris normalization techniques. For the processing of *CASIA* images, we use a custom normalization software applying Canny edge detection and Hough circle detection to localize inner and outer pupil boundaries. This localization is followed by Daugman's rubber sheet model [6] using a circular boundary model to transform the iris texture into a rectangular 512 × 64 pixel area. Finally, the texture is enhanced using blockwise brightness estimation.

While the first method was tuned to deliver accurate segmentation for the *CASIA* dataset, we obtained the open iris recognition software OSIRIS[3] (version 2.01) in order to segment images of the *ICE* dataset (OSIRIS also comes

---

[1] The Center of Biometrics and Security Research, CASIA Iris Image Database, http://www.sinobiometrics.com

[2] National Institute of Standards and Technology, Iris Challenge Evaluation (ICE) 2005, http://iris.nist.gov/ice/

[3] BioSecure Project, Open Source for IRIS reference system, http://svnext.it-sudparis.eu/svnview2-eph/ref_syst/Iris_Osiris

with an official evaluation on ICE-2005). The OSIRIS segmentation module uses again a cicular Hough transform, but also an active contour approach to detect the contours of iris and pupil. In order to get similar input textures for the feature extraction module, we employed adaptive histogram equalization after normalization on the 512×64 pixel sized textures with a window size of 32 pixels.

3.3   Feature Extraction

The first feature to be extracted is a custom implementation of the iris-code version by *Ma* et al. [14]. This algorithm extracts 10 one-dimensional horizontal signals averaged from pixels of 5 adjacent rows from the upper 50 pixel rows. Using dyadic wavelet transform, each of the 10 signals is analyzed, and from a total of 20 subbands (2 fixed bands per signal), local minima and maxima above a threshold define alternation points where the bitcode changes between successions of 0 and 1 bits. Finally, all 1024 bits per signal are concatenated and yield the resulting 10240 bits code.

The second applied feature is based on row-wise convolution with Log-Gabor filters, following an implementation by *Masek*[4] resembling Daugman's feature extraction approach. Again, rows are averaged to form 10 signals of 512 pixels each. The bit-code results from quantizing the phase angle of the complex response with two bits and sticking together all 10240 bits.

3.4   Matching and Decision

In order to assess the performance of our adapted version of the Levenshtein Distance (*Constrained LD*), we compare its recognition accuracy as well as average matching time requirements with the traditional Hamming Distance (*Minimum HD*), as used in commercial systems today [1]. The latter employs the fractional HD (i.e., the number of disagreeing bits of both codes divided by the total number of bits) over a number of bit shifts, and returns the minimum as a degree of dissimilarity between iris codes. Finally, the decision module compares the outcome with a threshold to classify the match as either genuine or imposter.

4   Experimental Results

The following subsections will cover each a specific research question. Unless otherwise noted, LD refers to the in Sect. 2 introduced constrained version in our tests and HD refers to the minimum fractional HD over a fixed number of shifts. Experiments are carried out in verification mode using the equal error rate (EER) and receiver operating characteristics (ROC) as the main performance indicators.

---

[4] L. Masek: Recognition of Human Iris Patterns for Biometric Identification, Master's thesis, University of Western Australia, 2003.

4.1   Does LD Enhance Iris Recognition Accuracy Compared to Traditional Minimum HD?

We have tested *Constrained LD* and *Minimum HD* on *CASIA* and *ICE* datasets using two different algorithms: *Ma* and *Masek*. From the obtained EERs summarized in Table 1 we can see, that LD clearly outperforms HD in all tested combinations. All rates refer to a fixed maximum amount of 20 shifts in order to ensure a fair comparison between LD and HD with respect to rotational tolerance. However, it seems that the feature extraction algorithm influences the amount of improvement. For *Ma*, relative EER improvements were more pronounced than for *Masek* on both databases: up to 40 percent improvement (from 8.6% to 4.96% EER in case of *ICE*) could be achieved. A possible explanation of this behaviour can be found in the manner this algorithm defines its iris-code bits: the alternating zero and one chains seem ideally suited for nonlinear LD alignment. Still, even for *Masek* relative EER improvements of up to 10 percent (from 6.08% to 5.49% EER in case of *ICE*) could be reported for LD. It is remarkable that even in case of very high HD accuracy on *CASIA* (0.58% EER for *Ma*, 0.89% EER for *Masek*) LD can push forward recognition rates (to 0.44% EER for *Ma*, 0.81% for *Masek*). Whereas the EER reflects only a single point of operation, the better performance of LD over HD becomes even more visible, if we take a closer look at the ROC curves for *ICE* and *CASIA* datasets in Figs. 2, 3. Almost all LD curves are clearly superior to HD, except the one

Table 1. EERs of HD versus LD (20 shifts)

EER (%)	ICE		CASIA	
	Ma	Masek	Ma	Masek
Minimum HD	8.60	6.08	0.58	0.89
Constrained LD	4.96	5.49	0.44	0.81

Table 2. Average Matching Times of HD versus LD (20 shifts)

AMT (ms)	ICE		CASIA	
	Ma	Masek	Ma	Masek
Minimum HD	0.73	0.73	0.99	0.97
Constrained LD	4.07	3.71	4.04	3.66

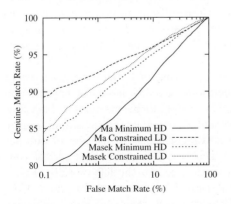

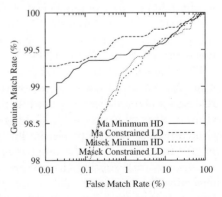

Fig. 2. HD versus LD: ROC for *Ma* and *Masek* (20 shifts) on *ICE* dataset

Fig. 3. HD versus LD: ROC for *Ma* and *Masek* (20 shifts) on *CASIA* dataset

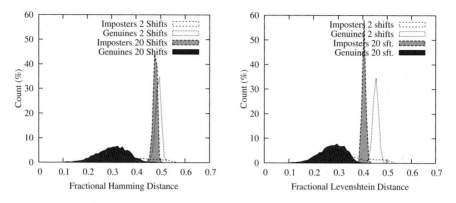

Fig. 4. Impact of max. HD shifts on Ma's score distributions (*CASIA* dataset)

Fig. 5. Impact of max. LD shifts on Ma's score distributions (*CASIA* dataset)

for Ma on *CASIA*, which also depends on the selected maximum shift count, as will be investigated in the next research question. Table 2 lists matching times.

4.2 Which Tradeoff Exists between the Maximum Number of Shifts, Time Complexity and Recognition Accuracy?

From Table 2 we can see, that LD requires additional matching time, however it theoretically lies in the same complexity class like HD. LD is on average 4-5 times slower than HD over different combinations of algorithms and datasets (results refer to the execution on a single processor at 2.8 GHz). While a single match using HD takes less than 1 ms, LD-based matching needs approximately 4 ms in case of 20 bit shifts. While assessing the performance of LD we noticed, that the number of shifts has an important impact on both recognition accuracy and certainly average matching time (due to additional comparisons). From the implementations it is easy to derive, that for both HD and LD in case of small bit shifts, doubling the number of shifts also results in approximately twice as much matching time. From the Figs. 4, 5 we can see, how larger maximum shifts cause inter-class (imposter) as well as intra-class (genuine) distributions to shift to the left, for both LD and HD.

The number of shifts is essential in order to cope with angular displacement of two iris textures to be matched. Furthermore, as shifts should be executed on iris-codes and not on iris-textures in order to avoid multiple extraction of features with resulting time overhead, it is important to consider the bit sampling rate with respect to texture pixels. In our tested algorithms the number of bits per texture row is equal to the number of pixels per row. A bit shift of one thus corresponds with shifting the texture a single pixel to the left or right. Fig. 6 illustrates the impact of shift count on recognition accuracy for the *CASIA* dataset. It is worth noticing, that for all algorithms a significant improvement of performance is achieved at around 7 and 14 shifts. Furthermore, we can see that for very few bit shifts, LD performs worse than HD, which makes sense, since

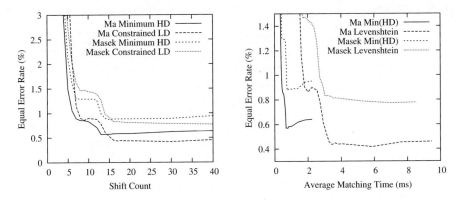

Fig. 6. Shifts-EER tradeoff for HD and LD on *CASIA* dataset

Fig. 7. Time-EER tradeoff for HD and LD on *CASIA* dataset

in order to benefit from the better non-linear alignment, we need at least an amount of shifts in the order of angular displacement of genuine pairs. Finally, Fig. 7 illustrates the resulting tradeoff between EER and average matching time.

5   Conclusion

In this paper we presented an adapted version of LD as a novel matching technique for iris recognition. Mislocation of pupil and iris centers cause significant and irrecoverable mapping distortions, which can not be overcome with simple HD-based matching. Even with accurate segmentation algorithms, inaccuracies are likely to occurr in unconstrained biometrics. Due to its transparent integration, LD is a useful matching technique to tolerate non-linear deformations of iris textures and has been shown to reduce recognition rates drastically (from 8.6% to 4.96% EER for Ma in case of the challenging ICE dataset). Given a sufficient number of bit shifts, all tested algorithms and datasets reported a superior performance of LD over HD. With the introduction of alignment constraints, LD is only 4-5 times slower than HD with an equal amount of shifts. Existing work in unconstrained iris biometrics has mainly concentrated on normalization issues so far, while alternative matching techniques have largely been neglected. We demonstrated and highlighted the ability to tolerate distortions at the matching stage in this work.

Acknowledgment

This work has been supported by the Austrian Federal Ministry for Transport, Innovation and Technology (FIT-IT Trust in IT-Systems, project no. 819382).

References

1. Bowyer, K.W., Hollingsworth, K., Flynn, P.J.: Image understanding for iris biometrics: A survey. Computer Vision and Image Understanding 110, 281–307 (2008)
2. Matey, J.R., Naroditsky, O., Hanna, K., Kolczynski, R., LoIacono, D.J., Mangru, S., Tinker, M., Zappia, T.M., Zhao, W.Y.: Iris on the move: Acquisition of images for iris recognition in less constrained environments. Proceedings of the IEEE 94, 1936–1947 (2006)
3. Wheeler, F.W., Perera, A.G., Abramovich, G., Yu, B., Tu, P.H.: Stand-off iris recognition system. In: Proceedings of the IEEE Second International Conference on Biometrics: Theory, Applications and Systems (BATS), pp. 1–7 (2008)
4. Proença, H., Alexandre, L.A.: Iris recognition: Analysis of the error rates regarding the accuracy of the segmentation stage. Elsevier Image and Vision Computing 28, 202–206 (2010)
5. Matey, J.R., Broussard, R., Kennell, L.: Iris image segmentation and sub-optimal images. Image and Vision Computing 28, 215–222 (2010)
6. Daugman, J.: How iris recognition works. IEEE Transactions on Circiuts and Systems for Video Technology 14, 21–30 (2004)
7. Cha, S.H., Tappert, C., Yoon, S.: Enhancing binary feature vector similarity measures. Journal of Pattern Recognition Research 1, 63–77 (2006)
8. Choi, S., Cha, S., Tappert, C.: A survey of binary similarity and distance measures. Journal of Systemics, Cybernetics and Informatics 8, 43–48 (2010)
9. Levenshtein, V.I.: Binary codes capable of correcting deletions, insertions, and reversals. Soviet Physics Doklady 10, 707–710 (1966)
10. Yuan, X., Shi, P.: A non-linear normalization model for iris recognition. In: Li, S.Z., Sun, Z., Tan, T., Pankanti, S., Chollet, G., Zhang, D. (eds.) IWBRS 2005. LNCS, vol. 3781, pp. 135–142. Springer, Heidelberg (2005)
11. Wyatt, H.J.: A 'minimum-wear-and-tear' meshwork for the iris. Vision Research 40, 2167–2176 (2000)
12. Schimke, S., Vielhauer, C., Dittmann, J.: Using adapted levenshtein distance for on-line signature authentication. In: Proceedings of the 17th International Conference on Pattern Recognition (ICPR 2004), Washington, DC, USA, vol. 2, pp. 931–934. IEEE Computer Society, Los Alamitos (2004)
13. Myers, C.S., Rabiner, L.R.: A comparative study of several dynamic time-warping algorithms for connected word recognition. The Bell System Technical Journal 60, 1389–1409 (1981)
14. Ma, L., Tan, T., Wang, Y., Zhang, D.: Efficient iris recognition by characterizing key local variations. IEEE Transactions on Image Processing 13, 739–750 (2004)

A Mobile-Oriented Hand Segmentation Algorithm Based on Fuzzy Multiscale Aggregation

Ángel García-Casarrubios Muñoz, Carmen Sánchez Ávila,
Alberto de Santos Sierra, and Javier Guerra Casanova

Group of Biometrics, Biosignals and Security, GB2S, Centro de Domótica Integral
Universidad Politécnica de Madrid
{agarcia,csa,alberto,jguerra}@cedint.upm.es

Abstract. We present a fuzzy multiscale segmentation algorithm aimed at hand images acquired by a mobile device, for biometric purposes. This algorithm is quasi-linear with the size of the image and introduces a stopping criterion that takes into account the texture of the regions and controls the level of coarsening. The algorithm yields promising results in terms of accuracy segmentation, having been compared to other well-known methods. Furthermore, its procedure is suitable for a posterior mobile implementation.

1 Introduction

Nowadays, one area of interest which is undergoing a continuous development is Biometrics [1]. Within the field of Biometrics, one of the techniques most widely used is related to the recognition of a person through their hand [2], since, as fingerprint, this physical characteristic is different for each person. Then, in order to carry out this identification, it will be necessary to separate first the required object (hand) from the background of the image. In that regard, image segmentation is the branch of image processing theory [18] that studies different techniques with the main aim of separating an object from the background within a digital image [3]. Specifically, this article presents an image segmentation algorithm aimed at segmenting hand images acquired by a mobile phone, being part of a whole biometric technique oriented to mobile devices for daily applications (bank account access, pin codes and the like).

In the literature, there are a wide number of algorithms that address the segmentation problem from different mathematical approaches [4]. One family of segmentation algorithms that have experienced a great development in recent years are those based on multiscale aggregation [5]. These methods try to segment an image by finding the objects directly, that is, finding the regions that compose the image instead of focusing on detecting the possible existing edges. Moreover, recent results obtained by these algorithms have shown improvement compared to other methods [11], like the Normalized Cuts method [12] and Mean-Shift method [13].

Within the family of multiscale aggregation methods, there are also a wide range of algorithms that differ in the type of mathematical operations applied to image pixels. The algorithms based on Segmentation by Weighted Aggregation (SWA [6]) have shown good results by means of similarities between intensities of neighboring pixels

G. Bebis et al. (Eds.): ISVC 2010, Part I, LNCS 6453, pp. 479–488, 2010.

[7] and with measurements of texture differences and boundary integrity [8]. Other approaches include more complicated operations, such as Gradient Orientations Histograms (GOH [9]), or more straightforward grouping techniques based on the intensity contrast between the boundary of two segments and the inside of each segment [10]. Moreover, thanks to its definite structure, SWA methods may be even used in conjunction with other approaches [20].

Even though our algorithm presents a structure based on SWA methods, there are several differences that must be taken into account:

- SWA algorithms identify each segment by one representative node [9], so that all the pixels which form the segment are strongly connected to that representative node. However, in our algorithm segments are identified by a number of characteristic measures. These measures gather information of each segment by computing the average intensity, variance and position centroid.
- SWA methods start the aggregation process by selecting a few seed nodes [9], satisfying the condition that all pixels are strongly connected to at least one seed node. In contrast, our algorithm builds the first scale by the direct formation of the first segments, so that they are formed by a few pixels similar to each other.
- SWA methods transfer information between scales through the interpolation coefficients p_{ik}, where p_{ik} represents the probability that node i belongs to the segment represented by the node k [19]. However, in our algorithm there is no need to compute any interpolation coefficient, because the segments in each scale collect the information of the segments from the previous scales. In addition, we use Delaunay triangulation [14] to include spatial information, so that it provides in each scale the segments that are more likely to be grouped. This considerably reduces the computational cost, since the number of weights to calculate is much smaller than the number of interpolation coefficients used in SWA methods.

2 Problem Statement

Given an image I containing $M \times N$ pixels, a graph $G = (V, E, W)$ is constructed, where nodes $v_i \in V$ represent pixels in I, edges $e_{ij} \in E$ connect pixels v_i and v_j according to a defined neighborhood system (namely a 4-connected structure) and weights w_{ij} are associated with each edge e_{ij}, indicating to what extent two nodes v_i and v_j are similar. This implementation proposes the use of fuzzy logic for w_{ij} definition.

The idea of segmentation consists of dividing image I into T segments. Let $s_t \subseteq V$ (with $t = 1, 2, ..., T$) be a segment gathering a set of pixels with similar properties in terms of w_{ij}. Concretely, the aim of this document consists of dividing the image I into $T = 2$ segments, distinguishing hand from background. However, sometimes it is difficult to achieve a result of $T = 2$ segments, and thereby it is important to isolate completely the hand in one only segment, even though the background may be formed by several segments.

Therefore, the problem is stated as follows: divide a given image I into segments s_t with $t = 1, 2, ..., T$, so that the hand is completely isolated in one segment from the background. Next section 2.1 will explain how nodes from V are assigned to segments s_t through different scales.

2.1 The Algorithm

Before starting with the main procedure of segmentation, some considerations on image I must be stated. Image I is obtained from a mobile device (section 3) and it represents a color image constructed by the RGB color space [15]. However, since the purpose of our algorithm is segmenting hand images, it is evident that the use of techniques for human skin detection might be useful. In that regard, we propose the use of the Lab color space (abbreviation for *CIE 1976 L*, a*, b**) [16]. Empirically, we found that the best results are obtained with the use of layer b, so $I \leftarrow b$. Notice that the calculation of the layer b from a RGB image is a complicated process that is beyond the scope of this article, see [16] for more information about this conversion.

In Fig. 1 we present the steps followed in the construction of the algorithm's structure and they will be explained below.

1. If we consider $G^{[s]}$ the representation of the image in the scale s, then $G^{[0]} = I$. In this first scale, weights w_{ij} are calculated according to the following Gaussian-like fuzzy distribution (Eq. 1):

$$w_{ij} = \exp\left(-\frac{(I_i - I_j)^2}{2\sigma^2} \right) \qquad (1)$$

where I_i and I_j are the image intensities of I at positions i and j, and $\sigma > 0$ is a parameter responsible for accuracy in terms of segmentation, because if $\sigma \rightarrow \infty$, then $w_{ij} \rightarrow 1$. In our algorithm, for the sake of accuracy, this parameter is fixed to $\sigma = 0.01$.

Once the graph G is obtained, the aggregation algorithm proceeds by sorting every pair of nodes (namely v_i and v_j) according to their weight w_{ij} in descending order. In other words, first pairs (those with higher values regarding w_{ij}) are more similar, and therefore they deserve to belong to the same segment. Taking into account that an unassigned segment s_i is denoted by $s_i = 0$, the aggregation process is carried out until every node in V is assigned to a segment, according to Eq. 2.

$$(s_i, s_j) = \begin{cases} (s_q, s_q) & \text{if } s_i = s_j = 0 \\ (s_i, s_i) & \text{if } s_i \neq 0, s_j = 0 \\ (s_j, s_j) & \text{if } s_i = 0, s_j \neq 0 \end{cases} \qquad (2)$$

where labels s_i and s_j are assigned to nodes v_i and v_j indicating to which segment they belong, according to three possible situations:

- If both nodes have not been assigned yet, then both nodes are assigned to the new segment s_q, being $q \leq M \times N$.
- If in a given edge e_{ij}, only one of the nodes has been assigned to a certain segment label (s_i or s_j depending on the case), then both nodes are assigned to that segment.
- If in a given edge e_{ij}, both nodes have been assigned to a segment label, then both nodes remain assigned to the corresponding segment.

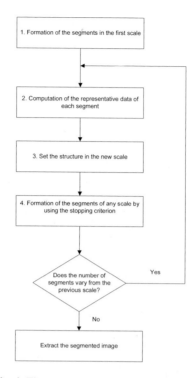

Fig. 1. Flow chart of the algorithm's structure

2. As it was mentioned in section 1, our algorithm does not identify each segment by a unique representative node. In contrast, it carries out that identification through measures that collect the characteristic information of each segment. Those measures are: position centroid ξ (Eq. 3), average intensity μ (Eq. 4) and variance σ^2 (Eq. 5). The centroid position is calculated based on the physical position of nodes within image I and is used later to find the neighbors of every segment. Considering that a node v_i has two components (x_{vi}, y_{vi}), the centroid is computed as follows:

$$\xi = \left(\overline{x}_{s_i}, \overline{y}_{s_i} \right) \tag{3}$$

where x_{si} and y_{si} represent the coordinate locations of nodes within segment s_i.

Similarly, average intensity is obtained by averaging the node intensities in v_{si}, as follows:

$$\mu_i = \left(\overline{I}_{v_{s_i}} \right) \tag{4}$$

where v_{si} indicates the set of nodes within segment s_i.

Finally, the variance provides information about the texture of the segment, as it takes into account its dispersion. The variance is calculated, based on previous Eq. 4:

$$\sigma_i^2 = \frac{1}{N_{s_i}} \sum_{i=1}^{N_{s_i}} \left(I_{v_{s_i}} - \mu_i \right)^2 \tag{5}$$

where N_{si} represents the number of nodes from V in segment s_i.

Therefore, from the second scale, a segment s_i is defined as a three-vector component $s_i = (\xi_i, \mu_i, \sigma_i^2)$, facilitating in subsequent scales the recursive procedure, where the representative data of the segments in a scale s become the nodes in the new scale $s+1$.

3. Due to the aggregation process shown before, the 4-neighbor structure is not conserved, so it is necessary to build a new structure for each scale. This new structure is created by the Delaunay triangulation [17], which establishes a network based on the centroids ξ_i computed in the previous scale. Although the number of neighbors cannot be fixed for each node of the new scale, this algorithm provides a quick way to set every node's neighbors based on their proximity.

4. In order to calculate the weights associated to the new edges, the same Gaussian-like fuzzy distribution shown before (Eq. 1) is used, although in this case it computes the average intensities of the corresponding segments, μ_i and μ_j.

One problem that is encountered when using the aggregation process mentioned above is that it does not stop aggregating nodes until it reaches the end of the list of edges, which means that nodes connected by weak edges (at the bottom of the list) would be added. To solve this problem, we propose the introduction of a stopping criterion that takes into account the texture of the segments. So, if this stopping criterion is not satisfied by a pair of nodes, they will not be aggregated under the same segment. Thus, two nodes v_i and v_j of any scale will be aggregated only if the following condition is satisfied:

$$\sigma_{ij}^2 \leq \sqrt{\sigma_i^2 \cdot \sigma_j^2} + k \tag{6}$$

where σ_{ij}^2 represents the variance of the likely aggregate from nodes v_i and v_j, σ_i^2 and σ_j^2 represent the cumulative variances of the nodes v_i and v_j, and k is a constant set based on empirical results (section 4) that controls the level of coarsening, i.e. the number of remaining segments at the end of the process.

This condition will yield better results regarding segmentation, although k must be tuned to obtain $T = 2$ segments, since the algorithm get blocked in a point where no more segments are aggregated, and therefore $T \geq 2$, in general. As future work, this requirement must be refreshed dynamically in order to obtain $T = 2$ segment labels automatically.

3 Database Acquisition

An important aspect of a segmentation algorithm regards the data involved to validate the proposed approach. Since this document presents a segmentation algorithm oriented for mobile devices, it is obvious that the data must be acquired with the selected device. In other words, the data acquisition consists of images acquired by an iPhone Mobile device. The dimensions of each image are 2 Mega-pixels and 1600 x 1200 pixels, with a resolution of 72 dpi. Furthermore, each image was taken with the hand open, considering a distance of 15-30 cm between the device and the hand. The database gathers 50 users of a wide range of ages (from 16 to 60) and containing different races. Moreover, the images were taken under uncontrolled illumination settings and hands were not required to be placed on a platform. The image acquisition procedure requires no removal of objects such as rings, bracelets, watches and so forth, so that the image is taken non-invasively. The database is publicly available at http://sites.google.com/site/engb2s/databasehand.

4   Results

The evaluation of a segmentation method is a difficult task. In Fig. 2 the reader may observe the degree of accuracy achieved when segmenting the hand from the background.

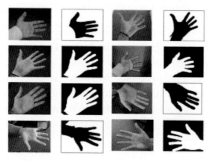

Fig. 2. Eight visual examples of the algorithm's performance

Apart from the obvious way of a direct observation, some quantitative evaluation methods have been used in order to validate our algorithm.

First, the performance of the segmentation has been assessed based on the ratio F-Measure [11] (Eq. 7), which compares our algorithm with ground truth segmentations. These ground truth segmentations were obtained by manually segmenting the images into two classes, foreground and background.

$$F = \frac{2RP}{R+P} \tag{7}$$

where R (Eq. 8) and P (Eq. 9) are the Recall and Precision values of a particular segmentation, according to [11].

$$R = \frac{\text{number of true positives}}{\text{number of true positives} + \text{number of false negatives}} \tag{8}$$

$$P = \frac{\text{number of true positives}}{\text{number of true positives} + \text{number of false positives}} \tag{9}$$

In Fig. 3 (left) the results of the F-Measure for different images and the values of the constant k (stopping criterion) are provided. Notice the high scores achieved, validating the accuracy of the algorithm.

In addition, we have simulated the blurry effect that occurs sometimes when taking a photograph (Fig. 3 - right). For that, the original images were eroded by motion filters that approximate, once convolved with an image, the linear motion of a camera by L pixels, with an angle of θ degrees. Notice that, in spite of the erosion, the scores are over 96%. Moreover, if the original images are eroded by a Gaussian filter of variance $\sigma^2 \leq 20$, the scores are still over 90%.

Furthermore, it is provided a comparison with two well-known methods (Fig. 4), "Normalized cuts and image segmentation" (implementation in MATLAB available[1]) [12] and "Efficient graph-based image segmentation" (according to our implementation) [10], resulting in an outperformance by our approach. It is remarkable that our algorithm allows the implementation of a light software program, which is proved by the fact that it performs considerably well with relatively large images (800x600 pixels computed in 36 seconds), whereas the software implementations of other algorithms, like Normalized Cuts, only perform well with smaller images (160x160 pixels in 13 seconds). In fact, computing a small image (160x160 pixels) with our method took only 4 seconds, meaning that our algorithm is over 3 times faster than the Normalized Cuts algorithm. The implementation of this algorithm has been carried out on a PC Computer @1.8 GHz, with MATLAB 7 R14. A mobile oriented implementation remains as future work.

Image	F-Measure	Constant k
	0.99 ± 0.013	0.003
	0.99 ± 0.016	0.003
	0.99 ± 0.011	0.001
	0.99 ± 0.012	0.001

Image	F-Measure	L	θ	Constant k
	0.99 ± 0.015	20	10	0.003
	0.97 ± 0.016	30	20	0.003
	0.96 ± 0.013	40	20	0.003
	0.97 ± 0.015	40	30	0.003

Fig. 3. Accuracy of the algorithm (left) and performance when eroding the original image with a filter that approximates the linear motion of a camera by L pixels, with an angle of θ degrees (right)

[1] MATLAB implementation at http://www.cis.upenn.edu/~jshi/

Original Image   Our Method   Eff. Graph-Based Segm.   Normalized Cuts

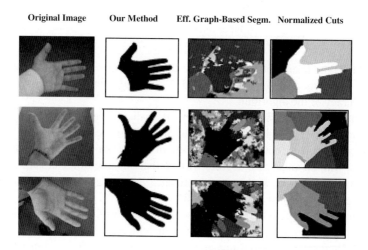

Fig. 4. Results of applying our method compared to other state of the art algorithms

5   Conclusions

We have introduced a multiscale algorithm for image segmentation and aimed for biometric purposes.

The algorithm uses a process of recursive aggregation in order to group the pixels and segments which share the most number of properties. In this article we have focused on the gray intensity and the Lab color space, but the structure of the method is designed to introduce as many features as the user may like, such as second order statistical measures. These properties are computed recursively by weights between pixels and segments, taking into account that fuzzy techniques are used for their conformation. Moreover, the algorithm is flexible enough to allow the user to introduce new fuzzy functions in order to calculate the weights.

We have also presented a stopping criterion that computes the texture of each segment and controls the level of coarsening by adjusting the parameter k.

In addition, one remarkable feature of the algorithm presented is the lack of interpolation matrix, which contributes to the lightness and fastness of the method. This lack of the interpolation matrix is in certain way compensated by the introduction of a Delaunay Triangulation algorithm, which is responsible for the assignment of neighbors in every scale. Moreover, we have introduced the concept of using the centroid of a segment as a way of representing its spatial properties and, therefore, allowing the Triangulation algorithm to create every new graph properly.

We have also introduced a new fast grouping technique that works with the representative properties of each segment. This algorithm receives the graph obtained by the Triangulation method and assigns every segment of the current scale to a new one of the next scale, taking into account that the weight between a pair of segments determines the likelihood of those segments to be grouped.

Finally, we have evaluated the efficiency of our quasi-linear algorithm and we have compared it to the "Normalized Cuts" algorithm and to the "Efficient-graph based segmentation" method, turning out that our algorithm drew better results.

It is remarkable that our algorithm allows the implementation of a light software program, which is proved by the fact that it performs considerably well with relatively large images (800x600 pixels computed in 36 seconds), whereas the software implementations of other algorithms, like Normalized Cuts, only perform well with smaller images (160x160 pixels in 13 seconds). In fact, computing a small image (160x160 pixels) with our method took only 4 seconds, meaning that our algorithm is over 3 times faster than the Normalized Cuts algorithm.

Acknowledgment

This research has been supported by the Ministry of Industry, Tourism and Trade of Spain, in the framework of the project CENIT-Segur@, reference CENIT-2007 2004. (https://www.cenitsegura.es).

References

1. Li, Y., Xu, X.: Revolutionary Information System Application in Biometrics. In: International Conference on Networking and Digital Society, ICNDS 2009, May 30-31, vol. 1, pp. 297–300 (2009)
2. Fong, L.L., Seng, W.C.: A Comparison Study on Hand Recognition Approaches. In: International Conference of Soft Computing and Pattern Recognition, SOCPAR 2009, December 4-7, pp. 364–368 (2009)
3. Shirakawa, S., Nagao, T.: Evolutionary image segmentation based on multiobjective clustering. In: IEEE Congress on Evolutionary Computation, CEC 2009, May 18-21, pp. 2466–2473 (2009)
4. Kang, W.-X., Yang, Q.-Q., Liang, R.-P.: The Comparative Research on Image Segmentation Algorithms. In: First International Workshop on Education Technology and Computer Science, ETCS 2009, March 7-8, vol. 2, pp. 703–707 (2009)
5. Sharon, E., Galun, M., Sharon, D., Basri, R., Brandt, A.: Hierarchy and adaptivity in segmenting visual scenes. Macmillan Publishing Ltd., Basingstoke (2006)
6. Son, T.T., Mita, S., Takeuchi, A.: Road detection using segmentation by weighted aggregation based on visual information and a posteriori probability of road regions. In: IEEE International Conference on Systems, Man and Cybernetics, SMC 2008, October 12-15, pp. 3018–3025 (2008)
7. Sharon, E., Brandt, A., Basri, R.: Fast multiscale image segmentation. In: IEEE Conference on Computer Vision and Pattern Recognition, Proceedings, vol. 1, pp. 70–77 (2000)
8. Sharon, E., Brandt, A., Basri, R.: Segmentation and boundary detection using multiscale intensity measurements. In: Proceedings of the 2001 IEEE Computer Society Conference on Computer Vision and Pattern Recognition, CVPR 2001, vol. 1, pp. I-469 – I-476 (2001)
9. Rory Tait Neilson, B.N., McDonald, S.: Image segmentation by weighted aggregation with gradient orientation histograms. In: Southern African Telecommunication Networks and Applications Conference, SATNAC (2007)
10. Felzenszwalb, P.F., Huttenlocher, D.P.: Efficient graph-based image segmentation. Int. J. Computer Vision 59, 167–181 (2004)
11. Alpert, S., Galun, M., Basri, R., Brandt, A.: Image segmentation by probabilistic bottom-up aggregation and cue integration. In: IEEE Conference on Computer Vision and Pattern Recognition, CVPR 2007, pp. 1–8 (June 2007)

12. Shi, J., Malik, J.: Normalized cuts and image segmentation. IEEE Transactions on Pattern Analysis and Machine Intelligence 22, 888–905 (2000)
13. Comaniciu, D., Meer, P., Member, S.: Mean shift: A robust approach toward feature space analysis. IEEE Transactions on Pattern Analysis and Machine Intelligence 24, 603–619 (2002)
14. Dyer, R., Zhang, H., Möller, T.: Delaunay mesh construction. In: Proceedings of the Fifth Eurographics Symposium on Geometry Processing, SGP 2007, Aire-la-Ville, Switzerland, pp. 273–282. Eurographics Association (2007)
15. Vassili, V.V., Sazonov, V., Andreeva, A.: A survey on pixel-based skin color detection techniques. In: Proc. Graphicon 2003, pp. 85–92 (2003)
16. Hunter, R.S.: Photoelectric Color-Difference Meter. Proceedings of the Winter Meeting of the Optical Society of America, JOSA 38(7), 661 (1948)
17. de Berg, M., van Kreveld, M., Overmars, M., Schwarzkopf, O.: Computational Geometry: Algorithms and Applications, 3rd edn., Springer, Heidelberg (April 2008)
18. Gonzalez, R.C., Woods, R.E.: Digital Image Processing. Addison-Wesley Longman Publishing Co., Inc., Boston (1992)
19. Meirav, G., Eitan, S., Basri, R., Brandt, A.: Texture segmentation by multiscale aggregation of filter responses and shape elements. In: Proceedings of the Ninth IEEE International Conference on Computer Vision, ICCV 2003, Washington, DC, USA, p. 716. IEEE Computer Society, Los Alamitos (2003)
20. Xiao, Q., Zhang, N., Gao, S., Li, F., Gao, Y.: Segmentation based on shape prior and graph model optimization. In: 2nd International Conference on Advanced Computer Control (ICACC), March 27-29, vol. 3, pp. 405–408 (2010)

Analysis of Time Domain Information for Footstep Recognition

R. Vera-Rodriguez[1,2], J.S.D. Mason[2], J. Fierrez[1], and J. Ortega-Garcia[1]

[1] Biometric Recognition Group - ATVS, Universidad Autónoma de Madrid,
Avda. Francisco Tomás y Valiente, 11 - 28049 Madrid, Spain
[2] Speech and Image Research Group, Swansea University,
Singleton Park SA1 8PP, Swansea, UK
{ruben.vera,julian.fierrez,javier.ortega}@uam.es,
j.s.d.mason@swansea.ac.uk

Abstract. This paper reports an experimental analysis of footsteps as
a biometric. The focus here is on information extracted from the time
domain of signals collected from an array of piezoelectric sensors. Results
are related to the largest footstep database collected to date, with almost
20,000 valid footstep signals and more than 120 persons, which is well
beyond previous related databases. Three feature approaches have been
extracted, the popular ground reaction force (GRF), the spatial average
and the upper and lower contours of the pressure signals. Experimental
work is based on a verification mode with a holistic approach based on
PCA and SVM, achieving results in the range of 5 to 15% EER depending
on the experimental conditions of quantity of data used in the reference
models.

1 Introduction

Footstep signals have been used in different applications including medicine [1],
surveillance [2], smart homes [3] and multimedia [4]. Footstep recognition was
first suggested as a biometric in 1977 [1], but it was not until 1997 when the first
experiments were reported [5]. Since then the subject has received relatively little
attention in the literature compared to other biometrics, even though it possesses
some worthwhile benefits: unobtrusive, unconstrained, robust, convenient for
users, etc.

Different techniques have been developed using different sensors, features and
classifiers as described in [6]. The identification rates achieved of around 80-
90% are promising and give an idea of the potential of footsteps as a biometric
[7,8]. However, these results are related to relatively small databases in terms of
number of persons and footstep signals, typically around 15 people and perhaps
20 footsteps per person [5]; this is a limitation of the work to date.

A database is an essential tool to assess any biometric; therefore, this paper
reports experimental results of footsteps as a biometric on the largest footstep
database to date, with more than 120 people and almost 20,000 signals, enabling
assessment with statistical significance.

G. Bebis et al. (Eds.): ISVC 2010, Part I, LNCS 6453, pp. 489–498, 2010.
© Springer-Verlag Berlin Heidelberg 2010

Regarding the sensors employed to capture the footstep signals, two main approaches have been followed in the literature: switch sensors [9,10,11] have been used with a relatively high sensor density (ranging from 50 to 1024 sensors per m^2) in order to detect the shape and position of the foot. On the other hand, different types of sensors that capture transient pressure [5,12,13,14,15] have been used with relatively low sensor density (typically 9 sensors per m^2), more focused in the transient information of the signals along the time course.

The capture system considered here uses a high density of approximately 650 piezoelectric sensors per m^2 which gives a good spatial information and measures transient pressure.

This paper is focused on the analysis of the temporal information of the footstep signals. In this sense the most popular feature extracted in the related works is the ground reaction force (GRF), in some cases used in a holistic manner [5], and in other cases geometric measurements are extracted from the GRF [12,15,16]. In our previous works [8,17,18] geometric and holistic features were compared obtaining in all cases better results for the holistic approach. In this paper, the GRF profiles are compared with other features in a holistic manner and also a fusion of them is carried out obtaining verification results in the range of 5 to 15% of EER depending on the experimental conditions. The experimental protocol is focused on the study of the influence of the quantity of data used in the reference models.

The paper is organized as follows. Section 2 describes the footstep signals used and the feature extraction process, focused on time information. Section 3 describes briefly the database, Section 4 presents the experimental results; and finally conclusions and future work are presented in Section 5.

2   Description of the Signals and Feature Extraction

2.1   Footstep Signals

As mentioned above, the capture system considered here uses piezoelectric sensors with a relatively high density, and therefore footstep signals collected contain information in both time and spatial domains. This is in contrast to previous related works, e.g. [12,10,11]. In fact, footstep signals collected here contain information in four dimensions namely: pressure, time, and spatial positions X and Y. The sensors are mounted on a large printed circuit board and placed under a conventional mat. There are two such mats positioned appropriately to capture a typical (left, right) stride footstep. Each mat contains 88 piezoelectric sensors in an area of 30 × 45 cm.

Figure 1 shows three different 3D plots for an example of a footstep signal reflecting its three stages: Figure 1(a) shows the differential pressure for an instant in the first stage of the footstep, i.e. when the heel strikes the sensor mat, Figure 1(b) shows the same but for an instant in the second stage of the footstep, i.e. when the whole foot rests over the sensors, and Figure 1(c) the same but for an instant in the third stage of the footstep, i.e. when the heel leaves the surface and the toes push off the sensor mat. It is worth noting that the output of the

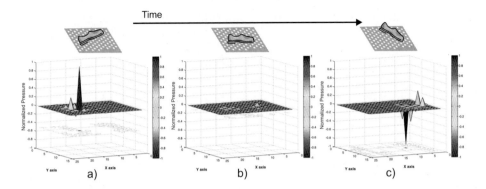

Fig. 1. Spatio-temporal footstep signal in the different stages. a) The derivative of the pressure against the position X and Y at the first stage of footstep. b) The same but for second stage of footstep signal. c) The same but for third stage of the footstep signal.

piezoelectric sensors is the differential pressure in time; thus, it can be seen in Figure 1(c) that there are negative values.

In this paper, the focus is on the analysis of the information of the footstep signals contained in the time domain, leaving the analysis of the spatial domain for further work.

2.2 Feature Extraction and Matching

This section describes the time domain features that are used to assess footsteps as a biometric. Figure 2(a) shows an ensemble of signals from a single footstep. Each signal represents the differential pressure against time for each of the 88 sensors in one mat. An energy detector across the 88 sensors is used to obtain the beginning of each footstep to align the signals.

The most popular time domain feature in related works is the ground reaction force (GRF) [5,7,12,15,16]. Figure 2(b) shows the GRF profile for the example footstep considered here. In this case, as the piezoelectric sensors provide the differential pressure, the GRF is obtained by accumulating for each sensor signal across the time, and then an average of the 88 single profiles is computed to provide a global GRF. Formally, let $s_i[t]$ be the output of the piezoelectric sensors i, where t are the time samples being $t = 1,...,T_{max}$ and i are the sensors $i = 1,...,88$. Then the global GRF (GRF_T) is defined by:

$$GRF_T[t] = \frac{1}{88}\sum_{i=1}^{88}(\sum_{\tau=0}^{t}(s_i[\tau])) \tag{1}$$

Apart from the GRF, two other feature approaches are studied here. The first comes from a spatial average [13,19] of the 88 sensors of the mat to produce a single profile. An example is shown in Figure 2(b).

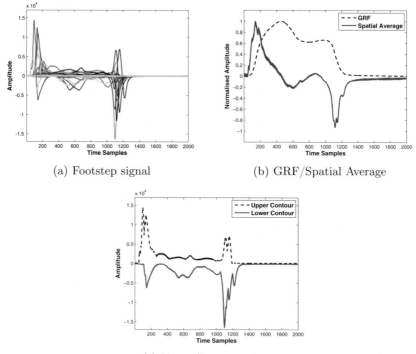

(a) Footstep signal (b) GRF/Spatial Average

(c) Upper/Lower contours

Fig. 2. Feature extraction in time domain for a footstep signal. (a) Differential pressure against time for the 88 sensors. (b) Normalised ground reaction force profile from (a) as defined in Equation 1, and normalised spatial average of the 88 sensors as defined in Equation 2. (c) Upper and lower contour profiles from (a) as defined in Equations 3 and 4 respectively.

$$s_{ave}[t] = \frac{1}{88}\sum_{i=1}^{88}(s_i[t]) \tag{2}$$

The second approach uses the upper and lower contour coming from the maxima and minima of the sensors for each time sample, as shown in Figure 2(c). These two signals are then concatenated into one contour signal.

$$s_{up}[t] = \max_{i=1}^{88}(s_i[t]) \tag{3}$$

$$s_{lo}[t] = \min_{i=1}^{88}(s_i[t]) \tag{4}$$

Equations 1 to 4 lead to a high dimensionality in the time domain with a vector of 8000 samples per footstep. Data dimensionality is further reduced using principal component analysis (PCA), retaining more than 96% of the original information by using the first 120 principal components for each feature approach. Regarding the classifier, a support vector machine (SVM) was adopted with a radial basis function (RBF) as the kernel, due to very good performance in previous studies in this area [7,8].

3   Database and Experimental Protocol

The database collected, apart from footsteps, contains another three biometric modes: speech, face and gait. These modes were included in order to assist in the labelling of the footstep signals, as the collection was an unsupervised process. The speech mode was used to carry out an automatic labelling of the database. A novel iterative process was developed using an identification strategy, labelling the data with the highest confidence first and leaving the data with less confidence for the last iterations [20].

Regarding the experimental protocol followed to assess footsteps as a biometric, special attention has been paid to the partitioning of the data into three sets, namely Reference data and two test sets. The first test set, called Development was used to set the parameters of the system such as the features, the PCA components and the SVM classifier. Then the unseen Evaluation test set is comprised of the last 5 signals collected from each person. It is worth noting that in this paper the data used in the different sets keeps the chronological time of the collection. Therefore, for each user the reference data is comprised of the first data provided, and the data used in the Evaluation set is the last collected. This is a realistic approach reflecting actual usage in contrast to previous related works [8,9,10].

The influence of the quantity of data used to train and test the system is a key factor in any performance assessment; while common in more established biometric modes this aspect is not considered in many cases of footstep studies, for example in [5,12,13], due to limited numbers of data per person in the databases. Different applications can be simulated using different quantities of data in the reference models. In the present work we simulate two important applications: smart homes and access control scenarios. In the case of a smart home there would be potentially a very large quantity of reference data available for a small number of persons, while in security access scenarios such as a border control, limited reference data would be available, but potentially for a very large group of people.

A characteristic of the database considered here is that it contains a large amount of data for a small subset of people (>200 signals for 15 people) and a smaller quantity of data for a larger group of people (>10 signals for 60 people). This reflects the mode of capture which was voluntary and without reward. The assessment of the system is carried out in several points or benchmarks considering different amounts of reference data.

For example, Table 1 shows the quantity of data used in benchmark B1 (using 40 signals in the reference models and 40 models) for the different data sets of Development and Evaluation. Each signal from the test sets is matched against

Table 1. Database configuration for benchmark B1 (40 signals per reference model)

Benchmark B1	Reference Data	Test	
		Development	Evaluation
Clients	P1 – P40	P1 – P40	P1 – P40
Footsteps per client	40	170 (8-650)	5
Total for clients	1,600	6,697	200
Out of class users	P41 - P127	P41 - P78	P41 - P110
Total out of class	763	380	350
Total set	2,363	7,077	550
Total		9,990	

all the 40 reference models defined. As can be seen in the table, the total number of stride signals in the database is 9,990, i.e. 19,980 single (right and left) signals in total. As a result, the number of genuine matchings is 6,697 and 200 for Development and Evaluation respectively; and the number of impostor matchings are 276,383 and 21,800 for Development and Evaluation respectively. Similarly, other benchmark points have been defined with different number of models and signals per model. Profile results of these other benchmark points can be seen in Figure 3.

4   Experimental Results

This section describes the assessment of the time domain features described in Section 2 following the protocols defined in Section 3.

Figure 3(a) shows the EER against the different quantities of stride footstep signals used to train the reference models, bearing in mind that the number of reference models decreases as defined by the top abscissa axis. For example, the points on the left of the figure relate to 75 reference models using only 1 footstep signal to train each model; whereas points on the right relate to 5 reference models with 500 signals to train each model.

The figure shows EER results for the three feature approaches, i.e. the GRF, the spatial average, and the contour. Also a fourth plot in the figure shows the result of the fusion at the feature level of the three approaches, carried out concatenating the features of the single approaches after PCA. These results are generated for stride footsteps, which are comprised of concatenated right and left footstep signals.

All four plots have a similar overall shape with (i) an initial steep fall from approximately 35% EER to 15-20% EER when using 1 to 10 footsteps for training, (ii) a smooth knee curve when increasing the number of signals used in the reference models from 20 to 80 where the error rates change less rapidly from 18 to 13% for the cases of GRF and spatial average, from 15% to 11% for the case of the contour and from 13% to 9% for the case of the fusion of the three approaches; and (iii) relatively flat profiles where error rates are around 10% for

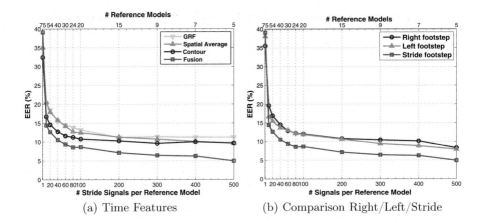

(a) Time Features   (b) Comparison Right/Left/Stride

Fig. 3. (a) Four plots of EER against number of stride signals used to train the reference models for concatenated (stride) footsteps. (b) EER against number of signals used to train the reference models with the fusion of the three feature approaches in the time domain for single (right and left) and stride footsteps.

the three feature approaches (11% for the GRF) and around 5% EER for the case of the fusion when using 500 signals in the reference models.

This shows that in all cases the performance saturates when the number of signals per model exceeds approximately 80 footsteps. Also, errors as low as 5% are viable, especially with further system optimisation. It should be emphasized that: (i) these results relate to features extracted from the time domain only, no spatial information is considered here, and (ii) the number of trials varies along the abscissa axis.

It is interesting to note that the GRF and the spatial average features give similar performance, while the contour features provide the best results of the three approaches more accentuated in the left part of the figure, i.e. when using 10 to 100 signals to train the reference models. Also, the fusion outperforms the three single approaches. This approach provides the best results for footstep recognition using time domain features. The following results of this section relate to the fusion of the three feature approaches in the time domain.

Figure 3(b) shows the EER against different quantities of reference data for the case of the single footsteps (right and left) and the stride for the fusion of the three time domain feature approaches. As can be seen the three plots follow the same trend, but there is a significant improvement of performance when using the stride compared to the single footsteps (right and left), reducing the EER by an average of 3%.

Figure 4 analyses in more detail the case of benchmark B1 (i.e. using 40 signals per reference models and 40 models) comparing results obtained for the Development (Fig. 4(a)) and Evaluation (Fig. 4(b)) sets. In both cases there is a superior performance for the case of the stride footstep with an average relative improvement of 25%. It is interesting to see such a significant performance degradation

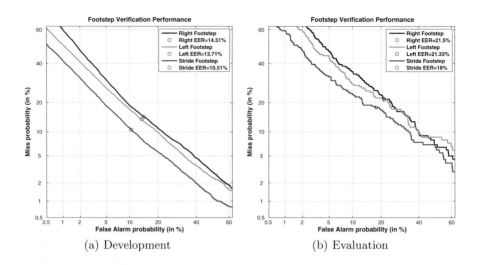

(a) Development   (b) Evaluation

Fig. 4. DET curves for fusion of the three features approaches. (a) Results for benchmark B1 for the Development set, and (b) for the Evaluation set.

between the Development and the Evaluation. As described in Table 1, in both datasets there is a common reference data. As described, in these experiments the time sequence of the collection is kept, i.e. data used in the test sets was collected later in time than data used to train the reference models. Therefore data used in the Development set is closer in time to the reference than data used in the Evaluation set, and therefore more likely to be more similar. This explains the degradation observed in the Evaluation set.

Results achieved here are better compared to those obtained in the related works. Also, it is worth noting that the experimental setup here is the most realistic at least in two factors: (i) it considers the largest footstep database to date, and (ii) it keeps the time lapse between reference and test data, in contrast to most previous works, for example [8,9,10], which randomize the time sequence of the data in the experiments. The randomization makes reference and test datasets more similar and therefore it is possible to achieve artificially good results.

5   Conclusions and Future Work

This paper studies footstep signals as a biometric based on the largest footstep database to date with more than 120 people and almost 20,000 signals. Footstep signals collected contain information in both time and spatial domains, in contrast with previous related works.

This paper focuses on the analysis of the time information of the signals. Features such as the popular ground reaction force, together with two others approaches named the spatial average and the contour are compared and fused following a holistic approach with PCA and SVM.

The experimental protocol is designed to study the influence of the quantity of data used in the reference models, simulating conditions of possible extreme applications such as smart homes or border control scenarios. Results in the range of 5 to 15% EER are achieved in the different conditions for the case of the stride footstep for the fusion of the three feature approaches, which are better than previous works, and with a much more realistic experimental setup.

The time gap between reference data and test is an important point to consider in further work as we have observed a significant degradation of the performance in the Evaluation set which is comprised of the last data collected in the database.

Also, the analysis of the spatial information of the footstep signals and a fusion with the time domain information are very interesting lines for further research.

Acknowledgements

R.V.-R., J.F. and J.O.-G. are supported by projects Contexts (S2009/TIC-1485), Bio-Challenge (TEC2009-11186), TeraSense (CSD2008-00068) and "Cátedra UAM-Telefónica". The multimodal footstep database was collected with support from UK EPSRC and in this context the Authors wish to acknowledge the major contributions of Nicholas W.D. Evans and Richard P. Lewis.

References

1. Pedotti, A.: Simple Equipment Used in Clinical Practice for Evaluation of Locomotion. IEEE Trans. on Biomedical Engineering 24(5), 456–461 (1977)
2. Shoji, Y., Takasuka, T., Yasukawa, H.: Personal Identification Using Footstep Detection. In: Proc. ISPACS, pp. 43–47 (2004)
3. Liau, W.H., Wu, C.L., Fu, L.C.: Inhabitants Tracking System in a Cluttered Home Environment Via Floor Load Sensors. IEEE Trans. on Automation Science and Engineering 5(1), 10–20 (2008)
4. Srinivasan, P., Birchefield, D., Qian, G., Kidane, A.: A Pressure Sensing Floor for Interactive Media Applications. In: Proc. ACM SIGCHI (2005)
5. Addlesee, M.D., Jones, A., Livesey, F., Samaria, F.: The ORL Active Floor. IEEE Personal Communications, 4, 235–241 (1997)
6. Vera-Rodriguez, R., Evans, N., Mason, J.: Footstep Recognition. In: Encyclopedia of Biometrics. Springer, Heidelberg (2009)
7. Suutala, J., Roning, J.: Methods for person identification on a pressure-sensitive floor: Experiments with multiple classifiers and reject option. Information Fusion 9(1), 21–40 (2008)
8. Vera-Rodriguez, R., Lewis, R., Mason, J., Evans, N.: Footstep recognition for a smart home environment. International Journal of Smart Home 2, 95–110 (2008)
9. Yun, J.S., Lee, S.H., Woo, W.T., Ryu, J.H.: The User Identification System Using Walking Pattern over the ubiFloor. In: Proc. ICCAS, pp. 1046–1050 (2003)
10. Middleton, L., Buss, A.A., Bazin, A.I., Nixon, M.S.: A floor sensor system for gait recognition. In: Proc. AutoID, pp. 171–176 (2005)

11. Suutala, J., Fujinami, K., Röning, J.: Gaussian process person identifier based on simple floor sensors. In: Roggen, D., Lombriser, C., Tröster, G., Kortuem, G., Havinga, P. (eds.) EuroSSC 2008. LNCS, vol. 5279, pp. 55–68. Springer, Heidelberg (2008)
12. Orr, R.J., Abowd, G.D.: The Smart Floor: A Mechanism for Natural User Identification and Tracking. In: Proc. Conference on Human Factors in Computing Systems (2000)
13. Cattin, C.: Biometric Authentication System Using Human Gait. PhD Thesis (2002)
14. Suutala, J., Pirttikangas, S., Riekki, J., Roning, J.: Reject-optional LVQ-based Two-level Classifier to Improve Reliability in Footstep Identification. In: Ferscha, A., Mattern, F. (eds.) PERVASIVE 2004. LNCS, vol. 3001, pp. 182–187. Springer, Heidelberg (2004)
15. Gao, Y., Brennan, M.J., Mace, B.R., Muggleton, J.M.: Person recognition by measuring the ground reaction force due to a footstep. In: Proc. RASD (2006)
16. Suutala, J., Roning, J.: Combining classifiers with different footstep feature sets and multiple samples for person identification. In: Proc. ICASSP, vol. 5 (2005)
17. Vera-Rodriguez, R., Evans, N.W.D., Lewis, R.P., Fauve, B., Mason, J.S.D.: An experimental study on the feasibility of footsteps as a biometric. In: Proc. EUSIPCO, pp. 748–752 (2007)
18. Vera-Rodriguez, R., Mason, J., Evans, N.: Assessment of a Footstep Biometric Verification System. In: Handbook of Remote Biometrics. Springer, Heidelberg (2009)
19. Stevenson, J.P., Firebaugh, S.L., Charles, H.K.: Biometric Identification from a Floor Based PVDF Sensor Array Using Hidden Markov Models. In: Proc. SAS 2007 (2007)
20. Vera-Rodriguez, R., Mason, J., Evans, N.: Automatic cross-biometric footstep database labelling using speaker recognition. In: Tistarelli, M., Nixon, M.S. (eds.) ICB 2009. LNCS, vol. 5558, pp. 503–512. Springer, Heidelberg (2009)

Shaped Wavelets for Curvilinear Structures for Ear Biometrics

Mina I.S. Ibrahim, Mark S. Nixon, and Sasan Mahmoodi

ISIS, School of Electronics and Computer Science, University of Southampton, UK
{mis07r,msn,sm3}@ecs.soton.ac.uk

Abstract. One of the most recent trends in biometrics is recognition by ear appearance in head profile images. Determining the region of interest which contains the ear is an important step in an ear biometric system. To this end, we propose a robust, simple and effective method for ear detection from profile images by employing a bank of curved and stretched Gabor wavelets, known as banana wavelets. A 100% detection rate is achieved here on a group of 252 profile images from XM2VTS database. The banana wavelets technique demonstrates better performances than Gabor wavelets technique. This indicates that the curved wavelets are advantageous here. Also the banana wavelet technique is applied to a new and more challenging database which highlights practical considerations of a more realistic deployment. This ear detection technique is fully automated, has encouraging performance and appears to be robust to degradation by noise.

1 Introduction

Biometrics concerns the recognition of individuals based on a feature vector extracted from their anatomical and/or behavioral characteristic, and plays a vital role in security and surveillance systems. Any automatic biometric system needs detection and partitioning process to extract the region of interest from the background.

Ear as a biometric identifier has attracted much attention in the computer vision and biometric communities in recent years. Ear, which is characterized by the appearance of the outer ear, lobes and bone structures is frequently used in biometric. Ear identification has some advantages over other biometric technologies for various reasons. An ear contains a large number of specific and unique features that assist in human identification. It contains a rich and stable structure that does not change significantly over time [1]. An ear can be remotely captured without any knowledge or consent of the person under examination. It also does not suffer from changes in facial expression. These properties make ears very attractive as a biometric identifier. As a result, the ear biometric is suitable for security, surveillance, access control and monitoring applications.

Iannarelli [2] performs two early studies suggesting ears are unique to individuals and supporting the use of an ear as a biometric modality. There are some studies which show how the ear can be used for recognition, using 2D and 3D images [1, 3].

G. Bebis et al. (Eds.): ISVC 2010, Part I, LNCS 6453, pp. 499–508, 2010.

The 2D approaches use the ear as a planar structure affixed to the head. Alternatively, 3D approaches can be used and this has so far been achieved with range scan data. Ear detection is the most important step in an ear recognition system, and the detection quality will therefore affect directly the performance of the whole recognition system.

Recent approaches mostly focus on ear recognition without a fully automated method for ear detection [1, 3]. However automated schemes have recently been proposed for ear detection prior to recognition. Some researchers have focused on 2D [4, 5, 6] and 3D [7, 8] ear detection. The two most sophisticated approaches in 2D ear detection are proposed in [4] and [5]. Islam et al. [4] modifies the cascaded AdaBoost approach to detect the ear from 2D profile images in a learning method by using a training data of ear images. They report good results on large size databases. However if the ear image is rotated with the respect to the training data or if its appearance is different from the ears in the training data, their method could fail to detect the ear, because the training data does not contain example test data in such cases. Forming a database of rotated ears will require much more storage than that required for a technique which is inherently immune to change in feature orientation. Arbab-Zavar et al. [5] propose an ear detection algorithm based on the elliptical shape of the ear by using a Hough transform. Their method is robust with occlusion; however their ear detection algorithm only works under some specific conditions applied to the images of the database to avoid errors caused by the presence of nose and/ or spectacles.

We contend that it is prudent to continue investigating approaches which consider the ear as a planar surface. This will allow for application in access control and surveillance, and for acquisition from documents. The ear plane is not aligned to that of the head, but it is chosen as such we can consider the ear to be on a flat surface. For ear detection, we shall need to locate the ear in a profile image automatically with an algorithm which is robust at the presence of noise. It therefore appears appropriate to investigate a technique which depends on the general structure of the ear.

We employ a bank of banana wavelets, which are generalized Gabor wavelets, to extract curvilinear structures. In addition to the frequency and orientation, banana wavelets are also characterized with properties associated with the bending and curvature of the filter. The ear is an image structure mainly contains features which are similar to those of banana wavelets. These features then appear well matched to the general structure of the ear which has many curvilinear structures, particularly in the region of the helix (the uppermost part of the ear) and the tragus (which are the lower parts).

This paper is structured as follows. Section 2 gives a brief background on banana wavelet filters. Section 3 describes our new technique to detect the ear. The extraction results are provided in section 4. Finally, the conclusions are presented in section 5.

2 Banana Wavelets

Banana wavelets are a generalization of Gabor wavelets and are localized filters derived from a mother wavelet [9], particularly suited to curvilinear structures.

A banana wavelet $B^\mathbf{b}$ is parameterized by a vector $\mathbf{b}$ of four variables, i.e. $\mathbf{b} = (f, \alpha, c, s)$ where f, α, c and s are frequency, orientation, curvature, and size respectively. This filter is built from a rotated and curved complex wave function $F^\mathbf{b}(x, y)$ and a Gaussian $G^\mathbf{b}(x, y)$ function rotated and curved in the same way as $F^\mathbf{b}(x, y)$ [9]:

$$B^\mathbf{b}(x, y) = \gamma \cdot G^\mathbf{b}(x, y) \cdot \left(F^\mathbf{b}(x, y) - DC^\mathbf{b}\right) \tag{1}$$

where $G^\mathbf{b}(x, y) = \exp\left(-\frac{f^2}{2} \cdot \frac{1}{\sigma_x^2} \cdot \left(x_c + c \cdot x_s^2\right)^2 + \frac{1}{\sigma_y^2} \cdot \frac{1}{s^2} \cdot x_s^2\right)$, $x_c = x \cdot \cos\alpha + y \cdot \sin\alpha$,

$x_s = -x \cdot \sin\alpha + y \cdot \cos\alpha$, $F^\mathbf{b}(x, y) = \exp\left(i \cdot f \cdot \left(x_c + c \cdot x_s^2\right)\right)$, γ is a constant, chosen empirically, σ_x and σ_y are the scales of the Gaussian filter in x and y directions respectively, and $DC^\mathbf{b} = e^{-\frac{\sigma_x}{2}}$ is the bias of the banana wavelets.

Any image can be represented by the banana wavelet transform allowing the description of both spatial frequency structure and spatial relations. The convolution of the image with complex banana filters with different frequencies, orientations, curvatures, and sizes, captures the local structure points of an object.

3 Ear Detection

We argue that any ear contains curvilinear structures such as helix, anti-helix and inter-tragic notch. The essence of our ear detection technique is to initially calculate the magnitude of the filter responses $AI(\mathbf{x}_0, \mathbf{b})$ by convolving a banana wavelet $B^\mathbf{b}$ with an image I and then to find the positions where this magnitude has local maxima, i.e.:

$$AI(\mathbf{x}_0, \mathbf{b}) = \arg\left(\max_{\mathbf{x}_0}\left(\|FI(\mathbf{x}_0, \mathbf{b})\|\right)\right) \tag{2}$$

where $FI(\mathbf{x}_0, \mathbf{b}) = \left(B^\mathbf{b} * I\right)(\mathbf{x}_0)$ and $\mathbf{x}_0$ is the position of a pixel in an input image I. A banana wavelet $B^\mathbf{b}$ produces a strong response at pixel position $\mathbf{x}_0$ when the local structure of the image at that pixel position is similar to $B^\mathbf{b}$. An input ear image and the response magnitudes, which are calculated by convolving the input image with the filters depicted in Fig. 1, are shown in Fig. 2. In this figure, white pixels represent high values in the response magnitudes. Therefore there are local maxima (highlighted) at those positions where ear has similar curvature, orientation, and size to those of the corresponding banana wavelet.

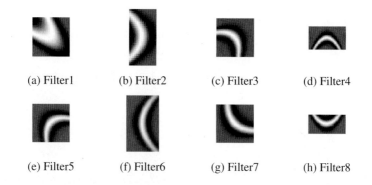

(a) Filter1   (b) Filter2   (c) Filter3   (d) Filter4

(e) Filter5   (f) Filter6   (g) Filter7   (h) Filter8

Fig. 1. (a)-(h) 8 filters used in this work

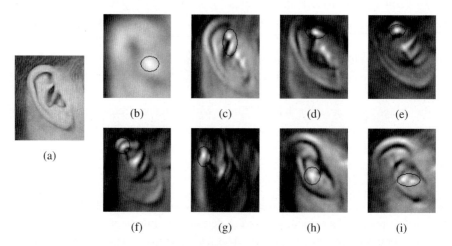

Fig. 2. (a) Input image, and (b)-(i) after convolution with 8 banana filters

A position of interest is selected by considering two conditions: i) the response magnitude has to represent a local maximum ($Q1$) and ii) its value should be greater than a certain threshold ($Q2$):

$$Q1: AI(\mathbf{x}_0,\mathbf{b}) \geq AI(\mathbf{x}_i,\mathbf{b}) \qquad \mathbf{x}_i \in N(\mathbf{x}_0) \qquad (3)$$

$$Q2: AI(\mathbf{x}_0,\mathbf{b}) > \lambda \cdot T(\mathbf{x}_0) \qquad (4)$$

where λ is a constant, $T(\mathbf{x}_0)=0.25 \cdot (E(I)+E(I,\mathbf{x}_0))$, $E(I)=\sum\limits_{\mathbf{x} \in I}\sum\limits_{\mathbf{b} \in B} AI(\mathbf{x},\mathbf{b})$, $E(I,\mathbf{x}_0)=\sum\limits_{\mathbf{x} \in N(\mathbf{x}_0)}\sum\limits_{\mathbf{b} \in B} AI(\mathbf{x},\mathbf{b})$, and $N(\mathbf{x}_0)$ represents a square window with center $\mathbf{x}_0$ and length of side w.

In addition to the conditions (Q1) and (Q2), the spatial arrangement of the positions of local maxima should match a template representing the ear structure (Fig. 3

illustrates this template). Locations of the local structure points in this template are general for any ear. For example, the convolution of filter 4 with any ear, produces a strong response at the top middle part as illustrated in Fig. 2-e.

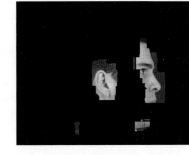

Fig. 3. The ear template   **Fig. 4.** Regions of interest

To reduce computational burden, the ear detection process starts with a coarse search by applying the banana wavelets to the whole image to extract the regions of interest containing curved lines in order to perform a finer search for the ear within these regions (Fig. 4). These regions are much smaller than the whole image and therefore the fine search performed by applying the banana wavelets to these regions, requires less computational demand.

In the fine search, the regions of interest are divided into a group of smaller neighborhoods and convolved with a bank of banana filters (8 filters in this paper are chosen, as shown in Fig. 1) to calculate positions corresponding to local maxima in each neighborhood. The neighborhood with maximum number of positions matching the ear template and meeting conditions (Q1) and (Q2) is considered as the neighborhood containing the ear. In the case that many overlapping neighborhood windows are detected, only one region is selected which contains maximum percentage of overlapped windows. Our technique is generic and applicable to any database. The parameters of the 8 filters are chosen by experiments (Table 1). R and C in the table are the number of rows and columns of the banana wavelet filters, respectively, which are used in the convolution process between banana wavelet filters and the image.

Table 1. Parameter Settings for the Banana Wavelets

	f	α	c	s	R	C
Filter 1	0.05	$\pi/4$	0.1	1	50	50
Filter 2	0.28	$\pi/2$	0.05	1	30	15
Filter 3	0.28	$3\pi/4$	0.05	1	30	30
Filter 4	0.28	π	0.05	1	30	50
Filter 5	0.28	$5\pi/4$	0.05	1	30	30
Filter 6	0.28	$3\pi/2$	0.02	1	50	30
Filter 7	0.28	$7\pi/4$	0.03	1	30	30
Filter 8	0.28	2π	0.05	1	20	40

4 Results

Our primary purpose is to evaluate success in ear detection. The efficiency of banana wavelet technique is tested using a database of 2D images selected from the XM2VTS face profile database [10]. Our database consists of 252 images from 63 individuals with four images per person collected during four different sessions over a period of five months to ensure the natural variation between the images of the same person. The images selected are those where the whole ear is visible in a 720×576 24-bit image. The ears in the database are not occluded by hair but there are few images with some occlusion by earrings. This is the same subset of the XM2VTS face profile database used by Hurley et al. [11] and Arbab-Zavar et al. [5].

The new technique correctly detects all the ears in the images in the database (the detection rate was 100%). Some results of detection using banana wavelets are shown in Fig. 5. The system is fully automatic and it does not require any manual interference for ear detection. As such the approach appears suitable for real time biometric applications. The parameters used in these results are $\sigma_x = 1$, $\sigma_y = 2\pi$, $\gamma = 0.08$, $\lambda = 1.8$, and $w = 7$.

Banana wavelet filters can capture the curved structures better than Gabor wavelet filters. To show this, Gabor wavelet technique is applied to the same subset of the XM2VTS face profile database with the same filter sizes, orientations and frequencies as those of banana wavelet filters (the same parameters in Table 1, except that curvature c was set to zero). The detection rate obtained by Gabor wavelets is 97.2%.

Banana wavelet technique is robust to degradation of images such as the motion blur, partly shown in Fig. 5-a. It is also accurate and robust to some rotations (note the large subject rotation in Fig. 5-b). In addition to that, the technique is robust at presence of earrings and/or glasses, as shown in Fig. 5-c.

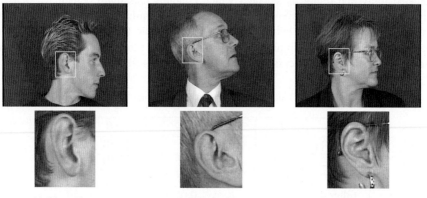

(a) Blurred ear   (b) Rotated ear   (c) Ear occluded by earrings

Fig. 5. Samples of ear detection using our technique

It also appears robust to noise. The accuracy of detection in presence of noise is more than 98% when the noise standard deviation σ is quite high. These results are illustrated in Fig. 6. Here, images are contaminated by additive zero mean Gaussian noise with various noise variances. Here, the technique is successful until $\sigma = 500$ in which case a region containing the eye is erroneously selected (see Fig. 6-f).

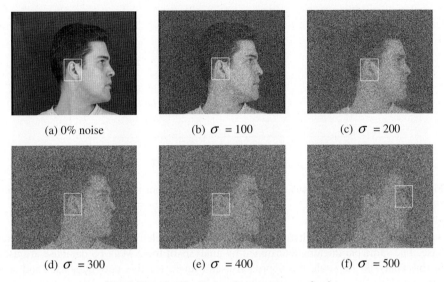

| (a) 0% noise | (b) $\sigma = 100$ | (c) $\sigma = 200$ |
| (d) $\sigma = 300$ | (e) $\sigma = 400$ | (f) $\sigma = 500$ |

Fig. 6. Samples for the results at presence of noise

The results of the noise analysis are provided in Fig. 7. Here, for noise-free images the detection rate is 100%. As expected, the detection rate drops with increasing noise. The analysis of the database is shown in Fig. 7 where at $\sigma = 100$ the recognition rate is still above 98% but this drops to under 80% when $\sigma = 200$. This is actually quite a severe level of noise, as shown in Fig. 6-c. In much worse cases, i.e., beyond the level experienced in surveillance video footage, as shown in Fig. 6-f detection rate drops to about 10%. The graph also shows the performance of the Gabor wavelet. As the noise increases, the advantages associated with using curvature become masked by the noise and in cases of severe noise the Gabor wavelet is more successful than the banana wavelets.

We also apply banana wavelet technique to detect the ears from a new database [12] from which a selection of images is shown in Fig. 8. The advantage of this database is that, it has a lot of variations of ear orientation, size, color skin, and lighting condition, allowing investigation of the performance of our technique on a data acquired in a more realistic scenario (see Fig. 8). The database is acquired as subjects walk past a camera triggered by a light beam signal, and other biometrics are acquired at the same time. As the acquisition is largely uncontrolled, subjects sometimes present the whole head without occlusion and other combination with partial or large occlusion, and partial or sometimes the head is even absent.

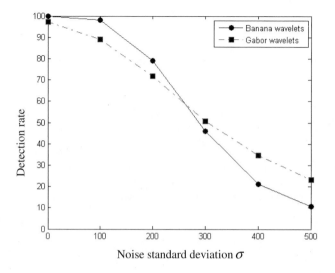

Fig. 7. Detection rate for banana wavelets technique and Gabor wavelets technique in the presence of noise

Table 2 shows the results of applying banana wavelet technique to the new database. The same parameters determined by analysis of the XM2VTS database are used in this analysis. As such, the approach is not tuned for this new database (and scenario) and it is likely that these results could be improved further. We do believe that the structure of the results will remain similar, in that some subjects' ears will remain concealed by hair, and that in a walk-through scenarios it is difficult to acquire images which consistently capture the whole ear.

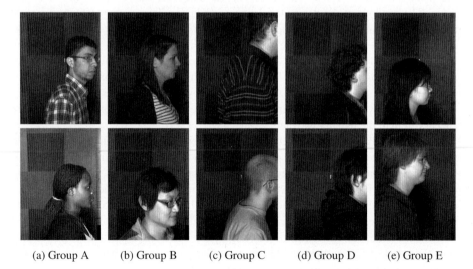

(a) Group A   (b) Group B   (c) Group C   (d) Group D   (e) Group E

Fig. 8. Samples from the new database: (a)-(e) show the groups according to Table 2

Table 2. Results of applying banana wavelet filters to the new database

	Number of samples	Success ear detection
Group A: whole head, no occlusion	885	85%
Group B: whole head, small occlusion	308	65.3%
Group C: partial head, no occlusion	653	79%
Group D: partial head, small occlusion	208	44.7%
Group E: large occlusion	383	17.8%

5 Conclusions

This paper demonstrates how banana wavelets can be used to find the ear from head profile images for biometric purposes. The complexity of the task has been reflected in the fact that ear images can vary in appearance under different viewing and illumination conditions. The experiments show that the system is effective for ear detection, which the proposed technique correctly detects all the test images selected from the XM2VTS database. The technique accrues advantages of noise tolerance and relative immunity to noise. It does not depend on a controlled lighting conditions or skin color; it therefore appears suitable for general applications. The technique proposed here is applied to a more complex database which the acquisition of this database is largely uncontrolled. The result of the ear detection for the new database is good enough according to the uncontrolled conditions, and shows the expected performance in occlusion. The performance of the banana wavelets technique is compared with that of Gabor wavelets technique which shows that banana wavelets can capture the curved structures better than Gabor wavelets. Finally, the technique proposed here is fully automated and does not require any help to detect the ear. The success of our technique relies on the fact that the selected curvilinear structures are general for any ear. We look forward to using this new approach as a primer for recognition purposes.

References

1. Hurley, D.J., Arbab-Zavar, B., Nixon, M.S.: The ear as a biometric. In: Jain, A., Flynn, P., Ross, A. (eds.) Handbook of Biometrics (2008)
2. Iannarelli, A.: Ear Identification. Paramount Publishing Company, Freemont (1989)
3. Bhanu, B., Chen, H.: Human Ear Recognition by Computer. Springer, Heidelberg (2008)
4. Islam, S.M.S., Bennamoun, M., Davies, R.: Fast and Fully Automatic Ear Detection Using Cascaded AdaBoost. In: Proc. of IEEE Workshop on Application of Computer Vision (WACV 2008), USA, pp. 1–6 (January 2008)
5. Arbab-Zavar, B., Nixon, M.S.: On shape mediated enrolment in ear biometrics. In: Bebis, G., Boyle, R., Parvin, B., Koracin, D., Paragios, N., Tanveer, S.-M., Ju, T., Liu, Z., Coquillart, S., Cruz-Neira, C., Müller, T., Malzbender, T. (eds.) ISVC 2007, Part II. LNCS, vol. 4842, pp. 549–558. Springer, Heidelberg (2007)
6. Bustard, J.D., Nixon, M.S.: Robust 2D Ear Registration and Recognition Based on SIFT Point Matching. In: BTAS (2008)
7. Yan, P., Bowyer, K.W.: Biometric recognition using 3d ear shape. IEEE Transactions on Pattern Analysis and Machine Intelligence 29(8), 1297–1308 (2007)

8. Chen, H., Bhanu, B.: Human ear recognition in 3d. IEEE Transactions on Pattern Analysis and Machine Intelligence 29(4), 718–737 (2007)
9. Krüger, N., Pötzsch, M., Peters, G.: Principles of Cortical Processing Applied to and Motivated by Artificial Object Recognition. In: Information Theory and the Brain. Cambridge University Press, Cambridge (2000)
10. Messer, K., Matas, J., Kittler, J., Luettin, J., Maitre, G.: XM2VTSDB: The Extended M2VTS Database. In: Proc. AVBPA 1999, Washington D.C. (1999)
11. Hurley, D.J., Nixon, M.S., Carter, J.N.: Force field feature extraction for ear biometrics. Computer Vision and Image Understanding 98, 491–512 (2005)
12. Samangooei, S., Bustard, J., Nixon, M.S., Carter, J.N.N.: On Acquisition and Analysis of a Dataset Comprising of Gait, Ear and Semantic Data. In: Bhanu, B., Govindaraju, V. (eds.) Multibiometrics for Human Identification, CUP (2010) (in press)

Face Recognition Using Sparse Representations and Manifold Learning

Grigorios Tsagkatakis[1] and Andreas Savakis[2]

[1] Center for Imaging Science
[2] Department of Computer Engineering
Rochester Institute of Technology, NY, 14623

Abstract. Manifold learning is a novel approach in non-linear dimensionality reduction that has shown great potential in numerous applications and has gained ground compared to linear techniques. In addition, sparse representations have been recently applied on computer vision problems with success, demonstrating promising results with respect to robustness in challenging scenarios. A key concept shared by both approaches is the notion of sparsity. In this paper we investigate how the framework of sparse representations can be applied in various stages of manifold learning. We explore the use of sparse representations in two major components of manifold learning: construction of the weight matrix and classification of test data. In addition, we investigate the benefits that are offered by introducing a weighting scheme on the sparse representations framework via the weighted LASSO algorithm. The underlying manifold learning approach is based on the recently proposed spectral regression framework that offers significant benefits compared to previously proposed manifold learning techniques. We present experimental results on these techniques in three challenging face recognition datasets.

Keywords: Face recognition, manifold learning, sparse representations.

1 Introduction

Dimensionality reduction is an important initial step when extracting high level information from images and video. An inherent assumption is that data from a high dimensional space can be reliably modeled using low dimensional structures. Traditional techniques such as Principal Components Analysis (PCA) and Linear Discriminant Analysis (LDA) use linear subspaces as the low dimensional structures for signal modeling [12]. However, certain limitations of linear subspace methods suggest the investigation of non-linear dimensionality reduction techniques.

Manifolds are non-linear structures that have gained considerable attention in recent years. In manifold learning, the objective is to identify the intrinsic dimensionality of the data and project it into a low dimensional space that can reliably capture important characteristics while preserving various geometric properties, such as the geodesic distances or the local neighborhood structure. In general, we may think of a manifold as a low dimensional surface embedded in an ambient higher dimensional space. Manifolds arise when there is a smooth variation of some key parameters that

G. Bebis et al. (Eds.): ISVC 2010, Part I, LNCS 6453, pp. 509–518, 2010.

define the system's degrees of freedom, and are usually far less in number than the dimension of the space where the signal is initially described. In other words, the parameters that characterize the generation of the manifold are sparse with respect to the support set of the signal [1]. Manifold learning has been successfully applied in various computer vision problems such as face recognition [14], activity recognition [18], pedestrian detection [19] and structure-from-motion [20] among others.

Sparse representations (SRs) are another type of signal representation that has received considerable attention in the past few years. SRs were initially applied in signal reconstruction, where it was shown that signals, such as images and audio, can be naturally represented in fixed bases using a small number of coefficients [15], [16]. As far as computer vision applications are concerned, SRs have been very successful in extracting useful information based on the assumption that although images are high dimensional by nature, collections of images, e.g. human faces in various poses, can be reliably represented by a small number of training examples [11].

From the above discussion, it becomes evident that manifold learning and sparse representations share similar motivation in the context of computer vision, although they are different in terms of signal reconstruction. It is therefore natural to ask whether these two approaches can be combined in order to obtain better results. The contribution of this paper lays in the investigation of the connections between the sparse representations and the manifold learning in terms of face recognition accuracy. More specifically, we examine the application of the sparse representation framework for both the generation of the weight matrix as well as the subsequent classification. Although these two approaches have been explored independently [11] and [9], they have not been examined in combination. In addition, we also examine a weighting scheme that introduces a distance based parameter on the sparse representation framework. The sparse representation framework is evaluated using the spectral regression framework that offers better modeling capabilities for manifold learning compared to linearized techniques [9].

The rest of the paper is organized as follows. The concepts of manifold learning and sparse representations are presented in Sections 2 and 3 respectively. The notion of sparsity and how it is applied for the weight matrix construction is explored in Section 4. Experimental results and discussion are provided in Section 5. The paper is concluded in Section 6.

2   Manifold Learning

Manifold learning techniques can be divided in two categories: non-linear and linear approximations. Non-linear dimensionality reduction, including Isometric Feature Embedding (Isomap) [3], Local Linear Embedding (LLE) [4] and the graph Laplacian Eigenmap (LE) [5], reduce the dimensionality of the data while preserving various geometric properties such as the geodesic distances or the local neighboring structure. Most algorithms of this nature represent the training data through a distance matrix (based on the adjacency graph) and achieve dimensionality reduction by applying eigenanalysis on this distance matrix.

A critical issue related to non-linear embedding methods such as Isomap, LLE and LE, is the out-of-sample extension problem, where the lack of a straightforward extension of the mapping to new data (testing) limits the applicability of the methods.

More recent methods try to overcome the out-of-sample problem by constructing a linear approximation of the manifold. The benefits of linear approximations are mostly expressed in terms of savings in computational time, although in some cases very promising results have been reported. Methods in this category include locality preserving projections (LPP) [7] and neighborhood preserving embedding (NPE) [8].

A major limitation of these methods is the requirement for solving a large scale eigen-decomposition which makes them impractical for high dimensional data modeling. In the spectral regression framework (SR) [23] Cai et al. proposed an efficient approach that tackles this problem by recasting the learning of the projection function into a regression framework. Formally, given a data set $\{x_i\}_{i=1}^{m} \in \mathbb{R}^n$, the first step of SR is the construction of a weight matrix $W_{ij} = d(x_i, x_j)$, where each data sample is connected (given a non-zero weight) to another data sample if it is one of its nearest neighbors or belongs to the surrounding ϵ ball. In the second step, the diagonal degree matrix $D_{ii} = \sum_{j \neq i} W_{ij}$ and the Laplacian $L = D - W$ are calculated. The optimal low dimensional projection y of the high dimensional *training* data points x is given by the maximization of the following eigen-problem

$$Wy = \lambda Dy \tag{1}$$

Once the low dimensional representation of the input data is found, new testing data are embedded by identifying a linear function such that

$$x \to y = S^T x = [s_0, \dots, s_{k-1}]^T x \tag{2}$$

where the linear projection function S is a matrix whose columns are the eigenvectors obtained by solving the following eigen-problem

$$XLX^T s = \lambda XDX^T s \tag{3}$$

In SR however, the last step, which is the most expensive, is replaced by the solution of a regression problem given by

$$s = arg \max_{s} \sum_{i=1}^{m} (s^T x_i - y_i)^2 \tag{4}$$

In SR the optimal solution to Eq. (4) is given by the regularized estimator

$$s = (XX^T + \gamma I)^{-1} Xy \tag{5}$$

Once the embedding function has been learned, classification of new test points is usually performed using the k-nearest neighbors (*k*NN) classifier. In addition to the desirable general properties of the *k*NN, such as the guarantees in the error rate [2], *k*NN has been widely adopted by the manifold learning community because of two main reasons. First, it is closely related to the local linear assumption used in the generation of the manifold embedding and, second, *k*NN is an instance-based classifier that does not require training and thus can be used in unsupervised settings.

Furthermore, approximate *k*NNs have been proposed [13] that can deal with large datasets with moderate requirements in terms of classification speed and memory.

Despite these benefits, *k*NN presents a number of drawbacks that may compromise the classification accuracy. The major limitation stems from the choice of *k* which is typically selected via cross-validation. Nevertheless, even if a globally optimal value for *k* is found, the lack of an adaptive neighborhood selection mechanism may result in a poor representation of the neighborhood structure.

3 Sparse Representations

Sparse representations (SRs) were recently investigated in the context of signal processing and reconstruction in the emerging field of compressed sensing [15], [16]. Formally, assume a signal $x \in \mathbb{R}^m$ is known to be sparsely represented in dictionary $D \in \mathbb{R}^{m \times n}$ i.e. $x = Da$ where $\|a\|_0 \leq k$ and $a \in \mathbb{R}^n$. The goal is to identify the non-zero elements of a that participate in the representation of x. An example could be the case where x represents the face of a particular individual and D is the dictionary containing the representation of all faces in the database as in [11]. The individual may be identified by locating the elements of the dictionary that are used for reconstruction, or equivalently the non zero elements of a. The solution is obtained by solving the following optimization problem

$$\hat{a} = \arg min \|a\|_o \quad \text{s.t.} \quad x = Da \tag{6}$$

This problem is NP-hard and therefore difficult to solve in practice. In the pioneering work of Donoho [16] and Candes [15], it was shown that if the solution satisfies certain constraints, such as the sparsity of the representation, the solution of the problem in Eq. 6 is equivalent to the solution of the following problem known as LASSO in statistics

$$\hat{a} = \arg min \|a\|_1 \quad \text{s.t.} \quad x = Da \tag{7}$$

This Sparse Representations approach (SRs) has been recently applied in various computer vision problems including face recognition [10] and image classification [17]. When noise is present in the signal, a perfect reconstruction using Eq. (7) may not be feasible. Therefore, we require that the reconstruction is within an error constant and Eq. (7) is reformulated as

$$\hat{a} = \arg min \|a\|_1 \quad \text{s.t.} \quad \|x - Da\|_2 \leq \epsilon \tag{8}$$

The LASSO algorithm can be efficiently applied for solving the problem in Eq. (8), but it assumes that all elements are equally weighted. This approach, called first order sparsity, only deals with the question of how to sparsely represent a signal given a dictionary. However, other means of information could also be used in order to increase the performance of LASSO or adjust it towards more desirable solutions. One such case is the weighted LASSO [25]. In the weighted LASSO, each coefficient is weighted differently according to its desired contribution. Formally, the optimization of the weighted LASSO is similar to Eq. (8) and it is given by

$$\hat{a} = \arg min \|Pa\|_1 \quad s.t. \quad \|x - Da\|_2 \leq \epsilon \qquad (9)$$

where P is a vector of weights. In this paper, we examine the effects of the solution of Eq. (9) when P corresponds to the distance between the sample x and the individual dictionary elements $d_i \in D$.

Regarding the use of the SRs for classification, the notion of sparsity was proposed in [11] for face recognition based on a well known assumption that the training images of an individual's face span a linear subspace [12]. Although this approach achieved excellent results, it required that the class labels were available during classification. In an unsupervised setting, the class label information is not available and therefore an alternative approach has to be applied. In this paper we use the coefficient with the maximum value as the indicator of the training sample that is most similar to the test example analogous to a nearest neighbor approach. Formally the class of an unknown sample u is given by $class(u) = class(d_i)$ where d_i is the dictionary element found in Eq. (9) i.e. $d_i \equiv max_i(\|a\|_2))$.

4 Sparse Graphs

As discussed in Section 2, the first step in manifold learning is the generation of the adjacency graph. Recently, the ℓ^1-graph was proposed, which employs the concept of sparse representations during graph construction. The objective in ℓ^1-graph is to connect a node with the nodes associated with the data points that offer the sparsest representation. One could select the weights of the edges connecting x_i to other vertices by solving the following ℓ^1 minimization problem

$$\hat{a}_k = \arg min \|a\|_1, \quad s.t. \|x_i - D_i a\|_2 \leq \epsilon \qquad (10)$$

where $D_i = [x_1, \dots x_{i-1}, 0, x_{i+1}, \dots x_j]$. In this case, the weights of the graph correspond to the coefficients of the linear approximation of each data point with respect to the rest of the training set. ℓ^1-graphs offer significant advantages compared to the typical nearest neighbor graphs, the most important of which is that there is no need to specify k, the number of neighbors. The adaptive selection of neighborhood size can more accurately represent the neighborhood structure compared to nearest neighbors. In addition, the ℓ^1-graph is more robust to noise and outliers, since it is based on linear representations that have shown promising results under difficult scenarios such as illumination variation. Furthermore, ℓ^1-graphs encode more discriminative information, especially in the case where class label information is not available.

In [9], L. Qiao et al. applied the ℓ^1- graph construction approach in a modified version of the NPE and reported higher recognition accuracy compared to state-of-the-art manifold learning algorithms using typical weight graphs for face recognition. The approach was later extended to semi-supervised manifold learning in [6]. A similar approach was also presented in [10] where the authors applied the ℓ^1-graph in subspace learning and semi-supervised learning. In all three works the nearest neighbor was used for the classification scheme of the test data.

In addition to the simple sparsity constraint that only deals with the cardinality of the solution, we propose the application of the weighted LASSO in order to take distances into account. We therefore propose to replace Eq. (10) by Eq. (9) where each weighting coefficient is given by $P_i = dist(x, d_i)$ where d_i is the dictionary element (training example) associated with the coefficient a_i. We investigated different choices for the distance metric and report the results using the Euclidean distance.

5   Experimental Results

The goal of the experimental section is to investigate how the SRs framework can be used in conjunction with manifold learning for face recognition as discussed in the previous sections. In order to evaluate the classification accuracy that is achieved by this combination, we performed a series of experiments on three publicly available face recognition datasets: the Yale, the AT&T and the Yale-B. The YALE dataset [22] contains 165 face images of 15 individuals, 11 images per individual. These images contain faces in frontal poses with significant variation in terms of appearance (expressions, glasses etc). The second one is the AT&T dataset [21] which contains 400 images of 40 individuals. The images included in the AT&T dataset exhibit variation in expression, facial details and head pose (20 degrees variation). The Yale-B [24] dataset contains 21888 images of 38 persons under various pose and illumination conditions. We used a subset of 2432 of nearly frontal face images in this experiment.

We note that all images were resized to 32x32 pixels and the pixel values were rescaled to [0,1]. The results are presented in Table 1-11 and correspond to the average classification accuracy after a 10-fold cross validation. In these tables, columns indicate the method used for the generation of the weight matrix using Eq. (1). These techniques are the typical nearest neighbor graph (NN-Graph) using 2 neighbors (as used in [9]), the sparse representation technique for weight matrix construction (ℓ^1-Graph) and the weighted sparse representations (w ℓ^1-Graph). The rows indicate the method used for classification. These methods are the 1-nearest neighbor (NN), the sparse representation classification (SRC) using Eq. (8) and the weighted sparse representations using the weighted LASSO (wSRC) using Eq. (9).

Tables 1-4 present the classification results on the Yale dataset. Based on these results a number of observations can be made. First, we observe that regarding the method used for graph construction, the ℓ^1-Graphs and the wℓ^1-Graphs achieve significantly higher accuracy compared to the NN-Graph, especially when the NN is used as the classifier. The increase in accuracy observed using the NN classification ranges from 13% to 22% depending on the number of training examples available. This indicates that using either the ℓ^1-Graph or the wℓ^1-Graph can provide significant benefits, especially when computational constraints prohibit the application of the SRC or the wSRC classifiers during testing. Regarding classification, we observe that the SRC and the wSRC achieve much higher recognition accuracy compared to the NN classifier, particularly when the NN-Graph is used for the weight matrix generation. As for the weighting extension of the sparse representations, we observe that the results are similar to the ones obtained without the weighting scheme.

Table 1. Classification results on Yale with 2 training examples/class

YALE - 2	NN-Graph	ℓ^1-Graph	w ℓ^1-Graph
NN	39.39	44.90	44.80
SRC	47.68	**47.46**	47.43
wSRC	47.14	47.00	46.97

Table 2. Classification results on Yale with 3 training examples/class

YALE - 3	NN-Graph	ℓ^1-Graph	w ℓ^1-Graph
NN	42.83	50.45	50.53
SRC	53.15	**52.80**	52.68
wSRC	52.65	52.83	52.56

Table 3. Classification results on Yale with 4 training examples/class

YALE - 4	NN-Graph	ℓ^1-Graph	w ℓ^1-Graph
NN	44.78	54.53	54.64
SRC	57.13	**57.90**	57.67
wSRC	57.52	57.77	57.39

Table 4. Classification results on Yale with 5 training examples/class

YALE - 5	NN-Graph	ℓ^1-Graph	w ℓ^1-Graph
NN	46.13	56.68	56.60
SRC	60.73	60.86	60.71
wSRC	60.68	60.86	**60.88**

The results for the AT&T dataset are presented in Tables 5-8 for the cases of 2, 4, 6 and 8 training examples per individual. We observe that similarly to the Yale dataset, using either the ℓ^1-Graph or the wℓ^1-Graph can provide significant increase in accuracy, especially for the case of the NN classifier. We further notice that the wℓ^1-Graph performs better than the ℓ^1-Graph, although the increase in accuracy is minimal.

Table 5. Classification results on AT&T with 2 training examples/class

AT&T - 2	NN-Graph	ℓ^1-Graph	w ℓ^1-Graph
NN	57.90	69.93	68.45
SRC	77.14	76.12	**76.34**
wSRC	75.00	74.76	74.76

Table 6. Classification results on AT&T with 4 training examples/class

AT&T - 4	NN-Graph	ℓ^1-Graph	w ℓ^1-Graph
NN	74.06	82.20	82.77
SRC	89.54	89.31	**89.55**
wSRC	89.10	89.00	89.06

Table 7. Classification results on AT&T with 6 training examples/class

AT&T - 6	NN-Graph	ℓ^1-Graph	w ℓ^1-Graph
NN	84.84	89.34	89.40
SRC	92.43	93.25	93.40
wSRC	92.96	93.37	**93.56**

Table 8. Classification results on AT&T with 8 training examples/class

AT&T - 8	NN-Graph	ℓ^1-Graph	w ℓ^1-Graph
NN	91.43	93.62	93.81
SRC	95.31	95.93	95.87
wSRC	95.41	95.93	**96.00**

The classification results for the YaleB dataset are shown in Tables 9-11 for 10, 20 and 30 training examples per individual. We note that the YaleB dataset is more demanding due to its larger size. Regarding the performance, we again observe the superiority of the ℓ^1-Graph and the wℓ^1-Graph for the weight matrix construction and the SRC and wSRC for the classification. However, we notice that there is a significant increase in terms of accuracy when the weighted sparse representation is used. We can justify this increase in accuracy by the fact that the larger number of training examples offers better sampling of the underlying manifold in which case the use of distances provide more reliable embedding and classification.

Table 9. Classification results on YALE-B with 10 training examples/class

YALEB - 10	NN-Graph	ℓ^1-Graph	w ℓ^1-Graph
NN	74.92	84.90	84.95
SRC	82.10	85.84	86.18
wSRC	82.39	86.23	**86.33**

Table 10. Classification results on YALE-B with 20 training examples/class

YALEB - 20	NN-Graph	ℓ^1-Graph	w ℓ^1-Graph
NN	84.76	87.60	87.96
SRC	87.06	89.17	90.10
wSRC	87.90	90.62	**91.05**

Table 11. Classification results on YALE-B with 30 training examples/class

YALEB - 30	NN-Graph	ℓ^1-Graph	w ℓ^1-Graph
NN	87.83	89.56	90.03
SRC	89.01	90.26	90.50
wSRC	90.50	**91.99**	91.75

6 Conclusions

In this paper, we investigated the use of manifold learning for face recognition when the sparse representations framework was utilized in two key steps of manifold learning: weight matrix construction and classification. Regarding the weight matrix construction, we examined the benefits of the sparse representation framework instead of the traditional nearest neighbor approach. With respect to classification, we compared the recognition accuracy of a typical classification scheme, the k-nearest neighbors, and the accuracy achieved by the sparse representation classifier. In addition, we investigated the effects of introducing a distance based weighting term in the sparse representation optimization and examined its effects on the weight matrix construction and the classification.

Based on the experimental results, we can make the following suggestion regarding the design of a manifold based face recognition system. When sufficient computational resources are available during the training stage, using the sparse representation framework will always lead to significantly better results especially when resource limitations during the testing phase prohibit the application of the more computationally demanding sparse representation framework for classification. On the other hand, when the available processing power during testing is adequate, then the sparse representation classification outperforms the nearest neighbor while the method used for graph construction plays a less significant role.

References

1. Baraniuk, R.G., Cevher, V., Wakin, M.B.: Low-dimensional models for dimensionality reduction and signal recovery: A geometric perspective. To appear in Proceedings of the IEEE (2010)
2. Cover, T.M.: Estimation by the nearest neighbor rule. IEEE Trans. on Information Theory 14(1), 50–55 (1968)
3. Tenenbaum, J.B., Silva, V., Langford, J.C.: A global geometric framework for nonlinear dimensionality reduction. Science 290, 2319–2323 (2000)
4. Saul, L.K., Roweis, S.T.: Think globally, fit locally: unsupervised learning of low dimensional manifolds. Journal of Machine Learning Research 4, 119–155 (2003)
5. Belkin, M., Niyogi, P.: Laplacian eigenmaps for dimensionality reduction and data representation. Neural Computation 15(6), 1373–1396 (2006)
6. Qiao, L., Chen, S., Tan, X.: Sparsity preserving discriminant analysis for single training image face recognition. Pattern Recognition Letters (2009)
7. He, X., Niyogi, P.: Locality preserving projections. In: Advances in Neural Information Processing Systems, vol. 16, pp. 153–160 (2003)
8. He, X., Cai, D., Yan, S., Zhang, H.J.: Neighborhood preserving embedding. In: IEEE Int. Conf. on Computer Vision, pp. 1208–1213 (2005)
9. Qiao, L., Chen, S., Tan, X.: Sparsity preserving projections with applications to face recognition. Pattern Recognition 43(1), 331–341 (2010)
10. Cheng, B., Yang, J., Yan, S., Fu, Y., Huang, T.: Learning with L1-Graph for Image Analysis. IEEE Transactions on Image Processing (2010) (accepted for publication)
11. Wright, J., Yang, A.Y., Ganesh, A., Sastry, S.S., Ma, Y.: Robust face recognition via sparse representation. IEEE Trans. on Pattern Analysis and Machine Intelligence 31(2), 210–227 (2009)

12. Belhumeur, P., Hespanda, J., Kriegman, D.: Eigenfaces versus Fisherfaces: Recognition Using Class Specific Linear Projection. IEEE Trans. Pattern Analysis and Machine Intelligence 19(7), 711–720 (1997)
13. Andoni, A., Indyk, P.: Near-optimal hashing algorithms for approximate nearest neighbor in high dimensions. Communications of the ACM 51(1), 117–122 (2008)
14. He, X., Yan, S., Hu, Y., Niyogi, P., Zhang, H.J.: Face recognition using laplacianfaces. IEEE Trans. on Pattern Analysis and Machine Intelligence, 328–340 (2005)
15. Candès, E., Romberg, J., Tao, T.: Robust uncertainty principles: Exact signal reconstruction from highly incomplete frequency information. IEEE Trans. on Information Theory 52(2), 489–509 (2006)
16. Donoho, D.: Compressed sensing. IEEE Trans. on Information Theory 52(4), 1289–1306 (2006)
17. Mairal, J., Bach, F., Ponce, J., Sapiro, G., Zisserman, A.: Supervised dictionary learning. In: Advances in Neural Information Processing Systems, vol. 21 (2009)
18. Elgammal, A., Lee, C.S.: Inferring 3D body pose from silhouettes using activity manifold learning. In: IEEE Conf. on Computer Vision and Pattern Recognition, vol. 2 (2004)
19. Tuzel, O., Porikli, F., Meer, P.: Pedestrian detection via classification on Riemannian manifolds. IEEE Trans. on Pattern Analysis and Machine Intelligence, 1713–1727 (2008)
20. Rabaud, V., Belongie, S.: Linear Embeddings in Non-Rigid Structure from Motion. In: IEEE Conf. on Computer Vision and Pattern Recognition (2008)
21. AT&T face database, http://www.cl.cam.ac.uk/research/dtg/attarchive/facedatabase.html
22. Yale Univ. Face Database, http://cvc.yale.edu/projects/yalefaces/yalefaces.html
23. Cai, D., He, X., Han, J.: Spectral regression for efficient regularized subspace learning. In: Proc. Int. Conf. Computer Vision, pp. 1–8 (2007)
24. Georghiades, A.S., Belhumeur, P.N., Kriegman, D.J.: From Few to Many: Illumination Cone Models for Face Recognition under Variable Lighting and Pose. IEEE Trans. Pattern Analysis and Machine Intelligence 23(6), 643–660 (2001)
25. Zou, H.: The adaptive Lasso and its oracle properties. Journal of the American Statistical Association 101(476), 1418–1429 (2006)

Face Recognition in Videos Using Adaptive Graph Appearance Models

Gayathri Mahalingam and Chandra Kambhamettu

Video/Image Modeling and Synthesis (VIMS) Lab.
Department of Computer and Information Sciences,
University of Delaware, Newark, DE, USA

Abstract. In this paper, we present a novel graph, sub-graph and super-graph based face representation which captures the facial shape changes and deformations caused due to pose changes and use it in the construction of an adaptive appearance model. This work is an extension of our previous work proposed in [1]. A sub-graph and super-graph is extracted for each pair of training graphs of an individual and added to the graph model set and used in the construction of appearance model. The spatial properties of the feature points are effectively captured using the graph model set. The adaptive graph appearance model constructed using the graph model set captures the temporal characteristics of the video frames by adapting the model with the results of recognition from each frame during the testing stage. The graph model set and the adaptive appearance model are used in the two stage matching process, and are updated with the sub-graphs and super-graphs constructed using the graph of the previous frame and the training graphs of an individual. The results indicate that the performance of the system is improved by using sub-graphs and super-graphs in the appearance model.

1 Introduction

Face recognition has long been an active area of research, and numerous algorithms have been proposed over the years. For more than a decade, active research work has been done on face recognition from still images or from videos of a scene [2]. A detailed survey of existing algorithms on video-based face recognition can be found in [3] and [4]. The face recognition algorithms developed during the past decades can be classified into two categories: holistic approaches and local feature based approaches. The major holistic approaches that were developed are Principal Component Analysis (PCA) [5], combined Principal Component Analysis and Linear Discriminant Analysis (PCA+LDA) [6], and Bayesian Intra-personal/Extra-personal Classifier (BIC) [7].

Chellappa *et al.* [8] proposed an approach in which a Bayesian classifier is used for capturing the temporal information from a video sequence and the posterior distribution is computed using sequential importance sampling. As for the local feature based approaches, Manjunath and Chellappa [9] proposed a feature based approach in which features are derived from the intensity data without assuming

G. Bebis et al. (Eds.): ISVC 2010, Part I, LNCS 6453, pp. 519–528, 2010.
© Springer-Verlag Berlin Heidelberg 2010

any knowledge of the face structure. Topological graphs are used to represent relations between features, and the faces are recognized by matching the graphs. Fazl Ersi and Zelek [10] proposed a feature based approach in which Gabor histograms are generated using the feature points of the face image and are used to identify the face images by comparing the Gabor histograms using a similarity metric. Wiskott *et al.* [11] proposed a feature based approach in which the face is represented as a graph with the features as the nodes and each feature described using a Gabor jet. A similar framework was proposed by Fazl-Ersi *et al.* [12] in which the graphs were generated by triangulating the feature points.

Video-based face recognition has the advantage of using the temporal information from each frame of the video sequence. Zhou *et al.* [13] proposed a probabilistic approach in which the face motion is modeled as a joint distribution, whose marginal distribution is estimated and used for recognition. Li [14] used the temporal information to model the face from the video sequence as a surface in a subspace and performed recognition by matching the surfaces. Kim *et al.* [15] fused pose-discriminant and person-discriminant features by modeling a Hidden Markov Model (HMM) over the duration of a video sequence. Stallkamp *et al.* [16] used K-nearest neighbor model and Gaussian mixture model (GMM) for classification purposes. Liu and Chen [17] proposed an adaptive HMM to model the face images. Lee *et al.* [18] represented each individual by a low dimensional appearance manifold in the ambient image space. Park and Jain [19] used a 3D model of the face to estimate the pose of the face in each frame and then matching is performed by extracting the frontal pose from the 3D model.

In this paper, we propose a novel adaptive graph based approach that uses graphs, sub-graphs, and super-graphs for spatially representing the faces for face recognition in a image-to-video scenario. The graphs, sub-graphs and super-graphs are constructed using the facial feature points as vertices which are labeled by their feature descriptors. An adaptive probabilistic graph appearance model is built for each subject, which captures the temporal information. Adaptive matching is performed using the probabilistic model in the first stage and a graph matching procedure in the second stage. The appropriate appearance model is updated with the results of recognition from the previous frame of the video sequence, and the associated graph model set is updated with the sub-graphs and super-graphs generated using the graph of the previous frame and the model graphs.

2   Face Image Representation

In this section, we describe our approach in representing the face images. In our approach, the face image is represented by a graph which is constructed using the facial feature points as vertices. The vertices are labeled by their corresponding feature descriptors which are extracted using the Local Binary Pattern (LBP) [20], [21]. Every face is distinguished not by the properties of individual features, but by the contextual relative location and comparative appearance of these features. Hence, it is important to identify those features that are conceptually common in every face such as eye corners, nose, mouth, etc. The feature points are extracted by using a similar approach as [1], where the authors extract the features points using a modified Local Feature Analysis (LFA) [22] which constructs kernels that spatially represent a pixel in the image. A subset of kernels are extracted that correspond to discriminative facial features using Fisher scores. Figure 1 shows the feature points extracted from the image and a frame of the video sequences. The images are ordered according to their resolution from high to low.

Fig. 1. First 150 Feature points extracted from the training image (first pair of images) and the testing video frames (second & third pair of images)

2.1 Feature Description with Local Binary Pattern

A feature descriptor is constructed for each feature point extracted from an image using Local Binary Pattern (LBP).

The original LBP operator proposed by Ojala *et al.* [20] labels the pixels of an image by thresholding the $n \times n$ neighborhood of each pixel with the value of the center pixel, and considering the result value as a binary number. The histogram of the labels of the pixels is used as a texture descriptor. The LBP operator with P sampling points on a circular neighborhood of radius R is given by,

$$LBP_{P,R} = \sum_{p=0}^{P-1} s(g_p - g_c)2^p. \tag{1}$$

where

$$s(x) = \begin{cases} 1 \text{ if } x \geq 0 \\ 0 \text{ if } x < 0 \end{cases} \tag{2}$$

The LBP operators with at most two bitwise transitions from 0 to 1 or vice versa were called as uniform patterns by Ojala *et al.* which reduced the dimension of LBP significantly. In our experiments, we use $LBP_{8,2}^{u2}$ which represents an uniform LBP operator with 8 sampling points in a radius of 2 within a window of 5×5 around the pixel which give a 59 element vector.

3 Adaptive Graph Appearance Model

An adaptive appearance model is constructed for each subject using the set of feature points and their descriptors from all the images of the subject. The appearance of a graph is another important distinctive property and is described using the feature descriptors of the vertices of the graph. In our approach, we construct a graph appearance model by modeling the joint probability distribution of the appearance of the vertices of the graphs of an individual. The probabilistic appearance model is constructed using the feature descriptors from all the images of a subject which makes it easy to adapt to the changes in the size of the training data. The model can easily be adapted to the changes in the training set as it is constructed using the feature descriptors. The adaptation is performed at the matching stage where the result of recognition from each frame is adapted to the appropriate appearance model. Given N individual and M training face images, the algorithm to learn the model is described as follows:

1. Initialize N empty model sets.
2. For each individual i with M_i images
   a. For each image I_i^j, (j^{th} image of the ith individual)
      * Extract feature points and corresponding feature descriptors (subsection 2.1).
      * Construct image graphs (subsection 3.1) and add it to the ith model set.
   b. For each pair of graphs in the ith model set
      * Extract the feature points for sub-graph and super-graph (subsection 3.2).
      * Construct the sub-graph and super-graph using the extracted feature points (subsection 3.2) and add it to the ith model set.
   c. Construct the appearance model for the ith individual using the ith model set.

The appearance model denoted as Φ_n is constructed by estimating the joint probability distribution of the appearance of the graphs which is modeled using Gaussian Mixture Model (GMM) [23]. GMMs can efficiently represent heterogeneous data and capture dominant patterns in the data using Gaussian components. Mathematically, a GMM is defined as:

$$P(F|\Theta) = \sum_{i=1}^{K} w_i N(X|\mu_i, \sigma_i) \qquad (3)$$

where

$$N(X|\mu_i, \sigma_i) = \frac{1}{\sigma_i \sqrt{2\pi}} \exp^{-\frac{(X-\mu_i)^2}{2\sigma^2}} \qquad (4)$$

and $\Theta = {w_i, \mu_i, \sigma_i^2}_{i=1}^{K}$ are the parameters of the model, which includes the weight w_i, the mean μ_i, and variance σ_i^2 of the K Gaussian components. In order to maximize the likelihood function $P(F|\Theta)$, the model parameters are re-estimated using the Expectation-Maximization (EM) technique [24]. For more details about the EM algorithm see [24].

3.1   Image Graph Construction

The most distinctive property of a graph is its geometry, which is determined by the way the vertices of the graph are arranged spatially. Graph geometry plays an important role in discriminating the graphs of different face images. In our approach, the graph geometry is defined by constructing a graph with constraints imposed on the length of the edges between a vertex and its neighbors. We propose a graph generating procedure that generates a unique graph with the given set of vertices for each face image. At each iteration, vertices and edges are added to the graph in a Breadth-first search manner and considering a spatial neighborhood distance for each vertex. This generates a unique graph given a set of feature points. The following proof illustrates the uniqueness property of the graph generated.

Theorem 1. *Given a set of vertices V, the graph generation procedure generates a unique graph $G(V, E)$.*

Proof. Proof by contradiction. Let there be two graphs $G_1(V, E_1)$ and $G_2(V, E_2)$ generated by the graph generation procedure, such that $G_1 \neq G_2$. In other words, $E_1 \neq E_2$. Without loss of generality, let us assume that there exists an edge $e \in E_1$ which connects two vertices u and v, where $u, v \in V$, and $e \notin E_2$. This implies that the Euclidean distance between u and v is greater than the threshold, and hence $e \notin E_1$ as well. Hence, $E_1 = E_2 \Rightarrow G_1 = G_2$. Hence the proof.

3.2   Common Sub-graph and Super-graph

The graph representation effectively represents the inherent shape changes of a face and also provides a simple and powerful matching technique. Including the shape changes and the facial deformations caused by pose changes in the model improves the recognition rate. In our approach, we capture these shape changes due to change in pose of the face by constructing a common sub-graph and super-graph using the set of graphs of an individual. The common sub-graph and super-graph are defined in our system as follows;

Definition 1. *Given two graphs G_1 and G_2, the sub-graph H of G_1 and G_2 is defined as*

$$H = \{v | v \in G_1 \cap G_2, \ni \cos(f_1(v), f_2(v)) \approx 1, f_1(v) \in G_1, f_2(v) \in G_2\} \quad (5)$$

where v is the vertex, e is the edge, and $f(v)$ is the feature descriptor of v.

The sub-graph includes those vertices that have spatial similarity and vertex similarity in G_1 and G_2.

Definition 2. *Given two graphs G_1 and G_2, the super-graph H of G_1 and G_2 is defined as*

$$H = \{v | v \in G_1 \cup G_2\} \quad (6)$$

where v is the vertex, e is the edge, and $f(v)$ is the feature descriptor of v.

The sub-graphs and super-graphs are constructed for each pair of graphs of the images of a subject using a similar approach to construct the graph of an image. The sub-graph and super-graph essentially capture the craniofacial shape changes and the facial deformations due to various poses of the face.

4   Adaptive Matching and Recognition

We use a two stage adaptive matching procedure to match every frame of the video with the trained models and graphs. The first stage of matching involves the computation of a Maximum a Posterior probability using the test graph $G(V, E, F)$ with vertex set V and set of feature vectors F and is given by,

$$P_k = \max_n P(G|\Phi_n). \tag{7}$$

where P_k is the MAP probability of G belonging to model set k.

The MAP solution is used to prune the search space for the second stage of matching in which we use a simple deterministic algorithm that uses cosine similarity measure and spatial similarity constraints to compare the test graph with the training graphs. The appropriate GMM is adapted by the result of recognition and is used for matching subsequent frames. The recognition result is considered correct if the difference between the highest score and the second highest score is greater than a threshold. This measure of correctness is based on the idea proposed by Lowe [25], that reliable matching requires the best match to be significantly better than the second best match. The appropriate model set and the GMM is updated with the result of recognition from each frame. The update involves adding the graph of the frame to the model set, along with the sub-graphs and super-graphs generated using the graph of the frame and the graphs in the model set. The entire matching procedure is given as follows;

1. For *each frame f in the video sequence*
   a. Construct the image graph G using the extracted feature points and their descriptors.
   b. Compute the MAP solution for G belonging to each appearance model and select k model sets (10% in our experiments) with highest probability.
   c. Compute similarity scores between G and the graphs from k model sets using cosine similarity measure.
   d. Update the appropriate GMM and the model set with G using the likelihood score and similarity scores.
2. Select the individual with the maximum number of votes from all the frames.

An iterative procedure is used to find the similarity between graphs. Given two graphs G and H with $|H| \leq |G|$, we use spatial similarity (spatial location of a vertex in H and G) and vertex similarity (vertices with similar feature descriptors) to match H with a subgraph of G that maximizes the similarity score. At each iteration, vertex $u \in H$ is compared with $v \in G$ such that u and v have high spatial and vertex similarity. The procedure is repeated with neighbors of u and v. The spatial constraint imposed on the vertices reduces the number of vertex comparisons and allows for faster computation.

5   Experiments

In order to validate the robustness of the proposed technique, we used the UTD database [26]. The UTD database consists of a series of close and moderate range, videos of 315 subjects and also their high resolution images in various poses. The neutral expression close range videos and the parallel gait videos were used in our experiments. The high resolution images of each subject were used as training set. Figure 2 shows sample video frames of both the close-range and moderate-range videos from the UTD database.

The preprocessing steps include extracting the face region and resizing it to 72×60 pixels. We extracted 150 feature points from each image and their corresponding feature descriptors were computed using 5×5 window around each point. The dimension of the feature vectors are reduced using PCA from 59 to 20 retaining 80% of the non-zero eigenvalues. Graphs including sub-graphs and super-graphs are generated for the images of each individual. The maximum spatial neighborhood distance of each vertex was set to 30 pixels. A GMM with 10 Gaussian components is constructed for each individual using the set of graphs. K-means clustering is used for initializing the GMM.

(a) Sample video frames from UTD dataset in close-range

(b) Sample video frames from UTD database in moderate-range

Fig. 2. Sample video frames from the UTD video datasets

During the testing stage, we randomly selected a set of frames from the videos of a subject. A graph is generated for each frame after preprocessing the frame. The likelihood scores are computed for the test graph and the GMMs and the training graphs are matched with the test graph to produce similarity scores, and the appropriate GMM is updated using the similarity and likelihood scores. The threshold is determined by the average of the difference in likelihood scores and similarity scores between each class of data. Though the threshold value is data dependent, the average proves to be an optimum value. The performance of the algorithm is compared with video-based recognition algorithm in [17] (denoted as HMM) which handles video-to-video based recognition. In addition to this, we compare the performance by considering the effects of temporal information and spatial information individually and when combined. We denote the above two approaches as AGMM and Graphs respectively, and the proposed approach as AGMM+Graphs. The results are tabulated in the Table 1. Figure 3 shows the Cumulative Match Characteristic curve obtained for various algorithms (AGMM+Graphs, AGMM and HMM).

Table 1. Comparison of the error rates with different algorithms

	HMM	AGMM	Graphs	AGMM+Graphs
UTD Database (close-range)	24.3%	24.1%	21.2%	16.1%
UTD Database (moderate-range)	31.2%	31.2%	26.8%	19.4%

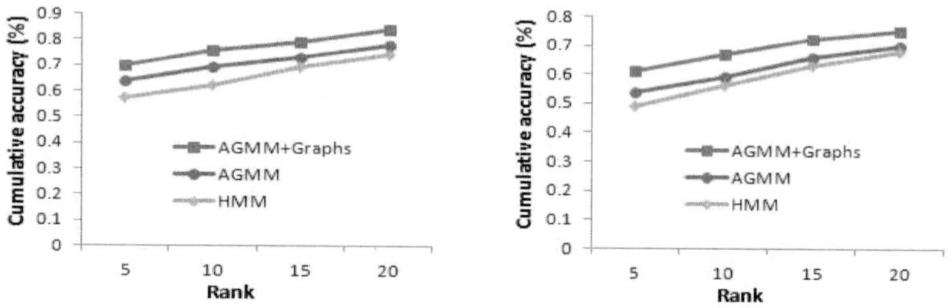

(a) CMC curve for close-range videos of UTD database

(b) CMC curve for moderate-range videos of UTD database

Fig. 3. Cumulative Match Characteristic curves for close-range and moderate-range videos

A few observations were made from the error rates and the CMC curves. The first observation is that the recognition performance is improved by the spatial representation using the sub-graph, and super-graph representations. It is evident from the results that the account of spatial and temporal information together improves the performance of the system in case of matching high resolution images with that of low resolution videos. This observation can be made from the error rates of the HMM approach and our approach. The inclusion of spatial information in addition to the temporal information provided by the HMM or AGMM improves the performance of the system. The second observation is that the close-range videos of the UTD database has lower error rates than the moderate-range videos. This is due to the fact that the frame of the video sequence mostly contains the face region thus gathering more details of the facial features than the moderate-range videos. The third observation is that the adaptive appearance model along with the update to the graph model sets improves the performance significantly from our previous work [1]. This is due to adding graphs, sub-graphs and super-graphs to the model set and the appearance model that is spatially similar to those generated for each frame of the individual. Also, the chance of updating the incorrect appearance model is low due to the abundant spatial information available from the graphs. The fourth observation is that the performance of the system is affected by the amount of training data given for each individual. The lack of sufficient training images of a subject affect the performance of the system. This eventually leads us to the conclusion that the system's performance can be improved in the case of video-video based face recognition where the training set is a set of videos which has more number of frames with the wealth of spatial and temporal information.

The effect of various parameters on the performance was also tested. From our experiments, we observed that the parameters do not significantly affect the performance of the system. For example, increasing the maximum Euclidean distance between two vertices of a graph to a value greater than the width or length of the image will have no effect as this does not change the spatial neighborhood of a vertex in the graph. Hence, a lower threshold value of half the value of the width of face region was set to ensure a connected graph. The Gaussian components of a GMM represented heterogeneous data of the training set which are basically various facial features (e.g. eyes, nose, mouth, etc.). Hence, the 10 Gaussian components were sufficient to represent the heterogeneous facial features.

6   Conclusion

In this paper, we proposed a graph based face representation for face recognition from videos. The spatial characteristics are captured by constructing graphs for each face image and extracting the common sub-graphs and super-graphs from the set of graphs of each subject. An adaptive graph appearance model is generated that incorporates the temporal characteristics of the video sequence. A modified LFA and LBP were used to extract the feature points and feature descriptors, respectively. A two stage adaptive matching procedure that exploits the spatial and temporal characteristics is proposed for efficient matching. The experimental results show that graph based representation is robust and gives better performance. As a future work, we would like to test the system on video-video based face recognition and other standard databases with benchmarks.

References

1. Mahalingam, G., Kambhamettu, C.: Video based face recognition using graph matching. In: The 10th Asian Conference on Computer Vision (2010)
2. Chellappa, R., Wilson, C.L., Sirohey, S.: Human and machine recognition of faces: a survey. Proceedings of the IEEE 83, 705–741 (1995)
3. Wang, H., Wang, Y., Cao, Y.: Video-based face recognition: A survey. World Academy of Science, Engineering and Technology 60 (2009)
4. Zhao, W., Chellappa, R., Phillips, P.J., Rosenfeld, A.: Face recognition: A literature survey (2000)
5. Turk, M., Pentland, A.: Eigenfaces for recognition. Journal of Cognitive Neuroscience 3, 71–86 (1991)
6. Etemad, K., Chellappa, R.: Discriminant analysis for recognition of human face images. Journal of the Optical Society of America 14, 1724–1733 (1997)
7. Moghaddam, B., Nastar, C., Pentland, A.: Bayesian face recognition using deformable intensity surfaces. In: Proceedings of Computer Vision and Pattern Recognition, pp. 638–645 (1996)
8. Zhou, S., Krueger, V., Chellappa, R.: Probabilistic recognition of human faces from video. Computer Vision and Image Understanding 91, 214–245 (2003)
9. Manjunath, B.S., Chellappa, R., Malsburg, C.: A feature based approach to face recognition (1992)

10. Ersi, E.F., Zelek, J.S.: Local feature matching for face recognition. In: Proceedings of the 3rd Canadian Conference on Computer and Robot Vision (2006)
11. Wiskott, L., Fellous, J.M., Kruger, N., Malsburg, C.V.D.: Face recognition by elastic bunch graph matching. IEEE Trans. on Pattern Analysis and Machine Intelligence 19, 775–779 (1997)
12. Ersi, E.F., Zelek, J.S., Tsotsos, J.K.: Robust face recognition through local graph matching. Journal of Multimedia, 31–37 (2007)
13. Zhou, S., Krueger, V., Chellappa, R.: Face recognition from video: A condensation approach. In: Proc. of Fifth IEEE International Conference on Automatic Face and Gesture Recognition, pp. 221–228 (2002)
14. Li, Y.: Dynamic face models: construction and applications. Ph.D. Thesis, University of London (2001)
15. Kim, M., Kumar, S., Pavlovic, V., Rowley, H.A.: Face tracking and recognition with visual constraints in real-world videos. In: CVPR (2008)
16. Stallkamp, J., Ekenel, H.K.: Video-based face recognition on real-world data (2007)
17. Liu, X., Chen, T.: Video-based face recognition using adaptive hidden markov models. In: CVPR (2003)
18. Lee, K.C., Ho, J., Yang, M.H., Kriegman, D.: Visual tracking and recognition using probabilistic appearance manifolds. Computer Vision and Image Understanding 99, 303–331 (2005)
19. Park, U., Jain, A.K.: 3d model-based face recognition in video (2007)
20. Ojala, T., Pietikainen, M., Harwood, D.: A comparative study of texture measures with classification based on feature distributions. Pattern Recognition, 51–59 (1996)
21. Ojala, T., Pietikainen, M., Maenpaa, T.: A generalized local binary pattern operator for multiresolution gray scale and rotation invariant texture classification. In: Second International Conference on Advances in Pattern Recognition, Rio de Janeiro, Brazil, pp. 397–406 (2001)
22. Penev, P., Atick, J.: Local feature analysis: A general statistical theory for object representation. Network: Computation in Neural Systems 7, 477–500 (1996)
23. McLachlam, J., Peel, D.: Finite mixture models (2000)
24. Redner, R.A., Walker, H.F.: Mixture densities, maximum likelihood and the em algorithm. SIAM Review 26, 195–239 (1984)
25. Lowe, D.G.: Distinctive image features from scale-invariant keypoints. International Journal of Computer Vision 60, 91–110 (2004)
26. O'Toole, A., Harms, J., Hurst, S.L., Pappas, S.R., Abdi, H.: A video database of moving faces and people. IEEE Transactions on Pattern Analysis and Machine Intelligence 27, 812–816 (2005)

A Spatial-Temporal Frequency Approach to Estimate Cardiac Motion

Marco Gutierrez[1,*], Marina Rebelo[1], Wietske Meyering[1], and Raúl Feijóo[2]

[1] Heart Institute, University of São Paulo Medical School
Av. Dr. Enéas de Carvalho Aguiar 44, CEP 05403-000, São Paulo, SP, Brazil
[2] National Laboratory for Scientific Computation
Av. Getúlio Vargas 333, CEP 25651-075, Petrópolis, RJ, Brazil
{marco.gutierrez,marina.rebelo,wietske.meyering}@incor.usp.br
rualfeijoo@gmail.com

Abstract. The estimation of left ventricle motion and deformation from series of images has been an area of attention in the medical image analysis and still remains an open and challenging research problem. The proper tracking of left ventricle wall can contribute to isolate the location and extent of ischemic or infarcted myocardium. This work describes a method to automatically estimate the displacement fields for a beating heart based on the study of the variation in the frequency content of a non-stationary image as time varies. Results obtained with this automated method in synthetic images are compared with traditional gradient based method. Furthermore, experiments involving cardiac SPECT images are also presented.

1 Introduction

Left ventricular contractile abnormalities can be an important manifestation of coronary artery disease. Wall motion changes may represent ischemia or infarction of the myocardium [1]. Quantifying the extent of regional wall motion abnormality may aid in determining the myocardial effects of coronary artery disease. It would also simplify the analysis of wall motion changes after diagnostic and therapeutic interventions and permit comparison of different imaging techniques to assess their diagnostic accuracy. For this reason, the proper tracking of left ventricle wall can contribute to isolate the location and extent of ischemic or infarcted myocardium and constitutes a fundamental goal of medical image modalities, such as Nuclear Medicine.

Cardiac Single-Photon-Emission Computerized Tomography (SPECT) provides the clinician a set of 3D images that enables the visualization of the distribution of radioactive counts within the myocardium and surrounding structures. Defects on the distribution of some radionuclides such as ^{201}TI and ^{99m}Tc-MIBI in the myocardium indicate a muscle hypoperfusion [2]. Electrocardiographic gating of SPECT (gated-SPECT) images provides the additional ability to track the wall motion and wall

---

* Corresponding author.

G. Bebis et al. (Eds.): ISVC 2010, Part I, LNCS 6453, pp. 529–538, 2010.
© Springer-Verlag Berlin Heidelberg 2010

thickening associated with myocardial infarcts by acquiring sets of volumes in different phases of the cardiac cycle [3].

The tracking of structures in time series images has been studied by the computer vision community, especially in the areas of non-rigid motion, segmentation and surfacing mapping. The goal is to obtain a displacement field that establishes a correspondence between certain points in the structure at time t and time t+1. Generally speaking, the common methods to obtain velocity vector fields lie within feature matching, gradient, and spatio-temporal frequency techniques. The matching technique is not suitable when the sequence of images involves deformable structures.

This work describes a method to automatically estimate the displacement fields for a beating heart based on the study of the variation in the frequency content of a non-stationary medical image as time varies. Results obtained with this automated method in synthetic images are compared with a traditional gradient based method. Furthermore, preliminary results with cardiac SPECT images are also presented. The remainder of this paper is organized as follows. Section 2 offers an overview of methods for the tracking of structures in time series images. A new spatial-temporal frequency approach to estimate cardiac motion is also presented in this section. In Section 3 we present experimental evaluation with numerical phantoms and real images. This paper concludes with Section 4.

2   Methods

In this section some methods for the tracking of structures in time series images are presented. Particular attention is given to gradient-based method and spatial-temporal frequency based approach.

2.1   Feature Matching Approach

The feature matching technique is useful for the determination of shape and surface orientation from motion and for tracking applications. However, it is very sensitive to ambiguity amongst the structures to be matched. This is particularly true for structures submitted to non-rigid motion, as in the case of medical images, where the geometrical relationships between object points may be distorted. Moreover, the matching process involved in verifying the correspondences among structures in the correspondence model can suffer from combinatorial explosion [4].

Recent methods to track features in images estimate their temporal behavior by stochastic filters and in each image the predicted state of each feature is matched with the segmented feature's data. Consequently, to successfully accomplish the tracking of features along images the following steps should be employed: a stochastic filter; a matching strategy; a methodology to segment the features from each image; an efficient model to perform the management of the tracked features; and a stochastic approach to learn the dynamic model of the features from the tracked motion [5,6].

2.2 Gradient-Based Approach

The gradient-based method or the differential method was introduced by Fennema and Thompson [7] and developed by Horn [8] as the Optical Flow (OF) equation. This method is based on the assumption that the intensity of image elements is conserved between the images. The equation formulated in the continuum is the well-known motion constraint equation:

$$E_x u + E_y v + E_t = 0.$$ (1)

where, E_x, E_y and E_t are the image derivatives in x, y and t directions, u and v are the components of the local velocity vector $\vec{v}$ along the directions x and y, respectively.

In cardiac SPECT images using a radionuclide such as ^{201}TI and ^{99}Tc^m-MIBI, the principal structure is the myocardium, therefore it is a reasonable assumption that the total intensity of voxels within the ventricle is conserved between successive frames, and the motion of connected tissue within the myocardium should be smooth.

Eq. (1) is usually called the Optical Flow Constraint (OFC) and in order to evaluate the flow velocity $\vec{v}$, Horn introduced a smoothness constraint in addition to the fundamental constraint. Using this method and the extension to three dimensions, the velocity flow field can be obtained by minimizing the functional:

$$\min \iiint \left[\begin{array}{l} \left(E_x u + E_y v + E_z l + E_t\right)^2 + \\ \alpha^2 \left(u_x^2 + u_y^2 + u_z^2 + v_x^2 + v_y^2 + v_z^2 + l_x^2 + l_y^2 + l_z^2 \right) \end{array} \right] dxdydz.$$ (2)

where $u, u_x, u_y, u_z, v, v_x, v_y, v_z, l, l_x, l_y,$ and l_z are the components of the local velocity vector $\vec{v}$ and its partial derivatives along the directions x, y, and z respectively. The first term is the OFC, the second is a measure of the Optical Flow field smoothness, and α is a weighting factor that controls the influence of the smoothness constraint. The functionally Eq. (2) is minimized by using the calculus of variations [9], which leads to a system of three coupled differential equations from the Euler-Lagrange equations:

$$\nabla^2 u = \frac{E_x}{\alpha^2} \left(E_x u + E_y v + E_z l + E_t\right)$$
$$\nabla^2 v = \frac{E_y}{\alpha^2} \left(E_x u + E_y v + E_z l + E_t\right)$$ (3)
$$\nabla^2 l = \frac{E_z}{\alpha^2} \left(E_x u + E_y v + E_z l + E_t\right)$$

These equations can be easily decoupled, and an interactive solution can be defined using discrete approximation of the Laplacian operator with a finite difference method. Therefore, the following set of equations is used to estimate the 3D OF components at each time instant

$$
\begin{cases}
u^{n+1} = \bar{u}^n - \dfrac{E_x\left(E_x\bar{u}^n + E_y\bar{v}^n + E_z\bar{l}^n + E_t\right)}{\alpha^2 + E_x^2 + E_y^2 + E_z^2} \\[4mm]
v^{n+1} = \bar{v}^n - \dfrac{E_y\left(E_x\bar{u}^n + E_y\bar{v}^n + E_z\bar{l}^n + E_t\right)}{\alpha^2 + E_x^2 + E_y^2 + E_z^2} \\[4mm]
l^{n+1} = \bar{l}^n - \dfrac{E_z\left(E_x\bar{u}^n + E_y\bar{v}^n + E_z\bar{l}^n + E_t\right)}{\alpha^2 + E_x^2 + E_y^2 + E_z^2}
\end{cases}
\tag{4}
$$

where n is the iteration index; E_x, E_y, E_z are the partial derivatives of the image intensity in the directions x, y and z, respectively; E_t is the partial derivative in time; $\bar{u}, \bar{v}$ and $\bar{l}$ are the mean velocities in each direction for the voxels in a neighborhood of a given voxel; and α is a weighting factor. Velocity components in the x, y and z directions for each voxel are computed as the solution of a linear algebraic system of equations whose coefficients are determined by the spatial and temporal derivatives given by Eq. (4). The linear system described in Eq. (4) can be solved by methods like Conjugate Gradient methods [10] or Algebraic Reconstruction techniques [11].

2.3 Spatial-Temporal Frequency Based Approach

The spatio-temporal frequency (STF) based approach to optical flow derivation encompasses all methods which are based upon some underlying spatio-temporal frequency image representation [12,13]. The major motivation for considering the use of the STF image representation as a basis for computing optical flow comes from the literature on mammalian vision. In particular, recent investigations have demonstrated that many neurons in various visual cortical areas of the brain behave as spatio-temporal frequency bandpass filters [14,15].

In the field of non-stationary signal analysis, the Wigner-Ville Distribution (WVD) [16] has been used for the representation of speech and image. Jacobson and Wechsler [17] first suggested the use of the WVD for the representation of shape and texture information. In particular, they formulated a theory for invariant visual pattern recognition in which the WVD plays a central role. Given a time-varying image $f(x, y, t)$, its WVD is a 6-dimensional function defined as:

$$
W_f(x, y, t, w_x, w_y, w_t) = \int\!\!\!\int\!\!\!\int_{-\infty}^{+\infty} R_f(x, y, t, \alpha, \beta, \tau)\, e^{-j(\alpha w_x + \beta w_y + \tau w_t)}\, d\alpha\, d\beta\, d\tau.
\tag{5}
$$

where $x, y,$ and t are the pixel position in space and time and $w_x, w_y,$ and w_t are the frequency components in $x, y,$ and t directions, respectively. In Eq. (5) the term on the left is the pseudo-correlation function, where * denotes complex conjugation.

$$
R_f(x, y, t, \alpha, \beta, \tau) = f(x+\alpha, y+\beta, t+\tau) . f^*(x-\alpha, y-\beta, t-\tau).
\tag{6}
$$

For the special case where a time-varying image is uniformly translating at some constant velocity $\bar{v}$ with components u and v along the directions x and y, respectively, the image sequence can be expressed as a convolution between a static image and a translating delta function.

$$f(x, y, t) = f(x, y) * \delta(x - ut, y - vt).$$ (7)

Using the convolution and windowing properties of the WVD, we obtain:

$$W_f(x, y, t, w_x, w_y, w_t) = \delta(uw_x + vw_y + w_t)W_f(x - ut, y - vt, w_x, w_y).$$ (8)

The WVD of a linearly translating image with constant velocity $\vec{v}$ is everywhere zero except in the plane defined by $\{(x, y, t, w_x, w_y, w_t) : uw_x + vw_y + w_t) = 0\}$, for fixed u and v. Equivalently, for an arbitrary pixel at x, y and t, each local spatial and temporal frequency spectrum of the WVD is zero everywhere except on the plane defined by $\{(w_x, w_y, w_t) : uw_x + vw_y + w_t) = 0\}$.

From Eq. (5) the WVD assigns a three-dimensional spatio-temporal frequency spectrum to each pixel over which the image is defined. However, the WVD assigns a 3D spectrum with interference due to cross correlation when more than one frequency is present [16].

In order to smooth the spectrum of WVD a filter must be introduced. In this work we adopted a Hanning filter to smooth the spectrum.

$$h = 0.5 * \left[1 - \cos\left(\frac{2\pi n}{N} \right) \right] \quad \text{for} \quad 0 \le n \le N - 1.$$ (9)

To reduce the effects of the cross-terms we also used the Choi-Williams distribution (CWD) [16]. The CWD was introduced with the aim of controlling the cross-terms encountered in Wigner-Ville distribution. The exponential kernel introduced by Choi and Williams is defined as:

$$cw = \frac{1}{\sqrt{4\pi \tau^2 \sigma}} \exp\left(\frac{-(\mu - t)^2}{4\tau^2 \sigma} \right).$$ (10)

From the Eq. (10), if a small σ is chosen, the Choi-Williams distribution approaches the Wigner-Ville distribution, since the kernel approaches to one. On the other hand, for large σ more cross-terms are suppressed and auto-terms are affected.

The spatial orientation of the smoothed 3D frequency spectrum is completely governed by the pixel velocity, whose components can be obtained through a simple multiple linear regression model [18].

$$w_t = b + uw_x + vw_y + e.$$ (11)

Eq. (11) is a linear regression extension where w_t is a linear function of two independent variables w_x and w_y. The values of the coefficients b, u and v are achieved by solving the following linear system:

$$
\begin{bmatrix}
n & \sum_{k=1}^{n} w_{x_k} & \sum_{k=1}^{n} w_{y_k} \\
\sum_{k=1}^{n} w_{x_k} & \sum_{k=1}^{n} w_{x_k}^2 & \sum_{k=1}^{n} w_{x_k} w_{y_k} \\
\sum_{k=1}^{n} w_{y_k} & \sum_{k=1}^{n} w_{x_k} w_{y_k} & \sum_{k=1}^{n} w_{y_k}^2
\end{bmatrix}
\cdot
\begin{bmatrix} b \\ u \\ v \end{bmatrix}
=
\begin{bmatrix}
\sum_{k=1}^{n} w_{t_k} \\
\sum_{k=1}^{n} w_{x_k} w_{t_k} \\
\sum_{k=1}^{n} w_{y_k} w_{t_k}
\end{bmatrix}.
\tag{12}
$$

where n is the number of pixels, w_{x_k}, w_{y_k} and w_{t_k} are the frequency components in each direction and u and v are the velocity components on x and y directions, respectively.

3 Experimental Evaluation

3.1 Numerical Phantom Simulations

The velocity vector along three frames of 3D images was computed by the gradient and frequency-based methods at each voxel of a mathematical phantom. The phantom consists of a cylinder with 74 pixels in diameter and 5 pixels in length. Each cylinder's cross-section comprises voxels with intensity function described below:

$$
E(x, y, z) = \gamma + \beta[\sin(\omega_1 x) + \sin(\omega_2 y)].
\tag{13}
$$

where $E(x, y, z)$ is the intensity of the voxel in the spatial position x, y, z of the voxel space, γ and β are constants, and ω_1 and ω_2 are the spatial frequencies.

The cylinder was submitted to translation and rotation with known velocities and the Root Mean Square Error (RMSE) was used as a measure of error between the estimated and the real velocities. In the computation of Eq. (4) and Eq. (10), α and σ parameters where fixed with values 10 and 1, respectively. These values where maintained fixed for the experiments with simulated and real images.

$$
RMSE = 100 \frac{\sqrt{\sum_{i}^{M} \sum_{j}^{N} (u_{i,j} - \hat{u}_{i,j})^2 + (v_{i,j} - \hat{v}_{i,j})^2}}{\sqrt{\sum_{i}^{M} \sum_{j}^{N} (u_{i,j})^2 + (v_{i,j})^2}}.
\tag{14}
$$

Tables 1 and 2 show the results obtained after applying translation to the cylinder using different velocities between each image frame. In these Tables, u and v are the actual velocities (pixels/frame) in x and y directions respectively, $\hat{u}$ and $\hat{v}$ are the mean estimated velocities (pixels/frame) and ε_{rms} is the RMSE expressed as percentages.

Table 3 shows the results obtained after applying rotation to the same phantom using the gradient and frequency-based methods. In Table 3, ω and $\hat{\omega}$ are the real and estimated angular velocities (degree/frame), respectively, and ε_{rms} is the RMSE expressed as percentages. These results show that the method has a satisfactory performance for translation and rotation when the velocities applied to the phantom are less than 2 pixels per frame, corresponding to speeds around 200 mm/s which are higher than typical heart wall speeds.

Table 1. Results obtained after translation in a section of the cylinder using the gradient-based method (velocities in pixels/frame)

u	v	$\hat{u}$	$\hat{v}$	ε_{rms} (%)
1	0	0.9308	0.0002	7.13
2	0	1.9741	0.0012	11.10
1	1	0.9747	0.9585	6.86
1	2	1.1635	1.9095	14.13

Table 2. Results obtained after translation in a section of the cylinder using the frequency-based method and two different distributions (velocities in pixels/frame)

u	v	Wigner-Ville			Choi-Williams		
		$\hat{u}$	$\hat{v}$	ε_{rms} (%)	$\hat{u}$	$\hat{v}$	ε_{rms} (%)
1	0	0.9012	0.0000	11.87	0.9482	-0.0007	7.12
2	0	1.8022	-0,0004	12.29	1.8874	0.0002	8.21
1	1	0.9018	0.8562	15.32	0.9448	0.9266	8.96
1	2	0.9050	1.7209	16.39	0.9467	1.8714	8.73

Table 3. Results obtained after rotation movement in a section of the cylinder by the gradient and frequency based methods and two different distributions (velocities in degrees/frame)

ω	Gradient-based method		Frequency-based method			
			Wigner-Ville		Choi-Williams	
	$\hat{\omega}$	ε_{rms} (%)	$\hat{\omega}$	ε_{rms} (%)	$\hat{\omega}$	ε_{rms} (%)
2	1.9441	8.09	1.7900	23.80	1.8600	25.97
5	5.1011	9.56	4.4500	21.45	4.8050	14.71
7	6.9944	20.27	6.1950	20.94	6.8200	12.43

3.2 Real Images

The proposed method was applied to gated perfusion studies MIBI-^{99m}Tc obtained from a dual-head rotating gamma camera (ADAC Vertx+ with a LEAP Collimator). The acquisition process is synchronized with the electrocardiogram and the cardiac

cycle can be divided into 8 or 16 frames per cycle. A total of 64 projections were obtained over a semi-circular 180° orbit. All projection images were stored using a
64x64, 16 bits matrix. All transverse tomograms were reconstructed with a thickness
of one pixel per slice (6.47 mm). The volume of transverse tomograms was re-
oriented, and sets of slices perpendicular to the long axis (obliques) and of slices parallel to the long axis (coronals and sagittals) were created.

The dense displacement fields were obtained from a series of 2D gated SPECT
slices. Fig.1 depicts the results obtained for one oblique SPECT slice at systole and
diastole.

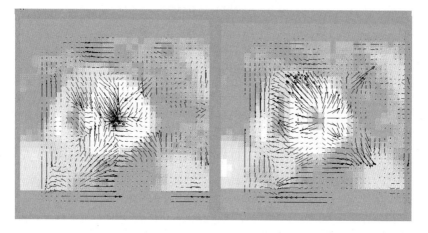

Fig. 1. An oblique slice of a SPECT study at systole (left) and diastole (right). Both images
include the superimposition of the dense displacement field that delineates the non-rigid motion
performed by the cardiac structures, estimated by the Spatial-temporal frequency based
approach

4   Conclusions

This work described a method based on a spatial-temporal frequency analysis to estimate dense displacement fields from time-varying images. The method was applied in
sequences of synthetic and gated SPECT images to obtain the dense displacement
field that delineates the non-rigid motion of cardiac structures.

The majors motivations for considering this kind of motion representation were: (1)
in non-rigid motion, the image frequency components change with time; (2) some investigations on mammalian vision have demonstrated that many neurons in various
visual cortical areas of the brain behave as spatio-temporal frequency band-pass filters.

The synthetic phantom experiments have shown that the proposed method has a satisfactory performance for translation and rotation when the velocities applied to the
phantom are less than 2 pixels per frame, values corresponding to speeds around 200
mm/s which are higher than typical heart wall speeds. A qualitative analysis with a
series of images from a SPECT study and its corresponding velocity vector field has
shown that the spatio-temporal frequency approach could detect complex motion such
as contraction and rotation.

Although this is a work in progress, we believe that the proposed method will allow a better understanding of the heart's behavior. Further research includes: 1) to establish a training phase to estimate the parameters α and σ in Eq. (2) and Eq. (10), respectively, to reduce the errors associated with the velocity measurements; 2) to extend the experimental evaluation with realistic phantoms; 3) to understand the information from a physiological viewpoint.

Acknowledgments

This work is supported by National Council for Scientific and Technological Development (CNPq, Grants 57.3710/2008-2 and 301735/2009-3), The National Institute for Science and Technology - Medicine Assisted by Scientific Computing (INCT-MAAC) and Zerbini Foundation.

References

1. Meier, G.D., Ziskin, M., Santamore, W.P., Bove, A.: Kinematics of the beating heart. IEEE Transaction on Biomedical Engineering 27, 319–329 (1980)
2. Blokland, J.A.K., Reiber, J.H.C., Pauwels, E.K.J.: Quantitative analysis in single photon emission tomography (SPET). Eur. J. Nucl. Med. 19, 47–61 (1992)
3. Garcia, E.V., Van Train, K., Maddahi, J., Prent, F., Areeda, J.: Quantification of rotational thallium-201 myocardial tomography. Journal of Nuclear Medicine 26, 17–26 (1985)
4. Laplante, P.A., Stoyenko, A.D.: Real-time imaging: theory, techniques and applications. IEEE Press, New York (1996)
5. Pinho, R., Tavares, J.: Tracking Features in Image Sequences with Kalman Filtering, Global Optimization, Mahalanobis Distance and a Management Model. Computer Modeling in Engineering & Sciences 46(1), 51–75 (2009)
6. Pinho, R., Tavares, J., Correia, M.: Efficient Approximation of the Mahalanobis Distance for Tracking with the Kalman Filter. Int. Journal of Simulation Modelling 6(2), 84–92 (2007)
7. Fennema, C.L., Thompson, W.B.: Velocity determination in scenes containing several moving objects. Computer Vision, Graphics and Image Processing 9, 301–315 (1979)
8. Horn, B.K.P.: Robot Vision. McGraw-Hill, New York (1996)
9. Sagan, H.: Introduction to the Calculus of Variations. Dover Publications, New York (1969)
10. Axelsson, O.: In Iterative solution methods. Cambridge University Press, Cambridge (1996)
11. Herman, G.T.: In Medical imaging systems techniques and applications: computational techniques. In: Leondes, C.T. (ed.). Gordon and Breach Science Publishers, Amsterdam (1998)
12. Beauchemin, S.S., Barron, J.L.: On the Fourier Properties of Discontinuous Motion. Journal of Mathematical Imaging and Vision 13(3), 155–172 (2000)
13. Langer, M.S., Mann, R.: Optical Snow. International Journal of Computer Vision 55(1), 55–71 (2003)
14. Adelson, E.H., Bergen, J.R.: Spatiotemporal energy models for the perception of motion. Journal of Optical Society of America A 12, 284–299 (1985)

15. Gafni, H., Zeevi, Y.Y.: A model for processing of movement in the visual system. Biological Cybernetics 32, 165–173 (1979)
16. Cohen, L.: In Time-frequency analysis. Prentice Hall PTR, New Jersey (1995)
17. Jacobson, L.D., Wechsler, H.: Joint spatial/spatial-frequency representation. Signal Processing 14, 37–68 (1988)
18. Chapra, S.C., Canale, R.P.: In Numerical methods for engineers with programming and software applications. McGraw-Hill, New York (1998)

Mitosis Extraction in Breast-Cancer Histopathological Whole Slide Images

Vincent Roullier[1,*], Olivier Lézoray[1],
Vinh-Thong Ta[2], and Abderrahim Elmoataz[1]

[1] Université de Caen Basse-Normandie, ENSICAEN, CNRS
GREYC UMR 6072 - Équipe Image
[2] LaBRI (Université de Bordeaux – CNRS) – IPB

Abstract. In this paper, we present a graph-based multi-resolution approach for mitosis extraction in breast cancer histological whole slide images. The proposed segmentation uses a multi-resolution approach which reproduces the slide examination done by a pathologist. Each resolution level is analyzed with a focus of attention resulting from a coarser resolution level analysis. At each resolution level, a spatial refinement by semi-supervised clustering is performed to obtain more accurate segmentation around edges. The proposed segmentation is fully unsupervised by using domain specific knowledge.

1 Introduction

Breast cancer is the second leading cause of cancer death for women. Its incidence increases substantially and continuously while the mortality rate remains high despite earlier detection and advances in therapeutic care. The identification and the use of reliable prognostic and therapeutic markers is a major challenge for decision-making regarding therapy. Proliferation has been shown to be the strongest prognostic and predictive factor in breast carcinoma, especially in patients lacking lymph node metastases [1]. This parameter is daily taken into account by the pathologist for establishing the histopathological grading of breast carcinomas, using enumeration of mitotic figures, through the lens of the microscope. The recent use of immunohistochemical staining of mitosis is able to facilitate their detection. Nevertheless, the visual counting method remains subjective and leads to reproducibility problems due to the frequent heterogeneity of breast tumors [2].

The recently introduced microscopical scanners allow recording large images of the whole histological slides and offer the prospect of fully automated quantification for a better standardization of proliferation rate appraisal. If the advent of such digital whole slide scanners has triggered a revolution in histological imaging, the processing and the analysis of breast cancer high-resolution histopathological images is a very challenging task. First, the produced images are relatively

---

* This work was supported under a research grant of the ANR Foundation (ANR-06-MDCA-008/FOGRIMMI).

G. Bebis et al. (Eds.): ISVC 2010, Part I, LNCS 6453, pp. 539–548, 2010.

huge and their processing requires computationally efficient tools. Second, the biological variability of the objects of interest makes their extraction difficult. As a consequence, few works in literature have considered the processing of whole slide images and most of these works rely only on machine learning techniques [3,4].

In this work, we present a graph-based multi-resolution segmentation and analysis strategy for histological breast cancer whole slide images. The proposed strategy is based on a top-down approach that mimics the pathologist interpretation under the microscope as a focus of attention. The proposed segmentation performs an unsupervised clustering at each resolution level (driven by domain specific knowledge) and refines the associated segmentation in specific areas as the resolution increases. The whole strategy is based on a graph formalism that enables to perform the segmentation adaptation at each resolution.

The paper is organized as follows. A description of the considered images is presented in Sect. 2. In this Section, we also describe the visual analysis process performed by pathological experts to evaluate mitotic figures proliferation and their inherent multi-resolution approach. Our graph-based formulation for image segmentation is presented in Sect. 3 and its integration into a multi-resolution segmentation strategy is detailed in Sect. 4. Sect. 5 presents visualization tools of extracted mitosis. Last Section concludes.

2   Image Description

2.1   Breast Cancer Histological Whole Slide Images

Breast cancer tissue samples are sectioned at 5 μm thickness and stained with an immunohistochemical (hematoxylin and eosin) method. A ScanScope CS ® (Aperio, San Diego, CA) digital microscopical scanner is then used to digitalize each slice at 20x magnification scale and the resulting digital images are compressed with a quality of 75% following the JPEG compression schema.

To facilitate the visualization and the processing, scanned samples acquired by the scanner are directly stored as an irregular pyramid where each level of the pyramid is an under resolved version of the highest resolution image (the pyramid base).

The usual size of a compressed whole slide image is about 100~500 Megabytes after compression. However, the resulting whole slide images are too large in size to be processed or visualized as a whole. Therefore, the whole slide image is tiled by the scanner to ease both its processing and visualization: each resolution level of the pyramid is split into image tiles in a non-overlapping layout.

2.2   Visual Analysis Process

Within the last decade, histologic grading has become widely accepted as a powerful indicator of prognosis in breast cancer. The majority of tumor grading systems currently employed for breast cancer combine nuclear grade, tubule formation and mitotic rate. In general, each element is given a score of 1 to 3 (1 being the best and 3 the worst) and the score of all three components are added together to give the breast cancer grading. The usual breast cancer grading scheme is the Elston-Ellis criterion [5] and is based on three separated scores:

- Gland (tubule) formation: one scores the proportion of whole carcinoma that forms acini (1: <75%; 2: 10-75%; 3: <10%).
- Nuclear pleomorphism: one scores the nuclear atypia according to size, shape and chromatin pattern (1: none; 2: moderate; 3: pronounced).
- Mitotic count: one scores the number of mitotic figures per 10 consecutive high power fields (1: 0-9 mitoses; 2: 10-19 mitoses; 3: > 19 mitoses).

The final grading is obtained by adding the three scores. The total score is in the range 3-9 and the final obtained grading is:

- **Grade 1** if total score is 3-5.
- **Grade 2** if total score is 6-7.
- **Grade 3** if total score is 8-9.

In this work, we are interested in helping pathologists to establish an accurate mitotic count. Indeed, with the Elston-Ellis criterion, a pathologist bases its scores only on ten consecutive high power fields. This can be lesser representative than having a score established according to the whole preparation which was of course too tedious for pathologists under a classical microscope until now. With the advent of fast whole slide image scanners, it is now possible [6]. Our study drives towards this direction.

2.3 Multi-resolution Approach

Whole slide images (WSI) are usually huge in size. Fortunately, they are stored as a pyramid of tiled images that enables to process them in a hierarchical way [7]. As a consequence, a multi-resolution segmentation method is a natural approach for segmenting whole slide images. Moreover, such a strategy reproduces the analysis done by the pathologists under the microscope: regions of interest are determined at low resolution while cellular classification is performed at high resolution. The proposed multi-resolution segmentation method is based on a top-down segmentation that mimics pathologist interpretation according to specific domain knowledge. Fig. 1 illustrates the identification (by a pathologist) of mitosis in breast cancer slides stained with hematoxyline and eosine.

3 Graph-Based Segmentation

3.1 Preliminaries on graphs

A graph is a structure used to describe a set of objects and the pairwise relations between those objects. The objects are called *vertices* and a link between two

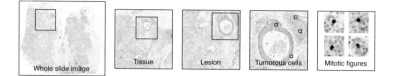

Fig. 1. Illustration of the visual analysis process performed by a pathologist expert. Each decision is performed at a higher resolution than the previous one in the region of interest. Each interior square in an image is magnified in a subsequent image. In the second square, the tissue is detected. In the third square, the pathologist separate tissue and lesion. The third square allows to detect tumourous cells and the last, mitotic figures.

objects is called an *edge*. A weighted graph $G = (V, E, w)$ is composed of a finite set $V = \{u_1, \ldots, u_N\}$ of N vertices, a set of edges $E \subset V \times V$, and a *weight function* $w : E \to \mathbb{R}^+$. An edge of E, which connects two *adjacent neighbor* vertices u and v, is noted (u, v). In the rest of this paper, the notation $v \sim u$ means that vertex v is an adjacent neighbor of vertex u. We assume that the graph G is simple, connected and undirected. This implies that the weight function w is symmetric i.e. $w(u, v) = w(v, u)$ if $(u, v) \in E$ and $w(u, v) = 0$ otherwise. Let $\mathcal{H}(V)$ be the Hilbert space of real valued functions on the vertices of a graph. Each function $f : V \to \mathbb{R}$ of $\mathcal{H}(V)$ assigns a real value $f(u)$ to each vertex $u \in V$. Similarly, let $\mathcal{H}(E)$ be the Hilbert space of real valued functions defined on the edges of the graph. For the case of images, nodes are pixels, edges connect neighbor pixels with 8-adjacency.

3.2 Discrete Operators on Graphs

Let us recall some basic definitions. We consider that a graph $G = (V, E, w)$ and a function $f \in \mathcal{H}(V)$ are given. The *weighted difference* $d_w : \mathcal{H}(V) \to \mathcal{H}(E)$ of a function f on an edge (u, v) linking two vertices $u, v \in V$ is defined as $(d_w f)(u, v) = \sqrt{w(u, v)}(f(v) - f(u))$. This operator leads us to define the *directional derivative* of f, over an edge (u, v), as $\partial_v f(u) = (d_w f)(u, v)$. Then, the *weighted gradient* $\nabla_w f$ of the function f, at a vertex $u \in V$, is defined as $(\nabla_w f)(u) = (\partial_{v_1} f(u), \ldots, \partial_{v_k} f(u))^T$. This operator corresponds to the *local variation* of the function f at the vertex u and measures the regularity of f in the adjacent neighborhood $v_1, \ldots, v_k$ of the vertex u. Hence, the $\mathcal{L}_2$-norm of the weighted gradient is $\|(\nabla_w f)(u)\|_2 = \left[\sum_{v \sim u} w(u, v)(f(v) - f(u))^2 \right]^{1/2}$. Then, the *weighted p-Laplacian* $(\Delta_w^p f)(u)$ at vertex u is defined as

$$(\Delta_w^p f)(u) = \sum_{v \sim u} \gamma_p(u, v)(f(v) - f(u)) \tag{1}$$

where $\gamma_p(u, v) = w(u, v)(\|(\nabla_w f)(u)\|_2^{p-2} + \|(\nabla_w f)(v)\|_2^{p-2})$. Clearly, in the case where $p = 1$ and $p = 2$, we have the definitions of the standard graph curvature

$\Delta_w^1 f = \kappa f$ and graph Laplace $\Delta_w^2 f = \Delta f$ operators. More details on these definitions can be found in [8].

In the following, we only consider the case where $p = 2$.

3.3   Discrete Regularization Framework

To regularize a function $f^0 \in \mathcal{H}(V)$ using the p-Laplacian (Eq. (1)), we consider the following general variational problem on graphs:

$$\min_{f \in \mathcal{H}(V)} \left\{ E_w(f, f^0, \lambda, p) = R_w(f, p) + \frac{\lambda}{2} \|f - f^0\|_2^2 \right\}. \tag{2}$$

The first term, $R_w(f, p)$, is the regularizer and is defined as, with $0 < p < +\infty$: $R_w(f, p) = \frac{1}{p} \sum_{u \in V} \|(\nabla_w f)(u)\|_2^p$. The second term is the fitting term. This optimization problem has a unique solution for $p = 1$ and $p = 2$ which satisfies, for all $u \in V$:

$$\frac{\partial E_w(f, f^0, \lambda, p)}{\partial f(u)} = (\Delta_w^p f)(u) + \lambda(f(u) - f^0(u)) = 0,$$

which is equivalent to

$$\left(\lambda + \sum_{v \sim u} \gamma(u, v) \right) f(u) - \sum_{v \sim u} \gamma(u, v) f(v) = \lambda f^0(u).$$

To approximate the solution of the minimization (2), we can linearize this system of equations and use the Gauss-Jacobi method to obtain the following iterative algorithm:

$$\begin{cases} f^{(0)}(u) = f^0(u) \\ f^{(n+1)}(u) = f^0(u) + \sum_{v \sim u} \lambda + \sum_{v \sim u} \gamma^{(n)}(u, v), \end{cases} \tag{3}$$

where $\gamma^{(n)}(u, v)$ is the γ function (in Eq. (1)) at the iteration step n. The interested reader can refer to [8] for more details on the formulation and the connections with other formalisms. The above algorithm enables to simplify functions living on graphs by a discrete diffusion process.

3.4   Discrete Semi-supervised Clustering

The previously presented discrete regularization framework can be naturally adapted to address discrete semi-supervised clustering problems. Let $V = \{u_1, \ldots, u_N\}$ be a finite set of data, where each data u_i is a vector of $\mathbb{R}^m$. Let $G = (V, E, w)$ be a weighted graph such that all vertices are connected by an edge of E. The semi-supervised clustering of the set V consists in grouping the set V into k classes where the number of k classes is given. For this, the set V is composed of labeled and unlabeled data. The objective is then to estimate the labels of unlabeled data from labeled ones. Let $C = \{c_i\}_{i=1,\ldots,k}$ the set of classes, L the set of labeled vertices and $V \setminus L$ be te initially unlabelled vertices (the whole set of vertices except the labeled ones). For each vertex of L, its classes is available with the function $\mathcal{L} : L \rightarrow C$.

This situation can be modeled by considering k initial label functions (one per class) $f_i^0 : V \rightarrow \mathbb{R}$, with $i = 1, \ldots, k$. For a given vertex u, if u is initially labeled ($u \in L$) then $f_i^0(u) = +1$ if $\mathcal{L}(u) \in c_i$ and $f_i^0(u) = -1$ otherwise. If u is initially unlabeled (i.e. $u \in V \setminus L$) then $f_i^0(u) = 0$. Then, the vertex clustering is accomplished by k regularization processes. This corresponds to estimate functions $f_i : V \rightarrow \mathbb{R}$ for each ith class using the discrete diffusion process (Eq. (3)). At the end of the label propagation processes, the final label of a given vertex $u \in V$ can be obtained by $\underset{i}{\operatorname{argmax}}\left\{ f_i(u) / \sum_{j=1,\ldots,k} f_j(u) \right\}$.

4 Multi-resolution Segmentation Approach

4.1 Principle

As it has been previously pointed out, a multi-resolution segmentation process is a natural approach to analyze whole slide images [7,9]. Indeed, we have seen in Sect. 2.2 that the whole slide image analysis visual process performed by pathologist experts is a multi-resolution process. An expert determines regions of interest at low resolution while cellular classification is performed at high resolution.

Our proposed multi-resolution segmentation process is based on a top-down segmentation that reproduces exactly the interpretation process performed by pathologist experts according to specific domain knowledge (expressed by the pathologists themselves). At a given resolution i, a clustering is performed by the following steps: the image is simplified by discrete regularization (Fig 2 (a)-(b)) and clustered by an unsupervised 2-means clustering (Fig. 2 (c)-(d)). The clustering is performed inside specific region that were segmented at the previous resolution. The obtained clustering is spread by pixel replication at a finer level of resolution (Fig. 2 (e)-(f)) and refined in specific region (according to domain knowledge) (Fig. 2 (g)-(i)). As clustering being performed in a feature space, it does not take into account spatial information, and the obtained segmentation is not accurate around image edges. In addition, the propagation of the labels across the different resolution levels is performed by plain pixel replication and the segmentation is coarse around edges. To alleviate both these effects, each obtained clustering is refined by our discrete semi-supervised clustering in a narrow band around the boundaries of the clusters. The whole segmentation strategy can be summarized by Algorithm 1 where I_i denotes an image at resolution level i.

At the last resolution level, mitotic figures are extracted (Fig. 2 (j)-(l)). Fig. 3 provides results for several whole slide images.

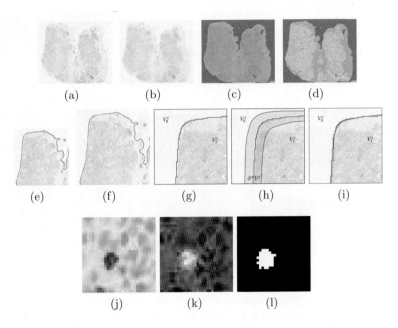

Fig. 2. Illustration of the multi-resolution segmentation process. (a) initial image, (b) regularized image. (c) and (d) clustered image at two level 1 and 2. (e) and (f) replication problem. (g)-(i) illustration of the spatial refinement clustering. (j)-(l) the mitotic figures extraction.

Algorithm 1. Multi-resolution WSI segmentation

1: $I_0^s = \text{Regularization}(I_0)$
2: $I_0^c = \text{2-MeansClustering}(I_0^s)$
3: $I_0^r = \text{SpatialClusteringRefinement}(I_0^c)$
4: **for** $i = 1$ to 3 **do**
5:     $I_i^p = \text{ReplicatePreviousResolutionClustering}(I_{i-1}^r)$
6:     $I_i^s = \text{Regularization}(I_i)$
7:     $I_i^c = \text{2-MeansClustering}(I_i^s)$ inside one class of I_i^p
8:     $I_i^r = \text{SpatialClusteringRefinement}(I_i^c)$
9: **end for**

---

5   Visualization of Mitotic Figures

Once all the mitosis have been extracted at the highest resolution using our top-down multi-resolution graph-based extraction algorithm, the pathologist has to visualize them to establish the mitotic score. Our proposal does not intend to compute the mitotic score but at helping the pathologist to do it. To do so, we provide for pathologists two visualization tools.

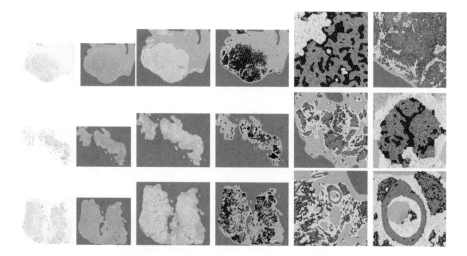

Fig. 3. Illustration of the multi-resolution clustering process. The first column presents original images. The second column is the segmentation obtained at the first resolution (background in pink, tissue in cyan), the third column is the segmentation obtained at the second resolution (lesion in yellow), the fourth column is the segmentation obtained at the third resolution (stroma in yellow and tumorous cells dark blue). The fourth column is the segmentation obtained at the fifth resolution (stroma in dark blue and tumorous cells in green). The sixth column is the segmentation obtained at the fifth resolution (mitotic figures in red).

The first one enables the pathologist to evaluate the global repartition of mitosis on the whole slide. This visual information is provided by superimposing a graph on the whole slide image. The graph is constructed as follows. Each detected mitosis is represented by a vertex of the graph. A Voronoi map is computed on the vertices coordinates and the associated Delaunay graph is obtained. This enables the pathologist to evaluate regions on the whole slide image where the mitotic activity is important by the superposition of either the Delaunay graph or the Voronoi distance map on the whole slide image (Fig. 4(a)-(c)).

The second visualization tool enables the pathologist to see the extracted mitosis altogether on a single 3D projection. Indeed, the first projection tool enables the pathologist to visualize each extracted mitosis on the whole image while appreciating the whole global distribution of mitosis. However, this does not allow to simultaneously visualize all the extracted mitosis of the slide to appreciate their aspect (e.g. to differentiate the different mitotic phases). Therefore, we propose a specific visualization tool that provides this information to the pathologist.

The proposed visualization tool is based on dimensionality reduction with Laplacian Eigenmaps [10]. Dimensionality reduction requires a distance measure to evaluate the similarity between two objects in the initial space. In the case of mitosis, since they can be in different mitotic phases, the most prominent information is texture. Therefore, a texture description of each mitotic figure is computed in the form of a Locally Binary Pattern (LBP) histogram (introduced in [11]) and this feature vector is used as an input for dimensionality reduction with a χ^2 histogram distance.

Once the dimensionality reduction has been performed, the pathologist can visualize simultaneously all the mitotic figures of a whole slide image in the form of a 3D projection where each mitosis is projected at coordinates defined by the projection. With this projection, the pathologist can appreciate the similarity of mitosis that are not necessarily spatially close in the whole slide image. Finally, performing a dimensionality reduction with Laplacian Eigenmaps provides also geometrical information on the projection. Indeed, the sign of the first eigenvector enables to partition the data into two sets. The obtained partitions correspond to the normalized cut criterion of the initial data [12]. As a consequence, this clustering information is also provided by coloring the bounding box of each mitosis in a color corresponding to one of the two clusters (red or green). This enables the pathologist to quickly distinguish mitosis with low or high textural content (Fig. 4(d)-(f)).

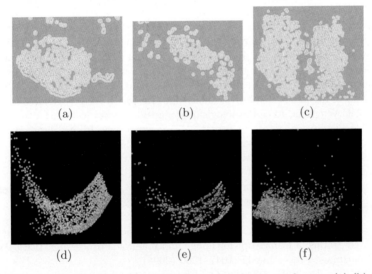

Fig. 4. Illustration of visualization tools of extracted mitotic figures. (a)-(b)-(c): Delaunay graph superimposed on the distance maps associated. (d)-(e)-(f): visualization of extracted mitotic figures by dimensionality reduction for the mitosis of (a)-(c) (see electronic version of the paper for better visualization).

6   Conclusion

In this paper, a multi-resolution image analysis strategy for automatic enumeration of mitotic figures on whole slide images is proposed. The whole classification process begins with the lowest resolution image and moves to higher resolution into regions of interest gradually identified. Graph-based regularization provides a unified formalism for both image simplification and spatial cluster refinement. Contrary to methods that can be found in literature, our method is completely unsupervised and has the advantage of reducing the amount of data to be processed at each resolution level by selecting regions of interest.

We also propose two methods for the visualization of mitotic figures. The first method allows to visualize the distribution of mitosis on the tissue samples. The second method groups mitosis according to texture parameters.

Future works will concern the automation of the other scores of the Elston-Ellis grading systems.

References

1. Tavassoli, F., Devilee, P. (eds.): Pathology and Genetics, Tumours of the Breast and Female Genital Organ. International Agency for Research on Cancer Press (2003)
2. Petushi, S., Garcia, F., Haber, M., Katsinis, C., Tozeren, A.: Large-scale computations on histology images reveal grade-differentiating parameters for breast cancer. BMC Medical Imaging 6, 1–11 (2006)
3. Ruiz, A., Kong, J., Ujaldon, M., Boyer, K., Saltz, J., Gurcan, M.: Pathological image segmentation for neuroblasma using the gpu. In: ISBI, pp. 296–299 (2008)
4. Signolle, N., Plancoulaine, B., Herlin, P., Revenu, M.: Texture-based multiscale segmentation: Application to stromal compartment characterization on ovarian carcinoma virtual slides. In: Elmoataz, A., Lezoray, O., Nouboud, F., Mammass, D. (eds.) ICISP 2008. LNCS, vol. 5099, pp. 173–182. Springer, Heidelberg (2008)
5. Elston, C.W., Ellis, I.O.: Pathological prognostic factors in breast cancer. I. The value of histological grade in breast cancer: experience from a large study with long-term follow-up. Histopathology 19, 403–410 (1991)
6. Dalle, J., Leow, W., Racoceanu, D., Tutac, A., Putti, T.: Automatic breast cancer grading of histopathological images. In: EMBS, pp. 3052–3055 (2008)
7. Kong, J., Sertel, O., Shimada, H., Boyer, K., Saltz, J., Gurcan, M.: Computer-aided evaluation of neuroblastoma on whole-slide histology images: Classifying grade of neuroblastic differentiation. Pattern Recognition 42, 1080–1092 (2009)
8. Elmoataz, A., Lézoray, O., Bougleux, S.: Nonlocal discrete regularization an weighted graphs: a framework for image and manifolds processing. IEEE Transactions on Image Processing 17, 1047–1060 (2008)
9. Roullier, V., Ta, V.T., Lézoray, O., Elmoataz, A.: Graph-based multi-resolution segmentation of histological whole slide images. In: ISBI, pp. 153–156 (2010)
10. Belkin, M., Niyogi, P.: Laplacian eigenmaps for dimensionality reduction and data representation. Neural Computation 15, 1373–1396 (2003)
11. Ojala, T., Pietikäinen, M., Harwood, D.: A comparative study of texture measures with classification based on feature distributions. Pattern Recognition 29, 51–59 (1996)
12. Shi, J., Malik, J.: Normalized cuts and image segmentation. IEEE Transactions on Pattern Analysis and Machine Intelligence 22, 888–905 (1997)

Predicting Segmentation Accuracy for Biological Cell Images[*]

Adele P. Peskin[1], Alden A. Dima[2], Joe Chalfoun[2], and John T. Elliott[2]

[1] NIST, Boulder, CO 80305
[2] NIST, Gaithersburg, MD 20899

Abstract. We have performed segmentation procedures on a large number of images from two mammalian cell lines that were seeded at low density, in order to study trends in the segmentation results and make predictions about cellular features that affect segmentation accuracy. By comparing segmentation results from approximately 40000 cells, we find a linear relationship between the highest segmentation accuracy seen for a given cell and the fraction of pixels in the neighborhood of the edge of that cell. This fraction of pixels is at greatest risk for error when cells are segmented. We call the ratio of the size of this pixel fraction to the size of the cell the extended edge neighborhood and this metric can predict segmentation accuracy of any isolated cell.

1 Introduction

Cell microscopy is being used extensively to monitor cellular behavior under experimental settings. The common use of CCD cameras and the availability of microscopes with excellent optics, light sources, and automated stages and filter wheels allows collection of quantitative multiparameter image sets of large numbers of cells [1]. When combined with image analysis procedures, these image sets can provide several measurements of cellular behavior under the experimental conditions.

One of the most common image analysis procedures for cellular images is segmentation of an image object from the remainder of the image. For example, for images of cells stained with a fluorescent dye that covalently attaches to cellular proteins [2], segmentation procedures can be used to identify image pixels that are associated with the cell and separate them from the background. The results of this type of segmentation on images of cells that are at low density (i.e. minimal cell-cell contact) can be used to generate metrics such as spreading area, cell shape, and edge perimeter and provide high quality information about the morphological shape of the cells. This information is characteristic of cell phenotype or state and provides measurements that can be used to compare experimental conditions [3].

---

[*] This contribution of NIST, an agency of the U.S. government, is not subject to copyright.

G. Bebis et al. (Eds.): ISVC 2010, Part I, LNCS 6453, pp. 549–560, 2010.

There is extensive literature on procedures used to segment whole cells. Depending on the properties of the imaged cells, automated segmentation algorithms can provide differential results even when applied to the same image. New sophisticated algorithms such as level sets and active contours can segment particular cell features, but they are not necessarily readily available for conventional users of image analysis software. Automated segmentation routines based on histogram analysis or simple gradient-based edge detection routines are more common in most image analysis software. Although these methods can provide appropriate cell segmentation under many conditions, these segmentation routines often fail to adequately segment certain cells. In this study, we segmented a large number of cells from cellular images with various algorithms to identify what features in a cell object can influence segmentation outcome for a particular cell. A metric that evaluates the staining properties at the edge of a cell was developed to score the cell edge properties. This metric, called the edge neighborhood fraction, was found to be predictive of segmentation accuracy under several experimental conditions.

Because image segmentation is critical to biological image analysis, many segmentation methods have been published, including histogram-based, edge-detection-based, watershed, morphological, and stochastic techniques [4]. There are few examples of systematic comparisons of image analysis algorithms for cell image data; these include a comparison of cell shape analysis, a recent report comparing segmentation algorithms [5] and a study on predicting confidence limits in segmentation of cells [6]. An ongoing study in our group compares nine different segmentation techniques to manually segmented cells on a small number of cell images [7]. The study presented here evaluates cells from images of two different cell lines under five different sets of imaging conditions. Overall, we evaluated over 40000 cells from both NIH3T3 fibroblast and A10 smooth muscle cells in 9000 images. The 40000 cells represent 4 replicate wells of each of 2000 unique cells imaged under five different settings which varied edge quality. A study on this scale was large enough to produce statistically reliable results about the accuracy of the segmentation methods evaluated here and form predictive information for individual cells in different cell lines over a range of representative imaging conditions.

2   Data Description

The data used in this study examine two cell lines and five imaging conditions. These images consist of A10 rat smooth vascular muscle cells and NIH3T3 mouse fibroblasts stained with a fluorescent Texas Red-C2-maleimide cell body stain [2]. The overall geometric shape of the cell lines differ. A10 cells are well spread large cells. NIH3T3 cells are smaller fibroblasts with a spindly shape. The five imaging conditions varied in both the exposure time and the filter settings. These settings resulted in varying the line per mm resolution and the signal to noise ratio. Multiple cells are present on most images.

Both cell lines were maintained as previously described [8]. Cells were seeded at 1200 (NIH3T3) or 800 (A10) cells/cm^2, in 6 well tissue culture PS plates and incubated overnight. The cells were fixed with 1% PFA in PBS, and stained with Texas-Red maleimide and DAPI as previously described [2]. Images of the stained cells were collected with an Zeiss automated microscope with automated filter wheels controlled by Axiovision software. For optimal filter conditions, the stained cells were visualized with a Texas Red filter set (Chroma Technology, Excitation 555/28, #32295; dichroic beamsplitter #84000; Emission 630/60, #41834). For non-optimal filter conditions, the cells were imaged with Texas Red excitation filter (Chroma Technology, Excitation 555/28 filter dichroic beam-splitter #84000; Emission 740lp, #42345). These imaging conditions result in reduced intensity signal to noise ratios and introduce blurring.[1] Exposure times were selected to use either 1/4, full, or partially saturated dynamic range of the CoolSnap HQ2 camera. The five imaging conditions are summarized in Table 1.

To segment each of the cells and generate reference data that closely mimics human drawn segmentations, we used a procedure that was developed based on the analysis of manual segmentation processes. The procedure is described in reference [9]. This algorithm was applied to images with the highest contrast (i.e. conditions 3, Table 1) and edge pixels were identified as pixels with at least one neighbor pixel with an intensity less than 0.7 of its adjacent value. This intensity gradient feature has been shown to correlate well with manually selected edge pixels. Figure 1 shows a typical image and the reference data cell masks.

Table 1. The five sets of imaging conditions

Image	Exposure time(s) A10	Exposure time(s) NIH3T3	Filter type
1	0.015	0.01	optimal filter
2	0.08	0.05	optimal filter
3	0.3	0.15	optimal filter
4	1.0	1.0	non-optimal filter
5	5.0	5.0	non-optimal filter

3   Extended Edge Neighborhood

The apparent cell edges vary widely in clarity and sharpness across the five different images of the same cells. In particular, the images vary in terms of the number of pixel lengths (distance between pixels) needed to represent the thickness of the edge regions of the cells. We have previously quantified this thickness with a metric we call the cell edge quality index (QI) [10]. In the next section we describe how we find and quantify the fraction of pixels that are at risk for inaccuracy during a segmentation, using the quality index and the cell geometry.

---

[1] Certain trade names are identified in this report only in order to specify the experimental conditions used in obtaining the reported data. Mention of these products in no way constitutes endorsement of them. Other manufacturers may have products of equal or superior specifications.

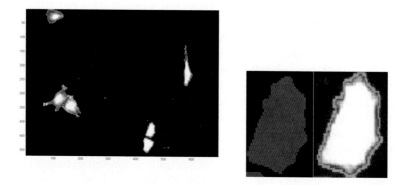

Fig. 1. The outlines in green of masks resulting from our semi-automated method for mimicking the manual segmentation process; A close-up of one cell near the bottom of the image, with the mask in red and the cell outline in green

In an accompanying paper, we compare manual segmentation masks on a set of 16 of the images used here, with type 3 imaging conditions from Table 1. We find that the our manual segmentation sets differ from one another, and that the extent of misalignment from one set to another depends upon the cell geometry [7] [9]. We observed that smaller, less round cells were more at risk for error in the selection. The smaller the cell, the less likely that two people hand selecting a mask would pick a large fraction of exactly the same pixels for the mask: if most of the pixels are near an edge, a cell is more at risk for any kind of segmentation error. In addition to cell size and shape, the gradient of the pixel intensity at the cell edge also plays a large role in determining whether a cell image can be segmented properly. We combine these concepts into a single quantity that can be calculated quickly for each cell. The metric represents the size of the extended edge neighborhood of pixels and is a fraction derived from the ratio of pixels at risk to the total area of the cell. The pixels at risk are determined by multiplying the cell perimeter by a factor determined from the quality index (QI), that represents the physical thickness (Th) of the cell edge.

4   Quality Index and Edge Thickness Calculation

For each cell in an image, we evaluate the pixel intensities within an isolated region containing the cell and background pixels. The quality index is calculated as follows [10]:

1. Identify the 3-component Gaussian mixture, whose components correspond to background (B), edge (E), and cell (C) pixels, via the EM (Expectation-Maximization) algorithm; x_B, x_E, and x_C denote the means of each component [11], [12].
2. Find the average gradient magnitude at each intensity between x_B and x_E.

3. Smooth the gradient in this region to fill in any gaps, and denote the resulting function by G(Intensity).
4. Find the intensity, Intensity A, at which the smoothed gradient magnitude is maximized.
5. Find the expected neighboring pixel to a pixel with Intensity A and denote this intensity as B; i.e., Intensity B = A - G(A)*(1 pixel unit).
6. Find the expected neighboring pixel to a pixel with Intensity B; i.e., Intensity C = B - G(B)*(1 pixel unit) = A - G(A)*(1 pixel unit) - G(A-G(A)*(1 pixel unit))*(1 pixel unit).
7. Compute the quality index as QI = (A - C)/(A - x_B).

The quality index ranges from 0.0 to 2.0, with a perfectly sharp edge at a value of 2.0. The edge thickness is defined as Th = 2.0/QI to scale a perfectly sharp edge to be equal to 1.0 pixel unit. We approximate the number of pixels at the edge by multiplying the edge thickness, Th, by the cell perimeter, and then define our new metric, the ratio of pixels at the edge to the total number of pixels, the extended edge neighborhood (EEN), as:

$$EEN = (P \times Th)/area \tag{1}$$

This value is effectively the fraction of the total cell area that makes up the cell edge. We determine the cell perimeter-to-area ratio from a 5-means clustering segmentation mask. This is a simple, fast segmentation, which we know from our ongoing study [7] is consistently a good performing algorithm over a wide range of extended edge neighborhoods. Because we need only the ratio of perimeter to area and not an accurate measurement of each, we can use this simple method.

The extended edge neighborhood is influenced by the intensity contrast of the image, through the calculation of the gradient at the cell edge, and by the overall geometry of the cell, the ratio of cell perimeter to cell area. This metric can vary from 0.0 to 3.0 or more depending upon the geometric features of the cell, although most images of cells have values between 0.0 and 1.0. Figure 2 shows three cells with very different extended edge neighborhoods. If the edge is very thick because the image is blurry, as in the third picture of Figure 2, the extended edge neighborhood can be larger than 1.0. If a cell is very thin but the cell edges are very sharp, as in the middle picture of Figure 2, it still has a higher extended edge neighborhood than a larger cell with similar edges, as in the first picture of Figure 2. This metric can be calculated for each cell in the absence of good segmentation, because knowledge of the cell edge is not required for either the edge thickness or cell geometry estimates. The calculation for each cell can be performed efficiently even for very large datasets.

5   Testing 40000 Cells

Figure 3 shows a histogram plot of the EEN metric for 40000 cells. It shows that the A10 cells have lower EEN values on average than the NIH3T3 cells. To determine if the extended edge neighborhood metric is predictive of accuracy

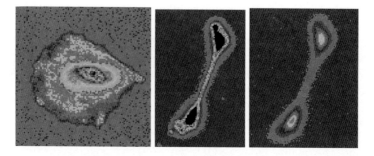

Fig. 2. 3 cells are colored according to pixel intensity, with the full range shown divided into 40 different colors: a large, round cell with low extended edge neighborhood; a small, thin cell similar edges but a higher extended edge neighborhood; the same cell but a blurrier image, where the extended edge neighborhood is greater than 1.0

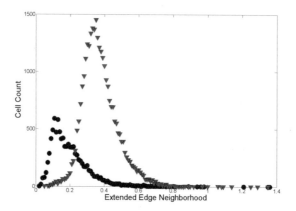

Fig. 3. Numbers of cells as a function of extended edge neighborhood for the A10 cells in red, and the NIH3T3 cells in blue

for segmentation algorithms, we studied the accuracy of four segmentation algorithms that use different methods to identify cell edges as a function of extended edge neighborhood for each cell. The algorithms tested were 3-means clustering, 4-means clustering, 5-means clustering, and a Canny edge method. Segmentation masks were generated from the k-means clustering algorithms by assuming that the cluster with the lowest centroid represents the background and the remaining clusters belong to the cells. To determine an accuracy metric for each cell that is segmented by an automated segmentation, we compared the results of the algorithm to that of a reference segmentation data set derived with an a computer assisted manual segmentation and expert visual inspection, using bivariate similarity metrics, previously described in [7] [9]. Definitions of these metrics and a justification for their use are summarized in the next section.

6   Bivariate Similarity Index

Various similarity metrics have been used to evaluate segmentation algorithm performance. The commonly used Jaccard similarity index [13], for example, compares a reference data set, T, with another set of estimates, E, defined by:

$$S = |T \cap E|/|T \cup E|, \tag{2}$$

where $0.0 \leq S \leq 1.0$. If an estimate matches the truth, $T \cap E = T \cup E$ and $S = 1$. If an algorithm fails, then $E = 0$ and $S = 0$. However, S cannot discriminate between certain underestimation and overestimation cases. For example, if the true area = 1000, then both the underestimated area of 500, and the overestimated area of 2000 yield the same value for the similarity index $S = 500/1000 = 1000/2000 = 0.5$. Here we used a set of bivariate similarity indices that can distinguish between underestimation and overestimation.

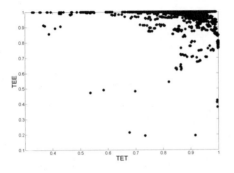

Fig. 4. Plot of TET vs. TEE for 3,000 A10 cells

We define these indices as follows, to compare the reference pixel set T, with a segmentation mask, pixel set E:

$$TET = |T \cap E|/|T|, 0.0 \leq TET \leq 1.0 \tag{3}$$

$$TEE = |T \cap E|/|E|, 0.0 \leq TEE \leq 1.0 \tag{4}$$

Each similarity metric varies between 0 and 1. If the estimate matches the reference mask, both TET and TEE = 1.0. TET and TEE were constructed to be independent and orthogonal and divides performance into four regions: Dislocation: TET and TEE are small; Overestimation: TET is large, TEE is small; Underestimation: TET is small, TEE is large; and Good: both TET and TEE are large. Figure 5 illustrates the use of the indices to compare a group of approximately 3000 A10 cells, segmentated with a 5-means clustering segmentation in Figure 4. This plot shows the tendency of 5-means clustering segmentations to underestimate cell edge compared to the manually segmented data.

In some situations, segmentation comparisons may be facilitated by combining the bivariate indices into a univariate metric. For these purposes we define a metric called the segmentation distance as the Euclidean distance from the point corresponding to the TET and TEE values to the point corresponding to perfect segmentation ($TET = 1.0, TEE = 1.0$). The univariate metric does not contain information about over- or undersegmentation, but it does provide a general measure of segmentation accuracy and can be used to evaluate correlations with the extended edge neighborhood metric.

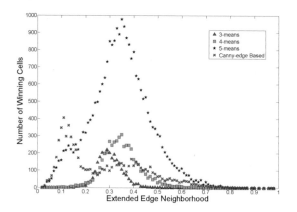

Fig. 5. Number of cells for which a technique worked best: 3-means clustering (red), 4-means clustering (green), 5-means clustering (blue), Canny edge (purple), as a function of extended edge neighborhood

7   Results

To begin, we evaluate which segmentation algorithm worked best for each individual cell. Figure 5 shows how many of the 40000 cells scored best for each method, as a function of our new metric, the extended edge neighborhood. Often several of the methods gave similar results, but this plot counts the number of times each method gave the best results, regardless of whether another method came close. If two methods gave identical results for a given cell, results for that cell were not included in the plot. The 5-means clustering gives the best results for this cell image dataset and the cells most likely to be segmented best using this method tended to have a mean EEN value greater than 0.2. Interestingly, the Canny edge segmentation method works very well in only a small region of the extended edge neighborhood curve, at low extended edge neighborhoods between 0.0 and approximately 0.15, which represents larger cells with sharp edges. In this region the Canny algorithm segmentation results are more similar to manual segmentation than 5-means clustering, 4-means clustering, or 3-means clustering. 3-means clustering and 4-means clustering methods did best for only

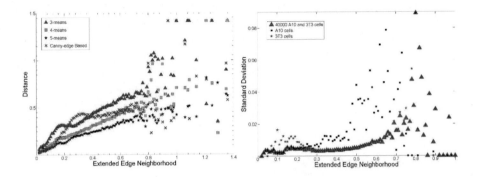

Fig. 6. Averaged segmentation distance for each group of cells with the same extended edge neighborhood for 3-means clustering (red), 4-means clustering (green), 5-means clustering (blue), and Canny edge (purple); Standard deviations of 5-means clustering averaged segmentation distance results as a function of extended edge neighborhood for the A10 cells (blue); for the NIH3T3 cells (red); for all 40000 cells (purple)

a small number of cells in the extended edge neighborhood region between 0.2 and 0.4, where the accuracy using any method was not very high.

Figure 6 shows all of the segmentation results for each of four methods over the whole range of our extended edge neighborhood metric. The results are presented in terms of the averaged segmentation distance for groups of cells with the same extended edge neighborhoods. We can draw a number of conclusions from this data. In general, the accuracy of these methods varies monotonically with extended edge neighborhood, and the 5-means clustering results are on average always better than the 4-means clustering, which are always better on average than the 3-means clustering. An occasional cell has better results for 3-means clustering than 4-means clustering or 5-means clustering. However, the results show that in general, the extended edge neighborhood metric predicts the accuracy of k-means algorithms in segmenting these cell images. The standard deviations of segmentation distances averaged over a group of cells with the same extended edge neighborhoods are low, suggesting a high level of predictability for most cells. As an example, the standard deviation results for the 5-means clustering data are shown in the second plot of Figure 6. The data show that for A10 smooth muscle cells, the standard deviation in segmentation accuracy is low for the EEN range 0.0 to 0.2. The standard deviation for the NIH3T3 cells is low for the range 0.2 to 0.5.

The Canny edge segmentation results are similar to the 5-means clustering method at low extended edge neighborhoods. Above we saw that more cells were segmented accurately in this extended edge neighborhood than with 5-means clustering, but the variability of the Canny edge results produce a similar plot on data averaged over all of the cells, in the region between 0.0 and 0.15 extended edge neighborhood. Overall, we see a fairly linear trend in the best averaged segmentation as a function of extended edge neighborhood.

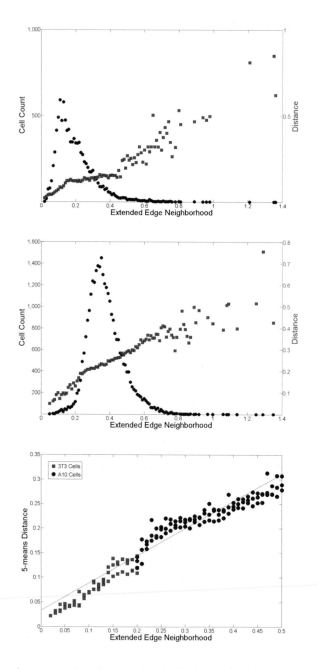

Fig. 7. Averaged segmentation distance for the A10 cells (red), along with a plot of cell numbers as a function of extended edge neighborhood (blue); Same for the NIH3T3 cells; Results from the A10 cells (red) and NIH3T3 cells (blue) from the first two plots. A straight line is fitted to this data.

To investigate further the relationship between the EEN metric, the extended edge neighborhood, and the best averaged segmentation results, we look only at data for which there are a large number of cells. We find the region on the extended edge neighborhood curves for each cell line that includes 90 % of the cell data in each cell line. The first two plots of Figures 7 overlay plots of cell counts for the A10 and NIH3T3 cell lines respectively with the segmentation distances from a 5-means clustering for each cell line. In the third plot of Figure 7, we graph the A10 cells in the extended edge region from 0.02 to 0.2, and the NIH3T3 results in the extended edge region 0.2 to 0.5, where 90% of the cells from each cell line occur. Both of these data sets are fitted with a linear model, which is also plotted in Figure 7: predicted distance $= 0.051 + \text{EEN} \times 0.477$, with a correlation coefficient of 0.9815.

8   Conclusions and Future Work

From this large scale test, we define a method to pre-process images and determine their vulnerability to segmentation error. The accuracy that is possible from any given segmentation technique is directly proportional to the extended edge neighborhood of each individual cell within an image. Rounder, larger cells have a lower extended edge neighborhood than smaller less round cells, and segmentation will more closely align with manual segmentation for these cell images. Our results suggest that of the four segmentation methods tested here, a 5-means clustering segmentation is the most reliable. The Canny edge segmentation method performs best with cells within a very small extended edge neighborhood range that is less than approximately 0.15. We can now use the methods outlined in this paper to look at a wider range of segmentation algorithms for identifying more accurate segmentation techniques. We have written a software segmentation pre-processor that calculates extended edge neighborhood for each cell in an image and then provides the best technique and expected accuracy for the segmentation of each cell based on these four algorithms, which can evolve as we compare more algorithms. We believe that this processor will be of great use for optimizing the segmentation of cells seeded at low density and stained as described here. The EEN can also be used as a metric to rank cells for most-likely best segmentation. Measurements of cell function can be weighted by this ranking to potentially improve the measurement robustness in a cell-based assay. The EEN metric will have significant value in determining which cells in a data set are most at risk during a segmentation procedure.

Acknowledgements

James Filliben of NIST's Statistical Engineering Division, Anne Plant and Michael Halter, of NIST's Biochemical Science Division, provided helpful discussion.

References

1. Plant, A.L., Elliott, J.T., Tona, A., McDaniel, D., Langenbach, K.J.: Tools for Quantitative and Validated Measurements of Cells. In: Taylor, L., Giuliano, K., Haskins, J. (eds.) High Content Screening: A Powerful Approach to Systems Cell Biology and Drug Discovery. Humana Press, Totowa (2006)
2. Elliott, J.T., Tona, A., Plant, A.L.: Comparison of reagents for shape analysis of fixed cells by automated fluorescence microscopy. Cytometry 52A, 90–100 (2003)
3. Elliott, J.T., Woodward, J.T., Langenbach, K.J., Tona, A., Jones, P.L., Plant, A.L.: Vascular smooth muscle cell response on thin films of collagen. Matrix Biol. 24(7), 489–502 (2005)
4. Zhou, X., Wong, S.T.C.: High content cellular imaging for drug development. IEEE Signal Processing Magazine 23(2), 170–174 (2006)
5. Coelho, L.P., Shariff, A., Murphy, R.F.: Nuclear Segmentation in Microscope Cell Images: A Hand-Segmented Dataset and Comparison of Algorithms. In: ISBI (2009)
6. Cardinale, J., Rauch, A., Barral, Y., Szkely, G., Sbalzarini, I.F.: Bayesian image analysis with on-line confidence estimates and its application to microtubule tracking. In: IEEE International Symposium of Biomedical Imaging, pp. 1091–1094 (June 2009)
7. Dima, A., Elliott, J.T., Filliben, J.J., Halter, M., Peskin, A., Bernal, J., Stotrup, B.L., Kociolek, M., Brady, M.C., Tang, H.C., Plant, A.L.: Comparison of segmentation algorithms for flourescence microscopy images of cells. Cytometry Part A (submitted)
8. Langenbach, K.J., Elliott, J.T., Tona, A., Plant, A.L.: Evaluating the correlation between fibroblast morphology and promoter activity on thin films of extracellular matrix proteins. BMC-Biotechnology 6(1), 14 (2006)
9. Chalfoun, J., Dima, A., Peskin, A.P., Elliot, J., Filliben, J.J.: A Human Inspired Local Ratio-Based Algorithm for Edge Detection in Fluorescent Cell Images. In: 6th International Symposium on Visual Computing 2010 (2010)
10. Peskin, A.P., Kafadar, K., Dima, A.: A Quality Pre-Processor for Biological Cells. In: 2009 International Conference Visual Computing (2009)
11. Hastie, T., Tibshirani, R., Friedman, J.H.: The Elements of Statistical Learning: Data Mining, Inference, And Prediction. Springer, New York (2001)
12. Peskin, A.P., Kafadar, K., Santos, A.M., Haemer, G.G.: Robust Volume Calculations of Tumors of Various Sizes. In: 2009 International Conference on Image Processing. Computer Vision, and Pattern Recognition (2009)
13. Rand, W.M.: Objective criteria for the evaluation of clustering methods. Journal of the American Statistical Association 66(336), 846–850 (1971)

Multiscale Analysis of Volumetric Motion Field Using General Order Prior

Koji Kashu[1], Atsushi Imiya[2], and Tomoya Sakai[2]

[1] School of Advanced Integration Science, Chiba University
[2] Institute of Media and Information Technology, Chiba University
Yayoi-cho 1-33, Inage-ku, Chiba, 263-8522, Japan

Abstract. We introduce variational optical flow computation involving the prior with the fractional order differentiations. The fractional order differentiation is a typical tool in signal processing and image analysis. The zero crossing of a fractional order Laplacian yields a good performance for edge detection. As a sequel of edge detection with the fractional order differentiations, we deal with variational optical flow computation involving the fractional order differentiations on optical flow vectors. The method allows us to detect discontinuity of optical flow using linear operations.

1   Introduction

In this paper, we introduce a motion segmentation method using variational optical flow computation with the prior involving fractional order differentiations [1][1] for variational optical flow computation [2].

Using the Fourier transform pair

$$F(\xi, \eta, \zeta) = \frac{1}{2\pi^{3/2}} \int_{\mathbf{R}^3} f(x, y, z)e^{-i(x\xi+y\eta+z\zeta)} dxdydz, \tag{1}$$

$$f(x, y, z) = \frac{1}{2\pi^{3/2}} \int_{\mathbf{R}^3} F(\xi, \eta, \zeta)e^{i(x\xi+y\eta+z\zeta)} d\xi d\eta d\zeta, \tag{2}$$

we define the operation Λ as

$$\Lambda f(x, y, z) = \frac{1}{2\pi} \int_{\mathbf{R}^3} (\sqrt{\xi^2 + \eta^2 + \zeta^2}) F(\xi, \eta)e^{i(x\xi+y\eta+z\zeta)} d\xi d\eta d\zeta. \tag{3}$$

Furthermore, we have the equality

$$\int_{\mathbf{R}^3} |\nabla f|^2 dxdydz = \int_{\mathbf{R}^3} |\Lambda f|^2 dxdydz, \tag{4}$$

since

$$\int_{\mathbf{R}^3} |f|^2 dxdydz = \int_{\mathbf{R}^3} |F|^2 d\xi d\eta d\zeta. \tag{5}$$

---

[1] Fractional order differentiations satisfies the relations $\frac{d^{1/2}}{dx^{1/2}}x = \frac{1}{\sqrt{2\pi}}x^{1/2}$ and $\frac{d^{1/2}}{dx^{1/2}} \frac{1}{\sqrt{2\pi}}x^{1/2} = 1$.

G. Bebis et al. (Eds.): ISVC 2010, Part I, LNCS 6453, pp. 561–570, 2010.
© Springer-Verlag Berlin Heidelberg 2010

The mathematical properties of eq. (4) on the operator Λ allows us to focus on variational image analysis in the form

$$J_\alpha(f) = \int_{\mathbf{R}^3} \{M(f) + \kappa |\Lambda^\alpha f|^2\} dxdydz, \tag{6}$$

where $M(f)$ is the model term for image analysis for $\kappa \geq 0$ and $\alpha = 1 + \varepsilon$ where $0 \leq \varepsilon < 1$ as a generalization of the energy function such that

$$J(f) = \int_{\mathbf{R}^3} \{M(f) + \kappa |\nabla f|^2\} dxdydz. \tag{7}$$

Since $\Lambda = \Lambda^*$, the Euler-Lagrange equation of eq. (6) is

$$\Lambda^{2\alpha} f + \frac{1}{\kappa} M(f) = 0. \tag{8}$$

Fractional order differentiations are typical tools in signal processing [4] and is applied to the edge detection of images [5]. In edge detection, a zero-crossing set of a fractional order Laplacian derives good performance [5].

Total variational (TV) regularization [3] is a successful method of optical flow computation of an image with a discontinuity of the gray values and the optical flow field. TV regularization uses the total variation of optical flow field as the prior, although the classical Horn-Schunck method [6,7] uses the L_2 norm of the gradient of flow field. TV regularization optical flow computation [3] derives a nonlinear elliptic partial differential equation as the Euler-Lagrange equation of the energy functional of the problem.

The generalization of the order of differentiation in a Horn-Schunck type prior is another modification of the original Horn-Schunck regularization, since this generalization yields a linear Euler-Lagrange equation. There are two types of generalization of the differentiations in priors; the first one is to deal with higher-order differentiations, and the second one is to deal with fractional order differentiations. We focus on the second type of generalization, that is, we deal with the variational optical flow computation whose prior term involves a fractional order differentiation of optical flow vectors.

Since fractional differentiations are linear operations, the fractional order regularization for optical flow computation [2] derives a linear fractional order elliptic partial differential equation as the Euler-Lagrange equation of the energy functional. Therefore, we can numerically solve the problem using the same strategy that is used to solve the Horn-Schunck method.

2   Optical Flow Computation

For a spatio-temporal image $f(x, y, z, t)$, the total derivative with respect to the time argument t is given as

$$\frac{d}{dt} f = \frac{\partial f}{\partial x} \frac{dx}{dt} + \frac{\partial f}{\partial y} \frac{dy}{dt} + \frac{\partial f}{\partial z} \frac{dz}{dt} + \frac{\partial f}{\partial t} \frac{dt}{dt} \tag{9}$$

where $\boldsymbol{u} = (u, v, w)^\top = (\dot{x}, \dot{y}, \dot{z})^\top = (\frac{dx}{dt}, \frac{dy}{dt}, \frac{dz}{dt})^\top$ is the motion of each point $\boldsymbol{x} = (x, y, z)^\top$. Optical flow consistency [6,7,8] $\frac{d}{dt}f = 0$ implies that the motion $\boldsymbol{u} = (u, v, w)^\top$ of the point $\boldsymbol{x} = (x, y, z)^\top$ is the solution of the singular equation,

$$f_x u + f_y v + f_z w + f_t = 0. \tag{10}$$

The mathematical properties of eq. (4) on the operator Λ allows us to focus on variational optical flow computation in the form

$$J_\alpha(\boldsymbol{u}) = \int_{\mathbf{R}^2} (\nabla f^\top \boldsymbol{u} + \partial_t f)^2 dx dy dz + \kappa E_\alpha(\boldsymbol{u})$$

$$E_\alpha(\boldsymbol{u}) = \int_{\mathbf{R}^2} (|\Lambda^\alpha u|^2 + |\Lambda^\alpha v|^2 + |\Lambda^\alpha w|^2) dx dy dz, \tag{11}$$

for $\kappa \geq 0$ and $\alpha = n + \varepsilon$ where $0 \leq \varepsilon < 1$ as a generalization of the energy functional of the Horn-Schunck method [6] such that

$$J(\boldsymbol{u}) = \int_{\mathbf{R}^2} (\nabla f^\top \boldsymbol{u} + \partial_t f)^2 dx dy dz + \kappa E_1(\boldsymbol{u})$$

$$E_1(\boldsymbol{u}) = \int_{\mathbf{R}^2} (|\nabla u|^2 + |\nabla v|^2 + |\nabla w|^2) dx dy dz. \tag{12}$$

These energy functionals lead to the following definition.

Definition 1. *We call the minimizer of eq. (11) the α optical flow*

Since $\Lambda = \Lambda^*$, the Euler-Lagrange equation of eq. (11) is

$$\Lambda^{2\alpha} \boldsymbol{u} + \frac{1}{\kappa}(\nabla f^\top \boldsymbol{u} + \partial_t f)\nabla f = 0. \tag{13}$$

Specially, for $\alpha = 1, \frac{3}{2}, 2$, the Euler-Lagrange equations are

$$\Delta \boldsymbol{u} - \frac{1}{\kappa}(\nabla f^\top \boldsymbol{u} + \partial_t f)\nabla f = 0, \tag{14}$$

$$\Delta \Lambda \boldsymbol{u} - \frac{1}{\kappa}(\nabla f^\top \boldsymbol{u} + \partial_t f)\nabla f = 0, \tag{15}$$

$$\Delta^2 \boldsymbol{u} + \frac{1}{\kappa}(\nabla f^\top \boldsymbol{u} + \partial_t f)\nabla f = 0. \tag{16}$$

since $\Lambda^2 = -\Delta$, $\Lambda^3 = -\Delta\Lambda$, and $\Lambda^4 = \Delta^2$.

The solution involving the lth-order prior is

$$f(x, y, z) = \sum_{i,j,k=0}^{l-1} a_{ij} x^i y^j z^k \tag{17}$$

for nonnegative integers i, j, k, that is, the solution is locally a $(k-1)$th-order polynomial of x, y, and z. This property implies that the priors involving the first- and second- order differentiations derive a piecewise linear and affine optical flow, respectively.

3   Numerical Examples

The associated diffusion equation

$$\frac{\partial u}{\partial \tau} = -\Lambda^{2\alpha} u - \frac{1}{\kappa}(\nabla f^\top u + \partial_t f)\nabla f. \tag{18}$$

of eq. (13) derives the semi-implicit discretization

$$\frac{u^{(n+1)} - u^{(n)}}{\Delta\tau} = -\Lambda^{2\alpha} u^{(n+1)} - \frac{1}{\kappa}(\nabla f^\top u^{(n)} + \partial_t f)\nabla f. \tag{19}$$

From eq. (19) we have the iteration form

$$(I + \frac{\Delta\tau}{\kappa} S_{kmn}) u_{kmn}^{(l+1)} = u_{kmn}^{(l)} + \Delta\tau(-\Lambda^{2\alpha}) u_{kmn}^{(l)} - \frac{\Delta\tau}{\kappa} c_{kmn}, \ l \geq 0, \tag{20}$$

for the numerical computation of α optical flow.

Equation (20) is expressed as

$$A u^{(l+1)} = P^\top B P u^{(l)} + c, \ \ B = I + \Delta\tau(-L)^\alpha \tag{21}$$

where L is the discrete Laplacian such that

$$L = diag(l, l, l), \ \ l = D \otimes I \otimes I + I \otimes D \otimes I + I \otimes I \otimes D \tag{22}$$

$$D = \begin{pmatrix} -1 & 1 & 0 & \cdots & 0 & 0 \\ 1 & -2 & 1 & 0 \cdots & & \\ \vdots & \vdots & \ddots & \vdots & & \\ 0 & 0 & \cdots & 0 & 1 & -1 \end{pmatrix}, \tag{23}$$

and $P = diag(Q, Q, Q)$ for the permutation matrix Q which transforms the collection of triplets (k, m, n) $1 \leq k, m, n \leq N$ from the lexicographical order to the reverse lexicographical order.

Setting Φ and Σ to be the discrete cosine transform matrix and its eigenmatrix, the matrix L^α is rewritten as[2]

$$L^\alpha = (\Phi \otimes \Phi \otimes \Phi)^\top (\Sigma^\alpha \otimes \Sigma^\alpha \otimes \Sigma^\alpha)(\Phi \otimes \Phi \otimes \Phi) \tag{24}$$

Therefore, we can compute Λ^α numerically using discrete cosine transform.

Figures 1 (a) and (b) show biological and medical images, respectively Figure 2 shows Motion field of chromosome segmentation in binary fission process for $\alpha = 0, 5, 1.0, 1.5, 2.0, 2.5, 3.0$. The image sequence is dynamically measured using a confocal laser microscopy. If the optimal order of the points is between n and $n+1$, the points are viscoelastically moving [9] on an image. These results shows that, in the segmentation process, the ribosome vicoustically moves and that the motions of nucleolus and nucleus are inactive. Figure 3 show the results

---

[2] For a positive definite matrix A and a real number α the eigenvalues of A^α is λ^α for $Au = \lambda u$, where u is the eigenvector of A.

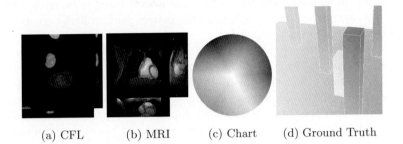

| (a) CFL | (b) MRI | (c) Chart | (d) Ground Truth |

Fig. 1. Original images on on the sagittal, coronal, and transverse planes and the color chart for planar motion field. (a) Dynamic confocal laser image of chromosome segmentation in binary fission process. (b) MRI image of a beating heart. (c) Color chart for the planar vector. (d) Color expression of a planar motion field.

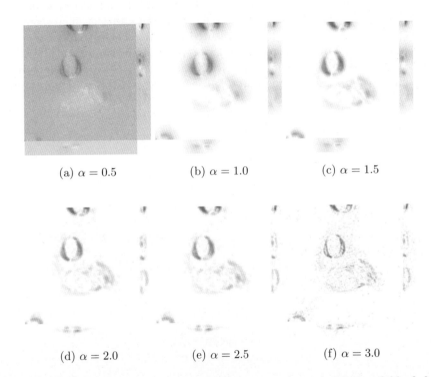

| (a) $\alpha = 0.5$ | (b) $\alpha = 1.0$ | (c) $\alpha = 1.5$ |

| (d) $\alpha = 2.0$ | (e) $\alpha = 2.5$ | (f) $\alpha = 3.0$ |

Fig. 2. Computational results for a biological image sequence. Motion field of chromosome segmentation in binary fission process. The optical flow fields for $\alpha = 0.5, 1.0, 1.5, 2.0, 2.5, 3.0$ show that the continuity order of flow field is non-uniform on the screen.

a beating heart image sequence, respectively for $\alpha = 0, 5, 1.0, 1.5, 2.0, 2.5, 3.0$. These results separates the heart wall based on the continuity orders.

In these examples, optical flow field is computed using volumetric method, and on the sagittal, coronal, and transverse planes, motion fields are expressed using color chart of Fig. 1.

The examples show that our method extracts motion boundary, and that using hierarchical analysis of motion fields with respect to α, it is possible to classify the motion fields using the continuity order of the optical flow vectors. As shown in results, if α is non-integer, the algorithm extracts the motion boundary. Furthermore, since the detected flow fields depend on the parameter α. Moreover, by tuning the parameter α, the algorithm detects the position of discontinuity of the boundary.

Figures 4(a) and 4(d) show the average and variance curves of residuals

$$R(\alpha) = \text{average}\left\{\int_{\mathbf{R}^3} r(\alpha)d\mathbf{x}\right\}, \quad r(\alpha) = (\nabla f^\top \mathbf{u}_\alpha + f_t)^2 + \kappa|\Lambda^\alpha \mathbf{u}_\alpha|^2 \quad (25)$$

for three-dimensional volumetric images, where $\mathbf{u}_\alpha$ is the numerical solution of the problem for each α for the image sequence of confocal laser microscopy and volumetric beating heart sequence measured by MRI. These results show that the volumetric motion boundary of HeLa moves viscoelastically and that these of volumetric beating heart moves elastically, since the $R(\alpha)$ curve of HeLa is flat and that of volumetric beating curve is convex and has the minimum around

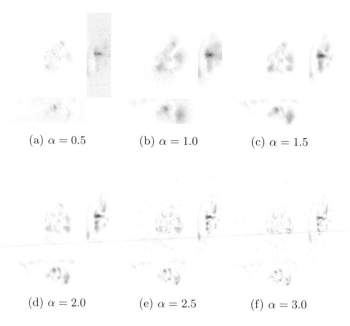

(a) $\alpha = 0.5$ (b) $\alpha = 1.0$ (c) $\alpha = 1.5$

(d) $\alpha = 2.0$ (e) $\alpha = 2.5$ (f) $\alpha = 3.0$

Fig. 3. Computational results for a medical image sequence. Wall motion of a beating heart. The optical flow fields for $\alpha = 0.5, 1.0, 1.5, 2.0, 2.5, 3.0$ show that the continuity order of flow field is non-uniform on the screen.

$\alpha = 2$. Figures 4(c) and 4(d) show optical flow for $\boldsymbol{u}^*$. Since each image of the heart image sequence is produced as the average of many images measured in the different observations, The background of each image is assumed to be the average of white noise with various levels and the motion vectors in the background are average of various motion vectors. Therefore, it is not so easy to extract discontinuity of motion field from the background.

For Λ we obtain the relations

$$\Lambda^{2\alpha} = (-\Delta)^n(-\Delta)^{\varepsilon}, \ \Lambda^{2\alpha} = (-\Delta)^{n+1}(-\Delta)^{-\bar{\varepsilon}} \tag{26}$$

for $\alpha = n + \varepsilon = (n+1) - \bar{\varepsilon}$ where $0 < \varepsilon, \bar{\varepsilon} < 1$ and $\bar{\varepsilon} + \varepsilon = 1$. Since symbolically we have the relation $(-\Delta)^{-\bar{\varepsilon}}(-\Delta)^{\bar{\varepsilon}} f = f$, the operation $(-\Delta)^{-\bar{\varepsilon}} f$ is a low-pass filter in the Fourier domain.

This decomposition can be read that $\Lambda^{2\alpha} f$ is achieved by applying the harmonic operation $(-\Delta)^{n+1}$ to $g = (-\Delta)^{-\bar{\varepsilon}} f$, which is achieved by convolution between the function f and the Riesz potential [10]. The operation $(-\Delta)^{-\bar{\varepsilon}}$ possesses smoothing effect to the optical flow field in each step of iteration, since $(-\Delta)^{\bar{\varepsilon}}$ is a high-pass operation for $\bar{\varepsilon} > 0$. Therefore, our numerical scheme derived generates a smoothed optical flow before applying the harmonic operation,

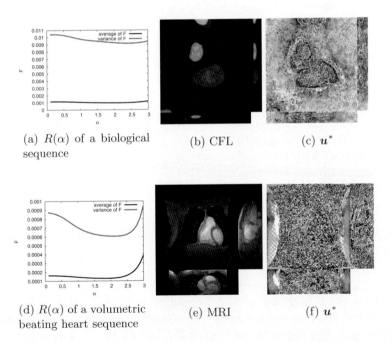

(a) $R(\alpha)$ of a biological sequence

(b) CFL

(c) $\boldsymbol{u}^*$

(d) $R(\alpha)$ of a volumetric beating heart sequence

(e) MRI

(f) $\boldsymbol{u}^*$

Fig. 4. The residual curves of a volumetric image and the optimal motion. (a) Residual of optical flow vector $\boldsymbol{u}_\alpha$ of HeLa (b) Dynamic confocal laser image of chromosome segmentation in binary fission process. (c) $\boldsymbol{u}^*$ is optical flow for $r(\boldsymbol{x}, \alpha)$. (d) Residual of optical flow vector $\boldsymbol{u}_\alpha$ of beating heart. (e) MRI beating heart image. (f) $\boldsymbol{u}^*$ is optical flow for $r(\boldsymbol{x}, \alpha)$.

which is the main part of the prior for selection of model in optical flow computation. This pre-smoothing property of the numerical scheme yields a better performance for $\alpha = n + \varepsilon$ such that $0 < \varepsilon < 1$.

4   Concluding Remarks

The order of differentiation in the prior decides the continuity order of the optical flow filed. Therefore, our results show that the orders between one and two are preferable to detect discontinuity optical flow vectors, which appear on the motion boundary.

The Nagel-Enckelmann [8] and TV regularization [2] methods are image and flow driven optical flow computation methods, respectively. On the other hand, our method is an operator driven. In Nagel-Enckelmann method, the structure tensor, which is the local moment of the gradient image, controls the local coordinate of optical flow computation, although the numerical scheme is linear. In TV regularization, the direction of the flow vector itself controls the direction to compute local average and the numerical scheme is nonlinear.

Setting T to be a shift-invariant linear operation to images, the operation T is expressed as a convolution operation such that

$$Tf(x,y,z) = \int_{\mathbf{R}^2} t(x - x', y - y', z - z') f(x', y', z') dx' dy' dz'. \qquad (27)$$

We can deal with the variational equation

$$G(u) = \int_{\mathbf{R}^2} \left\{ (\nabla f^\top u + \partial_t f)^2 + \kappa(|Tu|^2 + |Tv|^2 + |Tw|^2) \right\} dx dy dz \qquad (28)$$

for the optical flow computation. We showed that if T is the differentiation operator of the order α, such that, $\alpha = 1 + \varepsilon$ for $0 < \varepsilon < 2$, we can detect the motion boundary with motion discontinuity.

The Euler-Lagrange equation of eq. (28) is

$$T^*Tu + \frac{1}{\kappa}(\nabla f^\top u + \partial_t f)\nabla f = 0, \qquad (29)$$

where T^* is the conjugate operation of T. The associated diffusion equation of eq. (29) is

$$\frac{\partial}{\partial \tau} u = Fu + g, \qquad (30)$$

where $F = (T^*T + S)$ for $g = f_t \nabla f$ and $S = \nabla f \nabla f^\top$. Therefore, for the selection of T, the operator theory [11,12] is helpful. Setting T to be the Fourier transform of t, the Fourier transform of Tf is TF. If T is a real function of $\rho = \sqrt{\xi^2 + \eta^2 + \zeta^2}$, the relation $T^* = T$ is satisfied. For example, if $T(\xi, \eta, \zeta) = -\rho^2 = (\xi^2 + \eta^2 + \zeta^2)$, T is the Laplacian, which is linear and shift-invariant operation. For the semi-explicit discretization of

$$\frac{\partial u}{\partial \tau} = T^*Tu + \frac{1}{\kappa}(\nabla f^\top u + \partial_t f)\nabla f, \qquad (31)$$

using the operator splitting

$$u + \frac{\Delta\tau}{\kappa}Su = u + \Delta\tau(-T^*T)u - \frac{\Delta\tau}{\kappa}\partial_t\nabla f, \tag{32}$$

where $S = \nabla f\nabla f^\top$, eq. (29) derives the iteration form

$$(I + \frac{\Delta\tau}{\kappa}S)u^{(l+1)} = u^{(l)} + \Delta\tau(-T^*T)u^{(l)} - \frac{\Delta\tau}{\kappa}\partial_t\nabla f, \, l \geq 0, \tag{33}$$

which converges to

$$u = \left(T^*T + \frac{1}{\kappa}S\right)^{-1}\left(-\frac{1}{\kappa}\partial_t f\nabla f\right) = Cu_n, \; u_n = -\frac{\partial_t f}{|\nabla f|^2}\nabla f. \tag{34}$$

if $\rho(I + \frac{\Delta\tau}{\kappa}S) > 1$ and $\rho(I - \Delta\tau T^*T) < 1$. Equation. (34) implies that the operator T decides the linear operation A which derives the optical flow vector from the normal flow u_n at each point, although an image and optical flow field decides the operator A in the Nagel-Enckelmann method and TV regularization. The selection of T is expected to derive remarkable results for optical flow computation.

The second term of the right-hand side of eq. (33) is numerically computed as

$$(-T^*Tu)_{kmn} = \sum_{ijq} t_{ijp}u_{k-i\,m-j,n-q} \tag{35}$$

since T^*T is a shift-invariant operation, which is efficiently computed by FFT.

This research was supported by "Computational anatomy for computer-aided diagnosis and therapy: Frontiers of medical image sciences" funded by Grant-in-Aid for Scientific Research on Innovative Areas, MEXT, Japan, Grants-in-Aid for Scientific Research founded by Japan Society of the Promotion of Sciences and Grant-in-Aid for Young Scientists (A), NEXT, Japan.

References

1. Tadjeran, C., Meerschaert, M.M.: A second-order accurate numerical method for the two-dimensional fractional diffusion equation. J. of Computational Physics 220, 813–823 (2007)
2. Papenberg, N., Bruhn, A., Brox, T., Didas, S., Weickert, J.: Highly accurate optic flow computation with theoretically justified warping. IJCV 67, 141–158 (2006)
3. Yin, W., Goldfarb, D., Osher, S.: A comparison of three total variation based texture extraction models. J. Visual Communication and Image Representation 18, 240–252 (2007)
4. Tseng, C.-C., Pei, S.-C., Hsia, S.-C.: Computation of fractional derivatives using Fourier transform and digital FIR differentiator. Signal Processing 80, 151–159 (2000)
5. Mathieu, B., Melchior, P., Oustaloup, A., Ceyral, C.: Fractional differentiation for edge detection. Signal Processing 83, 2421–2432 (2003)

6. Horn, B.K.P., Schunck, B.G.: Determining optical flow. Artificial Intelligence 17, 185–204 (1981)
7. Beauchemin, S.S., Barron, J.L.: The computation of optical flow. ACM Computer Surveys 26, 433–467 (1995)
8. Nagel, H.-H.: On the estimation of optical flow:Relations between different approaches and some new results. Artificial Intelligence 33, 299–324 (1987)
9. Momani, S., Odibat, Z.: Numerical comparison of methods for solving linear differential equations of fractional order. Chaos Solitons and Fractals 31, 1248–1255 (2007)
10. Ortiguera, M.D.: Riesz potential operations and inverses via fractional centred derivatives. International Journal of Mathematics and Mathematical Sciences, Article ID 48391, 1–12 (2006)
11. Yosida, K.: Functional Analysis. Springer, Berlin (1980)
12. Kaku, T.: Perturbation Theory for Linear Operator(Reprint). Springer, Berlin (1995)

Appendix

The Riemann-Liouville fractional differentiation

$$\frac{d^\alpha}{dx^\alpha} f(x) = \frac{1}{\Gamma(n-\alpha)} \frac{d^n}{dx^n} \int_0^t (t-\tau)^{n-\alpha-1} f(\tau) d\tau,$$

for $n = [\alpha] + 1$. involves the Cauchy integral formula, which is unstable to numerical implementation, since the formula involves a singular integral.

Let f_n and F_n for $0 \le n \le (N-1)$ be the discrete Fourier transform pair such that

$$F_n = \frac{1}{\sqrt{N}} \sum_{m=0}^{N-1} f_m \exp\left(-2\pi i \frac{mn}{N}\right), \quad f_n = \frac{1}{\sqrt{N}} \sum_{m=0}^{N-1} F_m \exp\left(2\pi i \frac{mn}{N}\right).$$

Since

$$\frac{1}{2}(f_{n+\frac{1}{2}} - f_{n-\frac{1}{2}}) = \frac{1}{\sqrt{N}} \sum_{m=0}^{N-1} i \sin\left(\pi \frac{m}{N}\right) F_m \exp\left(2\pi i \frac{mn}{N}\right),$$

we can compute

$$(D^\alpha f)_n = \frac{1}{\sqrt{N}} \sum_{m=0}^{N-1} \left(i \sin\left(\pi \frac{m}{N}\right)\right)^\alpha F_m \exp\left(2\pi i \frac{mn}{N}\right),$$

for the discrete difference operation D.

A Multi-relational Learning Approach for Knowledge Extraction in in Vitro Fertilization Domain

Teresa M.A. Basile, Floriana Esposito, and Laura Caponetti

Università degli Studi di Bari, Dipartimento di Informatica, Bari, Italy
{basile,esposito,laura}@di.uniba.it

Abstract. In the field of assisted reproductive technologies, ICSI fertilization is a medically-assisted reproduction technique, enabling infertile couples to achieve successful pregnancy. In this field crucial points are: the analysis of clinical data of the patient, aimed at adopting an appropriate stimulation protocol to obtain an adequate number of oocytes, and the selection of the best oocytes to fertilize. In this paper we would provide a framework able to extract useful morphological features from oocyte images that combined with the provided clinical data of the patients can be used to discover new information for defining therapeutic plans for new patients as well as selecting the most promising oocytes.

1 Introduction

Many assisted reproductive techniques have been designed to overcome the always more frequent problem of infertility. One of these techniques is the intracytoplasmic sperm injection (ICSI) in which a single sperm is directly injected into an oocyte. The fertilized oocyte grows in a laboratory for one to five days, then it is placed in the woman's uterus. Due to ethical and medical reasons a specified number of embryos have to be selected and hence transferred in woman's uterus. As a consequence, even the number of oocytes to fertilize could be under such a restriction and clinicians prefer to appropriately select the most promising oocytes among all the oocytes taken from the woman.

Generally, the oocytes selection is manually done by non-invasive examination of the morphology and dynamics of the oocyte. Indeed, a set of morphological parameters to be examined are present in medical literature such as oocyte/cytoplasm dimension, perivitelline space and zona pellucida thickness, first polar body conformation, and more subtle abnormalities of cytoplasm such as central granularity, inclusions and vacuolation.

However, these variables are not the unique and independent parameters involved in the process. Indeed, in general, before the ICSI procedure, a hormone stimulation protocol of the female patient, consisting of a set of pharmacological treatments, is carried out in order to ensure the development of multiple preovulatory follicles to obtain multiple oocytes to aspire. In this phase, the couples' health conditions have to be taken into account as well.

G. Bebis et al. (Eds.): ISVC 2010, Part I, LNCS 6453, pp. 571–581, 2010.

In this work we introduce a multi-relational learning approach able to deal with clinical data and relevant features extracted from oocyte images. The goal is to discover new information useful to support the clinicians both in the definition of the stimulation protocol for new unseen patients and in the selection of oocytes from new unseen oocytes. At this aim we have developed a system that exploits image processing techniques to extract useful morphological features from oocyte images and machine learning methods, that work on such features and combine them with the patients clinical data. Due to the presence of strong relationships among different stages of the process, multi-relational learning techniques that are able to take into account the relationships existing among all the entities involved in the process seem to be the most suitable approaches in this medical application domain. The approach consists of a multi-relational clustering followed by a multi-relational rule induction. The induced rules are represented in a easily comprehensible form and can be used as an advisor to the clinicians during their work in order to help them in determining what knowledge sources are relevant for a treatment plan.

Some works faced the problem of designing systems to support clinicians in this domain. Some approaches work with low level features extracted from oocyte images to assess their quality [1–3]. Other approaches work with higher level characteristics such as the clinical data of the patients with the aim of grasping structural patterns that define the peculiarities of the patient. Few approaches are presented in literature that work with both these kinds of information, but they assume that all the information on both features extracted from the images and the clinical data are available [4–6]. Furthermore, they exploit an attribute-value description of the data thus losing the relationships existing between oocytes and patients data. Indeed, an important aspect and commonly neglected by these approaches is that each set of variables, i.e. patient data and image features, cannot be considered as a stand-alone set since relationships between such sets of data can occur [7]. For example, clinical data of the patient are related both to the oocyte quality and to the implantation success rate; the oocyte quality, intended as its maturity, plays a fundamental role in the embryo development. For these reasons, multi-relational learning techniques seem to be the most suitable approaches in this application domain.

2   The Framework

The general framework we propose, depicted in fig. 1, is made up of a module devoted to image-based features extraction - based on mathematical morphology - and a module for knowledge extraction from both clinical and image features data - based on multi-relational learning techniques.

2.1   Image-Based Features Extraction

The features extraction module is oriented to extract some relevant morpho-structural features from oocyte images, such as the measures of oocyte and

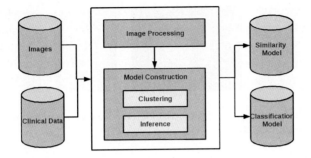

Fig. 1. Schematic representation of the proposed framework

cytoplasm diameters. This can be addressed as an image segmentation problem. Since we are interested in extracting the shape of the oocyte from the image, we employ a segmentation method better suited for shape analysis, that is mainly based on mathematical morphology [8].

Basic concept of mathematical morphology is the structuring element: given a two-dimensional binary image $X \subset Z^2$, a structuring element is a particular set $B \subset Z^2$, that gets translated over X and whose relations with X are studied at each location. In the following, we denote B_x the translation of B by x.

The basic operations of mathematical morphology are dilation and erosion. The dilation of an image $X \subset Z^2$ by a structuring element B, denoted by $\oplus$, is the set of points $x \in Z^2$ such that the translation of B by x has a non-empty intersection with set X: $X \oplus B = \{x \in Z^2 \mid X \cap B_x \neq 0\}$. The erosion of X by a structuring element B, denoted by $\ominus$, is the set of points $x \in Z^2$ such that the translation of B by x is included in X: $X \ominus B = \{x \in Z^2 \mid B_x \subseteq X\}$.

From the erosion and dilation operators, two fundamental morphological operations can be derived as follows: the opening of X by B, denoted by $\circ$, is the union of all the translations of the structuring element that fit inside the image X, i.e. $X \circ B = \bigcup\{B_x \mid B_x \subset X\} = ((X \ominus B) \oplus B)$. The dual operation of the opening is the closing, denoted by $\bullet$, which is defined as: $X \bullet B = ((X \oplus B) \ominus B)$.

On such operators we designed a procedure able to extract the region containing the oocyte, and its diameter, along with a good approximation of the cytoplasm diameter. Specifically, the proposed procedure works as follows.

Oocyte region detection. Firstly, an edge detection and a binarization steps are performed. After the binarization, elements that are not of interest surrounding the image borders, such as the holding and injection pipettes, have to be taken out. This is done by selecting a point p in the border region and, starting form it, by finding the connected components. This step uses an *extraction of connected components* algorithm [8] based on dilation and intersection of the set of pixels of the binary image.

At this point, the binary image shows segments of high contrast that do not quite delineate the outline of the object of interest. Indeed gaps in the segments surrounding the object are evident. These gaps will disappear as soon as the image is dilated twice using circular structuring elements.

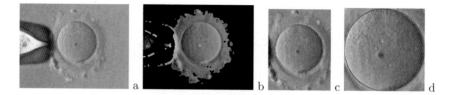

Fig. 2. Image Processing. (a) Original. (b) Detected region. (c) Oocyte region extraction. (d) Cytoplasm detection.

The dilated image shows the outline of the object quite nicely, but there are still holes in the interior of the object. The filling of these holes is performed by starting from a point p in the region to process and iteratively dilating it and intersecting the resulting dilation with the complement of the starting image [8]. Finally, in order to make the segmented object look natural, the region is smoothed by an opening-closing operation with a circular structuring element.

Now, by subtracting the obtained image from the original one, the region of the oocyte on a black-background is achieved (fig. 2b). Finally, in order to obtain the smallest rectangle that contains the object, the center of mass of the oocyte region is calculated and, starting from it, the 4-directional Euclidean distances, until a pixel background is encountered, are computed. The mean of these values represents the diameter of the oocyte and the minimum and maximum x and y coordinates of the 4-directional Euclidean distances are the starting points from which to extract the bounding rectangular region containing the oocyte (fig. 2c).

Cytoplasm region detection. As according to medical literature the cytoplasm dimension is about the 66% of the oocyte dimension [9], this value can be used to approximate the cytoplasm diameter. A more accurate measure has been obtained by considering that the shape of the cytoplasm can be approximated by a circumference. To this aim the Hough transform is applied to the binary image so as to detect the best circle fitting the shape of the oocyte cytoplasm. This has been done by searching for circles of radius r, varying from $(d/2 - \delta)$ to $(d/2 + \delta)$, where δ has been chosen equal the 10% of the oocyte dimension d. The resulting cytoplasm detection is shown in fig. 2d.

2.2   Knowledge Extraction

The knowledge extraction step involves the representation of both clinical and image-based data, extracted from oocyte images, and the application of the multi-relational learning approach to build a model able to solve the issues concerning the identification of similarities among situations such as stimulation protocols under specific patients'health conditions and, hence, the predictivity of the goodness of the oocytes to select as the most promising for the fertilization.

As to the data representation, the description language reported in Table 1 was exploited: for each entity involved in the domain, a set of descriptive attributes are reported along with the existing relationships. Specifically, the general information on a patient and the clinical data about the couple diseases is

Table 1. Attributes (bold) and Relations (italic) descriptors used to represent entities (patient P, protocol PR, hormone H, oocyte O, component C) and relationships

General information and clinical data	
age(P,val)	val: integer
bmi(P,val)	val: real
basal_FSH(P, val)	val: real
basal_LH(P, val)	val: real
male_infertility(P,inf)	inf: oligospermia, azoospermia, teratospermia
female_infertility(P,inf)	inf: tubaric, pcos, thyroid, uterine, endometrial
Stimulation Protocol Data	
stimulation_protocol(P,PR)	PR: nominal (performed protocol identification)
hormone_stimulation(PR, H)	H: agoGnRH,antagoGnRH,rFH,rLH,HMG,uFSH
dose(H,val)	val: real (provided dose)
timing(H,val)	val: real (duration of the stimulation type)
duration_of_stimulation(PR,val)	val: real (duration of the stimulation protocol)
estradiol_level(PR,value)	val: real (estradiol at the HCG injection day)
HCG_dose(PR,val)	val: real
aspiration_timing(PR,val)	val: real (hours from HCG injection to aspiration)
Oocyte data	
part_of(P,O)	O: nominal (oocyte identification)
dimension(O, val)	val: real (μm)
part_of(O, C)	C: cytoplasm
dimension(C, val)	val: real (μm)

followed by the data describing the ovarian stimulation protocol. Then, the data about the oocyte aspiration phase are introduced: For each patient a set of n (a value varying form one patient to another) oocytes is obtained and each oocyte is described according to the own features extracted from the images.

As to the knowledge extraction phase, we apply two submodules. The first one concerns the application of clustering techniques to identify similarities among patients. Indeed, the aggregation of patients that show a similar behavior could be useful to better understand the conditions under which a pregnancy could be obtained. Once the clustering has been taken place, for each cluster a set of rules are induced that will be able to identify relations between stimulation protocol and health conditions or between number and quality of oocytes obtained.

Due to the complexity of data in our application domain, multi-relational techniques were exploited. In particular, we use APAM [10] as it is very robust with respect to the existence of outliers. This is a fundamental characteristic for our application domain as clinicians can adopt very different stimulation protocols according to their experience and, more importantly, according to the patients' health conditions. Furthermore, the APAM algorithm is based on an approximate evaluation of the clustering membership thus allowing to tackle the uncertainty in the data. On the other hand, the induction process on the clustered data was performed by means of the incremental multi-relational inductive logic system [11] as its incremental capability makes it able to learn a satisfiable model even with few examples and, more importantly, to revise the learned rules as new examples are provided without restart the learning step from scratch.

In the following the multi-relational techniques exploited are briefly reported.

Multi-Relational Clustering. Clustering is an unsupervised learning technique used to find a partition of a set of objects into clusters so that the objects within each cluster are similar to each other. The similarity between objects can be determined using various distance measures. Relational clustering works on relational data and uses distance measures that are generally more complex than those used in the case of attribute-value representations. Indeed, the generic Euclidean distance cannot be applied to relational representations of the data as they are not represented by a feature vector of a fixed number of measurements.

Here we use the distance function and the modification of a partitional clustering algorithm introduced in [10] and here briefly reported.

As to the distance function, an adaptation of the Tanimoto metric to relational descriptions is exploited. Specifically, the Tanimoto metric adaptation to define the distance between two multi-relational descriptions D_1 and D_2 is:

$$d_{T_\cap}(D_1, D_2, \alpha) = \frac{|D_1| + |D_2| - 2s_\cap(D_1, D_2, \alpha)}{|D_1| + |D_2| - s_\cap(D_1, D_2, \alpha)},$$

where $|D_i|$ is the number of descriptors in D_i and $s_\cap(D_1, D_2, \alpha)$, the number of descriptors in common between D_1 and D_2, is approximated by the mean of the number of common descriptors in D_1 and D_2 for each of the α renamings of D_2. In this setting a renaming $R(D)$ of a multi-relational description D is a ground description obtained by firstly turning constants into variables in D and then applying a substitution (i.e., a mapping from variables onto a new set of constants) to the result. The set of renamings is generated randomly choosing k renamings of D onto the set of constants C.

As to the partitional clustering, the following generic schema is considered:

1. randomly choose k representatives for clusters;
2. iteratively improve these initial representatives until the change in the objective function from one iteration to the next drops below a given threshold:

    (a) assign each object to the cluster it "fits best" in the current clustering
    (b) compute new cluster representatives using these new assignments

One of the most well-known and commonly used partitioning method is the k-*medoids* clustering algorithm. Traditional k-medoids clustering algorithm seeks to find k medoids among the objects in the data set minimizing, for a given clustering solution $\mathcal{C}$, the following objective function:

$$tightness(\mathcal{C}) = \frac{1}{n} \sum_{i=1,\dots,n} d(\mathbf{x}_i, \mu_i)),$$

where μ_i is the medoid of the cluster $\mathbf{x}_i$ belongs to and $d(\cdot, \cdot)$ is the distance.

The k-medoids clustering algorithm PAM on which APAM is based starts with a set of clusters containing the medoids of the complete data set, and greedily inserts new objects into this set of clusters while minimizing the above objective function. Then, it tries to improve the previously obtained clustering by exploring all possible replacements of medoids by non-medoids picking the replacement that enhances the fitness function. If no such fitness improving replacement can be found, the procedure terminates.

APAM, the approximate relational clustering variant of PAM, uses the following objective function:

$$\mathcal{J}_{tightness}(\mathcal{C}, \alpha) = \frac{1}{n} \sum_{i=1,\ldots,n} d_{T_\cap}(\mathbf{x}_i, \mu_i, \alpha).$$

Similarly to PAM, it starts by randomly selecting k medoids and finding the first clustering solution $\mathcal{C}$ by associating each non-medoid instance to the cluster whose medoid is more similar. Then, it iteratively tries to swap a medoid with a non-medoid object, exploring all possible replacements, in order to minimize the value of the objective function. It terminates if no replacement can be found that leads to a clustering with a better (lower) objective value.

Multi-Relational Rule Induction. Rule induction is a supervised learning technique concerning the extraction of a set of formal rules from a set of labelled observations. One of the rule induction paradigms able to deal with multi-relational representation language is the Inductive Logic Programming (ILP) framework [12]. In this setting, given a background knowledge and a set of labelled (positive/negative) observations, the aim is to derive a set of rules (or theory) which entails all the positive and none of the negative observations. The rules induction is performed by exploring the lattice-based concepts by means of some operators such as least general generalisation and inverse resolution.

In this work we adopted the ILP system INTHELEX [11] to process the results obtained by the multi-relational clustering technique. It is a learning system for the induction of theories from positive and negative observations. It is fully and inherently incremental. This means that, in addition to the possibility of taking as input a previously generated version of the theory, learning can also start from an empty theory and from the first available observation; moreover, at any moment the theory is guaranteed to be correct with respect to all of the observations encountered thus far. Indeed, the correctness is checked on any new example and, in case of failure, a revision process is activated.

In the theory revision process the system exploits a previous theory (if any) and a memory of all the past (positive/negative) observations that led to the current theory. The new observations are exploited incrementally to modify incorrect hypotheses according to a data-driven strategy. In particular, when a positive observation is not covered, a revision of the theory to restore its completeness is performed as follows:

- replacing a rule in the theory with one of its least general generalizations against the problematic observation;
- adding a new rule to the theory, obtained by properly turning constants into variables in the problematic example;
- adding the problematic observation as a positive exception.

On the other hand, when a negative observation is covered, the system revises the theory to restore consistency by performing one of the following actions:

- adding *positive* information able to characterize all the past positive observations (and exclude the problematic one) to the rule that covers the example;
- adding *negative* information to discriminate the problematic observation from all the past positive ones to the rule that covers the problematic observation;
- adding the problematic observation as a negative exception.

3   Evaluation

The overall framework was tested on a preliminary set of data collected by the Department of Endocrinology and Molecular and Clinical Oncology of the University Federico II of Naples including clinical data of the patients along with the corresponding light microscope images of the oocytes. The dataset consisted of about 30 patients and 120 oocytes images.

The image processing devoted to the extraction of morpho-structural features from the oocyte images was able to correctly extract the region of interest (oocyte and cytoplasm) in the 90% of the cases. On about the 10% of the processed images the procedure doesn't work well due to very noisy images in which the background is confused whit the oocyte region. In the cases when the procedure correctly was able to extract the oocyte/cytoplasm region and hence the diameter, the standard deviation of the difference between the automatically extracted and the real manually measured diameter was about 11 μm, i.e. it represents about 0.06% of the real oocyte/cytoplasm measure.

As to concern the experimental outcomes of the applications of the multi-relational techniques, they revealed some interesting features that correlate health conditions to the stimulation protocol, that could confirm in some cases the medical literature. In particular the clustering was setted so to generate two clusters in order to differentiate good from not good practice. In particular, the first cluster, that we labelled as *protocol stimulation practice in order to obtain a greater number of oocytes,* has put together couples characterized by few female infertility conditions and many/severe male infertility ones. For such couples mainly long stimulation protocol was carried out resulting in a production of a mean of 6 oocytes for patient. This cluster characterization was confirmed by the rule induction step that exactly was able to grasp the concept as above reported by inferring rules such the one reported in fig. 3. This rule says that couples with some female infertility factors and severe male infertility factors, for which the patients were subjected to a *long* stimulation protocol, a great number of oocytes was obtained. Furthermore, it gives further information characterizing the obtained oocytes, i.e. that in such conditions almost all of them have medium oocyte/cytoplasm dimension.

On the other hand, the other cluster has aggregated couples characterized by many female infertility conditions and male infertility conditions of different seriousness. For such couples prevalently short stimulation protocols were carried out and a lesser number of oocytes were obtained (3 in average). This can be labelled as *protocol stimulation practice in order to obtain a lesser number of oocytes.* Even in this case, the rule induction phase was able to learn the concept as described in the found rule reported in the follows:

```
cluster1(Patient):-
    age(Patient,[26,30[), basal_fsh(Patient, [4,6[), oocyte_aspired(Patient, [5,7]),
    female_infertility(Patient,thyroid),
    stimulation_protocol(Patient,Protocol),
    hormon_stimulation(Protocol,agoGnRH),   specification(agoGnRH,long),
    hormon_stimulation(Protocol,antagoGnRH),specification(antagoGnRH,none),
    hormon_stimulation(Protocol,rFH),       specification(rFH,yes),
    hormon_stimulation(Protocol,uFSH),      specification(uFSH,none),
    hcg_dose(Protocol,[10000,12000]),
    aspiration_timing(Protocol,[35.5,36]),
    is_oocyte_of(Patient,Ovo1), dimension_ovo(Ovo1,[156,164[),
    is_oocyte_of(Patient,Ovo2), dimension_ovo(Ovo2,[156,164[),
    is_oocyte_of(Patient,Ovo3), dimension_ovo(Ovo3,[156,164[),
    is_oocyte_of(Patient,Ovo4), dimension_ovo(Ovo4,[156,164[).
```

Fig. 3. Sample learned rule

```
cluster2(Patient):-
    age(Patient, [35,40]),  oocyte_aspired(Patient, [2,4]),
    male_infertility(Patient,oligo), specification(oligo,normal),
    stimulation_protocol(Patient,PR_ID),
    hormone_stimulation(PR_ID,agoGnRH), specification(agoGnRH,short),
    hcg_dose(PR_ID,[10000,12000]), aspiration_timing(PR_ID,[35.5,36]).
```

This rule says that: *a patient belongs to the cluster2 iff the patient is between 35 and 40 years old, in the couple there is a male infertility problem, specifically an oligospermia characterized as normal, the patient was subjected to a stimulation protocol in which a short treatment of agoGnRH hormone was carried out, and an hcg dose in [10000,12000] UI was injected during the treatment. For these patients the aspiration time, i.e. the hours between the hcg injection and the oocytes aspiration is between 35.5 and 36 hours and a number of oocytes ranging form 2 to 4 was obtained.*

More interestingly, an in deep analysis of the the oocytes quality in the clustered data was performed according to the further information provided by the clinicians about the oocyte fertilization and growing. This analysis revealed that most of the data about oocytes in the first cluster concern poor quality oocytes as they do not grown in the following days after fertilization or they present an high fragmentation rate (> 10) at the embryo stage (fragmentation is a process where portions of the embryo's cells have broken off and are separate from the nucleated portion of the cell. According to the medical literature, little or no fragmentation are preferable as embryos with more than 25% fragmentation have a low implantation potential). On the contrary, the data in the second cluster regard good quality oocytes that in almost all of case grows in embryos and with a low rate of fragmentation (≤ 10) at embryo stage.

4   Conclusion and Future Work

In this paper an automatic tool to support ICSI applications is presented. The existing approaches work at different level according to the data starting point, i.e. images or clinical data. On the contrary, the proposed framework involves both data with the aim of putting together the automatically extracted morpho-structural data of the image and the clinical data with the aim of further elaboration steps devoted to discovery relationships among data.

Future work will concern the extension of the clinical data to elaborate by considering more parameters in both the stimulation protocol and in the definition of health conditions, and the extension of the image processing module in order to extract more features from the oocyte images and from other images that follow the oocyte development after fertilization, i.e. zygote and embryo images. Finally, an exhaustive experimental phase is planned to be carried out.

Acknowledgment

This work is partially supported by the italian project "MIUR-FAR 08-09: The Molecular Biodiversity LABoratory Initiative (MBLab-DM19410)."

References

1. Griffin, J., Emery, B.R., Huang, I., Peterson, M.C., Carrell, D.T.: Comparative analysis of follicle morphology and oocyte diameter in four mammalian species. Journal of Experimental & Clinical Assisted Reproduction 3, 2 (2006)
2. Losa, G., Peretti, V., Ciotola, F., Cocchia, N., De Vico, G.: The use of fractal analysis for the quantification of oocyte cytoplasm morphology. In: Fractals in Biology and Medicine, pp. 75–82 (2005)
3. Montag, M., Schimming, T., Kster, M., Zhou, C., Dorn, C., Rsing, B., van, d.V.H., Ven, d.V.K.: Oocyte zona birefringence intensity is associated with embryonic implantation potential in icsi cycles. Reprod. Biomed. Online 16, 239–244 (2008)
4. Morales, D.A., Bengoetxea, E., Larrañaga, P., García, M., Franco, Y., Fresnada, M., Merino, M.: Bayesian classification for the selection of in vitro human embryos using morphological and clinical data. Computer Methods and Programs in Biomedicine 90, 104–116 (2008)
5. Uyar, A., Ciray, H.N., Bener, A., Bahceci, M.: 3P: Personalized pregnancy prediction in ivf treatment process. In: Weerasinghe, D. (ed.) eHealth, Social Informatics and Telecommunications Engineering. Lecture Notes of the Institute for Computer Sciences, vol. 1, pp. 58–65. Springer, Heidelberg (2009)
6. Morales, D.A., Bengoetxea, E., Larranaga, P.: XV-Gaussian-Stacking Multiclassifiers for Human Embryo Selection. In: Data Mining and Medical Knowledge Management: Cases and Applications, IGI Global Inc. (2009)
7. Rjinders, P., Jansen, C.: The predictive value of day 3 embryo morphology regarding blastocysts formation, pregnancy and implantation rate after day 5 transfer following in-vitro fertilisation or intracytoplasmic sperm injection. Hum. Reprod. 13, 2869–2873 (1998)
8. Gonzalez, R.C., Woods, R.E.: Digital Image Processing, 3rd edn. Prentice-Hall, Inc., Upper Saddle River (2006)
9. Veek, L.L.: An Atlas of Human Gametes and Conceptuses: An Illustrated Reference for Assisted Reproductive Technology. Parthenon (1999)

10. Di Mauro, N., Basile, T., Ferilli, S., Esposito, F.: Approximate relational reasoning by stochastic propositionalization. In: Ras, Z.W., Tsay, L.S. (eds.) Advances in Information and Intelligent Systems. Studies in Computational Intelligence, vol. 265, pp. 81–109. Springer, Heidelberg (2010)
11. Esposito, F., Ferilli, S., Fanizzi, N., Basile, T.M.A., Di Mauro, N.: Incremental learning and concept drift in INTHELEX. Intell. Data Anal. 8, 213–237 (2004)
12. Muggleton, S., De Raedt, L.: Inductive logic programming: Theory and methods. Journal of Logic Programming 19(20), 629–679 (1994)

Reconstruction of Spectra Using Empirical Basis Functions

Jakob Bärz, Tina Hansen, and Stefan Müller

University of Koblenz, Germany

Abstract. Physically-based image synthesis requires measured spectral quantities for illuminants and reflectances as part of the virtual scene description to compute trustworthy lighting simulations. When spectral distributions are not available, a method to reconstruct spectra from color triplets needs to be applied. A comprehensive evaluation of the practical applicability of previously published approaches in the context of realistic rendering is still lacking. Thus, we designed three different comparison scenarios typical for computer graphic applications to evaluate the suitability of the methods to reconstruct illumination and reflectance spectra. Furthermore, we propose a novel approach applying empirical mean spectra as basis functions to reconstruct spectral distributions. The mean spectra are derived from averaging sets of typical red, green, and blue spectra. This method is intuitive, computationally inexpensive, and achieved the best results for all scenarios in our evaluation. However, reconstructed spectra are not unrestrictedly applicable in physically-based rendering where reliable synthetic images are crucial.

1 Introduction

In physically-based image synthesis, the natural appearance of a virtual scenario is predicted by simulating the distribution of light. To guarantee reliable and trustworthy results, a spectral rendering algorithm with measured radiometric input data for light sources and materials is obligatory. Especially in product design as a typical field of application of physically-based rendering, decision-makers rely on photometrically and colorimetrically consistent images. However, when spectral data is not available due to expensive measurement devices, legacy RGB data or efficiency reasons, e.g. measuring textures or environment maps using RGB cameras, spectra need to be derived from colors in 3D color spaces.

A number of promising approaches to compute spectra from color triplets have been presented in the past. The key problem when converting colors into spectra is that an infinite number of metameric spectra [1] with the same color coordinates exist. But, in order to apply reconstructed spectra in a physically-based rendering context, not only the luminance and chromaticity values of these spectra are to be indistinguishable from the original spectra. Also, using reconstructed spectral distributions for light sources or materials needs to exhibit a reliable photometric and colorimetric appearance of the reflected spectra resulting from interactions between light and matter. A systematic evaluation of existing methods addressing this issue has not yet been published.

G. Bebis et al. (Eds.): ISVC 2010, Part I, LNCS 6453, pp. 582–591, 2010.

We present a novel approach to reconstruct spectra from RGB triplets using empirical basis functions derived from averaged sets of typical red, green, and blue spectra. In a comprehensive evaluation, we compared our approach with different previously proposed methods. To account for better practical applicability in physically-based rendering, three scenarios with diverse foci were investigated. The first scenario is a direct comparison of ground truth spectra and reconstructed spectra. The input RGB triplets are derived from the ground truth by performing a spectrum to XYZ conversion followed by converting the XYZ tristimuli to RGB values. In the second scenario, a reflectance spectrum is illuminated by different light spectra applying both ground truth and reconstructed spectra before comparing the resulting spectra to evaluate reliability with respect to varying lighting conditions. The third scenario was designed to investigate the applicability of reconstructed spectra as spectral radiance distributions of light sources. To generate perceptually meaningful results, all comparisons were carried out by computing the 1976 CIELAB difference between the original and reconstructed spectra.

The remainder of the paper is organized as follows. Section 2 reviews existing approaches to reconstruct spectra and related problems. Our method is outlined in section 3 and the evaluation scenarios are described in section 4. The results follow in section 5, the conclusion in section 6.

2   Related Work

A general approach to create spectra from RGB values is to transform the color triplet into a color space, where the basis functions are known. Let $\bar{x}(\lambda)$, $\bar{y}(\lambda)$, and $\bar{z}(\lambda)$ be the 1931 CIE XYZ color matching functions and M_{rgb_xyz} the matrix transforming values from a well-defined RGB color space into the CIE XYZ color space, then the color coordinates $W = (w_1, w_2, w_3)^T$ in the 3D color space defined by arbitrary linearly independent basis functions $f_1(\lambda)$, $f_2(\lambda)$, and $f_3(\lambda)$ are derived from RGB value $C = (r, g, b)^T$ by

$$\begin{bmatrix} w_1 \\ w_2 \\ w_3 \end{bmatrix} = \begin{bmatrix} \int f_1(\lambda)\bar{x}(\lambda)d\lambda & \int f_2(\lambda)\bar{x}(\lambda)d\lambda & \int f_3(\lambda)\bar{x}(\lambda)d\lambda \\ \int f_1(\lambda)\bar{y}(\lambda)d\lambda & \int f_2(\lambda)\bar{y}(\lambda)d\lambda & \int f_3(\lambda)\bar{y}(\lambda)d\lambda \\ \int f_1(\lambda)\bar{z}(\lambda)d\lambda & \int f_2(\lambda)\bar{z}(\lambda)d\lambda & \int f_3(\lambda)\bar{z}(\lambda)d\lambda \end{bmatrix}^{-1} M_{rgb\_xyz} \begin{bmatrix} r \\ g \\ b \end{bmatrix} \quad (1)$$

Weighting the basis functions by the coordinates yields spectrum $S(\lambda)$ with color sensation C:

$$S(\lambda) = w_1 f_1(\lambda) + w_2 f_2(\lambda) + w_3 f_3(\lambda) \quad (2)$$

A number of different basis functions have been suggested in the past. To generate spectra from designer input or existing libraries in RGB color space, Glassner [2] defines a three-dimensional space of monochromatic line spectra with an orthonormal base of axis given by the delta functions $f_1(\lambda) = \delta_u(\lambda)$, $f_2(\lambda) = \delta_v(\lambda)$, and $f_3(\lambda) = \delta_w(\lambda)$, with $\lambda_{max} \geq \lambda_u > \lambda_v > \lambda_w \geq \lambda_{min}$.

The resulting spectra may contain physically implausible negative values and are empty except for three discrete wavelengths. A solution for the latter is a multiple execution of the procedure, applying a different orthonormal base choosing varying nonzero wavelengths for the delta functions each. Averaging the resulting spectra yields a fuller spectrum, which is a metamer of its components. Alternatively, Glassner proposes the primaries of a monitor [3] or the first three functions from a Fourier basis as basis functions. In 1985, Wandell [4] proved that equal energy spectrum $f_1(\lambda) = 1.0$, a single cycle of sine $f_2(\lambda) = sin(2\pi(\lambda - \lambda_{min})/(\lambda_{max} - \lambda_{min}))$, and a single cycle of cosine $f_3(\lambda) = cos(2\pi(\lambda - \lambda_{min})/(\lambda_{max} - \lambda_{min}))$ suffice to match a number of 462 Munsell color chips with an RMS error of about 0.02.

The conversion of a camera RGB color signal into a spectrum has been addressed by Drew et al [5] in 1992. Four different sets of basis functions were investigated. Firstly, the 1931 CIE standard observer color-matching functions as in Horn [6] and secondly the Fourier functions [4] were applied as basis functions. Thirdly, Drew at al [5] used basis function products of illumination $E_i(\lambda)$ and reflectance $S_i(\lambda)$ functions: $f_1(\lambda) = E_1(\lambda)S_1(\lambda)$, $f_2(\lambda) = E_1(\lambda)S_2(\lambda)$, and $f_3(\lambda) = E_2(\lambda)S_1(\lambda)$. As basis set for illumination, Judd et al's daylight functions [7], determined by a principal component analysis of 622 measured samples, were employed. The reflectance functions were derived from the Krinov catalogue [8] of 370 natural reflectances applying the Karhunen-Loève [9, p. 275] analysis. Lastly, a fourth set of basis functions was computed from a large collection of 1710 natural color signals being synthesized by multiplying five typical daylight spectra from Judd et al [7] with 342 reflectance spectra from Krinov [8]. The three most important vectors accounting for the highest variability in the whole set were again determined by the KLT. According to Drew at al [5], the best results were delivered by the last two methods, but the error increased when applying the methods to a set of color signals not being included in the PCA.

Sun et al [10] presented a method to derive spectra from colors in the context of rendering the wavelength-dependent phenomenon of light interference. Since previous approaches were unable to generate spectra of rich shape as exist in nature, this method applies Gaussians as basis functions that are centered at the wavelengths of red, green, and blue with their width depending on the saturation of the colors to reproduce. Specifically, the three Gaussian basis functions are given by $f_i(\lambda) = e^{-ln2[2(\lambda - \lambda_{c,i})/w_i]^2}$ with $i \in 1, 2, 3$ and $\lambda_{c,i}$ representing the center and w_i the width of the ith Gaussian function. While the centers are chosen at 680, 550, and 420nm, the widths depend on the local saturation of the color to be reproduced. Sun et al compared the Gaussian to the Fourier basis functions by computing the difference between the input color and the color resulting from applying the respective color matching functions to the reconstructed spectrum. Sun reports a lower relative error for the Gaussian basis functions, especially for highly saturated colors.

In 1999, Smits [11] published a solution to convert existing RGB data into plausible reflectance spectra for spectral-based rendering systems. It is emphasized that creating metameric spectra with the exact RGB values is a well-defined problem, but illuminating these spectra with non-equal energy light yields indirect spectra, which are usually no metamers. Thus, Smits formulates a number of constraints for good metamers: The reflectance spectra are piecewise constant functions defined from 360 to 800nm and lie between 0 and 1 to be physically plausible. They should be smooth and have a minimum amount of variation for computational reasons. Smits [11] proposes to express spectra as a combination of as much white as possible, followed by as much of a secondary color out of cyan, magenta, or yellow as possible, and one primary color out of red, green, or blue. The smoothness of the basis functions is guaranteed by minimizing the differences between adjacent samples.

Another approach to generate realistic spectra for reflectances was proposed by Wang et al [12] in 2004. The authors measured a set of 1400 physical reflectances and created a number of more saturated colors applying Bouguer's law [13]. These spectra serve as basis functions to reconstruct any color P by trilinear interpolation. Therefore, eight subspaces in the color space of P are constructed by computing three planes that pass through P and are parallel to the xy-, xz-, and yz-plane, respectively. For each subspace, the color point closest to P is selected. Four of these eight points serve to build the smallest tetrahedron containing P. Finally, trilinear interpolation of the four spectra corresponding to the corner points of the tetrahedron yields the desired spectrum for P.

Bergner et al [14] presented a tool to design reflected spectra as a product of light and reflectance as a part of the scene description. The design principles are: A combination of light and reflectance should produce the color sensation requested by the user, the second order difference of the resulting spectrum should be minimized to guarantee smoothness, and the spectrum should be positive. The design problem is expressed by $Q \operatorname{diag}(E) W = C$, where E is the illumination spectrum and $\operatorname{diag}(E)$ the diagonal matrix containing E on the main diagonal. W is the reflectance producing the desired color sensation C under illumination E. Matrix Q is the product of the hardware-specific matrix Q_{rgb_xyz} to transform RGB to XYZ values and matrix Q_{xyz_spec} with the rows containing the CIE XYZ color matching functions. Asking for multiple lights E_k to produce colors C_k is solved by vertically concatenating the multiple $Q diag(E_k)$ to Q^{E_k} and the multiple C_k to Y. The solution for W is found by minimizing $Q^{E_k} W - Y$.

Recently, Rump et al [15] emphasized the need of spectral data for exact color reproduction in lighting simulations. Given a dense RGB and a sparse spectral sampling from a line scanner of a texture or environment map, their approach is able to infer full spectral data. The authors derive the spectral image by minimizing the euclidean distance between the desired and the measured spectrum for each pixel as well as between the corresponding RGB color of the desired spectra and the measured RGB data. Furthermore, neighboring spectra with respect to intensity-free euclidean distance in RGB color space are forced to be close to each other. To evaluate the resulting spectral data, the environment maps were applied to illuminate the Munsell ColorChecker chart and the textures were illuminated by ten different light sources prior to computing the 1994 CIELAB

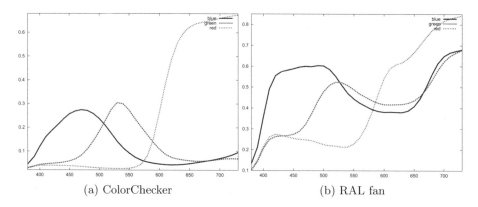

(a) ColorChecker   (b) RAL fan

Fig. 1. Mean spectra derived from the ColorChecker chart and RAL fan

color difference between the approximated and the ground truth spectra. The ΔE^*_{94} error was acceptable selecting reasonable line scans.

3   Approach

This work was motivated by research on physically-based lighting simulation in the context of product design where a predictive appearance of the synthetic images is crucial. Since spectral rendering is required to achieve photometric and colorimetric consistency, the problem of incorporating existing RGB data for environment maps or textures needed to be addressed. To choose the best state-of-the-art method, a systematic evaluation of previous approaches to reconstruct spectra from color triplets was still lacking. Furthermore, existing algorithms either delivered physically implausible and significantly negative results [2] [4], required a priori knowledge of a set of spectra and costly pre-processing [5] [12] or included unintuitive computation [10]. Our approach was motivated by the simplicity of Smits [11] in combination with the basis function model [2] and the idea of interpolating physical basis functions [12].

Contrary to [2], [10], and [4], where arbitrary base functions were applied, we used empirically determined mean spectra for red, green, and blue. A subset of spectra for red, green, and blue was selected subjectively from the measured spectra of the Munsell Digital ColorChecker SG chart with 140 patches. Averaging each subset yields the mean spectrum serving as the base spectrum for the corresponding color. This process was repeated with a RAL fan containing 126 spectra, to obtain a set of different base spectra to evaluate the dependency of the results from the training data. The base spectra are displayed in figure 1. Let $r^d_i(\lambda)$, $g^d_j(\lambda)$, and $b^d_k(\lambda)$ be subjectively selected sets of measured red, green, and blue spectra from data set $d \in \{\text{ColorChecker}, \text{Ral fan}\}$, then the basis functions are given by

$$R^d(\lambda) = \frac{1}{I}\sum_{i=1}^{I} r^d_i(\lambda), \ G^d(\lambda) = \frac{1}{J}\sum_{j=1}^{J} g^d_j(\lambda), \ B^d(\lambda) = \frac{1}{K}\sum_{k=1}^{K} b^d_k(\lambda) \qquad (3)$$

Setting $f_1(\lambda) = R^d(\lambda)$, $f_2(\lambda) = G^d(\lambda)$, and $f_3(\lambda) = B^d(\lambda)$ in equation 1 converts a given RGB tristimulus C into color W in the 3D space defined by the basis functions. Applying the weights W to the basis functions as in equation 2 yields the desired spectrum $S(\lambda)$ with color sensation C.

4   Evaluation

To evaluate the different approaches to reconstruct spectra, we investigated three scenarios with diverse foci. The scenarios were designed to determine the practical applicability of the resulting spectral distributions for light sources or materials in a lighting simulation. In each scenario, we applied the same test cases to ground truth spectra and reconstructed counterparts for better comparability. To generate perceptually meaningful results, we transformed both ground truth and reconstructed spectra into the CIELAB color space and computed the 1976 CIELAB color difference [16] between the resulting colors. The ΔE_{ab}^* distance is a measure for the difference between two stimuli perceived by the same observer. The human eye is unable to recognize the difference of two colors with a distance below the *just noticeable difference (JND)* of 2.3 [17]. The evaluation included our approach applying empirical basis functions by averaging typical spectra from the ColorChecker chart (in the following denoted as Emp_{CC}) and from the RAL fan with 126 spectra (Emp_{RAL}). We compared our approach to Smits [11] and the methods using Fourier [4], Gaussian [10], and CIE XYZ color matching functions [6] as basis functions. All spectra were represented as vectors containing 71 samples at 5nm spacing from 380 to 730nm.

In the first scenario, we compared the reconstructed with the ground truth spectra directly. For each spectrum of the Munsell Digital ColorChecker SG chart and the two different RAL fans, we determined the respective sRGB value serving as input for the different reconstruction methods. The reconstructed spectrum and the ground truth spectrum were compared by computing the ΔE_{ab}^* error. As reference white, illuminant E was applied in this scenario. The second scenario investigated the quality of a reflected spectrum after illuminating a reconstructed material spectrum with a reconstructed spectrum of a natural light source. The spectra for the light source (D50, D65, or E) and the material were reconstructed as described in Scenario I. Afterward, the reconstructed material was illuminated by the reconstructed light and the ground truth material by the ground truth light, respectively. Between the two resulting reflected spectra, ground truth and reconstructed, the ΔE_{ab}^* error was calculated. The reference white was set to the respective light spectrum.

The third scenario was designed to determine the applicability of reconstructing illuminant spectra. For instance, this situation occurs when an environment map captured by an RGB camera is reconstructed and serves as light source for the scene. The reconstructed light spectra were applied to illuminate the reconstructed reflection spectra. For the resulting reflected spectra, we calculated the ΔE_{ab}^* error to determine the color difference to the ground truth counterparts. Notably, we chose a D65 spectrum as reference white in this scenario to

Table 1. CIELAB differences for the three scenarios

	data set	error	Fourier	Gauss	Smits	XYZ	Emp$_{CC}$	Emp$_{RAL}$
Scenario I	CC140	mean	1.23	9.52	34.28	1.55	0.76	0.45
		max	17.00	89.25	86.28	21.26	8.03	8.65
		min	0	0	3.42	0	0	0
	RAL	mean	0.08	5.16	19.35	0.83	1.30	0.08
		max	4.32	49.81	59.75	11.78	7.90	0.83
		min	0	0	2.76	0	0	0
	RAL-K5	mean	1.53	12.24	37.04	2.16	0.60	0.53
		max	38.25	104.28	91.16	32.30	19.89	17.17
		min	0	0	4.31	0	0	0
Scenario II	CC140	mean	1.69	10.12	36.50	5.45	3.16	1.44
		max	17.59	90.08	90.16	26.63	9.95	11.31
		min	0	0	7.83	0.29	0.38	0.06
	RAL	mean	0.75	6.17	22.91	5.11	3.59	1.35
		max	5.62	52.77	64.64	16.70	10.02	3.84
		min	0	0.01	7.12	0.34	0.76	0.04
	RAL-K5	mean	1.89	12.90	38.22	5.98	2.93	1.50
		max	37.47	107.82	94.59	40.80	21.31	15.40
		min	0	0.01	8.45	0.15	0.20	0.06
Scenario III	CC140	mean	12.39	28.25	35.75	15.41	14.00	11.97
		max	156.82	242.86	119.19	186.76	136.11	152.43
		min	0	0	1.19	0.17	0.23	0.06
	RAL	mean	11.21	22.07	23.15	14.89	15.09	11.38
		max	75.82	121.80	71.61	76.73	71.78	65.40
		min	0.02	0.04	0.58	0.21	0.57	0.08
	RAL-K5	mean	15.09	33.37	34.61	18.61	16.26	14.31
		max	143.05	223.22	146.87	126.61	114.10	113.99
		min	0	0	0.85	0.06	0.09	0.01

account for the fact that a real setup will most likely incorporate many different light sources. Thus, the resulting error is higher as though applying the actual illuminant as reference white.

5   Results

The results in this section were calculated for the three different scenarios described in section 4. For each scenario, we applied three different sets of spectral data. The first data set was the Munsell Digital ColorChecker SG chart with 140 color patches. The other two data sets were measured from the RAL fan containing 126 and the RAL-K5 fan with 211 spectra, respectively. For each method and data set the minimum and maximum ΔE_{ab}^*, and the mean ΔE_{ab}^* distance are displayed in table 1. For the results of scenario I, the ΔE_{ab}^* errors for the basis function approaches is expected to be zero by definition. However, since the reconstructed spectra were forced to be physically plausible by cropping

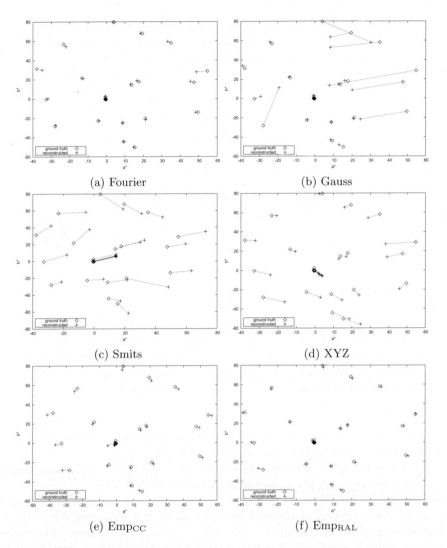

Fig. 2. Results for Scenario II projected onto the a*b* plane

negative values to zero and values greater one to one, the mean error is a good measure for the plausibility of the reconstruction methods. While the Emp_{CC}, Emp_{RAL}, Fourier, and XYZ basis functions delivered good overall results for all data sets, the Gauss function's mean error was too high, and Smits could not compete. Notably, Emp_{RAL} outperforms Emp_{CC} in terms of mean error even for the ColorChecker data which served as training set for the latter. Even for this simple scenario, the maximum error of all methods for nearly every data set lies above the *JND*. The only exception is Emp_{RAL}, applied on its training data, despite just a few representative red, green, and blue spectra were selected to derive the mean functions.

In scenario II, where the reconstructed spectra were illuminated by the reconstructed counterparts of the smooth natural standard light sources D65, D50 and E, the ΔE^*_{ab} error tends to be slightly higher overall. In this scenario, only the mean values of Fourier and Emp_{RAL} are still below the *JND* threshold with the latter performing better on average. No method was able to reproduce every spectrum of the data sets faithfully. Again, the choice of the training data does not seem to influence the outcome of our approach. For better comparability, the results of this scenario were projected onto the a*b* plane in figure 2. The results of the third scenario reveal that none of the methods was able to deliver a mean error below the *JND* for reflectance spectra illuminated by arbitrary lights. Still, Emp_{RAL} performs best, followed by Fourier, Emp_{CC}, and XYZ. Gauss and Smits delivered the worst results. Arguably, the reconstruction approaches are not applicable in scenarios with arbitrary illumination or multiple reflections when a trustworthy appearance is crucial. For a decent estimation, Emp_{RAL} exhibits a mean value below fifteen for all data sets even in this worst case setup.

6   Conclusion

Motivated by the issue of incorporating existing RGB data for illuminants and reflectances into physically-based image synthesis for product design, this work focused on comparing previously published methods to reconstruct spectra from color triplets. We conducted a comprehensive evaluation based on practical applicability in a spectral rendering context. Thus, we designed three typical scenarios for computer graphics: A direct comparison of ground truth and reconstructed spectra, an evaluation of reconstructed reflectances illuminated by reconstructed standard light sources and a scenario to rate the suitability of reconstructing arbitrary light sources to illuminate reconstructed reflectances.

Furthermore, the evaluation included our novel approach to reconstruct spectra applying mean red, green, and blue spectra as basis functions. These spectra are averaged from representative sets of typical red, green, and blue spectra from the Munsell Digital ColorChecker SG chart or a RAL fan, for instance. Our method is simple, intuitive, and computationally inexpensive. The evaluation revealed a good overall performance for this new method when selecting reasonable spectra to derive the basis functions. Notably, the results indicate no significant dependency between the choice of the training data and the test data.

In general, reconstructing spectra from color triplets is not unrestrictedly applicable in physically-based lighting simulation setups. When reconstructing spectra for both arbitrary light sources and reflectances, the resulting reflected spectra are not guaranteed to be indistinguishable from the reflected spectra computed with ground truth spectra for light and material. Thus, radiometrically measured spectral data is required as input to generate truly predictive synthetic images that decision-makers can rely on. Future work might focus on selecting good sets of representative red, green, and blue spectra to generate basis functions yielding lower errors on average. Investigations should include the evaluation whether the reconstructed spectra relate positively with the training data for specific applications, e.g. choosing the mean spectra from sets of car paints when rendering automobiles.

References

1. Hunt, R.: The Reproduction of Color. Wiley & Sons, New York (1975)
2. Glassner, A.S.: How to derive a spectrum from an rgb triplet. IEEE Comput. Graph. Appl. 9, 95–99 (1989)
3. Glassner, A.S.: Principles of Digital Image Synthesis. Morgan Kaufmann Publishers Inc., San Francisco (1994)
4. Wandell, B.A.: The synthesis and analysis of color images. IEEE Trans. Pattern Anal. Mach. Intell. 9, 2–13 (1985)
5. Drew, M.S., Funt, B.V.: Natural metamers. CVGIP: Image Underst 56, 139–151 (1992)
6. Horn, B.K.P.: Exact reproduction of colored images. Computer Vision, Graphics, and Image Processing 26, 135–167 (1984)
7. Judd, D.B., MacAdam, D.L., Wyszecky, G., Budde, H.W., Condit, H.R., Henderson, S.T., Simonds, J.L.: Spectral distribution of typical daylight as a function of correlated color temperature. J. Opt. Soc. Am. 54, 1031–1036 (1964)
8. Krinov, E.: Spectral reflectance properties of natural formations. Technical Translations (1947)
9. Tou, J.T., Gonzalez, R.C.: Pattern recognition principles [by] Julius T. Tou [and] Rafael C. Gonzalez. Addison-Wesley Pub. Co., Reading (1974)
10. Sun, Y., Fracchia, F.D., Calvert, T.W., Drew, M.S.: Deriving spectra from colors and rendering light interference. IEEE Comput. Graph. Appl. 19, 61–67 (1999)
11. Smits, B.: An rgb-to-spectrum conversion for reflectances. J. Graph. Tools 4, 11–22 (1999)
12. Wang, Q., Xu, H., Sun, Y.: Practical construction of reflectances for spectral rendering. In: Proceedings of the 22th International Conference in Central Europe on Computer Graphics, Visualization and Computer Vision, pp. 193–196 (2004)
13. Evans, R.M.: An introduction to color. J. Wiley, New York (1961)
14. Bergner, S., Drew, M.S., Möller, T.: A tool to create illuminant and reflectance spectra for light-driven graphics and visualization. ACM Trans. Graph. 28, 1–11 (2009)
15. Rump, M., Klein, R.: Spectralization: Reconstructing spectra from sparse data. In: Lawrence, J., Stamminger, M. (eds.) SR 2010 Rendering Techniques, Saarbruecken, Germany, Eurographics Association, pp. 1347–1354 (2010)
16. Wyszecki, G., Stiles, W.S.: Color Science: Concepts and Methods, Quantitative Data and Formulae, 2nd edn. Wiley Series in Pure and Applied Optics. Wiley-Interscience, Hoboken (2000)
17. Mahy, M., Van Eycken, L., Oosterlinck, A.: Evaluation of uniform color spaces developed after the adoption of cielab and cieluv. Color Res. Appl. 19, 105–121 (1994)

Experimental Study on Approximation Algorithms for Guarding Sets of Line Segments

Valentin E. Brimkov[1], Andrew Leach[2], Michael Mastroianni[2], and Jimmy Wu[2]

[1] Mathematics Department, SUNY Buffalo State College, Buffalo, NY 14222, USA
brimkove@buffalostate.edu
[2] Mathematics Department, University at Buffalo, Buffalo, NY 1426-2900, USA
{ableach,mdm32,jimmywu}@buffalo.edu

Abstract. Consider any real structure that can be modeled by a set of straight line segments. This can be a network of streets in a city, tunnels in a mine, corridors in a building, pipes in a factory, etc. We want to approximate a minimal number of locations where to place "guards" (either men or machines), in a way that any point of the network can be "seen" by at least one guard. A guard can see all points on segments it is on (and nothing more). As the problem is known to be NP-hard, we consider three greedy-type algorithms for finding approximate solutions. We show that for each of these, theoretically the ratio of the approximate to the optimal solution can increase without bound with the increase of the number of segments. Nevertheless, our extensive experiments show that on randomly generated instances, the approximate solutions are *always* very close to the optimal ones and often are, in fact, optimal.

Keywords: guarding set of segments, art gallery problem, approximation algorithm, set cover, vertex cover.

1 Introduction

In recent decades, and especially in the last one, security issues of diverse nature are becoming increasingly important. In particular, in many occasions 24 hour surveillance is carried out for various purposes. It can be secured either through observers (e.g., policemen, guards, etc.) or by technology (including observing cameras or other sensors). Both ways are costly, therefore it is desirable to optimize the number of people or devices used. In this paper we consider a model of such an optimization problem.

Consider any real structure that can be modeled by a set of straight line segments. This can be a network of streets in a city, tunnels in a mine or corridors in a building, pipes in a factory, etc. We want to find a minimal (or close to the minimal) number of locations where to place "guards" (either men or machines), in a way that any point of the network can be "seen" by at least one guard. Alternatively, we can view this problem as finding a minimal number of devices

G. Bebis et al. (Eds.): ISVC 2010, Part I, LNCS 6453, pp. 592–601, 2010.

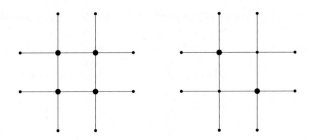

Fig. 1. *Left:* Any minimal vertex cover of the given plane graph requires four vertices. One of them is marked by thick dots. *Right:* Two vertices can guard the same graph. One optimal solution is exhibited.

to place so that a user has access (or a connection) to at least one device from any location. In terms of our mathematical model, we look for a minimum (or close to the minimum) number of points on the given segments, so that every segment contains at least one of them. We call this problem *Guarding a Set of Segments* (GSS).

We remark that GSS belongs to the class of the *art gallery problems*. A great variety of such problems have been studied for at least four decades. Related studies have started much earlier by introducing the concepts of starshapedness and visibility (see [3,7]). The reader is referred to the monograph of Joseph O'Rourke [8] and the more recent one of Jorge Urrutia [10]. See also [1,6] and the bibliography therein for a couple of examples of art-gallery problems defined on sets of segments.

GSS is germane to the set cover (SC) and vertex cover (VC) problems, which are fundamental combinatorial problems that play an important role in complexity theory. GSS can be formulated as a special case of the set cover problem (see Section 2) and, under certain conditions, as a vertex cover problem, as well. However, in general, GSS and VC are different, as Figure 1 demonstrates. It is well-known that both SC and VC are NP-complete [5,4]. In a recent work [2] we proved that GSS is NP-complete as well. This suggests that it is unrealistic to expect any efficient algorithm for finding the exact optimal solution. Therefore, in the present paper we study GSS' approximability, both theoretically and through massive experimentation.

To this end, we consider three greedy-type algorithms for finding approximate solutions of GSS. We show that for each of these, theoretically the ratio of the approximate to the optimal solution can increase without bound with the increase of the number of segments. However, our extensive experiments demonstrate that on randomly generated instances the approximate solutions are *always* very close to the optimal ones, and often are, in fact, optimal.

The paper is organized as follows. The next section includes certain preliminaries useful for understanding the rest of the paper. In Section 3 we present the three approximation algorithms and study their theoretical performance.

Section 4 presents our experimental study. We conclude with some final remarks in Section 5.

2   Preliminaries: Vertex Cover, Set Cover, and GSS

Given a universe set U and an arbitrary family of subsets $F \subseteq \mathcal{P}(U)$, the optimization *set cover* problem looks for a minimum cover $C \subseteq F$ with $\bigcup C = U$.

Given a graph G, a *vertex cover* of G is a set C of vertices of G, such that every edge of G is incident to at least one vertex of C. The optimization vertex cover problem has as an input a graph $G = (V, E)$, and one looks for a vertex cover with a minimum number of vertices.

Now let $S = \{s_1, s_2, \ldots, s_n\}$ be a set of segments in the plane. Denote by $\bar{S}$ the set of all points of segments in S and by V the set of all intersection points of segments of S. The elements of V will be called *vertices* of $\bar{S}$.[1] If $A \subseteq V$, let S_A denote the set of segments which contain the vertices in A and and $\bar{S}_A$ denote the set of points on segments in S_A. We consider as degenerate the case of intersecting collinear segments since it is trivial to discover such segments and merge them into one. In terms of the above notations, our problem can be formulated as follows.

Guarding a Set of Segments (GSS)
Find a minimally sized subset of vertices $\Gamma \subseteq V$ such that $\bar{S}_\Gamma = \bar{S}$.

In other words, one has to locate a minimum number of guards at the vertices of $\bar{S}$ so that every point of $\bar{S}$ is seen by at least one guard.

W.l.o.g, we will assume throughout that the set $\bar{S}$ is connected.

It is easy to see that the requirement to locate guards at vertices is not a restriction of the generality: every non-vertex point on a segment s can see the points of s only, while each of the vertices on s can see s and other segments.

In view of the above remark, GSS admits formulation in terms of a set cover problem, as follows.

Set Cover Formulation of GSS
Let S be a set of segments and V the set of their intersections, called vertices *of S. Find a minimally sized subset of vertices $\Gamma \subseteq V$ such that $S_\Gamma = S$.*

3   Approximation Algorithms for GSS

The four approximation algorithms we consider can all be classified as greedy algorithms adapted from the related set cover and vertex cover problems.

G1 Set Cover Simplest Greedy (SC)
On each iteration, this algorithm simply chooses a vertex of greatest degree ($\deg(v) = |S_v|$), removes it and all incident segments from the set of segments, and repeats until there are no longer any segments to cover.

---

[1] A vertex can be an intersection point of arbitrarily many segments. Segments may intersect at their endpoints.

G2 Vertex Cover-Inspired Matching Greedy (VC)

The approximation algorithm for the vertex cover problem finds a maximal matching[2] for the graph and selects both endpoints of each edge of the matching. Because in GSS there can be multiple intersections along a segment, we needed to make some changes before this algorithm can be correctly applied to GSS. For this algorithm, "both endpoints of each edge" was translated to "all intersections along each segment."

On each iteration of G2 it chooses a segment with fewest intersections on it, adds all intersections on the segment to the cover, and removes all incident segments from the GSS, repeating until there are no longer any segments to cover. The segments chosen on each iteration form a matching.

G3 Improved Vertex Cover Matching Greedy (IVC)

This algorithm is the same as G2 with the following modification: the intersections along each chosen segment are added to the cover in order of greatest degree first, ignoring intersections that do not contribute any new segments to the cover.

G4 Improved Set Cover Greedy Hybrid (ISC)

This algorithm is a heuristic that uses G2 as its base but abandons the idea of a matching. For each segment chosen by G2, G4 simply chooses one intersection of greatest degree along that segment and ignores the rest.

3.1 Theoretical Performance

By *performance*, or *approximation ratio* of an approximation algorithm on a problem instance we will mean the ratio $\frac{g_{approx}}{g_{opt}}$ where g_{approx} and g_{opt} are the respective values of the approximate and the optimal solution on the instance. Then the *(relative) error* of the approximation algorithm is defined as $\frac{g_{approx} - g_{opt}}{g_{opt}} = \frac{g_{approx}}{g_{opt}} - 1$.

To show that the approximation error of algorithm G1 can be arbitrarily large with respect to the optimal solution, we use the approach similar to a known estimation of a greedy solution to the set and vertex cover problems (see, e.g., [9]).

We define the placement of line segments in the plane as follows. Fix an $m \in \mathbb{N}$, $m > 3$, $a \in \mathbb{R}$, $a > 0$, and place the m points $a_1 = (1, a), (2, a), \ldots, (m, a) = a_m$. Let $n = \sum_{i=2}^{m} \lfloor \frac{m}{i} \rfloor$. Place the n points $b_{2,0} = (1, 0), (2, 0), \ldots, (n, 0) = b_{m,0}$ partitioned into groups of $\lfloor \frac{m}{2} \rfloor, \lfloor \frac{m}{3} \rfloor, \ldots, \lfloor \frac{m}{m-1} \rfloor, \lfloor \frac{m}{m} \rfloor$ points where each partition is appended to the previous partition along the line $y = 0$. Let $b_{i,j}$ be the kth point from the origin on the x-axis, where $k = \left(j + 1 + \sum_{r=1}^{i-1} \lfloor \frac{m}{r} \rfloor - m \right)$. This way i is the index of the partition $(i = 2 \ldots m)$ and j is an index within each partition

---

[2] A *matching* in a graph is any set of edges without common vertices. A matching is *maximal* if it is not a proper subset of any other matching in the graph.

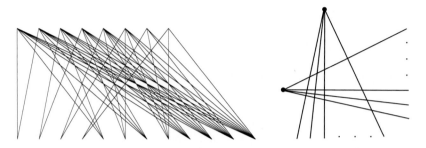

Fig. 2. *Left:* Illustration to the theoretical construction on which G1 obtains a solution that is far from the optimal. *Right:* If each of the two sheaves of segments contains $n/2$ segments (for any positive even $n \geq 4$), then algorithm G3 finds an approximate solution of $n/2+1$ segments. Clearly, the optimal solution comprises the two vertices marked by thick dots. Note that G1 is optimal here.

$(j = 0 \ldots m-1)$. We have one row of points on $y = a$ which is placed on integer coordinates in the following order:

$$a_1, a_2, \ldots, a_m$$

and one row on $y = 0$ which is placed on integer coordinates in the following order:

$$b_{2,0}, b_{2,1}, \ldots, b_{2,\lfloor \frac{m}{2} \rfloor -1}, b_{3,0}, b_{3,1}, \ldots, b_{3,\lfloor \frac{m}{3} \rfloor -1}, \ldots, b_{m-1,\lfloor \frac{m}{m-1} \rfloor -1}, b_{m,0}$$

Connect $b_{i,j}$ to all points in the set $\{a_{ji+1}, a_{ji+2}, \ldots, a_{ji+i}\}$ forming i segments for each $b_{i,j}$. In this way, $b_{i,j}$ lies on exactly i segments. Figure 2 (left) illustrates the construction for $m = 8$.

One can show that the optimal solution to the so-constructed GSS problem is $A = \{a_1, \ldots, a_m\}$ while the set cover greedy algorithm always finds the solution $B = \{b_{2,0}, b_{2,2}, \ldots, b_{2,\lfloor \frac{m}{2} \rfloor -1}, \ldots, b_{m-1,\lfloor \frac{m}{m-1} \rfloor -1}, b_{m,0}\}$ (i.e. $\{b_1, \ldots, b_n\}$). This makes the approximation ratio

$$\frac{g_{\text{approx}}}{g_{\text{opt}}} \geq \frac{n}{m} = \frac{\sum_{i=2}^{m} \lfloor \frac{m}{i} \rfloor}{m} \approx \frac{m \log m}{m} = \log m,$$

The technical details of the proof, although not too difficult, are verbose and therefore omitted due to page limit.

Figure 2 (right) illustrates that a solution for a GSS problem on n segments found by the Improved Vertex Cover Matching Greedy (G3) can be $(n/2+1)/2$ times the optimal.

Despite the above theoretical results, the experimental results of the next section show that on "random" GSS instances the tested approximation algorithms have excellent performance.

4   Experimental Studies

4.1   Generators

We currently have two GSS generators designed to generate random sets of segments. Both are written in Java.

Segment Generator. The *Segment Generator* was originally designed to simply generate random segments. Because the results consisted almost entirely of degree two intersections, some modifications including two new methods were added to ensure more vertices of higher degree.

Parameters

- Number of segments to generate
- Size of square grid (length along a side)
- Probabilities
    - Random Intersecting Segment
    - Connect two random vertices
    - Extend segment
- Output files: GSS (line segment coordinates), Matrix (incidence matrix), Intersections (intersection coordinates)

Algorithm. The *Segment Generator* repeats a simple loop that runs until the requested number of segments has been generated. First it creates four random connected segments to create some intersections. Then, for each remaining segment it chooses among the three following segment generation methods according to the probability distribution given in the parameters: random intersecting segment, random vertex to vertex segment, and random vertex to random point segment (probability $= 1-$ sum of the other two). The two methods that use an intersection as an endpoint also have a chance of extending the segment past the intersection, using the given probability parameter. The three methods are described in further detail below.

Random Intersecting Segment randomly chooses two arbitrary lattice points under the constraint that the resulting segment must be connected to the already existing generated segments.

Random Vertex to Vertex Segment randomly chooses two intersection points from the existing segment intersections and connects them to form a new segment.

Random Vertex to Random Point randomly chooses an intersection point and an arbitrary lattice point and connects them to form a new segment.

Vertex Generator. The *Vertex Generator* is designed to generate segment sets with a large number of high-degree vertices fairly quickly.

Parameters

- Number of segments to generate
- Size of square grid (Length along a side)
- Probability: Connect two random vertices
- Output files: GSS, Matrix, Intersections

Algorithm. The *Vertex Generator* repeats a simple loop that runs until the requested number of segment has been generated. First it creates a random segment as a base. Then, for each remaining segment it chooses one of two segment generation methods based on the probability distribution specified in the parameters: random vertex to vertex segment, and random vertex to random point.

Random Vertex to Vertex Segment randomly chooses two distinct endpoints from different segments out of the existing set and connects them to form a new segment.

Random Vertex to Random Point randomly chooses an endpoint from an existing segment and a random lattice point and connects the two to form a new segment.

Exact and Approximation Algorithms

In order to study the performance of the algorithms, we needed to have the optimal solution computed for each problem instance. This is implemented in Mathematica using the software's integer linear programming (ILP) algorithm. Note that ILP itself is NP-complete, which limited the number of segments for which an exact solution can be found within a reasonable amount of time. We therefore restrict our analysis to sets of segments small enough to make this computation feasible.

For all the approximation algorithms, we represent the *GSS* problem as a 0-1 segment-vertex incidence matrix and find the matrix row/column cover. A side effect is that these algorithms can solve any covering problem that can be represented as a 0-1 matrix. These algorithms are implemented in C++.

Data Analysis

Once we had generated *GSS* instances and found solutions using the various algorithms, we saved the results in tabular form within a text file. Further statistical analysis (e.g. finding means, plotting correlations, etc.) was done with GNU R.[3]

Additionally, we wrote a script that uses gnuplot[4] to graph a *GSS* instance along with the solutions found using any of the algorithms for visual inspection. Illustration is omitted due to page limit but can be found in [2].

4.2   Experimental Performance

In the following presentation algorithm G2 is ignored from evaluation and discussion, as the solutions it produced were always significantly inferior to the solutions found by all other algorithms.

Throughout the following analysis, mention of probability parameters refers to the probabilities that the given generator will use the corresponding methods to generate the next line segment. These parameters are adjusted for various experiments in order to create sets of segments that exhibit desired properties.

---

[3] http://www.r-project.org/
[4] http://www.gnuplot.info/

Table 1. Experiments on sets of different cardinality

Number of segments	5	10	20	40	80	160
mean ILP	2.94	4.97	9.73	19.45	38.57	76.29
mean SC	2.96	5.39	10.48	20.63	40.32	78.91
mean IVC	3.17	6.81	14.98	32.34	68.92	143.92
mean ISC	2.94	5.08	9.98	19.75	39.4	78.19
Approximation Ratio						
mean SC	1.006667	1.084333	1.077333	1.061020	1.045344	1.034336
mean IVC	1.080000	1.370667	1.541556	1.663740	1.787754	1.887043
mean ISC	1.000000	1.022000	1.027139	1.015798	1.021741	1.025017
max SC	1.333333	1.250000	1.300000	1.157895	1.128205	1.078947
max IVC	1.500000	1.600000	1.888889	1.944444	1.945946	2.000000
max ISC	1.000000	1.200000	1.222222	1.105263	1.081081	1.053333
SC: % incorrect	2	42	63	84	90	100
IVC: % incorrect	23	98	100	100	100	100
ISC: % incorrect	0	11	24	29	67	97
% SC better than ISC	0	5	8	0	9	18
% SC better than IVC	22	86	100	100	100	100
SC-ISC Concurrent approximation ratio						
mean	1.000000	1.012000	1.018333	1.015798	1.019073	1.021578
max	1.000000	1.200000	1.222222	1.105263	1.054054	1.053333

Random Intersecting Segment Sets

The sets of segments used for this data table were generated using Segment Generator with parameters of 1, 0, 0, which means that line segments are only created using the Random Intersecting Segment method. This resulted in an average of 98100 sets of segments were generated for each of the following sizes of segment sets: 5, 10, 20, 40, 80 and 160. All constructions were solved using ILP (the optimal), the set cover greedy algorithm, the improved vertex cover greedy algorithm, and the improved set cover greedy algorithm.

Generalized results are presented in Table 1 and it is to a great extent self-explanatory. The main conclusion one can make on the basis of the obtained data is that all approximation algorithms perform well. The approximation ratio over all 600 problems of different dimensions is never greater than 2.000. Moreover, at least one of the approximation algorithms always does better than 0.222 error at worst, and better than 0.022 error on average.

It is clear from the data that for the improved vertex cover algorithm, the mean and maximum approximation ratios strictly increase with the number of segments. On the other hand, the mean approximation ratios of set cover and improved set cover do not display any positive growth correlation. In fact, they tend to decrease as the number of line segments increases which is rather unexpected. Counter-intuitively, the percent of optimal solutions given by all studied approximation algorithms actually decreases despite the increasing accuracy of approximation.

Table 2. Results of experiments on 49,902 sets of 12 segments

12 segments	All GSS (49902)	IVC < SC (1057)	IVC < ISC (183)
mean ILP	4.669071	4.577105	4.415301
mean SC	4.925412	5.703879	4.901639
mean IVC	6.031161	4.686850	4.469945
mean ISC	4.824195	4.686850	5.480874
Approximation Ratios			
mean SC	1.056135	1.249306	1.113661
mean IVC	1.294305	1.023825	1.012568
mean ISC	1.034772	1.025181	1.244536
max SC	1.600000	1.600000	1.500000
max IVC	2.500000	1.400000	1.250000
max ISC	1.500000	1.500000	1.500000
SC: % incorrect	24.860730	100.000000	48.087430
IVC: % incorrect	83.319300	10.879850	5.464481
ISC: % incorrect	15.398180	10.879850	100.000000
% SC < ISC	6.671075	0.000000	54.644810
% SC < IVC	73.323153	N/A	2.185792
% IVC ≠ ISC	78.624103	15.61	N/A
SC-ISC concurrent approximation ratio			
mean	1.019562	same as ISC	1.113661
max	1.500000	same as ISC	1.500000

Further Experiments

The motivation for this test was to obtain the most variety possible among segment sets of a fixed size. For this purpose, 49902 sets of 12 segments were created using Segment Generator with probabilities of .3, .4, .3, yielding a mean of 68% degree two intersections. The maximum number of intersections created in one of these pictures was 53 and the minimum was 4.

The obtained results, summarized in Table 2 show that even over greatly varied arrangements of line segments, the studied approximation algorithms still perform extremely well. In fact, if we take the minimum of the approximation ratios among set cover and improved set cover for each case, the mean and maximum approximation ratios are 1.02 and 1.5 respectively. This suggests that the two algorithms complement each other in a concurrent manner as set cover performs better than improved set cover 6.67% of the time. Thus it would be interesting to answer the question: when improved set cover performs poorly, does set cover perform well and vice versa?

Five Segment Experiment

For human inspection, 3 groups of 10000 instances of *GSS* with 5 segments were generated with the Segment Generator using 3 varied probability profiles. The most notable results were:

- ISC performed better than IVC in all instances
- either ISC or SC was optimal in every instance

Vertex Generator Experiments

Experiments were carried out using the Vertex Generator, but the results were very similar to those produced by the Segment Generator. This is reassuring as it suggests that the accurate experimental approximations are not just an artifact of the sets of line segments that Segment Generator creates.

5   Concluding Remarks

In this paper we considered an art-gallery problem defined on a set of segments and three approximation algorithms for it. In theory, there exist sets of segments in which the algorithms can be made to perform as poorly as desired. However, in practice the approximate solutions are near optimal. Work in progress is aimed at providing a theoretical explanation of this phenomenon.

Acknowledgements

The authors thank the three anonymous referees for a number of useful remarks and suggestions.

This work was supported by NSF grants No. 0802964 and No. 0802994.

References

1. Bose, P., Kirkpatrick, D., Li, Z.: Worst-case-optimal algorithm for guarding planar graphs and polyhedral surfaces. Computational Geometry: Theory and Applications 26(3), 209–219 (2003)
2. Brimkov, V.E., Leach, A., Mastroianni, M., Wu, J.: Guarding a set of line segments in the plane. Theoretical Computer Science (2010), doi:10.1016/j.tcs.2010.08.014
3. Brunn, H.: Über Kerneigebiete. Matt. Ann. 73, 436–440 (1913)
4. Garey, M., Johnson, D.: Computers and Intractability. W.H. Freeman & Company, San Francisco (1979)
5. Karp, R.: Reducibility among combinatorial problems. In: Miller, R.E., Thatcher, J.W. (eds.) Complexity of Computer Computation, pp. 85–103. Plenum Press, New York (1972)
6. Kaucic, B., Žalik, B.: A new approach for vertex guarding of planar graphs. J. of Computing and Information Technology - CIT 10(3), 189–194 (2002)
7. Krasnoselśkii, M.A.: Sur un Critère pour qu'un Domain Soit Étoilé. Mat. Sb. 19, 309–310 (1946)
8. O'Rourke, J.: Art Gallery Theorems and Algorithms. Oxford University Press, Oxford (1987)
9. Papadimitriou, C., Steiglitz, K.: Combinatorial Optimization. Prentice-Hall, New Jersey (1982)
10. Urrutia, J.: Art Gallery and Illumination Problems. In: Sack, J.-R., Urrutia, J. (eds.) Handbook of Computational Geometry, ch. 22. North Holland, Amsterdam (2000)

Toward an Automatic Hole Characterization for Surface Correction

German Sanchez T.[1] and John William Branch[2]

[1] University of Magdalena, Faculty of Engineering, Systems Program, Santa Marta, Colombia
gsanchez@unimagdalena.edu.co
[2] National University of Colombia, Faculty of Minas, Systems and Informatic School, Medellin, Colombia
jwbranch@unal.edu.co

Abstract. This paper describes a method for Automatic hole characterization on 3D meshes, avoiding user intervention to decide which regions of the surface should be corrected. The aim of the method is to classify real and false anomalies without user intervention by using a contours irregularity measure based on two geometrical estimations: the torsion contour's estimation uncertainty, and an approximation of geometrical shape measure surrounding the hole.

1 Introduction

The shape reconstruction process requires estimating a mathematical representation of an object's geometry using a measured data-set from the object [1]. The purpose is to convert a big amount of data points coming from the surface of a real object acquired by special devices namely 3D scanners, into a digital model preserving the geometrical features like volume and shape.

There are many measuring drawbacks in the acquisition step: topological characteristics of the objects, sensor structure, physical properties of the object's material, illumination conditions and among others. Those represent the main source of anomaly generation, distorting the measured data and avoid estimating an accurate representation of the geometrical characteristic. These inadequacies represent the main source of anomaly generation, and must be repaired in order to create a valid digital model [2].

In this paper the anomalies can be classified into three groups: Noise, Holes artifacts and Redundancy. Noisy points are points which location may be perturbed by unknown levels of noise [4]. Holes artifacts or incomplete sampling is caused by not well-sampled set which need user efforts. Typically, these anomalies are repaired in a phase called integration [3].

The classical reconstruction methods need to apply a post-processing procedure after the final stage of surface-fitting. This is mainly due to the difficulty in differentiating the nature of the discontinuity, that is, whether or not it belongs to the actual surface. One of the main desirable features in surface reconstruction methods is the ability to fill holes or to complete missing regions. Lack of information is caused

G. Bebis et al. (Eds.): ISVC 2010, Part I, LNCS 6453, pp. 602–611, 2010.
© Springer-Verlag Berlin Heidelberg 2010

mainly by the reflective properties of the material, or by occlusion problems in regions inaccessible to the sensor.

Some methods make an implicit correction during the fitting phase by means of global reconstruction [1] [5]. This approach has two disadvantages: first, it does not permit to keep false holes, *i.e.* those belonging to the object, and second, the quality of the portion generated depends on the technique used and not on the analysis of the intrinsic geometry of the object. While taking the surface as continuum and complete, these techniques reproduce visually appropriate solutions. However, the correction of these anomalies is still limited to particular cases when objects are closed.

The hole-filling procedure is based on surface segment generation, for which different techniques have been proposed [6-14]. There are two general trends in the group of geometry-based techniques: repair based on triangle meshes, and repair based on voxels. Wei [15] proposed an algorithm for filling holes that starts with a hole identification phase and then applies a triangulation of the hole region using the Advancing Front Method. Finally, by solving a Poisson equation, the vertex of the generated triangles is adjusted. Although adequate visual results are obtained with this procedure, it is time costly and depends on the size of the hole.

Voxel-based approaches estimate an initial surface volumetric representation by voxel set. These voxel units are marked with a sign according to their relative position to the surface, that is, inside or outside the surface. Different techniques have emerged to fill the hole in the volumetric space. Curless [16] proposed a method based on volumetric diffusion consisting of a distance function estimation --which is used to mark the voxel--, and then diffusion is applied through the volume to find the zero set that defines the surface. Ju [17] proposed a method of contour surface reconstruction by marking the voxel using an Octree data structure. The procedure is able to fill small gaps, taking into account geometric characteristics. The main limitation is that the hole size must be smaller than the relative surface size. Chun [18] describes a two-phase approach to 3D model repair. In the first phase, a radial basis function interpolation is used to fill the region inside the hole; the second one is a post-processing stage to refine the geometrical detail. In the refinement stage the normals are adjusted to produce adequate visual results.

In this paper, we propose a metric for robust hole characterization in 3D models. This metric intends to characterize holes through geometric features measures. Our hypothesis is based on a widely accepted definition [19]: free-form objects are smooth except in regions that represent specific geometric details. Hence, if there are not any problems in the acquisition stage, a false contour anomaly should not have large geometric variations; otherwise, it could be caused by problems in the acquisition stage and constitute an anomaly to be repaired. Thus, if there were any problems during the acquisition process then the data is altered introducing distortion that should not be equal for the segments that define the anomaly. That is, acquisition problems introduce some degree of "contour distortion". The characterization of each anomaly is based on the quantification of this distortion, which for this particular case is approximated by a quantification of the entropy in the boundary geometry.

The paper is organized as follows: section 2 introduces the topic of anomalies classification; section 3 describes the proposed contour's irregularity; and section 4 presents the experimental design and results.

2 Anomalies Classification

The key difference between 3D scattered data and other data types such as images or video is that 3D scattered data are typically irregularly sampled. The points' distribution of vertices across the surface is not uniform, so to quantify a measure it must be robust under different distribution of points.

In order to get the most accurate estimation of the irregularity of a hole, we propose a metric that measures the hole´s most important geometrical features: the contour curve irregularity measure from the torsion and curvature entropy.

3 Irregularity Measure

In this step we are interested in measuring the geometrical characteristic of the contour curve, i.e. its curvature and its torsion. The aim of these estimations is to quantify its irregularity by means of the uncertainty using the entropy measure.

Entropy Measure of contour
The two fundamental characteristics of a curve are its curvature and torsion; these allow to measure how a curve bends in 3D space, therefore, it constitutes a curve's particular characteristic. Often, we assumed that discontinuous contour curves in smooth objects without acquisition problems are smooth too.

Contour bends give us a measure of irregularity. However, estimating the accurate torsion value of a sample curve, defined by a piecewise linear approximation through an ordered finite collection of points $\{p_i\}$, is not a trivial task since noise is present. i.e., the points p_i stay too close to the curve, but not necessarily lie on it.

In order to approximate a correct classification of contour curves, we used the torsion measure. For a spatial curve torsion is defined by $B'(s) = \tau(s)N(s)$ where $N(s) = r''(s)/\|r''(s)\|$ is the normal vector, s is the arc length from a specific position $r(t_0)$ given by a parametric curve r, to a close position $r(t_1)$ and defined by $s(t_1) = \int_{t_0}^{t_1}\|r'(u)\|\,du$. For a non arc-length parameterized $r(t)$, $\tau(s)$ is thus estimated:

$$\tau(s) = -\frac{(r' \times r'') \cdot r'''}{\|r' \times r''\|^2} \tag{6}$$

To estimate the torsion we adopt the weighted least squares approach and local arc-length approximation [2] [3] [4]. It considers a samples-set $\{p_i\}$ from a spatial

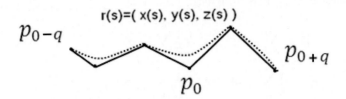

Fig. 1. Weight least square curve

curve. The estimation of derivates of r at p_0 is performed with a point-subset P of $2q + 1$ points such that (see Figure 1):

Then, a parametric curve $(\hat{x}(s), \hat{y}(s), \hat{z}(s))$ is fitted locally, assuming $p_{0=}r_0$ and an arc-length s_i value associated to the samples p_i:

$$\hat{x}(s) = x_0 + x_0' \cdot s_i + \frac{1}{2}x_0'' \cdot s_i^2 + \frac{1}{6}x_0''' \cdot s_i^3$$

$$\hat{y}(s) = y_0 + y_0' \cdot s + \frac{1}{2}y_0'' \cdot s_i^2 + \frac{1}{6}y_0''' \cdot s_i^3 \tag{7}$$

$$\hat{z}(s) = z_0 + z_0' \cdot s + \frac{1}{2}z_0'' \cdot s_i^2 + \frac{1}{6}z_0''' \cdot s_i^3$$

Taking $\hat{x}$ coordinate, the derivates x_0', x_0'', x_0''' are obtained minimizing [21]:

$$E_x(x_0', x_0'', x_0''') = \sum_{i=-q}^{q} w_i \left(x_i - x_0's_i - \frac{1}{2}x_0''(s_i)^2 \frac{1}{2}x_0''(s_i)^2 - \frac{1}{6}x_0'''(s_i)^3 \right)^2 \tag{8}$$

Where $w_i = 1, s_i = \sum_{k=0}^{i-i} \sqrt{((p_k - p_{k+1})^2)}, p_i \in \mathbb{R}^3$. It can be written in terms of matrix inversion:

$$A \cdot x = b \tag{9}$$

A similar approach is used to estimate the x and y derivates getting the vectors:

$$Y = \begin{bmatrix} y_0' \\ y_0'' \\ y_0''' \end{bmatrix} \qquad z = \begin{bmatrix} z_0' \\ z_0'' \\ z_0''' \end{bmatrix}$$

From the equations system:

$$\begin{cases} A \cdot x = b \\ A \cdot y = b \\ A \cdot z = b \end{cases} \tag{10}$$

Where,

$$\begin{bmatrix} a_1 & a_2 & a_4 \\ a_2 & a_3 & a_5 \\ a_4 & a_5 & a_6 \end{bmatrix} \cdot \begin{bmatrix} x_0' & y_0' & y_0' \\ x_0'' & y_0'' & y_0'' \\ x_0''' & y_0''' & y_0''' \end{bmatrix} = \begin{bmatrix} b_{x,1} & b_{y,1} & b_{z,1} \\ b_{x,2} & b_{y,2} & b_{z,2} \\ b_{x,3} & b_{y,3} & b_{z,3} \end{bmatrix}$$

The a_i values and $b_{x,i}$ are defined thus:

$$a_1 = \sum_{i=-q}^{q} w_i s_i^2 \qquad a_2 = \frac{1}{2}\sum_{i=-q}^{q} w_i s_i^3 \qquad a_3 = \frac{1}{4}\sum_{i=-q}^{q} w_i s_i^4$$

$$a_4 = \frac{1}{6}\sum_{i=-q}^{q} w_i s_i^4 \qquad a_5 = \frac{1}{12}\sum_{i=-q}^{q} w_i s_i^5 \qquad a_6 = \frac{1}{36}\sum_{i=-q}^{q} w_i s_i^6$$

$$b_{x,1} = \sum_{i=-q}^{q} w_i s_i x_i \qquad b_{x,2} = \frac{1}{2}\sum_{i=-q}^{q} w_i s_i^2 x_i \qquad b_{x,3} = \frac{1}{6}\sum_{i=-q}^{q} w_i s_i^3 x_i$$

$$b_{y,1} = \sum_{i=-q}^{q} w_i s_i y_i \qquad b_{y,2} = \frac{1}{2}\sum_{i=-q}^{q} w_i s_i^2 y_i \qquad b_{y,3} = \frac{1}{6}\sum_{i=-q}^{q} w_i s_i^3 y_i$$

$$b_{z,1} = \sum_{i=-q}^{q} w_i s_i z_i \qquad b_{z,2} = \frac{1}{2}\sum_{i=-q}^{q} w_i s_i^2 z_i \qquad b_{z,3} = \frac{1}{6}\sum_{i=-q}^{q} w_i s_i^3 z_i$$

Finally, it defines:

$$r_0' = \begin{bmatrix} x_0' \\ y_0' \\ z_0' \end{bmatrix} \qquad r_0'' = \begin{bmatrix} x_0'' \\ y_0'' \\ z_0'' \end{bmatrix} \qquad r_0''' = \begin{bmatrix} x_0''' \\ y_0''' \\ z_0''' \end{bmatrix}$$

The computation of $\tau(s)$ is straightforward, thus:

$$\tau(p_0) = -\frac{(r_0' \times r_0'') \cdot r_0'''}{\|r_0' \times r_0''\|^2}$$

Due to their nature, hole-characterization problems suggest solutions based on inference, since it needs a process of drawing conclusions from available information that is partial, insufficient and that does not allow to reach an unequivocal, optimal and unique solution. Then we need to make inferences from the available data assuming it is noisy. Specifically, the topic of hole characterization constitutes a highly ambiguous example to take decisions because there are many possible configurations of irregular contours. Both aspects, noise and ambiguity, imply taking uncertainty into account.

The adequate way to deal with the presence of uncertainty, related to lack of information, is to introduce assumptions about the problem's domain or *a priori* knowledge about the data, by means of the notion of degrees of belief. It should be treated using the classical rules of calculus of probabilities. The rules of probability theory allow us to assign probabilities to some "complex" propositions on the basis of the probabilities that have been previously assigned to other, perhaps more "elementary" propositions. However, in order to estimate a measure to characterize contours, we are not interested in just probability estimation about a geometrical characteristic, but also in its variability. High variability could be measured through entropy. Specifically, conditional entropy is used.

Given two variables x and y, the $S_{x|y}$ quantity that measures the amount of uncertainty about one variable x when we have some limited information about another variable y is conditional entropy [5]. It is obtained by calculating the entropy of x as if the precise value of y were known and then taking the expectation over the possible values of y.

$$S_{x|y} = -\sum_y p_y \, S[p_{x|y}] = -\sum_y p_y \sum_x p_{x|y} \log\,(p_{x|y})$$

In a similar way,

$$S_{x|y} = -\sum_{x,y} p_{xy} \log\,(p_{x|y}) \tag{11}$$

Given a sequence of n points $P: p_i \in \mathbb{R}^3$ forming the contour of a 3D curve defining an anomaly, and a set φ of geometrical characteristics measures associated to each one in P. We want to measure the irregularity in φ from a prior knowledge of some geometrical characteristic measure. It means that the certainty of a point p_i is estimated by taking an l -set $\psi_i: \{\varphi_k : i - l - 1 < k < i - 1\}$ over a sorted sequence of points used to estimate the next value. The certainty or, in inverse form, the unpredictability of all ψ_i is related to entropy.

$$S_\varphi = \sum_i^n p(\varphi_i, \psi_i)\log\,(p(\varphi_i|\psi_i)) \tag{12}$$

Where $\psi_i: \{\varphi_{i-(l+1)}, \varphi_{i-l}, \ldots, \varphi_{i-2}, \varphi_{i-1}\}$, and:

$$\varphi_i = \tau(p_i) = -\frac{(r' \times r'') \cdot r'''}{\|r' \times r''\|^2} \;.$$

Contours' curvature Measure
The principal limitation of a descriptor based on torsion measure is to deal with planar curves. Planar curves may appear as a result of occlusions; although these are less common and torsion based decision is inappropriate. However, in order to attain completeness, our metric takes into account those cases and uses a tangent vector variability measure for planar cases like irregularity.

For anomalies in planar cases, tangent variability is usually high; conversely, real holes show smooth changes between tangent angles (see Figure 2).

To estimate this measure, we take the weighted least squares approach and local arc-length approximation made in section 4. Tangent vector is defined as $T(t) = N(s) \times B(s)$, or in derivate terms $T(t) = \frac{r'(t)}{\|r'(t)\|}$. We estimate the entropy S_T of angle between successive tangents with equation 12, by replacing the torsions distributions by the angle between tangents distribution. And finally, we quantify the global entropy of the contour S_c by:

$$S_c = S_\varphi + S_T \tag{13}$$

Finally,

$$Irregularity = \|\overline{SI}\|(S_\varphi + S_T)$$

For undefined cases of $\overline{SI}$, the S_t measure is an accurate estimation of irregularity.

a)   b)

Fig. 2. Tangent vectors variability of both real and false holes respectively

4 Experiment Design and Result

All tests were performed using a computer with 3.0GHz Intel processor, 3.0GB RAM running under Microsoft Windows XP operating system. The implementations of the models were made in C++ for identification and mesh handling and MATLAB for entropy estimation, as well as programming a OpenGL graphics engine in order to obtain the graphical representation of the images. Most data used were obtained with the non-invasive 3D Digitizer VIVID 9i Konica Minolta range scanner of the National University of Colombia, Manizales.

In order to estimate the p_{xy} and $p_{x|y}$ quantity given a continuous variable of torsion measure, we used a bin size r to discrete the domain of an experimental set of hole-contour configurations. The experimental set was obtained from 10 images with real hole anomaly, like Figure 3a,b and 10 contours with false hole anomalies like Figure 3c-f, in partial and complete 3d models' range data. Some images were scales to maintain a fixed scale. It was done by subtracting the mean and dividing by the standard deviation. The irregularity was estimated with equation 13. Figure 4 shows the irregularity estimated for both sets. It shows that the irregularity measure is highly sensitive to the irregularities in the contour. Determining if an anomaly is real or false is straightforward because the values are sufficiently separated. The midpoint of the range of separation is 3.1. The experimental threshold for classification was estimated in this value.

However, our goal is to quantify the irregularity of contour. We are interested in extending the method using a Bayesian classifier for the anomalies classification as well as the correction by the generation of the missing segment of surface.

The irregularity is increased when the separation of the data is greater. The method is highly sensitive to noise; small changes in the regularity of the contour show an equal increase in the estimation of entropy. This method can be used as an initial step in the process of correcting anomalies. We aim to complement the method with the filling process to propose an automatic robust method to correct anomalies.

The proposed method allows estimating a metric for the automatic classification of anomalies in range images. The purpose of the method is to automate a process that has traditionally required user intervention. The method estimates the most relevant geometric characteristics of curves and surfaces to describe them.

The anomalies used as the working set were mostly generated by occlusion. We are interested in extending the study to holes generated by surface brightness.

(a)   (b)

(c)   (d)

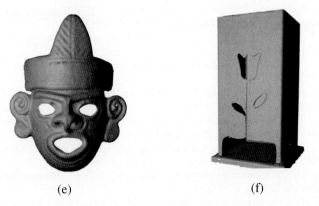
(e)   (f)

Fig. 3. Examples of surface discontinuity: a-b) false discontinuities, c) real discontinuity of the object

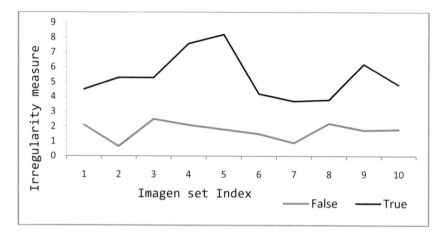

Fig. 4. Irregularity values of false and real anomalies

References

[1] Koenderink, J.J., van Doorn, A.J.: Surface shape and curvature scales. Image Vision Comput. 10(8), 557–565 (1992) ISSN:0262-8856

[2] Lewiner, T., Gomes Jr., J.D., Lopes, H., Craizer, M.: Curvature and torsion estimators based on parametric curve fitting. Computers & Graphics 29(5), 641–655 (2005) ISSN 0097-8493

[3] Lancaster, P., Salkauskas, K.: Surfaces generated by moving least squares methods. Math. Comp. 37(155), 141–158 (1981)

[4] Lancaster, P., Salkauskas, k.: Curve and Surface Fitting: An Introduction. Academic Press, London (2002)

[5] Ariel, C.: Lectures on Probability, Entropy, and Statistical Physics. Al-bany: Department of Physics, University at Albany (2008); 0808.0012

[6] Sun, X., et al.: Noise in 3D laser range scanner data. In: IEEE International Conference on Shape Modeling and Applications, SMI 2008, pp. 37–45 (2008)

[7] Turk, G., Levoy, M.: Zippered polygon meshes from range images, pp. 311–318. ACM, New York (1994)

[8] Curless, B.: New Methods for Surface Reconstruction from Range Images. Stanford University (1997)

[9] Kumar, A., et al.: A Hole-filling Algorithm Using Non-uniform Rational B-splines, pp. 169–182. Springer, Heidelberg (2007)

[10] Besl, P.J.: The free-form surface matching problem, in Machine Vision for Three-Dimensional Scenes. Academic Press, London (1990)

[11] Pauly, M., Gross, M., Kobbelt, L.P.: Efficient simplification of point-sampled surfaces, pp. 163–170. IEEE Computer Society, Los Alamitos (2002)

[12] do Carmo, M.: Differential geometry of curves and surfaces. Prentice Hall, Englewood Cliffs (1976)

[13] Dorai, C., et al.: Registration and Integration of Multiple Object Views for 3D Model Construction. IEEE Transactions on Pattern Analysis and Machine Intelligence 20, 83–89 (1998)

[14] Curless, B., Levoy, M.: A volumetric method for building complex models from range images, pp. 303–312. ACM Press, New York (1996)

[15] Carr, J.C., et al.: Smooth surface reconstruction from noisy range data, pp. 119–126. ACM Press, New York (2003)

[16] Davis, J., et al.: Filling Holes in Complex Surfaces Using Volumetric Diffusion, pp. 428–438 (2002)

[17] Wang, J., Oliveira, M.M.: A hole-filling strategy for reconstruction of smooth surfaces in range images. In: XVI Brazilian Symposium on Computer Graphics and Image Processing, SIBGRAPI 2003, pp. 11–18 (2003)

[18] Podolak, J., Rusinkiewicz, S.: Atomic volumes for mesh completion, p. 33. Eurographics Association (2005)

[19] Bischoff, S., Pavic, D., Kobbelt, L.: Automatic restoration of polygon models. ACM Trans. Graph. 24, 1332–1352 (2005)

[20] Guo, T., et al.: Filling Holes in Meshes and Recovering Sharp Edges. In: IEEE International Conference on Systems, Man and Cybernetics, SMC 2006, vol. 6, pp. 5036–5040 (2006)

[21] Bendels, G.H., Schnabel, R., Klein, R.: Fragment-based Surface Inpainting. In: Desbrun, M., Pottmann, H. (eds.) The Eurographics Association (2005)

[22] Branch, J., Prieto, F., Boulanger, P.: Automatic Hole-Filling of Triangular Meshes Using Local Radial Basis Function, pp. 727–734. IEEE Computer Society, Los Alamitos (2006)

[23] Liepa, P.: Filling holes in meshes, pp. 200–205. Eurographics Association (2003)

[24] Zhao, W., Gao, S., Lin, H.: A robust hole-filling algorithm for triangular mesh. Vis. Comput. 23, 987–997 (2007)

[25] Ju, T.: Robust repair of polygonal models, pp. 888–895. ACM, New York (2004)

[26] Chen, C.-Y., Cheng, K.-Y.: A Sharpness-Dependent Filter for Recovering Sharp Features in Repaired 3D Mesh Models. IEEE Transactions on Visualization and Computer Graphics 14, 200–212 (2008)

A Local-Frame Based Method for Vector Field Construction on Raw Point Cloud

Xufang Pang[1,2], Zhan Song[1,2], and Xi Chen[1,2]

[1] Shenzhen Institutes of Advanced Technology, Chinese Academy of Sciences, China
[2] The Chinese University of Hong Kong, Hong Kong, China
{xf.pang,zhan.song,xi.chen}@sub.siat.ac.cn

Abstract. Direction fields are an essential ingredient in controlling surface appearance for applications ranging from anisotropic shading to texture synthesis and non-photorealistic rendering. Applying local principal covariance analysis, we present a simplistic way for constructing local frames used for vector field generation on point-sampled models. Different kinds of vector fields can be achieved by assigning different planar vectors to the local coordinates. Unlike previous methods, in the proposed algorithm, there is no need of user constraints or any extra smoothing and relaxation process. Experimental results in the isotropic remeshing and texture synthesis are used to demonstrate its performance.

1 Introduction

Vector fields have been used as the fundamental information for controlling the appearance of surface in many applications, for instance, example-driven texture synthesis makes use of a vector field to define local texture orientation and scale [1-3]. In non-photorealistic rendering, vector fields are used to guide the orientation of brush strokes [4] and hatches [5]. In fluid simulation, the external force is a vector field which need not correspond to any physical phenomenon and can exist on synthetic 3D surfaces [6]. For quadrilateral remeshing, edges are aligned in accordance with the direction tensor [7-8].

Most of the recent algorithms for constructing vector fields are based on user given constraints and some time-consuming operators. In [1], Praun et al. specify the desired orientation and scale over the mesh as a tangential vector field. Firstly, vectors at a few faces need to be defined, and then vectors at the remaining faces are interpolated using Gaussian radial basis functions, where the radius is defined as distance over the mesh, as computed using Dijkstra's algorithm. In [2], a multi-resolution representation of the mesh and employs a push-pull technique to interpolate sparse user information about the vector field is introduced. For every level representation of meshes, tangent planes of each vector have been estimated in the context of processing operations. In [3], a relaxation technique to generate smooth vector field is presented. The method can be described as a diffusion process where the desired values

G. Bebis et al. (Eds.): ISVC 2010, Part I, LNCS 6453, pp. 612–621, 2010.

are smoothly propagated from seed points to the rest of the surface. The energy function is used to represent different types of symmetries. In [9], the vector field is defined as parametric domain. In [10], each point in the point set is defined as a direction that conforms to user defined or automatically detected directions so that the discrepancy of close point is minimized. This operator needs to assign at least one point a direction manually. If no preferred directions are specified, random directions of the point set need to be smoothed iteratively by repeatedly averaging directions of the points. In [11], a discrete version of the surface gradient is proposed by the use of K-nearest neighbors. Gradient information is used to define the parameterization as an optimization problem. The solution minimizes a measure of distortion of the gradient along the surface. The local parameterizations are used for local operations on the geometry or texture mapping operations. In [12], the vector field is formulated as a linear problem by using intrinsic, coordinate-free approach based on discrete differential forms and the associated Discrete Exterior Calculus (DEC) [13]. In [14], a smooth tangent vector fields is defined on non-flat triangle meshes. Their principal design tool is the placement of singular basis vector fields, which are mapped to the surface from their planar domain via the use of polar geodesic maps and parallel transport. For removing singularities, they use a posteriori harmonic smoothing.

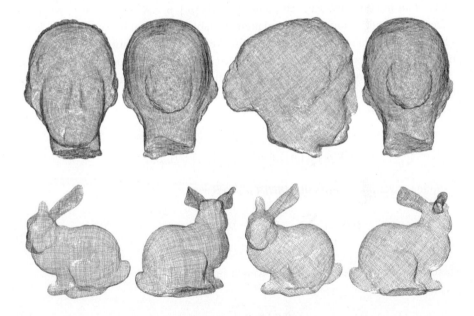

Fig. 1. Some salient results generated via the proposed vector construction algorithm

In this paper, a new method for computing direction fields on raw point cloud is presented. For point clouds without normal information, surface normal at each point is firstly calculated by performing local Principle Component Analysis (PCA). After adjusting the surface normal orientation, a local coordinate for each surface point can be built. Then different kinds of vector fields can be constructed by assigning different planar vectors to the local coordinates, and the generated vector fields can be edited by rotating the assigned planar vector as shown in Fig. 1.

2   Local-Frame Based Vector Field Generation

In the proposed algorithm, local-frame vector field is generated by firstly building a local coordinate for each point of the point-sampled model. And then different kinds of tangent vector fields can be constructed by assigning different planar vectors to local coordinates.

2.1   Local Frame Coordinate Construction

For point cloud with correct normal information, we can easily build local frame for each point by assuming that the normal direction is the Z axis of local coordinate. However, if the normals are not given, our algorithm needs extra computation for calculating the corresponding normal of each point in point cloud [15-16].

Given an input of unorganized point cloud $P=\{p_i\}, p_i \in R^3, i \in \{1,...,N\}$, where no surface normal and connection information available. The point set surface is required to be manifold, it may be of arbitrary genus. We calculate the normal of each point in point cloud by performing principal covariance analysis over its local neighborhoods.

Neighborhoods of each point p_i in a local region with radius of r are denoted as $NBHD(p_i)$ (For point cloud which is uneven sampled, we use k-nearest neighbors), where r is set to be 2-5 times of the average station distance of the point set [17], and the neighborhoods are defined as:

$$NBHD(p_i) = \{p_j\}, \parallel p_j - p_i \parallel \leq r, j = 0,...,k \tag{1}$$

Assuming covariance matrix of $NBHD(p_i)$ to B as:

$$B = \sum_{p_j \in NBDH(p_i)} \theta(p_j - p_i)(p_j - p_i)^T \tag{2}$$

where θ refers to a radials Gaussian weight function as:

$$\theta(d) = e^{-\frac{d^2}{h^2}} \tag{3}$$

where $d = \parallel p_j - p_i \parallel$ and h is the average station distance of the point set. Eigen values and their corresponding eigenvectors of B are represented by $\lambda_0 < \lambda_1 < \lambda_2$, and V_0, V_1, V_2. Reference domain H_i, which can be seen as the tangent plane of point set surface at p_i, is defined by V_1, V_2 and p_i, and the surface normal n_i at p_i is set to be V_0, $\parallel n_i \parallel = 1$.

Since the normal obtained by this method are pointing to the different sides of point set surface, so we get each normal to point in direction 'out' consistently by performing normal adjustment algorithm as described in [19].

A local coordinate system on H_i can be defined as (p_i, u, v, n_i), where p_i is the origin, and n_i is the Z-axis. The coordinate transformation matrixes between object frame and local frame are defined as follows:

Matrix T_1 for transformation of object frame to local frame can be expressed as:

$$T_1 = \begin{bmatrix} 1 & 0 & 0 & -p_i.x \\ 0 & 1 & 0 & -p_i.y \\ 0 & 0 & 1 & -p_i.z \\ 0 & 0 & 0 & 1 \end{bmatrix} \begin{bmatrix} 1 & 0 & 0 & 0 \\ 0 & \cos 1 & \sin 1 & 0 \\ 0 & -\sin 1 & \cos 1 & 0 \\ 0 & 0 & 0 & 1 \end{bmatrix} \begin{bmatrix} \cos 2 & 0 & -\sin 2 & 0 \\ 0 & 1 & 0 & 0 \\ \sin 2 & 0 & \cos 2 & 0 \\ 0 & 0 & 0 & 1 \end{bmatrix} \quad (4)$$

Matrix T_2 for transformation of object frame to local frame is defined as:

$$T_2 = \begin{bmatrix} \cos 2 & 0 & \sin 2 & 0 \\ 0 & 1 & 0 & 0 \\ -\sin 2 & 0 & \cos 2 & 0 \\ 0 & 0 & 0 & 1 \end{bmatrix} \begin{bmatrix} 1 & 0 & 0 & 0 \\ 0 & \cos 1 & -\sin 1 & 0 \\ 0 & \sin 1 & \cos 1 & 0 \\ 0 & 0 & 0 & 1 \end{bmatrix} \begin{bmatrix} 1 & 0 & 0 & p_i.x \\ 0 & 1 & 0 & p_i.y \\ 0 & 0 & 1 & p_i.z \\ 0 & 0 & 0 & 1 \end{bmatrix} \quad (5)$$

where,

$$v = \sqrt{n_i.z^2 + n_i.y^2}$$

$$\cos 1 = n_i.z / v$$

$$\sin 1 = -n_i.y / v$$

$$\cos 2 = v$$

$$\sin 2 = -n_i.x$$

2.2 Vector Field Construction

Some regular kinds of vector fields can be obtained by assigning a constant planar vector (vectors on reference domain) to the local frame coordinate of each point. For example, a common used isotropic tensor can be achieved by simply assigning two orthogonal planar vectors to each coordinate of point set model. Fig. 2 shows the direction fields resulting from two given orthogonal vectors $D_1 = (1.0, 0.0, 0.0)$ and $D_2 = (0, 1.0, 0.0)$. These kinds of orthogonal vector fields are very suitable for isotropic remeshing as shown in Fig. 3. Fig. 4 shows that the proposed local-frame based vector field can be changed by rotating the assigned planar vectors of each local coordinate.

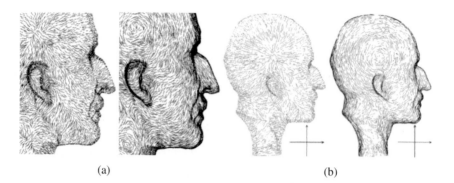

Fig. 2. Vector fields constructed by assigning two orthogonal constant planar vectors to each local frames. (a) Partial vector fields of Max-Planck model; (b) Simplified vector fields.

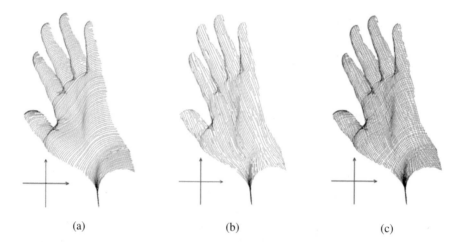

(a)   (b)   (c)

Fig. 3. Isotropic flow lines generated according to the direction fields constructed by assigning two agreed orthogonal planar vectors to each local coordinate of existing points in point cloud. (a) Flow lines generated via assigned vector $D_1 = (1.0, 0.0, 0.0)$ to each local coordinate. (b) $D_2 = (0, 1.0, 0.0)$. (c) Flow lines generated from these two orthogonal vector fields.

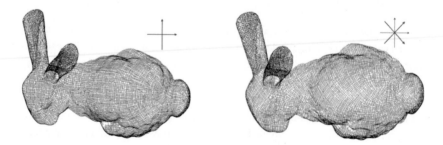

Fig. 4. Vector fields (flow lines) can be changed by rotating the assigned planar vectors

Another two point clouds with direction flows with $60°$ rotation are also conducted to evaluate the algorithm performance. For the purpose of remeshing and multi-resolution modeling, previous works like [20] usually adopt Laplacian fields to construct the vector fields on polygon meshes. In the proposed approach, the results are generated from unorganized point cloud with less computation as shown in Fig. 5.

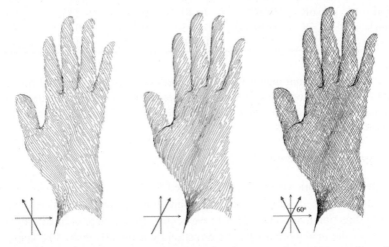

Fig. 5. Assigning two vectors with 60 degree rotation to each local coordinate; Two different groups of direction flows are obtained which are suitable for triangulation tasks

3　Experiments and Discussion

To evaluate the performance of the proposed local-frame based vector field generating method, we make an analysis on consuming-time and memory requirement of the procedure, as performed over different point clouds with no normal information. This section also demonstrates the feasibility of the algorithm by comparing the work with some previous works.

3.1　Run Time Performance and Memory Requirement

Table 1 shows the run times for each stage in the proposed algorithm for point-sampled models with different scales, which includes times for computing normal, adjusting normal orientation and generating vector fields of two different directions. The total execution time and maximum memory used are also reported in the table.

The experimental data show that the most time-consuming part of our implementation is calculating oriented normal for point cloud, and it takes very little time for generating vector fields. So for point clouds with correct normal information, the proposed approach can provide vector fields with extremely high efficiency.

Table 1. Run time performance (in seconds) and maximal memory requiremnt (in Megabytes) of the proposed method, as measured on a 2.7GHz AMD Athlon processor

Model	# of points	Normal estimation	Normal orientation	Vector Field	Total time	Maximal memory
Bunny	35.9K	0.390	0.985	0.015	1.390	54
Venus	134.4K	1.484	3.312	0.047	4.843	100
Hand	6.6K	0.079	0.140	0.016	0.235	35
Planck	25.4K	0.297	0.703	0.015	1.015	48
Dino	56.3K	0.766	1.656	0.016	2.438	62

The entry of maximum memory in Table 1 demonstrates that our local-frame based algorithm performed with low memory requirements. According to the algorithm procedures, we can also derive that it needs almost no extra memory during the stage of constructing vector fields.

3.2 Comparison and Limitation

As shown in Fig. 6(a), vector field is obtained by firstly manually position one or more seeds on point set surface, then any point within a sphere of a user defined radius gets a direction that points either to the center of the sphere or away from it. The rest points of the model are assigned a direction by using a Dijkstra-like approach; if no seed point is given, successive relaxation passes is performed to smooth a random set of directions to a certain degree, Fig. 6(b) shows the result of this procedure.

Vector field represented in Fig. 6(c) is a tangential vector field, specifying vectors at a few faces, the remaining directions are computed using an RBF scattered data interpolation approach, where distances are determined using Dijkstra's algorithm on the edge graph. Vector field depicted in Fig. 6(d) is achieved by firstly creating a hierarchy of mesh from low to high density over a given surface, and then the user specifies a vector field over the surface that indicates the orientation of the texture. The mesh vertices on the surface are then sorted in such a way that visiting the points in order will follow the vector field and will sweep across the surface from one end to the other.

Algorithms used in Fig. 6(a-d) are all based on user given constraints and some time-consuming operators, such as relaxation process, Dijkstra's approach, hierarchy interpolation, Gaussian smoothing and so on. Instead of using manually given constraints and some iterative smoothing operators, the proposed algorithm constructs vector fields over raw point cloud in a much simple manner, and high quality results are obtained from one-shot computation as shown in Fig. 6(e) and 6(f).

The experimental results demonstrate that the proposed algorithm build different kinds of reasonable vector fields which are particularly suitable for isotropic quadrilateral remeshing or example-driven texture synthesis without any manually given constraints or time-consuming iterative smoothing operators.

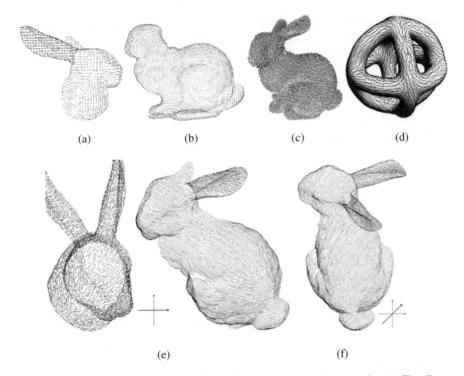

Fig. 6. Comparison between the proposed algorithm and some previous works. (a) The direction field resulting from the first pass of the algorithm starting in one, respectively two, user specified discontinuities [10]. (b) Starting from a set of random directions, this point set is smoothed iteratively by repeatedly averaging directions of the points [10]. (c) A tangential vector field is defined on triangle mesh by using RBF and Dijkstra's algorithm [1]. (d) The interpolated orientation field based on user defined constraint [2]. (e), (f) The direction fields constructed by our algorithm which can be changed by rotating the given local planar vector.

For point clouds with no surface normal information, the construction of the local frame is dependent on the local sampling rate, which may not be suitable for sparse or uneven sampled model. However, for uniformly sampled point clouds, especially for point clouds which provide correct normal information, satisfied vector field can be obtained via the proposed algorithm.

Currently, our algorithm hasn't taken the effects of features into account, but uses the singularities closely related to the initial pose of the model. So it has to change the model pose for getting a satisfying vector field. For most of the models of symmetric objects and other models without extremely complicated features, the proposed algorithm is applicable.

4 Conclusion and Future Work

In this paper, we have presented a local-frame based method to establish direction fields over point-sampled geometry, which is both efficient and easy to implement.

Vector fields generated by our algorithm are applicable for example-driven texture synthesis, isotropic remeshing and other graphic applications. For the present, our algorithm hasn't taken features into consideration of our generated vector fields. This work is concluded as a possible future work. Moreover, its application can also be extended to many other point cloud based applications such as quadrilateral meshing and triangulation etc.

Acknowledgments

The work described in this article was partially supported by NSFC (Project no. 61002040) and Knowledge Innovation Program of the Chinese Academy of Sciences (Grant no. KGCX2-YW-156).

References

1. Praun, E., Finkelstein, A., Hoppe, H.: Lapped textures. In: Proc. ACM SIGGRAPH, pp. 465–470 (2000)
2. Turk, G.: Texture synthesis on surfaces. In: Proc. SIGGRAPH 2001, pp. 347–354 (2001)
3. Wei, L.-Y., Levoy, M.: Texture synthesis over arbitrary manifold surfaces. In: Proc. SIG-GRAPH 2001, pp. 355–360 (2001)
4. Hertzmann, A.: Painterly rendering with curved brush strokes of multiple sizes. In: Proc. 25th Annual Conference on Computer Graphics and Interactive Techniques, pp. 453–460 (1998)
5. Hertmann, B., Zorin, D.: Illustrating smooth surfaces. In: Proc. ACM SIGGRAPH 2000, pp. 517–526 (2000)
6. Stam, J.: Flows on surfaces of arbitrary topology. ACM Trans. Graph. 22, 724–731 (2003)
7. Dong, S., Kircher, S., Garland, M.: Harmonic functions for quadrilateral remeshing of arbitrary manifolds. Computer Aided Geometric Design 22, 392–423 (2005)
8. Canas, G.D., Gortler, S.J.: Surface remeshing in arbitrary codimensions. The Visual Computer 22, 885–865 (2006)
9. Stam, J.: Flows on surfaces of arbitrary topology. ACM Trans. Graph. 22, 724–731 (2003)
10. Alexa, M., Klug, T., Stoll, C.: Direction fields over point-sampled geometry. WSCG 11 (2003)
11. Zwicker, M., Pauly, M., Knoll, O., Gross, M.: Pointshop 3d: An interactive system for point-based surface editing. ACM Transactions on Graphics 21, 322–329 (2000)
12. Fisher, M., Schröder, P.: Design of tangent vector fields. ACM Trans. Graph. 26, 56–63 (2007)
13. Desbrun, M., Kanso, E., Tong, Y.: Discrete Differential Forms for Computational Modeling. In: Grinspun, et al. (eds.) (2006)
14. Zhang, E., Mischaikow, K., Turk, G.: Vector Field Design on Surfaces. ACM Trans. Graph. 25, 1294–1326 (2006)
15. Mitran, J., Nguyen, A.: Estimating surface normals in noisy point cloud data. In: Proc. Nineteenth Annual Symposium on Computational Geometry, pp. 322–328 (2003)
16. Dey, T.K., Li, G., Sum, J.: Normal estimation for point clouds: a comparison study for a voronoi based method. In: Proc. Eurographics 2005, pp. 39–46 (2005)

17. Pang, X.F., Pang, M.Y.: An algorithm for extracting geometric features from point cloud. In: Proc. Information Management, Innovation Management and Industrial Engineering 2009, vol. 4, pp. 78–83 (2009)

18. Alexa, M., Behr, J., Cohen-Or, D., Fleishman, S., Levin, D., Silva, C.T.: Computing and rendering point set surfaces. IEEE Transactions on Visualization and Computer Graphics (S1077-2626) 9, 3–15 (2003)

19. Hoppe, H., Derose, T., Duchamp, T., Mcdonald, J., Stuetzie, W.: Surface reconstruction from unorganized points. In: Proc. Computer Graphics 1992, vol. 26, pp. 71–78 (1992)

20. Xiong, Y.H., Li, G.Q., Han, G.Q., Peng, L.: Remeshing and multi-resolution modeling method based on the glow line. Software, 131–142 (2008)

Preprocessed Global Visibility for Real-Time Rendering on Low-End Hardware

Benjamin Eikel, Claudius Jähn, and Matthias Fischer

Heinz Nixdorf Institute, University of Paderborn

Abstract. We present an approach for real-time rendering of complex 3D scenes consisting of millions of polygons on limited graphics hardware. In a preprocessing step, powerful hardware is used to gain fine granular global visibility information of a scene using an adaptive sampling algorithm. Additively the visual influence of each object on the eventual rendered image is estimated. This influence is used to select the most important objects to display in our approximative culling algorithm. After the visibility data is compressed to meet the storage capabilities of small devices, we achieve an interactive walkthrough of the Power Plant scene on a standard netbook with an integrated graphics chipset.

1   Introduction

State of the art graphics hardware for desktop computers is capable of rendering millions of triangles in real-time. By contrast, netbooks and mobile devices, which are optimized for low power consumption, lack comparably powerful graphics adapters. Nevertheless do many people use a laptop as a replacement for a desktop computer especially in a professional environment. Therefore tasks, which used to be reserved for desktop computers, have to be completed on mobile devices. For instance for business trips or informal sales meetings it may be desirable to present complex 3D scenes to customers on a mobile device. These 3D scenes may originate from computer-aided design (CAD) applications or 3D scans of real objects. Fortunately our approach allows the real-time rendering of complex scenes on high-performance workstations as well as on low-end hardware with weak graphics capabilities. The same preprocessed 3D data can be used for rendering on different devices, therefore several data preparations are no longer necessary.

We exploit the observation, that at most positions in a virtual scene, only a small fraction of the whole geometry contributes almost all visible pixels of the final image. Our global visibility preprocessing determines the importance of the geometry for the visual impression in the rendered image. The resulting visible sets are small enough to be rendered in real-time on a strong graphics machine. Due to the importance estimations of the objects in the scene, the presented approximative culling provides an adjustable frame rate while providing reasonable visible quality even when rendering complex scenes on a netbook.

G. Bebis et al. (Eds.): ISVC 2010, Part I, LNCS 6453, pp. 622–633, 2010.

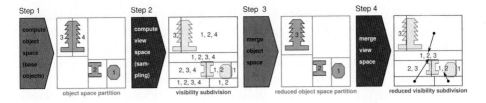

Fig. 1. Preprocessing steps to build the data structures and reduce their storage space. In the reduced object space partition after Step 3 the two objects 3 and 4 have been merged. Thereby references to Object 4 in the visibility subdivision can be saved. In the reduced visibility subdivision, cells with the same visible objects share the same reference set (indicated by the arrows), instead of storing it twice. Additively, the cells with similar reference sets (e.g. $\{1, 2, 3\}$ and $\{2, 3\}$) could also be merged.

2 Outline

We give a short outline of our visibility preprocessing and rendering during runtime. The basic idea of our data structure is a separated partition of view space and object space. A partition of the 3D model's triangles into objects is called *object space OS*. A partition of the view space $VS \subset \mathbb{R}^3$ into view cells is called *visibility subdivision VS*.

Computation of object space and view space take place in four steps (see Figure 1). The first two steps compute objects and partition the view space into cells. The last two steps reduce the space requirement of the computed data structures of the first two steps.

1. First we compute an initial object space partition using a kD tree. The leaves of the kD tree are *base objects* used for computing visibility information in the second step (confer Figure 1). Other spatial data structures producing relatively compact objects with adjustable sizes should be applicable as well. The complexity of the base objects is the first parameter used to adjust the preprocessing costs on the one hand and the quality of the final global visibility information and therefore the image quality on the other hand.
2. Secondly, the initial visibility subdivision is computed with an adaptive sampling approach (see Section 4.1). Each cell of the partition provides a list of objects visible from (almost) all positions of the cell and the estimated influence of the object on the final image.
3. The third step merges the object space's objects for two reasons: Firstly, objects consisting of too few triangles should be merged in order to obtain reasonable batch sizes for the graphics pipeline. Secondly, merging objects is the main mechanism to reduce the data structure's memory consumption (see Section 4.2). By merging two objects, only one object's meta information has to be stored and the number of visibility references inside the view cells is reduced.
4. The fourth step merges the view space's cells. This step also reduces the memory consumption by decreasing the number of cells and visibility references (see Section 4.2).

Step 1 and 2 result in an object space and a view space with finest granularity. Starting with coarser granularity would lead to more imprecise visibility computations. An initial memory reduction of the data structure would cause more imprecise results. Therefore, we start with a fine granularity and reduce the data structure such that the size fits the target device's memory in the end.

The rendering during the scene's walkthrough is quite straightforward. Based on the capabilities of the used hardware and the desired frame rate, a triangle budget is introduced. Until this budget is exhausted, only the most relevant objects of the current cell are rendered – that are the objects, which are expected to produce the most visible pixels weighted by the object's rendering costs. In this manner, even a comparatively small triangle budget leads to a good visual quality. Using the same data structure on stronger hardware with a large budget, all objects that were classified visible can be rendered.

3   Related Work

Culling The literature offers many kinds of different culling algorithms. Our method can be classified by the taxonomy from Cohen-Or et al. [1] as *from-region occlusion culling* algorithm. We apply *image precision* visibility in our *sampling strategy*. Our preprocessing stage can cope with *generic scenes* composed of triangles. Our rendering algorithm limits the number of triangles displayed and therefore applies *approximate visibility*. The geometry of the whole model can be used for occlusion (*all occluders, generic occluders, occluder-fusion*).

Visibility Preprocessing Instead of analyzing the scene geometry's visibility at run-time (on-line algorithm), it can be analyzed in a preprocessing step (off-line algorithm). The preprocessing can be carried out on a powerful machine that stores the global visibility results. These results can be used by a low-end machine to render only the scene's visible parts.

The aspect graph [2] is helpful to theoretically explain the high complexity of exact visibility. It is a subdivision of the view space into regions, in which the object's topological appearance stays the same. For perspective projection in $\mathbb{R}^3$ the aspect graph contains $\mathcal{O}(n^9)$ regions, where n is the number of facets. Due to its high complexity it is not suitable for practical usage. To be able to use preprocessed visibility for practical sizes of 3D scenes, we use a less exact definition of visibility. We are only interested in whether an object is visible or not. Furthermore we do not determine the exact visibility of triangles but approximate the visibility of objects.

Beside the classical approaches based on potentially visible sets (PVS) [3, 4], there exist several methods based on a sampled preprocessed visibility. One approach is presented by Nirenstein and Blake [5]. Here the term aggressive visibility is introduced, meaning that the calculated potentially visible sets are real subsets of the exact visible sets. We do not determine the visibility of geometry on triangle level but on object level, in order to be able to cope with larger scenes. Therefore it is possible that objects classified as visible by our algorithm contain triangles, which are in fact not visible from the tested region. Hence our approach is approximate and not aggressive. Furthermore, the solution from the aforementioned article focuses on an adaptive subdivision of the view space and not the object space. Mattausch et al. [6] present a solution for the integrated partitioning of the object space and the view space. We pick up this idea and also subdivide both spaces. By merging visibility cells and objects respectively we are able to reduce the storage space needed by the partitions. To gain visibility information they use ray casting which yields exact geometric visibility. Because of the uniform ray casting and computational costly calculations – for finding out which object or region to subdivide next – their solution is not well suited for large scenes.

A much faster method is presented by Bittner et al. [7]. By using a casted ray as information for all cells intersected by the ray, global visibility information can be gained much faster. The method is able to preprocess very complex scenes with millions of triangles, but it classifies much more triangles as visible than our approach. The diagram [7] shown for the Power Plant model states that about 1.4 million triangles are visible on average, whereas our sampling approach classifies only 600,000 triangles as visible. Because our goal is rendering on low-end hardware, we are interested in as few visible triangles as possible.

Van de Panne and Stewart [8] present compression techniques that efficiently store the visibility information as a table of single bits. This also allows efficient merging of objects and cells with similar visibility. Because our method stores the number of visible pixels for each object and each cell it is visible in, these compression techniques cannot be applied directly. Furthermore we amply reduce the data structures' storage space to be loadable on netbooks and eschew the additional computation demand involved with compression techniques.

Similar to our approach the HDoV tree [9] uses information about an object's impact on the image displayed on screen. This information is stored in a tree structure together with the geometry and its different levels of detail. The rendering algorithm can use the objects' visual importance to assign triangle budgets to the tree nodes and fitting approximations can be chosen.

Rendering on Mobile Devices There are different classes of rendering algorithms for mobile devices. One class uses a remote server to render an image and sends this image to the mobile device for display (e.g. [10]). Another class uses image-based rendering [11, 12]. A third class of solutions [13–16] renders the geometry on the mobile device directly, whereas the scenes are specifically designed and prepared for mobile rendering. On the one hand, we aim at using PC-like systems that are not as small as some devices utilized in these articles. On the other hand, we want to visualize standard scenes that are normally displayed on CAD workstations.

4 Preprocessing

In Section 4.1 we present our global visibility sampling algorithm. Section 4.2 describes the steps to reduce the data structures' storage space. In Section 4.3 the influences on the rendered image's visual quality are identified.

4.1   Sampling Visibility

The goal of the sampling process is to get a fine granular partition of the view space, where each cell holds references to all base objects that are visible from a point inside the cell, with a rating of the object's importance. To sample the visibility at one position in the scene, the base objects are rendered on the six sides of a cube around that position. For each object it is determined whether it is fully occluded or how many pixels are visible by using hardware assisted occlusion queries. The ratio of the object's number of visible pixels to its number of triangles forms its *importance value*. This value models the influence of the object to the resulting image weighted by the costs for rendering the object.

In order to allow the handling of large scenes with a reasonable number of samples, the samples are not distributed uniformly but are adapted to changes in the visibility throughout the scene. This works as follows: At first, the bounding box of the scene is inserted into a priority queue with a *quality value* of zero. The quality of a cell is a measure of how similar the identified visible objects are at each sampled position in relation to the number of samples and the size of the cell. Then, until the overall number of sampled positions reaches the chosen limit, the cell with the lowest quality value is extracted from the queue. This aims at distributing the sampled positions among those cells, where the most benefit of additional subdivisions is expected, first. This cell is split along each axis into eight equal sized parts (like an octree) and all prior results associated with the extracted cell are distributed over the new cells, which cover the corresponding positions. For each of these new cells, new measurements are performed. At first, all corners are explored – if no corresponding result is yet available from a prior explored cell. Then some additional evenly distributed random positions inside the bounding box are sampled. The number of additional samples is chosen linearly in the diameter of the box, where the factor can be used to express the scaling of the scene. From experience, about five samples are sufficient for a house sized area. The importance value of the objects in the cell is set to the maximum value of all sample points in the cell. In order to calculate the estimated quality of a cell two sets of objects are determined: One containing the objects visible from every sampled position inside the cell and one containing the objects visible from at least one position. The difference in the accumulated triangle counts of these sets is an indication for the uniformity of the visibility inside this cell and is called $diffPoly(cell)$. The quality of a region is defined by: $numberOfSamples(cell)/(diameter(cell) \cdot diffPoly(cell))$. Hence, if many samples inside a small cell indicate nearly the same visible objects, it is very unlikely that a further subdivision of that cell increases the overall quality of the visibility subdivision.

The duration of the sampling process and the quality of the subdivision can mainly be controlled by the parameter for the number of sampled points. This leads to a high reliability and easy usage of this sampling technique.

4.2   Trading Space for (Rendering) Time

The sampling process returns a fine granular and memory-intensive partition of the scene, where each cell references the visible base objects with their expected projected visible size. In the next step, this initial partition is further processed in order to massively reduce the memory requirements and to gain objects of a higher complexity that allow an efficient rendering by the graphics hardware. The basic idea is to merge objects likely to be visible simultaneously and to merge cells from within nearly the same objects are visible. On the one hand, the merging reduces the number of references to store, but on the other hand, additional visible objects may be added to the cells. Obviously there is a trade-off between storage space and rendering time.

The merging is performed in two steps: In a first step, the base objects are merged until all objects consist of a reasonable number of triangles and the number of objects reaches a fixed value. Every merging step leads to a benefit in form of a memory consumption reduction and produces additional costs during rendering by the number of additionally rendered triangles weighted by the volume of the cell where this geometry is now rendered unnecessarily. The goal is to always merge those two objects, which result in the best cost-benefit ratio. Due to the large number of objects, it is impractical to determine this exactly and we apply a heuristic. We do the calculations only in a small constant working set of the smallest objects and merge the most promising candidates. When the termination conditions are met, the vertex ordering for all objects is optimized for vertex cache locality to improve rendering efficiency (see [17]).

In a second step, visibility cells are merged until their number also reaches a fixed value. This is done analogously to the merging of objects, where the costs are again defined by the additional rendering costs and the benefit by the amount of saved memory. If the boundaries of the merged cells do not form a box, the boxes of both cells are preserved and same visibility information is shared.

Through the merging, the resulting partitions have now been adapted to the memory size of the target machine.

4.3   Adjusting the Quality

When approximating global visibility there are basically two kinds of errors that can occur: If visible geometry is classified as invisible, it leads to missing parts in the rendered images. If occluded geometry is classified as visible, it leads to an increased rendering time and hence a lower frame rate – or if we use a fixed triangle budget, also in a decreasing visual quality. Another source of error in the presented technique is that inaccurate importance values of objects may result in a decreased image quality if less visible parts are rendered while actually more important parts are left out. To what extend these errors occur can be adjusted at several stages by different parameters: The larger the size of the base objects is chosen, the more triangles are classified as visible when only a fraction of the object is visible. On the other hand, larger objects reduce the possibility for visible geometry not to be identified as such, due to larger objects being easier to find from a random position. During the sampling process, accuracy is mainly to be controlled by the number of samples. The more samples are used, the better the objects are classified and the more accurate the importance values are gained. In order not to miss small objects, an adequate screen resolution, which should correspond to that of the target system, is necessary. Finally, during the merging step, the more objects and regions are merged, the more invisible geometry is classified as visible, while the amount of unidentified visible geometry may decrease. The overall quality of the partitions can be controlled according to the available time and memory for the preprocessing. The data structures can be adapted to the hardware resources of a wide range of target systems with different configurations.

5   Rendering

The presented technique allows a fluent, high quality interactive visualization of scenes within the realms of possibility, even when the complexity of all visible objects exceeds the capabilities of the used hardware. During the walkthrough, we utilize the graphics hardware to render the parts of the scene that have the greatest impact on the resulting image. When rendering a frame the visibility cell containing the observer's position is determined. At first frustum culling is applied to the set of objects from that cell. The remaining visible objects are rendered in descending order of importance. The number of drawn triangles is accumulated and the current frame ends when a triangle limit has been reached. If the observer does not move in the next frame, the rendering can continue at this point, until all referenced visible objects have been displayed.

If the graphics card is capable of rendering all objects in real time, this allows a very efficient rendering of visible geometry without any overhead for complex operations like occlusion tests. If the graphics card cannot handle all objects in real time, the ordering of the objects according to their importance still leads to a reasonable visual quality while preserving an adjustable frame rate.

6   Results

In this evaluation section, we analyze the number of rendered triangles, the rendering time and the visual quality of our algorithm, compared to a standard frustum culling and a conservative occlusion culling algorithm. Here we use an implementation of the coherent hierarchical culling algorithm (CHC) [18] with a kD tree with a leaf size of 20 k triangles as data structure, which yields good average running times.

The important constituents of the evaluation are the time needed for preprocessing a complex scene and the rendering efficiency and quality on a device with weak graphics adapter.

Systems. We used a Dell Precision T3500 workstation (Intel Xeon W3503 CPU (2×2.4 GHz), 12 GiBytes main memory, NVIDIA GeForce 9800 GT (1 GiByte memory) graphics adapter) for the preprocessing. As low-end device we used a Dell Latitude 2100 netbook (Intel Atom N270 CPU (2×1.6 GHz), 1 GiByte main memory, Intel 945GSE integrated graphics chipset (8 MiBytes memory)). Hardware assisted occlusion queries are not available on the netbook, which prevents the application of the CHC algorithm on this system.

Scene and Camera Path. For the evaluation of our algorithms we used the Power Plant model [19] with 12.7 million triangles. It is an architectural CAD model, which offers a pleasantly high amount of occlusion by the single parts of the building. The model was partitioned using a kD tree with a maximum of 150 triangles per leaf cell. The initial object space consists of nearly 150,000 objects.

Our camera path begins inside the model's main building with the camera pointing upwards where many tubes are visible. Then the camera moves outside, while the amount of geometry inside the frustum decreases. Beginning with the 380th waypoint, the main building containing nearly all objects is again contained in the viewing frustum, while the camera circles around the scene. Then the camera slowly moves away during the last hundred waypoints.

6.1   Sampling Efficiency

Our sampling approach was used with the object space, described in the previous section, to create a visibility subdivision. The scene has to be rendered six times for each sampled position. Therefore the overall time needed for the sampling process is mainly determined by the graphic capabilities of the device used for preprocessing. One limitation is introduced by the size of the base objects, which are typically too small to fully exploit the capacities of the graphics card (as the batch sizes are too small).

We created samples for 10,000 positions with a resolution of 720×720 pixels for each side of the sampling cube. The originating visibility subdivision consists of about 5,000 visibility cells that are containing nearly 35 million references to visible objects. On the test system, the overall sampling process took 3.5 hours.

6.2   Adapting to the Memory Requirements

After sampling, the overall memory consumption for the visibility subdivision and the base objects is over three gigabytes of data. In order to obtain a data set of practicable size, we reduce the *initial partitions* to two target states: One *intermediate state* that can be used for rendering on a high-performance computer with at least 2 GiBytes main memory and one *fully reduced state* that can be used for rendering on a low-performance computer with 1 GiByte main memory. For the intermediate state only small objects were merged until there were no objects with less than 200 triangles. The fully reduced state was created by merging the objects until there were no objects with less than 600 triangles and merging the visibility cells until there were at most 2000 cells. Here, the object's average size is about 1100 triangles. The time needed for reaching the fully reduced state was about 2.6 hours, using a working set of size 400. The data structures of the fully reduced state can be stored in 760 MiBytes memory and therefore can be loaded on our test system with 1 GiByte main memory.

Summarizing, we performed the complete preprocessing of a scene with 12.7 million triangles in about six hours.

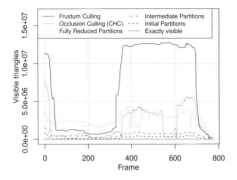
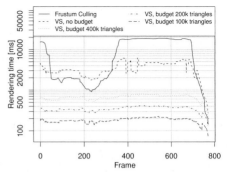

Fig. 2. Number of triangles that are classified visible by different rendering methods while traversing a camera path

Fig. 3. Comparison of rendering times on our test netbook. Mean frame times: Frustum Culling 9.7 s, VS (no budget) 3.5 s, VS (400k) 665 ms, VS (200k) 347 ms, VS (100k) 183 ms.

6.3   Rendering Performance and Quality

The diagram in Figure 2 shows the number of triangles that are classified as visible and thereby rendered when using different rendering algorithms while traversing the camera path. When determining the exact visibility – in our notion, that a triangle is visible if it contributes at least one pixel to the resulting image – only some hundred to a few thousand triangles are visible (see the gray line at the bottom). Rendering the original scene with frustum culling (the black curve at the top) only, naturally leads to the highest amount of geometry, whenever many objects lie in the frustum. The CHC algorithm (shown by the light gray curve) performs unevenly. When, for example, most of the geometry is hidden behind the chimney (around waypoint 600), it performs very well. But when the outer objects can be seen through the borders of the main building (around waypoint 680), much occluded geometry is classified as visible.

The other three curves in Figure 2 show our proposed rendering algorithm while using the initial, the intermediate and the fully reduced partitions respectively. By using the initial partitions, the algorithm only sends a small fraction of triangles inside the frustum to the rendering pipeline and yields better results in most of the cases compared to the CHC algorithm. Not much more triangles are classified as visible than that are exactly visible, which supports our sampling approach. Furthermore this diagram shows the trade-off between storage space and accuracy of the visibility data. The smaller the visibility subdivision gets in terms of storage space the more triangles are visible from the visibility cells. But even the reduced visibility subdivision achieves very good results compared to the frustum culling approach and is in some cases at least as good as the CHC algorithm. These results can be achieved with greatly reduced storage space.

Figure 3 shows the *rendering times* for the Power Plant model achieved by our algorithm with a storage space reduced visibility subdivision on the netbook. With frustum culling rendering, one frame may take over 18 s to complete. But even in this restricted setting it is possible to visualize such large scenes on mobile devices by using our proposed rendering algorithm. When restricting the maximum number of triangles rendered in one frame to 100,000, about five frames per second can be achieved almost constantly over the camera path.

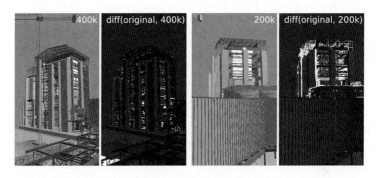

Fig. 4. Screenshots of the approximated rendering with a budget of 400 k triangles (1st image), 200 k triangles (3rd image) and pixel difference between the approximated rendering and rendering of the original model highlighted in white (2nd and 4th image)

To examine the *visual quality* (percentage of equally colored pixels) we compare the images generated by rendering the original Power Plant model with the images generated by our approximative culling algorithm while traversing the camera path. The Figure 4 gives an impression of how the visual quality is perceived by the user. The diagram in Figure 5 shows the distribution of the visual quality for the three different states of the partitions and for five different triangle budgets. If no budget is used only very few visible objects are not classified as visible due to sampling errors, leading to minor image errors in some cells on the camera path. The less triangles are rendered per frame and the stronger the data structure is compressed, the lower the visible quality becomes. If the budget is limited to only 100 k triangles using the fully reduced partitions, the rendered images still show the most important parts of the scene. An interactive navigation through the scene is still possible, but in order to observe smaller details, one has to stop the camera movement and wait a few frames until all details are shown.

So the overall performance of the presented method, in speed and image quality, depends on several parameters and properties:

- The invested preprocessing time,
- the design of the virtual scene (number of primitives, amount of occlusion),

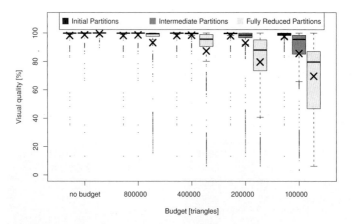

Fig. 5. Distributions of the visual quality for the initial partitions using the approximative culling rendering algorithm with different triangle budgets (boxes range from lower to upper quartile, black line shows the median, black cross shows the mean)

- the memory size of the target machine
- and finally on the capabilities of the target machine's graphics hardware.

7   Conclusion

We have presented a solution to process complex generic scenes and prepare them for the rendering on different kinds of devices. By using the Power Plant scene we were able to demonstrate that the preprocessing can be carried out with acceptable resources. The data structures can be reduced in storage space to a size that fits in a netbook's memory. Our approximate culling algorithm is finally able to provide real-time walkthroughs on low-end graphics hardware with a reasonable quality.

In order to further reduce the amount of rendered geometry necessary for achieving good visual results, the presented approximative culling method is an ideal starting point for the development of new LOD (level of detail) methods. An approach similar to the one of El-Sana et al. [20] could be applicably. Instead of omitting less important objects, the importance estimation can be used to select an appropriate level of detail for each visible object. This extension should allow an interactive walkthrough of complex CAD-generated scenes even on smartphones.

Acknowledgement. Partially supported by DFG grant Me 872/12-1 within SPP 1307 and Research Training Group GK-693 of the Paderborn Institute for Scientific Computation (PaSCo). Benjamin Eikel holds a scholarship of the International Graduate School Dynamic Intelligent Systems.

References

1. Cohen-Or, D., Chrysanthou, Y., Silva, C.T., Durand, F.: A survey of visibility for walkthrough applications. IEEE Trans. Vis. Comput. Graph. 9, 412–431 (2003)
2. Plantinga, H., Dyer, C.R.: Visibility, occlusion, and the aspect graph. International Journal of Computer Vision 5, 137–160 (1990)
3. Airey, J.M., Rohlf, J.H., Brooks Jr., F.P.: Towards image realism with interactive update rates in complex virtual building environments. Computer Graphics 24, 41–50 (1990)
4. Teller, S.J., Séquin, C.H.: Visibility preprocessing for interactive walkthroughs. Computer Graphics 25, 61–70 (1991)
5. Nirenstein, S., Blake, E.H.: Hardware accelerated visibility preprocessing using adaptive sampling. In: Eurographics Symposium on Rendering, pp. 207–216 (2004)
6. Mattausch, O., Bittner, J., Wonka, P., Wimmer, M.: Optimized subdivisions for preprocessed visibility. In: Proc. of GI 2007, pp. 335–342 (2007)
7. Bittner, J., Mattausch, O., Wonka, P., Havran, V., Wimmer, M.: Adaptive global visibility sampling. ACM Transactions on Graphics 28, 1–10 (2009)
8. van de Panne, M., Stewart, A.J.: Effective compression techniques for precomputed visibility. In: Rendering Techniques, pp. 305–316 (1999)
9. Shou, L., Huang, Z., Tan, K.L.: The hierarchical degree-of-visibility tree. IEEE Transactions on Knowledge and Data Engineering 16, 1357–1369 (2004)
10. Yoon, I., Neumann, U.: Web-based remote rendering with IBRAC (image-based rendering acceleration and compression). Computer Graphics Forum 19, 321–330 (2000)
11. Chang, C.F., Ger, S.H.: Enhancing 3D graphics on mobile devices by image-based rendering. In: Chen, Y.-C., Chang, L.-W., Hsu, C.-T. (eds.) PCM 2002. LNCS, vol. 2532, pp. 1105–1111. Springer, Heidelberg (2002)
12. Boukerche, A., Feng, J., de Araujo, R.B.: A 3D image-based rendering technique for mobile handheld devices. In: Proc. of WOWMOM 2006, pp. 325–331 (2006)
13. Nurminen, A.: m-LOMA - a mobile 3D city map. In: Proc. of Web3D 2006, pp. 7–18 (2006)
14. Rodrigues, M.A.F., Barbosa, R.G., Mendonça, N.C.: Interactive mobile 3D graphics for on-the-go visualization and walkthroughs. In: Proc. of SAC 2006, pp. 1002–1007 (2006)
15. Nurminen, A.: Mobile, hardware-accelerated urban 3D maps in 3G networks. In: Proc. of Web3D 2007, pp. 7–16 (2007)
16. Silva, W.B., Rodrigues, M.A.F.: A lightweight 3D visualization and navigation system on handheld devices. In: Proc. of SAC 2009, pp. 162–166 (2009)
17. Sander, P.V., Nehab, D., Barczak, J.: Fast triangle reordering for vertex locality and reduced overdraw. ACM Transactions on Graphics 26, Article No.: 89 (2007)
18. Bittner, J., Wimmer, M., Piringer, H., Purgathofer, W.: Coherent hierarchical culling: Hardware occlusion queries made useful. Computer Graphics Forum 23, 615–624 (2004)
19. The Walkthru Group: Power plant model. Internet page University of North Carolina at Chapel Hill (2001), http://gamma.cs.unc.edu/POWERPLANT/
20. El-Sana, J., Sokolovsky, N., Silva, C.T.: Integrating occlusion culling with view-dependent rendering. In: Proc. of VIS 2001, pp. 371–378 (2001)

A Spectral Approach to Nonlocal Mesh Editing

Tim McGraw and Takamitsu Kawai

Department of Computer Science and Electrical Engineering,
West Virginia University

Abstract. Mesh editing is a time-consuming and error prone process
when changes must be manually applied to repeated structures in the
mesh. Since mesh design is a major bottleneck in the creation of computer games and animation, simplifying the process of mesh editing is an
important problem. We propose a fast and accurate method for performing region matching which is based on the manifold harmonics transform.
We then demonstrate this matching method in the context of *nonlocal
mesh editing* - propagating mesh editing operations from a single source
region to multiple target regions which may be arbitrarily far away. This
contribution will lead to more efficient methods of mesh editing and
character design.

1 Introduction

Mesh editing can be a time-consuming task when similar changes must be made
to multiple regions of the mesh. Consider the problem of editing the limbs of
an animal, for example the hooves of horse. If an artist wishes to change the
shape of the hooves, the editing procedure must be manually applied to all four
hooves. This process is time-consuming and prone to error. In the case of meshes
representing inorganic structures, such as buildings, there may be dozens of
repeated structures, such as windows, making the problem even more complex.
We present a framework for nonlocal mesh editing and a method for quickly
identifying similar substructures in a mesh. We demonstrate our technique on
several meshes representing organic characters and inorganic structures.

Nonlocal approaches to image processing address the redundant nature of
many natural images due to periodicity and symmetry. This redundancy occurs,
for example, when the texture of a wall contains many similar bricks, or two eyes
are visible on the same face. The drawback of these methods is the search time
required to find similar regions. The classical image processing assumption of
smoothness is a local property, and may be assessed using differential operators
or local neighborhood searches. Nonlocal methods typically involve searches for
similarity over the entire image domain, and typically show improved results
over local methods at the expense of speed.

Many meshes also have the property of nonlocal self-similarity. The structure
of many living creatures is characterized by bilateral (left-right) symmetry. There

G. Bebis et al. (Eds.): ISVC 2010, Part I, LNCS 6453, pp. 634–643, 2010.

is geometric similarity between pairs of arms, legs, and facial features. Many non-living objects are also characterized by repetition : a car may have multiple wheels, a building may have periodic architectural details, such as windows and doors.

We propose to simplify mesh editing by identifying such redundancies and automatically propagating mesh editing operations on one structure to other similar structures. We quantify similarity of mesh regions by using a region spectrum determined from a frequency domain decomposition of the mesh.

2   Background

Our approach to propagating editing operations is inspired by the diverse concepts of nonlocal image processing and manifold harmonics.

2.1   Nonlocal Image Processing

Nonlocal image denoising was introduced by Buades et al. [1] in 2005. Many previous approaches to image denoising were based on local processing (such as local averaging). Nonlocal methods are based on the observation that due to texture and repetition, there may be nonlocal redundancy in an image which can be exploited for denoising. The idea was extended to denoising of video sequences [2], and also applied to other image processing problems, such as image demosaicing [3]. Nonlocal redundancy can be used as a regularizer for inverse problems, such as super-resolution [4] which has commonly used local smoothness as a standard constraint.

Searching over the entire image domain for similar regions can make these algorithms slow. Much recent work has focused on improving the efficiency of nonlocal methods. One approach is to restrict the search to large neighborhoods rather than the entire image.

2.2   Manifold Harmonics

The manifold harmonic basis (MHB) is analogous to the sinusoidal basis implied by the Fourier transform - both are eigenfunctions of Laplacian operators defined on their respective spaces. The Fourier transform operates on functions of real variables, but the manifold harmonic transform (MHT) operates on functions whose domain is a graph representing the connectivity of the mesh.

Vallet and Lévy [5] presented a method to convert the geometry of a mesh into frequency space for the purposes of mesh smoothing. The eigenfunctions of the Laplace-Beltrami operator are used to define Fourier-like basis functions called manifold harmonics. High frequency noise and other details can be removed by low-pass filtering. The manifold harmonics are described in further detail in section 4.

Spectral methods have been useful in many area of mesh processing, including segmentation [6] and deformation [7]. The mesh spectrum has previously been applied to matching and correspondence problems [8,9]. In contrast to earlier work, we are not considering the problem of comparing multiple meshes, but comparing different parts of the same mesh. This permits us to quickly compute an approximation of the spectrum, as we will demonstrate.

3 Our Approach to Nonlocal Mesh Editing

We utilize the MHT to estimate the frequency spectrum for the region being edited. The spectrum has several properties which make it efficient for region comparison:

- **Location invariance:** The spectrum is unchanged by translating the region.
- **Rotation invariance:** The spectrum is unchanged by changing the orientation of the region. This is important because similar regions may be rotated due to the pose of the mesh.
- Can be made **scale invariant:** It is not addressed in this work, but the matching can be made scale invariant by normalization of the spectra.
- **Conciseness:** A spectrum of 1000 coefficients is sufficient to represent most meshes for reconstruction purposes. However, by smoothing and subsampling the spectrum we can represent each region with a few hundred scalar values for matching applications.

Regions with similar spectra are candidates for more accurate matching using mesh registration. This step yields an error metric and a transformation matrix which can be used to propagate mesh edits.

Our mesh editing framework consists of the following steps

- Compute MHT for entire mesh (precomputed offline)
- Select region of interest (ROI)
- Estimate the spectrum for the ROI
- Estimate the spectrum for all other regions
- Perform mesh registration on regions whose spectra match. This step yields a set of target regions, and transformation matrices.
- Transform editing operations from the ROI to all target regions.

4 Implementation Details

In order to perform matching using the local spectra it is necessary to compute the MHB for the entire mesh. This involves computing the mesh Laplacian matrix, and the eigenvalue decomposition of this matrix. This computation can be performed once and stored in a file rather than recomputing it for every editing operation. See Vallet and Lévy [5] for more details about the numerical methods.

4.1 Manifold Harmonic Transform

The manifold harmonic basis (MHB) is analogous to the sinusoidal basis implied by the Fourier transform - both are eigenfunctions of Laplacian operators defined on their respective spaces. The Fourier transform operates on functions of real variables, but the manifold harmonic transform (MHT) operates on functions whose domain is a graph representing the connectivity of the mesh.

Given a triangulated mesh with n vertices we may compute m basis functions H^k, $k = 1...m$ called the manifold harmonic basis. Each basis function is assumed to be a piecewise linear function represented by its value H_i^k at each vertex, i. These basis functions, in matrix form, can be used to transform the vertices back and forth between their native geometric space and the frequency domain. Let $x = [x_1, x_2, ...x_n]$, $y = [y_1, y_2, ...y_n]$, and $z = [z_1, z_2, ...z_n]$ be the geometric coordinates and $\tilde{x} = [\tilde{x}_1, \tilde{x}_2, ...\tilde{x}_m]$, $\tilde{y} = [\tilde{y}_1, \tilde{y}_2, ...\tilde{y}_m]$, $\tilde{z} = [\tilde{z}_1, \tilde{z}_2, ...\tilde{z}_m]$ be the frequency coordinates.

The frequency domain vertices can be computed using the manifold harmonic transform

$$\tilde{x} = xDH, \tag{1}$$

where D is a lumped mass matrix which depends on triangle areas within the mesh. Similar expressions hold for $\tilde{y}, \tilde{z}$.

The mesh may be reconstructed from the frequency domain vertices by

$$x = \tilde{x}H', \tag{2}$$

and similar expressions can be written for y, z.

Spatial frequencies, ω_i, of the basis functions are related to the eigenvalues, λ_i, associated with each eigenvector by $\omega_i = \sqrt{\lambda_i}$. Plots of several MHB functions for a horse mesh are shown in Figure 1.

4.2 Local Spectrum Estimation

The key to our fast approach to local spectrum estimation is to not compute the MHT for the region around each vertex, but to reuse the global mesh harmonic basis functions to estimate the local MHT. Since the MHT involves computing eigenvalues of a large matrix, this approximation saves much computation time.

The local spectrum is estimated by restricting the basis functions to the selected region, R, by $G = H_j^k$ for $k \in [0, k_{max}]$ and $j \in R$. Let E denote a similar restriction of D. Then the approximate MHT of the vertices in R, x_R, is given by

$$\tilde{x_R} = x_R EG. \tag{3}$$

Similar expressions hold for the y and z coordinates.

A rotationally invariant spectrum is obtained by combining the frequency domain coefficients

$$\tilde{r} = \sqrt{\tilde{x_R}^2 + \tilde{y_R}^2 + \tilde{z_R}^2}. \tag{4}$$

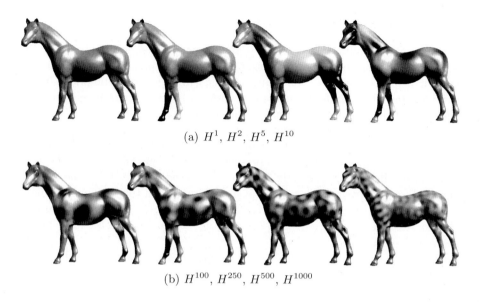

(a) H^1, H^2, H^5, H^{10}

(b) H^{100}, H^{250}, H^{500}, H^{1000}

Fig. 1. Selected manifold harmonic basis functions computed for the horse mesh

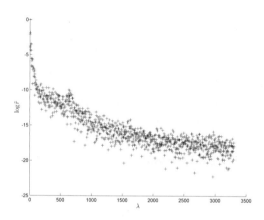

Fig. 2. Rotationally invariant spectrum $\tilde{r} = \sqrt{\tilde{x}^2 + \tilde{y}^2 + \tilde{z}^2}$ computed for the horse mesh and plotted on a semilog scale against eigenvalue, λ

A plot of the rotationally invariant spectrum of the horse mesh is shown in Figure (2). The coefficients tend to decay exponentially with increasing frequency.

4.3   Defining Regions

We define mesh regions as sets of vertices within a threshold geodesic distance from a selected center point. Given the center point we use the fast marching method [10] to find all of the vertices belonging to the region of interest. It may not be the case that all of the vertices in the ROI will be edited, but defining the regions in this way allows us to exploit the rotation invariance of the region spectrum.

4.4 Spectrum Similarity Metric

There has been much previous work on matching frequency domain spectra in application areas such as content-based audio retrieval and image analysis. Some common spectrum similarity measures are based on L_1 (Manhattan) and L_2 (Euclidean) distances [11], cross correlation [12], and information theoretic approaches [13]. After experimenting with these approaches we determined that the most robust region matching was obtained with a combination of Euclidean distance and cross correlation. Our measure, $d(a, b)$ is given by

$$d_E(a, b) = \sqrt{\sum_{i=1}^{n}(a_i - b_i)^2}$$

$$d_{cc}(a, b) = \frac{\sum_{i=1}^{n}(a_i - \bar{a})(b_i - \bar{b})}{(n - 1)s_a s_b}$$

$$d(a, b) = d_E(a, b) - d_{cc}(a, b) \tag{5}$$

where $\bar{a}$ is the mean of spectrum a, and s_a is the sample standard deviation of a.

Plots of the spectral similarity given several meshes and selected vertex sets are shown in Figure (3). Darker colors indicate a better match. Note that the selected left ear of the bunny matches the right ear, the selected finger of the hand matches the other 4 fingers, and the selected foot of the horse matches the other 3 feet. Also note that the matching function changes slowly. For large meshes we sparsely sample the matching function over the mesh, then find local maxima using hill-climbing. We then cluster the local maxima using the mean-shift algorithm [14], and take the vertex with best matching of the matching function in each cluster as the center point of a target region. The results of clustering and center point determination are shown in Figure 4. The axes draw in each region have their origin at the center point, and the axis directions correspond to the coordinate transformation determined by the mesh registration procedure describe in the following section.

The hand and bunny results demonstrate the rotation invariance of our spectral matching technique. Even though the fingers of the hand and the ears of the rabbit have different orientations we can still detect matches in these regions.

4.5 Mesh Registration

It is known that meshes may be *cospectral* (i.e. share the same frequency spectrum) but not be identical [15]. Due to this fact we must perform a geometrical comparison between regions whose spectra are similar. The iterative closest

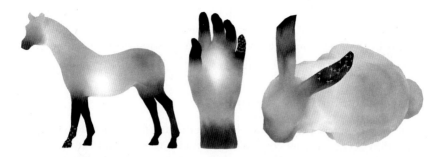

Fig. 3. Matching function for 3 different meshes. The ROI vertices are marked with '+'. Darker colors indicate better match.

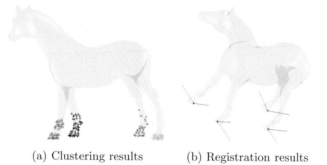

(a) Clustering results   (b) Registration results

Fig. 4. Clustering of the matching function (left) reveals 4 regions of similarity. After ICP registration of the ROI with these 4 regions the bilateral symmetry of the mesh is revealed in the plots of the coordinate frames of each region.

point (ICP) algorithm [16] is a method for mesh registration which minimizes an error metric. The error metric has value zero when two meshes are perfectly aligned. The ICP algorithm also yields a transformation matrix which allows us to transform coordinates from one region to another. We use a variant of ICP called *stochastic ICP* since it is less likely to become stuck in local minima of the energy functional during optimization. Mesh registration results are shown in Figure 4.

4.6   Propagating Editing Operations

Similar regions of the mesh may not be triangulated in the same way, so propagating changes in vertex location is problematic. One solution is to parameterize [17] the source and target regions to same domain. Then the target region may be retriangulated, or the vertex deformation may be interpolated.

In our experiments we use Laplacian mesh editing [18] to apply translations, rotations and scaling transformations to group of vertices with no need for exact vertex correspondence. Similar mesh editing techniques, such as differential

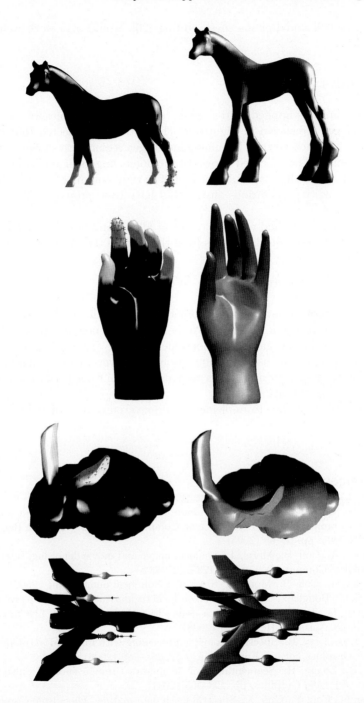

Fig. 5. Original meshes with ROI vertices plotted with '+', and automatically detected target regions in white (left). Results after ROI deformations have been automatically propagated to all target regions (right).

coordinates [19] and Poisson mesh editing [20], would also work well in this framework.

5 Results

Our algorithm was applied to the horse, hand, and bunny meshes, and the resulting images are shown in Figures 5. The initial user-selected ROI is shown with an '+' at each vertex. The matching target regions are shown in white in the left column. The repeated hooves of the horse, fingers of the hand, and ears of the bunny were all successfully located. The right column of images shows the results after editing operations on the ROI have been propagated to all target regions.

In all cases our method was able to detect all relevant matches and accurately propagate the vertex displacements from the ROI to the target regions.

6 Conclusions and Future Work

We have presented a framework for nonlocal mesh editing, and a fast method for performing mesh region matching. We demonstrated nonlocal mesh editing on several meshes, both organic and inorganic in nature. The approach we present is flexible enough to be used in conjunction with several mesh editing methods. The nonlocal mesh editing process relieves the user from performing repeated mesh editing operations.

In the future we plan to implement scale invariance into the editing process and investigate our proposed spectral matching method in the context of mesh denoising.

References

1. Buades, A., Coll, B., Morel, J.: A non-local algorithm for image denoising. In: IEEE Computer Society Conference on Computer Vision and Pattern Recognition, CVPR 2005, pp. 60–65 (2005)
2. Buades, A., Coll, B., Morel, J.: Nonlocal image and movie denoising. International Journal of Computer Vision 76, 123–139 (2008)
3. Buades, A., Coll, B., Morel, J., Sbert, C.: Non local demosaicing (2007)
4. Peyré, G., Bougleux, S., Cohen, L.: Non-local regularization of inverse problems. In: Forsyth, D., Torr, P., Zisserman, A. (eds.) ECCV 2008, Part III. LNCS, vol. 5304, pp. 57–68. Springer, Heidelberg (2008)
5. Vallet, B., Lévy, B.: Spectral geometry processing with manifold harmonics. Computer Graphics Forum (Eurographics) 27, 251–260 (2008)
6. Liu, R., Zhang, H.: Segmentation of 3d meshes through spectral clustering. In: Proc. of Pacific Graphics, pp. 298–305 (2004)
7. Rong, G., Cao, Y., Guo, X.: Spectral mesh deformation. The Visual Computer 24, 787–796 (2008)
8. Reuter, M., Wolter, F.E., Peinecke, N.: Laplacian-Beltrami spectra as 'shape-DNA' for shapes and solids. Computer Aided Geometric Design 38, 342–366 (2006)

9. Jain, V., Zhang, H.: Robust 3d shape correspondence in the spectral domain. In: Proc. IEEE Int. Conf. on Shape Modeling and Applications, pp. 118–129 (2006)
10. Kimmel, R., Sethian, J.: Computing Geodesic Paths on Manifolds. Proceedings of the National Academy of Sciences of the United States of America 95, 8431–8435 (1998)
11. Clark, B., Frost, T., Russell, M.: UV spectroscopy: techniques, instrumentation, data handling. Springer, Heidelberg (1993)
12. Van der Meer, F., Bakker, W.: CCSM: Cross correlogram spectral matching. International Journal of Remote Sensing 18, 1197–1201 (1997)
13. Chang, C.: An information-theoretic approach to spectral variability, similarity, and discrimination for hyperspectral image analysis. IEEE Transactions on Information Theory 46, 1927–1932 (2000)
14. Comaniciu, D., Meer, P.: Mean shift: A robust approach toward feature space analysis. IEEE Transactions on pattern analysis and machine intelligence 24, 603 (2002)
15. van Dam, E., Haemers, W.: Which graphs are determined by their spectrum? Linear Algebra and its Applications 373, 241–272 (2003)
16. Besl, P., McKay, N.: A method for registration of 3-D shapes. IEEE Transactions on pattern analysis and machine intelligence, 239–256 (1992)
17. Floater, M., Hormann, K.: Surface parameterization: a tutorial and survey. In: Advances in Multiresolution for Geometric Modelling, pp. 157–186 (2005)
18. Sorkine, O., Cohen-Or, D., Lipman, Y., Alexa, M., Rössl, C., Seidel, H.: Laplacian surface editing. In: Proceedings of the 2004 Eurographics/ACM SIGGRAPH Symposium on Geometry Processing, pp. 175–184. ACM, New York (2004)
19. Lipman, Y., Sorkine, O., Cohen-Or, D., Levin, D., Rössl, C., Seidel, H.: Differential coordinates for interactive mesh editing. In: Proceedings of Shape Modeling International, pp. 181–190 (2004)
20. Yu, Y., Zhou, K., Xu, D., Shi, X., Bao, H., Guo, B., Shum, H.: Mesh editing with poisson-based gradient field manipulation. ACM Transactions on Graphics (TOG) 23, 644–651 (2004)

Markov Random Field-Based Clustering for the Integration of Multi-view Range Images

Ran Song[1], Yonghuai Liu[1], Ralph R. Martin[2], and Paul L. Rosin[2]

[1] Department of Computer Science, Aberystwyth University, UK
{res,yyl}@aber.ac.uk
[2] School of Computer Science & Informatics, Cardiff University, UK
{Ralph.Martin,Paul.Rosin}@cs.cardiff.ac.uk

Abstract. Multi-view range image integration aims at producing a single reasonable 3D point cloud. The point cloud is likely to be inconsistent with the measurements topologically and geometrically due to registration errors and scanning noise. This paper proposes a novel integration method cast in the framework of Markov random fields (MRF). We define a probabilistic description of a MRF model designed to represent not only the interpoint Euclidean distances but also the surface topology and neighbourhood consistency intrinsically embedded in a predefined neighbourhood. Subject to this model, points are clustered in aN iterative manner, which compensates the errors caused by poor registration and scanning noise. The integration is thus robust and experiments show the superiority of our MRF-based approach over existing methods.

1   Introduction

3D surface reconstruction from multi-view 2.5D range images is important for a wide range of applications, such as reverse engineering, CAD and quality assurance, etc. Its goal is to estimate a manifold surface that approximates an unknown object surface using multi-view range images, each of which essentially represents a sample of points in 3D Euclidean space. These samples of points are usually described in local, system centred, coordinate systems and cannot offer a full coverage of the object surface. To reconstruct a complete 3D surface model, we need to register a set of overlapped range images into a common coordinate frame and then integrate them to fuse the redundant data contained in overlapping regions while retain enough data sufficiently representing the correct surface details. However, to achieve both is challenging due to its ad hoc nature. Scanning noise such as unwanted outliers and data loss typically caused by self-occlusion, large registration errors and connectivity relationship loss among sampled points in acquired data often lead to a poor integration. As a result, the reconstructed surface may include holes, false connections, thick and non-smooth or over-smooth patches, and artefacts. Hence, a good integration should be robust to inevitable registration errors and noise. Once multiple registered overlapping range images have been fused into a single reasonable point cloud, many techniques [1–3] can be employed to reconstruct a watertight surface.

G. Bebis et al. (Eds.): ISVC 2010, Part I, LNCS 6453, pp. 644–653, 2010.

2   Related Work

Existing integration methods can be classified into four categories: volumetric method, mesh-based method, point-based method and clustering-based method. The volumetric method [4–7] first divides the space around objects into voxels and then fuses the data in each voxel. But the comparative studies [8, 9] show they are time-consuming, memory-hungry and not robust to registration errors and scanning noise, resulting in poor reconstructed surfaces. The mesh-based method [10–13] first employs a step discontiuity constrained triangulation and then detects the overlapping regions between the triangular meshes derived from successive range images. Finally, it reserves the most accurate triangles in the overlapping regions and reconnects all remaining triangles subject to a certain objective function. Since the number of triangles is usually much larger than that of the sampled points, the mesh-based methods are computationally more expensive. Some mesh-based methods employ a 2D triangulation in the image plane to estimate the local surface connectivity as computation in a 2D sub-space is more efficient. But projection from 3D to 2D may lead to ambiguities if the projection is not injective. The mesh-based methods are thus highly likely to fail in non-flat areas where no unique projection plane exists. The point-based method [14, 15] produces a set of new points with optimised locations. But its integration result is often over-smooth and cannot retain enough surface details in non-flat areas. The clustering-based method [8, 9] employs classical clustering methods to minimise dissimilarity objective functions. It surpasses previous methods as it is more robust to noise and registration errors. Also, the iterative clustering optimises the locations of points and thus generates much fewer ill-shaped triangles. However, this clustering, only based on Euclidean distance, does not consider local surface topology and neighbourhood consistency, leading to errors in non-flat areas. For instance, in Fig.1(a), although point A is closer to B and thus the clustering-based methods wrongly group them together, we would rather group A with C or D to maintain the correct surface topology.

In this paper, we propose a novel integration method. A MRF model is designed based on both statistical and structural information that clustering-based methods neglect. This model is then converted into a specific description

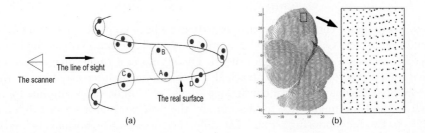

(a)   (b)

Fig. 1. (a) Local topology has a significant effect on the point clustering in non-flat areas (b) X-Y projection of point clouds from reference and registered range images where the gray points are the raw data and the black points are the registered ones

minimised in a clustering manner. The new method retains the advantages of clustering-based methods and is more robust as it uses more information from the input data. The integration is thus reliable in both flat areas and non-flat areas. It is worth mentioning that our method does not require a regular image grid whereas some state-of-the-art techniques [16, 17] rely heavily on it. The registered range images used as the input in this work are actually 3D unstructured point clouds due to large registration errors (Fig.1(b)). Generally, our method can also cope with the more general input–multiple 3D unstructured point clouds.

3   Markov Random Field-Based Clustering

MRF describes a system by local interaction and is able to capture many features of the system of interest by simply adding appropriate terms representing spatial or contextual dependencies into it. For this work, we denote a set of sites $s = \{1, \ldots, n\}$ representing the primitive points and define the label assignment $x = \{x_1, \ldots, x_n\}$ to all sites as a realisation of a family of random variables defined on s. We also define the feature space that describes the sites and understand it as a random observation field with its realisation $y = \{y_1, \ldots, y_n\}$. We also denote a label set $L = \{1, \ldots, m\}$, where each label corresponds to a class centroid.

An optimal labeling should minimise the posterior energy $U(x|y) = U(y|x) + U(x)$. Under the MRF assumption that the observations are mutually independent, the likelihood energy $U(y|x)$ can be computed as $U(y|x) = \sum_{i \in S} V(y_i|x_i)$. In this work, $V(y_i|x_i)$ is the Euclidean distance between point and class centroid:

$$V(y_i|x_i) = \|y_i - C_{x_i}\| = |y_i - C_{x_i}| \tag{1}$$

It can be understood in this way: given a label set C and an observation field y which has an observation y_i at point i, whether i should obey an assignment x_i which assigns it to a centroid C_{x_i} ($C_{x_i} \in C$) depends on the distance between them. The smaller the distance, the larger the probability for such an assignment.

Once we define a neighbourhood $N(i)$ for point i, the prior energy $U(x)$ can be expressed as the sum of different types of clique energies:

$$U(x) = \sum_i V_1(x_i) + \sum_i \sum_{i' \in N(i)} V_2(x_i, x_{i'}) + \sum_i \sum_{i' \in N(i)} \sum_{i'' \in N(i)} V_3(x_i, x_{i'}, x_{i''}) + \cdots \tag{2}$$

Here, single-point cliques $V_1(x_i)$ are set to 0 as we have no preference which label should be better. Our MRF model is inhomogenous and anisotropic due to the specific definition of the neighbourhood. Furthermore, a range image does not have such a discrete property attached to each point as intensity. Considering the inhomogeneous sites with continuous labels, we cannot use a simplified form such as the Ising model [18] to 'discourage' the routine from assigning different labels to two neighbouring points. The difference between the normals is used to evaluate the binary clique energy representing the neighbourhood consistency:

$$U(x) = \sum \sum_{i' \in N(i)} V_2(x_i, x_{i'}) = \sum \sum_{i' \in N(i)} w|\mathbf{A}_{C_{x_i}} - \mathbf{A}_{C_{x_{i'}}}| \tag{3}$$

where $\mathbf{A}_{C_{x_i}}$ and $\mathbf{A}_{C_{x_{i'}}}$ are the unit normal vectors of the centroids C_{x_i} and $C_{x_{i'}}$ respectively. The neighbourhood consistency constraint is thus based on the assumption that point normals should not deviate from each other too much in a small neighbourhood. The minimisation combining two types of energies arises from different value ranges. The range of the likelihood energy is usually dependent on the number and the positions of class centroids, whereas the range of the clique energy depends on the definition and the size of the neighbourhood system and the measurement of the difference between the normals. One weighting parameter w is thus necessary to balance the two kinds of energies.

We define a normal deviation parameter s related to local surface topology:

$$s_i = std(a_{i'}|i' \in N(i)), \quad a_{i'} = \cos\theta_{i'} = (\mathbf{A}_{i'} \cdot \mathbf{z})/|\mathbf{A}_{i'}||\mathbf{z}|, \quad i' \in N(i) \quad (4)$$

where std is the standard deviation function. We choose $\mathbf{z}$ (z axis) as the reference direction because the scanning accuracy of a point depends on the including angle between its normal and the line of sight from the scanner which is along $\mathbf{z}$ [13].

For most MRF applications in computer vision [19–22], 4 neighbouring pixels are chosen to produce a neighbourhood $N(i)$ for a pixel i. But to define $N(i)$ in a 3D unstructured point cloud is more difficult since the concept 'pixel' does not exist here. The simplest method is to use the k-nearest-neighbours algorithm (k-NN) to find the k points closest to i. But this method has its drawbacks. First, to employ k-NN needs extra computational cost. Second, more importantly, the neighbours produced by k-NN cannot deliver surface topological information. Due to incorrect registration and noise, some of the neighbours produced by k-NN may not be on the surface of point cloud. The following calculation of their normals will thus be inaccurate and makes the whole algorithm unreliable. In this work, before the integration, we first do a triangulation for the points in the overlapping regions and find the neighbouring triangles of point i. The vertices of these triangles, excluding i, are defined as the neighbours of i. The collection of all neighbours of i is defined as the neighbourhood of i, written as $N(i)$. Fig.2 shows a neighbourhood. The advantages of this definition are: (1) it reflects the local topology and makes it possible to evaluate the neighbourhood consistency defined as the difference between the normals in the MRF model; (2) there is no extra computational cost as the triangulation has been done at the beginning of the algorithm. Since points may have different number of neighbours, it is necessary to add a normalisation parameter to measure the clique energy. Also, the clique energy defined in this way is not relevant to scanning resolution, but the likelihood energy, described by the distance measurement is directly related to it. We thus use a constant c to balance the different magnitudes between the

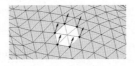

Fig. 2. A neighbourhood and the normalised normal vectors attached to the neighbours

two types of energies. In a general form, $w = s_i \times c \times R/n_i$, where n_i is the number of neighbours of i and R is the scanning resolution of the input images.

If point i lies in a non-flat area such as a crease edge, s_i will probably be a large value and the clique energy will have more weight. Thus the error illustrated in Fig.1 will probably be avoided. If i is located in a flat area, s_i will be quite small or even equal to 0 (in a planar area). In this case, the algorithm is actually downgraded to the classical k-means clustering. This is desired as the k-means clustering works well in flat areas. In other words, whether a point can be assigned to a certain centroid depends upon not only the distance between them but also the local topology and neighbourhood consistency.

In image segmentation where MRF has already been widely used [19–21, 23], the label set is usually defined using an intensity set and its domain remains unchanged in the algorithm. This is a 'pure' MRF labeling problem and can be solved by some well-known methods such as graph cuts and loopy belief propagation [24]. But these methods cannot gurantee a reconstructed mesh with well-shaped triangles. In contrast, the clustering-based methods can achieve that due to the averaging for the calculation of new centroids in the clustering [8, 9]. We thus solve this combinatorial minimisation problem in a clustering manner. In our method, both the label set and its domain are changed iteratively. A double optimisation is thus achieved. One is done by MRF under a given label set and the other one is done by the iteration analogous with the routine of the k-means clustering. The changing label set is the centroids. Once one point i is assigned to a centroid C_{x_i}, it is labeled as C_{x_i} and its normal is also labeled as $\mathbf{A}_{C_{x_i}}$. Analogously, the normal of i' should be labeled using $\mathbf{A}_{C_{x_{i'}}}$. But estimating $\mathbf{A}_{C_{x_{i'}}}$ needs n_i extra nearest neighbour searches for each point and that would significantly slow down the whole algorithm. We thus assume $\mathbf{A}_{C_{x_{i'}}} = \mathbf{A}_{i'}$, so

$$U(x|y) = \sum_i \sum_{i' \in N(i)} \frac{s_i}{n_i} \times c \times R \times |\mathbf{A}_{C_{x_i}} - \mathbf{A}_{i'}| + \sum_i |y_i - C_{x_i}| \qquad (5)$$

4   Implementation

Fig.3 shows the workflow of the new algorithm. We employ the method proposed in [8] to define correspondences and detect overlapping areas. Then, the points in the overlapping areas are triangulated to compute normals and find neighbours.

The initialisation is vital to the integration result and the speed of convergence. Each point in the overlapping area is shifted along its normal towards

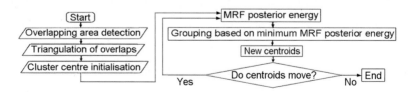

Fig. 3. The workflow of the MRF-based clustering algorithm

its corresponding point by half of the distance between them. For each point shifted from a point of the reference image, a sphere with a radius $r = m \times R$ (m is a parameter controlling the density of the output point cloud) is defined. If some points fall into this sphere, then their original points (without shifting) and normals are retrieved. Their averages are then used to initialise cluster centroids and their normals. The centroid yielding the lowest energy is chosen to label the point. All points labeled with the same centroid are grouped as one class. The centroid of a new class and its normal can be recalculated by computing the mean values of the coordinates of the class members and their normals. The iteration ends when centroids do not change any more. Finally, the new point set consisting of the new centroids and the original points not in the overlapping areas is used to reconstruct a triangular mesh and a watertight surface.

In practice, we use a speeded-up scheme to find the centroid C_{x_i}. We first perform k-NN to find the k centroids closest to i. Then we just search the one minimising $U(x|y)$ out of the k centroids. k is set to 5 in this work to ensure that the one searched can always minimise $U(x|y)$ among all centroids. Otherwise it means that the balance constant c is not set to a reasonable value and the clique energy has too much weight. Thus the new method has the same computational complexity as the k-means clustering. The computational complexity of our algorithm is $O(n_1 \log_2 n_2)$ where n_1 and n_2 are the numbers of points in the overlapping regions from both images and the reference image respectively.

5   Experimental Results and Performance Analysis

In our experiments, the input range images are all downloaded from the OSU Range Image Database (http://sampl.ece.ohio-state.edu/data/3DDB/RID/index.htm). On average, each 'Bird' image has 9022 points and each 'Frog' image has 9997 points. We employed the method proposed in [25] for pairwise registration. Inevitably, the images are not accurately registered, since we found the average registration error (0.30mm for the 'Bird' images and 0.29mm for the 'Frog' images) is as high as half the scanning resolution of the input data. A large registration error causes corresponding points in the overlapping area to move away from each other and the overlapping area detection will thus be more difficult. The final integration is likely to be inaccurate accordingly. Fig.4 and 5 show the comparative integration results produced by existing methods and our method.

Fig.6(a) illustrates the convergence performance of the new method. The new method achieves a high computational efficiency in terms of iteration number required for convergence. We also compute the proportion of the points really affected by clique energy. Here, 'really affected' means one point was not labeled with its closest centroid due to the effect from clique energy. Fig.6(b) shows the statistics over the integration of two Bird images ($0°$ and $20°$) using different parameters. Clique energy gains more weight when the product of c and R is larger. So, Fig.6(b) shows more points are really affected by clique energy when $c \times R = 80$. But it does not mean a better integration can be achieved as shown in Fig.6(c) and (d). We can see that it is important to choose an appropriate c.

Fig. 4. Integration results of 18 'Bird' images. Top left: volumetric method [5]. Top middle: mesh-based method [13]. Top right: fuzzy-c means clustering [9]. Bottom left: k-means clustering [8]. Bottom middle: the triangular mesh produced by the new method. Bottom right: the final integration result produced by the new method.

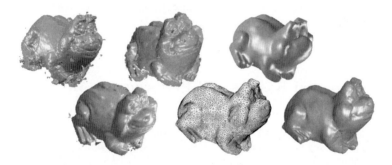

Fig. 5. Integration results of 18 'Frog' images. Top left: volumetric method [5]. Top middle: mesh-based method [13]. Top right: fuzzy-c means clustering [9]. Bottom left: k-means clustering [8]. Bottom middle: the triangular mesh produced by the new method. Bottom right: the final integration result produced by the new method.

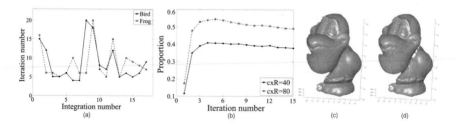

Fig. 6. (a): The convergence performance of the new mehtod. (b): The proportion of the points really affected by clique energy (c): Integration result when $c \times R = 40$ (d): Integration result when $c \times R = 80$. Please note the holes around the neck.

Fig. 7. Integration results of 18 'Bird' images produced by different clustering methods. Left: MRF-based clustering; Middle: fuzzy-c means; Right: k-means.

Fig. 8. Integration re0.7ts (back view) of 18 'Bird' images produced by different clustering methods. From left to right: k-means clustering, fuzzy-c means clustering, the reconstructed surface and triangular mesh produced by MRF-based clustering.

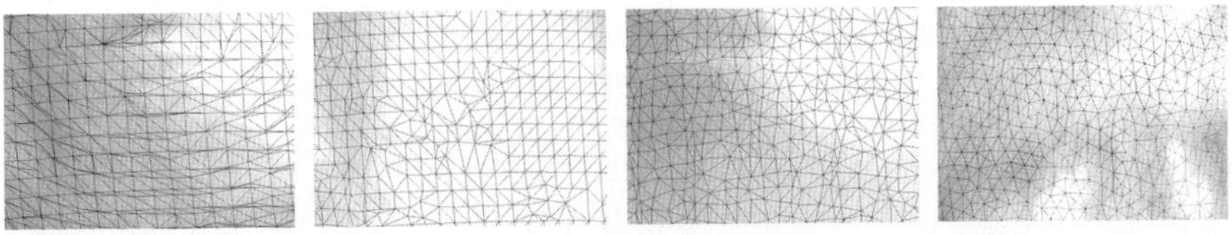

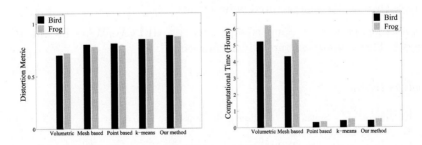

Fig. 9. Triangulated meshes of the integrated surface using different methods. From left to right: volumetric[5], mesh-based method[13], point-based method[15], our method.

Fig. 10. Different performance measures of integration algorithms. Left: distortion metric. Right: computational time.

Due to the different objective functions, it is very difficult to define a uniform metric such as the integration error [8, 9] for comparison because we reject the idea that the closest centroid is the best choice. However, Fig.7 and 8 highlight the visual difference of the integration results produced by classical clustering methods and our method. It can be seen that our method performs better, particularly in the non-flat regions such as the neck and the tail of the bird.

Even so, for a fair comparison, we introduce some measurement parameters widely used but not relevant to the objective function: 1) The distribution of interior angles of triangles. The angle distribution shows the global optimal degree of triangles. The closer the interior angles are to $60°$, the more similar the triangles are to equilateral ones; 2) Average distortion metric [26]: the distortion metric of a triangle is defined as its area divided by the sum of the squares of the lengths of its edges and then normalised by a factor $2\sqrt{3}$. The value of distortion metric is in [0,1]. The higher the average distortion metric value, the higher the quality of a surface; 3) The computational time; Fig.9 and 10 show that the MRF-based algorithm performs better in the sense of the distribution of interior angles of triangles, the distortion metric and the computational time. All experiments were done on a Pentium IV 2.40 GHz computer. Additionally, because there is no specific segmentation scheme involved in the MRF-based clustering, our algorithm saves computational time compared with the techniques using some segmentation algorithms as a preprocessing before the integration [9].

6   Conclusion

Clustering-based methods proved superior to other existing methods for integrating multi-view range images. It has, however, been shown that classical clustering methods lead to significant misclassification in non-flat areas as the local surface topology are neglected. We develop a MRF-based method to tackle this problem and produce better integration results. It does not only focus on minimising the integration errors defined by the Euclidean distance, but also considers the effect from local topology and neighbourhood consistency. The reconstructed surfaces are geometrically realistic since the new method essentially uses more information contained within the input data. Also, it is applicable to more general data sources such as 3D unstructured point clouds.

Acknowledgments. Ran Song is supported by HEFCW/WAG on the RIVIC project. This support is gratefully acknowledged.

References

1. Katulakos, K., Seitz, S.: A theory of shape by space carving. Int. J. Comput. Vision 38, 199–218 (2000)
2. Dey, T., Goswami, S.: Tight cocone: a watertight surface reconstructor. J. Comput. Inf. Sci. Eng. 13, 302–307 (2003)
3. Yemez, Y., Wetherilt, C.: A volumetric techinque for surface reconstruction from silhouette and range data. Comput. Vision Image Understanding 105, 30–41 (2007)

4. Curless, B., Levoy, M.: A volumetric method for building complex models from range images. In: Proc. SIGGRAPH, pp. 303–312 (1996)
5. Dorai, C., Wang, G.: Registration and integration of multiple object views for 3d model construction. IEEE Trans. Pattern Anal. Mach. Intell. 20, 83–89 (1998)
6. Rusinkiewicz, S., Hall-Holt, O., Levoy, M.: Real-time 3d model acquisition. In: Proc. SIGGRAPH, pp. 438–446 (2002)
7. Sagawa, R., Nishino, K., Ikeuchi, K.: Adaptively merging large-scale range data with reflectance properties. IEEE Transactions on PAMI 27, 392–405 (2005)
8. Zhou, H., Liu, Y.: Accurate integration of multi-view range images using k-means clustering. Pattern Recognition 41, 152–175 (2008)
9. Zhou, H.: et al: A clustering approach to free form surface reconstruction from multi-view range images. Image and Vision Computing 27, 725–747 (2009)
10. Hilton, A.: On reliable surface reconstruction from multiple range images. Technical report (1995)
11. Turk, G., Levoy, M.: Zippered polygon meshes from range images. In: Proc. SIG-GRAPH, pp. 311–318 (1994)
12. Rutishauser, M., Stricker, M., Trobina, M.: Merging range images of arbitrarily shaped objects. In: Proc. CVPR, pp. 573–580 (1994)
13. Sun, Y., Paik, J., Koschan, A., Abidi, M.: Surface modeling using multi-view range and color images. Int. J. Comput. Aided Eng. 10, 137–150 (2003)
14. Li, X., Wee, W.: Range image fusion for object reconstruction and modelling. In: Proc. First Canadian Conference on Computer and Robot Vision, pp. 306–314 (2004)
15. Zhou, H., Liu, Y.: Incremental point-based integration of registered multiple range images. In: Proc. IECON, pp. 468–473 (2005)
16. Bradley, D., Boubekeur, T., Berlin, T., Heidrich, W.: Accurate multi-view reconstruction using robust binocular stereo and surface meshing. In: CVPR (2008)
17. Christopher, Z., Pock Thomas, B.H.: A globally optimal algorithm for robust tv-l1 range image integration. In: Proc. ICCV (2007)
18. Li, S.: Markov random field modelling in computer vision. Springer, Heidelberg (1995)
19. Wang, X., Wang, H.: Markov random field modelled range image segmentation. Pattern Recognition Letters 25, 367–375 (2004)
20. Suliga, M., Deklerck, R., Nyssen, E.: Markov random field-based clustering applied to the segmentation of masses in digital mammograms. Computerized Medical Imaging and Graphics 32, 502–512 (2008)
21. Felzenszwalb, P., Huttenlocher, D.: Efficient belief propagation for early vision. International Journal of Computer Vision 70, 41–54 (2006)
22. Diebel, J., Thrun, S.: An application of markov random fields to range sensing. In: Proc. NIPS (2005)
23. Jain, A., Nadabar, S.: Mrf model-based segmentation of range images. In: Proc. Third International Conference on Computer Vision, pp. 667–671 (1990)
24. Szeliski, R., et al.: A comparative study of energy minimization methods for markov random fields with smoothness-based priors. IEEE Transactions on PAMI 30, 1068–1080 (2008)
25. Liu, Y.: Automatic 3d free form shape matching using the graduated assignment algorithm. Pattern Recognition 38, 1615–1631 (2005)
26. Lee, C., Lo, S.: A new scheme for the generation of a graded quadrilateral mesh. Computers and Structures 52, 847–857 (1994)

Robust Wide Baseline Scene Alignment Based on 3D Viewpoint Normalization[*]

Michael Ying Yang[1], Yanpeng Cao[2], Wolfgang Förstner[1], and John McDonald[2]

[1] Department of Photogrammetry, University of Bonn, Bonn, Germany
[2] Department of Computer Science, National University of Ireland,
Maynooth, Ireland

Abstract. This paper presents a novel scheme for automatically aligning two widely separated 3D scenes via the use of viewpoint invariant
features. The key idea of the proposed method is following. First, a number of dominant planes are extracted in the SfM 3D point cloud using
a novel method integrating RANSAC and MDL to describe the underlying 3D geometry in urban settings. With respect to the extracted 3D
planes, the original camera viewing directions are rectified to form the
front-parallel views of the scene. Viewpoint invariant features are extracted on the canonical views to provide a basis for further matching.
Compared to the conventional 2D feature detectors (e.g. SIFT, MSER),
the resulting features have following advantages: (1) they are very discriminative and robust to perspective distortions and viewpoint changes
due to exploiting scene structure; (2) the features contain useful local
patch information which allow for efficient feature matching. Using the
novel viewpoint invariant features, wide-baseline 3D scenes are automatically aligned in terms of robust image matching. The performance of
the proposed method is comprehensively evaluated in our experiments.
It's demonstrated that 2D image feature matching can be significantly
improved by considering 3D scene structure.

1 Introduction

Significant progress has recently been made in solving the problem of robust
feature matching and automatic Structure from Motion (SfM). These advances
allow us to recover the underlying 3D structure of a scene from a number of
collected photographs [1], [2], [3]. However, the problem of automatically aligning two individual 3D models obtained at very different viewpoints still remains
unresolved. Since the captured images are directly linked to the 3D point cloud
in the SfM procedure, 3D points can be automatically related in terms of the
matching of their associated 2D image appearances. Previously a number of successful techniques [4], [5], [6], [7], [8], [9] have been proposed for robust 2D image
matching - a comprehensive review was given in [10]. However the performances
of these techniques are limited in that they only consider the 2D image texture
and ignore important cues related to the 3D geometry. These methods cannot

---

[*] The first two authors contributed equally to this paper.

G. Bebis et al. (Eds.): ISVC 2010, Part I, LNCS 6453, pp. 654–665, 2010.
© Springer-Verlag Berlin Heidelberg 2010

produce reliable matching results of features extracted on wide baseline image pairs. In this paper our goal is to integrate recent advances in 2D feature extraction with the concept of 3D viewpoint normalization to improve the descriptive ability of local features for robust matching over largely separated views.

In predominantly planar environments (urban scenes), fitting a scene with a piecewise planar model has become popular for urban reconstruction [11], [12]. In this paper, we proposed a novel approach to extract a number of dominant planes in the 3D point cloud by integrating RANSAC and MDL. The derived planar structures are used to represent the spatial layout of a urban scene. The 2D image features can be normalized with respect to these recovered planes to achieve viewpoints invariance. The individual patches on the original image, each corresponding to an identified 3D planar region, are rectified to form the front-parallel views of the scene. Viewpoint invariant features are then extracted on these canonical views for further matching. The key idea of the proposed method is schematically illustrated in Fig. 1. Knowing how everything looks like from a front-parallel view, it becomes easier to recognize the same surface from different viewpoints. Compared with some previous efforts on combining 2D feature with 3D geometry [13], [14], our method exploited the planar characteristics of man-made environment and extracted a number of dominant 3D planes to represent its 3D layout. Viewpoint normalization can be performed *w.r.t.* the planes to achieve better efficiency and robustness.

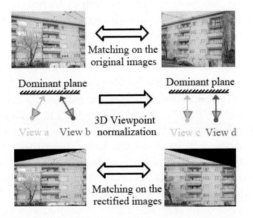

Fig. 1. The major procedure of generating and matching viewpoint invariant features

The remainder of the paper is organized as follow. Section 2 reviews some existing solutions for robust feature matching and 3D model alignment. The proposed method for 3D dominant plane extraction is presented in Section 3. In Section 4, we explain the procedures of 3D viewpoint normalization and propose an effective scheme to match the resulting viewpoint invariant features. In

Section 5, the performance of the proposed method is comprehensively evaluated. We finally conclude with a brief summary in Section 6.

2   Related Work

Automatic 3D scene alignment is a key step in many computer vision applications including large scale 3D modelling, augmented reality, and intelligent urban navigation. Given two sets of 3D points obtained at different viewpoints, the task is to estimate an optimal transformation between them. The most popular class of method for solving this problem is the Iterative closest point (ICP) based techniques [15], [16], [17]. They compute the alignment transformation by iteratively minimizing the sum of distances between closest points. However, the performances of ICP-based methods reply on a good estimation initialization and require good spatial configuration of 3D points. Recently many researchers proposed to enhance the performances of 3D point cloud alignment by referring to their associated 2D images. In [18], an effective method was presented for automatic 3D model alignment via 2D image matching. [19] presented a general framework to align 3D points from SfM with range data. Images are linked to the 3D model to produce common points between range data. [20] presented an automated 3D range to 3D range registration method that relies on the matching of reflectance range image and camera image. In [21], a flexible approach was presented for the automatic co-registration of terrestrial laser scanners and digital cameras by matching the camera images against the range image. These techniques work well for the small observation changes. To produce satisfactory registration results of 3D points clouds captured at significantly changed viewpoints, we need an effective image feature scheme to establish reliable correspondences between wide baseline image pairs.

A large number of papers have reported on robust 2D image feature extraction and matching, cf. [10] for a detailed review. The underlying principle for achieving invariance is to normalize the extracted regions of interest so that the appearances of a region will produce the same descriptors (in an ideal situation) under the changes of illumination, scale, rotation, and viewpoint. Among them the Scale-invariant feature transform (SIFT) [7] is the best scale-invariant feature scheme and the Maximally Stable Extremal Regions (MSER) [6] shows superior affine invariance. In [22], the authors conducted a comprehensive evaluation of various feature descriptors and concluded that the 128-element SIFT descriptor outperforms other descriptor schemes. Robust 2D feature extraction techniques have been successfully applied to various computer vision tasks such as object recognition, 3D modelling, and pose estimation. However, the existing schemes cannot produce satisfactory feature matching over largely separated views because perspective effects will add severe distortions to the resulting descriptors. Recently, many researchers have considered the use of 3D geometry as an additional cue to improve 2D feature detection. A novel feature detection scheme, Viewpoint Invariant Patches (VIP) [13], based on 3D normalized patches was proposed for 3D model matching and querying. In [14], both texture and depth information were exploited for computing a normal view onto the surface. In this way they kept the descriptiveness of similarity invariant features (e.g. SIFT) while achieving extra invariance against perspective distortions. However these methods directly make use of the preliminary 3D model from SfM. Viewpoint normalization with respect to the local computed tangent planes are prone to errors occurred in the process of 3D reconstruction. For predominantly planar scenes (urban scenes), piece-wise planar 3D models are more robust, compact, and efficient for viewpoint normalization of cameras with wide baselines.

3   3D Dominant Plane Extraction

One of the most widely known methodologies for plane extraction is the RANdom SAmple Consensus (RANSAC) algorithm [23]. It has been proven to successfully detect planes in 2D as well as 3D. RANSAC is reliable even in the presence of a high proportion of outliers.

Based on the observation that RANSAC may find wrong planes if the data has a complex geometry, we introduces a plane extraction method by integrating RANSAC and minimum description length (MDL).

Here, we apply MDL for plane extraction, similar to the approach of [24]. Given a set of points, we assume several competing hypothesis, here namely, outliers (O), 1 plane and outliers (1P+O), 2 planes and outliers (2P+O), 3 planes and outliers (3P+O), 4 planes and outliers (4P+O), 5 planes and outliers (5P+O), 6 planes and outliers (6P+O), ect..

Let n_0 points x_i, y_i, z_i be given in a 3D coordinate and the coordinates be given up to a resolution of ϵ and be within range R. The description length for the n_0 points, when assuming outliers (O), therefore is

$$\#bits(points \mid O) = n_0 \cdot (3lb(R/\epsilon)) \tag{1}$$

where $lb(R/\epsilon)$ bits are necessary to describe one coordinate.

If we now assume n_1 points to sit on a plane, n_2 points to sit on the second plane, and the other $\bar{n} = n_0 - n_1 - n_2$ points to be outliers, we need

$$\#bits(points \mid 2P + O) = n_0 + \bar{n} \cdot 3lb(R/\epsilon) + 6lb(R/\epsilon) + n_1 \cdot 2lb(R/\epsilon)$$
$$+ n_2 \cdot 2lb(R/\epsilon) + \left[\sum_{i=1}^{n_1+n_2} \left\{ \frac{1}{2ln2} \cdot (\mathbf{v}_i)^T \Sigma^{-1} (\mathbf{v}_i) + \frac{1}{2} lb(|\Sigma|/\epsilon^6) + \frac{k}{2} lb2\pi \right\} \right] \tag{2}$$

where the first term represents the n_0 bits for specifying whether a point is good or bad, the second term is the number of bits to describe the bad points, the third term is the number of bits to describe the parameters of two planes, which is the number of bits to describe the model complexity, a variation of [25]. We assumed the n_1 good points to randomly sit on one plane which leads to the fourth term, and the n_2 good points to randomly sit on the other plane which leads to the fifth term, and to have Gaussian distribution $\mathbf{x} \sim N(\mu, \Sigma)$ which leads to the sixth term.

$\#bits(points \mid 1P + O)$, $\#bits(points \mid 3P + O)$, $\#bits(points \mid 4P + O)$, $\#bits(points \mid 5P+O)$, and $\#bits(points \mid 6P+O)$, and so on, can be deducted in a similar way. RANSAC is applied to extract planes in the point cloud. The MDL principle, deducted above, for interpreting a set of points in 3D space,is employed to decide which hypothesis is the best one.

4   3D Viewpoint Invariant Features

In this step we perform normalization with respect to extracted dominant 3D planes to achieve viewpoint invariance. Given a perspective image of a world plane the goal is to generate the front-parallel view of the plane. This is equivalent to obtaining an image of the world plane where the camera viewing direction is parallel to the plane normal. It's well known that the mapping between a 3D world plane and its perspective image is a homography function. Since we know the 3D positions of the points shown in the scene and their corresponding image, we can compute the homography relating the plane to its image from four (or more) correspondences. The computed homography H enables us to warp the original image to a normalized front-parallel view where the perspective distortion is removed. Fig. 2 shows some examples of such viewpoint normalization. Within the normalized front-parallel views of the scene, the viewpoint invariant features are computed in the same manner as the SIFT scheme [7]. Potential keypoints are identified by scanning local extreme in a series of Difference-of-Gaussian (DoG) images. For each detected keypoint $\mathbf{x}$, appropriate scale $\mathbf{s}$ and orientation θ are assigned to it and a 128-element SIFT descriptor $\mathbf{f}$ is created based upon image gradients of its local neighbourhood.

Fig. 2. Some examples of viewpoint normalization. *Left*: Original images; *Right*: Normalized front views. Note the perspective distortions are largely reduced in the warped front-parallel views of the building walls (e.g. a rectangular window in the 3D world will also appear rectangular in the normalized images).

Given a number of features extracted on the canonical views, we applied the criterion described in [26] to generate the putative feature correspondences. Two features are considered matched if the cosine of the angle between their descriptors $\mathbf{f}_i$ and $\mathbf{f}_j$ is above some threshold δ as:

$$\cos(\mathbf{f}_i, \mathbf{f}_j) = \frac{\mathbf{f}_i \cdot \mathbf{f}_j}{\|\mathbf{f}_i\|_2 \|\mathbf{f}_j\|_2} > \delta \tag{3}$$

where $\|\cdot\|_2$ represents the $L2$-norm of a vector. This criterion establishes matches between features having similar descriptors and does not falsely reject potential correspondences extracted on the images of repetitive structures which are very common in man-made environments.

After obtaining a set of putative feature correspondences based on the matching of their local descriptors, we impose certain global geometric constraints to identify the true correspondences. The RANSAC technique [23] is applied for this task. The number of samples M required to guarantee a confidence ρ that at least one sample is outlier free is given in Table 1. When the fraction of outliers is significant and the geometric model is complex, RANSAC needs a large number of samples and becomes prohibitively expansive.

Table 1. The theoretical number of samples required for RANSAC to ensure 95% confidence that one outlier free sample is obtained for estimation of geometrical constraint. The actual required number is around an order of magnitude more.

Outlier ratio	40%	50%	60%	70%	80%
Our method (1 point)	4	5	6	9	14
H-matrix (4 point)	22	47	116	369	1871
F-matrix (7 point)	106	382	1827	13696	234041

The geometrical model can be significantly simplified via the use of these novel features, and thus, lead to a more efficient matching method. Since the effects of perspective transformation are not compensated in the standard SIFT scheme, only the 2D image coordinates of SIFT features can be used to generate geometric constraints (F-Matrix or H-Matrix). Therefore, a number of SIFT matches are required to compute F-Matrix (7 correspondences) or H-matrix (4 correspondences). In comparison, the viewpoint invariant features are extracted on the front-parallel views of the same continuous flat building facade, taken at different distances and up to a camera translation and rotation around its optical axis. Every feature correspondence provides three constraints: scale (camera distance), 2D coordinates on the canonical view (camera translation), and dominant orientation (rotation around its optical axis). Therefore, a single feature correspondence is enough to completely define a point-to-point mapping relation between two canonical views. Consider a pair of matched features $(\mathbf{x}_1^m, \mathbf{s}_1^m, \theta_1^m)$ and $(\mathbf{x}_2^n, \mathbf{s}_2^n, \theta_2^n)$ both extracted on the normalized front-parallel views, a 2D similarity translation hypothesis is generated as follows:

$$\begin{bmatrix} x_1 - x_1^m \\ y_1 - y_1^m \\ 1 \end{bmatrix} = \begin{bmatrix} \Delta s & 0 & 0 \\ 0 & \Delta s & 0 \\ 0 & 0 & 1 \end{bmatrix} \begin{bmatrix} \cos \Delta\theta & -\sin \Delta\theta & 0 \\ \sin \Delta\theta & \cos \Delta\theta & 0 \\ 0 & 0 & 1 \end{bmatrix} \begin{bmatrix} x_2 - x_2^m \\ y_2 - y_2^m \\ 1 \end{bmatrix} \qquad (4)$$

where $\Delta s = s_1^m/s_2^n$ is the scale ratio and $\Delta\theta = \theta_1^m - \theta_2^n$ is the orientation difference. Our experimental evaluations in Section 5.2 show that for all ground true correspondences the scale ratios and orientation differences are equal up to a very small offset. It means that the information of patch scale and dominant orientation associated with the viewpoint invariant features are robust enough to generate geometrical hypothesis, which is impossible in the SIFT scheme. Using this simplified geometric model, a much smaller number of samples are needed to guarantee the generation of the correct hypothesis (c.f. Table 1 for comparison). The correspondences consistent with each generated hypothesis (e.g. the symmetric transfer error is less than a threshold) are defined as its inliers. The hypothesis with the most supports is chosen and its corresponding inliers are defined as true matches.

5   Experimental Results

We conducted experiments to evaluate the performance of the proposed method on urban scenes, with focus on the building facade images.

5.1   Point Cloud Sets Generation

We have taken 15 pairs of images over largely separated views with a calibrated camera. Each pair consists of 10 images, of which 5 images represent left view, the other 5 images represent right view. Only one pair is exceptional, as shown in Fig. 7 (a), of which 10 images represent left view, the other 10 images represent right view. We intend to use this pair for further comparison *w.r.t.* multi-view image numbers. Then, we applied orientation software AURELO [27] to achieve full automatic relative orientation of these multi-view images. And we used the public domain software PMVS (patch-based multi-view stereo) [28] for deriving a dense point cloud for each view of image pairs. It provides a set of 3D points with normals at those positions where there is enough texture in the images. The algorithm starts by detecting features in each image, matches them across multiple images to form an initial set of patches, and uses an expansion procedure to obtain a denser set of patches, before using visibility constraints to filter away false matches. An example for a point cloud derived with this software is given in Fig. 3 *Middle*. Finally, 5 dominant planes were extracted from each point cloud, while the rest planes were removed. One example demonstrating dominant planes extraction is shown in Fig. 3 *Right*.

5.2   Performance Evaluations

After extracting dominant planes, we perform normalization *w.r.t.* these planes to achieve viewpoint invariance. After viewpoint normalization, corresponding scene elements will have more similar appearances. The resulting features will suffer less from the perspective distortions and show better descriptiveness. We tested our method on two wide baseline 3D point clouds, as shown in Fig. 4, to demonstrate such improvements. It's noted that both 3D point clouds covered a same dominant planar structure which can be easily related through a

Fig. 3. *Left*: One of three images taken for a building facade scene. *Middle*: A snap-shot image of corresponding 3D point cloud generated by PMVS. *Right*: The five dominant planes automatically extracted from the point cloud.

homography. A number of SIFT and viewpoint invariant features were extracted on the original images and on the normalized front-parallel views, respectively. Then we followed the method described in [22] to define a set of ground truth matches. The extracted features in the first image were projected onto the second one using the homography relating the images (we manually selected 4 well conditioned correspondences to calculate the homography). A pair of features is considered matched if the overlap error of their corresponding regions is minimal and less than a threshold [22]. We adjusted the threshold value to vary the number of resulting feature correspondences.

Fig. 4. Two 3D point clouds and their associated images captured at widely separated views

Our goal is to evaluate how well two actually matched features relate with each other in terms of the Euclidean distance between their corresponding descriptors, their scale ratio, and their orientation difference. Given a number of matched features, we calculated the average Euclidean distance between their descriptors. The quantitative results are shown in Fig. 5 *Left*. It's noted that the descriptors of corresponding features extracted on the front-parallel views become very similar. It's because the procedure of viewpoint normalization will compensate the effects of perspective distortion, thus the resulting descriptors are more robust to the viewpoint changes. For each pair of matched features, we also computed the difference between their dominant orientations and the ratio between their patch scales. The results are shown in Fig. 5 *Middle* and Fig. 5 *Right*, respectively. On the normalized front-parallel views, the viewing direction is normal to the extracted 3D plane. The matched features extracted on such normalized views have similar dominant orientations and consistent scale ratio. It means that the information of patch scale and dominant orientation associated with the

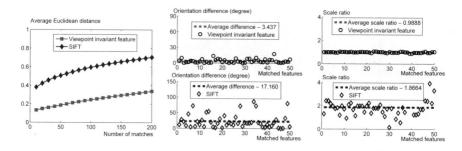

Fig. 5. Performance comparison between SIFT and Viewpoint invariant features. *Left*: The average Euclidean distances between the descriptors of matched features; *Middle*: The orientation differences between matched features; *Right*: The scale ratios between matched feature. The matched feature extracted on the normalized front-parallel views show better robustness to viewpoint changes.

Fig. 6. A number of matched features are shown. *Left*: on the original images; *Right*: on the front parallel views. Their scales and orientations are annotated. The feature matches on the viewpoint normalized views have very similar orientations and consistent scale ratios.

viewpoint invariant features are robust enough to determine camera distance and camera rotation around it optical axis, respectively. To qualitatively demonstrate the improvements, a number of matched features are shown on the original images (cf. Fig. 6 *Left*) and on the normalized images (Fig. 6 *Right*).

5.3   Wide Baseline Alignment

Next we demonstrate the advantages of the proposed feature matching scheme by applying it to some very difficult wide baseline alignment tasks. First, we extracted a number of viewpoint invariant features and establish putative correspondences according to Eq. 3 (the threshold δ was set at 0.9). Then, we applied the RANSAC algorithm impose the global geometric constraint (Eq. 4) to identify inliers. The number of inlier correspondences and correct ones were counted manually. For comparison, we applied SIFT and MSER for the same task. A set of putative matches were firstly established, among them the inlier correspondences were selected by imposing the homography constraint. In many cases, SIFT and MSER cannot generate enough correctly matched features to

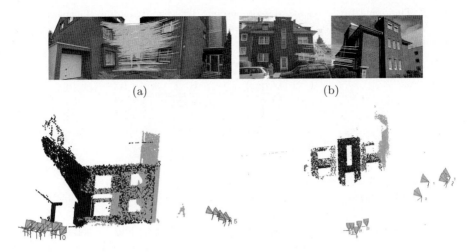

Fig. 7. Two example results of wide baseline 3D scene matching. Significant viewpoint changes can be observed on the associated image pairs shown on the top.

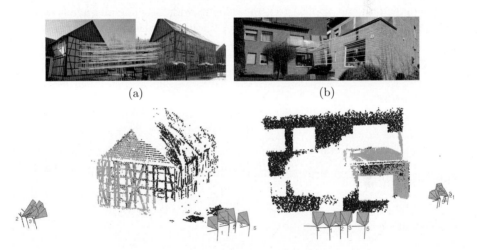

Fig. 8. Some other example results of wide baseline 3D scene matching. Our technique successfully aligned 3D scenes with very small overlap.

compute the correct H-matrix for identifying inlier correspondences due to the large viewpoint changes. Some matching results are shown in Fig. 7 and Fig. 8 with the quantitative comparisons provided in Tab. 2.

Table 2. The quantitative results of wide baseline 3D scene matching. (I - the number of initial correspondences by matching descriptors, N - the number of inliers correspondences returned by the RANSAC technique, T - the number of correct ones.)

Scene	SIFT			MSER			Our method		
	T	N	I	T	N	I	T	N	I
7a	70	89	7117	303	550	4511	420	421	8165
7b	0	13	704	3	13	512	23	23	658
8a	19	28	901	7	16	690	79	80	901
8b	0	10	640	4	19	412	41	41	804

6   Conclusions

We have proposed an intuitive scheme for aligning two widely separated 3D scenes via the use of viewpoint invariant features. To achieve this, we extracted viewpoint invariant features on the normalized front-parallel views *w.r.t.* 3D dominant planes derived from point cloud of a scene. This enables us to link the corresponding 3D points automatically in terms of wide baseline image matching. We evaluated the proposed feature matching scheme against the conventional 2D feature detectors, and applied to some difficult wide baseline alignment tasks of a variety of urban scenes. Our evaluation demonstrates that viewpoint invariant features are an improvement on current methods for robust and accurate 3D wide baseline scene alignment.

Acknowledgment

The work was funded by a Strategic Research Cluster grant (07/SRC/I1168) by Science Foundation Ireland under the National Development Plan, and Deutsche Forschungsgemeinschaft (German Research Foundation) FO 180/14-1 (PAK 274). The authors gratefully acknowledge these supports.

References

1. Hartley, R., Zisserman, A.: Multiple View Geometry in Computer Vision. Cambridge University Press, Cambridge (2003)
2. Snavely, N., Seitz, S.M., Szeliski, R.: Modeling the world from Internet photo collections. IJCV 80(2), 189–210 (2008)
3. Pollefeys, M., Van Gool, L., Vergauwen, M., Verbiest, F., Cornelis, K., Tops, J., Koch, R.: Visual modeling with a hand-held camera. IJCV 59(3), 207–232 (2004)
4. Bay, H., Ess, A., Tuytelaars, T., Van Gool, L.: Speeded-up robust features (SURF). Comput. Vis. Image Underst. 110(3), 346–359 (2008)
5. Tuytelaars, T., Van Gool, L.: Matching widely separated views based on affine invariant regions. IJCV 59(1), 61–85 (2004)
6. Matas, J., Chum, O., Urban, M., Pajdla, T.: Robust wide baseline stereo from maximally stable extremal regions. In: BMVC (2002)

7. Lowe, D.G.: Distinctive image features from scale-invariant keypoints. IJCV 60(2), 91–110 (2004)
8. Donoser, M., Bischof, H.: Efficient maximally stable extremal region (MSER) tracking. In: CVPR, pp. 553–560 (2006)
9. Mikolajczyk, K., Schmid, C.: Scale and affine invariant interest point detectors. IJCV 60(1), 63–86 (2004)
10. Mikolajczyk, K., Tuytelaars, T., Schmid, C., Zisserman, A., Matas, J., Schaffalitzky, F., Kadir, T., Van Gool, L.: A comparison of affine region detectors. IJCV 65(1-2), 43–72 (2005)
11. Sinha, S., Steedly, D., Szeliski, R.: Piecewise planar stereo for image-based rendering. In: ICCV, pp. 1881–1888 (2009)
12. Furukawa, Y., Curless, B., Seitz, S., Szeliski, R.: Manhattan-world stereo. In: CVPR, pp. 1422–1429 (2009)
13. Wu, C., Clipp, B., Li, X., Frahm, J., Pollefeys, M.: 3d model matching with viewpoint-invariant patches (VIP). In: CVPR, pp. 1–8 (2008)
14. Koeser, K., Koch, R.: Perspectively invariant normal features. In: ICCV, pp. 14–21 (2007)
15. Besl, P., McKay, N.: A method for registration of 3-d shapes. PAMI 14(2), 239–256 (1992)
16. Zhao, W., Nister, D., Hsu, S.: Alignment of continuous video onto 3d point clouds. PAMI 27(8), 1305–1318 (2005)
17. Pottmann, H., Huang, Q., Yang, Y., Hu, S.: Geometry and convergence analysis of algorithms for registration of 3d shapes. IJCV 67(3), 277–296 (2006)
18. Seo, J., Sharp, G., Lee, S.: Range data registration using photometric features. In: CVPR II, pp. 1140–1145 (2005)
19. Liu, L., Stamos, I., Yu, G., Zokai, S.: Multiview geometry for texture mapping 2d images onto 3d range data. In: CVPR II, pp. 2293–2300 (2006)
20. Ikeuchi, K., Oishi, T., Takamatsu, J., Sagawa, R., Nakazawa, A., Kurazume, R., Nishino, K., Kamakura, M., Okamoto, Y.: The great buddha project: Digitally archiving, restoring, and analyzing cultural heritage objects. IJCV 75(1), 189–208 (2007)
21. Gonzalez Aguilera, D., Rodriguez Gonzalvez, P., Gomez Lahoz, J.: An automatic procedure for co-registration of terrestrial laser scanners and digital cameras. IS-PRS Journal of Photogrammetry and Remote Sensing 64(3), 308–316 (2009)
22. Mikolajczyk, K., Schmid, C.: A performance evaluation of local descriptors. PAMI 27(10), 1615–1630 (2005)
23. Fischler, M., Bolles, R.: Random sample consensus: A paradigm for model fitting with applications to image analysis and automated cartography. Comm. of the ACM 24(6), 381–395 (1981)
24. Pan, H.: Two-level global optimization for image segmentation. ISPRS Journal of Photogrammetry and Remote Sensing 49, 21–32 (1994)
25. Rissanen, J.: Modelling by shortest data description. Automatica 14, 465–471 (1978)
26. Zhang, W., Košecká, J.: Hierarchical building recognition. Image Vision Comput. 25(5), 704–716 (2007)
27. Läbe, T., Förstner, W.: Automatic relative orientation of images. In: Proceedings of the 5th Turkish-German Joint Geodetic Days (2006)
28. Furukawa, Y., Ponce, J.: Accurate, dense, and robust multi-view stereopsis. PAMI (2009)

Modified Region Growing for Stereo of Slant and Textureless Surfaces[*]

Rohith MV[1], Gowri Somanath[1], Chandra Kambhamettu[1],
Cathleen Geiger[2], and David Finnegan[3]

[1] Video/Image Modeling and Synthesis (VIMS) Lab.
Department of Computer and Information Sciences,
University of Delaware, Newark, DE, USA
http://vims.cis.udel.edu
[2] Department of Geography, University of Delaware, Newark, DE, USA
[3] U.S. Army Corps of Engineers, Cold Regions Research and Engineering
Laboratory. Hanover, NH, USA

Abstract. In this paper, we present an algorithm for estimating disparity for images containing large textureless regions. We propose a fast and efficient region growing algorithm for estimating the stereo disparity. Though we present results on ice images, the algorithm can be easily used for other applications. We modify the first-best region growing algorithm using relaxed uniqueness constraints and matching for sub-pixel values and slant surfaces. We provide an efficient method for matching multiple windows using a linear transform. We estimate the parameters required by the algorithm automatically based on initial correspondences. Our method was tested on synthetic, benchmark and real outdoor data. We quantitatively demonstrated that our method performs well in all three cases.

1 Introduction

3D reconstruction using stereo cameras is being explored in the areas of urban reconstruction [1], robot navigation [2] and object recognition [3]. The algorithms being used for matching are as varied as the content encountered in the applications. In this paper, we present an algorithm for estimating disparity for images containing large textureless regions. This work is part of our effort to develop a fast and efficient method for capturing sea-ice surface topography using stereo cameras mounted on an icebreaker during a cruise. Obtaining the surface height and other characteristics are key factors in determining the volume of sea-ice and also for interpreting satellite data [4,5]. We propose a fast and efficient region growing algorithm for estimating the stereo disparity and demonstrate the working of our algorithm on synthetic and benchmark stereo datasets. We validate quantitatively our results on real scenes using data collected from a 3D

[*] This work was made possible by National Science Foundation (NSF) Office of Polar Program grants, ANT0636726 and ARC0612105.

G. Bebis et al. (Eds.): ISVC 2010, Part I, LNCS 6453, pp. 666–677, 2010.

Fig. 1. Reconstruction of a scene from a 4 MP image. Width of the scene is 20 meters and depth is 60 mts.

laser scanner. Though we present results on sea-ice images, the algorithm can be easily used for other applications.

A brief discussion on the nature of data and the requirements of the stereo algorithm will motivate our approach. The images acquired by the stereo system are large (4 megapixels each) and there are up to 500 disparity levels. Since the images mostly contain snow covered regions or ice, there is little color variation present. The terrain being imaged is viewed from an oblique angle due to the height of the system and hence the surface being reconstructed may not always be parallel to the baseline (violating the fronto-parallel assumption).

The first step in many stereo algorithms is the calculation of the disparity volume, that is, the cost associated with assigning various disparity values to pixels in the image. This would not only require computing the matching score of every pixel with corresponding feasible pixels, but also storing this information in an array whose size is the product of image area and disparity range. Our image size and disparity range preclude algorithms employing such a step. We propose an algorithm which grows regions out of sparse correspondences and hence its complexity does not depend significantly on disparity range, but only on image size. Our method also takes advantage of large planar areas in the image. Such areas are not matched in the stereo process, but are interpolated to fill in the disparity values. Compared to other region growing algorithms, our approach differs in the following aspects: (1) we provide dense disparity as output, (2) we explicitly handle change in appearance of a region due to slant, (3) we adaptively estimate most of the parameters necessary for our algorithm from the initial correspondences, (4) we also provide an efficient method for computing the similarity measure between windows being matched by discretizing the allowed angles of slant and precomputing the transforms.

We briefly review some related methods in Section 2, formulate the problem in Section 3 and discuss the details of our approach in Section 4. We present results on synthetic, benchmark and real outdoor datasets in Section 5 and conclude in Section 6.

2   Related Work

In this section, we discuss a few stereo methods which are similar to proposed method or present results on similar data. For a general discussion on stereo algorithms and their taxonomy, we refer the reader to [6]. One of the schemes involves classification of stereo algorithms based on the density of the estimated disparity map into sparse (only in some *feature* locations), dense (everywhere) and quasi-dense (almost everywhere) schemes. The relative merit of each scheme is discussed in [7]. One of the earliest region growing methods [8] relied on adaptive least squares correlation algorithm and provided dense correspondences. There is a recent extension [9] which attempts to increase the performance of the algorithm using parallel processing. [10] extends region growing method to multiview-cases. However, the discussion below concerns two other methods which are closer to the proposed method.

Lhullier and Quan [7] proposed a method to obtain surface reconstruction from uncalibrated images. The method involved obtaining and validating a sparse set of correspondences. These are then used to grow regions using a best-first strategy without epipolar constraint. The resulting disparity map is resampled using local homographies to obtain quasi-dense correspondences. These correspondences are used to estimate the fundamental matrix and then initial seeds are grown again using epipolar constraints. This process is repeated till the solution converges. Though we use a similar strategy for growing the seeds, there are some key differences. First, our system is calibrated and our images are rectified, so we do not estimate fundamental matrix, but employ epipolar constraints directly. They employ a combination of gradient cues and confidence measures to limit the propagation. We use statistics derived from the initial matches to decide when the propagation needs to stop. Though results are presented on planes oriented at different angles, their method does not explicitly handle fractional disparities and slanted surfaces in its formulation.

Growing correspondence seeds (GCS) [11] was proposed as a fast method for obtaining disparity of regions with random initial seeds. The method involved growing overlapping components in disparity space and then optimally matching those components. A distortion model for slant was suggested but not discussed in detail. It is noted that a simple region growing scheme does not perform well on its own and hence the components found are matched using maximum strict sub-kernel algorithm. However, the output of the algorithm is then filtered to provide only a quasi-dense estimate of the disparity. In our method, after growing the regions (using relaxed uniqueness constraints), we use the estimate as the initialization of an optimization algorithm that seeks to minimize a global energy function.

There are other approaches using belief propagation for slant regions [12] and texture-less regions [13]. However, these require computation of the entire disparity volume and are not suitable for large images with large disparity range.

3   Our Approach

Our approach is outlined below. As pointed out in [11], region growing alone does not provide a good estimate of the disparity. Hence, we attempt to minimize the following objective:

$$E(d) = \sum_{(x,y)} (1 - \lambda(x,y))[n(I(x,y), I'(x + d(x,y), y)) + \alpha \nabla^2 d] + \lambda(x,y)$$

Here, I and I' are the left and right images, d is the disparity map. λ is an indicator variable ($\lambda \in \{0, 1\}$). If $\lambda = 1$, then the pixel is occluded or lies on a discontinuity. We call a pixel (x,y) *matchable* if $\lambda(x,y) = 0$. It can be seen that if n is normalized, and $\nabla^2 d$ is small, then minimizing E will lead to minimizing the occluded regions in the image. However, minimizing this objective function is not a well-posed problem as there are multiple minima near textureless regions. Hence we propose to solve it using the following steps:

- Obtain sparse correspondences between left and right images.
- Grow the correspondences without entering textureless regions.
- Interpolate the obtained disparity to get an initial solution to the minimization problem.
- Solve the minimization by solving Euler-Lagrange equations at points where gradient is high.

The motivation for this approach is that region growing cannot cross regions where uniqueness constraint is violated. However, if we assume that each matchable region (a region consisting entirely of matchable pixels) contains at least one seed, the regions that remain after the growth step are either occluded or textureless. These can be interpolated according to the energy minimization. We use diffusion to achieve interpolation, though techniques using finite element methods may also be utilized [14].

4   Details

4.1   Sparse Correspondences

We use Harris detectors to find features for sparse correspondences and normalized cross correlation to match them. This is similar to the approach followed in [11]. Although various detector/descriptor combinations may be used, we have found that owing to lack of change in scale and orientation, Harris and correlation yield reliable matches. Since the images are already rectified, we filter those matches that do not follow the epipolar constraint. We will refer to coordinates of the resultant points as l_x and l_y (similarly r_x and r_y), each of which is $N \times 1$ vector (N is the number of points found).

Algorithm 4.1. Region growing algorithm

---

1. $\forall_{(x,y)} filled(x,y) = 0$.
2. for $i = 1$ TO N, insert $[l_x(i)\ l_y(i)\ l_x(i) - r_x(i)\ l_x(i) - r_x(i)\ l_x(i) - r_x(i)]$ into *queue* Q.
3. While Q is not empty, repeat steps 4 to 10
4. Pop first element of Q into c.
5. If $filled(c_x, c_y)$ then Goto step 3.
6. Get errors $e(d)$ of pixel (c_x, c_y) for disparities in range of c_{min} to c_{max}.
7. If $\min e < \theta$ and [Number of disparities d such that $e(d) < 2 * e_{min}$] $< M$, then continue, else Goto step 3.
8. $d(c_x, c_y) = \underset{d}{\mathrm{argmax}}\, e(d) + \alpha * |d - c_p|$.
9. $filled(x, y) = 1$.
10. For each neighbor (x, y) of (c_x, c_y): Insert $[x\ y\ d(c_x, c_y) - w\ d(c_x, c_y) + w\ c_p]$ into Q. The disparity range is restricted for left and right neighbor to avoid overlap.

4.2   Region Growing

We perform region growing iteratively using Algorithm 4.1. The matching measure used is sum of squared differences (SSD). Points which are to be processed are inserted into a FIFO queue Q. Each item c in Q has the following terms:

$$c = [c_x\ c_y\ c_{min}\ c_{max}\ c_p]$$

This item in the queue corresponds to the pixel at location (c_x, c_y). Its disparity will be searched over the range $[c_{min}, c_{max}]$. The last term c_p indicates the disparity of the initial correspondence which led to the current pixel being inserted into queue. c_p is used to regularize the disparity. Note that since noise levels may vary between images, we do not enforce a strict uniqueness constraint (each pixel has only one match). Instead we check to see if the number of disparities within twice the minimum error range is less than M. Choosing $M = 1$ would be close to enforcing uniqueness constraint. Also, there is a factor θ which is the maximum error allowed for a match. We choose M and θ adaptively based on initial matches (discussed in section 4.4). Though uniqueness is weakly enforced, it is sufficient to prevent growth in textureless regions. This may be observed in the synthetic results in section 5.1. We may also include gradient or segmentation cues if available to decide when a propagation needs to be stopped.

4.3   Sub-pixel Matching and Slant

In step 6 of Algorithm 4.1, we find errors of the current pixel with pixels in the disparity range $[c_{min}, c_{max}]$. This interval is divided into fractional intervals corresponding to sub-pixel disparities. We assume that scan lines of the image are independent, and hence we linearly interpolate the images for sub-pixel matching. This is similar to the method in [15]. Also, the appearance of the pixel in the right image may have changed due to slant. In stereo, slant refers to a plane

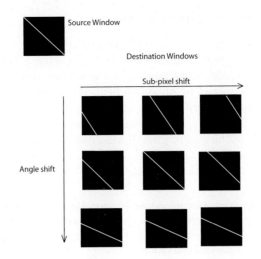

Fig. 2. Illustration of matching windows with varying sub-pixel value and angle. Matching with multiple angles helps decrease error in slanted regions.

whose normal is at significantly different angles with respect to the viewing directions of the two cameras. If the plane is very inclined or the scene is very close, the image of the plane is distorted in one of the images. To handle these cases, we compress and dilate the matching window corresponding to discrete angles. Figure 2 summarizes the matching performed for every window. For this paper, we have chosen 2 levels between every successive disparity $(0,0.25,0.5,1)$ and a total of 5 angles $(-60°, -30°, 0°, 30°, 60°)$. Note that each of these distortions (sub-pixel+angle distortion) is a linear interpolation of values of image. Hence, these operators can be pre-computed. When a window needs to be matched, the operators are applied through matrix multiplication and the result compared to the destination window. This is much more efficient than computing arbitrary affine parameters for each window and provides reasonable results if the slant is limited. An example of slant plane can be seen in Figure 3.

4.4 Parameter Estimation

We used parameters θ, M, w and α in Algorithm 4.1. Two of these parameters can be estimated from the initial set of correspondences. Given the correspondences, a tentative estimate of minimum and maximum disparity for the image is calculated as the minimum and maximum value of $l_x - r_x$. SSD errors are then calculated for each of the initial correspondence. θ is set to the average error of all correspondences. Similarly, the average number of disparity values with error less than $2e_{min}$ is chosen as M. The choice of α controls the smoothness and w decides the largest change in the gradient within a continous region (depends on the image size). For our experiments, we chose $w = 3$.

4.5   Interpolation and Optimization

Once the regions are grown, the resulting disparity map contains many pixels where disparity is not assigned. We fill them using bilinear interpolation. Euler-Lagrange equations corresponding to the energy in section 3 is evaluated iteratively on pixels sorted by their gradient. This forces the optimization to concentrate on regions where large gradients exist, thus smoothing them out. Since most of the pixels in the image are matchable, their disparity is already correctly assigned by the region growth algorithm. The step of optimization is useful when errors due to incorrect seeds or noise may have crept into the growth process (see Teddy example in Results).

Table 1. Change in number of iterations for growth algorithm with disparity range for image size of 200x200. Number of iterations decrease slightly with disparity range because occlusion increases as strip moves farther from the background. Occluded pixels are not handled by the growth algorithm.

Disparity Range	Iterations	Disparity volume
1	70985	40000
5	68276	200000
25	66964	1000000
50	60208	2000000

5   Results

The algorithm was implemented in MATLAB. The vectorization described in section 4.3 is exploited to perform parallel interpolations across rows and channels. The running times are reported on a PC with Intel Centrino 2 processor with 4GB RAM.

5.1   Synthetic Results

We present results on two synthetic images shown in Figure 3. In the first image, the white strip is moved laterally by 25 pixels. Note that the region growth algorithm assigns values only on the outline of the region and not inside where the matching is ambiguous. In the interpolation and optmization step, the disparity is filled inside the strip and the boundary smoothed out by diffusion. In the second image, it is moved and rotated, so that the disparity on the left side is 25 and right side is 10. Notice that the matching does not fail on the slant edge. Although the exact angle of the slant is not included in the search, the approximation is close enough to provide complete disparity on the boundary of the strip. We also conducted an experiment to measure the effect of disparity range on the number of matches performed. The strip in the first image was

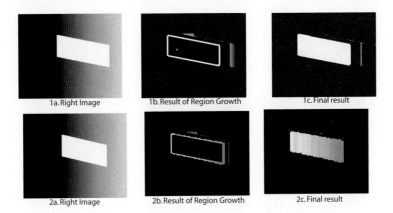

Fig. 3. Results on synthetic dataset. First row (1a-1c) corresponds to flat plane in front of a background leading to constant disparity. Second row (2a-2c) corresponds to a slanted plane and hence contains a gradient in disparity.

moved at steps and disparity calculated, the number of matches performed by the algorithm were noted. The findings are shown in Table 1. It can be seen that even though the disparity range changes by a large factor, the number of matches performed does not increase significantly. For image sizes of 200x200, the algorithm takes 12-14 seconds to converge.

5.2 Benchmark Results

We tested our algorithm on Middlebury stereo dataset. Results on two of the images are shown in Figure 4. Note that although large regions of Venus image are not assigned disparity in the growth step, the values are interpolated successfully in the next step. In the Teddy image, notice that the region on the roof is assigned wrong disparity in the first step. However, this is corrected in the subsequent steps. The errors on all the bench mark images are in table 2. Although the errors are not comparable to segmentation based techniques, it does however out-perform several global methods such as graph-cuts [16].

Table 2. Error in results for Middlebury dataset [6]. Bad pixels are those whose disparity error is greater than 1.

Image	Percentage of bad pixels
Tsukuba	1.23
Venus	1.57
Teddy	10.98
Cones	5.11

1a. Left Image with Inital
Correspondences

1b. Result of region growth

1c. Final Disparity

2a. Left Image with Inital
Correspondences

2b. Result of region growth

2c. Final Disparity

Fig. 4. Results on Middlebury dataset. First row (1a-1c) corresponds to Teddy image and second row (2a-2c) contains Venus image.

5.3 Real Outdoor Results

We performed tests on outdoor ice-scape to see the effectiveness of the algorithm on large images. The scene was river bank 15-25 meters far from the camera. It was covered with ice blocks and snow. The images were captured using a 16MP SLR camera and downsampled to 4MP. The scene was also measured using a 3D laser scanner that provides panaromic range scan of the scene with 1 mm precision. The scene was captured in 3 image pairs. Figure 5 shows reconstruction from one pair and corresponding section of laser data. 1,337,079 points were reconstructed from the stereo pair and a corresponding 100,000 points from the laser scanner were aligned with it for comparison. To compare the two surfaces, for each point obtained by the laser scanner, we compute the closest point on the stereo reconstructed surface and calculate the euclidean distance between them to estimate the reconstruction error. Table 3 shows mean error for all the points on the surface in centimeters for various values of α. Figure 1 shows ice blocks floating in river. The dimensions of rendered scene are 20 meters wide and 60 meters deep.

Table 3. Change in mean error between the stereo and 3D laser scanned surfaces when smoothness parameter α is varied

α	Mean Error (in cm)
0.1	19.34
0.3	18.88
0.8	18.62
2.0	19.15

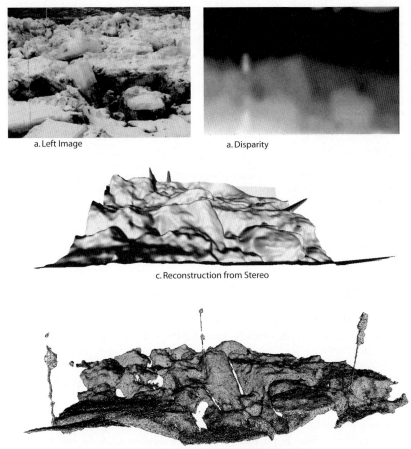

Fig. 5. Results on ice/snow surface. Comparison of surface reconstructed from stereo and 3D laser scanner. (a) Left image from stereo image pair. (b) Disparity estimated from our algorithm. (c) Reconstruction from stereo (mesh with 1,337,079 vertices). (d) Reconstruction from laser scanner (100,000 vertices).

6   Conclusion

To enable fast and accurate capture of surface topography, we proposed a stereo matching algorithm based on region growing. Since region growing alone cannot provide good results, we augmented it with an optimization step to correct the estimate. We modified the first-best region growing algorithm using relaxed uniqueness constraints and matching for sub-pixel values and slant surfaces and provided an efficient method for matching multiple windows using a linear transform. We estimated the parameters required by the algorithm automatically based on initial correspondences. We tested our method on synthetic, benchmark and real outdoor data. We quantitatively demonstrated that our method performs well in all three cases and showed that the number of iterations required does not grow significantly with disparity range, but only depends on the image size. Comparison with laser scanned data suggests that the algorithm performs well even in regions with little or no texture. In future work, we will combine the region growing technique with segmentation based methods to provide a robust disparity estimate.

References

1. Akbarzadeh, A., Frahm, J.-M., Mordohai, P., Clipp, B., Engels, C., Gallup, D., Merrell, P., Phelps, M., Sinha, S., Talton, B., Wang, L., Yang, Q., Stewenius, H., Yang, R., Welch, G., Towles, H., Nister, D., Pollefeys, M.: Towards urban 3d reconstruction from video. In: 3DPVT 2006: Proceedings of the Third International Symposium on 3D Data Processing, Visualization, and Transmission (3DPVT 2006), Washington, DC, USA, pp. 1–8. IEEE Computer Society Press, Los Alamitos (2006)
2. Murray, D., Little, J.: Using real-time stereo vision for mobile robot navigation. In: Autonomous Robots 2000 (2000)
3. Sumi, Y., Tomita, F.: 3d object recognition using segment-based stereo vision. In: Chin, R., Pong, T.-C. (eds.) ACCV 1998. LNCS, vol. 1352, pp. 249–256. Springer, Heidelberg (1997)
4. Worby, A.P., Massom, R., Allison, I., Lytle, V.I., Heil, P.: East antarctic sea ice: a review of its structure, properties and drift. Antarctic Sea Ice Properties, Processes and Variability. AGU Antarctic Research Series, pp. 41–67 (1998)
5. Worby, A.P., Geiger, C., Paget, M., Woert, M.V., Ackley, S., DeLiberty, T.: The thickness distribution of antarctic sea ice. J. Geophys. Res., 158–170 (2000)
6. Scharstein, D., Szeliski, R.: A taxonomy and evaluation of dense two-frame stereo correspondence algorithms. International Journal of Computer Vision 47, 7–42 (2002)
7. Lhuillier, M., Quan, L.: A quasi-dense approach to surface reconstruction from uncalibrated images. IEEE Transactions on Pattern Analysis and Machine Intelligence 27, 418–433 (2005)
8. Otto, G.P., Chau, T.K.W.: 'region-growing' algorithm for matching of terrain images. Image Vision Comput. 7, 83–94 (1989)
9. Shin, D., Muller, J.-P.: An explicit growth model of the stereo region growing algorithm for parallel processing. ISPRS Commission V: Close Range Image Measurement Techniques (2010)
10. Furukawa, Y., Ponce, J.: Accurate, dense, and robust multi-view stereopsis. In: IEEE Conference on Computer Vision and Pattern Recognition, CVPR 2007, pp. 1–8 (2007)
11. Čech, J., Šára, R.: Efficient sampling of disparity space for fast and accurate matching. In: CVPR Workshop Towards Benchmarking Automated Calibration, Orientation and Surface Reconstruction from Images, BenCOS 2007. IEEE, Los Alamitos (2007)
12. Ogale, A.S., Aloimonos, Y.: Shape and the stereo correspondence problem. Int. J. Comput. Vision 65, 147–162 (2005)

13. Rohith, M.V., Somanath, G., Kambhamettu, C., Geiger, C.A.: Stereo analysis of low textured regions with application towards sea-ice reconstruction. In: International Conference on Image Processing, Computer Vision, and Pattern Recognition (IPCV), pp. 23–29 (2009)
14. Rohith, M., Somanath, G., Kambhamettu, C., Geiger, C.: Towards estimation of dense disparities from stereo images containing large textureless regions. In: 19th International Conference on Pattern Recognition, ICPR 2008, pp. 1–5 (2008)
15. Birchfield, S., Tomasi, C.: A pixel dissimilarity measure that is insensitive to image sampling. IEEE Transactions on Pattern Analysis and Machine Intelligence 20, 401–406 (1998)
16. Kolmogorov, V., Zabih, R.: Computing visual correspondence with occlusions using graph cuts. In: Proceedings of Eighth IEEE International Conference on Computer Vision, ICCV 2001, vol. 2, pp. 508–5152 (2001)

Synthetic Shape Reconstruction Combined with the FT-Based Method in Photometric Stereo

Osamu Ikeda

Faculty of Engineering, Takushoku University
815-1 Tate, Hachioji, Tokyo, Japan

Abstract. A novel method of synthetic shape reconstruction from photometric stereo data combined with the FF-based method is presented, aiming at obtaining more accurate shape. First, a shape is reconstructed from color images using a modified FF-based algorithm. Then, with the shape as initial value, a more accurate shape is synthetically reconstructed based on the Jacobi iterative method. The synthesis is realized as follows: the reconstruction is sequentially made in each of small image subareas, using the depths in the neighboring subareas as boundary values, which is iterated until the overall shape converges. The division to image subareas enables us to synthesize large shapes.

1   Introduction

Several shape reconstruction methods have reported in the field of photometric stereo. One is to convert the surface normal map obtained from three images [1] to the depth map. The conversion is not straightforward but it needs sophisticated methods, as reported by Frankot and Chellappa [2] and Agrawal, Raskar and Chellappa [3]. One is a direct method that solves a set of relations between the surface normal vector and the depth, as presented by Basi and Jacobs [4]. The method needs a number of images to obtain a good shape and it is quite time consuming. Another direct method uses the Jacobi iterative method as shown by Ikeda [5]. The method appears mathematically sound, but it requires a huge space of memory. As a result, it is time consuming and reconstruction is limited in dimension of shape.

Future applications in computer vision may require accurate and detailed visual object information, such as shape, color, albedo and spectral reflection characteristics for recognition and/or identification purposes. The requirement for accurate shape, however, may not necessarily be met by the existing methods.

In this paper, based on the method [5], a novel synthetic reconstruction method is presented in combination with the FT-based method, to obtain accurate large-sized shapes. In the new method, the whole image area is divided into small subareas. In the reconstruction, one iteration of shape reconstruction is made sequentially in each of the subareas, using the depths in the neighboring subareas as the boundary condition, which is iterated until the entire shape converges. This results in the synthesis of the shape.

G. Bebis et al. (Eds.): ISVC 2010, Part I, LNCS 6453, pp. 678–687, 2010.
© Springer-Verlag Berlin Heidelberg 2010

To shorten the computation time, a shape is reconstructed using the FT-based method [2] and it is used as the initial shape. The method in its basic form is inferior to the iterative method [5] in accuracy, and it is applicable only to monochrome images. The algorithm is modified and optimized so that it is applicable to color images.

2   Reconstruction Using Jacobi Iterative Method

It is shown that the reconstruction using the Jacobi iterative method requires the amount of memory of $4N^4$ for N by N pixel images and that it also requires boundary values.

Referring to [5], the method is based upon the consistency between the reflectance functions $R_l(p, q)$ and color images $I_{c,l}(x,y)$ for three directions of the light source:

$$\alpha_c(x, y)R_l(p,q) = I_{c,l}(x, y), \ c= r, g, b, l = 1,2,3, x,y=1,\dots,N \tag{1}$$

where α_c are albedos that can be derived from the images [7] and (p, q) denote the surface tilts. For the case of Lambertian reflection, $R_l(p, q)$ are given by

$$R_l(p,q) = \mathbf{P} \bullet \mathbf{S}_l, \quad l = 1,2,3 \tag{2}$$

where $\mathbf{P}$ is the surface normal vector of the shape, and $\mathbf{S}$ is the illuminant vector. The simplest approximations of p and q may be given from the depth map, $z(x,y)$, as

$$p = \begin{cases} z(x-1, y)- z(x, y) \\ z(x, y)- z(x+1, y) \end{cases} \quad q = \begin{cases} z(x, y-1)- z(x, y) \\ z(x, y)- z(x, y+1) \end{cases} \tag{3}$$

So we can have four combinations for (p, q).

Defining the functions $f_{a,c,l}(x,y)$ as

$$f_{a,c,l}(x, y) \equiv J_{a,c,l}(x, y)- \alpha_c(x, y)R_{a,l}(p,q), \quad a=1-4 \tag{4}$$

where a denotes the combination of the approximations, and $J_{a,c,l}$ are $I_{a,c,l}$ with the lateral shifts in correspondence with the combination. The Jacobi iterative method gives the iterative relations for one of the four combinations as

$$- f_{1,c,l}(x, y)^{(n-1)} = \left(\frac{\partial f_{1,c,l}(x, y)}{\partial z(x, y)}\right)^{(n-1)} \left(z(x, y)^{(n)} - z(x, y)^{(n-1)}\right)+ \left(\frac{\partial f_{1,c,l}(x, y)}{\partial z(x-1, y)}\right)^{(n-1)}$$
$$\times \left(z(x-1, y)^{(n)} - z(x-1, y)^{(n-1)}\right)+ \left(\frac{\partial f_{1,c,l}(x, y)}{\partial z(x, y-1)}\right)^{(n-1)} \left(z(x, y-1)^{(n)} - z(x, y-1)^{(n-1)}\right) \tag{5}$$

where n is the number of iterations. Those relations can be rewritten in matrix form as

$$- \mathbf{f}_{a,c,l}^{(n-1)} = \mathbf{g}_{a,c,l}^{(n-1)} (\mathbf{z}^{(n)} - \mathbf{z}^{(n-1)}), \ a=1\text{-}4, c=r, g, b, l=1\text{-}3, n=1,2,\dots, \tag{6}$$

where $\mathbf{f}_{a,c,l}$ and $\mathbf{z}$ are vectors having of N^2 elements, and $\mathbf{g}_{a,c,l}$ are matrices having N^2 by N^2 elements.

12 of these relations are linearly combined in a least squares sense to reconstruct three color component shapes, and 36 of them for three colors combined shapes. We obtain the following relation for the combined case:

$$- \mathbf{F}^{(n-1)} = \mathbf{G}^{(n-1)} \left(\mathbf{z}^{(n)} - \mathbf{z}^{(n-1)} \right), \; n=1,2,\dots, \tag{7}$$

where $\mathbf{G}$ is the matrix having $36N^2$ by N^2 elements. The depth map is estimated following

$$\mathbf{z}^{(n)} = \mathbf{z}^{(n-1)} - \left(\mathbf{G}_2^{(n-1)} \right)^{-1} \mathbf{F}_2^{(n-1)}, \; n = 1,2,\dots, \tag{8}$$

$$\mathbf{G}_2 = \mathbf{G}^T \mathbf{G}, \, \mathbf{F}_2 = \mathbf{G}^T \mathbf{F} \tag{9}$$

To make the structure of $\mathbf{G}_2$ uniform, the elements on the four borders are excluded from the reconstruction. This enhances numerical stability. Hence, $\mathbf{G}_2$ results in $(N\text{-}2)^2$ by $(N\text{-}2)^2$ matrix and the reconstruction area is reduced to $N\text{-}2$ by $N\text{-}2$.

It is seen from Eqs. (8) and (9) that $\mathbf{G}_2$ occupies most of the required memory in the computation. For a case where available memory is limited to 2GB, N is limited to 150. The value may not be large enough for applications.

3   Procedure of Synthetic Shape Reconstruction

It is seen from Eq. (3) that depths just outside the reconstruction area are required to determine the surface tilts of the elements on the four borders. This means that when the whole image area is divided to small subareas, depths in subareas contribute to shape reconstruction in the neighboring subareas. If we carry out the reconstruction sequentially in each of the subareas, then it may be able to reconstruct the whole shape at every iteration in a similar fashion to the case where just one area is used. This may end up synthesizing the shape over a large area while repeating the reconstruction in small subareas.

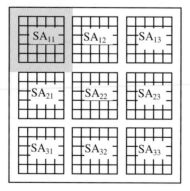

Fig. 1. Division of the reconstruction area into subareas

Let's consider the 3 by 3 subarea division shown in Fig. 1. The initial map of depth value is set to null. In the first iteration, one iteration of shape reconstruction is made in the subarea SA_{11} using the null depths in the subarea and those in the neighboring ones. The depth map is updated with the resulting depths, where the depths on the left of SA_{11} and those above SA_{11} are replaced with those at the nearest elements. One iteration of the reconstruction is made next in SA_{12} using the updated depth map, where part of the depth values in SA_{11} are used. The depth map is then updated in the same way. This is repeated through the subarea SA_{33}. In the second iteration, too, the sequential reconstruction begins with the subarea SA_{11} and ends with SA_{33}, while updating the depth map. The sequential reconstruction is repeated until the depth map converges.

4   Reconstruction Using FT

The FT-based shape reconstruction algorithm [2] is not so accurate but fast. So we use the shape as the initial one for the new method. The existing algorithm is for monochromatic color, so let's modify the algorithm in order to be applicable to color images.

Using $\mathbf{P} = (n_x, n_y, n_z)^T$ and $\mathbf{S} = (s_x^l, s_y^l, s_z^l)^T$, the following relation holds:

$$\frac{1}{\alpha_c}\begin{pmatrix} I_c^1 \\ I_c^2 \\ I_c^3 \end{pmatrix} = \begin{pmatrix} s_x^1 & s_y^1 & s_z^1 \\ s_x^2 & s_y^2 & s_z^2 \\ s_x^3 & s_y^3 & s_z^3 \end{pmatrix}\begin{pmatrix} n_{x,c} \\ n_{y,c} \\ n_{z,c} \end{pmatrix}, \quad c = r,g,b \tag{10}$$

From Eq. (10) the components of $\mathbf{P}$ are given by

$$\begin{pmatrix} n_{x,c} \\ n_{y,c} \\ n_{z,c} \end{pmatrix} = \frac{1}{\alpha_c}\begin{pmatrix} s_x^1 & s_y^1 & s_z^1 \\ s_x^2 & s_y^2 & s_z^2 \\ s_x^3 & s_y^3 & s_z^3 \end{pmatrix}^{-1}\begin{pmatrix} I_c^1 \\ I_c^2 \\ I_c^3 \end{pmatrix}, \quad c = r,g,b \tag{11}$$

Then, they are averaged over the three colors with weights dependent upon the image brightness:

$$n_i = \frac{\displaystyle\sum_{c=r,g,b} n_{i,c}\left(\sum_{l=1}^{3} I_c^l\right)}{\displaystyle\sum_{c=r,g,b}\sum_{l=1}^{3} I_c^l}, \quad i = x,y,z \tag{12}$$

The surface tilts (p, q) are given by

$$p = \frac{n_x}{n_z}, \quad q = \frac{n_y}{n_z} \tag{13}$$

where p and q vary depending on (x, y). Let $P(\omega_u, \omega_v)$ and $Q(\omega_u, \omega_v)$ be the Fourier transforms of p and q, respectively, and let the approximations, $\partial z/\partial x = z(x, y) - z(x-1, y)$ and $\partial z/\partial y = z(x, y) - z(x, y-1)$, be used. Then, the depth map is reconstructed as

$$z_r(x, y) = \mathrm{Re}\left[F^{-1}\left\{ \frac{\{1 - \exp(j\omega_u)\}P(\omega_u,\omega_v) + \{1 - \exp(j\omega_v)\}Q(\omega_u,\omega_v)}{|1 - \exp(-j\omega_u)|^2 + |1 - \exp(-j\omega_v)|^2} \right\} \right] \qquad (14)$$

where F^{-1} denotes the inverse Fourier transform. In its implementation using FFT, (ω_u, ω_v) are replaced with $(2\pi/M)(u, v)$, $u, v = 0, 1, \ldots, M\text{-}1$.

5   Experiments

Two objects were used in experiments. One is a plastic toy penguin, which is 36.8 mm in width, 48.1 mm in height, and 16.3 mm in depth. The other is a ceramic toy horse, which is 68 mm in width, 66.5 mm in height, and 19.2 mm in depth. These objects were illuminated with an incandescent light from three directions to capture images in RAW format. The color temperature of the light source is estimated to be around 3000K. One optical polarization filter was placed in front of the light source and the other in front of camera lens to get rid of specular reflection components in the images.

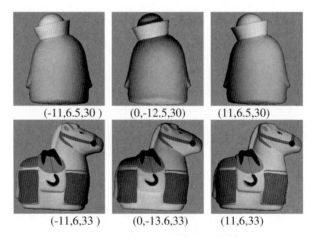

(-11,6.5,30)   (0,-12.5,30)   (11,6.5,30)

(-11,6,33)   (0,-13.6,33)   (11,6,33)

Fig. 2. Images of two objects used in experiments, where (S_x, S_y, S_z) denotes the illuminat vector

With the RAW development software, the captured images were made "linear" in brightness, and the color temperature was set to 4000K so that the resulting images agreed in color with our observation. Then, the three color component images were modulated as

$$I_c^l = \left(i_c^l\right)^{g_c}, \quad c = r, g, b, l = 1, 2, 3 \qquad (15)$$

The modulation may make the reflection of the objects Lambertian [8]. We reconstructed shapes and estimated (g_r, g_g, g_b) to be (0.679, 0.648, 0.656) for the horse and (0.889, 0.915, 0.880) for the penguin. The resulting color images are shown in Fig. 2.

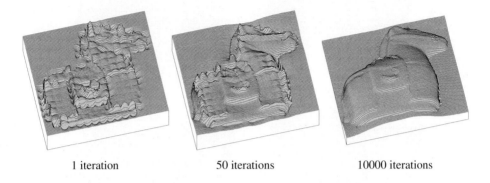

| 1 iteration | 50 iterations | 10000 iterations |

Fig. 3. Synthetic shape reconstruction with increasing number of iterations for the case of $M = 128$. In this example the reconstruction begins with null depth map.

The synthetic shape reconstruction is demonstrated in Fig. 3 for the case of $M = 128$ image, where the subarea size is 8 by 8 pixels and the initial depth map is $\mathbf{z}^{(0)}=\mathbf{0}$. It was observed that the subarea is optimal in the speed of convergence among subareas of N by N pixels, where $N = 4, 8, 16, 32, 64$, and 128, and that the reconstructed shapes for those different subareas are very similar to each other.

The initial 256 by 256 shape of the penguin, which was obtained using the modified FT algorithm, and the converged 256 by 256 shape, which was obtained using the synthetic reconstruction, are compared in Fig. 4. It is seen that the initial shape is inaccurate on the whole with some rotations; that is, the foot part is lower than the actual. It is also seen that the inaccuracy is serious in the dark image regions, as expected.

The modified FT algorithm assumes that surfaces of the brightest image parts face the lights normally for each color, but this does not necessarily hold valid. This may account for the inaccuracies in Fig. 4. The synthetic method, on the other hand, is accurate; since, the albedos are derived taking the light directions into account.

It's harder to know the surface tilts as the images are darker, and it becomes impossible to know them in the shadow regions. These lead to inaccurate shapes when using the FT-based algorithm. The synthetic method, on the other hand, is designed to interpolate the shapes in the shadow regions from the shapes in the bright regions.

The initial and converged shapes, reconstructed from the red, green and blue component images, are compared in Fig. 5. It can be observed that the hat in the converged shape for the red images is slightly lower than those for the green and blue images. This comes from the fact that there exist fringes in the body part in the red images, as shown in Fig. 5, which has the effect of enhancing the image contrast in the body part. The fringes are non-existent in the green and blue images, as shown in Fig. 5.

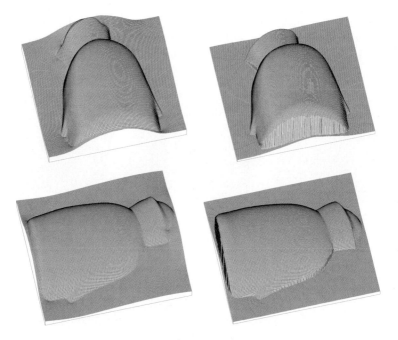

Fig. 4. Left: shape reconstructed using the modified FT algorithm, Right: shape reconstructed using the synthetic shape reconstruction. The top left shape was used as the initial depth map to reconstruct the shape using the synthetic method.

The existence of the fringes also affects the value of g_r in Eq. (15). It may be larger than 0.915 in the absence of the fringes, but its estimate is 0.889, which is smaller than 0.915 to reduce the enhanced image contrast.

On the whole, the three initial shapes look very similar. On the other hand, some differences are noticeable among the three converged shapes. Hence, the synthetic shape tends to reveal the color dependency more correctly.

These are the case of the horse, too, as shown in Fig. 6. On the whole, the converged shape is more accurate than the initial shape. Especially in the dark image regions and in the shadow regions, the converged shape is much better than the initial shape.

In this object case, there are no fringes in the red images but there are some fringes in the green images and, to a lesser extent, in the blue ones, in the hip and chest parts on the side of the body, appearing red in Fig. 2. They result in enhancing the image contrasts, and as a result, it is likely that g_r (0.648) has been made smaller than g_b (0.656).

The depth profiles as a function of number of iterations are shown in Fig. 7. In the case of the penguin, its dimensions in the 256 by 256 images are 193 in width and 252 in height, so that correspondingly the overall depth should be reconstructed to be 85.6. In the case of the horse, its dimensions are 252 in width and 246 in height, so that the overall depth should be reconstructed to be 71.1.

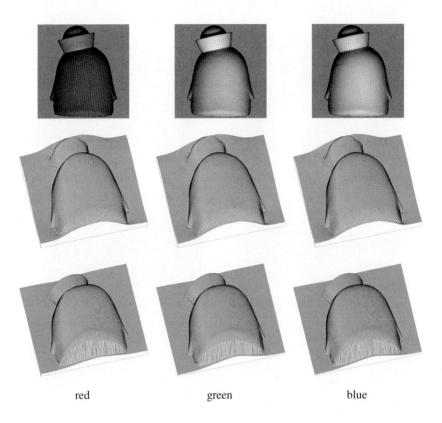

red   green   blue

Fig. 5. Top: from left, red, green and blue images for the direction (0, -12.5, 30), intermediate: shape reconstructed using the modified FT-based algorithm, bottom: shape reconstructed using the synthetic shape reconstruction

It is noticeable if we check the reconstructed shapes that the reconstruction is carried out first in the image regions with significant brightness, which is the portion marked A in Fig. 7. Then, it is done in the dark image regions or shadow regions, which is the portion marked B. The significant reconstruction completes in the portion A, and virtually no significant shape changes occur during the portion B. It is seen from the results in Fig. 7 that the portion A becomes longer as the object is more complex.

Three image-mapped shapes are shown in Fig. 8, where the object's shape was segmented and the images in Fig. 2 were mapped on it and made slightly brighter.

The new method was successfully applied to images of up to 512 by 512 pixels. We can know the shape in more details as we increase the dimensions. However, the existence of shadows in the images, which does not affect the resulting shape so much as shown in the results in Fig. 6, affects the computation time as implied in Fig. 7. Hence, not having dark image regions or shadow regions is important for fast convergence.

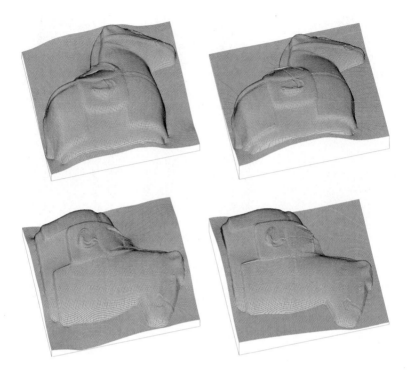

Fig. 6. Left: shape reconstructed using the modified FT-based algorithm, right: shape reconstructed using the synthetic shape reconstruction

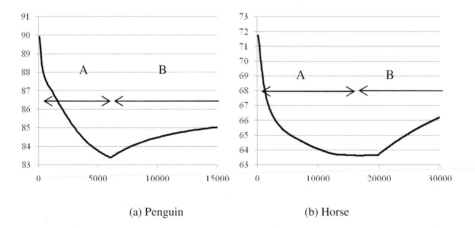

(a) Penguin   (b) Horse

Fig. 7. Overall depth profies in the iterative reconstruction, where object parts with sgnificantly bright images are reconstructed in the portion A and the remaining object parts with dark images or shadows are reconstructed mostly in B. B is much larger than A in terms of numbe rof iterations.

Fig. 8. Three images-mapped reconstructed shapes

6   Conclusions

A synthetic shape reconstruction method in combination with the FT-based method was presented. This method enables us to obtain accurate shapes from three color images, with no limit in dimension. Experimental results were given to examine the method, showing the usefulness of the method. There still are some factors that affect the accuracy such as shadows, fringes, and the conversion of the non-Lambertian reflection to Lambertian reflection. These will be studied in more details.

References

1. Woodham, R.J.: Photometric Method for Determining Surface Orientation from Multiple Images. In: Shape from Shading, ch. 17. MIT Press, Cambridge (1989)
2. Frankot, R.T., Chellappa, R.: A Method for Estimating Integrability in Shape from Shading Algorithms. IEEE Trans. PAMI 10, 439–451 (1988)
3. Agrawal, A., Raskar, R., Chellappa, R.: What is the Range of Surface Reconstruction from a Gradient Field? In: Leonardis, A., Bischof, H., Pinz, A. (eds.) ECCV 2006. LNCS, vol. 3951, pp. 578–591. Springer, Heidelberg (2006)
4. Basri, R., Jacobs, D.: Photometric stereo with General, Unknown Lighting. In: CVPR, vol. 2, pp. 374–381 (2001)
5. Ikeda, O.: Photometric Stereo Using Four Surface Normal Approximations and Optimal Normalization of Images. In: Proceedings ISSPIT, pp. 672–679 (2006)
6. Brown, M.Z., Burschka, D., Hager, G.D.: Advances in Computational Stereo. IEEE Trans. PAMI 25, 993–1007 (2003)
7. Barsky, S., Petrou, M.: The 4-Source Photometric Stereo Technique for Three-Dimensional Surfaces in the Presence of Highlights and Shadows. IEEE Trans. PAMI 25, 1239–1252 (2003)
8. Ikeda, O., Duan, Y.: Color Photometric Stereo for Directional Diffuse Object. In: Proceedings of Workshop on Applications of Computer Vision (2009)

Lunar Terrain and Albedo Reconstruction of the Apollo 15 Zone

Ara V. Nefian[1], Taemin Kim[2], Zachary Moratto[3],
Ross Beyer[4], and Terry Fong[2]

[1] Carnegie Mellon University
[2] NASA Ames Research Center
[3] Stinger-Ghaffarian Technologies Inc.
[4] SETI Institute

Abstract. Generating accurate three dimensional planetary models is becoming increasingly important as NASA plans manned missions to return to the Moon in the next decade. This paper describes a 3D surface and albedo reconstruction from orbital imagery. The techniques described here allow us to automatically produce seamless, highly accurate digital elevation and albedo models from multiple stereo image pairs while significantly reducing the influence of image noise. Our technique is demonstrated on the entire set of orbital images retrieved by the Apollo 15 mission.

1 Introduction

High resolution, accurate topographic and albedo maps of planetary surfaces in general and Lunar surface in particular play an important role for the next NASA robotic missions. More specifically these maps are used in landing site selection, mission planing, planetary science discoveries and as educational resources. This paper describes a method for topographic and albedo maps reconstruction from the Apollo era missions. The Apollo metric camera flown on an orbit at approximately 100km above the Lunar surface was a calibrated wide field (75°) of view orbital mapping camera that photographed overlapping images (80%). The scans of these film images recently made available [1,2] capture the full dynamic range and resolution of the original film resulting in digital images of size 22,000 ×

Fig. 1. Apollo Metric images from Orbit 33

G. Bebis et al. (Eds.): ISVC 2010, Part I, LNCS 6453, pp. 688–697, 2010.

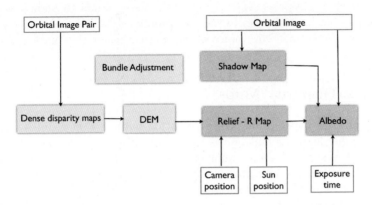

Fig. 2. The overall system for albedo reconstruction

22,000 pixels representing a resolution of 10 m/pixel. Figure 1 shows the images of one Lunar orbit captured by the Apollo 15 mission. Our method for geometric stereo reconstruction and photometric albedo reconstruction is illustrated in Figure 2. Each component of our system will be described in more detail in the following sections.

2 Bundle Adjustment

The Apollo-era satellite tracking network was highly inaccurate by today's standards with errors estimated to be 2.04-km for satellite station positions and 0.002 degrees for pose estimates in a typical Apollo 15 image [3]. Such errors propagate through the stereo triangulation process, resulting in systematic position errors and distortions in the resulting DEMs. These errors are corrected using bundle adjustment techniques. Our bundle adjustment solution uses SURF feature points [4]. Our bundle adjustment approach follows the method described in [5] and determines the best camera parameters that minimize the projection error given by $\epsilon = \sum_k \sum_j (I_k - I(C_j, X_k))^2$ where I_k are feature locations on the image plane, C_j are the camera parameters, and X_k are the 3D positions associated with features I_k. $I(C_j, X_k)$ is an image formation model (i.e. forward projection) for a given camera and 3D point. The optimization of the cost function uses the Levenberg-Marquartd algorithm. Speed is improved by using sparse methods described in [6]. Outliers are rejected using the RANSAC method and trimmed to 1000 matches that are spread evenly across the images. To eliminate the gauge freedom inherent in this problem, we add two addition error metrics to this cost function to constrain the position and scale of the overall solution. First, $\epsilon = \sum_j (C_j^{initial} - C_j)^2$ constrains camera parameters to stay close to their initial values. Second, a set of 3D *ground control points* are added to the error metric as $\epsilon = \sum_k (X_k^{gcp} - X_k)^2$ to constrain these points to known locations in the lunar coordinate frame. In the cost functions discussed above, errors are weighted by the inverse covariance of the measurement that gave rise to the constraint.

3   Dense Disparity Maps

This section describes our subpixel correlation method that reduces the effect of the scanning artifacts and refines the integer disparity map to sub-pixel accuracy. We investigated a large number of stereogrammmetric systems that can provide dense stereo matching from orbital imagery [7,8,9,10,11,12]. Our subpixel refinement method uses the statistical method described in [13].

In our approach the probability of a pixel in the right image is given by the following Bayesian model:

$$P(I_R(m,n)) = \prod_{(x,y) \in W} \mathcal{N}(I_R(m,n)|I_L(i + \delta_x, j + \delta_y), \frac{\sigma_p}{\sqrt{g_{xy}}})P(z = 0) + \quad (1)$$
$$+ \mathcal{N}(I_R(m,n)|\mu_n, \sigma_n)P(z = 1)$$

where W is a matching window,

$$I_L(i + \delta_x, j + \delta_y) \approx I_L(i,j) + \delta_x \frac{dI_L}{d_x}(i,j) + \delta_y \frac{dI_L}{d_y}(i,j) \quad (2)$$

and δ_x and δ_y are the local sub-pixel displacements given by

$$\delta_x(i,j) = a_1 i + b_1 j + c_1$$
$$\delta_y(i,j) = a_2 i + b_2 j + c_2. \quad (3)$$

The first mixture component ($z = 0$) is a normal density function with mean $I_L(i + \delta_x, j + \delta_y)$ and variance $\frac{\sigma_p}{\sqrt{g_{xy}}}$:

$$P(I_R(m,n)|z = 0) = \mathcal{N}(I_R(m,n)|I_L(i + \delta_x, j + \delta_y), \frac{\sigma_p}{\sqrt{g_{xy}}}) \quad (4)$$

The $\frac{1}{\sqrt{g_{xy}}}$ factor in the variance of this component has the effect of a Gaussian smoothing window over the patch. The second mixture component ($z = 1$) in Equation 4 models the image noise using a normal density function with mean μ_n and variance σ_n. With the assumption of independent observations within the matching window W the goal of the subpixel refinement algorithm becomes to maximize

$$P(\mathbf{I}_R(m,n)) = \prod_{(x,y) \in W} P(I_R(x,y)). \quad (5)$$

Figure 3 shows an example of a stereo image pair captured by the Apollo Metric Camera and used to generate a DEM of the Apollo 15 landing site.

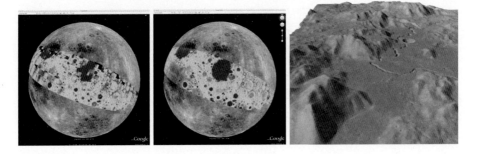

Fig. 3. Apollo 15 zone elevation model before (left) and after (center) bundle adjustment and oblique view of the Apollo 15 landing site (right)

4   Photometric Reconstruction

Each pixel of the Apollo Metric Camera images was formed by a combination of many factors, including albedo, terrain slope, exposure time, shadowing, and viewing and illumination angles. The goal of albedo reconstruction is to separate contributions of these factors. This is possible in part because of redundancy in the data; specifically, the same surface location is often observed in multiple overlapping images.

To do the albedo reconstruction, we include all of the factors in a image formation model. Many of the parameters in this model such as digital terrain slopes, viewing angle, and sun ephemeris are known. To reconstruct albedo, we first model how the Metric Camera images were formed as a function of albedo, exposure time, illumination and viewing angles, and other factors. Then we can formulate the albedo inference problem as a least-squares solution that calculates the most likely albedo to produce the observed image data.

Starting with the first images from the Apollo missions a large number of Lunar reflectance models were studied [14,15,16]. In this paper the reflectance is computed using the Lunar-Lambertian model [14,17]. As shown in Figure 4, we define the following unit vectors: $\mathbf{n}$ is the local surface normal; $\mathbf{l}$ and $\mathbf{v}$ are directed at the locations of the Sun and the spacecraft, respectively, at the time when the image was captured. We further define the angles $\mathbf{i}$ separating $\mathbf{n}$ from $\mathbf{l}$, $\mathbf{e}$ separating $\mathbf{n}$ from $\mathbf{v}$, and the phase angle α separating $\mathbf{l}$ from $\mathbf{v}$. The Lunar-Lambertian reflectance model is given by

$$F = AR = A \left[(1 - L(\alpha)) \cos(i) + 2L(\alpha) \frac{\cos(i)}{\cos(i) + \cos(e)} \right] \tag{6}$$

where A is the intrinsic albedo and $L(\alpha)$ is a weighting factor between the Lunar and Lambertian reflectance models [18] that depends on the phase angle and surface properties. R is a photometric function that depends on the angles α, i and e. The image formation model begins as follows. Let I_{ij}, A_{ij}, R_{ij} be the

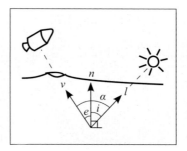

Fig. 4. Illumination and viewing angles used by the Lunar-Lambertian reflectance model

pixel value, albedo and R function at image location (i, j), and T be a variable proportional to the exposure time of the image. Then

$$I_{ij} = TA_{ij}R_{ij}. \tag{7}$$

Note that the image formation model described in Equation 7 does not take into consideration the camera transfer function since the influence of the non-linearities of the camera transfer function plays a secondary role in the image formation model [18]. From Equation 7 it can be seen that when the observed pixel value, exposure time, and R value are known, the image formation model in Equation 7 provides a unique albedo value. However, these values are subject to errors arising from measurement (exposure time), scanning (image value) or stereo modeling errors (reflectance), resulting in imprecise albedo calculations. The method proposed here mitigates these errors by reconstructing the albedo of the Lunar surface from *all* the overlapping images, along with their corresponding exposure times and DEM information. The albedo reconstruction is formulated as the least squares problem that minimizes the following cost function $\mathbf{Q}$:

$$\mathbf{Q} = \sum_{k} \sum_{ij} \left[(I_{ij}^k - A_{ij}T^k R_{ij}^k)^2 S_{ij}^k w_{ij}^k \right] \tag{8}$$

where super script k denotes the variables associated with the kth image and S_{ij}^k is a shadow binary variable. $S_{ij}^k = 1$ when the pixel is in shadow and 0, otherwise. The weights w_{ij}^k are chosen such that they have linearly decreasing values from the center of the image ($w_{ij}^k = 1$) to the image boundaries ($w_{ij}^k = 0$). The choice of these weights insures that the reconstructed albedo mosaic is seamless. As shown by Equation 8 and illustrated in Figure 2 the steps of our photometric reconstruction method are the computation of the shadow and relief map followed by albedo reconstruction. These steps are described next.

4.1   Shadow Map Computation

Discarding unreliable image pixels that are in shadow and for which the DEM and the reflectance models are unreliable plays an important role in accurate

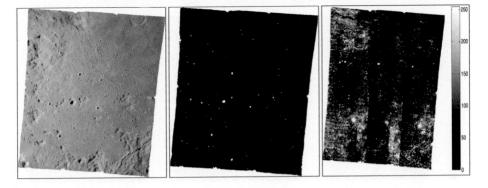

Fig. 5. Orbital image: (left) input image, (middle) binary shadow map with shadow regions shown in white, (right) DEM confidence map (brighter areas have higher estimated error)

albedo estimation [19,20]. Figure 5(left, middle) shows an input image together with its binary shadow map; shadowed areas are indicated in white.

4.2 Relief Map Computation

The geodetically aligned local DEM determine multiple values for the same location on the Lunar surface. A simple average of the local DEM value determines the value used in computing the local slopes and the reflectance value. The average DEM has the following benefits for albedo reconstruction:

- It is essential to the computation of a coherent "R map", since each point of the Lunar surface must have a unique DEM value.
- The statistical process produces more accurate terrain models by reducing the effect of random errors in local DEMs and without blurring the topographical features. Figure 6 shows the R map of a subregion of the orbital image in Figure 5 before and after the DEM averaging and denoising process. It can be seen that the noise artifacts in the original DEM are reduced in the denoised DEM while the edges of the large crater and mountain regions are very well preserved.
- The statistical parameters of the DEM values at each point are instrumental in building a confidence map of the Apollo coverage DEM. Figure 5(right) shows the error confidence map for the orbital image illustrated in Figure 5(left). The values shown in this error map are the 0.05×the variance values of the DEM expressed in meters.

This step of the algorithm computes the values of the photometric function R described by Equation 6 corresponding to every pixel in the image. We denote the set of R values as the "R map" of the image. The accurate DEM calculation influences the R values through the effect of surface normals on the angles i and e that appear in Equation 6.

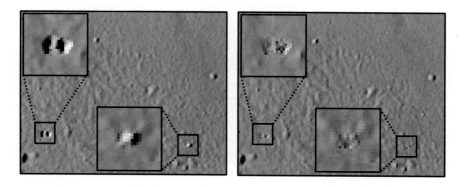

Fig. 6. R maps generated using (left) single local DEM and (right) denoised DEM derived from multiple overlapping local DEM. Our denoising approach preserves structure while reducing the artifacts shown in the insets.

4.3 Albedo Reconstruction

The optimal albedo reconstruction [21] from multi view images and their corresponding DEM is formulated as a minimization problem of finding

$$\{\tilde{A}_{ij}, \tilde{T}^k\} = \arg \min_{A_{ij}, T^k} \mathbf{Q} \tag{9}$$

for all pixels ij and images k, where $\mathbf{Q}$ is the cost function in Equation 8. An iterative solution to the above least square problem is given by the Gauss Newton updates described below.

- **Step 1:** Initialize the exposure time with the value provided in the image metadata. Initialize the albedo map with the average value of the local albedo observed in all images.

$$A_{ij} = \sum_k \frac{I_{ij}^k w_{ij}^k}{R_{ij}^k T^k} \tag{10}$$

- **Step 2:** Re-estimate the albedo and refine the exposure time using

$$\tilde{A}_{ij} = A_{ij} + \frac{\sum_k (I_{ij}^k - T^k A_{ij} R_{ij}^k) T^k R_{ij}^k S_{ij}^k w_{ij}^k}{\sum_k (T^k R_{ij}^k)^2 S_{ij}^k w_{ij}^k} \tag{11}$$

$$\tilde{T}^k = T^k + \frac{\sum_{ij} (I_{ij}^k - T^k A_{ij} R_{ij}^k) A_{ij} R_{ij}^k S_{ij}^k w_{ij}^k}{\sum_{ij} (A_{ij} R_{ij}^k)^2 S_{ij}^k w_{ij}^k} \tag{12}$$

- **Step 3:** Compute the error cost function $\mathbf{Q}$ (Eqn. 8) for the re-estimated values of the albedo and exposure time.

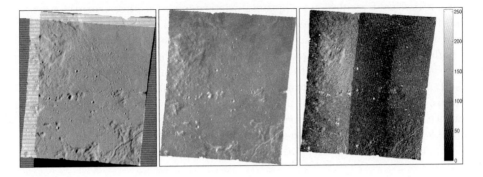

Fig. 7. Albedo reconstruction: (left) R map, (middle) reconstructed albedo, (right) albedo confidence map (brighter areas have higher estimated error)

- **Convergence:** If the convergence error between consecutive iterations falls below a fixed threshold then stop iterations and the re-estimated albedo is the optimal reconstructed albedo surface. Otherwise return to step 2.

Figure 7 shows the R map, the albedo map and the albedo reconstruction error map, respectively, for the original orbital image in Figure 5. The albedo reconstruction error map is computed as the absolute difference between the original image I_{ij}^k and the reconstructed image $T^k A_{ij} R_{ij}^k$. For display, the error values were multiplied by a factor 10 in Figure 7(right). Figure 8 illustrates the reconstructed albedo for one orbit of the Apollo mission data overlayed over previous low resolution Clementine imagery. The Clementine mission images were captured under incidence and emission angles close to zero, therefore capturing images that describe the relative Lunar albedo. It can be seen that the reconstructed albedo de-emphasizes the brightness variations shown in the original imagery (Figure 1) between images and produces a seamless albedo mosaic.

Fig. 8. Albedo reconstruction of orbit 33 of the Apollo 15 mission

5 Conclusions

This paper presents a novel approach for topographic and albedo maps generation from orbital imagery. The method for sub-pixel disparity maps uses a novel statistical formulation for optimally determining the stereo correspondence and reducing the effect of image noise. Our approach outperforms existing robust methods based on Lucas Kanade optical flow formulations at the cost of a higher computational complexity. The derived topographic maps are used to determine the albedo maps from an image formation model that incorporates the Lunar-Lambertian reflectance model. The optimal values of albedo and exposure time are learned from multiple image views of the same area on Luna surface. Further research will be directed towards a joint estimation of the topographic and albedo information using shape from shading techniques specific for the Lunar reflectance model and scanned image properties.

References

1. Robinson, M., Eliason, E., Hiesinger, H., Jolliff, B., McEwen, A., Malin, M., Ravine, M., Roberts, D., Thomas, P., Turtle, E.: LROC - Lunar Reconaissance Orbiter Camera. In: Proc. of the Lunar and Planetary Science Conference (LPSC), vol. XXXVI, p. 1576 (2005)
2. Lawrence, S.J., Robinson, M.S., Broxton, M., Stopar, J.D., Close, W., Grunsfeld, J., Ingram, R., Jefferson, L., Locke, S., Mitchell, R., Scarsella, T., White, M., Hager, M.A., Bowman-Cisneros, E., Watters, T.R., Danton, J., Garvin, J.: The Apollo Digital Image Archive: New Research and Data Products. In: Proc of the NLSI Lunar Science Conference, p. 2066 (2008)
3. Cameron, W.S., Niksch, M.A.: NSSDC 72-07: Apollo 15 Data User's Guide (1972)
4. Bay, H., Ess, A., Tuytelaars, T., Gool, L.V.: SURF: Speeded Up Robust Features. Computer Vision and Image Understanding (CVIU) 110, 346–359 (2008)
5. Triggs, B., Mclauchlan, P., Hartley, R., Fitzgibbon, A.: Bundle adjustment – a modern synthesis (2000)
6. Hartley, R.I., Zisserman, A.: Multiple View Geometry in Computer Vision, 2nd edn. Cambridge University Press, Cambridge (2004)
7. Cheng, L., Caelli, T.: Bayesian stereo matching. In: CVPRW 2004 Conference on Computer Vision and Pattern Recognition Workshop, p. 192 (2004)
8. Nehab, D., Rusinkiewicz, S., Davis, J.: Improved sub-pixel stereo correspondences through symmetric refinement. In: IEEE International Conference on Computer Vision, vol. 1, pp. 557–563 (2005)
9. Menard, C.: Robust Stereo and Adaptive Matching in Correlation Scale-Space. PhD thesis, Institute of Automation, Vienna Institute of Technology, PRIP-TR-45 (1997)
10. Nishihara, H.: PRISM: A Practical real-time imaging stereo matcher. Optical Engineering 23, 536–545 (1984)
11. Szeliski, R., Scharstein, D.: Sampling the Disparity Space Image. IEEE Transactions on Pattern Analysis and Machine Intelligence (PAMI) 26, 419–425 (2003)
12. Stein, A., Huertas, A., Matthies, L.: Attenuating stereo pixel-locking via affine window adaptation. In: IEEE International Conference on Robotics and Automation, pp. 914–921 (2006)
13. Nefian, A., Husmann, K., Broxton, M., To, V., Lundy, M., Hancher, M.: A Bayesian formulation for sub-pixel refinement in stereo orbital imagery . In: International Conference on Image Processing (2009)
14. McEwen, A.S.: Photometric functions for photoclinometry and other applications. Icarus 92, 298–311 (1991)

15. McEwen, A.S.: A precise lunar photometric function. In: Conf. 27th Lunar and Planet. Sci. (1996)
16. Minnaert, M.: The reciprocity principle in lunar photometry. Journal of Astrophysics (1941)
17. McEwen, A.S.: Exogenic and endogenic albedo and color patterns on Europa. Journal of Geophysical Research 91, 8077–8097 (1986)
18. Gaskell, R.W., Barnouin-Jha, O.S., Scheeres, D.J., Konopliv, A.S., Mukai, T., Abe, S., Saito, J., Ishiguro, M., Kubota, T., Hashimoto, T., Kawaguchi, J., Yoshikawa, M., Shirakawa, K., Kominato, T., Hirata, N., Demura, H.: Characterizing and navigating small bodies with imaging data. Meteoritics and Planetary Science 43, 1049–1061 (2008)
19. Arévalo, V., González, J., Ambrosio, G.: Shadow detection in colour high-resolution satellite images. Int. J. Remote Sens. 29, 1945–1963 (2008)
20. Matthies, L., Cheng, H.: Y.: Stereo vision and shadow analysis for landing hazard detection. In: IEEE International Conference on Robotics and Automation, pp. 2735 – 2742 (2008)
21. Yuille, A., Snow, D.: Shape and albedo from multiple images using integrability. In: CVPR 1997: Proceedings of the 1997 Conference on Computer Vision and Pattern Recognition (CVPR 1997), Washington, DC, USA, vol. 158, IEEE Computer Society, Los Alamitos (1997)

Super-Resolution Mosaicking of Unmanned Aircraft System (UAS) Surveillance Video Using Levenberg Marquardt (LM) Algorithm

Aldo Camargo[1], Richard R. Schultz[1], and Qiang He[2]

[1] Department of Electrical Engineering, University of North Dakota,
Grand Forks ND 58202 USA
[2] Department of Mathematics, Computer and Information Sciences, Mississippi Valley
State University, Itta Bena MS 38941 USA

Abstract. Unmanned Aircraft Systems (UAS) have been used in many military
and civil applications, particularly surveillance. One of the best ways to use the
capacity of a UAS imaging system is by constructing a mosaic of the recorded
video. This paper presents a novel algorithm for the construction of super-
resolution mosaicking. The algorithm is based on the Levenberg Marquardt
(LM) method. Hubert prior is used together with four different cliques to deal
with the ill-conditioned inverse problem and to preserve edges. Furthermore,
the Lagrange multiplier is compute without using sparse matrices. We present
the results with synthetic and real UAS surveillance data, resulting in a great
improvement of the visual resolution. For the case of synthetic images, we obtained a PSNR of 47.0 dB, as well as a significant increase in the details visible
for the case of real UAS frames in only ten iterations.

Keywords: Mosaic, Super-Resolution, UAS, Huber-Regularization, Ill-Posed
Inverse Problems. Steepest –Descent, Levenberg Marquardt, Nonlinear Multivariable Optimization, Unmanned Aircraft Systems, Video Surveillance.

1 Introduction

Multi-frame super-resolution refers to the particular case where multiple images of a
particular scene are available [3]. The idea is to use the low-resolution images containing motion, camera zoom, focus, and out of focus blur to recover extra data to
reconstruct an image with a resolution above of limits of the camera. The super-
resolved image should have more detail that any of the low resolution images. Mosaicking is the alignment or stitching of two or more images into a single composition
representing a 3D scene [5]. Generally, the mosaics are used to create a map which is
impossible to visualize with only one video frame.

Super-resolution mosaicking combines both methods, and it has number of applications when surveillance video from UAS or satellite is used. One clear application

G. Bebis et al. (Eds.): ISVC 2010, Part I, LNCS 6453, pp. 698–706, 2010.
© Springer-Verlag Berlin Heidelberg 2010

is the surveillance of certain areas even during night with the use of an infrared (IR) imaging system. The UAS can fly over areas of interest and generate super-resolved mosaics that can be analyzed at the ground control station. Other important applications involve the supervision of high voltage transmission lines, oil pipes, and the highway system. NASA also uses super-resolution mosaics to study the surface of Mars, the Moon, and other planets.

Super-resolution mosaicking has been studied by many researchers; Zomet and Peleg [6] use the overlapping area within a sequence of video frames to create a super-resolved mosaic. In this method, the SR reconstruction technique proposed in [7] is applied to a strip rather than a whole image. This means that the resolution of each strip is enhanced by the use of all the frames that contain that particular strip. The disadvantage is that this method is computationally expensive. Ready and Taylor [8] use a Kalman filter to compute the super-resolved mosaic. They add unobserved data to the mosaic using Dellaert's method. Basically, they construct a matrix that relates the observed pixels to estimated mixel values. This matrix is constructed using the homography matrix and the point spread function (PSF). The problem is that this matrix is extremely large, so they use a Kalman filter and diagonalization of the covariance matrix to reduce the amount of storage and computation required. The drawback of this algorithm is the use of the large matrix, and the best results with synthetic data obtain a PSNR of 31.6 dB. Simolic and Wiegand [9] use a method based on image warping. In this method, each pixel of each frame is mapped into the SR mosaic, and its gray level value is assigned to the corresponding pixel in the SR mosaic within a range of ± 0.2 pixel units. The drawback of this method is that it requires that the motion vectors and homography must be highly accurate, which is difficult when dealing with real surveillance video from UAS. Wang, Fevig, and Schultz [10] use the overlapped area within five consecutive frames from a video sequence. Then they use sparse matrices to model the relationship between the LR and SR frames, which is solved using maximum *a posteriori* estimation. To deal with the ill-posed problem of the super-resolution model, they use hybrid regularization. The drawback of this method is that it has to be used every five frames, which means that every five frames several sparse matrices has to be built. Therefore, this method does not seem to be appropriate to deal with a real video sequence which has thousand of frames. Pickering and Ye [1] proposed an interesting model for mosaicking and super-resolution of video sequences, using the Laplacian operator to find the regularization factor. The problem with the use of the Laplacian factor is that forces spatial smoothness. Therefore, noise and edge pixels are removed in the regularization process, eliminating sharp edges [11]. Arican and Frossard [17] use the Levenberg Marquardt algorithm to compute the SR of onmidirectional images. Chung [18] proposed different Gauss Newton methods to compute the SR of images, the disadvantage is that works only for small images.

Our method combines the ideas of most of these techniques, but it also inserts a different way to deal with the super-resolution mosaicking that does not require the construction of sparse matrices. Therefore, it is feasible to apply the algorithm to a relative large image sequence and obtain a video mosaic. Also, we use Huber regularization which preserves high frequency pixels, so sharp edges are preserved, and makes the super-resolution problem convex [20] that helps in the converge of our proposed algorithm.

2   Observation Model

Assuming that there are K frames of LR images available, the observation model can be represented as

$$y_k = DB_k W_k R[x]_k + \eta_k = H_k x + \eta_k \tag{1}$$

Here, y_k ($k = 1,2, ..., K$), x, and η_k represent the kth LR image, the part of the real world depicted by the super-resolution mosaic, and the additive noise, respectively. The observation model in (1) introduces $R[x]_k$ which represents the reconstruction of the kth warped SR image from the original high-resolution data x [1]. The geometric warp operator and the blur matrix between x and the kth LR image, $y_{k,}$ are represented by W_k and B_k, respectively. The decimation operator is denoted by D.

3   Robust Super-Resolution Mosaicking

The estimation of the unknown SR mosaic image is not only based on the observed LR images, but also on many assumptions such as the blurring process and additive noise. The motion model is computed as a projective model using the homography between frames; the blur is considered only optical. The additive noise η_k is considered to be independent and identically distributed white Gaussian noise. Therefore, the problem of finding the maximum likelihood estimate of the SR mosaic image $\hat{x}$ can be formulated as

$$\hat{x} = \arg \min{}_x \left\{ \left\| \sum_{k=1}^{K} y_k - DB_k W_k R[x]_k \right\|_2^2 \right\} \tag{2}$$

In this case, $\| \ \|_2$ denotes the Euclidean norm. As the SR reconstruction is an ill-posed inverse problem, we need to add another term for regularization which must have prior information of the SR mosaicking. This regularization term helps to convert the ill-posed problem into a well-posed problem. We use Huber regularization, because it preserves edges and high frequency information [2][3]. The equation :

$$\hat{x} = \arg \min{}_x \left\{ \left\| \sum_{k=1}^{K} y_k - DB_k W_k R[x]_k \right\|_2^2 + \lambda \sum_{g \in G_x} \rho(g, \alpha) \right\} \tag{3}$$

where $\mathbf{G}$ is the gradient operator over the cliques, and $\lambda^{(n)}$, the regularization operator can be computed as

$$\lambda^{(n)} = \left(\frac{\sum_{k=1}^{K} \left\| \mathbf{y}_k - \mathbf{DB}_k \mathbf{W}_k \mathbf{R}[\hat{\mathbf{x}}^{(n)}]_k \right\|_2}{K \sum_{g \in Gx} \rho(g,\alpha)} \right)^2 . \tag{4}$$

The Huber function is defined as

$$\rho(x,\alpha) = \begin{cases} x^2, & \text{if} \quad |x| \le \alpha, \\ 2\alpha|x| - \alpha^2, & \text{otherwise.} \end{cases} \tag{5}$$

4   Levenberg Marquardt Method

The Levenberg Marquardt (LM) method was proposed by [15] as a new method to solve nonlinear problems. This algorithm shares with gradient methods their ability to converge from an initial guess which may be outside of the region of convergence of other methods. Based on the Levenberg Marquardt method for minimizing (3), we have

Defining $f(x)$ as:

$$f(x) = \left\| \sum_{k=1}^{K} y_k - DB_k W_k R[x]_k \right\|_2^2 + \lambda \sum_{g \in G_x} \rho(g,\alpha) \tag{6}$$

Where J is the Jacobian matrix given by:

$$J = R^T \left[W_k^T B_k^T D^T (y_k - DB_k W_k R[\hat{x}^{(n)}]_k \right]_{k=1}^{K} - \lambda^{(n)} G^T \rho'(G\hat{x}^{(n)}, \alpha) \tag{7}$$

Levenberg Marquardt method is iterative: Initiated at the start point x_0. The method requires to find $\delta\square$ that minimizes $\left\| -F(\square + \delta\square) \right\| \approx \left\| \square - F(\square) - \square\delta\square \right\| = \left\| \varepsilon - \square\delta\square \right\|$
$\delta\square$ is found solving a non-linear-least square problem: The minimum is attained when $\square\delta\square - \varepsilon$ is orthogonal to the column space of J. This leads to:

$$H^* \delta x = J^T \varepsilon \tag{8}$$

and the pseudo Hessian is given by [15]:

$$H^* = J^T J \tag{9}$$

LM solves the equation (8), adding a damping term to the diagonal elements of H^* Therefore, the LM equation is

$$(H^* + cI)\delta x = J^T \varepsilon \tag{10}$$

then,

$$\hat{x}^{(n+1)} = \hat{x}^{(n)} + \delta x \qquad (11)$$

where c is the Levenberg Marquardt damping term that determines the behavior of the gradient in each iteration. If is close to zero then the algorithm behaves like a Gauss Newton method (GN), but if $c \to \infty$, then the algorithm behaves like steepest descent method (SD). The values of c during the iterative process are chosen in the

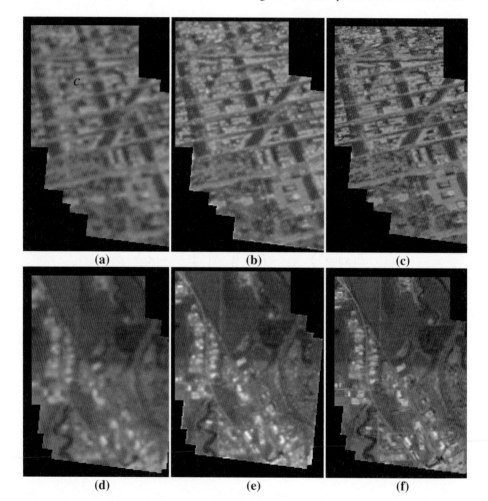

(a)   **(b)**   **(c)**

(d)   **(e)**   **(f)**

Fig. 1. Results of SR mosaicking for synthetic frames using the Levenberg Marquardt method. The mosaic was constructed using five frames. Figures (a) and (d) show the mosaic for the first and second sets of synthetic frames, respectively. These mosaics are the input to the algorithm. (b) and (e) are the super-resolved mosaics applying the LM method to (a) and (d), respectively. These mosaics are the output of the proposed algorithm. Figures (c) and (f) show the ground truth mosaics.

following way: at the beginning of the iterations, c is set to a large value, so the LM method use the robustness of SD, so the initial guess of the solution of (10) can be chosen with less caution. It is necessary to save the value of the errors for each iteration, and do the comparison between two consecutive errors. In the case that $error_{(k)}$ < $error_{(k-1)}$ decreases   by a certain amount so LM behaves like GN to take advantage of the speed up to converge, otherwise c increases to a to large value (increasing the searching area), which means that the LM behaves like SD.

5   Experimental Results for LM Method

We created synthetic LR frames from a single high resolution image. These LR frames where created using different translations (18 to 95 pixels), rotations ($5°$ to $10°$), and scales (1 to 1.5); we blurred the frames with a Gaussian kernel of size 3x3. With these LR frames we compute the LR mosaic which is the input to the proposed LM algorithm.

Figure 1 (d) and (f) shows the result for the proposed LM algorithm. The results SR mosaic shows a great improvement in the quality of the image and also looks close to the ground truth. Figure 2, shows the result for a real set of UAS frames, the SR mosaic has more details and is less cloudy. For the case of real UAS IR video frames there is some artifacts introduced by the LM algorithm, the reason of that is because the solution of (10) deals with close to singular matrices.

The PSNR obtained for the LM SR mosaic is 47.45 dB, using only 5 LR frames. Table 1 shows some results for two different of synthetic data, computed in ten iterations. Table 2 shows result for two different set of video frames captured in 2007 by the UASE team at the University of North Dakota.

Table 1. Results of the computation of super-resolution mosaicking using Levenberg Marquardt method for two different set of color synthetic frames

Test	PSNR (dB)	Final error $\left\|\hat{\mathbf{X}}_{k+1} - \hat{\mathbf{X}}_k\right\|_2 \Big/ \left\|\mathbf{X}_k\right\|_2$	Total Processing Time on CPU (sec)
First set of five synthetic color frames.	43.77	0.002833	5.109
Second set of five synthetic color frames	47.45	0.002505	4.281

Table 2. Results of the computation of super-resolution mosaicking using the proposed Levenberg Marquardt algorithm for three different sets real frames from UAS

Test	Final error $\left\| \hat{\mathbf{x}}_{k+1} - \hat{\mathbf{x}}_k \right\|_2 \Big/ \left\| \mathbf{x}_k \right\|_2$	Total Processing Time on CPU (sec)
Test #1: IR frames of buildings	0.003514	11.766
Test #2: IR frames of a forest.	0.004298	11.391

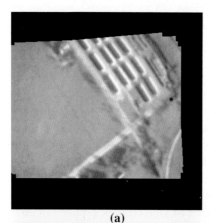

(a)

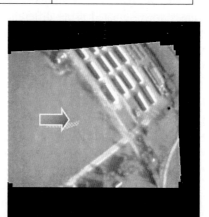

(b)

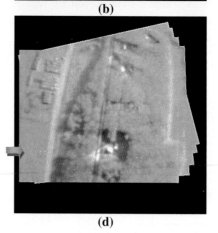

(c)

(d)

Fig. 2. Results of the SR mosaic for real frames from UAS using the proposed Levenberg Marquardt algorithm. The mosaic was constructed using five frames. Figures (a) and (c) show the mosaic for the first and second set of frames, respectively. These mosaics are the input to the algorithm. (b) and (d) are the super-resolved mosaics of (a) and (c) respectively. These mosaics are the output of the proposed algorithm. The arrows show the artifacs introduced by the algorithm.

6   Conclusions

We showed the construction of SR mosaic from input LR frames. For that, we first compute the LR mosaic which becomes the input to the SR LM algorithm The results for synthetic and real frames from UAS show a great improvement in the resolution. The algorithms work with IR and visible (RGB) frames. It is not necessary to use sparse matrices, so the processing time is reduced, and can be even faster if using GPU (Graphics Processing Unit) processors.

7   Future Work

This paper only works with a handful of five frames from a video; we are working to improve the algorithm to take a whole video and create the SR video mosaicking. Finally, we are considering the use of parallel programming by using GPU processors to speed up the computation of the super-resolution mosaic.

Acknowledgements

This research was supported in part by the FY2006 Defense Experimental Program to Stimulate Competitive Research (DEPSCoR) program, Army Research Office grant number 50441-CI-DPS, Computing and Information Sciences Division, "Real-Time Super-Resolution ATR of UAV-Based Reconnaissance and Surveillance Imagery," (Richard R. Schultz, Principal Investigator, active dates June 15, 2006, through June 14, 2010). This research was also supported in part by Joint Unmanned Aircraft Systems Center of Excellence contract number FA4861-06-C-C006, "Unmanned Aerial System Remote Sense and Avoid System and Advanced Payload Analysis and Investigation," as well as the North Dakota Department of Commerce grant, "UND Center of Excellence for UAV and Simulation Applications." Additionally, the authors would like to acknowledge the contributions of the Unmanned Aircraft Systems Engineering (UASE) Laboratory team at the University of North Dakota.

References

[1] Pickering, M., Ye, G., Frater, M., Arnold, J.: A Transform-Domain Approach to Super-Resolution Mosaicing of Compressed Images. In: 4th AIP International Conference and the 1st Congress of the IPIA. Journal of Physics: Conference Series, vol. 124 (2008); 012039

[2] Shultz, R.R., Meng, L., Stevenson, R.L.: Subpixel motion estimation for multiframe resolution enhancement. Visual Communication and Image Processing, 1317–1328 (1997)

[3] Pickup, L.C.: Machine Learning in Multi-frame Image Super-resolution. Ph.D. Dissertation, University of Oxford (2007)

[4] Borman, S.: Topic in Multiframe Superresolution Restoration. Ph.D. Dissertation, University of Notre Dame, Notre Dame, Indiana (2004)

[5] Capel, D.P.: Image Mosaicing and Super-resolution. University of Oxford, Ph. D. Dissertation. University of Oxford (2001)

[6] Zomet, A., Peleg, S.: Efficient Super-resolution and Applications to Mosaics. In: Proc. of International Conference of Pattern Recognition (September 2000)

[7] Irani, M., Peleg, S.: Improving resolution by image registration. Graph. Models Image Process. 53, 231–239 (1991)

[8] Ready, B.B., Taylor, C.N., Beard, R.W.: A Kalman-filter Based Method for Creation of Super-resolved Mosaicks. In: Robotics and Automation, UCRA 2006 (2006)

[9] Smolic, A., Wiegand, T.: High-resolution video mosaicing

[10] Wang, Y., Fevig, R., Schultz, R.R.: Super-resolution Mosaicking of UAV Surveillance Video. In: ICIP 2008, pp. 345–348 (2008)

[11] Farsiu, S., Robinson, D., Elad, M., Milanfar, P.: Fast and Robust Multi-Frame Super-resolution. IEEE Transaction on Image Processing 13(10), 1327–1344 (2004)

[12] Brown, M., Lowe, D.G.: Recognising Panoramas. In: Proceedings of the Ninth IEEE International Conference on Computer Vision, ICCV, vol. 2, p. 1218 (2003)

[13] Mikolajczyk, K., Schmid, C.: A Performance Evaluation of Local Descriptors. IEEE Transactions of Pattern Analysis and Machine Intelligence 27(10) (2005)

[14] Plaza, A., Plaza, J., Vegas, H.: Improving the Performance of Hyperspectral Image and Signal Processing Algorithms Using Parallel, Distributed and Specialized Hardware-Based Systems. Journal Signal Processing Systems, doi:10.1007/s11265-010-0453-1

[15] Marquardt, D.W.: An Algorithm for the Least-Squares Estimation of Nonlinear Parameters. SIAM Journal of Applied Mathematics 11(2), 431–441 (1963)

[16] Chung, J., Nagy, J.G.: Nonlinear Least Squares and Super Resolution. Journal of Physics: Conference Series 124 (2008); 012019

[17] Arican, Z., Frossard, P.: Joint Registration and Super-resolution with Omnidirectional Images. IEEE Transactions on Image Processing (2009)

[18] http://www.ics.forth.gr/~lourakis/levmar/

[19] Press, W.H., Teukolsky, S.A., Vetterling, A.W.T., Flannery, B.P.: Numerical Recipes in C: The Art of Scientific Computing. Cambridge University Press, New York (1992)

[20] Schultz, R.R.: Multichannel Stochastic Image Models: Theory, Applications and Implementations. PhD dissertation. University of Notre Dame, Indiana

Computer-Generated Tie-Dyeing Using a 3D Diffusion Graph

Yuki Morimoto* and Kenji Ono

RIKEN
yu-ki@riken.jp

Abstract. Hand dyeing generates artistic representations with unique and complex patterns. The aesthetics of dyed patterns on a cloth originate from the physical properties of dyeing in the cloth and the geometric operations of the cloth. Although many artistic representations have been studied in the field of non-photorealistic rendering, dyeing remains a challenging and attractive topic. In this paper, we propose a new framework for simulating dyeing techniques that considers the geometry of the folded cloth. Our simulation framework of dyeing in folded woven cloth is based on a novel dye transfer model that considers diffusion, adsorption, and supply. The dye transfer model is discretized on a 3D graph to approximate the folded woven cloth designed by user interactions. We also develop new methods for dip dyeing and tie-dyeing effects. Comparisons of our simulated results with real dyeing demonstrate that our simulation is capable of representing characteristics of dyeing.

1 Introduction

Since ancient times, dyeing has been employed to color fabrics in both industry and arts and crafts. Various dyeing techniques are practiced throughout the world, such as wax-resist dyeing (*batik dyeing*), hand drawing with dye and paste (*Yuzen dyeing*), and many other techniques [1,2]. Tie-dyeing produces beautiful and unique dyed patterns. The tie-dyeing process involves performing various geometric operations on a support medium, then dipping the medium into a dyebath. The process of dipping a cloth into a dyebath is called dip dyeing.

The design of dye patterns is complicated by factors such as dye transfer and cloth deformation. Professional dyers predict final dye patterns based on heuristics; they tap into years of experience and intimate knowledge of traditional dyeing techniques. Furthermore, the real dyeing process is time-consuming. For example, clamp resist dyeing requires the dyer to fashion wooden templates to press the cloth during dyeing. Templates used in this technique can be very complex. Hand dyed patterns require the dyer's experience, skill, and effort, which are combined with the chemical and physical properties of the materials. This allows the dyer to generate interesting and unique patterns. There are no other painting techniques that are associated with the deformation of the

---

* Corresponding author.

G. Bebis et al. (Eds.): ISVC 2010, Part I, LNCS 6453, pp. 707–718, 2010.

| Our simulated results along the time step | Real dyed pattern [3] | Magnified images |

Fig. 1. Comparisons of our simulated results with a real dyed pattern

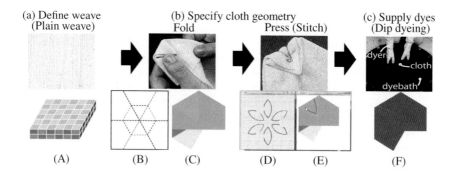

(a) Define weave (Plain weave)   (b) Specify cloth geometry — Fold — Press (Stitch)   (c) Supply dyes (Dip dyeing)

(A)   (B)   (C)   (D)   (E)   (F)

Fig. 2. The general steps for a real dyeing process (top row [3]) and our dyeing framework (bottom row) using the Chinese flower resist technique. (A) The woven cloth has a plain weave; the blue and yellow cells indicate the warp and the weft. (B) The unfolded cloth with user specified fold lines, the red and blue lines indicate ridges and valleys, and folds in the cloth. (C) The corresponding folded cloth in (B). (D, E) The interfaces representing user drawings on an unfolded and folded cloth. The gray lines indicate a user-specified boundary domain; these will be the dye resist regions. (F) The folded cloth with the red region indicating the exterior surfaces.

support medium. In contrast to hand dyeing, dyeing simulation allow for an inexpensive, fast, and accessible way to create dyed patterns. We focus on folded cloth geometry and dye transfer phenomenon. Figure 1 shows the simulated results obtained using our physics-based dyeing framework and a real dyed pattern.

Related work. Non-photorealistic rendering (NPR) methods for painting and transferring pigments on paper have been developed for watercolor and Chinese ink paintings [4,5,6,7,8,9]. These methods are often based on fluid mechanics: Kunii et al. [5] used Fick's second law of diffusion to describe water spreading on dry paper with pigments; Curtis et al. [9] developed a technique for simulating watercolor effects; Chu and Tai [7] presented a real-time drawing algorithm based on the lattice Boltzmann equation ; and Xu et al. [8] proposed a generic pigment model based on rules and formulations derived from studies of adsorption and diffusion in surface chemistry and the textile industry.

Several studies have also investigated dyeing methods: Wyvill et al. [10] proposed an algorithm for rendering cracks in *batik* that is capable of producing realistic crack patterns in wax; Shamey et al. [11] used a numerical simulation of the dyebath to study mass transfer in a fluid influenced by dispersion; and Morimoto et al. [12] used a diffusion model that includes adsorption to reproduce the details of dyeing characteristics such as thin colored threads by considering the woven cloth, based on dye physics.

However, the above methods are insufficient for simulating advanced dyeing techniques, such as tie-dyeing. Previous methods are strictly 2D, and are not designed to handle the folded 3D geometry of the support medium. There is no other simulation method that considers the folded 3D geometry of the support medium which clearly affects real dyeing results. In addition, the above methods cannot be used in tandem with texture synthesis [13] to generate dyed patterns, because dyeing is a time variant physical phenomenon. Because of the large number of dyeing factors (chemical and physical phenomenon with the dyer's design, skill, and experience), only physics-based models will be capable of simulating sensitive dyeing processes. We extend the physical 2D dyeing model (Morimoto et al. [12]) to consider the folded cloth geometry.

Real dyeing process and framework overview. Figure 2 depicts the framework with a corresponding dyeing process. In the real process, the cloth is prepared, folded, and pressed to form the cloth geometry that allows resist dyeing to occur in some parts of the cloth; tie-dyeing techniques (folding, stitching, clamping, etc) are coordinated with folding and pressing. The cloth is put into a dyebath, where dye transfer occurs. The cloth is removed from the dyebath, opened, and dried to complete dye transfer. Our framework is summarized in the bottom of Figure 2. First, we define the weave structure. Then, we model the dyeing techniques with user-specified folding and pressing. Next, we calculate the distribution of the dye supply and simulate dye transfer.

Our approach. We propose a novel simulation framework that simulates dyed patterns produced by folding the cloth. This dyeing framework is based on a new dye transfer model; it is implemented using a discrete 3D diffusion graph that approximates the folded cloth. Our dye transfer model is formulated using an evolutionary system of PDEs that accounts for diffusion with adsorption and dye supply from the dyebath.

We model the effects of pressing in tie-dyeing [1] via dye capacity maps. We model dip dyeing from a dye supply map. These maps are generated from distance fields of the exterior domain and dye resist regions of the folded cloth, respectively. The graph structure is constructed as follows. The graph vertices are sampled from a 2D cloth patch. The cloth patch is then folded along the fold lines specified by the user. The graph edges connect adjacent vertices in the folded cloth. We also incorporated the two-layered cellular cloth model of Morimoto et al. [12] to represent woven cloths.

Contributions and benefits. Our main contribution is a physically based dyeing simulation framework that accounts for the folded cloth geometry associated with various dyeing techniques. The technical contributions are a novel dye transfer model and its discretization in a 3D graph to simulate tie-dyeing and dip dyeing.

Our framework allows us to produce new stylized depictions for computer graphics. We expect the framework to have applications in real dyeing processes as a computer-aided design tool. Even dyeing neophytes would be able to generate and predict dyed patterns with minimal material and labor costs. By easing the design process, we hope to advance dyeing design and help spread the dyeing tradition around the world. Our 3D diffusion graph can be applied to simulate diffusion effects in layered objects.

2 Dye Transfer Model

Our dye transfer model accounts for the diffusion, adsorption, and supply terms of the dye as described in an equation (1). The diffusion and adsorption terms describe the dye behavior based on Fick's second law [14] and dyeing physics, respectively. The supply term enables arbitrary dye distribution for dip dyeing and user drawings.

Let $f = f(\mathbf{x}, t) \in (0, 1]$ be a dye concentration function with position vector $\mathbf{x} \in \Re^3$ and time parameter $t \geq 0$, where t is independent of $\mathbf{x}$. The dyeing model is formulated by the following evolutionary system of PDEs.

$$\frac{\partial f(\mathbf{x}, t)}{\partial t} = \mathrm{div}(D(\mathbf{x})\nabla f) + s(\mathbf{x}, f) - a(\mathbf{x}, f), \tag{1}$$

where $D(\mathbf{x})$ is the diffusion coefficient function, $\mathrm{div}(\cdot)$ and $\nabla(\cdot)$ are the divergence and gradient operators respectively, and $s(\cdot, \cdot)$ and $a(\cdot, \cdot)$ are the source and sink terms that represent dye supply and adsorption, whereas $\mathrm{div}(D(\mathbf{x})\nabla f)$ is the diffusion term. We model the functions $s(\cdot, \cdot)$ and $a(\cdot, \cdot)$ as follows.

$$s(\mathbf{x}, f) = \begin{cases} \alpha M_s(\mathbf{x}) & \text{if } M_s(\mathbf{x}) > f(\mathbf{x}, t) \text{ and } M_{cd}(\mathbf{x}) > f(\mathbf{x}, t). \\ 0 & \text{Otherwise}, \end{cases}$$

$$a(\mathbf{x}, f) = \begin{cases} \beta f(\mathbf{x}, t) & \text{if } h(\mathbf{x}, t) < a_d(\mathbf{x}, f) \text{ and } M_{ca}(\mathbf{x}) > h(\mathbf{x}, t). \\ 0 & \text{Otherwise}, \end{cases}$$

where α is the user-specified dye concentration, $M_s(\mathbf{x}) \in [0, 1]$ is the dye supply map, $M_{cd}(\mathbf{x})$ and $M_{ca}(\mathbf{x})$ are the diffusion and adsorption capacity maps, $\beta \in [0, 1]$ is the user-specified adsorption rate, and $a_d(\mathbf{x}, f)$ is the adsorption capacity according to adsorption isotherms. The adsorption isotherm depicts the amount of adsorbate on the adsorbent as a function of its concentration at constant temperature. In this paper, we employ the Langmuir adsorption model [15], which is a saturation curve, to calculate $a_d(\mathbf{x}, f)$ in our simulation based on the model developed by Morimoto et al. [12]. The adsorbed dye concentration $h(\mathbf{x}, t) \in [0, 1]$ is given by the equation bellow and the evolution of $f(\mathbf{x}, t)$ and $h(\mathbf{x}, t)$ as $t \to \infty$ describes the dyeing process.

$$\frac{\partial h(\mathbf{x}, t)}{\partial t} = a(\mathbf{x}, f)\frac{M_{cd}(\mathbf{x})}{M_{ca}(\mathbf{x})}, \tag{2}$$

Two-layered cloth model and diffusion coefficient [12]. Since the cloth is woven in wefts and warps, the layers of the cloth model naturally reflect this, and consist of a layer of weft cells and a layer of warp cells. The positions of cells are defined along the cloth weave. The diffusion coefficient $D(\mathbf{x})$ is defined between graph vertices, and is calculated from the composition of the dyestuff and cloth fibre, the style of the weave, and the relationship of cloth fibers between cells, causing an anisotropic diffusion. When considering the x-y plane, diffusion occurs between adjacent cells in the same plane. In the z-direction, diffusion is dictated by cloth geometry, which is described in more detail in Section 3.

Boundary condition. Let $\mathbf{z} \in \mathcal{B}$ be the boundary domain for dye transfer. We use the Neumann boundary condition $\frac{\partial f(\mathbf{z},t)}{\partial \mathbf{n}(\mathbf{z})} = b(\mathbf{z}), \mathbf{z} \in \mathcal{B}$ where $\mathbf{n}(\mathbf{z})$ is a unit normal vector of the boundary. In tie-dyeing, parts of the cloth are pressed together by folding and pressing. We assume that the pressed region is the boundary domain because no space exists between the pressed cloth parts for the dye to enter. In our framework, the user specifies $\mathcal{B}$ by drawing on both the unfolded and folded cloth, as shown in Figure 2 (D) and (E), respectively. In the case of the folded cloth, shown in Figure 2 (E), we project $\mathcal{B}$ to all overlapping faces of the folded cloth. Here, the faces are the polygons of the folded cloth, as shown in Figure 2 (C).

Press function. We introduce a press function $P(\mathbf{x}, c) \in [0, 1]$ as a press effect from dyeing technique. Here c is a user-specified cut-off constant describing the domain of influence of the dye supply and the capacity maps, and represents the physical parameters. The extent of the pressing effect and dye permeation depend on the both softness and elasticity of the cloth and on the tying strength. The press function serves to;

- Limit dye supply (Pressed regions prevent dye diffusion).
- Decrease the dye capacity (Pressed regions have low porosity).

We approximate the magnitude of pressure using the distance field $dist(\mathbf{x}, \mathcal{B})$ obtained from the pressed boundary domain $\mathcal{B}$. Note that the press effect only influences the interior surfaces of the cloth, as only interior surfaces can press each other.

We define Ω as the set of exterior surfaces of the tied cloth that are in contact with the dye and $\mathcal{L}$ as the fold lines that the user input to specify the folds. We define Ω, $\mathcal{B}$, and $\mathcal{L}$ as fold features (Figure 3). We calculate the press function $P(\mathbf{x}, c)$ as described in Pseudocode 1. In Pseudocodes 1 and 2, $CalcDF()$ calculates the distance field obtained from the fold feature indicated by the first argument, and returns infinity on the fold features indicated by the second and third arguments. We use m as the number of vertices in the diffusion graph described in Section 3.

Dye supply map. The dye supply map $M_s(\mathbf{x})$ (Figure 4) describes the distribution of dye sources and sinks on the cloth, and is applied to the dye supply term in equation (1). In dip dyeing, the dye is supplied to both the exterior and interior surfaces of the folded cloth. The folded cloth opens naturally except in pressed regions, allowing liquid to enter the spaces between the folds (Figure 5). Thus, we assume that $M_s(\mathbf{x})$ is inversely proportional to the distance from Ω, as it is easier to expose regions that are closer to the liquid dye. Also, the dye supply range depends on the movement of the cloth in the dyebath. We model this effect by limiting $dist(\mathbf{x}, \Omega)$ to a cut-off constant c_Ω. Another cut-off constant $c_{\mathcal{B}_1}$ limits the influence of the press function $P(\mathbf{x}, c_{\mathcal{B}_1})$.

The method used to model the dye supply map $M_s(\mathbf{x})$ is similar to the method used to model the press function, and is described in Pseudocode 2. In Pseudocode 2, $dist_{max}$ is the max value of $dist(\mathbf{x}, \Omega)$ and $GaussFilter()$ is a Gaussian function used to mimic rounded-edge folding (as opposed to sharp-edge folding) of the cloth.

$dist(\mathbf{x}, \mathcal{B}) \leftarrow CalcDF(\mathcal{B}, \Omega, \mathcal{L})$
for i=1 to m **do**
  if $dist(i, \mathcal{B}) > c$ **then**
    $P(i, c) \leftarrow 1.0$
  else
    $P(i, c) \leftarrow dist(i, \mathcal{B})/c$
  end if
end for

Pseudocode 1. Press function.

$dist(\mathbf{x}, \Omega) \leftarrow CalcDF(\Omega, \mathcal{B}, \mathcal{L})$
for i=1 to m **do**
  $dist(i, \Omega) \leftarrow (dist_{max} - dist(i, \Omega))$
  if $dist(i, \Omega) > c_\Omega$ **then**
    $dist(i, \Omega) \leftarrow P(i, c_{\mathcal{B}_1})$
  else
    $dist(i, \Omega) \leftarrow P(i, c_{\mathcal{B}_1})dist(i, \Omega)/c_\Omega$
  end if
end for
$M_s(\mathbf{x}) \leftarrow GaussFilter(dist(\mathbf{x}, \Omega))$

Pseudocode 2. Dye supply map.

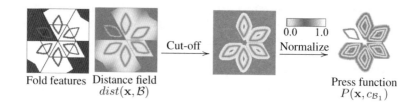

Fold features   Distance field
$dist(\mathbf{x}, \mathcal{B})$

Cut-off   0.0   1.0   Normalize

Press function
$P(\mathbf{x}, c_{\mathcal{B}_1})$

Fig. 3. An example for calculating the press function $P(\mathbf{x}, c_{\mathcal{B}_1})$ for the Chinese flower resist. In the fold features, the exterior surface Ω, pressed boundary domain $\mathcal{B}$, and the fold lines $\mathcal{L}$ are shown by red, gray, and blue colors.

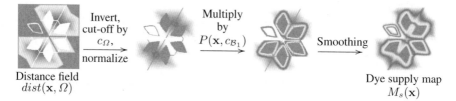

Distance field
$dist(\mathbf{x}, \Omega)$

Invert, cut-off by c_Ω, normalize

Multiply by $P(\mathbf{x}, c_{\mathcal{B}_1})$

Smoothing

Dye supply map
$M_s(\mathbf{x})$

Fig. 4. An illustration of the dye supply map $M_s(\mathbf{x})$ for the Chinese flower resist

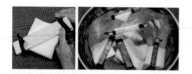

Basic capacity   Press function   Capacity map
          $P(\mathbf{x}, c_{\mathcal{B}_2})$ $M_{cd}(\mathbf{x})$

Fig. 5. Photographs of real dip dyeing. The left shows a folded cloth pressed between two wooden plates. The right shows folded cloths that have opened naturally in liquid.

Fig. 6. Illustration for our diffusion capacity map $M_{cd}(\mathbf{x})$ calculated by taking the product of the basic capacity and $P(\mathbf{x}, c_{\mathcal{B}_2})$

Capacity maps. The capacity maps indicate the dye capacities in a cloth; they define the spaces that dyes can occupy. We first calculate the basic capacities from a fibre porosity parameter [12]. The cut-off constant c_{B_2} limits the press function $P(\mathbf{x}, c_{B_2})$ used here. We then multiply these capacities by the press function to obtain the capacity maps $M_{cd}(\mathbf{x})$ and $M_{ca}(\mathbf{x})$ (see Figure 6).

3   Graph Diffusion

We approximate the user-specified cloth geometry by a 3D diffusion graph $\mathcal{G}$. Our transfer model is then discretized on the graph.

Diffusion graph construction. We construct multiple two-layered cells from the two-layered cellular cloth model by cloth-folding operations. The diffusion graph $\mathcal{G}$ is a weighted 3D graph with vertices $\mathbf{v}_i$, edges, and weights w_{ij} where $i, j = 1, 2, .., m$. The vertex $\mathbf{v}_i$ in the diffusion graph $\mathcal{G}$ is given by the cell center.

We apply the ORIPA algorithm [16], which generates the folded paper geometry from the development diagram, to the user-specified fold lines on a rectangular cloth. The fold lines divide the cloth into a set of faces as shown in Figure 2 (B). The ORIPA algorithm generates the corresponding vertex positions of faces on the folded and unfolded cloths and the overlapping relation between every two faces as shown in Figure 2 (C). We apply ID rendering to the faces to obtain the overlapping relation for multiple two-layered cells. We then determine the contact areas between cells in the folded cloth, and construct the diffusion graph $\mathcal{G}$ by connecting all vertices $\mathbf{v}_i$ to vertices $\mathbf{v}_j$ in the adjacent contact cell by edges, as illustrated in Figure 7.

Discretization in diffusion graph. The finite difference approximation of the diffusion term of equation (1) at vertex $\mathbf{v}_i$ of $\mathcal{G}$ is given by

$$\mathrm{div} D(\mathbf{x})\nabla f \approx \sum_{j \in N(i)} w_{ij} \frac{f(\mathbf{v}_j, t) - f(\mathbf{v}_i, t)}{|\mathbf{v}_j - \mathbf{v}_i|^2},$$

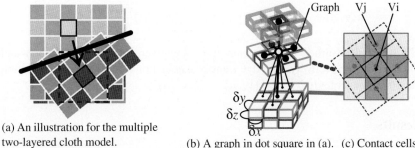

(a) An illustration for the multiple two-layered cloth model.

(b) A graph in dot square in (a).   (c) Contact cells.

Fig. 7. Graph construction from folded geometry of woven cloth. The bold line in (a) represents a fold line. The gray points in (c) are contact cell vertices $\mathbf{v}_j$ of the target cell vertex $\mathbf{v}_i$.

Fig. 8. From the left to right, five D_{ij} for x, y directions in the weft layer, z direction in both the layers, and x, y directions in the warp layer and two $D(\mathbf{v}_i)$ in the weft and warp layer in our results, where $D_{max} = 2.68e - 006$

where $w_{ij} = D_{ij}A_{ij}$, $N(i)$ is an index set of vertices adjacent to $\mathbf{v}_i$ in $\mathcal{G}$, and D_{ij} and A_{ij} are the diffusion coefficient and the contact area ratio between vertices $\mathbf{v}_i$ and $\mathbf{v}_j$, respectively (Figure 7). We calculate the distance between $\mathbf{v}_i$ and $\mathbf{v}_j$ from cell sizes, $\delta x, \delta y, \delta z$, and define $D_{ij} = (D(\mathbf{v}_i)+D(\mathbf{v}_j))/2$ between vertices connected by folding (Figure 8).

The following semi-implicit scheme gives our discrete formulation of equation (1):

$$(I - (\delta t)L)\mathbf{f}^{n+1} = \mathbf{f}^n + \delta t(\mathbf{s}^n - \mathbf{a}^n),$$

which is solved using the SOR solver [17], where I is the identity matrix, δt is the discrete time-step parameter, n is the time step, and

$$\mathbf{f}^n = \{f(\mathbf{v}_1, n), f(\mathbf{v}_2, n), .., f(\mathbf{v}_m, n)\},$$

$$\mathbf{s}^n = \{s(\mathbf{v}_1, f(\mathbf{v}_1, n)), s(\mathbf{v}_2, f(\mathbf{v}_2, n)), .., s(\mathbf{v}_m, f(\mathbf{v}_m, n))\},$$

$$\mathbf{a}^n = \{a(\mathbf{v}_1, f(\mathbf{v}_1, n)), a(\mathbf{v}_2, f(\mathbf{v}_2, n)), .., a(\mathbf{v}_m, f(\mathbf{v}_m, n))\},$$

L is the $m{\times}m$ graph Laplacian matrix [18] of $\mathcal{G}$. The element l_{ij} of L is given by

$$l_{ij} = \begin{cases} w_{ij}/|\mathbf{v}_j - \mathbf{v}_i|^2 & \text{if } i \neq j, \\ -\sum_{j \in N(i)} l_{ij} & \text{Otherwise.} \end{cases}$$

We then simply apply the forward Euler scheme to equation (2):

$$h(\mathbf{v}_i, n + 1) = h(\mathbf{v}_i, n) + \delta t a(\mathbf{v}_i, f(\mathbf{v}_i, n)) \frac{M_{cd}(\mathbf{v}_i)}{M_{ca}(\mathbf{v}_i)}$$

The diffused and adsorbed dye amounts at $\mathbf{v}_i$ are given by $f(\mathbf{v}_i, n)M_{cd}(\mathbf{v}_i)$ and $h(\mathbf{v}_i, n)M_{ca}(\mathbf{v}_i)$, respectively. The dye transfer calculation stops when the evolution of equation (1)converges to $\sum_{i=1}^{m} |f(\mathbf{v}_i, n)^n - f(\mathbf{v}_i, n)^{n-1}|/m \leq \epsilon$.

4   Results

Table 1 summarizes the parameters used in this study performed. For visualizations of the dyed cloth, we render the images by taking the product of the sum of dye (transferred and adsorbed) and its corresponding weft and warp texture.

Table 1. Parameters and timing for our results (Figure 9, 10)

Common parameters	(Intel Core i7 (3.2 GHz) PC with 3 GB RAM and a C++ compiler)			
Boundary condition value	$b(\mathbf{z}) = 0$	Initial dye condition $f(\mathbf{x}, 0) = h(\mathbf{x}, t) = 0$		
Discrete time step	$\delta t = 0.5$	Vertex number of $\mathcal{G}$ $m = 2\text{x}400\text{x}400$		
Stopping criterion	$\epsilon = 0.00001$	Cell size	$\delta x = \delta y = 0.0005,$	
Dye supply constant, adsorption rate $\alpha, \beta = 0.7, 0.1$			$\delta z = 0.005$	
Other parameters and timing	Figure 9	Seikaiha	Kumo shibori	Itajime
Cut-off constant $c_\Omega, c_{B_1}, c_{B_2}$	100.0, 16.0, 8.0	10.0, 16.0, 8.0	140.0, 16.0, 8.0	160.0, 16.0, 8.0
Computing time (sec/time step)	/	0.445	0.476	0.372

The simulated results obtained using both large and small cut-off constants are juxtaposed in Figure 9. From (a), we can observe that with large values of c_Ω, the dye is able to permeate into the interior of the folded cloth, while small values of c_Ω, only exterior surfaces receive dye due to the softness and elasticity of the cloth. From (b), large values of cut-off constant c_{B_1} limit dye supply throughout the cloth while small values of c_{B_1} only prevent dye supply to the pressed regions of the cloth, due to cloth properties mentioned above. Essentially, reducing the cut-off constant c_{B_1} affects the dye supply by reducing the space between the faces of the folded cloth. From (c), large values of cut-off constant c_{B_2} limit dye perfusion throughout the cloth while small values of c_{B_2} limit it in pressed regions only. Reducing cut-off constant c_{B_2} modifies the capacity maps by reducing the space between fibers. These results suggest that the cut-off constants can be manipulated to generate a wide variety of possible dyeing process. In future projects, we will aim to automate the selection of these cut-off constants.

Figure 10 shows real dyed cloth and our simulated results for a selection of tie-dyeing techniques. The dyeing results are evaluated by examining the color gradation, which stems from cloth geometry. Our framework is capable of generating heterogeneous dyeing results by visualizing the dye transferring process. While we cannot perform a direct comparison of our simulated results to real results, as we are unable to precisely match the initial conditions or account for other detailed factors in the dyeing process, our simulated results (a, b) corrspond well with real dyeing results (c). Since our dyeing

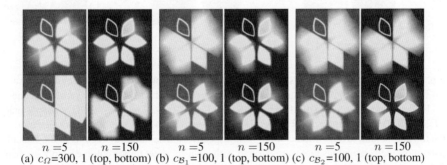

$n = 5$ $n = 150$ $n = 5$ $n = 150$ $n = 5$ $n = 150$

(a) $c_\Omega = 300, 1$ (top, bottom) (b) $c_{B_1} = 100, 1$ (top, bottom) (c) $c_{B_2} = 100, 1$ (top, bottom)

Fig. 9. Comparison of simulated results obtained using different cut-off constants while keeping the other parameters constant

framework enables dyeing simulations that account for the geometry of the folded cloth and advanced dyeing techniques, we can observe the characteristics of tie-dyeing as an interesting gradation in our simulated results.

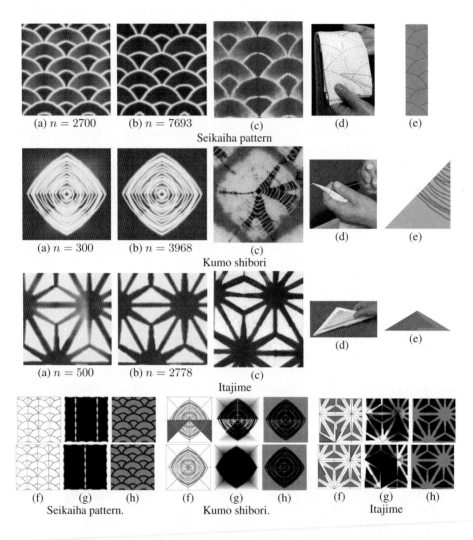

Fig. 10. Various tie-dyeing results and their corresponding conditions. (a) Our simulated results. (b) Our converged simulated results. (c) Real tie-dyeing results [3]. (d) Tie-dyeing techniques. (e) Folded cloths with user-specified boundary domains. (f) Fold features. (g) Dye supply maps $M_s(\mathbf{x})$. (h) Dye capacity maps $M_{cd}(\mathbf{x})$. In (f), (g), and (h), the top and bottom images indicate the top and bottom layers of the cloth model, respectively. The thin gray regions in the Seikaiha pattern (e) indicate regions that were covered with a plastic sheet to prevent dye supply on them; they are not the boundary domain. n is the number of time steps.

Discussion. Our tie-dyeing operation does not support unintentional deformations afterwards, such as wrinkles and stretching. The *Kumo shibori* in Figure 10 shows the importance of these effects. Our folding operation results in a triangular folded cloth (e) while real tied cloth looks like a stick (d), resulting in different dyed patterns. Our result show symmetry, but real results exhibit a warped geometry. In Figure 1, the real result has petals of varying sizes, with the smallest one appearing more blurred. On the other hand, our simulated result displays uniform petal size and blurring, save for the petal on the exterior. This is likely due to thickness of the cloth.

We should develop an interactive interface to edit the dye supply and capacity maps for an intuitive designing. Air pockets that are present between cloth layers in real folded cloths could be modeled in our system by varying the graph weights.

5   Conclusion

In this paper, we describe a novel dyeing framework based on a new dye transfer model and its discrete implementation in a 3D diffusion graph that approximates a folded cloth. The framework is able to generate a wide range of dyed patterns produced using a folded cloth, which are difficult to produce by conventional methods. Since our framework is based on dyeing physics and real dyeing processes, we expect that it is not only a new graphical tool for NPR, but it can also be used as a dyeing simulator for designing.

One limitation of our framework is designing multiple dye patterns using geometric operations on a single cloths; these operations can be quite complex and unintuitive with the current method. Thus our future work will include sketch-based design interface for dyeing. Cloth modeling using curved folds should be implemented for more complex dye patterns.

Acknowledgements

We would like to thank Prof. Hiromasa Suzuki, Prof. Yutaka Ohtake, Dr. Shin Yoshizawa, Dr. Gaku Hashimoto, and Mr. Adam Pan for their helpful comments. We also thank Asako Sakakibara and Jun Mitani. The images of real dyeing in this paper and ORIPA program are courtesy of Asako Sakakibara and Jun Mitani, respectively. This work was funded in part by a grant from the Japanese Information-Technology Promotion Agency (IPA).

References

1. Wada, Y.I.: Memory on Cloth (Shibori Now). Kodansha International Ltd (2002)
2. Polakoff, C.: The art of tie and dye in africa. African Arts, African Studies Centre 4 (1971)
3. Sakakibara, A.: Nihon Dento Shibori no Waza (Japanese Tie-dyeing Techniques), Shiko Sha (1999) (in Japanese)
4. Bruce Gooch, A.G.: Non-Photorealistic Rendering. A K Peters Ltd, Wellesley (2001)
5. Kunii, T.L., Nosovskij, G.V., Vecherinin, V.L.: Two-dimensional diffusion model for diffuse ink painting. Int. J. of Shape Modeling 7, 45–58 (2001)

6. Wilson, B., Ma, K.L.: Rendering complexity in computer-generated pen-and-ink illustrations. In: Proc. of Int. Symp. on NPAR, pp. 129–137. ACM, New York (2004)
7. Chu, N.S.H., Tai, C.L.: Moxi: real-time ink dispersion in absorbent paper. ACM Trans. Graph. 24, 504–511 (2005)
8. Xu, S., Tan, H., Jiao, X., Lau, W.F.C.M., Pan, Y.: A generic pigment model for digital painting. Comput. Graph. Forum 26, 609–618 (2007)
9. Curtis, C.J., Anderson, S.E., Seims, J.E., Fleischer, K.W., Salesin, D.H.: Computer-generated watercolor. Computer Graphics 31, 421–430 (1997)
10. Wyvill, B., van Overveld, C.W.A.M., Carpendale, M.S.T.: Rendering cracks in batik. In: NPAR, pp. 61–149 (2004)
11. Shamey, R., Zhao, X., Wardman, R.H.: Numerical simulation of dyebath and the influence of dispersion factor on dye transport. In: Proc. of the 37th Conf. on Winter Simulation, Winter Simulation Conference, pp. 2395–2399 (2005)
12. Morimoto, Y., Tanaka, M., Tsuruno, R., Tomimatsu, K.: Visualization of dyeing based on diffusion and adsorption theories. In: Proc. of Pacific Graphics, pp. 57–64. IEEE Computer Society, Los Alamitos (2007)
13. Kwatra, V., Wei, L.Y.: Course 15: Example-based texture synthesis. In: ACM SIGGRAPH 2007 courses (2007)
14. Fick, A.: On liquid diffusion. Jour. Sci. 10, 31–39 (1855)
15. Langmuir, I.: The constitution and fundamental properties of solids and liquids. part i. solids. Journal of the American Chemical Society 38, 2221–2295 (1916)
16. Mitani, J.: The folded shape restoration and the CG display of origami from the crease pattern. In: 13th International Conference on Geometry and Graphics (2008)
17. Press, W.H., Teukolsky, S.A., Vetterling, W.T., Flannery, B.P.: Numerical recipes in C the art of scientific computing, 2nd edn. Cambridge University Press, Cambridge (1992)
18. Chung, F.R.K.: Spectral Graph Theory. American Mathematical Society, Providence (1997); CBMS, Regional Conference Series in Mathematics, Number 92

VR Menus: Investigation of Distance, Size, Auto-scale, and Ray Casting vs. Pointer-Attached-to-Menu

Kaushik Das and Christoph W. Borst

University of Louisiana at Lafayette

Abstract. We investigate menu distance, size, and related techniques to understand and optimize menu performance in VR. We show how user interaction using ray casting and Pointer-Attached-to-Menu (PAM) pointing techniques is affected by menu size and distance from users. Results show how selection angle – an angle to targets that depends on menu size and distance – relates to selection times. Mainly, increasing selection angle lowers selection time. Maintaining a constant selection angle, by a technique called "auto-scale", mitigates distance effects for ray casting. For small menus, PAM appears to perform as well as or potentially faster than ray casting. Unlike standard ray casting, PAM is potentially useful for tracked game controllers with restricted DOF, relative-only tracking, or lower accuracy.

1 Introduction and Related Work

VR and immersive visualization involve widespread use of projection-based displays. For such displays, ray casting is the predominant pointing technique. We investigate VR menu properties related to menu size and distance and show how they affect user performance for pointing with ray casting and PAM [1]. Our work shows:

- Performance degrades with decrease in selection angle (a circular menu's center-to-target angle). We show the shape of the degradation both for increasing user-to-menu distance and for decreasing menu size with constant distance.
- Auto-scale mitigates the effect of menu distance on selection times for ray casting.
- PAM, a technique that allows separation of selection angle from distance and visual size, also has decreased performance with decreased selection angle, with an additional effect of visual size. For small menus, PAM may outperform ray casting.

This study complements our earlier work on menu performance with ray-casting pointing and PAM in projection-based 3D environments. We previously studied menu properties like layout and location and found that contextually-located pie layouts are promising [1], consistent with other work showing a benefit of pie menus over list menus, e.g., [2], [3]. Only one (projected) menu size was considered in our earlier work, using an auto-scale feature intended to mitigate distance effects. Other methods to deal with distance include 3D variations of marking menu [3] or the rapMenu [4].

G. Bebis et al. (Eds.): ISVC 2010, Part I, LNCS 6453, pp. 719–728, 2010.
© Springer-Verlag Berlin Heidelberg 2010

Our new study considers distance-size properties of contextual pie menus and includes an evaluation of the auto-scale technique. We also further investigate the PAM pointing technique (the orientation-only variant, PAMO [1]), considering distance and size effects. In the previous study, standard ray casting performed better overall than PAM, but PAM was better in some cases and reduced errors overall. A benefit of PAM is that it supports a broader range of controllers than ray casting, requiring as few as 2 degrees of freedom with only relative tracking.

2   Characteristics Related to Menu Distance and Size

2.1   Distance, Size, and Auto-scale

Ray-casting pointing faces a known problem of distant or precise pointing [5] due to perspective foreshortening and tracking or hand jitter that amplifies over distance. When pointing at pop-up menus, this problem might be mitigated by scaling menu size according to distance from user [1]. We name this mechanism "auto-scale", where menus at varying distances have constant projected size. It is important to know how auto-scale affects performance. As an alternative to auto-scale, menus could be placed at a fixed distance. However, for contextually-located menus, this would make menus appear at a different depth than the object on which the menu is invoked. This can lead to visual discordance in a stereo immersive environment. We have also considered PAM as a way to separate interaction motions from distance and size [1], but we did not previously study distance and size effects on PAM.

Auto-scale maintains constant projected size, but it is not known what sizes perform best. For pie menus with ray casting, no evaluative studies have been carried out, to our knowledge, to optimize size. Since ray casting has a known problem with small distant objects requiring overly precise pointing, and considering cases of larger interface elements outperforming small ones with ray casting [5], [6], we expect that larger menus would be faster, at least up to some optimal size.

2.2   Selection Angle

For ray casting, pie menu size can be described by selection angle: the angle a hand would rotate to move a ray-menu intersection point from menu center to a menu item. We expect this angle, instead of pre-projected menu radius, to be a suitable measure of required motion, due to perspective effects of distance. Considering the geometry in Fig. 1, selection angle is $\Phi = \arctan(radius/distance)$. For results reported in this paper, we note our menus allow selection at a threshold distance of 60% of menu radius, so users need not move through the entire selection angle.

If we increase pie radius at a fixed distance, selection angle increases. Selection angle also increases with decreasing menu distance for a fixed pie radius. The optimal angle would depend on characteristics of human limb and wrist motor movement for rapid aimed pointing tasks. We consider the angle as the required movement, since angular motion is usually predominant for ray-casting pointing.

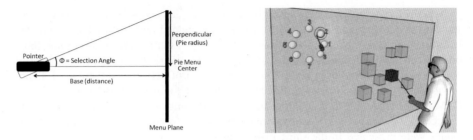

Fig. 1. *Left:* Side view of a pie menu and a pointer with selection angle. *Right:* PAM pointing: hand motions map to pointer (upper left) attached to a pie menu. The diagram, from [1], shows the menu at upper left, rather than contextually, for clarity.

2.3 PAM Selection Angle

PAM [1] is an indirect ray pointing technique that maps user motions to a ray selector that is local to the menu object. Specifically, as illustrated in Fig. 1 (right), PAM attaches a short menu-local pointer in front of the menu and maps wand motions to this attached pointer to aim a ray from attached ray origin to menu items. PAM separates menu visual size and distance from selection angle by calculating intersections disregarding menu visual size and distance. Selection angle can be controlled by changing a motion gain associated with the attached pointer – the higher the gain, the smaller the selection angle. This means that higher gain requires less angular motion to reach an item.

Changing selection angle in PAM by varying PAM gain might affect performance in a manner similar to changing selection angle for standard ray casting. If so, we could use PAM in a VR system where there is a restriction on menu visual size. PAM, with a (possibly optimal) large selection angle could then have faster performance than ray casting for small visual sizes or with high tracking jitter. However, a mismatch between visual angle and PAM selection angle may be distracting.

To avoid ambiguity, SRC angle (Standard Ray casting angle) is used to denote selection angle for ray casting. It also defines the visual size (screen-projected size) of a menu (auto-scaled or otherwise). For PAM pointing, SRC angle corresponds to projected visual size, not to PAM angle. PAM angle is used to denote PAM selection angle that depends on motion angle and gain but not on SRC angle. So, SRC angle can be used with a separate PAM angle for the same menu.

3 Pilot Study on PAM Angles

To estimate optimal PAM angle (that would be investigated further in the main study) we conducted a pilot study with varying PAM angles and visual sizes (SRC angles), on 16 subjects and a small number of trials (2 trials per PAM angle and SRC angle combination). The task and experiment settings were the same as what will be described in Section 4 (treatment type 2). Levels for SRC and PAM angles were also the same as in Section 4. Overall, the mean selection time with a PAM angle of 5° was best. Results, shown in Fig. 2, suggested that menu visual size (SRC angle) does have an effect in addition to the effect of PAM angle.

A subjective tuning task was also given. Subjects adjusted selection angle for both standard ray casting and PAM and picked a value judged best for selections. For ray casting, menu size was changed to tune SRC angle. A tuning task was repeated three times, each with a different initial size: small (1.3°), large (9.7°), and the average of two previous subject-chosen sizes. The overall mean of best angle (average of 3 chosen angles) was 4.4° (σ = 1.5°). For PAM, subjects controlled PAM gain to change PAM angle. Three different menu visual sizes (SRC angles) were presented randomly – large (9.7°), small (1.3°), and intermediate (4.3°). Within each SRC angle, PAM angles were presented in the same manner as before – large and small randomly ordered, then average. The mean of best PAM angle was 5.4° (σ = 2.7°). So, subjectively-tuned PAM angle was roughly consistent with the overall best PAM angle (5°). The following study will show how this (estimated) optimal PAM angle compares against ray casting for various SRC angles.

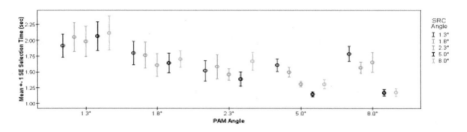

Fig. 2. Selection time (mean and SE) for different PAM and SRC angles from the pilot study

4 Main Study: Methods

Hypotheses: We are interested in five hypotheses (and independent variables):

1. *Distance*: As distance between a user and menu increases, we expect user performance would degrade due to increase in required precision for the pointing task. Distance is measured between pie menu center and the hand position.
2. *Auto-scale*: We expect effects with distance would not be found if the menu selection angle is kept constant with auto-scaling.
3. *SRC Angle*: We expect user performance would get better with increasing SRC angle and that this effect would be similar for increases resulting from reduced distance (not auto-scaled) and increase in specified auto-scale size.
4. *PAM Angle*: We expect that changing PAM angle by changing PAM gain would show similar effects as changing SRC angle for ray casting.
5. *Pointing Method:* For small SRC angles, we expect that PAM pointing with an estimated optimal PAM angle would be faster than standard ray casting.

We varied selection angle in three different ways:

1. Varying distance of menu (un-auto-scaled) from hand. (SRC angle)
2. Varying auto-scale size, specifying projected menu size independently from distance. (SRC angle)
3. Using PAM and varying PAM gain. (PAM angle)

For un-auto-scaled menus, we chose a fixed menu size along with a set of distances such that projected sizes had SRC angles that we were also evaluating with auto-scale. This allowed direct comparison of distance-based SRC angles to equivalent auto-scaled ones. Evaluated PAM angles were the same as evaluated SRC angles. For fair comparison of PAM to ray casting, the same set of menu distances was used in PAM conditions as in standard ray casting conditions.

Specifically, we evaluated SRC and PAM angles of 1.3°, 1.8°, 2.3°, 5.0°, and 8.0°. Hand-to-menu distances were 11m, 8m, 6.2m, 2.9m, and 1.8m in the 3D space. Minimum SRC angle was chosen so that menu labels were barely readable, although target item was indicated by color. Maximum size was large but did not cover the entire screen, to allow randomized menu position in a reasonable range.

Apparatus: We used a 1.5m x 1.1m rear-projection screen with its lower edge 0.7m from the floor. An InFocus DepthQ projector displayed stereo 800x600 pixel images at 120 Hz, which were viewed with StereoGraphics CrystalEyes glasses. A wired Intersense IS-900 Wand was the 6-DOF input device and its button was used to indicate selection of target boxes and menu items. Head tracking was also done with the IS-900. Subjects stood about 1.2m from the screen center as in Fig. 3 (left).

Subjects: There were 20 subjects, 6 female and 14 male, with age from 19 to 41 years. Two were undergraduate students and 18 were graduate students. Four were left handed. 12 subjects reported no VR experience, 6 reported experience with 3D motion game controllers, and 2 reported experience with VR systems.

Procedure: Subjects performed targeted menu item selection. A target box appeared at a random location but at a specific distance from the hand, based on current conditions. Subjects had to select this box with ray-casting pointing to pop up a contextual pie menu. Subjects had to select a red item amongst white items on the menu. Since distance is an independent variable here, additional depth cues were rendered. Target boxes were displayed on pedestals with shadows in a large enclosed space with textured walls. Subjects were instructed to select target menu items as quickly as possible while keeping errors low as well. However, accuracy was enforced but speed was not, to prevent subjects from achieving high speeds at the cost of accuracy. If an incorrect selection was made, an error sound was played, the menu disappeared, and subjects had to bring up the menu again. Furthermore, to explain the feature of pie menus that selections are possible by pointing in a direction, subjects were told that exact pointing at menu item spheres was not required. Sessions lasted for about 30 minutes.

Trials and Treatments: A trial consisted of selecting the menu item on a single-level 10-item menu. There were 10 trials per treatment, each with a unique target item. Target order was randomized within a treatment. Per treatment, 2 menus appeared at each of the 5 distances, but at an otherwise random screen position. Treatments, consisting of combinations of the independent variables, were presented in random order per subject. There were 24 treatments (240 trials, excluding practice), divided into the following five types, presented in random order:

1. Ray-casting pointing and auto-scaled menus. There were five treatments of this type, each with one of the five levels of SRC angle.

2. PAM pointing without auto-scaling (menu size was 0.25m before projection). There were five treatment of this type, each with one of the five levels of PAM angle.

3. PAM pointing and auto-scaled menus. There were five treatments of this type, each with one of the five levels of PAM angle with the matching SRC angle.

4. PAM pointing with (estimated) optimal PAM angle 5° and auto-scaled menus. There were four treatments of this type, each evaluated with one of four SRC angles. The optimal PAM angle matching SRC angle 5° occurred in treatment type 3.

5. Ray-casting pointing without auto-scaling (menu size was 0.25m before projection). There were five treatments of this type. Ray casting at specific distances is of interest, but to keep the presentation of all treatments similar, distance varies within the treatment as well. We later separate results per distance (SRC angle).

Dependent Variables: Dependent variables were selection times (appearance of menu to selection, including time spent correcting errors), error count, and movement. Due to space constraints, we focus mainly on selection times in this report.

 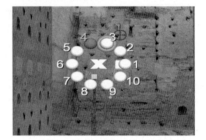

Fig. 3. *Left:* Experiment setup: rear-projection stereo display with 6-DOF head and wand tracking. *Right:* A screenshot of the experiment scene.

5 Results

Distance and Auto-scale: As seen in the leftmost box in Fig. 4, increasing distance in a un-auto-scaled menu with ray casting (treatment type 5) tends to raise selection times. A single-factor (distance) ANOVA on un-auto-scaled ray casting cases detects significant effect of distance on selection times ($F(4, 76)=46.102$, $p<0.001$). Pairwise comparisons with Bonferroni correction detect significance ($p<0.05$) except between the two closest distances. Similar effect of distance on selection times is not seen for menus auto-scaled to maintain specific SRC angle (Fig. 4, except leftmost box, treatment type 1). Note that auto-scaled conditions had 2 trials per distance, while there were 10 trials per distance in the un-auto-scaled condition. Single-factor ANOVAs for each auto-scaled SRC angle did not detect significant effect of distance on selection times. For auto-scaled menus at SRC angles of 1.3°, 2.3°, and 5°, the closest menu distance of 1.8 meters appears to take more time than further distances, but this was not detected significant overall.

SRC Angle: Fig. 5 shows how increasing SRC angle lowers selection times for ray-casting pointing. It also shows that SRC angle, changed either through varying menu distance without auto-scale (treatment type 5), or through changing menu size with auto-scaling (treatment type 1), has similar effect on user performance. A 2-factorial ANOVA on selection time with independent variables of SRC angle and auto-scale detects significant effect of SRC angle on selection time ($F(4, 76)=51.848$, $p<0.001$). All possible pairwise comparisons between different SRC angles, with Bonferroni correction, detect significance ($p<0.05$).

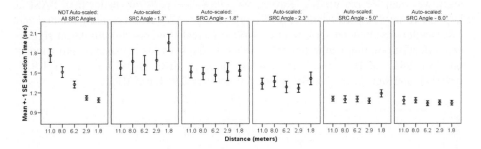

Fig. 4. Selection times (mean and SE) with ray casting at different menu distances without auto-scale in the leftmost panel (treatment type 5), and menus with auto-scale (except leftmost panel) having different SRC angles (treatment type 1)

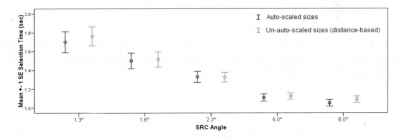

Fig. 5. Selection times (mean and SE) at different SRC angles clustered according to distance-based SRC angles and auto-scaled SRC angles. Distance-based SRC angles (right bar from each pair; treatment type 5) correspond to leftmost panel in Fig.4. Auto-scaled SRC angles (left bar from each pair; treatment type 1) are collapsed from error bars other than leftmost panel in Fig. 4 by averaging over menu distances.

PAM Angle: From Fig. 6 (right, treatment type 3) it seems that increasing PAM angle leads to lower selection times. If we compare 5° PAM angle cases (treatment type 4) to the SRC-angle-matched PAM angle cases (treatment type 3) at each of the four SRC angles, paired-sample t-tests detect significant effect of PAM angle at the smallest ($t(19)=2.278$, $p<0.05$) and largest ($t(19)=2.513$, $p<0.05$) SRC angle. It appears from Fig. 6 (left and right) that 5° PAM angle may perform better for smaller SRC angles (1.3°, 1.8°, and 2.3°) whereas matched PAM-SRC angles may perform better for larger SRC angles (5° and 8°). We can also see that increase of SRC angle leads to decrease in selection times (Fig. 6). An ANOVA on SRC angle for 5° PAM angle cases (treatment type 4), detects significant effect of SRC angle on selection times ($F(4, 76)=37.536$, $p<0.001$). Post-hoc tests with Bonferroni-adjusted pairwise comparisons show significance (p<0.05) between all pairs except for the two largest and two smallest SRC angles.

PAM vs. Ray casting: For smaller SRC angles, PAM seems to perform better with the estimated optimal PAM angle of 5° (treatment type 4) than ray casting (treatment type 1). It appears from Fig. 6 (left) that means for PAM at SRC angles of 1.3° and 1.8° are lower than for ray casting. Ray casting, however, performs better than PAM at larger SRC angles, particularly at SRC angle of 8.0°. A 2-factor (pointing method and SRC angle) ANOVA on selection times for auto-scaled menus did *not* detect significant effect of pointing method on times. Paired sample t-tests between pointing methods per SRC angle showed *near* significance at SRC angle 1.3° (t(19)=1.975, p<0.063) and significance at SRC angle 8.0° (t(19)= 2.914, p<0.01).

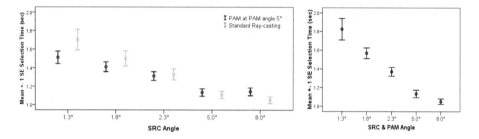

Fig. 6. Selection times (mean and SE) for auto-scaled sizes. *Left:* PAM with PAM angle 5° (treatment type 4) and standard ray casting at different SRC angles (treatment type 1). *Right:* PAM angle matched with SRC angle (treatment type 3).

6 Discussion

The hypotheses in Section 4 are largely supported by the observed results. An effect of menu visual size (SRC angle) in PAM pointing was also detected.

The basic effect of distance on menu selection with ray casting (Fig. 4, leftmost box) follows earlier studies of general object selection tasks [5]. Decreasing visual size with constant distance showed similar increase in selection times (Fig. 4, except leftmost box). Degradation of performance with increasing distance was mitigated by auto-scaling, which maintains a constant selection angle. Increasing selection angle decreased selection times, irrespective of how selection angle was varied – by changing auto-scaled size or by changing distance without auto-scale for ray-casting pointing, or by changing PAM gain for PAM pointing. These results show how selection angle can be used to understand user performance in ray-casting pointing.

Results suggest that larger menus should be used for faster pie menu selection with ray casting. However, from a practical standpoint, restrictions on size may be imposed by the display or application. For hierarchical menus, traveling through a sequence of large offset child menus could move a user's focus far from their object of interest. For the studied environment, we estimate optimal selection angle between 5° and 8° (for ray casting and PAM). The tuning study suggests users prefer selection angle closer to 5°. Note this optimal angle could vary depending on a selection threshold (60% of radius, in our study), as selection distance and area vary with it.

Higher selection times at small PAM angles may correspond to increased perceived sensitivity (by large C-D gain). Subjects may find it irritating to see the visual attached pointer move large amounts for small hand motions. In PAM pointing, visual size (SRC angle) has an effect on user performance, even when PAM angle is constant at 5°. From Fig. 2 and Fig. 6 we see that larger SRC angles typically led to lower selection times in PAM, especially for large PAM angles. Besides the performance increase with larger PAM angles, a likely reason that large SRC angles work well with large PAM angles is that subjects experience more consistency between visual effect and required motions.

Comparing PAM with estimated optimal PAM angle (5°) to ray casting at specific SRC angles (Fig. 6 left) suggests that PAM performs about as good as or better than ray casting at smaller SRC angles of 1.3° and 1.8° (ANOVA did not detect overall significance, but there was a near-significant t-test result at 1.3°, so we also consider overall plotted trends and note this result is less conservatively stated than others). PAM may be a good alternative to ray casting when menu visual sizes must be small, when jitter is large, or with limited-DOF tracking devices.

Selection time decreases with higher angular motion. That means users moved more rapidly towards a larger or a distant target. This follows from human motor performance for rapid aimed movements [7], where greater target distance results in faster motion. Also, as target size (pie-slice area) increases in a pie-menu with increasing distance to target, users need not spend much time on the slower corrective submovement [7] that occurs for precise pointing.

From results, we speculate that selection times may be modeled by a logarithmic function involving the inverse of selection angle [8]. However, such a model may not work well at very large or very small selection angles where additional factors such as tracking jitter become critical. A difference between visual target size (spheres drawn at visual menu circumference) and actual selection size (pie-slice with radius thresholds) may further complicate the model. A well-known logarithmic model for pointing is Fitts's law applied to 1D or 2D translation [9] or rotation [7], where an index of difficulty is a logarithmic function of the ratio of target distance to target width. Relating our results to Fitts's law may seem counterintuitive, as selection times decreased with an increase in distance, and our environment is 3D with interaction primarily involving rotation. However, both target distance and target width increase with increase in selection angle, and the effect of target width seemingly dominates. This may be explained by modeling target width as target area (e.g., resembling [9]), which would be proportional to the square of the pie radius [8]. Since distance-to-target is a fraction of the pie radius, this would predict a decrease in movement time for increase in pie radius (or selection angle).

7 Conclusion

We showed several properties related to menu distance and size and how these affect user performance in menu interaction. These findings can help UI designers optimize menu interaction in projection-based VEs. We confirmed that auto-scaling mitigates the effect of distance on menu pointing tasks. Auto-scaling could also work well with other user interface elements such as toolbars, list menus, etc. that afford auto-scaling (e.g., temporarily invoked widgets), where it could be used for more consistent user performance with widgets located at different distances. User performance was found to vary with selection angle: increasing selection angles lowered selection times. An estimated optimal selection angle of 5° is suggested for environments such as ours. We also observe that using PAM with a PAM angle of 5° can get performance as good as or faster than ray casting for small interface elements. In any case, inspection of results shows that performance differences between 5°-PAM and ray casting are not very large percentagewise, so PAM is a promising ray casting alternative for controllers that are supported by PAM but not ray casting.

References

1. Das, K., Borst, C.: An evaluation of menu properties and pointing techniques in a projection-based VR environment. In: IEEE Symposium on 3D User Interfaces (3DUI), pp. 47–50 (2010)
2. Chertoff, D.B., Byers, R.W., LaViola Jr, J.J.: An exploration of menu techniques using a 3D game input device. In: International Conference on Foundations of Digital Game, pp. 256–262 (2009)
3. Kurtenbach, G.P.: The design and evaluation of marking menus. PhD thesis, Toronto, Ont., Canada (1993)
4. Ni, T., McMahan, R.P., Bowman, D.A.: Tech-note: rapMenu: Remote menu selection using freehand gestural input. In: Proceedings of the 2008 IEEE Symposium on 3D User Interfaces, 3DUI 2008, Washington, DC, USA, pp. 55–58. IEEE Computer Society, Los Alamitos (2008)
5. Poupyrev, I., Weghorst, S., Billinghurst, M., Ichikawa, T.: Egocentric object manipulation in virtual environments: Evaluation of interaction techniques. Computers Graphics Forum 17(3), 41–52 (1998)
6. Kunert, A., Kulik, A., Lux, C., Fröhlich, B.: Facilitating system control in ray-based interaction tasks. In: 16th ACM Symposium on Virtual Reality Software and Technology, VRST 2009, pp. 183–186. ACM, New York (2009)
7. Meyer, D.E., Abrams, R.A., Kornblum, S., Wright, C.E., Smith, J.E.K.: Optimality in Human Motor Performance: Ideal Control of Rapid Aimed Movements. Psychological Review 95(3), 340–370 (1988)
8. Das, K.: Investigation of Menu Properties and Pointing Techniques in a Projection-based VR Environment. Master's thesis, University of Louisiana at Lafayette (2010)
9. MacKenzie, I.S., Buxton, W.: Extending Fitts' law to two-dimensional tasks. In: Proceedings of the SIGCHI Conference on Human Factors in Computing Systems, CHI 1992, pp. 219–226. ACM, New York (1992)

Contact Geometry and Visual Factors for Vibrotactile-Grid Location Cues

Nicholas G. Lipari and Christoph W. Borst

University of Louisiana at Lafayette

Abstract. Visual and haptic factors can affect a user's interpretation of vibrotactile cues communicating location of objects in a real or virtual environment. Identifying and understanding relevant factors will lead to better device and interface design, for example, through procedures that adjust for systematic error or per-user differences. We considered direct effects of hand-tactor contact geometry and a possible cross-modal effect of the visual interface. Our experiment examined contact geometry on a single row of tactors and presence of a visual border on a graphical region that mapped to the tactor array. We measured the relationship between vibrotactile array stimulus coordinates and user responses. Contact geometry that emphasized a certain tactor increased tendency for subjects to mark near it. Effects of visual borders were noticeable but subtle, acting more as a modulating factor.

1 Introduction

Multi-modal approaches for communicating position and direction can include the haptic sense as a supplementary or reinforcing data channel. Our work concerns a visual scene, rendered or real-life, containing an area that maps onto a haptic device (our work is not intended as an abstract perception study, but rather a human factors evaluation of the haptic device and visual interface design choices). Consider the conceptual diagram in Figure 1. In this example, an application renders a map for the purpose of navigation. The haptic device then renders a place of interest as a vibration pattern on the user's palm depending on point of view and a map-aligned region. Vibrotactile patterns may communicate additional information, for example, identity or status messages as intensity patterns. Based on this, the user navigates the environment in search of the intriguing location or feature. We are interested in errors due to discrepancies among haptic, visual, and multimodal stimuli.

We present an experiment regarding haptic device contact geometry, stimulus coordinate, and a visual interface property for users interpreting a vibrotactile location cue at the palm. Skin-tactor contact varies with body-site shape and variations between users. When marking a haptic stimulus's location in a graphical desktop environment, subjects marked locations nearer areas of emphasized contact with the palm (where contact pressure was generally higher). In this situation, a mapping is induced between vision and taction and may be skewed by

G. Bebis et al. (Eds.): ISVC 2010, Part I, LNCS 6453, pp. 729–738, 2010.

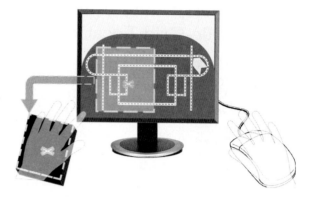

Fig. 1. Example of a location communicated haptically. The dashed border square is mapped to the tactile array and points of interest are relayed to the user as haptically rendered positions.

incorrect perception of tactile stimuli or variations in graphical interface parameters. We also investigated the parameter of borders in the visual scene. When borders of the response area were not present, the results suggested borders acted as a modulating factor. Such factors may influence the design of multimodal systems and calibration techniques that improve feedback by adapting to user trends.

2   Related Work

In a related experiment, Borst and Baiyya [1] investigated three parameters of a vibration pattern: position, direction, and profile. Each pattern had one of two possible shapes: point and line. The device, Figure 2 (left), activated adjacent tactors with varying intensities to provide the illusion of a point (or a point moving along a line) rendered somewhere on or between the tactors. After observing mean error approach zero at the center of the array, Borst and Baiyya postulated some systematic error could be present in the data.

We then conducted a preliminary experiment [2] to investigate systematic errors in position accuracy. The results demonstrated an effect of Visual Scale and Correct Answer Reinforcement on accuracy and suggested a metric to model systematic error. Radial error was intuitively defined as the distance between stimulus and response measured radially from the array's center. More analytically, this was the directed magnitude of the error vector projected onto the stimulus's normalized radial vector. Tendencies for positive radial error were called radial expansion and negative leanings were called contractions. We also proposed stimulus calibration based on this model. Assuming systematic error as radially symmetric, calibration adjusts the stimulus based on its radial distance. While the visual properties had a significant effect on radial error, we did not investigate how hand-tactor contact geometry contributed to radial error.

Fig. 2. The vibrotactile array used in previous experiments (left) and our current experiment (right). At left, six rows of five motors are mounted on a project box. Nylon washers and foam pads improve contact consistency across the palm. At right, one row of five pager motors mounted on a project box. Impostors take the place of the remaining five rows of motors. Shims are placed below foam pads to raise the height of individual motors, maintaining direct contact between tactors and skin.

The current experiment investigated two possible causes of error suggested in [2]. We considered the simplified comparison between a flat row of tactors, Figure 2 (right), and a raised middle tactor. These extremes of contact geometry emphasized the array's center. To supplement this, and provide some insight where most radial error occurred, we divided stimulus points into center and edge (of the palm). The experiment also varied the visibility of the interface's border to investigate subjects' tendencies at the borders of the array.

Although not a psychophysics study, this work was informed by several studies in applied psychology, haptic interfaces, and non-visual interaction techniques. In [3], mislocalization illusions regarding tactile perception were given. In the funneling effect, a tendency exists for simultaneous, adjacent stimuli to be felt as one stimulus. This may contribute to several haptic applications, including [4], [5], [6], and the current experiment, where attempts are made to combine signals from two or more tactors into one haptic point. Ryu et al. [4] have also developed a device with similar tactors and contact site as ours, the T-hive. Consisting of thirteen tactors mounted around a spherical handle, the T-hive was designed to provide directional information using multiple simultaneous vibrations for a six degree of freedom hand controller. Oakley et al. [5] suggest that a "spatial summing" occurs for some stimuli meeting the spatial and temporal specifications of [3]. Borst and Asutay [6] also use the technique of multiple, closely spaced tactors, but with an unweighted area sampling taken from graphics concepts. It appears that such configurations satisfy the psychophysical requirements listed by Hayward [3] for the funneling illusion.

In Lindeman et al. [7], eight tactors delivered directions at a subject's torso. Tasked with clearing a building, subjects located objects and avoided hazards in a virtual environment. Van Erp et al. demonstrated a similar apparatus in [8] for

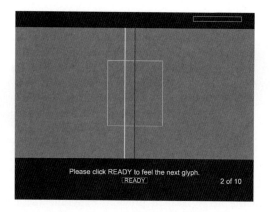

Fig. 3. A view of the data collection software for the current experiment with subject response (left) and reinforcement (right) vertical marker lines. A stimulus timer (top-right) and a session counter (bottom-right) indicated progress. Subjects were instructed to mark the horizontal position of the stimulus in the white rectangle (and in the same area for the Border Invisible case). After collecting the response, the experiment displayed the reinforcement marker.

piloting a helicopter and boat. These and other similar works have contributed to research in vibrotactile feedback devices geared toward specific applications.

3   Methods

Our repeated-measures experiment presented vibrotactile stimuli via a single-row array of tactors (Figure 2, right) and an abstracted visual environment (Figure 3) to users. The experiment interface represented a virtual environment with haptic location cues. We investigated these multi-modal components by haptically rendering horizontal positions and having subjects indicate corresponding positions in a visual region matching array size.

Hand shape may cause contact to vary across the array. By presenting haptic stimuli with various contact geometries, we investigate how this influences localization. Also, visual edges of a rectangle, as in Figure 3, may mentally anchor subject responses. This is relevant when an application provides a device-oriented context for visual feedback. Users may be biased toward, or constrained by, graphical boundaries of datasets or virtual rooms, or visual subregions mapped to a haptic device. A case without visible borders could show further expansion near the edges, illuminating their influence on the mental mapping of a haptic stimulus.

3.1   Design

We considered three Within-Subjects variables Height Pattern, Point Location, and Border Visibility. The two Height Patterns of tactors were Flat, with no

Fig. 4. Height Patterns used in the experiment and pilot study. a) Flat: all tactors have the same height. b) Exaggerated (Exag): The middle tactor is raised by two shims. c) Pilot Study: The middle tactor is raised by two shims and the adjacent tactors are raised by one. For the pilot study, (c) was also duplicated with washers above tactors.

raised tactors as in Figure 4a, and Exaggerated (Exag), having a raised center tactor as in Figure 4b. Point Location partitioned the stimulus coordinate set into Center and Edge subsets (our stimulus coordinates varied horizontally). Based on which stimulus coordinates involved the center tactor under the area sampling kernel of [6], the Center subset consisted of 41 coordinates, and the Edge had 39 coordinates. Center coordinates were those (from a random set) closer to the middle of the palm, and Edge coordinates were further. For the Border Invisible condition, the rectangle in Figure 3 was absent, and for the Border Visible case it was present. The dependent variables error magnitude and radial error were computed for each condition combination. As stimuli and responses were limited to one dimension, computation of each dependent variable was simplified to sign changes and absolute values of signed 1D error.

3.2   Apparatus

We modeled our array and haptic rendering technique after [2] and [6], but with modifications supporting our experiment. Seen in Figure 2 (left), the array used by Borst and Asutay [6] consisted of six rows of five tactors (14mm DC motors). Each tactor was placed on an 18 mm grid. Thus, 18 mm was one array unit. Borst and Baiyya [1] affixed nylon washers and foam pads to tactors to isolate vibrations and facilitate more consistent contact geometry for the palm. A controller board varied tactor levels according to an unweighted area sampling technique from [6].

   The device used here, Figure 2 (right), was a one dimensional version of the device in [2]. To vary tactor heights and examine the row of tactors with the most radial expansion from [2], we arranged one row of tactors in the same grid spacing with adjustable height. In other tactor locations, we placed impostors with nearly identical heights and diameters. A Measurement Computing USB-3114 controlled the five tactors with 16-bit precision analog voltage. The USB controller was commanded by the unweighted area sampling routine [6] and only minimal changes to driver and application code from [2] were required. To ensure consistent sensation between the experiments, we matched the controller output (voltage) from the USB device to those of [2].

While the device from [2] had nylon washers atop tactors to fit the palm, we selected Flat and Exaggerated patterns, shown in Figures 4a and 4b. Two implementation options were the placement of washers above the tactors or shims below. Since this could change vibration characteristics, a pilot study determined the best option with the geometry from Figure 4c. Four subjects completed eighty trials per day for two counterbalanced sessions. Regression smoothers for the device were similar to the response curves of [2]. Since the use of shims had a similar effect to washers and allowed direct tactor-skin contact, we chose shims for implementing the Height Patterns in this experiment.

3.3 Participants

Eight male subjects took part in this study. We considered them moderately experienced. Although levels of prior experience with the device varied, each subject had at least two hours of prior exposure to 2D vibrotactile palm arrays. Our experiment presented many trials, 160 per day, to the small number of subjects instead of fewer trials with many novice users, reducing the effects of learning and better representing a regular user. Two subjects were left-handed. The median age of subjects was 26 years, with a minimum age of 24 years and a maximum age of 37 years.

Table 1. Session Order Randomization. Rows correspond to sessions, with each subject exposed to two sessions (A and C or B and D) on separate days.

Label	Height	Border Sequence
A	Flat	BV BI BI BV
B	Exag	BV BI BI BV
C	Exag	BI BV BV BI
D	Flat	BI BV BV BI

3.4 Procedure

The experiment followed an open-response paradigm in which subjects received a vibrotactile stimulus on the left palm and indicated its horizontal position by moving a vertical line. Each session consisted of Demonstration, Training, and Testing stages. Starting in the Demonstration, subjects placed the left hand on the palm-array and felt a series of point vibrations (each lasting two seconds). We allowed, but did not instruct, subjects to look at the hand receiving stimuli, as they would be able to during normal use. The array rendered five point vibrations and the screen displayed horizontal positions with vertical lines. Subjects did *not* indicate position. During Training, subjects felt ten haptically rendered points. After each, they adjusted the position of a vertical line to indicate the perceived horizontal position on the interface shown in Figure 3 with the current experimental condition. Correct position was shown after the response. The Testing

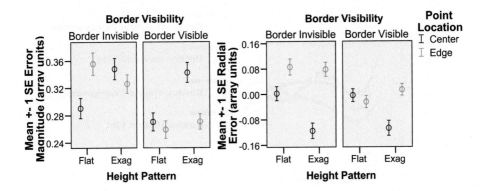

Fig. 5. Error magnitude and Radial Error against Height Pattern, Point Location, and Border Visibility

Stage followed; subjects again marked the horizontal position of haptically rendered points. Each subject completed two sessions on two non-consecutive days. Sessions consisting of one Height Pattern and both levels of Border Visibility lasted 20-30 minutes.

The order of conditions was randomized according to the following stipulations. All sessions had both levels of Border Visibility counterbalanced in four sub-sessions with 40 trials each, as seen in Table 1. To avoid the jarring effect of the BI case on the first testing session, all subjects began with the order (BV, BI, BI, BV). Height Patterns varied similarly across different sessions. Half of the subjects completed the A and C orders, half the B and D orders. Each session presented 160 trials, with 80 distinct horizontally varying points in a random order per (BV, BV) or (BI, BI) pair. We generated points prior to data collection; the set was split randomly within each Border Visibility condition.

4   Results

Figure 5 summarizes resulting error magnitude. We applied a 3-Way Repeated Measures ANOVA over the Within-Subjects variables Height Pattern, Point Location, and Border Visibility with dependent variable error magnitude. Border Visibility ($F(7,1) = 8.203, p < 0.05$) and Point Location ($F(7,1) = 7.007, p < 0.05$) had an effect. Post-hoc tests indicated the Border Invisible and Edge cases contributed more to error magnitude than Border Visible and Center, respectively. Height Pattern showed a *near* significant ($F(7,1) = 4.155, p < 0.085$) effect for error magnitude (due to a small number of subjects, p-values between 0.05 and 0.10 are noted as interesting and referred to as "near significant").

We also analyzed radial error from the same data. Height Pattern ($F(7,1) = 10.446, p < 0.05$) and Point Location ($F(7,1) = 19.504, p < 0.05$) both had a significant effect on radial error. Post-hoc tests with Bonferonni corrections

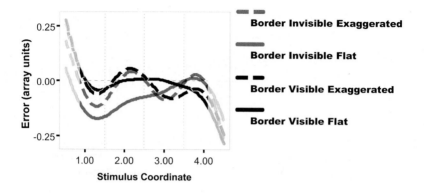

Fig. 6. Local Linear Regression Smoothers of Error (signed) over Border Visibility and Height Pattern. The horizontal axis is Stimulus Coordinate for a haptically rendered point varying horizontally. The horizontal dashed line is at zero radial error. The outer two vertical dashed lines are at the extents of the Center-Edge Point Locations. The middle vertical dashed line is the array's center tactor. Outer portions of the curves are faded to deemphasize points for which confidence diverges.

indicated the Flat and Center cases contributed more to radial error than Exaggerated and Edge, respectively. A *near* significant interaction existed between Border Visibility and Point Location ($F(7, 1) = 5.223, p < 0.06$).

5   Discussion

The above results demonstrate an effect of each independent variable for at least one metric. Height Pattern's significance in radial error, and near significance in error magnitude, signified an effect more apparent when investigating signed error, Figure 6. In Exaggerated (dashed) plots, responses had more oscillations, shifting left and right (negative and positive) of left and right Edge points, respectively. Then, subjects marked closer to the Exaggerated tactor for Center stimuli. In both of these cases, and for Border Visible Flat, mean error tended toward zero at the array's center (middle vertical dashed line). The relation to radial error can be seen by considering error relative to array center.

In the smoothed error plots of Flat Height Patterns (solid), less dramatic oscillations appear. Additionally, with the Border Visible Flat case (solid back), subjects performed better than both Exaggerated Height cases, suggesting the emphasized contact geometry increased perceived loudness of the middle tactor. Similarly, the slight negative tendencies for Border Visible Flat, and more drastically for Border Invisible Flat, occurred near the fleshy portion of the left palm where contact pressure with flat surfaces was likely highest. Note an interplay between contact pressure and vibration frequency may impact perceived location for the eccentric-mass motors. As these motors have a low cost and are available in many sizes, they are easily applicable to many situations. However, motor behavior regarding contact pressure, frequency, and vibration amplitude make it unclear which of these contributes most to the preference for areas of overly-conformant contact geometry.

Border Invisibility likely made subjects less confident, but its effects also vary depending on stimulus location. This was suggested by the near significant interaction of Border Visibility and Point Location. Both Border Invisible (grey) cases, Flat and Exaggerated, show how border invisibility served to amplify the error relative to the Border Visible cases. In both pairs of Height cases similar curves appeared, but with marked shifts away from the Border Visible (black) cases (erring away from the center) at the respective outer inflection points. A possible cause lies in the thought process for mapping a body part to a visual area. Since error at array center was near zero in both Exaggerated cases, subjects may have used the interface's center or response button for a reference point. If the point locations away from the center were judged relative to the center and nearest border, then the Border Invisible cases may have caused an over-estimated distance from the center reference point.

Considering contact geometry a factor in radial expansion, future devices should account for hand (or another stimulus site's) shape. This experiment indicated expected errors with inconsistent contact geometry. While our model of radial expansion provided insight into user performance, this effect was partly based on the structure of the hand, the composition of which varies from a fleshy perimeter to a denser, bony interior. This structure may contribute to the longitudinal asymmetry of the response curves, since a solid contact site in the center would make vibrations here more pronounced.

6   Conclusion

Contact geometry and visual references influence the mapping of haptic cues in a virtual environment. Results exhibited the least error near the array's center tactor. Error rose near the Center-Edge boundaries, at nearly identical coordinates in all cases. We have previously demonstrated [2] stimulus calibration using radial error. The highest points in radial error smoothers from [2] occurred at equivalent radial measures to here (1.2-1.25 array units from center). The peaks in [2] also occurred across three levels of Visual Scale, suggesting both effects originated with the devices but was modulated by visual stimuli. Involvement and understanding of haptic and visual conditions increases the power of stimulus calibration by allowing for their inclusion in calibration models.

Visual Borders also contribute to error as modulating factors, and the effects are relevant as design considerations for visualization applications. In our experiment, the background items were constant, yet the lack of Visual Borders increased subjects' error. In the case of a visualization system with an active scene and user interaction, the missing reference may also be detrimental.

A different body location could improve the contact between skin and tactors. With a more planar body-site, better conformance to a grid of tactors would be possible without height adjustments. While few applicable sites are as non-intrusive as the palm, the forearm and wrist do offer promising alternatives. These, however, have less area of glabrous (hairless) skin and lower tactile sensitivity than the palm, a change that would need to be investigated. Also, tactor response was dependent on contact pressure. Nonetheless, the questions raised remain relevant for any tactor type since variations in perceived loudness may cause irregular response shapes, as suggested here. A more definitive understanding of radial error gives insight into device design and modeling for stimulus calibration. Given the current trend of consumer grade virtual reality devices, haptic-enabled touch screens, and vibrational game controllers, we may see devices similar to ours in navigation, communication, or entertainment applications.

References

1. Borst, C.W., Baiyya, V.B.: A 2d haptic glyph method for tactile arrays: Design and evaluation. In: Third Joint EuroHaptics conference and Symposium on Haptic Interfaces for Virtual Environment and Teleoperator Systems, WHC 2009, Washington, DC, USA, pp. 599–604. IEEE Computer Society, Los Alamitos (2009)
2. Lipari, N.G., Borst, C.W.: Radial expansion and an effect of visual scale on the spatial perception of 2d vibrotactile position cues. In: IEEE VR 2009 Workshop on Perceptual Illusions in Virtual Environment, Lafayette, LA, USA, pp. 19–21. IEEE Computer Society, Los Alamitos (2009)
3. Hayward, V.: A brief taxonomy of tactile illusions and demonstrations that can be done in a hardware store. Brain Research Bulletin 75, 742–752 (2008); Special issue on "Robotics and Neuroscience"
4. Ryu, D., Yang, G.H., Kang, S.: T-hive: Vibrotactile interface presenting spatial information on handle surface. In: Proceedings of the IEEE International Conference on Robotics and Automation, Kobe, Japan, pp. 683–688. IEEE, Los Alamitos (2009)
5. Oakley, I., Kim, Y., Lee, J., Ryu, J.: Determining the feasibility of forearm mounted vibrotactile displays. In: Proceedings of the Symposium on Haptic Interfaces for Virtual Environment and Teleoperator Systems, HAPTICS 2006, Washington, DC, USA, pp. 27–34. IEEE Computer Society, Los Alamitos (2006)
6. Borst, C.W., Asutay, A.V.: Bi-level and anti-aliased rendering methods for a low-resolution 2d vibrotactile array. In: Proceedings of the First Joint Eurohaptics Conference and Symposium on Haptic Interfaces for Virtual Environment and Teleoperator Systems, WHC 2005, Washington, DC, USA, pp. 329–335. IEEE Computer Society, Los Alamitos (2005)
7. Lindeman, R.W., Sibert, J.L., Mendez-Mendez, E., Patil, S., Phifer, D.: Effectiveness of directional vibrotactile cuing on a building-clearing task. In: Proceedings of the SIGCHI Conference on Human Factors in Computing Systems, CHI 2005: pp. 271–280. ACM, New York (2005)
8. van Erp, J.B.F., van Veen, H.A.H.C., Jansen, C., Dobbins, T.: Waypoint navigation with a vibrotactile waist belt. ACM Trans. Appl. Percept. 2, 106–117 (2005)

Computer-Assisted Creation of 3D Models of Freeway Interchanges

Soon Tee Teoh

Department of Computer Science, San Jose State University

Abstract. Several existing procedural modeling systems are able to generate large 3D models of cities. However, none of these systems can automatically create 3D models of freeways and freeway interchanges, even though these are important features in 3D urban landscape. We have implemented a system that automatically creates 3D models of road surfaces, bridges, tunnels, freeways, and freeway interchanges, taking user design and preferences into account. While we allow the graphics designers to control the positions of the control points according to their aesthetic appeal, our system automatically generates 3D models of road surfaces and freeway connector ramps that are properly smoothed, banked and connected.

1 Introduction

While there exists several systems that generate large 3D models of cities, these systems do not automatically generate 3D models of freeways and freeway interchanges. This is a major limitation, because roads and freeways form an important part of an urban environment. A realistic, drivable 3D model of freeway and road networks needs to be produced, for example, if the output of the 3D city model is to be used for a drive-through.

We have implemented a system that takes control points input by the user, and generates 3D models of smoothed and banked road surfaces, heuristically creating bridges and tunnels in areas of rough terrain. The road surface generation algorithm can also be applied to procedurally-generated road paths or road paths generated by other systems. The output of our graphics model can be used in racing games, or as 3D models for urban landscapes used in games or movies.

While we allow the graphics designers to control the positions of the control points according to their aesthetic appeal, our system instantly automatically generates 3D models of road surfaces and freeway connector ramps that are properly smoothed, banked and connected. These calculations would be time-consuming for the human user to manually perform.

The main contribution of our work is the automatic creation of 3D models of freeway overpasses and connector ramps. The user can custom-design interchanges by setting the height of points on a freeway or connector ramp. The user can also determine the position, height, direction and number of lanes on

G. Bebis et al. (Eds.): ISVC 2010, Part I, LNCS 6453, pp. 739–750, 2010.

a ramp. Our system creates a 3D model of the freeway interchange, with appropriate smoothing and banking. Our system can also automatically detect all Freeway-Freeway and Road-Freeway intersections and create generic freeway interchanges in three common styles. Because we do not use rigorous civil engineering calculations, but use only approximate methods in our approach, the resulting roads do not match real-world roads, and are intended to be used in virtual environments such as computer games and movies.

2   Related Work

Procedural modeling systems that generate cities typically also create roads. For example. Weber et al.'s [1] system creates a network of major and minor streets, with varying widths, and they have to build geometry of these streets. Another example is CMPE, by Carrozzino et al. [2], which also creates a road network, and performs crossroad tessellation, curve tessellation, and texturing for the roads. Several other systems [3–5] also produce detailed road networks and geometry, however none of the above systems has examples of bridges and overpasses. On the other hand, Cabral et al. [6] have presented a method able to build roads, bridges and overpasses using example blocks. Galin et al. [7] have presented a system specifically for the procedural generation of roads, using the shortest path algorithm to find the best route for roads.

Some commercial software is also available for making road geometry. For example, like our system, Shape Magic Road Maker (www.shapemagic.com) is able create smooth roads and road intersections on rough terrain, but it does not include facilities to create freeway overpasses and interchanges. Cityscape (pixelactive3d.com) goes further in allowing users to create freeway overpasses and connectors. However, the user control over the appearance of the freeway overpass is limited. Our system allows users to create freeway interchanges with much more sophistication and realism.

3   Basic Road Surface Geometry

Our system loads a rectangular terrain map, with height information for each cell. The terrain map is displayed in 2D with color-coded heights. The user is able to click on the map to create control points for roads. A cubic curve is automatically created to fit through all the control points. Note that user-generated roads are not the main contribution of this paper, but are merely the input to the road-surface generation algorithms we describe in this paper. The input road routes can also come from procedural methods and other systems.

3.1   Joining Road with Terrain

A road specification consists of the control points of a curve, as well as the number of lanes each road segment has. Roads are typically level, except when they are banked. Sometimes roads are also smoothed and vary in height less than the underlying terrain, to make them more drivable. Given the height of the surface of a road, and the height-map of the underlying terrain, the program generates a triangle mesh that smoothly joins the road surface with the terrain.

First, the height of the road surface is encoded into RGB color. Then, the color-encoded road surface height for the road surface is rendered, up to two units outside the width of the road. The color buffer is read and stored in an array *ColorBuffer1*.

The program then renders the road with value 255 from the middle of the road until a distance of two units inside the edge of the road. From there, the value decreases linearly to 0 at a distance of two units outside the edge of the road. This is read to an array *ColorBuffer2*. The Marching Squares method is then performed to find the iso-contour of the value 128. This represents the edge of the road. Note that the road has to be at least 4 units wide for this method to work correctly.

Each cell that has an iso-contour is then triangulated. The points that are inside the road surface, as well as the points on the iso-surface, are set to the height of the road surface. The height of the road surface is read and decoded from *ColorBuffer1*. The height of points outside the road is set to the linear interpolation between the road surface height and the original terrain height: h = $(d/2)$*h1 + $(1-(d/2))$*h0, where d is the distance from the road edge, h0 is the height of the road surface (read from *ColorBuffer1*), and h1 is the original height of the terrain. For a distance of 2 units outside the road surface, the height of each point is found by linearly interpolate between road height and terrain height in the same way. For all terrain points completely within the road surface, their heights are set to the road surface height.

3.2   Triangulating Intersections

A polygon is created for an intersection between two or more roads. For each road intersection, the points of the intersection polygon are found by intersecting each outgoing road edge with all other outgoing road edges. For each outgoing road edge, the furthest intersection found forms part of the intersection polygon. For each outgoing road section, intersection points are found for both its road edges. An *Intersection Offset* is set to be the further of the two intersection points, and the road section is drawn only up to this point, so as not to overlap with the intersection polygon.

3.3   Automatic Bridges and Tunnels

The change in height of the terrain can sometimes be too steep for a road. When this occurs, a tunnel or bridge is built. For example, when a road needs to cross a deep ravine, a bridge is built over the ravine. An example is the Bixby Creek Bridge in California. Likewise, tunnels are dug in mountainous terrain, so that the road have a relatively gentle gradient compared to the terrain. Roads which are too steep make it dangerous to drive downhill, and some vehicle engines may not be powerful enough to climb a steep road.

After the initial height of each road section is set, according to the underlying terrain height at that point, the program makes one pass through all the section heights to detect a change in height of more than d, where d is a threshold set by the user to denote the maximum steepness of a road. For usual roads, $d = 0.2$. This point is set to be the start of the bridge/tunnel. From here, the program proceeds forward a distance of *MaxBridgeLength* to search for the first section that is the same height as the start. If this section is found, the point is set to be the end of the bridge/tunnel. Otherwise, the point which has the least slope from the start is chosen as the end.

Next, the program looks forward a small distance *BridgeLengthExtra* to see if a significantly better alternative end-point for the bridge/tunnel can be found. Suppose the difference in height between an end-point P1 and the start point S is d_1 and the difference in height between P2 and S is d_2. P1 is considered to be significantly better than P2 if $d_1 < d_2 + \delta$, where δ is a threshold value set in the program. If such an alternative end-point is found, this new end-point is set to be the end of the bridge, and the process repeats until no new better end-point can be found at the distance of *BridgeLengthExtra* from the bridge end. If this process of looking forward a small amount to find a better bridge/tunnel end-point is not performed, then in certain cases, a "silly" bridge will be built when a much better, slightly longer one can be built.

Figure 1 shows examples of some bridges and tunnels built by the program.

Fig. 1. Example of bridges, tunnels, and road banking at a sharp turn

3.4   Banking

Roads with high speed limits are often banked when they turn, so that cars can travel faster without the danger of flipping over. Banking refers to tilting the road surface perpendicular to the direction of travel. Suppose that the road makes an angle of theta with the horizontal, the speed at which a car can round a curve on an imaginary road without friction is v = sqrt(rg tan theta), where r is the radius of the turn, and g is gravitational force. This implies that the maximum allowable speed at a turn increases with the banking angle, and decreases with the sharpness of the turn. On a real-world road, calculation of banking angles depend on the friction of the road surface, types of vehicles traveling on the road, and the desired speed limit of the road.

Suppose that v0 and v1 are unit direction vectors of two consecutive sections of the road. Then, ϕ, the angle of curvature of the road is $\phi = v0.v1$. In our program, we set the banking angle $\theta = k\phi$, where k is a constant set by the user. In this way, the sharper the turn, the greater the banking angle. The coefficient k is set higher for roads with higher speed limits, in order to increase the banking angle. An example of road banking produced by the program is shown in Figure 1.

3.5   Lane Graphics

At each control point on a Road, the user sets the number of lanes. In the 3D view of the road, a shoulder is drawn on each side of the road, together with lanes, as well as lane dividers drawn as broken lines. If two adjacent control points have different number of lanes, the width of the road is linearly interpolated between the two points. A Road is set to be either a "Left" or "Right" road, indicating the side where a new lane is added.

4   Freeways

Freeways are common in the modern world and are especially important in computer games. Because of their large size, they are also prominent in cityscapes. Freeways have a few special characteristics which require them to be handled differently from ordinary roads. First, freeway do not have intersections, where one road crosses another. Instead, they have *interchanges* with overpasses so that cars can travel straight through without crossing, with connector ramps for cars to move to one freeway to another.

In our system, if the user designates a road as a "freeway", no intersection will be calculated for this freeway. Instead, we allow the user to make overpasses and underpasses, and also attach connector ramps to the freeway.

Also, freeways are divided roads; in other words, the road going one way is separated from the one going the opposite direction. At times, especially in mountainous environments, the two opposite directions can be separated by a significant distance. To define a freeway, the user first sets a series of control points as usual. These control points define the freeway median. Together with the median, two parallel Roads are created, one on each side of the median, going opposite directions. These two Roads also have their own control points, so that the user can move the control points to adjust their offsets from the median control points. If the user later moves a median control point, the two Roads attached to the control point move with the median control point.

4.1   Freeway Overpasses

To prevent interruption to the flow of traffic, a freeway does not have intersections. Instead, overpasses or underpasses are constructed. To create them in our system, the user clicks on an existing Road to make a new Height Point. The user then enters the desired height of the point. The number entered denotes the height offset from the ground at the point. For example, if the user creates a Height Point with height -2.0, the Road at this point would be depressed by a height of 2.0 from the ground height. If the user creates a Height Point with height 3.0, then at this point, the Road is a bridge with height 3.0 above the ground.

When the user creates multiple Height Points, the height of each point on the Road in between the Height Points is linearly interpolated. The user can also designate a Height Point as a "free" point. There are no restrictions of the height of the Road after a free point. Also, the user can designate a section of a Road to be "Bridge and Tunnel Free". Ordinarily, when there is an abrupt change in terrain height, a bridge or tunnel may be automatically created (described in Section 3.3). However, if a section is designated "Bridge and Tunnel Free", no bridge or tunnel would be automatically created. At freeway interchanges, no automatic bridges and tunnels should be created, so that they would not interfere with the user-created overpasses and connector ramps. If a user-defined bridge overlaps with an automatically-generated bridge, they are merged into one bridge, with each point taking the maximum height of the two bridges where they overlap. In between the Height Points set by the user, cubic interpolation is used to generate the heights for all points on the user-created bridge.

Pillars are drawn at regular intervals to support bridges. For overpasses, support pillars cannot be drawn over the road below. To prevent this from happening, the program takes a picture of the scene from the bottom, and renders all the road surfaces, each road having a unique ID, from top to bottom. In this way, each fragment in the final image contains the ID of the bottom-most road surface. A pillar for a Road A can be drawn at the location corresponding to a fragment if and only if the road ID of that fragment is the ID of Road A.

4.2   Freeway Connector Ramps

User can set road split point anywhere on one side of a freeway. A road split point can be defined to be incoming or outgoing, left or right. That defines the direction and side of the freeway connector ramp. The number of lanes before and after a road split point can also be determined by the user. If a road split point is created at the end of a road, it can either lead to one or two connector ramps. The different options are shown in Figure 2.

After defining split points, the user can create a freeway connector ramp by clicking on the end points of two split points. A freeway connector ramp is created to connect these two split points. A piece-wise cubic Hermite spline is created to define the path of the ramp. If no additional control points are set by the user, the two split points are the two control points of the cubic Hermite

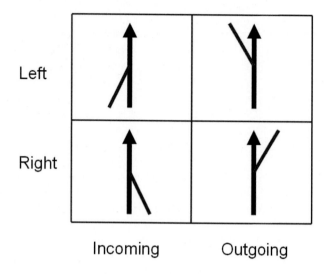

Fig. 2. The different types of split points on a freeway

curve. Note that the exact position of the end point has to be calculated, by finding the coordinates of the midpoint of the lane on the freeway that becomes the freeway ramp. The direction of the end-point of the Hermite curve is set to be the direction of the main freeway at that point. Similarly, the banking angle of the ramp is set to equal that of the freeway at the split point.

To create 3D surface geometry of the freeway connector ramp, the height at the end points are found from the freeway height at those points. If the freeway is a bridge at that point, then the freeway connector ramp also begins as a bridge. The ramp is then treated as an ordinary road, and the height of the road is determined by the usual algorithm, except for the additional bridge of the connector begins or ends as a bridge. If such an end bridge is required, the program searches for the other end point of the bridge by finding the point on the connector ramp that has the least height difference from the end point.

The user is able to customize each connector ramp, but pressing the "Subdivide" button. This will add an additional control point to the freeway ramp. The user can then use the mouse to move the control point to the desired position. The user can also optionally set a height for the control point. To attach another freeway ramp to this ramp, the user can set a control point as a "split point". Then, an additional ramp can be created from this split point. As usual, the user can set the direction of the split branch, as well as the number of lanes before and after the split.

Examples of freeway interchanges created by the user are shown in Figures 3, 4 and 5.

When a freeway ramp splits off from a freeway, it changes the *balance* of the road. Ordinarily, the balance of a road is zero, meaning that the control spline

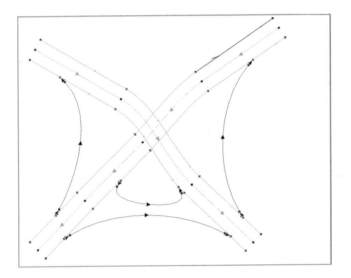

Fig. 3. User-defined "Partial Cloverleaf" freeway interchange

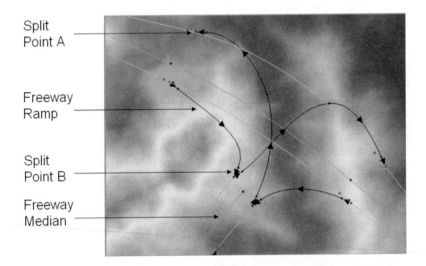

Fig. 4. User creates an intersection from an end of one freeway onto another freeway. Split Point A creates one ramp from a freeway side. Split Point B creates two ramps from the end of a freeway side. This type of freeway interchange is commonly know as the "Semi-Directional T".

Fig. 5. 3D model of a "Stack" freeway interchange

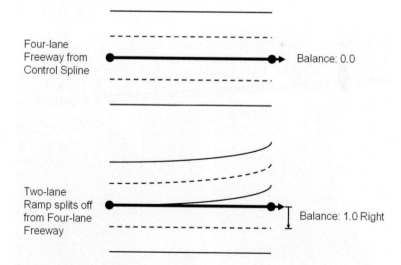

Fig. 6. The balance of a freeway changes after a ramp splits off. Before the split, the control spline is in the center of the freeway; after the split, the center of the freeway is one lane to the right of the control spline.

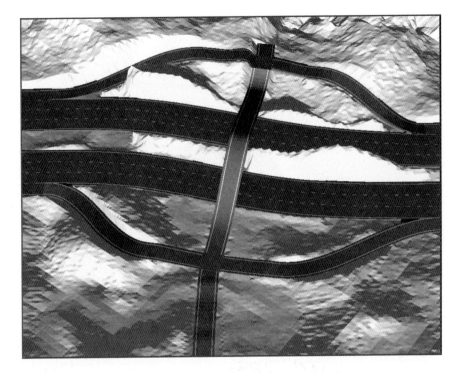

Fig. 7. 3D model of a "Diamond" interchange automatically created by the program

Fig. 8. 3D model of a "Cloverleaf" interchange automatically created by the program

is in the center of the road. However, when a freeway ramp splits off, this is changed. Figure 6 shows an example. In this example, before the split, a freeway has four lanes and the balance is zero. At the split, a two-lane freeway ramp exits the freeway, leaving two lanes on the main freeway. Immediately after the split, the center of the freeway road surface is now an offset of one to the right of the freeway control spline. The magnitude of the balance after a split is equal to half the remaining number of lanes on the freeway. The direction (left or right) of the balance depends on whether the ramp is to the left or right, and whether the ramp is incoming or outgoing. As 3D surface geometry is created for the road surface, our program keeps track of the balance so that the correct road surface is created. Our program also drifts the balance gradually back to zero after each split.

5   Automatic Freeway Interchanges

The method described by the previous section requires the user to manually create every freeway ramp, and set their height. While this allows the user great control over the appearance of each ramp, it is also rather time-consuming. If the user is only interested in creating generic interchanges, our system also provides an automatic tool so that with one single click, the user can request the program to automatically detect all Road-Freeway and Freeway-Freeway intersections, and automatically create the intersections. A classic "Diamond" interchange (Figure 7) is created for each Road-Freeway intersection, and either a "Full Cloverleaf" (Figure 8) or "Full Stack" interchange is created for each Freeway-Freeway intersection, depending on user preference. The program automatically creates these interchanges by automatically inserting split points, setting their directions, setting height points to lower one freeway, and raise the other freeway, as well as insert freeway ramps and set their heights. The user is allowed to manually change these automatically generated points if desired.

6   Conclusions

We have presented a system that automatically creates 3D models of roads and freeways, including freeway interchanges. First, the user specifies control points of a road, and at each control point, the user can also specify other parameters such as number of lanes and height. Our system automatically creates the geometry of a road surface, making sure that the road surface is smooth, curved, banked if necessary, and the road intersections are level. Over rough terrain, the program also heuristically creates tunnels and bridges to prevent roads from being too steep.

The major contribution of this paper is the automatic creation of freeway interchanges. Our tool allows the user to specify split points on the freeway for freeway connector ramps, as well as height points to create overpasses. Our system then generates a 3D model of the freeway interchange with smooth connections and proper banking. Finally, our program also allows the user with one click to automatically find all freeway intersections and automatically create a few common types of freeway interchanges.

References

1. Weber, B., Muller, P., Wonka, P., Gross, M.: Interactive geometric simulation of 4d cities. Computer Graphics Forum (Proceedings of the Eurographics Conference) 28, 481–492 (2009)
2. Carrozzino, M., Tecchia, F., Bergamasco, M.: Urban procedural modeling for real-time rendering. In: Proceedings of the 3rd ISPRS International Workshop 3D-ARCH (2009)
3. Aliaga, D., Vanegas, C., Benes, B.: Interactive example-based urban layout synthesis. ACM Transactions on Graphics (Proceedings of ACM SIGGRAPH Asia 2008) 27 (2008)
4. Chen, G., Esch, G., Wonka, P., Muller, P., Zhang, E.: Interactive procedural street modeling. ACM Trans. Graph. 27 (2008)
5. Thomas, G., Donikan, S.: Modelling virtual cities dedicated to behavioural animation. Computer Graphics Forum 19, 71–80 (2000)
6. Cabral, M., Lefebvre, S., Dachsbacher, C., Drettakis, G.: Structure preserving reshape for textured architectural scenes. Computer Graphics Forum (Proceedings of the Eurographics conference) 28, 469–480 (2009)
7. Galin, E., Peytavie, A., Marechal, N., Guerin, E.: Procedural generation of roads. Computer Graphics Forum (Proceedings of the Eurographics conference) 29 (2010)

Automatic Learning of Gesture Recognition Model Using SOM and SVM

Masaki Oshita and Takefumi Matsunaga

Kyushu Institute of Technology
680-4 Kawazu, Iizuka, Fukuoka, 820-8502, Japan
oshita@ces.kyutech.ac.jp, matsunaga@cg.ces.kyutech.ac.jp

Abstract. In this paper, we propose an automatic learning method for gesture recognition. We combine two different pattern recognition techniques: the Self-Organizing Map (SOM) and Support Vector Machine (SVM). First, we apply the SOM to divide the sample data into phases and construct a state machine. Next, we apply the SVM to learn the transition conditions between nodes. An independent SVM is constructed for each node. Of the various pattern recognition techniques for multi-dimensional data, the SOM is suitable for categorizing data into groups, and thus it is used in the first process. On the other hand, the SVM is suitable for partitioning the feature space into regions belonging to each class, and thus it is used in the second process. Our approach is unique and effective for multi-dimensional and time-varying gesture recognition. The proposed method is a general gesture recognition method that can handle any kinds of input data from any input device. In the experiment presented in this paper, we used two Nintendo Wii Remote controllers, with three-dimensional acceleration sensors, as input devices. The proposed method successfully learned the recognition models of several gestures.

Keywords: gesture recognition, automatic learning, SOM, SVM.

1 Introduction

Gesture recognition has many potential applications, such as gaming interfaces (for example Nintendo Wii, Microsoft Kinect, Sony Move), control interfaces for robots [1], and electronic devices (for example TVs, lights, air conditioning) [12], severance systems [5], and so on. In general, gesture recognition determines what kind of action the user is performing from various input signals such as body position, velocity and orientation. The input signals can be acquired from various devices, such as sensor-embedded remote controllers, pressure sensors and cameras.

There are many pattern recognition techniques. However, the problem with gesture recognition is that the inputs usually consist of multi-dimensional and time-varying data. If the input data were either multi-dimensional or time-varying, existing pattern recognition techniques could easily be applied. For example, the Support Vector Machine (SVM) and neural networks including the Self-Organizing Map (SOM) work well on multi-dimensional data, while the Hidden Markov Model (HMM) and Dynamic Programming (DP) are well-suited to time-varying data. Owing to the

G. Bebis et al. (Eds.): ISVC 2010, Part I, LNCS 6453, pp. 751–759, 2010.
© Springer-Verlag Berlin Heidelberg 2010

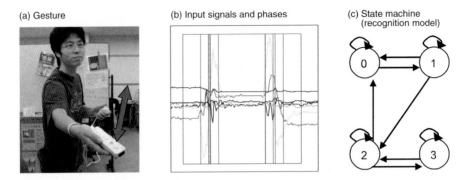

(a) Gesture

(b) Input signals and phases

(c) State machine (recognition model)

Fig. 1. Examples of gesture recognition using Wii remotes. (a) Performing a gesture. (b) Input signals and phases. The horizontal axis represents time. Each line represents one of the input values (in this case, accelerations), and each block a phase of the gesture. (c) State machine (recognition model). Each state represents a phase of the gesture and has conditions for transitions to connected states.

combination of multi-dimensional and time-varying data, gesture recognition introduces many problems.

A common approach to handling such data is to divide each gesture into short phases, and then to recognize each phase using a pattern recognition technique for multi-dimensional data [9]. For example, the gesture of raising and lowering a hand would contain phases such as stationary, moving upward, moving downward. By introducing a state machine, representing each phase as a state, and using a pattern recognition technique to determine the state transitions, we can construct a recognition model (Fig. 1). By preparing multiple recognition models for all gestures used in the application, the system can recognize any user gesture. The problem with this approach is that the state machine must be constructed manually. The designer must divide a gesture into several phases based on changes in the input signals. This requires much experience, effort, and trial and error processes.

The purpose of this research is to develop an automatic learning method for gesture recognition. We combine two different pattern recognition techniques: the Self-Organizing Map (SOM) [7] and Support Vector Machine (SVM) [1]. First, we apply the SOM to divide the sample data into phases and construct a state machine. Next, we apply the SVM to learn the transition conditions between nodes. An independent SVM is constructed for each node. Of the various pattern recognition techniques for multi-dimensional data, the SOM is suitable for categorizing data into groups, and thus it is used in the first process. On the other hand, the SVM is suitable for partitioning the feature space into regions belonging to each class, and thus it is used in the second process. Our approach is unique and effective for multi-dimensional and time-varying gesture recognition.

The proposed method is a general gesture recognition method that can handle any kinds of input data from any input device. In the experiment presented in this paper, we used two Nintendo Wii Remote controllers, with three-dimensional acceleration sensors, as input devices. The proposed method successfully learned the recognition models of several gestures.

The rest of this paper is organized as follows. In Section 2, we review related works. In Section 3, an overview of our method is presented, while Section 4 describes our method in detail. Experimental results and a discussion are presented in Section 5. Finally, Section 6 concludes the paper.

2 Related Work

There are many studies on gesture recognition techniques. As explained earlier, the problem is how to deal with multi-dimensional and time-varying input data.

The Hidden Markov Model (HMM) is a popular method for recognizing time-varying data. It represents a recognition model as a network of nodes that produce some symbols and transition probabilities between nodes. Since it handles input signals as a series of discrete symbols, in order to employ an HMM, the input signals must be reduced to a series of discrete symbols. This causes difficulty in handling multi-dimensional signals and prevents accurate recognition. Toyokura et al. [11] used manually specified conditions for discrimination and applied their approach to sign language recognition. Iuchi et al. [5] used an SOM for discretization. Liang et al. [8] used Principal Component Analysis (PCA) to discretize accelerations of multiple body parts.

Another method for recognizing time-varying data is dynamic programming. By matching the trajectory from the input signals and the sample trajectory from a gesture, the system can determine whether the gesture has been executed. However, to apply this approach, the trajectories must be projected onto a low-dimensional feature space. Moreover, an appropriate threshold must be specified to determine whether two trajectories are similar. Billon et al. [1] employed PCA for dimension reduction, while Yoshioka et al. [14] used manual discretization for each element of the feature vector.

Using a state machine is also a viable approach. Matsunaga et al. [9] used an SVM to learn the conditions for transitions, whereas Oshita [12] used manually described fuzzy-based rules. However, as explained above, the state machine must be created manually.

Even though some of these methods were designed for automatic learning of gesture recognition, certain parts or parameters need to be designed or specified for each gesture by a designer. On the contrary, our method realizes automatic learning.

3 System Overview

Our system consists of a learning and a recognition process. The learning process is an offline process, whereas the recognition process is an online process. The learning process takes input signals of sample data performed by a user as input, while the recognition process takes the current feature vector from the input device at each frame as input.

Each instance of the input signals from an input device is represented as a feature vector. Typically the dimension of the feature vector is between a few and several tens. In the experiment in this paper, we used six-dimensional feature vectors.

3.1 Recognition Model

Our recognition model is represented as a state machine (Fig. 1(c)). Each state in the state machine represents a phase of the gesture. As explained in Section 1, gestures are expressed by time-varying data and the conditions of the feature vector for a gesture vary from time to time. Therefore, we divide a gesture into phases.

Each state of the state machine has a connection to other states and its own SVM to determine to which state to transit based on the current feature vector.

The recognition model also has an initial state and a recognition path. The recognition path is a series of states. As the system transits through states according to the inputs, forming the recognition path, it determines that the gesture has been performed. In addition, motion speed can also be calculated from the intervals of the transitions [9].

Each recognition model recognizes only one type of gesture. Since a system usually uses multiple gestures, multiple recognition models must be constructed and processed for recognition in parallel. Each recognition model is trained independently for a specific gesture.

3.2 Learning Process

The learning process takes input signals of sample data performed by a user as input. A single sample data contains a series of feature vectors obtained during the performance of the gesture. To train a recognition model, a number of sets of sample data are required. For example, in our experiments, we used 10 sets of sample data. Since 16 feature vectors per second are recorded in our system and the length of a typical gesture is between a few seconds and 10 seconds, each sample data consists of a series of about 100 to 200 feature vectors.

Firstly, the system applies an SOM to categorize the feature vectors into a small number of groups (units). A state of the state machine is constructed from each group (unit). We use an SVM to represent the conditions for transitions in each node. Details of the learning process are explained in the next section.

3.3 Recognition Process

During the recognition process, the system takes the current feature vector from the input device at each frame as input. Based on the current feature vector, the system determines to which state to transit next or whether to stay in the current state. The system determines if a particular gesture has been performed based on the history of the state transitions and the recognition path as explained in Section 3.1.

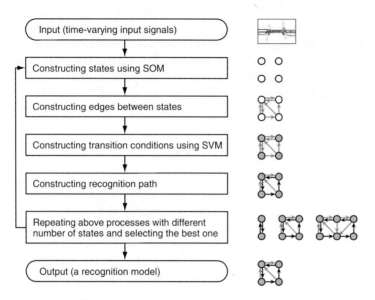

Fig. 2. Flow chart and data structures for the learning process

4   Automatic Learning of Recognition Model

In this section, we explain our automatic learning method, which involves constructing states using an SOM, edges between states, transition conditions using an SVM, and finally the recognition path (Fig. 2). In the following subsections, we explain each of these processes.

4.1   Constructing States Using an SOM

We use an SOM to construct the states of the state machine. The SOM is a pattern recognition technique incorporating unsupervised learning and a type of neural network. It is widely used in many applications for classifying various kinds of data such as images, voices, medical data, genes, etc. The SOM is suitable for categorizing sample data. It maps high-dimensional data into a number of units, which are usually arranged as a two dimensional grid. As result, it groups similar data into the same or a nearby unit.

We simply apply an SOM to all feature vectors from the sample data and then use each unit as a state in the state machine. Each state contains the feature vectors belonging to the corresponding unit and these are used in the following processes. Units with no data are ignored and no states are constructed from them.

In general, when applying an SOM, the number of units must be specified manually. If the number is too small, dissimilar data may be mapped to the same unit. On the other hand, if the number is too high, similar data may be mapped to different units causing too many states to be created in our system. This can cause recognition errors. In many applications, the number of units is chosen empirically. However, in our case, the number depends on the type of input and the gesture to be recognized. Therefore, we introduce automatic selection of the number of units. Our system constructs several state machines (learning models) with different numbers of units. It then evaluates these using the sample data. Finally, the best state machine that minimizes the following criterion is chosen.

$$e = e_1 + k\,e_2 \tag{1}$$

where e_1 is the error rate that correct inputs are not recognized, e_2 is the error rate that incorrect inputs are recognized, and k is a parameter denoting the weights, specified by the user, of e_1 and e_2. The appropriate value of k depends on the application. For example, in an application such as a gaming interface, a small e_1 is important and a small value is specified for k. On the other hand, in an application such as a security system, a small e_2 is important and thus, a large value is specified for k.

4.2 Constructing Edges between States

Once the states of the state machine have been created, our system determines possible transitions for each state. If a feature vector of state A is followed by a feature vector of state B in the input sample data, an edge from state A to state B is created. In addition, each state has an edge to itself and an edge to the initial state, as the user could terminate a gesture at any time and start from the beginning.

4.3 Constructing Transition Conditions Using an SVM

We use an SVM to learn the transition conditions for each state. The SVM is a popular pattern recognition technique with supervised learning. Since it divides the feature space for each class, the SVM can handle unknown data well, although it is not suited to grouping sample data. This is the reason that we combine two pattern recognition techniques, the SOM and SVM, with different characteristics.

To construct an SVM for each state, the sample data belonging to the state and those belonging to adjacent (possible transition) states are used as training data. In our implementation, we employ LibSVM [3], which supports multi-class recognition.

Once the SVM has been trained, the transition to the next state is determined based on the current feature vector. The transition could be to remain in the same state or to move to one of the adjacent states.

4.4 Constructing Recognition Path

A state machine is constructed from the above processes. Finally, the recognition path is determined. At this point, we have an initial state, to which the initial feature vectors belong, and a terminal state, to which the terminal feature vectors belong.

However, even if the system transits from the initial state to the terminal state, it does not necessarily mean that the particular gesture has been performed. Generally, a state machine is a next work of states. Therefore, there can be many paths from the initial state to the terminal state and not all paths mean that the gesture has been performed.

Our system determines a correct path for gesture recognition based on the input sample data. A sample data is represented by a series of states. The system stores the recognition path and uses it for recognition. When there are different recognition paths for different sample data, all recognition paths are used in the recognition process.

5 Experiment and Discussion

We have tested our method with various gestures. In this experiment, we used Wii Remote controllers as input devices. With these controllers, the user has one spatial acceleration sensor for each hand. An acceleration sensor detects three directional (X, Y and Z) accelerations. Therefore, the feature vector has a dimension of six. We used two kinds of gestures in the experiments. A simple gesture (Gesture A) consists of raising and then lowering the right hand while keeping the left hand still. A complex gesture (Gesture B) consists of pushing both hands forward, swinging both hands to the right, swinging both hands to the left, and extending both hands outwards. Note that Wii Remote controllers have only an acceleration sensor. Therefore, even these simple gestures are difficult to recognize, because the positions of the controllers cannot be used. We used 10 sets of sample data for training each recognition mode; that is, an individual performed each gesture 10 times and the input signals were recorded for the learning process.

As explained in Section 4.1, to determine the appropriate number of units for the SOM, the system repeats the learning and recognition processes with a different number of units: two, four, and six. Both gestures (A and B) achieved the best recognition rate with four units. Therefore, four units (four states) were used in the following experiment.

To evaluate the effectiveness of our automatic learning method, especially the effectiveness of using the SOM, we compared the results of our method and a manual method. In the manual method, a participant who is a graduate student with basic knowledge of gesture recognition manually labeled each of about 2,000 feature vectors of input sample data by checking the images on the screen. For the number of states, we used the number determined automatically by our method. Once the categorization was done, we used our automatic method for the later processes, since manually specifying transition conditions requires much trial and error and is not practical.

Results of the above experiment are shown in Table 1. Obviously, the manual categorization required a great deal of time compared to our automatic method, especially for complex gestures. On the other hand, the manual method showed slightly better results. This is because the classification using an SOM does not work well in the presence of noise. When a human finds noise in a series of feature vectors, he/she is able to categorize these according to the previous and following feature vectors.

Table 1. Comparison between the proposed automatic method and manual method

Method / Gesture	Time for learning [sec]	Recognition rate [%]
Automatic (Proposed) / Gesture A	510	100
Manual / Gesture A	4	98
Automatic (Proposed) / Gesture B	5100	84
Manual / Gesture B	8	80

However, since our method simply categorizes all feature vectors independently, a sudden transition to a wrong node could happen as a result of the noise. To solve this problem, an additional process to correct the categorization error could be introduced.

6 Conclusion

In this paper, we proposed a method for automatically learning a recognition model by incorporating both an SOM and an SVM. We also showed the efficiency thereof.

Gesture recognition has many potential applications. Applying our method to various actual applications with higher dimensions remains a future work. For most applications, camera-based sensors are suitable, since users do not have to hold or wear any input devices. However, in order to apply our method to input signals from cameras, a method for extracting feature vectors from images is needed. In addition, when applying our method to very high-dimensional data, for example, motion data acquired using motion capture equipment and which could have more than 40 dimensions, we would need a method for dimension reduction, such as PCA. Further research on these extensions as well as performing further experiments for comparison with alternative methods are also future works.

References

1. Billon, R., Nedelec, A., Tisseau, J.: Gesture recognition in flow based on PCA and using multiagent system. In: Proc. of Advances on Computer Entertainment, pp. 139–146 (2008)
2. Cristianini, N., Shawe-Taylor, J.: An Introduction to Support Vector Machines. Cambridge University, Cambridge (2000)
3. Chang, C.-C., Lin, C.-J.: 2002, LIBSVM: a Library for Support Vector Machines (2001), http://www.csie.ntu.edu.tw/~cjlin/libsvm/index.html
4. Irie, K., Wakamura, N., Umeda, K.: Construction of an Intelligent Room Based on Gesture Recognition -Operation of Electric Appliances with Hand Gestures. In: Proc. IEEE/RSJ Int. Conf. on Intelligent Robots and Systems, pp. 193–198 (2004)
5. Iuchi, H., Maeda, S., Tsuruta, N.: Gesture Recognition using Self-Organizing Maps and Hidden Markov Model. In: IPSJ SIG Notes, Computer Vision and Image Media, vol. 2001(36), pp. 127–134 (2001)
6. Joutou, T., Yanai, K.: Web Image Classification with Bag-of-keypoints. IPSJ SIG Notes, Computer Vision and Image Media 2007(42), 201–208 (2007)
7. Kohonen, T.: The self-organizing map. Proc. of the IEEE 78(9), 1464–1479 (1990)
8. Liang, X., Li, Q., Zhang, X., Zhang, S., Geng, W.: Performance-driven motion choreographing with accelerometers. Computer Animation and Virtual Worlds 20(2-3), 89–99 (2009)

9. Matsunaga, T., Masaki, O.: Recognition of Walking Motion Using Support Vector Machine. In: Proc. of ISICE 2007, pp. 337–342 (2007)
10. Nanri, T., Otsu, N.: Anomaly Detection in Motion Images Containing Multiple Persons. In: Proc. of PRMU 2004-77, vol. 104(291), pp. 583–588 (2004)
11. Noma, K., Nakai, M., Shimodaira, H., Sagayama, S.: Sequential-Pattern Recognition by Support Vector Machines using Dynamic Time-Alignment Kernels, Technical report of IEICE, vol. 100(507), pp. 63–68 (2000)
12. Oshita, M.: Motion-Capture-Based Avatar Control Framework in Third-Person View Virtual Environments. In: ACM SIGCHI International Conference on Advance in Computer Entertainment Technology (2006)
13. Yamada, T., Umeda, K.: Improvement of the Method of Operating a Mobile Robot by Gesture Recognition. In: JSME Conference on Robots and Mechatronics, vol. 2001, p. 49 (2001)
14. Yoshioka, T., Koga, H., Watanabe, T., Yokoyama, T.: Online Automatic Acquisition of Human Motion Models with Voting, Technical report of IEICE, vol. 105(302), pp. 119–124 (2005)
15. Toyokura, Y., Nankaku, Y., et al.: Approach to Japanese Sign Language Word Recognition using Basic Motion HMM. In: Proceedings of the Society Conference of IEICE, vol. 2006, p. 72 (2006)

Author Index

Printing: Mercedes-Druck, Berlin
Binding: Stein+Lehmann, Berlin